Recent Titles in This Series

(*Continued in the back of this publication*)

Differential Geometry:
Geometry in Mathematical
Physics and Related Topics

Proceedings of Symposia in
PURE MATHEMATICS

Volume 54, Part 2

Differential Geometry: Geometry in Mathematical Physics and Related Topics

Robert Greene
S. T. Yau
Editors

American Mathematical Society
Providence, Rhode Island

PROCEEDINGS OF THE SUMMER RESEARCH INSTITUTE
ON DIFFERENTIAL GEOMETRY
HELD AT THE UNIVERSITY OF CALIFORNIA, LOS ANGELES
LOS ANGELES, CALIFORNIA
JULY 8–28, 1990

with the support from the National Science Foundation
Grant DMS–8913610.

1991 *Mathematics Subject Classification.*
Primary 53A10, 49F20, 58E20, 58E12, 58G11 (Part 1)
81E13, 53C80, 53B50, 32C10, 53C15, 58F05, 83C75 (Part 2)
53C20, 53C40, 58C40, 58G25, 58F07, 58F11, 58F17 (Part 3).

Library of Congress Cataloging-in-Publication Data
Geometry in mathematical physics and related topics/Robert Greene and S. T. Yau, editors.
    p.   cm.—(Proceedings of symposia in pure mathematics; v. 54, pt. 2) (Differential geometry; pt. 2)
"Proceedings of a Summer Research Institute on Differential Geometry, held at the University of California, Los Angeles, July 8–28, 1990"—T.p. verso.
    ISBN 0-8218-1494-X (Part 1)
    ISBN 0-8218-1495-8 (Part 2)
    ISBN 0-8218-1496-6 (Part 3)
    ISBN 0-8218-1493-1 (set) (alk. paper)
    1. Complex manifolds—Congresses.   2. Mathematical physics—Congresses.   I. Greene, Robert Everist, 1943– .   II. Yau, Shing-Tung, 1949– .   III. Summer Research Institute on Differential Geometry (1990: University of California, Los Angeles)   IV. Series.   V. Series: Differential geometry; pt. 2.
QA641.D3833   1993   pt. 2
[QA331.7]
516.3′6 s—dc20   92-32831
[515′.9]   CIP

This publication was typeset using $\mathcal{A}_{\mathcal{M}}\mathcal{S}$-TEX,
the American Mathematical Society's TEX macro system.
            10 9 8 7 6 5 4 3 2 1   98 97 96 95 94 93

Dedication.

To Professor S.-S. Chern,

in appreciation of his formative influence on

modern differential geometry.

Contents

(∗ denotes one-hour survey lectures)

PART 2

PART 3

Preface

The 1990 American Mathematical Society Summer Institute on Differential Geometry took place at University of California, Los Angeles from July 9 to July 27, 1990. This was the largest AMS Summer Institute ever. There were 426 registered participants and 270 lectures. The organizing committee for the Institute consisted of Robert Bryant, Duke University; Eugenio Calabi, University of Pennsylvania; S. Y. Cheng, University of California, Los Angeles; H. Blaine Lawson, State University of New York, Stony Brook; H. Wu, University of California, Berkeley; and as co-chairmen, Robert E. Greene, University of California, Los Angeles, and S. T. Yau, Harvard University.

In the years since the previous AMS Summer Institute on Differential Geometry in 1973, the field has undergone a remarkable expansion, both in the number of people working in the area and in the number and scope of the topics under investigation. Even in the context of the rapid growth of mathematics as a whole during this period, the growth of geometry is striking.

It is our hope that the three volumes of these proceedings, taken as a whole, will provide a broad overview of geometry and its relationship to mathematics in toto, with one obvious exception; the geometry of complex manifolds and the relationship of complex geometry to complex analysis were the subject of one week of the (three-week) 1989 AMS Summer Institute on Several Complex Variables and Complex Geometry. While some topics in complex geometry arose naturally in the 1990 Summer Institute and are covered in these Proceedings, the coverage of this subject in 1989 justified a reduced emphasis in 1990.

Thus the reader seeking a complete view of geometry would do well to add the second volume on complex geometry from the 1989 Proceedings to the present three volumes.

Each week of the 1990 Summer Institute was given a general emphasis as to subject and the Proceedings volumes are organized in the same way. While overlap is natural, and indeed inevitable, the subjects of the volumes are as follows:

I. Partial differential equations on manifolds: harmonic functions and mapping, Monge-Ampre equation, differential systems, minimal submanifolds;

II. Geometry in mathematical physics and related topics: gauge theory, symplectic geometry, complex geometry, L^2 cohomology, Lorentzian geometry;

III. Riemannian geometry: curvature and topology, groups and manifolds, dynamical systems in geometry, spectral theory of Riemannian manifolds.

The articles in these Proceedings are also of several types. We requested broad-ranging surveys from the people who had given one-hour survey lectures at the Institute. These articles are marked with an asterisk in the table of contents. Such surveys were also encouraged from other participants. The remaining articles are either research papers in the usual sense or relatively brief announcements of results. But in these categories as well, we encouraged authors to provide more background information and references to related work than usual. Thus we hope that readers will find the volumes a source of broad perspectives on the rapidly expanding literature of geometry.

The editors themselves essayed two efforts in this direction. Volume I begins with a problem list by S. T. Yau, successor to his 1980 list [Seminar in Differential Geometry, Annals of Math. Studies, no. 102, Princeton University Press]. Volume III begins with an overview by R. E. Greene of some recent trends in Riemannian geometry, in the interests of identifying themes common to the remaining papers in that volume and in outlining certain topics that, as it happened, would not otherwise have been included.

An event such as the 1990 AMS Summer Institute involves the efforts of a great many people. We would like to thank all the participants, both for their participation in the Institute and for their prompt and abundant response to our call for papers for these Proceedings. We are indebted to the staff of the UCLA Mathematics Department for their cooperation during the Summer Institute and to the American Mathematical Society in general and in particular to Mr. Wayne Drady and Ms. Susan Blyth for their invaluable on-the-spot assistance during the Institute. Without the financial support of the National Science Foundation, the Summer Institute could not have had anything like the scope it in fact attained; we are particularly grateful for their willingness to support the participation of advanced graduate students, a willingness that we believe was a valuable investment in the future of geometry. Finally, we are indebted to the publication staff of the American Mathematical Society and particularly to Ms. Alison Buckser, Ms. Donna Harmon, and Ms. Christine Thivierge, whose unfailing patience and helpfulness made the preparation of these volumes a far easier task than it would have been otherwise.

Robert E. Greene
S. T. Yau

Proceedings of Symposia in Pure Mathematics
Volume **54** (1993), Part 2

Invariants of Gauss Maps of Theta Divisors

MALCOLM ADAMS, CLINT MCCRORY, TED SHIFRIN, AND ROBERT VARLEY

1. The general set-up

Let $\mathbb{A}^n$ be a principally polarized abelian variety of dimension n, and let $\Theta \subset \mathbb{A}^n$ be its theta divisor. The Gauss map of Θ, $\gamma : \Theta \to \mathbb{P}^{n-1*} = \mathbb{P}(T_0\mathbb{A}^n)^*$, is the obvious one: $\gamma(z)$ is the translation to $T_0\mathbb{A}^n$ of the tangent space $T_z\Theta$. (In the event that Θ has singularities, we use the Nash tangent bundle and the Nash Gauss map. For the purposes of this article, we avoid these complications.) The focus of this work is the structure of γ: what are its singularities, is it a stable map, and, ultimately, to what extent is there a Torelli theorem (i.e., can we recover the abelian variety from certain data involving γ)?

We will make reference to the Thom-Boardman singularity type

$$\Sigma^i(\gamma) = \{z \in \Theta : \text{corank } d\gamma_z = i\},$$

and if z is a smooth point of $\Sigma^i = \Sigma^i(\gamma)$, then

$$z \in \Sigma^{i,j}(\gamma) \iff \text{corank } d(\gamma|\Sigma^i)_z = j.$$

(The process continues inductively to symbols of length ≥ 3.) We will also make occasional reference to the Arnold symbol of a singularity of the Gauss map of a hypersurface: this is the singularity type of the tangent hyperplane section. The normal forms one should keep in mind (cf. [**4, 8**]) are shown in Table 1.

In the special case of a Jacobian variety, we have some beautiful projective geometry. Let C be a generic (nonhyperelliptic, smooth) curve of genus g, embedded as its canonical model in $\mathbb{P}^{g-1}$; let $J = J(C)$ be its (g-dimensional) Jacobian variety, and Θ the theta divisor of J. Recall that Θ

1991 *Mathematics Subject Classification.* Primary 14H40.

The first author was partially supported by a grant from the University of Georgia Faculty Research Grant; the last, by NSF Grant DMS-8803487.

This paper is in final form and no version will be submitted for publication elsewhere.

TABLE 1. Normal forms

Arnold symbol	$n = 3$	$n = 4$
A_k, $k \geq 2$	$x^2 + y^{k+1}$	$x^2 + y^2 + z^{k+1}$
D_k, $k \geq 4$	$x^{k-1} + xy^2$	$x^2 + y^{k-1} + yz^2$
E_6	$x^3 + y^4$	$x^2 + y^3 + z^4$
P_8		$x^3 + y^3 + z^3 + axyz$

is the divisor of Riemann's theta function ϑ,

$$(1) \qquad \vartheta(z, \Omega) = \sum_{n \in \mathbb{Z}^g} e^{\pi\sqrt{-1}\langle n, \Omega n\rangle + 2\pi\sqrt{-1}\langle n, z\rangle}, \qquad z \in \mathbb{C}^g,$$

where Ω is the period matrix of the abelian variety J. Choose a basis $\omega_1, \ldots, \omega_g$ for the holomorphic differentials on C; let $\vec{\omega} = (\omega_1, \ldots, \omega_g)$. Letting $C^{(g-1)}$ denote the $(g-1)$-fold symmetric product of C, the Abel-Jacobi map $\mu : C^{(g-1)} \to J(C)$ is given by

$$(2) \qquad \mu(P_1 + \cdots + P_{g-1}) = \int_{P_o}^{P_1} \vec{\omega} + \cdots + \int_{P_o}^{P_{g-1}} \vec{\omega} \quad \text{(mod periods)},$$

and is birational to its image, which is (a translate of) Θ. (Its fibers consist of linearly equivalent $(g-1)$-tuples of points on C, i.e., in the nontrivial case, the special divisors of degree $g-1$.)

Differentiating (2), the Gauss map of Θ may be interpreted (at a general point of Θ) in terms of the geometry of C by means of the equation

$$(3) \qquad \gamma(P_1 + \cdots + P_{g-1}) = \overline{P_1 \cdots P_{g-1}} = P_1 \wedge \cdots \wedge P_{g-1} \in \mathbb{P}^{g-1*}.$$

If we let $s_1, \ldots, s_{g-1}$ be local coordinates on C centered at (the distinct points) $P_1, \ldots, P_{g-1}$ respectively, then differentiating (3) yields

$$(4) \quad d\gamma_{P_1 + \cdots + P_{g-1}} = P_1' \wedge P_2 \wedge \cdots \wedge P_{g-1} \, ds_1 + \cdots + P_1 \wedge \cdots \wedge P_{g-1}' \, ds_{g-1}$$
$$\text{(mod })P_1 \wedge \cdots \wedge P_{g-1},$$

from which we infer that γ drops rank when the hyperplane $\overline{P_1 \cdots P_{g-1}}$ is tangent to C at one of the points P_i. Indeed, when the plane is bitangent at, say, P_1 and P_2, then γ has corank two, and when the plane is tritangent γ has corank three. Since $d\gamma$ is symmetric (mirroring the Lagrangian structure on the Gauss map), we "expect" γ to have corank two along a subvariety of codimension three, and to have corank three along a subvariety of codimension six.

One of the new ingredients in these considerations is the presence of a $\mathbb{Z}/2$ symmetry. The (-1)-map on $\mathbb{A}^n$ induces an involution ι on Θ, which in the Jacobian case can be interpreted in the following way: the canonical curve $C \subset \mathbb{P}^{g-1}$ has degree $2g - 2$; the general $(g-1)$-tuple of points $P_1 + \cdots + P_{g-1}$ spans a hyperplane H, and $\iota(P_1 + \cdots + P_{g-1})$ is the residual

$(g-1)$-tuple in $C \cdot H$. (In the language of divisors, ι associates to the effective divisor D of degree $(g-1)$ the residual divisor $K-D$.) Note that, by (3), in this case γ is ι-invariant, as the residual $(g-1)$-tuple spans the same hyperplane.

We will discuss here some of the less technical results for two- and three-dimensional theta-divisors (cf. [2, 9]). For the more classical case of projective surfaces and threefolds, cf. [6, 7, 8], and for some of the background on Lagrangian structures, cf. [1].

2. The two-dimensional Θ-divisor: a brief review

Let C be a nonhyperelliptic smooth curve of genus three, embedded in $\mathbb{P}^2$ as a quartic. In this case the Gauss map (3) is given by the secant map

$$\gamma : C^{(2)} \to \mathbb{P}^{2*}, \qquad \gamma(P+Q) = \overline{PQ}.$$

As we saw above, the parabolic curve $\Pi = \overline{\Sigma}^1$ corresponds to the point pairs $P+Q$ where $\overline{PQ}$ is tangent to C at P (or Q). (Cf. Figure 1.) It is not hard to see that Π has ordinary double points at the points $P+Q$ such that $\overline{PQ}$ is bitangent to C at P and Q. (E.g., let s and t be local coordinates on C centered at P and Q respectively; then a local equation for the Gauss map at such a point is $\gamma(s,t) = (s^2 + \cdots, t^2 + \cdots)$.) There are 28 lines bitangent to C, hence 28 points of $\Sigma^2(\gamma)$. (This can be computed by Chern classes, but also follows from Plücker formulas for plane curves, or by counting odd theta characteristics—see below.) While one expects to find Σ^2 in codimension three, the fact that these points are (isolated) fixed points of the $\mathbb{Z}/2$ action forces their existence.

A local equation for Π is $P \wedge P' \wedge Q = 0$. Then, by Lagrange multipliers,

$$\begin{aligned}
P+Q \in \Sigma^{1,1}(\gamma) &\iff d\gamma_{P+Q} \\
&= 0 \ (\mathrm{mod}\ P \wedge Q, \ d(P \wedge P' \wedge Q)) \iff P \wedge Q'\, dt \\
&= 0 (\mathrm{mod}\ P \wedge Q, \ d(P \wedge P' \wedge Q)) \iff P \wedge P'' \wedge Q = 0,
\end{aligned}$$

if and only if P is a flex point. (Here γ looks like $\gamma(s,t) = (s^3 + st + \cdots, t + \cdots)$.) The generic plane quartic has 24 flex points (its Weierstrass points). Now this computation is valid only away from the diagonal of $C^{(2)}$, and there are 24 further $\Sigma^{1,1}$ points on the diagonal, namely the images of these under ι (since γ is ι-invariant, so are its Thom-Boardman loci).

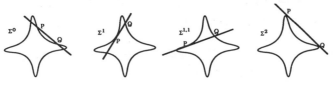

FIGURE 1

While γ is not stable as a (Lagrangian) map (due to the presence of singularities in the unexpected codimension), it is $\mathbb{Z}/2$-stable precisely when C has only normal Weierstrass points (i.e., no higher flexes). Cf. [9, 1].

3. The three-dimensional Θ-divisor

The generic three-dimensional abelian variety is a Jacobian variety. When we come to dimension four, the Jacobians comprise a hypersurface (dimension 9) in the space of all principally polarized abelian varieties (dimension 10).

The Jacobian case. Let C be a generic (nonhyperelliptic, smooth) curve of genus four; its canonical embedding is a smooth sextic in $\mathbb{P}^3$, got by intersecting a (smooth) quadric surface Q and a cubic surface F. It follows from Riemann-Roch that on C there are only two g_3^1's (linear systems of degree three and dimension one), given by intersecting F with the two families of rulings on Q. The Abel-Jacobi map μ blows down these two $\mathbb{P}^1$'s of linearly equivalent triples to double points of Θ (Riemann Singularity Theorem) and is elsewhere an isomorphism between $C^{(3)}$ and Θ.

Analogous to the genus three case, symmetry forces the existence of Σ^3 points. In fact, there are 120 tritangent planes to C. To count them (cf. also [10]), we observe that the divisors $D = P + Q + R$ so that $2D$ is a canonical divisor comprise the *odd theta characteristics* on C (note that $1 \le h^0(\mathcal{O}(D)) = h^0(\mathcal{O}(K - D)) \le 1$, by geometric Riemann-Roch, since P, Q and R cannot be collinear); on a curve of genus g, there are $2^{g-1}(2^g - 1)$ odd theta characteristics (cf. [5, p. 143]). For future reference, we remark that the parabolic surface $\Pi = \overline{\Sigma}^1$ has the local equation $xyz = 0$ at a Σ^3 point.

Perhaps more special is the presence of ∞^1 bitangent planes to C, so that $\Sigma^2(\gamma)$ contains at least a curve. (An argument for this is as follows: consider the image $\Gamma \subset \mathbb{G}(1, 3)$ of the Gauss map of C. The set $\sigma_1(l_o)$ of lines incident to a general line $l_o \in \mathbb{G}(1, 3)$ comprises the codimension-one Schubert cycle in $\mathbb{G}(1, 3)$, and hence cuts $\deg \Gamma$ ($= 18$) points on Γ. For the general point $P \in C$, specializing l_o to $T_P C$, one can check easily that l_o appears with multiplicity four in $\Gamma \cdot \sigma_1(l_o)$, and so we can find a point $Q \in C$, $Q \ne P$, so that $T_P C$ and $T_Q C$ are coplanar.) Since there is no nontrivial local $\mathbb{Z}/2$ action here, the unexpected dimension of the Σ^2 locus suggests the nongenericity (or nonstability) of γ as a $\mathbb{Z}/2$-Lagrangian map.

In the Jacobian case, there is an elaborate cast of characters, which we (partially) summarize in Table 2.

It follows that the curve $\overline{\Sigma}^{1,1}$ does not pass through the singular points of Π (since a hyperplane section of type $2P + 2Q + 2R$ is not the limit of those of type $3P + Q + R + S$); on the other hand, the curve $\overline{\Sigma}^2$ certainly

TABLE 2. SINGULARITIES OF THE GAUSS MAP OF Θ ON $J(C)$

hyperplane section type	divisor	Thom-Boardman symbol	Arnold symbol
$P+Q+R+S+T+U$	$P+Q+R$	Σ^0	A_1
$2P+Q+R+S+T$	$P+Q+R$	Σ^1	A_2
$3P+Q+R+S$	$P+Q+R, \quad 2P+S$	$\Sigma^{1,1}$	A_3
$4P+Q+R$	$P+Q+R, \quad 3P$	$\Sigma^{1,1,1}$	A_4
$2P+2Q+R+S$	$P+Q+R$	Σ^2	D_4
$4P+Q+R$	$2P+Q$	Σ^2	D_6
$3P+2Q+R$	$P+Q+R, \quad 2P+Q$	$\Sigma^{2,1}$	E_6
$2P+2Q+2R$	$P+Q+R$	Σ^3	P_8

does. $\Sigma^{1,1}$ is a 6-fold covering of C, ramified at the $\Sigma^{1,1,1}$ points. These are the 120 points $(P+Q+R, \quad 3P)$ corresponding to the 60 Weierstrass points P of C. Finally, it is of interest to note that $\gamma(\Pi) \subset \mathbb{P}^{3*}$ is a developable ruled surface, the tangent developable of the curve $\gamma(\overline{\Sigma}^{1,1})$ (the ruling consists of the pencil of planes containing $T_P C$; since each ruling intersects the infinitely near ruling—i.e., in the dual of the osculating plane to C at P—the surface is developable).

The generic abelian variety. In contrast with the Jacobian case, we have the following result.

THEOREM. *For the theta divisor Θ of a generic four-dimensional abelian variety, the Gauss map γ has rank ≤ 1 only at the 120 fixed points of the involution $\iota : \Theta \to \Theta$ (the Σ^3 points of γ).*

SKETCH OF PROOF.

Step 1. Using Harris-Tu (as in [8, p. 697]), one computes the homological $\Sigma^{\geq 2}(\gamma)$ locus to be a zero-cycle of degree 480. The 120 Σ^3 points are certainly contained therein, but each is counted with multiplicity 4 (in the universal setting, $\{0\} \subset \{$symmetric 3×3 complex matrices of rank $\leq 1\}$ has multiplicity four, since the latter variety is the cone over the Veronese embedding of $\mathbb{P}^2$ in $\mathbb{P}^5$). This accounts for the 480 and so the Σ^3 points comprise the whole $\Sigma^{\geq 2}$ locus ... provided the preimage $\mathrm{II}^{-1}(\{$symmetric matrices of rank $\leq 1\})$ under the second fundamental form II ($\leftrightarrow d\gamma$) is zero-dimensional.

Step 2. Let $\mathcal{H}_4$ denote genus four Siegel upper halfspace, the space of period matrices of four-dimensional principally polarized abelian varieties, and let $\Theta \to \mathcal{H}_4$ denote the universal theta divisor. Consider the relative second fundamental form $\mathrm{II} : \Theta \to \mathrm{Sym}^2(T^*\Theta) \otimes N_\Theta$. The issue becomes this: passing to local coordinates and viewing II as having values in $\{$symmetric 3×3 matrices$\}$, we claim that at a general Σ^2 point ξ of the theta divisor Θ_o of a Jacobian, II is transverse to $\{$symmetric matrices of rank $= 1\}$. The technical point is to compute $\mathbf{II}$ in terms of the hessian of the defining function ϑ of Θ, and then to observe that, because the Riemann theta function (1) satisfies the heat equation (cf. [3]),

$$\frac{\partial \vartheta}{\partial \Omega_{ij}} = \frac{1}{2\pi\sqrt{-1}(1 + \delta_{ij})} \frac{\partial^2 \vartheta}{\partial z_i \partial z_j},$$

the derivative of an entry of $\mathbf{II}$ (say, $\frac{\partial^2 \vartheta}{\partial z_1 \partial z_2}$) with respect to Ω_{ij} can be computed in terms of the fourth-order partial derivative $\frac{\partial^4 \vartheta}{\partial z_i \partial z_j \partial z_1 \partial z_2}$. But this is explicitly computable at points of the theta divisor of a Jacobian: we parametrize Θ_o near ξ by (2), and using the projective geometry of C, set up a convenient linear coordinate system on $\mathbb{C}^4$, letting $\ker \mathbf{II}_\xi$ be spanned by $\partial/\partial z_1$ and $\partial/\partial z_2$. The crucial point to check, then, is that $\frac{\partial^4 \vartheta}{\partial z_1 \partial z_2^2 \partial z_3}$ does not vanish at ξ. This implies that $\mathbf{II}$ is submersive at (ξ, Θ_o), as desired. $\square$

As we observed earlier, the parabolic surface Π of a Jacobian theta divisor has the local equation $xyz = 0$ at a Σ^3 point. Versally deforming γ as a $\mathbb{Z}/2$-*Lagrangian map* gives rise to a local normal form $x^3 + y^3 + z^3 + axyz$ for γ and a parabolic surface with equation $x^3 + y^3 + z^3 + bxyz = 0$, where $b = -(a^3 + 108)/3a^2$. The coefficient a is a modulus of the P_8 point, and the above analysis shows that, while $a = 0$ (and hence $b = \infty$) for the theta divisor of a Jacobian, a is not identically zero (for when $a = 0$ one has a curve of Σ^2 points). This modulus has been studied in the context of singularity theory (cf. [4]).

While we noted that in the Jacobian case the curve $\overline{\Sigma}^{1,1}$ does not pass through the Σ^3 points, in the generic case, it does, and has nine branches at each such point. (This suggests that the curve $\overline{\Sigma}^{1,1} \cup \overline{\Sigma}^2$ in the Jacobian case should be viewed as the specialization of the generic $\overline{\Sigma}^{1,1}$.) It is also the case that $\gamma(\Pi)$ is not ruled for the generic abelian variety $\mathbb{A}^4$ (this is a consequence, in fact for $\mathbb{A}^n$, $n \geq 4$, of recent work of Debarre); it would be quite interesting to provide a direct, differential geometric proof of this fact.

4. Homogeneity and the Lagrangian structure of the Gauss map

If $M = \{z_{n+1} = f(z_1, \ldots, z_n)\} \subset \mathbb{C}^{n+1}$ is the graph of a function, then there is a natural "gradient" Lagrangian structure on γ:

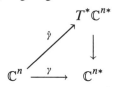

where $\tilde{\gamma}(z) = (df_z, z)$, and $\gamma = \pi \circ \tilde{\gamma}$. $T^*\mathbb{C}^{n*}$ is a (holomorphic) symplectic manifold, π is a Lagrangian submersion (i.e., its fibers are Lagrangian submanifolds), and the image of $\tilde{\gamma}$ is a Lagrangian submanifold of $T^*\mathbb{C}^{n*}$.

Thus, the Gauss map of any smooth complex hypersurface $M \subset \mathbb{C}^{n+1}$ locally has such a Lagrangian structure. The problem is that the local structures are not at all canonical.

REMARK. In the case of an oriented real hypersurface $M \subset \mathbb{R}^{n+1}$, using the Euclidean structure, Arnold gives the following canonical global Lagrangian structure on the Gauss map:

$$
\begin{array}{ccc}
 & & T^* S^n \cong TS^n \\
 & \nearrow^{\tilde{\nu}} & \downarrow \\
M & \xrightarrow{\ \nu\ } & S^n
\end{array}
$$

with $\tilde{\nu}(x) = (\nu(x), \mathrm{proj}_{T_x M}\, x)$.

The above local Lagrangian structure is not rigid; a simple related example is the following: the Lagrangian factorizations $x \rightsquigarrow (x^3, x + tx^2) \rightsquigarrow x^3$ are inequivalent for different values of t. Thus, if we represent M as a graph in different ways (even arbitrarily close in the appropriate topology), we may well obtain inequivalent Lagrangian structures. So we instead work with a homogeneous Lagrangian structure: let $N^*(M) \subset T^*\mathbb{C}^{n+1}$ be the conormal bundle of M; then γ lifts to the homogeneous Gauss map Γ,

$$
\begin{array}{ccc}
N^*(M) & \xrightarrow{\ \Gamma\ } & T_0^*\mathbb{C}^{n+1} \\
\downarrow & & \downarrow \\
M & \xrightarrow{\ \gamma\ } & \mathbb{P}^{n*}
\end{array}
$$

The same construction works for $\Theta \subset \mathbb{A}^{n+1}$. If we remove the zero-section of $T^*\mathbb{A}^{n+1}$, Γ is a conic Lagrangian map, and the $\mathbb{Z}/2$ symmetry is induced by the (-1)-map of $\mathbb{A}^{n+1}$. What is more, the $\mathbb{Z}/2$ action is symplectic in the homogeneous set-up, whereas it's antisymplectic in the local gradient set-up. (Note that there is an important difference from the usual "microlocal" game: we are projecting to the fiber, rather than to the base, of the cotangent bundle.)

QUERY. Is there a global (holomorphic) Lagrangian structure on the Gauss map of a complex hypersurface? (We can prove—nonconstructively—that the answer is yes for $n = 1$.)

REFERENCES

See especially the references in 1, 8, and 9.

1. M. Adams, C. McCrory, T. Shifrin, and R. Varley, *Symmetric Lagrangian singularities and Gauss maps of theta divisors*, Lecture Notes in Math., Singularity Theory and its Applications, Warwick, 1989 (D. Mond and J. Montaldi, eds.), Springer-Verlag, New York, pp. 1-26.
2. _____, *The Gauss map of a genus four theta divisor* (to appear).
3. E. Arbarello, M. Cornalba, P.A. Griffiths, and J. Harris, *Geometry of algebraic curves*. I, Springer-Verlag, New York, 1985.

4. V.I. Arnold, S.M. Gusein-Zade, and A.N. Varchenko, *Singularities of differentiable maps.* I, Birkhäuser, Boston, 1985.
5. C. H. Clemens, *A scrapbook of complex curve theory*, Plenum Press, New York, 1980.
6. V. S. Kulikov, *Calculation of singularities of an imbedding of a generic algebraic surface in projective space* $\mathbb{P}^3$, Funct. Anal. Appl. **17** (1983), 176-186.
7. C. McCrory and T. Shifrin, *Cusps of the projective Gauss map*, J. Differential Geom. **19** (1984), 257-276.
8. C. McCrory, T. Shifrin, and R. Varley, *The Gauss map of a generic hypersurface in* $\mathbb{P}^4$, J. Differential Geom. **30** (1989), 689-759.
9. _____, *The Gauss map of a genus three theta divisor*, Trans. Amer. Math. Soc. **331** (1992), 727–750.
10. R. Piene, *Cuspidal projections of space curves*, Math. Ann. **256** (1981), 95-119.

UNIVERSITY OF GEORGIA

Proceedings of Symposia in Pure Mathematics
Volume **54** (1993), Part 2

On Characteristics of Hypersurfaces
in Symplectic Manifolds

AUGUSTIN BANYAGA

1. Introduction

The goal of this paper is to bring more geometric flavour in the problem of existence of periodic solutions of hamiltonian systems on a prescribed energy hypersurface. We describe the geometric surrounding of the problem and survey some existence theorems obtained by geometric considerations.

2. Hypersurfaces in almost hermitian manifolds

An almost hermitian manifold is a smooth manifold P equipped with a riemannian metric G and an almost complex structure J such that $G \circ J = G$.

Let M be an oriented hypersurface in an almost hermitian manifold P with almost hermitian structure (G, J). The following is a classical construction due to Tashiro (see [22]): for $x \in M$, let $C_x \in T_x P$ be the unit outward normal vector to M. The vector field $C: x \mapsto C_x$ along M satisfies $G(C, i_* X) = 0$, for all vector field X tangent to M. We have $G(C, JC) = G(JC, J^2 C) = -G(JC, C) = -G(C, JC)$. Therefore $G(C, JC) = 0$, which means that JC is tangent to M. The nowhere vanishing vector field Z on M such that $JC = -i_* Z$ is called the *characteristic vector field* of M. Here $i: M \to P$ denotes the embedding of M into P. If M is compact, Z defines a one-dimensional foliation on M by the orbits of Z, called the characteristic foliation. The (unparametrised) orbits of Z are called the characteristics: the leaves of the characteristic foliation.

Let $\pi: TP \to TM$, $\pi^\perp: TP \to TM^\perp$ be the natural projections coming from the decomposition $TP = TM \oplus (TM)^\perp$ obtained via the metric G.

1991 *Mathematics Subject Classification.* Primary 58F05, 58F22, 58F25; Secondary 53C15, 53C57.

Supported in part by NSF grant DMS 90-01861.

This paper is in final form and no version of it will be submitted for publication elsewhere.

One defines a 1-1 tensor field $\varphi: TM \to TM$ and a one-form η on M as follows: for any vector field X, φX is defined as $i_*\varphi X = \pi(J i_* X)$ and $\eta(X)$ is the unique function on M such that $\pi^\perp(J i_* X) = \eta(X)C$, i.e., $J(i_* X) = i_*(\varphi X) + \eta(X)C$. The following are easy to check

$$\eta(Z) = 1, \qquad \varphi Z = 0, \qquad \varphi^2 X = -X + \eta(X)Z,$$
$$g(\varphi X, \varphi Y) = g(X, Y) - \eta(X)\eta(Y),$$

where $g = i^*G$ is the pull back of G on M. As a consequence of the above formulas, we have $\eta(X) = g(Z, X)$ for all vector field X. Moreover, $\Phi(X, Y) = g(X, \varphi Y)$ is a skew symmetric two-form of maximum rank, its kernel is just spanned by the characteristic field Z, i.e., $i(Z)\Phi = 0$. Here, if X is a vector field and ω is a p-form, $i(X)\omega$ is the $(p-1)$-form defined by $(i(X)\omega)(Y_1, Y_2, \ldots Y_{p-1}) = \omega(X, Y_1, \ldots, Y_{p-1})$. The two-form Φ is called the fundamental two-form. We have that $\eta \wedge \Phi^n$ is a volume form on M, here dimension of M is $(2n+1)$.

The couple (η, Φ) is an almost contact structure [4]. If $\Phi = d\eta$, then the almost contact structure (η, Φ) is a contact structure and η is a contact form. The characteristic field Z is usually called the Reeb vector field in that case. Observe that the characteristic field Z satisfies the equation $i(Z)(\eta \wedge \Phi^n) = \Phi^n$ which in turn characterises it completely since $\eta \wedge \Phi^n$ is a volume-form.

REMARK. The characteristic field Z appears as the kernel of a two-form of maximal rank on the odd dimensional manifold M.

In general, the flow of Z does not preserve the fundamental one-form η, i.e., we do not have in general that $L_Z \eta = 0$. However when this happens, the flow of Z is geodesible: the flow lines appear to be geodesics (in some riemannian metric, actually in the very riemannian metric involved here). Indeed, according to Sullivan, Gluck-Ziller or Wadsley (see [18, Lemma 2]), a nowhere vanishing vector field Z is geodesible if and only if there is a one-form η such that $\eta(Z) > 0$, and $i(Z) d\eta = 0$. For our η and Z, we know that $\eta(Z) = 1$, and since $L_Z \eta = i(Z) d\eta + d i(Z)\eta = 0$, $i(Z) d\eta = 0$.

PROPOSITION 1. *Suppose for any vector field X on M with $g(X, Z) = 0$, we have $g([X, Z], Z) = 0$, then the characteristic foliation is geodesible.*

PROOF. We want to prove that $L_Z \eta = d i(Z)\eta = 0$. Since $(i(Z) d\eta)(Z) = 0$, it is enough to prove that $(i(Z) d\eta)(X) = 0$ for all vector fields X such that $g(X, Z) = 0$. Now

$$(i(Z) d\eta)(X) = (d\eta)(Z, X) = Z \cdot \eta(X) - X \cdot \eta(Z) - \eta([X, Z])$$
$$= Z \cdot g(X, Z) - X(1) - g(Z[X, Z]) = 0. \quad \square$$

In general the fundamental two-form Φ is not a closed form. However, when this happens, the flow of Z has recurrence properties, more precisely, if M is compact, and Φ closed, then any point is nonwandering, since we have the following

PROPOSITION 2. *If the fundamental two-form* Φ *is closed, then the characteristic field* Z *preserves the volume form* $\eta \wedge \Phi^n$.

PROOF. We know that $i(Z)(\eta \wedge \Phi^n) = \Phi^n$. Hence $d(i(Z)(\eta \wedge \Phi^n)) = d\Phi^n = 0$. Also $d(\eta \wedge \Phi^n) = 0$ for dimension reason. Thus

$$L_Z(\eta \wedge \Phi^n) = d(i(Z)(\eta \wedge \Phi^n)) = 0. \quad \square$$

Among natural questions in studying foliations are the existence of compact leaves, the holonomy of those leaves, their stability, riemannian problems like the geodesibility, the existence of bundle-like metrics etc. For instance, in our case the problem of existence of compact leaves is just to find periodic orbits for Z .

QUESTION. Find geometric, topological (or analytical) conditions on the submanifold M on the ambient manifold P, and the embedding $i: M \to P$ (the submanifold geometry) to ensure the existence of periodic orbits for the characteristic vector field Z .

In order to hope to have reasonable answers, we have to put some restrictions. Our first restriction is to assume that the ambient manifold P is a symplectic manifold.

3. Hypersurfaces in symplectic manifolds

A symplectic manifold is a smooth $2m$-dimensional manifold P equipped with a closed two-form Ω of maximum rank, this means that Ω^m is a volume-form, or equivalently, that the mapping $X \mapsto i(X)\Omega$, from the vector space of vector fields on P into the vector space of one-forms on P, is an isomorphism of vector spaces.

A symplectic manifold (P, Ω) carries (many) almost hermitian structures (G, J) said adapted to the symplectic structure Ω, meaning by that the following holds $\Omega(X, Y) = G(X, JY)$, for all vector fields X and Y (see for instance [19]).

Now let M be a $(2n+1)$-dimensional submanifold of a $(2n+2)$-dimensional symplectic manifold (P, Ω). Choosing and fixing an adapted almost hermitian structure (G, J), gives raise to the characteristic vector field Z and the almost contact structure (η, Φ).

A Kähler manifold is a manifold equipped with a riemannian metric G and a complex structure J such that $G \circ J = G$ and $\Omega(X, Y) = G(X, JY)$ is a symplectic form.

When we are given a hypersurface M in a Kähler manifold P we use the given structures G, J, Ω to construct canonically an almost contact structure (η, Φ). For instance any oriented hypersurface in $\mathbb{C}P^n$ carries a natural almost contact structure coming from the usual Kähler structure of $\mathbb{C}P^n$.

PROPOSITION 3. (i) $i^*\Omega = \Phi$.

(ii) *For each* $x \in M$ *the kernel of* $(i^*\Omega)(x)$ *is a one-dimensional vector space spanned by* $Z(x)$. *In short we say* $\ker(i^*\Omega) = Z$.

COROLLARY. *The characteristic foliation is independent of the choice of an adapted almost hermitian structure* (G, J).

The fundamental two-form Φ *is closed, hence the characteristic vector field* Z *preserves the volume form* $\eta \wedge \Phi^n$. *If* M *is compact, under the flow of* Z *any point is a nonwandering point.*

PROOF. We compute $i^*\Omega$: $(i^*\Omega)(X, Y) = \Omega(i_*X, i_*Y) = G(i_*X, Ji_*Y)$. Here as usual $i: M \to P$ is the embedding of M into P. Since $Ji_*Y = i_*\varphi Y + \eta(Y)C$,

$$(i^*\Omega)(X, Y) = G(i_*X, i_*\varphi Y) + \eta(Y)G(i_*X, C) = G(i_*X, i_*\varphi Y)$$
$$= (i^*G)(X, \varphi Y) = g(X, \varphi Y) = \Phi(X, Y). \quad \square$$

REMARK. Since $i^*\Omega = \Phi$ is a closed two-form, which when restricted to vectors normal to Z i.e., $\{X$ such that $g(X, Z) = 0\}$, is nondegenerate we say that $i^*\Omega$ is a transverse symplectic structure.

4. Connection with hamiltonian systems

A smooth function $H: P \to \mathbb{R}$ on a symplectic manifold (P, Ω) defines a vector field X_H, called the hamiltonian vector field with hamiltonian $H: X_H$ is the unique vector field corresponding to the one-form $dH: i(X_H)\Omega = dH$. The system of ordinary differential equations $\dot{x} = X_H$ are called the Hamilton equations. Since $X_H H = dH(X_H) = \Omega(X_H, X_H) = 0$, it follows that H is constant under the trajectories of X_H (this is called in mechanics the principle of conservation of energy).

Let (G, J) be an adapted almost hermitian structure. Using the metric G we define the gradient γH of H as the unique vector field such that $G(\gamma H, Y) = (dH)(Y)$. Since $G(Y, JX_H) = \Omega(Y, X_H) = -(i(X_H)\Omega)(Y) = -(dH)(Y) = -G(Y, \gamma H)$, for all Y we conclude that $JX_H = -\gamma H$, and finally that $X_H = J\gamma H$.

Let $a \in \mathbb{R}$ be a regular value of $H: P \to \mathbb{R}$ i.e., $M_a = H^{-1}(a)$ contains no critical points, or dH does not vanish on M_a. In that case, M_a is a smooth manifold of codimension one. Moreover, since H is constant on M_a, $(dH)(Y) = 0$ for any vector field Y on M_a, i.e., $G(\gamma HY) = 0$ for all vector field on M_a, which means that γH is normal to M_a. Since $G(\gamma H, J\gamma H) = \Omega(\gamma H, \gamma H) = 0$ we see that X_H is tangent to M_a. Now let M stand for M_a and $i: M \to P$ be the natural inclusion. Let Y be any tangent vector to M,

$$(i^*\Omega)(Y, X_H) = \Omega(i_*Y, i_*X_H) = \Omega(i_*Y, X_H)$$
$$= G(i_*Y, JX_H) = G(i_*Y, J^2\gamma H) = -G(i_*Y, \gamma H) = 0$$

since γH is normal to M. This shows that X_H is a span of the kernel of $i^*\Omega$, hence it is proportional to the characteristic vector field Z obtained as $i_*Z = -JC$, where C is the unit normal vector along M, here $C = \gamma H / \|\gamma H\|$, where $\|\gamma H\|^2 = G(\gamma H, \gamma H)$.

Hence finding compact leaves of the characteristic foliation is exactly the same thing as finding periodic solutions to the Hamilton equations on the "fixed energy" level surface M.

In the euclidean space $\mathbb{R}^{2n}$ with its symplectic form $\Omega = \sum_{i=1}^{n} dp_i \wedge dq_i$, where $z = (p_i, q_i)$ denote the coordinates, G is the standard inner product and $J(p, q) = (-q, p)$ is the usual complex structure of $\mathbb{C}^n = \mathbb{R}^{2n}$. The Hamilton equations have the familiar look $\dot{p} = -H_q$, $\dot{q} = H_p$.

The first general result on the existence of periodic solutions of the Hamilton equations on an energy level is due to Seifert [15], who in 1948 proved a theorem containing this statement: let $H(p, q) = \|p\|^2 + \|q\|^2 + V(q)$, where V is a smooth convex function, then on any regular energy surface, Hamilton equations have a periodic solution.

The next important results came in 1978. Weinstein [20] proved that if $H: \mathbb{R}^{2n} \to \mathbb{R}$ is a C^2 function, then X_H has a closed orbit on any convex regular energy hypersurface. About the same time, Rabinowitz [13] proved the same result assuming only that the regular energy hypersurface is starshaped.

The proofs of the two results are completely different. Weinstein uses generalisations of geometric methods of Seifert and Luisternik; Rabinowitz, instead, uses a new approach. The classical action-functional on loops of the calculus of variations was known for a long time, but it was notorious for its bad properties. Rabinowitz developed a daring method (the so-called "minimax-method") to get around the difficulties and found critical points of the action functional on loop spaces in $\mathbb{R}^{2n}$ corresponding to periodic orbits of X_H on a fixed energy level hypersurface which is assumed to be starshaped. This method and its variants/improvements by among others, Conley-Zehnder, Hofer, Ekeland, Viterbo, Floer etc., have led to spectacular results in the field. I will give a brief comment on these results in §6. The interested reader is urged to refer to the very interesting surveys by Viterbo [17] and by Rabinowitz [14].

5. Geometric approach

In the main results of Weinstein and Rabinowitz, the geometric hypothesis are clear: convex means that the curvature is positive, and starshaped means that the space is obtained from a sphere (positive curvature one) by radial diffeomorphisms. The latter is however not a "symplectic" condition; this is why Weinstein tried to find a symplectically invariant condition and was led to his conjecture, stated below in this section.

Here we are going to explore some more sufficient geometric conditions to guarantee the existence of closed characteristics.

Let M be a hypersurface in an almost hermitian manifold P with almost hermitian structure (G, J) and $g = i^*G$ the induced metric on M where $i: M \to P$ is the embedding of M in P. Let $\nabla, \overline{\nabla}$, be the Levi-Civita connections of G and g respectively. The Gauss and Weingarten equations for the hypersurface are

$$\nabla_X Y = \overline{\nabla}_X Y + h(X, Y)C, \qquad \nabla_X C = -AX$$

where $h(X, Y)$ is the second fundamental form and A is the corresponding Weingarten map defined by $g(AX, Y) = h(X, Y)$.

One of the results we can derive is the following

THEOREM 1. *Let M be a compact simply connected hypersurface in a Kähler manifold P. If the Weingarten map commutes with the fundamental tensor φ then the characteristic foliation has a compact leaf.*

The situation described in Theorem 1 occurs for instance when the hypersurface is quasi-umbilical, i.e., when A has the form $AX = aX + b\eta(X)Z$, where a, b are smooth functions and η is the fundamental one-form.

EXAMPLE: THE GENERALIZED CLIFFORD HYPERSURFACES. Let

$$M'_{p,n,r} = \{(z_1, \ldots, z_{n+1}) \in \mathbb{C}^{n+1} \mid |z_1|^2 + \cdots + |z_{n+1}|^2 = 1, \text{ and}$$
$$|z_1|^2 + \cdots + |z_p|^2 = r(|z_{p+1}|^2 + \cdots + |z_{n+1}|^2),$$
$$1 \le p \le n - 1, \ r > 0\}.$$

The submanifolds $M'_{p,n,r}$ of S^{2n+1} are diffeomorphic with $S^{2p-1} \times S^{2n-(2p-1)}$ and under the Hopf map project onto hypersurfaces $M_{p,n,r}$ of the projective space CP^n. These, so-called generalized Clifford hypersurfaces, are compact, and simply connected moreover their Weingarten map commute with the fundamental tensor, as it is shown by Okumura [12].

For more examples of existence of closed characteristics derived from Theorem 1, see [2].

PROOF. According to Okumura [12], the condition $A \circ \varphi = \varphi \circ A$ implies that Z is a Killing vector field. Hence it defines a one-dimensional riemannian foliation [9]. Moreover, we know already that the foliation has a transverse symplectic structure. Extending Attiyah-Guillemin-Sternberg convexity theorem of the momentum map, Molino [10] has proved that provided that M is simply connected, there exists a momentum map $I: M \to \mathbb{R}^p$, for some $p > 0$, which is constant on the leaves, and such that $I(M)$ is a closed convex polytope whose vertices correspond to compact leaves. Since $I(M)$ has one or more vertices, the foliation has at least one compact leaf. □

DEFINITION [21]. A hypersurface M in a symplectic manifold (P, Ω) is said to be a *hypersurface of contact type* if there exists a contact form α on M such that $i^*\Omega = d\alpha$.

Led by examples known that at times, where periodic orbits have been shown to exist by variational methods, Weinstein formulated the following

CONJECTURE (1978). *Any compact simply connected hypersurface of contact type in any symplectic manifold has at least one closed characteristic.*

Let M be any oriented hypersurface in a symplectic manifold (P, Ω). We saw in §3 that when we choose an almost an almost hermitian structure

(G, J) adapted to Ω, we get an almost contact structure (η, Φ) such that $i^*\Omega = \Phi$. If $d\eta = \Phi$, then η is a contact form and the hypersurface M is a hypersurface of a contact type. Following Okumura [11], we call this hypersurface a *contact hypersurface.*

THEOREM 2. *Any compact contact hypersurface in a Kähler manifold of constant positive holomorphic sectional curvature carries at least two closed characteristics.*

DEFINITION. Let (M, α) be a contact manifold with Reeb field Z, we say that α is an R-contact form (or (M, α) is an R-contact manifold) if there exists a riemannian metric g_T on the codimension one subbundle $\ker \alpha$ of TM, whose fibre over x is $\{X \in T_x M | (\alpha_x)(X) = 0\}$, such that $L_Y g_T = 0$ for all vector Y multiple of Z. In short, an R-contact form is a contact form such that its Reeb field generates a riemannian foliation. The metric g_T is called a transverse metric.

REMARK. If (M, α) is an R-contact manifold, with Reeb field Z and transverse metric g_T then we can define a genuine riemannian metric on M by setting

$$g(X, Y) = g_T(X - \alpha(X)Z, Y - \alpha(Y)Z) + \alpha(X)\alpha(Y)$$

and observe that Z is Killing for this metric. However, we did not require g to be a "contact metric" (see Blair [4]), hence α may not be a "K-contact" form. But of course, a K-contact form is an R-contact form. The difference between these two concepts has to be clarified, since there is a lot of confusion about them in the literature.

In [3], we gave an elementary proof of the following

THEOREM 3. *The Reeb field of any compact R-contact manifold has at least two closed orbits.*

SKETCH OF THE PROOF. We show that the R-contact condition implies that Z is almost periodic: there exists $T > 0$ such that if ψ_t is the flow of Z, then for every point $x \in M$, x and $\psi_T(x)$ are within the injectivity radius of M. We define a smooth function S by: $S(x)$ is the integral of the contact form α on the unique geodesic from x to $\psi_T(x)$. Let V be the displacement vector of ψ_T. It is shown that at a point $p \in M$ at which V_p is not vertical, i.e., proportional to Z, then $(d_p S)(v) \neq 0$, where $v = (\varphi V)_p$ and φ is the 1-1 tensor field in the almost contact structure defined by α. This says that at a critical point of S, either V is vertical or zero, which in both cases means that the orbit through p is closed. See [3] for more details. □

PROOF OF THEOREM 2. The calculations of [12] of the Weingarten map of a contact hypersurface in a Kähler manifold of constant positive holomorphic sectional curvature, show that the Reeb field Z is Killing. Hence the manifold is an R-contact manifold. In fact it is a K-contact manifold. Now apply Theorem 3 to conclude. □

REMARK. Assuming M simply connected we would deduct from Theorem 1, only the existence of one closed orbit. Observe that in Theorem 1, M is not supposed to be a hypersurface of contact type. The only things we need are that Z generate a riemannian foliation which has a transverse symplectic structure. The simply connectivity condition is essential.

6. Variational methods

I will comment very briefly on recent important results obtained by analytical methods. In September 1986, Viterbo [16] proved the Weinstein conjecture for compact hypersurfaces (not necessary simply connected) in $\mathbb{R}^{2n}$ with its standard symplectic form. The idea was to translate the problem into finding periodic orbits for a well choosen hamiltonian system then translate it into terms of finding critical points of some action-functional on an infinite dimensional space of loops in $\mathbb{R}^{2n}$. In October 1986, Hofer and Zehnder [7] simplified Viterbo's arguments and proved a surprising result, implying Viterbo's.

THEOREM. *Let M be a compact regular energy level of a function $H \colon \mathbb{R}^{2n} \to \mathbb{R}$, $M = H^{-1}(1)$. Then for all $\delta > 0$ there exist ε such that $0 \leq |\varepsilon| < \delta$ and $M_{\varepsilon} = H^{-1}(1 + \varepsilon)$ contains a periodic orbit $z_{\varepsilon} = (p_{\varepsilon}, q_{\varepsilon})$ of the hamiltonian vector field X_H. Moreover, if $T_{\varepsilon} > 0$ is the period of z_{ε} then $\exists \beta > 0$,*

$$0 < \int_0^{T_{\varepsilon}} (p_{\varepsilon} \dot{q}_{\varepsilon}) \, dt < \beta.$$

This *almost-existence* phenomenon (i.e., existence of a periodic orbit on a nearby hypersurface) seems to be a very deep and important surprise for the future of the theory. More recently, Floer, Hofer and Viterbo have obtained such almost existence for compact hypersurfaces in a symplectic manifold of the form $P \times C^n$, where P is a symplectic manifold with symplectic form Ω whose cohomology class vanishes on $\pi_2(P)$, C^n has its symplectic form as $\mathbb{R}^{2n}$, and $P \times C^n$ has the "product" symplectic form. Here a new ingredient is entered in the picture: the theory of pseudo-holomorphic curves of Gromov [8]. For surveys of these fascinating results, see [14, 17].

Acknowledgment

I would like to thank the Forschungsinstitut für Mathematik, ETH, Zürich, for its support and hospitality while the final version of this paper was written.

References

1. A. Banyaga, *A note on Weinstein's conjecture*, Proc. Amer. Math. Soc. **109** (1990), 855–858.
2. ____, *On characteristics of hypersurfaces in Kaehler manifolds*, preprint.
3. A. Banyaga and P. Rukimbira, *Weak stability of almost regular contact foliations* (to appear).

4. D. E. Blair, *Contact manifolds in Riemannian geometry*, Lecture Notes in Math., vol. 509, Springer-Verlag,

5. D. E. Blair and D. D. Showers, *Almost contact manifolds with Killing structure tensors*, J. Differential Geom. **9** (1974), 577–582.

6. V. L. Ginzburg, *New generalizations of Poincaré's geometric theorem*, Funct. Anal. Appl. **21** (1987), 100–109.

7. H. Hofer and E. Zehnder, *Periodic solutions on hypersurfaces and a result of Viterbo*, Invent. Math. **90** (1987), 1–9.

8. H. Hofer, A. Floer, and C. Viterbo, *The proof of Weinstein conjecture in $P \times C$* , Math. Z. **203** (1990), 469–482.

9. P. Molino, *Riemannian foliations* Progress in Math. Birkhäuser, Boston, MA, 1984.

10. ____, *Reduction symplectique et feuilletages riemanniens, moment structural et theorème de convexite*, Séminaire Gaston Darboux de Géometrie et Topologie differentielle, 1987–1988 pp. 11–25; University of Montpellier, France.

11. M. Okumura, *Contact hypersurfaces in certain Kaehler manifolds*, Tôhoku Math. J. **18** (1966), 74–102.

12. ____, *On some real hypersurfaces of complex projective spaces*, Trans. Amer. Math. Soc. **212** (1975), 355–364.

13. P. H. Rabinowitz, *Periodic solutions of Hamiltonian systems*, Comm. Pure Appl. Math. **108** (1978), 507–518.

14. ____, *The prescribed energy problem for periodic solutions of hamiltonian systems*, Contemporary Math. **81** (1988), 183–191.

15. H. Seifert, *Periodische Bewegungen mechanischen Systeme*, Math. Z. **51** (1948), 337–357.

16. C. Viterbo, *A proof of Weinstein' conjecture in $\mathbb{R}^{2n}$* , Ann. Inst. Henri Poincaré **4** (1987), 337–356.

17. ____, *Recent progress in periodic orbits of autonomous hamiltonian systems and applications to symplectic geometry*, Nonlinear Functional Analysis, (P. S. Milojevic, ed.), Lectures in Pure and Applied Mathematics Series, Marcel Dekker Inc., NY, no. 121, 1990, pp. 227–250.

18. A. V. Wadsley, *Geodesic foliation by circles*, J. Differential Geom. **10** (1975), 541–549.

19. A. Weinstein, *Lectures on symplectic manifolds*; CBMS Regional Conf. Ser. in Math., vol. 53, Amer. Math. Soc., Providence, RI, 1970.

20. ____, *Periodic orbits for convex hamiltonians systems*, Ann. of Math. (2) **108** (1978), 507–518.

21. ____, *On the hypothesis of Rabinowitz' periodic orbit theorem*, J. Differential Equations **33** (1978), 353–358.

22. K. Yano and M. Kon, *CR submanifolds of Kaehlerian and Sasakian manifolds*, Progr. Math. **30** (1983), Birkhäuser Boston, MA.

PENNSYLVANIA STATE UNIVERSITY

Proceedings of Symposia in Pure Mathematics
Volume **54** (1993), Part 2

Disprisoning and Pseudoconvex Manifolds

JOHN K. BEEM

I. Introduction

A common assumption in the study of Riemannian (i.e., positive definite) manifolds is that the space be complete. This assumption is quite reasonable for positive definite manifolds and has several well-known implications for the geodesic systems of these spaces. For the investigation of manifolds with a linear connection and for the study of semi-Riemannian manifolds (i.e., manifolds with a nondegenerate but not necessarily positive definite metric tensor), the situation is somewhat different. In this more general setting, one uses geodesic completeness since there is usually no induced distance function with the usual distance properties. Here a geodesic is said to be *complete* if the domain of the geodesic is all of the real line. In this paper, geodesics will always be assumed to have been extended to their maximal domain. One difficulty with the assumption of geodesic completeness for spaces which are not Riemannian is that there exist a large number of important examples which fail to be geodesically complete. In particular, a number of important examples of exact solutions to Einstein's equations in General Relativity fail to be geodesically complete. Thus, completeness is a more restrictive assumption for the semi-Riemannian (or linear connection) case as compared to the Riemannian case. Another problem is that for manifolds which are not positive definite, the completeness assumption yields less control on the geodesic system. Thus, one faces the problem of finding reasonable alternatives to use either in place of geodesic completeness or in conjunction with this assumption.

A semi-Riemannian manifold (M, g) with a metric g of signature $(-, +, \ldots, +)$ is *Lorentzian* [**3**, **16**, **20**, **25**]. For these manifolds one has a causality condition known as global hyperbolicity [**16**]. This condition is

1991 *Mathematics Subject Classification.* Primary 53C50, 53C80, 53C22.
Supported in part by NSF grant DMS-8803511.
This paper is in final form and no version of it will be submitted for publication elsewhere.

a type of "internal" completeness. Unfortunately, this important tool does not exist for semi-Riemannian spaces which have more general metric signatures. In this paper, we will consider the implications for geodesic systems of two assumptions which are known as disprisonment and pseudoconvexity. The disprisoning property corresponds to real principal type used in the study of pseudodifferential equations [13, 22, 23]. Pseudoconvexity has been used in pseudodifferential equations as a condition on the bicharacteristics [13]. These conditions have also been useful in the investigation of semi-Riemannian manifolds [1, 2, 5, 6, 7, 8, 9, 10, 12, 19].

Let (M, ∇) be a smooth n-dimensional manifold M with a linear connection ∇ and corresponding geodesics. The inextendible geodesics are said to form a *disprisoning* system if each end of each geodesics leaves every compact subset K of M. More precisely, (M, ∇) is disprisoning if for each inextendible geodesic $\gamma:(a, b) \to M$ and any compact set K there are parameter values t_1 and t_2 with $a < t_1 < t_2 < b$ such that $\gamma(t) \in M - K$ for all $a < t < t_1$ and all $t_2 < t < b$. The geodesic system is *pseudoconvex*, if for each compact subset K of M, there is a second compact set H such that each geodesic segment with endpoints in K lies in H. These two properties have important implications for the geodesic structure. For example, the classical Hadamard-Cartan theorem may be extended to (M, ∇) when these properties hold and there are no conjugate points. One may make the geodesics of (M, ∇) into a topological space $G(M)$ in a natural way. Assuming that the original geodesics satisfy a certain nonreturning property, then $G(M)$ is a $(2n - 2)$-dimensional manifold iff (M, ∇) is pseudoconvex. When (M, g) is a semi-Riemannian manifold, then ∇ is the usual Levi-Civita connection and one may restrict these two conditions to certain classes of geodesics. Thus, one may just require these two properties hold for causal geodesics in the case of Lorentzian signature. If this is done, then one obtains a generalization of global hyperbolicity. One application in this setting is to show that completeness for causal geodesics is C^1-fine stable when the Lorentzian manifold is disprisoning and pseudoconvex for the class of causal geodesics. A corollary is that this same result holds for all globally hyperbolic Lorentzian manifolds.

2. Lorentzian manifolds and space-times

Let (M, g) be a Loretzian manifold. It is a *space-time*, if (M, g) may be *time oriented*. In other words, if there is a smooth vector field X on M such that X is everywhere *timelike* (i.e., $g(X, X) < 0$ for all points of M). A smooth curve $c:(a, b) \to M$ is *causal* if $g(c', c') \leq 0$ for all t. It is *future directed* if $g(c', X) < 0$ for all t in (a, b). The *chronological future* $I^+(p)$ of a point p consists of all points q in M such that there is a future directed timelike curve from p to q. The chronological past $I^-(p)$ of p is defined as all q such that $p \in I^+(q)$. The *causal future* $J^+(p)$ of p consists of p and of all points q in M such that there is some future directed causal curve from p to q. If $q \in J^+(p)$, then we say p is in the *causal past* of q and write $p \in J^-(q)$. Given any fixed point p, then both the chronological future and past of p are open sets. In general, the causal future and causal past of p need be neither open nor closed [3]. The space-time (M, g) is said to be *strongly causal* if, given any point p and any neighborhood $U(p)$, there is some smaller neighborhood $V(p)$ contained in $U(p)$ such that any causal curve which leaves $U(p)$ fails to ever return to $V(p)$ (see [16]). A strongly causal space-time is *globally hyperbolic* if for all $p, q \in M$ the set $J^+(p) \cap J^-(q)$ is compact. In a globally hyperbolic space-time one may always obtain a maximal length timelike geodesic joining two causally related points (see Seifert [21]). The causal curves are the ones of primary interest to physicists since they correspond to paths of particles moving at or below the speed of light. One may restrict the notions of disprisoning and pseudoconvexity to just the causal geodesics. Thus, one says that the space-time (M, g) is *causally disprisoning* if, for each inextendible causal geodesic $\gamma: (a, b) \to M$ and any compact set K, there are parameter values t_1 and t_2 with $a < t_1 < t_2 < b$ such that $\gamma(t) \in M - K$ for all $a < t < t_1$ and all $t_2 < t < b$. In other words, a causally disprisoning space-time is one that has no causal geodesic partially trapped in any compact set. The geodesic system is *causally pseudoconvex*, if for each compact subset K of M, there is a second compact set H such that each causal geodesic segment with endpoints in K lies in H. We now show that global hyperbolicity implies both causally pseudoconvex and causally disprisoning.

PROPOSITION 1. *If (M, g) is a globally hyperbolic space-time, then (M, g) is causally disprisoning and causally pseudoconvex.*

PROOF. Since (M, g) is strongly causal, there are no compact sets with partially trapped causal curves [16, p. 195]; consequently, no compact set has any partially trapped causal geodesic. Thus, (M, g) is causally disprisoning. To show that (M, g) is causally pseudoconvex, let K be a given compact subset of M. Cover K with open sets of the form $I^+(p)$ and take a finite subcover corresponding to $I^+(p_1), \ldots, I^+(p_k)$. Next take an open cover of K with open sets of the form $I^-(q)$ and then take a finite subcover $I^-(q_1), \ldots, I^-(q_r)$. Let H be the union of all sets of the form $J^+(p_i) \cap J^-(q_j)$ where $1 \leq i \leq k$ and $1 \leq j \leq r$. Since (M, g) is globally hyperbolic, each of the sets $J^+(p_i) \cap J^-(q_j)$ is compact. Using the fact that there are only a finite number of sets of the indicated form, it follows that H is compact. Given any causal geodesic segment $\gamma: [a, b] \to M$ with $\gamma(a) \in K$ and $\gamma(b) \in K$, let i and j be chosen such that $\gamma(a) \in I^+(p_i)$ and $\gamma(b) \in I^-(q_j)$. Then the image $\gamma[a, b]$ lies in $J^+(p_i) \cap J^-(q_j)$ and hence also in H, as desired. $\square$

Given M, let Semi(M) be the collection of all semi-Riemannian metrics on M. One defines the C^r-fine Whitney topologies on Semi(M) using a fixed countable covering $\{B_i\}$ of M by open sets where each B_i has compact closure lying in a coordinate chart of M and where the closures of the sets B_i form a locally finite cover of M. Assume $\varepsilon: M \to (0, +\infty)$ is a continuous function. Let $g, h \in \text{Semi}(M)$; then $|g - h|_r < \varepsilon$ if, for each $p \in M$, all of the corresponding coefficients and their mixed derivatives up to order r of the two metric tensors g and h are $\varepsilon(p)$-close at p when calculated in the coordinates of all B_i which contain p. The sets $U(g, \varepsilon) = \{h \in \text{Semi}(M) \mid |g - h|_r < \varepsilon\}$ for arbitrary $g \in \text{Semi}(M)$ and continuous $\varepsilon: M \to (0, +\infty)$ form a basis for the C^r-fine topology on $\text{Semi}(M)$. One defines a property to be C^r-fine stable if it holds for a subset of $\text{Semi}(M)$ which is open in the C^r-fine topology. It is clear from the definition that if r and s are nonnegative integers with $r < s$, then the C^s-fine topology is finer than the C^r topology. Thus, properties which are C^r-fine stable are also C^s-fine stable for all s larger than r. One may clearly restrict the C^r-fine topologies to certain subsets of $\text{Semi}(M)$. For example, one can use these topologies on the space $\text{Riem}(M)$ of all positive definite Riemannian metrics on M or one can use these topologies on the space $\text{Lor}(M)$ for all Lorentzian metrics on M. The following result of Beem and Parker [7] shows that if one considers the two properties of causal disprisonment and causal pseudoconvexity jointly, then they are stable in the C^1-fine topology.

THEOREM 2. *Let (M, g) be a Lorentzian manifold. If (M, g) is both causally disprisoning and causally pseudoconvex, then there is a C^1-fine open neighborhood $U(g)$ of g in $\text{Lor}(M)$ such that each $h \in U(g)$ is both causally disprisoning and causally pseudoconvex.*

3. Geodesic connectedness

Let (M, ∇) be a smooth connected manifold with linear connection ∇. The manifold (M, ∇) is said to be *geodesically connected* if for each pair of distinct points p, q of M, there is at least one geodesic segment from p to q. For positive definite Riemannian manifolds, the Hopf-Rinow Theorem guarantees that Cauchy completeness is equivalent to geodesic completeness and that either of these two properties implies geodesic connectedness. On the other hand, for semi-Riemannian manifolds the condition of geodesic completeness fails to yield geodesic connectedness (see O'Neill [20, p. 105]). In fact, it is quite common for geodesically complete Lorentzian manifolds to fail to be geodesically connected. Furthermore, one may construct compact and analytic Lorentzian manifolds which fail to be geodesically connected.

EXAMPLE 3. Let $M = S^1 \times S^1 = \{(t, s) \mid -4\pi \leq t \leq 4\pi$ and $0 \leq s \leq 1\}$ using the obvious identifications. Define the Lorentzian metric g on M by $g = (-\cos t)\, dt^2 + (2 \sin t)\, dt\, ds + (\cos t)\, ds^2$. It is easy to show that the vector field $X = \cos(t/2)\, \partial/\partial t - \sin(t/2)\, \partial/\partial s$ is timelike and hence time orients (M, g). It is not hard to check that all geodesics which start on the circle

$t = 0$ lie in the set $\{(t, s) \mid -5\pi/2 \leq t \leq 7\pi/2\}$. It follows that (M, g) is not geodesically connected.

The next result of Beem and Parker [9] is a generalization of the Hadamard-Cartan theorem [17]. Related results for future cones of space-times have been obtained by Flaherty [14, 15].

THEOREM 4. *Let (M, ∇) be both disprisoning and pseudoconvex. If (M, ∇) has no conjugate points, then (M, ∇) is geodesically connected. Furthermore, M is diffeomorphic to $\mathbf{R}^n$ and the exponential map at each point is a diffeomorphism from its domain onto M.*

4. Stability of geodesic completeness

Geodesic completeness is not a stable property for general space-times as is shown by the following example of Williams [24]. Let $M = S^1 \times S^1 = \{(x, y) \mid 0 \leq x \leq 2\pi$ and $0 \leq y \leq 2\pi\}$ using the obvious identifications. This manifold may be given the complete flat metric $g = dx\, dy$. it may also be given the geodesically incomplete metrics $g_n = dx\, dy + [\sin(x)/n]\, dy^2$. For all of these metrics, the circle $S\{(0, y) \mid 0 \leq y \leq 2\pi\}$ represents the image of a null geodesic. For the original flat metric this null geodesic is complete; however, for each of the metrics g_n, this image S represents an incomplete geodesic. For the g_n metrics, the tangent vector to this null geodesic does not return to itself after tracing out S. Then tangent vector returns to a multiple $(\neq 1)$ of its original value. It follows that S is incomplete for each g_n. On the other hand, the sequence of metrics $\{g_n\}$ converges to g in each of the Whitney C^r-fine topologies. Notice also that all of the metrics involved are analytic.

Beem and Ehrlich [5] have shown that causal geodesics completeness is stable for space-times which satisfy causal pseudoconvexity and causal disprisonment. It should be noted that these sufficient conditions for the stability of geodesic completeness are not necessary conditions. Recall that $\text{Lor}(M)$ denotes the set of all Lorentzian metrics on M.

THEOREM 5. *Let (M, g) be a space-time which is both causally pseudoconvex and causally disprisoning. If all causal geodesics of (M, g) are complete, then there is a fine C^1 neighborhood $U(g)$ of g in $\text{Lor}(M)$ such that each metric h in $U(g)$ is causally geodesically complete.*

COROLLARY 6. *Let (M, g) be a globally hyperbolic space-time. If (M, g) is causally geodesically complete, then there is a fine C^1 neighborhood $U(g)$ of g in $\text{Lor}(M)$ such that each metric h in $U(g)$ is causally geodesically complete.*

On the other hand, very little is known about the stability of geodesic incompleteness. The standard "big bang" models used in cosmology are examples of Robertson-Walker space-times. They are warped products of the form $(a, b) \times_f H$ where H is a positive definite Riemannian space of constant sectional curvature and the open interval (a, b) has the negative definite metric $-dt^2$. The warping function is generally chosen such that the resulting space-time is incomplete because of physical considerations (see [16]). Beem and Ehrlich [4] have shown that for this special class of space-times, one does have stability of incompleteness. The methods used rely very heavily on the symmetry of these space-times and it is not clear how these results may be generalized. Examples of Williams [24] show that for some space-times geodesic incompleteness is not stable. An interesting open problem is to find reasonable sufficient conditions for the stability of geodesic incompleteness for semi-Riemannian manifolds.

The stability of completeness and incompleteness for positive definite Riemannian manifolds is much easier. In the positive definite case one has the induced distance function $d: M \times M \to \mathbf{R}^1$ and the Hopf-Rinow Theorem guarantees that (M, g) is geodesically complete iff the corresponding distance function is Cauchy complete. The next result shows that one has stability of both completeness and incompleteness using the C^0-fine Whitney topology on the space of all Riemannian metrics on M. Recall that $\text{Riem}(M)$ represents the collection of all positive definite Riemannian metrics on M and that C^0-fine stable implies C^r-fine stable for all $r \geq 0$.

PROPOSITION 7. *Let (M, g) be a connected positive definite Riemannian manifold. There is a C^0-fine open neighborhood $U(g)$ of g in $\text{Riem}(M)$ such that either all $h \in U(g)$ are complete or else all $h \in U(g)$ are incomplete.*

PROOF. If M is compact, then all metrics in $\text{Riem}(M)$ are complete and the result follows using $U(g) = \text{Riem}(M)$. Assume M is not compact and choose any constant $0 < \alpha < 1$. Let $\{C_i\}$ be an expanding sequence of compact sets with $M = \bigcup C_i$. For each i, let $\varepsilon_i > 0$ be chosen such that if $|g - h|_0 < \varepsilon_i$ on $C_i - C_{i-1}$, then for each nontrivial tangent vector $X \in T(C_i - C_{i-1})$ one has

$$\alpha^2 h(X, X) < g(X, X) < \alpha^{-2} h(X, X).$$

Now choose any positive valued continuous function $\varepsilon: M \to \mathbf{R}^1$ with $0 < \varepsilon(p) < \varepsilon_i$ for all $p \in C_i - C_{i-1}$. Let $U(g) = \{h \in \text{Riem}(M) \mid |g - h|_0 < \varepsilon\}$. If $c: [a, b] \to M$ is any smooth curve in M, then the length $L_h(c)$ of c with respect to h is related to the length $L_g(c)$ of c with respect to g by $\alpha L_h(c) < L_g(c) < \alpha^{-1} L_h(c)$. It follows that the distance functions d_h and d_g are related by $\alpha d_h(p, q) < d_g(p, q) < \alpha^{-1} d_h(p, q)$. Thus, (M, g) is complete iff (M, h) is complete whenever $h \in U(g)$, as required. $\square$

5. The space of geodesics

Let (M, ∇) be a connected n-dimensional manifold with a linear connection ∇. An equivalence relation may be defined on the set of all inextendible geodesics of (M, ∇) by letting two geodesics be equivalent if they differ by a reparametrization. Two equivalent geodesics are then parameterizations of the same point set of M Let $[\gamma]$ denote the equivalence class of the geodesic γ and let $G(M)$ be the set of all equivalence classes of geodesics of (M, ∇). The space $G(M)$ of geodesics has been an important tool in the study of manifolds with closed geodesics of the same length (see Besse [11]). Low [18] has considered this construction for the space of all null geodesics of a Lorentzian manifold. He has also considered the space of causal curves and used pseudoconvexity [19]. One difference that should be mentioned in this paper is that we require manifolds to be Hausdorff and Low does not.

One may consider $G(M)$ to be a quotient space of the reduced tangent bundle $T'M$ (i.e., the tangent bundle less the zero vectors) with corresponding map $\pi: T'M \to G(M)$. The map π induces the quotient topology on $G(M)$. The map π is always an open map and never a closed map. Examples show that $G(M)$ need not be T_0. The space $G(M)$ is T_1 iff each geodesic has a closed image.

A sequence of geodesics $\{\gamma_n\}$ *converges tangentially* to the geodesic γ: $(a, b) \to M$ if there is some t_0 with $a < t_0 < b$ and a sequence of geodesics $\{\beta_n\}$ with $\beta_n \in [\gamma_n]$ for each n such that $\beta'_n(t_0) \to \gamma'(t_0)$. In other words, at some point $\gamma(t_0)$ of γ the original geodesics are converging to $\gamma(t_0)$ and their tangents are converging to the direction of the tangent to γ at that point. If $\{\gamma_n\}$ converges tangentially to the geodesic γ, then $\beta'_n(t) \to \gamma'(t)$ for all t in the domain (a, b) of γ. It follows that the image of γ lies in the Hausdorff lower limit of the images of the β_n (or γ_n). Note that the set of t such that $\beta_n(t)$ converges need not be a connected set. It turns out that tangential convergence is equivalent to convergence in the topology of $G(M)$. In general, a sequence of geodesics which converges tangentially to a geodesic may have images in M which fail to have a Hausdorff limit. On the other hand, the Hausdorff lower limit of these images will always exist and contain the image of the limit geodesic. The space $G(M)$ is a Hausdorff topological space iff the Hausdorff limit of the images of each tangentially convergent sequence of geodesics exists and is equal to the image of the limit geodesic, see Beem and Parker [10]. If $G(M)$ is Hausdorff, then for each fixed p in M the set of points which may be joined to p is a closed subset of M. When M has no conjugate points and $G(M)$ is Hausdorff, then each pair of points of M is joined by at least one geodesic [10]. Thus, (M, ∇) is geodesically connected. If each point of M has arbitrarily small neighborhoods such that every geodesic leaving such a neighborhood fails to return, then we say that the *nonreturning property* is satisfied. If (M, ∇) satisfies the nonreturning property, the following three conditions are equivalent: (1) $G(M)$ is Hausdorff; (2) $G(M)$ is a $(2n - 2)$-dimensional manifold; (3) (M, ∇) is pseudoconvex.

Acknowledgment

The author wishes to thank P. E. Ehrlich, R. J. Low and P. E. Parker for some helpful discussions.

REFERENCES

1. D. Allison, *Pseudoconvexity of Lorentzian doubly warped products*, Preprint.
2. J. K. Beem, *Pseudoconvexity and Lorentzian geometry*, Colloquia Mathematica Societatis János Bolyai, Proc. Differential Geom. Conf. 1989, 87–91.
3. J. K. Beem and P. E. Ehrlich, *Global Lorentzian geometry*, Marcel Dekker, New York, 1981.
4. ____, *Stability of geodesic incompleteness for Robertson-Walker space-times*, Gen. Relativity Gravitation **13** (1981), 239–255.
5. ____, *Geodesic completeness and stability*, Math. Proc. Cambridge Philos. Soc. **102** (1987), 319–328.
6. J. K. Beem and P. E. Parker, *Klein-Gordon solvability and the geometry of geodesics*, Pacific J. Math. **107** (1983), 1–14.
7. ____, *Whitney stability of solvability*, Pacific J. Math. **116** (1985), 11–23.
8. ____, *Pseudoconvexity and general relativity*, J. Geom. Phys. **4** (1987), 71–80.
9. ____, *Pseudoconvexity and geodesic connectedness*, Ann. Mat. Pura Appl. (4) **155** (1989), 137–142.
10. ____, *The space of geodesics*, Geometriae Dedicat38 (191991), 87–99.
11. A. L. Besse, *Manifolds all of whose geodesics are closed*, Springer-Verlag, Berlin, 1978.
12. L. Del Riego and P. E. Parker, *Pseudoconvex and disprisoning sprays*, Preprint.
13. J. J. Duistermaat and L. Hörmander, *Fourier integral operators*, II, Acta. Math. **128** (1972), 183–269.
14. F. Flaherty, *Lorentzian manifolds of nonpositive curvature*, Proc. Sympos. Pure Math., vol. 27, part 2, Amer. Math. Soc., Providence, RI, 1975, pp. 395–399.
15. ____, *Lorentzian manifolds of nonpositive curvature. II*, Proc. Amer. Math. Soc. **48** (1975), 199–202.
16. S. W. Hawking and G. F. R. Ellis, *The large scale structure of space-time*, Cambridge Univ. Press, New York, 1973.
17. S. Kobayashi, *Riemannian manifolds with conjugate points*, Ann. Math. Pura Appl. **53** (1961), 149–155.
18. R. J. Low, *The geometry of the space of null geodesics*, J. Math. Phys. **30** (1989), 809–811.
19. ____, *Spaces of causal paths and naked singularities*, Classical Quantum Gravity **7** (1990), 943–954.
20. B. A. O'Neil, *Semi-Riemannian Geometry*, Academic Press, New York, 1983.
21. H.-J. Seifert, *Global connectivity by timelike geodesics*, Z. Naturforsch. A **22** (1967), 1356–1360.
22. M. Taylor *Pseudodifferential operators*, Princeton Univ. Press, Princeton, NJ, 1981.
23. F. Treves, *Introduction to pseudodifferential and Fourier integral operators*. vol. 1: *Pseudodifferential operators*, Plenum Press, New York, 1980.
24. P. M. Williams, *Instability of geodesic completeness and incompleteness*, Preprint, Dept. of Math., Univ. of Lancaster, 1985.
25. J. Wolf, *Spaces of constant curvature*, 5th ed., Publish or Perish, Berkeley, CA, 1984.

UNIVERSITY OF MISSOURI

Proceedings of Symposia in Pure Mathematics
Volume **54** (1993), Part 2

On the Holonomy of Lorentzian Manifolds

L. BERARD BERGERY AND A. IKEMAKHEN

1. Introduction

The notion of the holonomy group of a Riemannian manifold was introduced by E. Cartan around 1920 in [**CN1**] and used in his study of Riemannian symmetric spaces [**CN2**]. The holonomy groups of Riemannian products were characterized locally by A. Borel and A. Lichnérowicz in [**B-L**] and globally by G. de Rham in [**DR**]. Then, in 1955, M. Berger showed in [**BR1**] that for irreducible nonlocally symmetric Riemannian manifolds, their (restricted) holonomy groups have to belong to a very short list. And recently, in [**BT**], R. Bryant was able to produce examples of manifolds whose holonomy groups are precisely the last two "exceptions" of Berger's list, so all the possible (restricted) holonomy groups are known in the Riemannian case. We will recall that briefly in §2.

In the pseudo-Riemannian case, the holonomy group is defined exactly in the same way than in the Riemannian case, but the corresponding classification problem is more difficult. We will recall below the main results of M. Berger [**BR1, BR2, WU1, WU2**]. But the corresponding "list" of all possible (restricted) holonomy groups is not known up to now. The main difference between the Riemannian case and the pseudo-Riemannian case is the possible existence (in the later case) of holonomy groups whose holonomy representations are not irreducible, but are not decomposable as a direct sum of pseudo-Riemannian holonomy representations. This was already pointed out by H. Wu, who exhibited many examples and counter-examples, showing the difficulty of the problem in the general case. Notice that even the classification of pseudo-Riemannian symmetric spaces is not known, except in the irreducible case M. Berger [**BR2**], in the Lorentzian case M. Cahen and

1991 *Mathematics Subject Classification.* Primary 53B30.

Parts of this paper were contained in the thesis of the second author, prepared at the University of Nancy 1, France, under the guidance of the first author.

This paper is in final form, and no version of it will be submitted for publication elsewhere.

N. R. Wallach [C-W], or (almost completely) in the case of signature $(2, m - 2)$ M. Cahen and M. Parker [C-P].

In this paper, we present some progress toward a classification in the Lorentzian case [i.e., signature $(1, m - 1)$]. In §3, we give in Theorem I the complete list of connected subgroups of the orthogonal group $O(1, m - 1)$ which are not irreducible, but do not leave invariant any nondegenerate proper subspace. This is not sufficient to give the solution to our holonomy problem, but restricts the possibilities. We give also in §4 a further obstruction, which is described in Theorem II. Up to now, we know no other obstruction. In §5, we indicate some examples; in particular we show explicit examples whose restricted holonomy groups are not closed (Theorem III).

2. Holonomy groups

We recall briefly the main definitions and theorems, and refer to classical textbooks like [BE, Chapter 10 or K-N tome 1, Chapter IV] for more details.

On any pseudo-Riemannian manifold (M, h) with dimension m and signature (p, q) [where p is the number of $+$, and $q = m - p$ the number of $-$], there is a unique torsion-free metric connection D, called the Levi-Civita connection, and it gives rise to a parallel transport τ along each curve. Given a point x in M, the holonomy group H_x is the subgroup of $O(T_xM, h_x)$ generated by parallel transports along all the closed curves starting at x. Since H_x is given as a linear group, we will often speak of the "holonomy representation" instead of the holonomy group. Now the restricted holonomy group is the subgroup H_x^0 of H_x generated by parallel transports along closed curves at x which are null-homotopic. Then H_x^0 is an arcwise-connected subgroup of $O(T_xM, h_x)$; in particular, it is a connected Lie subgroup of the connected component $SO_0(T_xM, h_x)$ of the identity. A theorem by W. Ambrose and I. M. Singer [A-S] asserts that the Lie algebra $\mathcal{H}$ of H_x^0 (and H_x) is generated by all $\tau_c^{-1} \circ R_y(X, Y) \circ \tau_c$, where c is any curve starting at x, the point y is the end of c, X, Y are any vectors in T_yM, and R is a curvature tensor of (M, h). Also, if c is any curve in M as above, then the holonomy group at the end point y is related to the holonomy group at the origin x by $H_y = \tau_c \circ H_x \circ \tau_c^{-1}$, and $H_y^0 = \tau_c \circ H_x^0 \circ \tau_c^{-1}$. In particular, if M is connected, the holonomy representations at the points of M are all isomorphic, and we may speak of "the" holonomy representation of M, up to conjugacy.

If (M, h) and (M', h') are two pseudo-Riemannian manifolds with signature (p, q) and (p', q') respectively, and holonomy groups H_x and $H_{x'}$ respectively at x of M and x' of M', then the product $(M \times M', h \oplus h')$ is a pseudo-Riemannian product with signature $(p + p', q + q')$. Now the holonomy group of $M \times M'$ at (x, x') is the product $H_x \times H_{x'}$, more precisely, in the canonical identification $T_{(x, x')}(M \times M') = T_xM \oplus T_{x'}M'$, then $H_{(x, x')} = (H_x \times \{1\}) \oplus (\{1\} \times H_{x'})$, that is the holonomy representation of

$M \times M'$ is the exterior direct sum of the holonomy representations of M and M'. In order to state the converse of that result, we set a few definitions.

DEFINITIONS 1. A (connected) pseudo-Riemannian manifold (M, h) is called

irreducible if its holonomy representation is irreducible;

indecomposable if its holonomy group H_x does not leave invariant any *nondegenerate* proper subspace of $T_x M$;

strictly irreducible if its restricted holonomy group H_x^0 is irreducible;

strictly indecomposable if its restricted holonomy group H_x^0 does not leave invariant any

nondegenerate proper subspace of $T_x M$.

These notions do not depend on the particular point x of M where the holonomy is computed. Of course, if (M, h) is strict irreducible (resp. strictly indecomposable), it is irreducible (resp. indecomposable). We recall also that the holonomy group of the universal covering of (M, h) coincides with the restricted holonomy group of (M, h).

A theorem by A. Borel and A. Lichnérowicz, generalized by H. Wu, reduces the study of "decomposable" holonomy groups to strictly indecomposable ones, at least locally.

THEOREM A (A. Borel and A. Lichnérowicz [**B-L**], Wu [**Wu1**]). *Let* (M, h) *be a connected pseudo-Riemannian manifold with restricted manifold holonomy group* H_x^0 *at the point* x *of* M. *Let* $T_x M = E_0 \oplus E_1 \oplus \cdots \oplus E_r$ *be an orthogonal decomposition into* H_x^0-*invariant nondegenerate subspaces, such that the holonomy representation of* H_x^0 *is trivial in* E_0, *and indecomposable in each* E_i *(*$i \geq 1$ *[i.e., does not contain any nondegenerate invariant proper subspace]. Then there exists* $r + 1$ *(immersed) submanifolds* $N_0, N_1, \ldots, N_r$ *of* M *such that a neighborhood of* x *in* M *is isometric to a product* $U_0 \times U_1 \times \cdots \times U_r$ *where* U_i *is an open subset of* N_i *with the induced metric,* $T_x N_i = E_i$, *the manifold* N_0 *is flat, each manifold* N_i *(*$i \geq 1$*) is strictly indecomposable, and the holonomy representation of* H_x^0 *is the exterior direct sum representation of the trivial representation in* E_0 *and the holonomy representations of the restricted holonomy groups of the manifolds* N_i *for* $i \geq 1$.

This local decomposition into a product is even a global one in the simply-connected complete case; this is the well-known de Rham's theorem, generalized by H. Wu to the pseudo-Riemannian case.

THEOREM B (G. de Rham [**DR**], H. Wu [**WU1**]). *Let* (M, h) *be a simply-connected complete pseudo-Riemannian manifold. Then* (M, h) *is isometric to a product of simply-connected complete* **indecomposable** *pseudo-Riemannian manifolds. More precisely, if* $T_x M = E_0 \oplus E_1 \oplus \cdots \oplus E_r$ *is an orthogonal decomposition of* $T_x M$ *in nondegenerate subspaces such that the holonomy representation of* $H_x = H_x^0$ *is trivial in* E_0, *and indecomposable in each* E_i

$(i \geq 1)$, *then there exists* r *simply-connected complete* **indecomposable** *pseudo-Riemannian manifolds* (N_i, h_i) $i = 1, \ldots, r$, [*with* (N_i, h_i) *different from* $(\mathbb{R}, \pm dt^2)$], *such that* (M, h) *is isometric to the pseudo-Riemannian product* $(\mathbb{R}^a, h_0) \times (N_1, h_1) \times \cdots \times (N_r, h)_r)$, [*where* a *is the dimension of* E_0 *and* h_0 *is a flat canonical pseudo-Riemannian metric*], *in such a way that* E_i *corresponds to the tangent space to* N_i (*for any* $i \geq 1$), *and that* E_0 *corresponds to* $\mathbb{R}^a$.

The list of all possible restricted holonomy groups of **strictly irreducible** pseudo-Riemannian manifolds was established by M. Berger in [**BR1, BR2**]. More precisely, Berger proved the following:

THEOREM C (M. Berger [**Br1**]). *Let* (M, h) *be a connected* **strictly irreducible** *pseudo-Riemannian manifold which is not locally symmetric. Then its restricted holonomy group* H_x^0 *is, up to conjugacy, one of the following*:

$SO_0(p, q)$ *with signature* (p, q),

$U(p, q)$ *or* $SU(p, q)$ *with signature* $(2p, 2q)$,

$Sp(1)Sp(p, q)$ *or* $Sp(p, q)$ *with signature* $(4p, 4q)$,

$SO(n, \mathbb{C})$ *with signature* (n, n),

$SO(n, \mathbb{H})$ *with signature* $(2n, 2n)$,

$G_2^{\mathbb{C}}, G_2$ *and* G_2^2 *with signature respectively* $(7, 7), (7, 0)$ *and* $(4, 3)$,

$Spin(7, \mathbb{C}), Spin(7), Spin(4, 3)$ *with signature respectively* $(8, 8), (8, 0)$ *and* $(4, 4)$.

As a matter of fact, the original list of [**BR1**] contains also $Spin(9)$ and two other similar groups. $Spin(9)$ was ruled out by D. V. Alekseevskii[**AI**], but it is possible to rule out these groups even by Berger's methods (see [**BT**, p. 535]).

For all those groups except $SO(n, \mathbb{H})$, explicit examples of metrics with the corresponding holonomy group are known; see in particular [**BT**], where R. Bryant solved the cases of G_2 and $Spin(7)$.

In the Riemannian case, the indecomposable (respectively strictly indecomposable) manifolds are irreducible (respectively strictly irreducible), and the decomposition into a product for a decomposable one gives rise only to Riemannian manifolds. Moreover, the Riemannian symmetric spaces were classified by E. Cartan in [**CN2**] and their holonomy coincides with the isotropy representation. Hence the problem of the determination of all possible restricted holonomy groups is solved in the Riemannian case. See [**BE**, Chapter 10] for more details and the discussion of what is known for the holonomy group.

For the pseudo-Riemannian case, the **irreducible** symmetric spaces were classified by M. Berger in [**BR2**]. [We will not reproduce here this long list, and we refer to [**C-L**] for the computation of the signatures in Berger's list.] Note that indecomposable pseudo-Riemannian spaces are not classified in the general case, except for Lorentzian ones (see below) and (almost completely) for the case of signature $(2, m - 2)$ M. Cahen and M. Parker [C-P].

In the Lorentzian case, we deduce from Theorem A that a Lorentzian manifold is locally the product of an indecomposable Lorentzian manifold (eventually with dimension 1) and s strictly irreducible manifolds $(N_i, -g_i)$, where g_i is a Riemannian metric. So we are reduced to the indecomposable case. The list of possible **irreducible** restricted holonomy representations may be extracted from Theorem C and Berger's general classification [**BR2**] for the symmetric case; we reproduce it here for completeness.

COROLLARY (TO BERGER'S THEOREMS). *Let (M, h) be a strictly irreducible Lorentzian manifold. Then either (M, h) is "trivial," [i.e., it is locally $(\mathbb{R}, -dt^2)$] or its restricted holonomy group at x is exactly $\mathrm{SO}_0(T_x M, h_x)$. In particular, if (M, h) is a strictly Lorentzian locally symmetric space, it is trivial or has nonzero constant curvature.*

REMARKS. We do not know any direct proof of this corollary. It would be very nice to have a direct proof that $\mathrm{SO}_0(1, m - 1)$ is the only possible irreducible restricted holonomy group in the nonsymmetric case [compare with the Riemannian case, where J. Simons gave in [**SS**] a direct proof that the restricted holonomy group has to be transitive on the sphere, which gives Berger's list up a few details]. Also, a direct proof that an irreducible Lorentzian symmetric space has constant curvature would be interesting.

There is a complete classification of Lorentzian symmetric spaces, due to M. Cahen and N. R. Wallach [**C-W**]. We reproduce it for completeness.

We give first an explicit construction of some of them.

Let $(\mathbb{R}^n, \langle \cdot, \cdot \rangle)$, $n \geq 2$, be the canonical Euclidean space and $\varphi : \mathbb{R}^n \to \mathbb{R}^n$ any symmetric invertible linear map. We consider the solvable Lie algebra $\mathfrak{G} = \mathbb{R}^{2n+2} = \mathbb{R}^n \times \mathbb{R}^n \times \mathbb{R} \times \mathbb{R}$, with the bracket

$$[(x, y, t, u), (x', y', t', u')]$$
$$= (u\varphi(y') - u'\varphi(y), ux' - u'x, \langle x, \varphi(y') \rangle - \langle x', \varphi(y) \rangle, 0).$$

Then $\mathscr{K} = \{(x, 0, 0, 0)\}$ is an abelian subalgebra of $\mathfrak{G}$ and $\mathscr{P} = \{(0, y, t, u)\}$ is an $\mathrm{ad}(\mathscr{K})$-invariant complement of $\mathscr{K}$ in $\mathfrak{G}$. The symmetric bilinear form

$$\eta((0, y, t, u), (0, y', t', u')) = tu' + t'u - \langle y, y' \rangle$$

is nondegenerate with signature $(1, n + 1)$ and $\mathrm{ad}(\mathscr{K})$-invariant. We denote by G the simply connected solvable subgroup generated by $\mathfrak{G}$ and by K the closed subgroup corresponding to $\mathscr{K}$. Then η induces a G-invariant Lorentzian metric h on $M = G/K$ and we have

THEOREM D (M. Cahen and N. R. Wallach [**C-W**]. *The Lorentzian manifolds (M, h) defined above are simply connected, indecomposable, nonstrictly irreducible, Lorentzian symmetric spaces with dimension $m = n + 2$, and any strictly indecomposable, nonstrictly irreducible, locally symmetric Lorentzian manifold is locally isometric to one of these.*

Notice that, in those examples, the group K is precisely the holonomy group at the origin, and that G is the group generated by the transvections. The isometry group may be larger, and depends on φ. Finally, any two such examples are isometric if and only if the corresponding maps φ are conjugate by orthogonal maps [so if we want uniqueness, we may choose the φ's to be diagonal, with nonzero diagonal elements $\lambda_1 \geq \lambda_2 \geq \cdots \geq \lambda_n$].

3. A problem in linear algebra

We are left with strictly indecomposable, nonstrictly irreducible, nonlocally symmetric Lorentzian manifolds, and their possible holonomy representations. In order to restrict the problem, we will consider first an induced problem in linear algebra.

If a Lorentzian manifold (M, h) is strictly indecomposable but not strictly irreducible, then the restricted holonomy representation H_x^0 leaves invariant to nondegenerate proper subspace of $T_x M$, but leaves invariant at least one **degenerate** subspace E of $T_x M$. Then H_x^0 leaves also invariant the orthogonal complement $E^\perp$ and the intersection $E \cap E^\perp$, which is a one-dimensional isotropic subspace. Hence H_x^0 is a connected subgroup of $SO_0(T_x M, h_x)$ which leaves invariant a null direction, but no nondegenerate invariant subspace. We classify all the possibilities for such a subgroup.

We consider on $\mathbb{R}^m$ the symmetric 2-form η with signature $(1, m-1)$ defined by

$$\eta((x_0, x_1, \ldots, x_{n+1}), (y_0, y_1, \ldots, y_{n+1})) = x_0 y_{n+1} + y_0 x_{n+1} - \sum_{i=1}^{n} x_i y_i,$$

where $n = m - 2$.

If $m = 2$, then the connected component $SO_0(\eta)$ in the orthogonal group $O(\eta)$ is not irreducible, since it leaves invariant any null direction. On the other hand, $SO_0(\eta)$ is one dimensional. Hence we will consider only the case $m \geq 3$.

If $e_0, \ldots, e_{n+1}$ is the canonical basis of $\mathbb{R}^m$, then the line R generated by e_0 is a null direction and its orthogonal $F^\perp$ is the hyperplane generated by $e_0, e_1, \ldots, e_n$. We denote by G the connected subgroup of $SO_0(\eta)$ generated by all the maps which have invariant the line F. Then the Lie algebra $\mathfrak{G}$ of G may be identified with the subalgebra of matrices in the following "block" form

$$\begin{pmatrix} a & {}^t X & 0 \\ 0 & A & X \\ 0 & 0 & -a \end{pmatrix},$$

where a is in $\mathbb{R}$, X in $\mathbb{R}^n$ is viewed as a $(1, n)$ column matrix, ${}^t X$ is the $(n, 1)$ line matrix transposed of X, and A is a (n, n) alternate matrix, i.e. ${}^t A + A = 0$ or A is in the Lie algebra $so(n)$.

In order to simplify the notations, we identify $\mathfrak{G}$ with $\mathbb{R} \oplus so(n) \oplus \mathbb{R}^n$ and we denote the above element by (a, A, X). Notice that $\mathcal{N} = \{(0, 0, X), X \in \mathbb{R}^n\}$ is an abelian ideal of $\mathfrak{G}$, that $\mathcal{K} = \{(0, A, 0), A \in so(n)\}$ is a subalgebra isomorphic to $so(n)$, and that $\mathcal{A} = \{(a, 0, 0), a \in \mathbb{R}\}$ commutes with $\mathcal{K}$. One sees easily that $\mathcal{N}$ and $\mathcal{A} \oplus \mathcal{N}$ leave invariant any subspace F_1 such that $F \subset F_1 \subset F^\perp$, and only these ones. Notice that those F_1 are the only degenerate subspaces containing F. Obviously, any subalgebra $\mathcal{H}$ such that $\mathcal{N} \subset \mathcal{H} \subset \mathfrak{G}$ leaves invariant F and does not leave invariant any nondegenerate subspace, but there are other algebras with the same property, at least if $m \geq 5$.

THEOREM I. *Let $\mathcal{H}$ be the Lie algebra of a connected subgroup H of $SO_0(\eta)$ which leaves invariant $F = \mathbb{R}e_0$ and does not leave invariant any nondegenerate proper subspace of $\mathbb{R}^m$. Then $\mathcal{H}$ is in $\mathfrak{G}$ and*

either $\mathcal{H}$ contains $\mathcal{N}$,

or *there exists a nontrivial decomposition $n = p + q$ and $\mathbb{R}^n = \mathbb{R}^p \oplus \mathbb{R}^q$, a nontrivial abelian subalgebra $\mathcal{T}$ of $so(p)$, a semisimple subalgebra $\mathcal{D}$ of $so(p)$ (eventually 0), commuting with $\mathcal{T}$ and a **surjective** linear map $\varphi: \mathcal{T} \to \mathbb{R}^q$ such that, up to conjugacy in $\mathfrak{G}$, $\mathcal{H}$ is the subalgebra of $\mathfrak{G}$ of the following "block" form*

$$\left\{ \begin{pmatrix} 0 & {}^tX & {}^t\varphi(A) & 0 \\ 0 & A+B & 0 & X \\ 0 & 0 & 0 & \varphi(A) \\ 0 & 0 & 0 & 0 \end{pmatrix} \text{ for all } A \text{ in } \mathcal{T}, \text{ all } B \text{ in } \mathcal{D} \text{ and all } X \text{ in } \mathbb{R}^p \right\}.$$

PROOF. We suppose that $\mathcal{H}$ is as in the hypothesis and does not contain $\mathcal{N}$. We denote $\mathcal{P} = \mathcal{H} \cap \mathcal{N}$, and $\mathcal{Q}$ the orthogonal complement of $\mathcal{P}$ in $\mathcal{N}$ (for the canonical scalar product in the identification of $\mathcal{N}$ with $\mathbb{R}^n$). Then $\mathcal{P}$ is an abelian ideal of $\mathcal{H}$ and $\mathcal{Q}$ is nonzero. We denote $\mathcal{L} = \{(a, A, X) \in \mathcal{H} \text{ such that } X \in \mathcal{Q}\}$, and $\mathcal{L}_1$ the projection of $\mathcal{L}$ and $\mathcal{A} \oplus \mathcal{K}$. By definition, there exists a linear map $\psi: \mathcal{L} \to \mathcal{N}$ such that $\mathcal{L} = \{(a, A, \psi(a, A)) \text{ such that } (a, A) \in \mathcal{L}_1\}$. Obviously, $\mathcal{H} = \mathcal{P} \oplus \mathcal{L}$. Moreover

LEMMA 1. *$\mathcal{L}$ is a subalgebra of $\mathcal{H}$.*

PROOF OF LEMMA 1. For any $(0, 0, Y)$ in $\mathcal{P}$ and any (a, A, X) in $\mathcal{L}$, we have $[(a, A, X), (0, 0, Y)] = (0, 0, AY)$. Notice that the action of $\mathcal{K}$ on $\mathcal{N}$ with the bracket is the same that the action of $\mathcal{K}$ on $\mathbb{R}^n$, and in the following, we identify $\mathcal{N}$ and $\mathbb{R}^n$. Hence $A\mathcal{P} \subset \mathcal{P}$. Since A is alternate, we deduce that $A\mathcal{Q} \subset \mathcal{Q}$. Now, if (a', A', X') is also in $\mathcal{L}$, we have $[(a, A, X), (a', A', X')] = (0, [A, A'], (aI+A)X' - (a'I+A')X)$, and $(aI + A)X'$ is in $\mathcal{Q}$.

From the proof of the lemma, we deduce moreover that ψ satisfies:

(1) $\psi(0, [A, A']) = (aI + A)\psi(a', A') - (a'I + A')\psi(a, A)$,

for any (a, A) and (a', A') in $\mathcal{L}_1$.

Also, if $\mathcal{K}_1$ is the subalgebra of $\mathcal{K}$ which leaves invariant $\mathcal{P}$ and annihilates $\mathcal{Q}$, and similarly, $\mathcal{K}_2$ is the subalgebra of $\mathcal{K}$ which leaves $\mathcal{Q}$ and annihilates $\mathcal{P}$, then $\mathcal{L}_1 \subset \mathcal{A} \oplus \mathcal{K}_1 \oplus \mathcal{K}_2$.

In order to prove the theorem, we consider first the case where $\mathcal{L}_1$ is not included in $\mathcal{K}$, that is there exists $(1, A_0, X_0)$ in $\mathcal{L}$. Since A_0 is alternate, $I + A_0$ is invertible. We set $X_1 = (I + A_0)^{-1} X_0$, and for any $(0, A)$ in $\mathcal{B} = \mathcal{L}_1 \cap \mathcal{K}$, we define $\theta(A) = \psi(0, A) - A X_1$.

We compute (1) for $(1, A_0)$ and $(0, A)$ and get

$$\theta([A_0, A]) + A_0 A X_1 - A A_0 X_1 = (I + A_0)\theta(A) + A X_1 + A_0 A X_1 - A X_0,$$

so

$$\theta([A_0, A]) = (I + A_0)\theta(A), \quad \text{and} \quad \theta((\operatorname{ad} A_0)^r A) = (I + A_0)^r \theta(A),$$

hence for any polynomial P in one variable, we get $\theta(P(\operatorname{ad} A_0)A) = P(I + A_0)\theta(A)$.

We choose for P the characteristic polynomial of $\operatorname{ad}(A_0)$. Then the first term is zero. Now the spectras of A_0 and $\operatorname{ad}(A_0)$ are purely imaginary, hence $P(I + A_0)$ is invertible, and $\theta(A) = 0$. Let E be the subspace of $\mathbb{R}^m = \mathbb{R} \oplus \mathbb{R}^n \oplus \mathbb{R}$ defined by $E = \{(-{}^t X_1 Z, Z, 0) \text{ for any } Z \text{ in } \mathcal{Q}\}$. One sees easily that E is a nontrivial nondegenerate subspace of $\mathbb{R}^m$. Now, for any Z in $\mathcal{Q}$, we have $(1, A_0, X_0)(-{}^t X_1 Z, Z, 0) = (-{}^t X_1 Z + {}^t X_0 Z, A_0 Z, 0) = (-{}^t X_1 A_0 Z, A_0 Z, 0) \in E$, then

$$(0, A, A X_1)(-{}^t X_1 Z, Z, 0) = ({}^t(A X_1)Z, A Z, 0)$$
$$= (-{}^t X_1 A Z, A Z, 0) \in E$$

for any $(0, A)$ in $\mathcal{B}$, and finally $(0, 0, Y)(-{}^t X_1 Z, Z, 0) = ({}^t Y Z, Z, 0) = (0, 0, 0)$ for any Y in $\mathcal{P}$. Hence $\mathcal{K}$ leaves invariant the nondegenerate proper subspace E.

We are left with the case where $\mathcal{L}_1$ is in $\mathcal{K}$, and more precisely in $\mathcal{K}_1 \oplus \mathcal{K}_2$. We define $\varphi: \mathcal{L}_1 \to \mathcal{Q} \subset \mathbb{R}^n$ by $\varphi(A) = \psi(0, A)$. Then (1) becomes

$$(2) \qquad\qquad \varphi([A, A']) = A\varphi(A') - A'\varphi(A).$$

LEMMA 2. $\mathcal{L}_1$ is in $\mathcal{K}_1$.

PROOF OF LEMMA 2. We define $\mathcal{C}_1 = \mathcal{L}_1 \cap \mathcal{K}_1$. This is an ideal of $\mathcal{L}_1$, and the orthogonal $\mathcal{C}_2$ of $\mathcal{C}_1$ in $\mathcal{L}_1$ (for the Killing form of $\mathcal{K}$) is also an ideal. We suppose that $\mathcal{C}_2 \neq 0$, and we will build a nondegenerate $\mathcal{K}$-invariant proper subspace of $\mathbb{R}^m$. We consider the projection $\mathcal{C}_3$ of $\mathcal{C}_2$ into $\mathcal{K}_2$. We denote by E_1 the intersection of all the kernels of the elements B_1 of $\mathcal{K}_2$, considered as operating in $\mathcal{Q}$, and by E_2 the orthogonal of E_1 in $\mathcal{G}$. Notice that E_2 is nonzero. For any element V in E_2, we may consider the subspace $E = \{({}^t V Z, Z, 0) \text{ for any } Z \text{ in } E_2\}$ of $\mathbb{R}^m$, and we will choose

V such that E is invariant by $\mathcal{H} = \mathcal{P} \oplus \mathcal{C}_1 \oplus \mathcal{C}_2$. First, any $(0, 0, Y)$ in $\mathcal{P} \subset \mathcal{H}$ sends E to 0.

If $(0, A, \varphi(A))$ is in $\mathcal{C}_1$ and $(0, B, \varphi(B)$ is in $\mathcal{C}_2$, then $[A, B] = 0$ since $\mathcal{C}_1$ and $\mathcal{C}_2$ are mutually commuting ideals, $A\varphi(B) = 0$ since $\varphi(B)$ is in $\mathcal{Q}$ and A in $\mathcal{K}_1$, so by (2) $B\varphi(A) = 0$. Since $\varphi(A)$ is in $\mathcal{Q}$, this means that $\varphi(A)$ is in the kernel of the projection B_1 of B in $\mathcal{K}_3$, so that $\varphi(A)$ is in E_1. Now $(0, A, \varphi(A))$ sends $({}^tVZ, Z, 0)$ of E to 0, since ${}^t\varphi(A)Z = 0$ and $AZ = 0$.

Finally, if $(0, B, \varphi(B))$ is in $\mathcal{C}_2$, it sends $({}^tVZ, Z, 0)$ of E to $({}^t\varphi(B)Z, B_1Z, 0)$, where B_1 is the projection of B in $\mathcal{K}_3$. Since this projection is an isomorphism between $\mathcal{C}_2$ and $\mathcal{C}_3$, we set $\theta(B - 1) = \varphi(B)$, and (2) gives, for any B_1 and B_1' in $\mathcal{C}_3$:

$$(3) \qquad \theta([B_1, B_1']) = B_1\theta(B_1') - B_1'\theta(B_1).$$

Let C be the Casimir element of the representation of $\mathcal{C}_3$ in E_2, for the invariant scalar product on $\mathcal{C}_3$ induced by the Killing form of $\mathcal{K}$. That is $C = \sum_i D_i^2$, where D_i is an orthonormal basis of $\mathcal{C}_3$ for that scalar product. One knows that C commutes with $\mathcal{C}_3$, and we remark that C is invertible since E_2 is the orthogonal of the intersection of the kernels of the D_i's. Now

$$\begin{aligned} C\varphi(B_1) &= \sum_i D_i^2\varphi(B_1) = \sum_i D_iD_i\varphi(B_1) \\ &= \sum_i D_i(\varphi([D_i, B_1]) + B_1\varphi(D_i)) \\ &= \sum_i D_i(\varphi([D_i, B_1]) + [D_i, B_1]\varphi(D_i) + B_1D_i\varphi(D_i)) \\ &= \varphi\left(\sum_i [D_i, [D_i, B_1]]\right) + B_1\left(\sum_i D_i\varphi(D_i)\right). \end{aligned}$$

Since C commutes with B_1, the first term is zero, and $\varphi(B_1) = B_1C^{-1}(\sum_i D_i\varphi(D_i))$. Now we choose $V = -C^{-1}(\sum_i D_i\varphi(D_i))$ and we get that ${}^t\varphi(B_1)Z = -{}^tV{}^tB_1Z = {}^tVB_1Z$. Hence E is $\mathcal{H}$-invariant. By contradiction, $\mathcal{C}_2$ has to be 0 and $\mathcal{L}_1$ is in $\mathcal{K}_1$.

END OF THE PROOF OF THE THEOREM. Now $\mathcal{H} = \{(0, A, Y + \varphi(A))$ for any A in $\mathcal{L}_1$ and Y in $\mathcal{P}\}$. We denote by E the orthogonal of the image of φ in $\mathcal{Q} \subset \mathbb{R}^n$. Then $\mathcal{H}E = 0$. Since E is nondegenerate, we deduce that φ must be surjective.

Applying (1) once again, we get $\varphi([A, A']) = 0$, for any A and A' in $\mathcal{L}_1$, and the derived Lie algebra of $\mathcal{L}_1$ is in $\text{Ker}\,\varphi$.

Finally, up to conjugacy, we may change coordinates in $\mathbb{R}^n = \mathbb{R}^p \oplus \mathbb{R}^q$ such that $\mathcal{P} = \mathbb{R}^p$ and $\mathcal{Q} = \mathbb{R}^q$, and we get $\mathcal{H}$ as in the theorem, where $\mathcal{T}$ is the center and $\mathcal{D}$ the derived algebra of the Lie algebra $\mathcal{L}_1$.

Conversely, let E be a $\mathscr{H}$-invariant nondegenerate subspace of $\mathbb{R}^m$. If E contains an element $(\alpha, \beta, 1)$ in $\mathbb{R}^m = \mathbb{R} \oplus \mathbb{R}^n \oplus \mathbb{R}$, then by applying twice an element $(0, 0, Y)$ of $\mathscr{P}$ with ${}^t YY = 1$ we get e_0, so E contains e_0 and the orthogonal of E for η is still nondegenerate, and is contained in $\mathbb{R} \oplus \mathbb{R}^n$. So we may consider only the case where $E = \{({}^t VZ, Z, 0) \in \mathbb{R} \oplus \mathbb{R}^n \oplus \mathbb{R}$, where Z is in some subspace E_1 of $\mathbb{R}^n\}$.

If we apply $(0, 0, Y)$ of $\mathscr{P}$ to $({}^t VZ, Z, 0)$, we get $({}^t YZ, 0, 0)$ proportional to e_0, so we must have E_1 in $\mathscr{Q}$. If we apply $(0, A, \varphi(a))$ of $\mathscr{L}$ to $({}^t VZ, Z, 0)$, we get $({}^t \varphi(A)Z, 0, 0)$ since A is in $\mathscr{H}_1$. But φ is surjective onto $\mathscr{Q}$, hence $Z = 0$ and there is no nondegenerate proper subspace. And Theorem I is proved.

4. Another obstruction

As before, let H_x^0 be the restricted holonomy group at x of a Lorentzian manifold (M, h), which is strictly indecomposable and leaves invariant a null line F. If we identify $(T_x M, h_x)$ with $(\mathbb{R}^m, \eta)$ in such a way that F becomes $\mathbb{R}e_0$, then the Lie algebra of H_x^0 is identified with a $\mathscr{H}$ as in Theorem I. But there is a further obstruction on $\mathscr{H}$. An example showing such an obstruction is given in [WU2, Example 4, p. 354]. Then a general formulation is the following:

THEOREM II. *Let H_x^0 be the restricted holonomy group at x of a strictly indecomposable Lorentzian manifold (M, h), and suppose that H_x^0 leaves invariant a null line F. Let $\mathscr{H}$ be the Lie algebra of H_x^0, where we identify $T_x M$ with $(\mathbb{R}^m, \eta)$ in such a way that F is identified with $\mathbb{R}e_0$. Then the projection $\mathscr{B}$ of $\mathscr{H}$ into $\mathscr{K}$ with respect to $\mathscr{A} \oplus \mathscr{N}$ satisfies the following property:*

There exists an orthogonal decomposition $\mathbb{R}^n = E_0 \oplus E_1 \oplus \cdots \oplus E_r$ and a corresponding decomposition into mutually commuting ideals $\mathscr{B} = \mathscr{B}_1 \oplus \mathscr{B}_2 \oplus \cdots \oplus \mathscr{B}_r$ such that $\mathscr{B}_i(E_j) = 0$ if $i \neq j$, $\mathscr{B}_i(E_i) = E_i$, $\mathscr{B}_i \subset so(E_i)$ and $\mathscr{B}_i$ is irreducible on E_i, i.e. the representation of $\mathscr{B}$ in $\mathbb{R}^n$ is the exterior direct sum representation of a trivial representation and r irreducible representations.

In some sense, $\mathscr{B}$ satisfies a "Borel-Lichnérowicz" type property, as in the Riemannian case.

PROOF. Suppose that $\mathscr{B}$ leaves invariant a decomposition $\mathbb{R}^n = \mathbb{R}^p \oplus \mathbb{R}^q$, with $p + q = n$. As before, we denote by $\mathscr{H}_1$ the subalgebra of $\mathscr{H}$ which leaves invariant $\mathbb{R}^p$ and annihilates $\mathbb{R}^q$, and similarly, we denote by $\mathscr{H}_2$ the subalgebra of $\mathscr{H}$ which leaves invariant $\mathbb{R}^q$ and annihilates $\mathbb{R}^p$. Let $\mathscr{B}_1 = \mathscr{B} \cap \mathscr{H}_1$ and $\mathscr{B}_2 = \mathscr{B} \cap \mathscr{H}_2$. We will prove that $\mathscr{B} = \mathscr{B}_1 \oplus \mathscr{B}_2$. More precisely, we denote by $\mathscr{H}_3$ the subalgebra of $\mathscr{H}$ whose projection on $\mathscr{K}$ is $\mathscr{B}_1 \oplus \mathscr{B}_2$. We will prove that $\mathscr{H}_3 = \mathscr{H}$.

We look first to the curvature R of (M, h) at $T_x M$, with the identification of $T_x M$ to $\mathbb{R}^m = \mathbb{R} \oplus \mathbb{R}^p \oplus \mathbb{R}^q \oplus \mathbb{R}$ as before. For any X in $\mathbb{R}^p$ and Y

in $\mathbb{R}^q$, and any U, V in $\mathbb{R}^m$, we have by hypothesis $\langle R(U, V)X, Y\rangle = 0$, since $R(U, V)$ is in $\mathscr{H}$. Hence $R(X, Y) = 0$. Now, if Y' is in $\mathbb{R}^q$, $\langle R(X, U)Y, Y'\rangle = -\langle R(U, Y)X, Y'\rangle - \langle R(Y, X)U, Y'\rangle = 0$, and similarly, if X' is in $\mathbb{R}^p$, then $\langle R(Y, U)X, X'\rangle = 0$. Now $R(U, V)e_0 \in \mathbb{R}e_0$ hence $R(e_0, X) = R(e_0, Y) = 0$ and $R(e_0, e_{n+1})$ is in $\mathscr{A} \oplus \mathscr{N}$. All this together implies that $R(U, V)$ belongs to $\mathscr{H}_3$ for any U and V.

We consider now the subspaces $E = \mathbb{R}e_0 \oplus \mathbb{R}^p$ and $E^\perp = \mathbb{R}e_0 \oplus \mathbb{R}^q$. They are $\mathscr{H}$-invariant, so there exists two parallel distributions of spaces on M (more precisely on the universal covering) generated by parallel transport of E and $E^\perp$. And the same calculation as above may be done on the curvature at each point. Using Ambrose-Singer Theorem, we deduce that $\mathscr{H} = \mathscr{H}_3$.

The proof of the theorem follows easily through coordinate changes and induction.

5. Some examples

Up to now, we do not know any other obstruction for a group to be the restricted holonomy group of a strictly indecomposable, nonstrictly irreducible, Lorentzian manifold. We will give below some examples.

Let $\mathscr{H}$ be a Lie subalgebra of $\mathscr{G}$ satisfying the hypothesis of both Theorems I and II.

We may distinguish four cases:

TYPE 1: $\mathscr{H}$ CONTAINS $\mathscr{A}$. Then $\mathscr{H} = \mathscr{A} \oplus \mathscr{B} \oplus \mathscr{N}$, where $\mathscr{B}$ is a subalgebra of $\mathscr{K}$ satisfying the conclusion of Theorem 2. Such examples appear already in dimension $m = 2$, where $\mathscr{K} = \mathscr{N} = 0$ (in dimension 3, we have $\mathscr{K} = 0$).

The connected subgroup A of $G \subset SO_0(1, m-1)$ generated by $\mathscr{A}$ is

$$A = \left\{ \begin{pmatrix} a & 0 & 0 \\ 0 & \mathrm{Id} & 0 \\ 0 & 0 & a^{-1} \end{pmatrix}, \quad \text{with } a > 0 \right\}$$

and is a closed one-dimensional subgroup. Similarly, the connected subgroup N of G generated by $\mathscr{N}$ is

$$N = \left\{ \begin{pmatrix} 1 & {}^t X & \frac{1}{2}{}^t XX \\ 0 & \mathrm{Id} & X \\ 0 & 0 & 1 \end{pmatrix}, \quad \text{with } X \text{ in } \mathbb{R}^n \right\}$$

and is a closed abelian subgroup, isomorphic to $\mathbb{R}^n$. The subgroup K generated by $\mathscr{K}$ is compact and isomorphic to $SO(n)$. Now the conclusion of Theorem II implies that the subgroup B of K generated by $\mathscr{B}$ is a product of irreducible subgroups, and it is known that such irreducible subgroups are compact. Hence B is compact, and the connected subgroup H generated by $\mathscr{H}$ is closed. So the restricted holonomy groups of type 1 are closed.

Suppose that B is the holonomy group of a Riemannian metric g on $\mathbb{R}^n$. Then the Lorentzian metric $h = 2dx_0 dx_{n+1} - g + kdx_{n+1}^2$ has $H = ABN$ as (restricted) holonomy group as soon as k is sufficiently generic (for example $k = \sum_{i=0}^{n} x_i^2$). It is also possible to construct examples where B is not the holonomy group of a Riemannian metric (see [**IN**]).

TYPE 2. $\mathscr{H}$ CONTAINS $\mathscr{N}$ AND IS CONTAINED IN $\mathscr{K} \oplus \mathscr{N}$. Then $\mathscr{H} = \mathscr{B} \oplus \mathscr{N}$, where $\mathscr{B}$ is a subalgebra of $\mathscr{K}$ satisfying the conclusion of Theorem II. Such examples appear also in dimension $m = 2$ $(\mathscr{K} = 0)$, 3 $(\mathscr{K} = 0)$ or more. Here again, the subgroup B generated by $\mathscr{B}$ is compact, and the connected subgroup H generated by $\mathscr{H}$ is closed. So the restricted holonomy groups of type 2 are closed.

As examples, we may take the same as above, but with k independent of x_0. Here again, there are examples where B is not the holonomy group of a Riemannian metric.

We recall that the restricted holonomy group of the strictly indecomposable, nonstrictly irreducible, Lorentzian locally symmetric spaces is N. But N is also the restricted holonomy group of nonlocally symmetric Lorentzian manifolds. It suffices to take k sufficiently generic.

TYPE 3. $\mathscr{H}$ CONTAINS $\mathscr{N}$ BUT IS NOT OF TYPE 1 OR 2. Let $\mathscr{B}$ be the projection of $\mathscr{H}$ into $\mathscr{K}$, which satisfies the conclusion of Theorem II. Then $\mathscr{B} = \mathscr{T} \oplus \mathscr{D}$, where $\mathscr{T}$ is abelian and $\mathscr{D}$ is semisimple. Now there exists a surjective linear map $\varphi: \mathscr{T} \to \mathscr{A}$ such that $\mathscr{H} = \mathscr{L} \oplus \mathscr{L} \oplus \mathscr{N}$, where $\mathscr{L}$ is the subalgebra of $\mathscr{A} \oplus \mathscr{T}$ defined by $\mathscr{L} = \{(\varphi(A), A) \in \mathscr{A} \oplus \mathscr{T}$, for any A in $\mathscr{T}\}$. Such examples appear in dimension 4 (with $\mathscr{D} = 0$) or more, and φ is injective in dimension 4 and 5. Here, the subgroup generated by B is still compact, and may be written $B = TD$, where T is a compact torus generated by $\mathscr{T}$ and D is a compact semisimple Lie group generated by $\mathscr{D}$. Now the subgroup H generated by $\mathscr{H}$ may be written $H = LDN$, where L is the subgroup generated by $\mathscr{L}$. But this subgroup L of AT is not necessarily closed, since its intersection with T is not necessarily compact. More precisely, L is closed if and only if the subalgebra $\mathscr{T}_1 = \operatorname{Ker} \varphi$ of $\mathscr{T}$ generates a compact Lie subgroup of T. And H is closed if and only if L is closed. Notice that if φ is injective, H is closed, so the first nonclosed example appears only in dimension 6.

We consider the Lorentzian metric

$$h = 2d_0 dx_5 - dx_1^2 - dx_2^2 - dx_3^2 - dx_4^2 + 2x_1 x_2^2 dx_1 dx_5 - 2x_1^2 x_2 dx_2 dx_5$$
$$+ 2x_3 x_4^2 dx_3 dx_5 - 2x_3^2 x_4 dx_4 dx_5 + 2x_0(x_1 x_2 + \alpha x_3 x_4) dx_5^2$$

on $\mathbb{R}^6$. Then a straightforward computation shows that its (restricted) holonomy group is of type 3 and **closed if and only if** α **is rational** (with $D = 0$).

TYPE 4. $\mathscr{H}$ DOES NOT CONTAIN $\mathscr{N}$. We have seen in Theorem I that $\mathscr{H} \subset \mathscr{K} \oplus \mathscr{N}$ and there exists a decomposition $\mathscr{N} = \mathscr{N}_1 \oplus \mathscr{N}_2$, a decomposition $\mathscr{B} = \mathscr{T} \oplus \mathscr{D}$ as above (of the projection $\mathscr{B}$ of $\mathscr{H}$ into $\mathscr{K}$), and a surjective linear map $\varphi: \mathscr{T} \to \mathscr{N}_2$ such that $\mathscr{H} = \mathscr{L} \oplus \mathscr{D} \oplus \mathscr{N}_1$, where $\mathscr{L}$ is the subalgebra of $\mathscr{T} \oplus \mathscr{N}_2$ defined by $\mathscr{L} = \{(A, \varphi(A)) \in \mathscr{T} \oplus \mathscr{N}_2$, for any A in $\mathscr{T}\}$. Such examples appear in dimension 5 (with $\mathscr{D} = 0$) or more, and φ is injective in dimension 5 or 6. Once again, the subgroup $B = TD$ generated by $\mathscr{B}$ is compact, as are T generated by $\mathscr{T}$ and D generated by $\mathscr{D}$. Also, the subgroups N_1 and N_2 generated respectively by $\mathscr{N}_1$ and $\mathscr{N}_2$ are closed. Now the subgroup H generated by $\mathscr{H}$ may be written LDN_1, where L is the subgroup of TN_2 generated by $\mathscr{L}$. As before, L is closed if and only if the subalgebra $\operatorname{Ker}\varphi$ of $\mathscr{T}$ generates a compact torus (which is satisfied if φ is injective). So the first nonclosed example appears in dimension 7.

We consider the Lorentzian metric

$$h = 2dx_0dx_6 - \sum_{i=1}^{5} dx_i^2 + 2x_1x_2^2dx_1dx_6 - 2x_1^2x_2dx_2dx_6 + 2x_3x_4^2dx_3dx_6$$

$$- 2x_3^2x_4dx_4dx_6 + 2(x_1x_2 + \alpha x_3x_4)x_6dx_5dx_6 + dx_6^2$$

on $\mathbb{R}^7$.

Then a straightforward computation shows that its (restricted) holonomy group is of type 4 and **closed if and only if** α **is rational** (with $D = 0$).

Summarizing all that, we get in particular

THEOREM III. *The restricted holonomy group of a Lorentzian metric in dimension ≤ 5 is closed. In dimension ≥ 6, there are examples of Lorentzian manifolds with nonclosed restricted holonomy group.*

The first assertion was already obtained by M. Cahen.

Notice that examples of nonclosed (restricted) holonomy groups of pseudo-Riemannian manifolds have been given already by H. Wu, for example in signature $(2, 4)$ (see Example 6, p. 355 in [WU2]).

We hope to be able to give other examples in a near future.

[The first author apologizes for having asserted erroneously the closedness in all cases at the AMS Summer Institute.] Both authors thank the C.N.R.S. for its support through UA 750, and the European Communities for their support through the program G.A.D.G.E.T.

REFERENCES

[AI] D. V. Alekseevskiĭ, *Riemannian manifolds with exceptional holonomy groups*, Funksional Anal. i Prilozhen. **2** (2) (1968), 1–10. (English translation: Functional Anal. Appl. **2** (1968), 97–105.

[A-S] W. Ambrose and I. M. Singer, *A theorem on holonomy*, Trans. Amer. Math. Soc. **79** (1953), 428–443.

[BE] A. L. Besse, *Einstein manifolds*, Springer-Verlag, Berlin-Heidelberg-New York, 1987

[B-L] A. Borel and A. Lichnerowicz, *Groupes d'holonomie des variétés riemanniennes*, C. R. Acad. Sci. Paris **234** (1952), 1835–1837.

[BR1] M. Berger, *Sur les groupes d'holonomie des variétés à connexion affine et des variétés riemanniennes*, Bull. Soc. Math. France **83** (1955), 279–330.

[BR2] ____, *Les espaces symétriques non compacts*, Ann. Sci. École Norm. Sup. **74** (1975), 85–177.

[BT] R. Bryant, *Metrics with exceptional holonomy*, Ann. of Math. (2) **126** (1987), 525–576.

[C-L] M. Cahen, J. Leroy, M. Parker, F. Tricerri, and L. Vanhecke, *Lorentz manifolds modelled on a Lorentz symmetric space*, J. Geom. and Phys. **7** (1990), 571–591.

[CN1] E. Cartan, *Les groupes d'holonomie des espaces généralisés*, Acta Math. **48** (1926), 1–42.

[CN2] ____, *Sur une classe remarquable d'espaces de Riemann*, Bull. Soc. Math. France **54** (1926), 214–264; **55** (1972), 114–134.

[C-P] M. Cahen and M. Parker, *Pseudo-Riemannian symmetric spaces*, Mem. Amer. Math. Soc. **229** (1980).

[C-W] M. Cahen and N. Wallach, *Lorentzian symmetric spaces*, Bull. Amer. Math. Soc. **79** (1970), 585–591.

[DR] G. de Rham, *Sur la réductibilité d'un espace de Riemann*, Comm. Math. Helv. **26** (1952), 328–344.

[HN] S. Helgason, *Differential geometry, Lie groups and symmetric spaces*, Academic Press, New-York and London, 1978.

[IN] A. Ikemakhen, *Sur l'holonomie des variétés lorentziennes*, Thèse de doctorat de l'Université de Nancy 1, 1990.

[K-N] S. Kobayashi and K. Nomizu, *Foundations of differential geometry*, Interscience Wiley, New York, vol. **1** 1963: vol. 2 1969.

[SS] J. Simons, *On the transitivity of holonomy systems*, Ann. of Math. (2) **76** (1962), 213–234.

[WU1] H. Wu, *On the de Rham decomposition theorem*, Illinois J. Math. **8** (1964), 291–311.

[WU2] ____, *Holonomy groups of indefinite metrics*, Pacific J. Math. **20** (1967), 351–392.

INSTITUT ELIE CARTAN, URA 750 CNRS UNIVERSITE DE NANCY 1, B.P. 239, 54506 VAN DOEUVRE CEDEX, FRANCE

Proceedings of Symposia in Pure Mathematics
Volume **54** (1993), Part 2

Spinors, Dirac Operators, and Changes of Metrics

JEAN-PIERRE BOURGUIGNON

In the last fifteen years, the Dirac operator has attracted a lot of attention from mathematicians because of the many pieces of topological information that can be drawn from its spectral data. Since it is a square root of the Laplacian, its principal symbol incorporates the metric via Clifford multiplication. This fact makes the Dirac operator an important invariant attached to a metric. (An excellent reference on this whole subject is the new book [L-M].)

Nevertheless, for reasons which are quite obscure, its dependence on the metric has never been studied thoroughly. It is precisely the aim of this talk, a report on joint work with Paul Gauduchon (cf. [B-G]), to successively compare orthonormal frames, spinors, and Dirac operators for two metrics. We close by giving the variation formula for the eigenvalues of the Dirac operator.

1. Comparison of orthonormal bases for two scalar products

We work over a fixed vector space V. A basis of V is viewed as an invertible map from $\mathbf{R}^n$ to V. We denote by BV the space of (linear) bases of V. (This is just $\mathbf{R}G\ell_n$ where one has "lost" the origin.) To any $f \in BV$, one can associate an inner product on V by setting $\pi(f) = (f^{-1})^*(e)$ where e denotes the standard Euclidean/Scalar product on $\mathbf{R}^n$. The map $\pi : BV \to$ Met V is an O_n-fibration whose fibre at $g \in$ Met V consists of the set $B_g V$ of g-orthonormal bases of V. The space Met V has a natural Riemannian metric with nonpositive sectional curvature (Met $V \simeq \mathbf{R}G\ell_n/O_n$).

We are looking for an O_n-equivariant process associating to any $f \in B_g V$ an h-orthonormal basis for $h \in$ Met V. If we set $H_g = g^{-1} \cdot h$ (i.e., H_g is the linear map associated with h via the duality defined by g, hence is

---

1991 *Mathematics Subject Classification.* Primary 53A50, 58G25, 58G30.
The final version of this paper has been submitted for publication elsewhere.

automatically invertible), then the map b_h^g

$$b_h^g(f) = (H_g^{-1/2}) \circ f$$

has the required property (cf. [B-P]). For later purposes, it is convenient to view b_h^g as follows. The bundle $\pi : BV \to \text{Met } V$ has a natural O_n-connection; namely, at $f \in B_g V$ the horizontal subspace is given by the space of maps from $\mathbf{R}^n$ to V which, in the canonical basis of $\mathbf{R}^n$ and the basis f of V, are represented by a symmetric matrix. Then, we have

PROPOSITION 1. *Let g, $h \in \text{Met } V$. The horizontal lift associated to this connection along any path from g to h lying in the flat subspace of $\text{Met } V$ determined by g and h coincides with b_h^g.*

2. Comparison of spinors for two different Riemannian metrics

With this new definition of b_h^g at hand, it becomes easy to carry over the construction to spinorial bases. We first suppose that V is oriented and we denote by B^+V the set of positive bases of V.

We introduce $\tilde{B}^+V$ the double cover of B^+V (a copy of the double cover $\widetilde{RG\ell}_n$ of $RG\ell_n$ where the identity has been lost). By composing the covering map with π, we obtain a Spin_n-fibration $\tilde{\pi} : \tilde{B}^+V \to \text{Met } V$ whose points in the fibre $\tilde{B}_g^+V$ can be called *spinorial bases*.

Our lifting construction carries over to this situation since the SO_n-connection can be lifted to a Spin_n-connection. It defines a Spin_n-equivariant map β_h^g from $\tilde{B}_g^+V$ to $\tilde{B}_h^+V$, associating an h-spinorial basis $\beta_h^g(\varphi)$ in a natural way to any g-spinorial basis φ.

Our notations have been chosen to fit nicely with the principal bundle formalism. For an orientable differentiable manifold M of dimension n, we denote by $G\ell^+M$ its positive linear frame bundle, and by P a $\widetilde{RG\ell}_n^+$-principal bundle covering $G\ell^+M$. This is our version of choosing a spin structure on M (which we of course suppose to exist). We denote by γ the simultaneous choice of P and a Riemannian metric g, and call it a *spin metric*. We denote the SO_n-bundle of positive g-orthonormal frames on M by $SO_g M$, and the Spin_n-bundle of γ-spinorial frames by $\text{Spin}_\gamma M$.

Let us take two Riemannian metrics g and h determining two spin metrics γ and η. By patching together the horizontal lifts in each tangent space, we obtain a Spin_n-equivariant map β_η^γ from $\text{Spin}_\gamma M$ to $\text{Spin}_\eta M$.

We can now compare spinors for the spin metrics γ and η. We fix a *spinor representation* of Spin_n in a vector space Σ (i.e., a restriction to Spin_n of a representation of the Clifford algebra $C\ell_n$), and we define the γ-spinors as being the elements of the associated bundle $\Sigma_\gamma M = \text{Spin}_\gamma M \times_{\text{Spin}_n} \Sigma$. Thanks to the Spin_n-equivariance of β_η^γ, we have a well-defined comparison of γ-spinors and η-spinors by setting $\beta_\eta^\gamma([\varphi, \psi]) = [\beta_\eta^\gamma(\varphi), \psi]$ where φ is a γ-spinorial basis and $\psi \in \Sigma$.

3. Comparison of Dirac operators for two spinor metrics

Recall that the Dirac operator $\mathscr{D}^\gamma$ acting on γ-spinor fields is defined as

$$\mathscr{D}^\gamma \psi = \sum_{i=1}^{n} e_i \cdot_\gamma D_{e_i}^\gamma \psi$$

where (e_i) denotes a g-orthonormal basis, $\cdot_\gamma$ Clifford multiplication, and D^γ the lift of the Levi-Civita connection of g to the spinor bundle $\Sigma_\gamma M$. The main issue is that Dirac operators for different spin metrics act on different (though isomorphic) spaces of sections.

For many purposes, in particular in order to make sense of the infinitesimal variation of the Dirac operator, one needs to work with operators acting on a fixed space of sections. This can be done using a gauge transform of the Dirac operator via the transformation β.

For $\psi \in \Gamma(\Sigma_\gamma M)$, we set $^\gamma\mathscr{D}^\eta \psi = \beta_\gamma^\eta(\mathscr{D}^\eta(\beta_\eta^\gamma(\psi)))$.

PROPOSITION 2. *The gauge transformed Dirac operator satisfies*

$$(1) \qquad ^\gamma\mathscr{D}^\eta \psi = \sum_{i=1}^{n} e_i \cdot_\gamma {}^\gamma D_{H_g^{-1/2}(e_i)}^\eta \psi \;.$$

Here, $^\gamma D^\eta$ is the gauge transformed Levi-Civita connection for the spin metric η. It is given by the formula

$$(2) \qquad ^\gamma D^\eta = D^\gamma + \frac{1}{2}\,(G_h^{-1/2} d^{D^h} G_h^{-1/2})\;.$$

REMARK. In connection with the almost conformal invariance of the Dirac operator, one recovers the classical formula relating the Dirac operators of two conformally related metrics g and $h = e^{2u}g$, namely,

$$^g\mathscr{D}^{e^{2u}g} = e^{-(n+1)u/2} \circ \mathscr{D}^g \circ e^{(n-1)u/2}\;.$$

From formulas (1) and (2), one can derive the infinitesimal variation of the Dirac operator, namely, for a curve $t \mapsto \gamma_t$ of spin metrics,

$$(3) \qquad \frac{d}{dt}(^\gamma\mathscr{D}^{\gamma_t})\big|_{t=0} = -\frac{1}{2}\sum_{i=1}^{n} e_i \cdot_\gamma D_{K_g(e_i)}^\gamma - \frac{1}{2}\,(\delta_g k + d(\mathrm{trace}_g k))\cdot_\gamma$$

where $k = \frac{d}{dt}g_t\big|_{t=0}$ and δ_g is the codifferential of symmetric 2-tensor fields defined by the Riemannian metric g.

4. Variations of the eigenvalues of the Dirac operator

For this section, we assume that the manifold M is compact. Then, as is well known, the Dirac operator has a discrete real spectrum going to infinity in both directions. We are interested in following the eigenvalues along an analytic curve $t \mapsto \gamma_t$ of spin metrics.

It is a classical computation that the infinitesimal variation $\dot{\lambda}$ of an eigenvalue λ of an operator A corresponding to the normalized eigenfunction ψ is given by

$$\dot{\lambda} = \langle \dot{A}\psi , \psi \rangle$$

where $\langle \, , \, \rangle$ denotes the global inner product on sections. In our case, we get

(4) $$\frac{d}{dt}\lambda_{g+tk}{}_{|t=0} = \frac{1}{2}\langle k , Q_\psi \rangle$$

where Q_ψ is the symmetric bilinear form defined, for $X \in TM$, as

$$Q_\psi(X, X) = \mathscr{R}e(X \cdot D_X\psi , \psi) \, .$$

Hence, the only remaining term in the infinitesimal variation of the eigenvalue of the Dirac operator comes from the change in the spinor bundle.

Formula (4) greatly simplifies for infinitesimal conformal deformations of the metric, say $k = 2ug$. In that case, the derivative of the eigenspinor disappears, and one gets $\dot{\lambda} = \lambda \langle u, |\psi|^2 \rangle$. (This reflects the conformal weight of the Dirac operator.)

REMARK. In particular we derive from (4) that λ is critical among metrics having a fixed volume element if ψ is a Killing spinor field.

References

[B-G] J. P. Bourguignon and P. Gauduchon, *Spineurs, opérateurs de Dirac et changement de métriques*, Comm. Math. Phys. **144** (1992), 581–599.

[B-P] E. Binz and R. Pferschy, *The Dirac operator and the change of the metric*, C. R. Math. Rep. Acad. Sci. Canada **5** (1983), 269–274.

[L-M] H. B. Lawson Jr. and M.-L. Michelsohn, *Spin geometry*, Princeton Univ. Press, Princeton, N.J., 1989.

CENTRE DE MATHÉMATIQUES, ECOLE POLYTECHNIQUE, PALAISEAU, FRANCE
E-mail address: jpb@cmep.polytechnique.fr

Proceedings of Symposia in Pure Mathematics
Volume **54** (1993), Part 2

The Hyperkähler Geometry of the ADHM Construction and Quaternionic Geometric Invariant Theory

CHARLES P. BOYER AND BENJAMIN M. MANN

ABSTRACT. In this paper we analyze various geometric structures on the moduli space of instantons over S^4. We study the hyperkähler geometry of the ADHM construction which is best expressed in Lie algebraic terms and examine its connections with hyperkähler Morse theory and geometric invariant theory. Topological implications are also discussed.

1. Introduction

The moduli spaces of instantons (self-dual connections with respect to a conformal class of metrics) associated to principal G bundles over Riemannian four-manifolds have proven to be fundamental objects in modern geometry. When the base manifold is taken to be S^4 with its standard flat conformal structure, these moduli spaces, which we denote by $\mathcal{M}_k(G)$, exhibit surprisingly rich geometrical and topological structures. First, on the topological side, we have shown in previous work [**10, 11, 13**] that the disjoint union of $\mathcal{M}_k(G)$ taken over all k have the homotopy structure of a four-fold loop space. On the geometrical side it had been observed by several people (Donaldson, Hitchin, Taubes) that the self-duality equations on S^4 have the interpretation of an infinite dimensional hyperkähler moment map with respect to the action of the based gauge group on the infinite dimensional affine space of G connections. Hence, the hyperkähler quotient procedure of [**22**] gives the moduli space $\mathcal{M}_k(G)$ the structure of a hyperkähler manifold. It was also observed by Nigel Hitchin (private communication) that the quadratic constraints of the well-known ADHM construction can be interpreted as the zero set of a hyperkähler moment map with respect to the natural $O(k)$

1991 *Mathematics Subject Classification*. Primary 81T13; Secondary 58E15.

Key words and phrases. Instantons, hyperkähler geometry and Morse theory, geometric invariant theory.

The first author was partially supported by NSF grants DMS-8815581, DMS-9004076, and the second author by NSF grant DMS-8901879.

This paper is in final form and no version of it will be submitted for publication elsewhere.

action that occurs there. Furthermore, it has now been proven by Kronheimer and Nakajima [25] and by Maciocia [30] that these two methods give equivalent hyperkähler structures on $\mathscr{M}_k(G)$.

In this paper we study in detail the hyperkähler geometry on $\mathscr{M}_k(G)$ obtained by hyperkähler reduction from the ADHM construction. In general, we shall restrict our discussion to the case $G = \mathrm{Sp}(n)$, but many of our results extend to general G, cf. [13]. First we show that the hyperkähler geometry of the ADHM construction is a special case of a more general theory. Let $\mathfrak{g}$ denote a real orthogonal symmetric Lie algebra, and $\mathbb{H}$ the algebra of quaternions. The tensor product over the reals $\mathfrak{g} \otimes \mathbb{H}$ has a natural algebraic structure determined by the Lie algebra on $\mathfrak{g}$ and the quaternionic algebra on $\mathbb{H}$. We refer to this algebraic structure as the *quaternionic adjoint map*. If $\mathfrak{p}$ denotes the plus one eigenspace in the symmetric space decomposition of $\mathfrak{g}$, then the diagonal restriction of the quaternionic adjoint map is a hyperkähler moment map on $\mathfrak{p} \otimes \mathbb{H}$ with respect to the adjoint action of the Lie group K generated by the minus one eigenspace of $\mathfrak{g}$. It then turns out that the ADHM construction is associated to the K module $\mathfrak{p}(\mathbb{H})$ obtained as the direct sum of the plus one eigenspaces of the symmetric Lie algebras $\mathfrak{gl}(k, \mathbb{R})$ and n copies of $\mathfrak{o}(k, 1)$ tensored with $\mathbb{H}$.

Although the quadratic constraints of the ADHM construction have a nice algebraic interpretation in terms of the diagonal of the quaternionic adjoint map, as well as their geometric interpretation as a hyperkähler moment map, from a strictly topological point of view they are still somewhat cumbersome. Recall Kirwan [24] showed that, under the right conditions, the symplectic quotient $\mu^{-1}(0)/G$ obtained from a moment map μ on a compact Kähler manifold M coincides with the geometric invariant theory quotient $M^{ss}/G^{\mathbb{C}}$ (here M^{ss} is a certain subset of well-behaved points of M). Motivated by Kirwan's work, we looked for a quaternionic analogue. However, in the quaternionic case a result such as that of Kirwan's cannot be hoped for since the "quaternionification" of a Lie group makes no sense. Indeed, the maximal compact subgroup of the quaternionic orthogonal group $O(k, \mathbb{H})$ is not $O(k)$. In view of this we should try for the next best thing, namely a homotopy equivalence. Thus, if $\mathfrak{p}^{(k)}(\mathbb{H})$ denotes the dense open subset of $\mathfrak{p}(\mathbb{H})$ where the rank condition of the ADHM construction holds, one should try to construct a homotopy equivalence between $\mathfrak{p}^{(k)}(\mathbb{H})/O(k)$ and $\mathscr{M}_k$. In this regard we have had only partial success, and have constructed such a homotopy equivalence when $k = 2$. In fact, we conjecture that it is true for general k, but so far this remains an open question.

To prove such a homotopy equivalence we construct an equivariant strong deformation retraction from $\mathfrak{p}^{(k)}(\mathbb{H})$ to the maximal strata of the zero set of the hyperkähler moment map μ. However, as is the case with many similar problems involving Morse theory on a noncompact manifold, the Palais-Smale condition C is violated, so a careful study of the topology of the ends of the moduli space is necessary. Moreover, as was pointed out to us by Cliff Taubes, that attempting to use the norm squared of the moment map as a Morse function does not give a map of sufficiently fast descent. We are greatly indebted to him for showing us how to construct a flow which gives the moment map an exponential descent. It turns out that this dissipative system on $\mathfrak{p}^{(k)}(\mathbb{H})$ can be written in terms of the quaternionic adjoint map and has a form reminiscent of an Euler flow on a Lie algebra. Although the vector field which generates this flow is not differentiable on $\mathfrak{p}(\mathbb{H})$, it is Lipschitz and the various strata are invariant sets for finite time t. This part of the construction holds for all k. However, the more delicate point is to show that the flow does not jump strata in the limit as $t \to \infty$. In fact, we show by example that the lower strata can indeed jump strata in the limit. However, we conjecture that this does not happen on the maximal strata.

We establish our conjecture for instanton number $k = 2$ by showing that the generalized Euler flow is completely integrable in the sense that there exist enough constants of the motion to integrate the system by quadratures. Using these constants of the motion, we show that, when restricted to the maximal strata, $\mathfrak{p}^{(k)}(\mathbb{H})$, the flow does not jump strata in the limit. This implies that $\mathfrak{p}^{(2)}(\mathbb{H})/O(2)$ is homotopy equivalent to $\mathscr{M}_2$. We conclude this paper by then discussing some topological consequences. Additionally, it is mentioned that many of the results of this paper were announced previously without complete proofs in [12].

During the course of our investigations we have had many conversations with many people. It is a pleasure to thank Fred Cohen, Nigel Hitchin, Jacques Hurtubise, Tom Lada, Jim Milgram, Rick Schoen, Jim Stasheff, and Cliff Taubes for helpful discussions and correspondence concerning this work.

NOTE ADDED IN PROOF. This paper was written at the time of the conference and represents our knowledge at that time. Recently there have been advances made in several directions discussed here. For example, in the topological direction, in joint work with J. Hurtubise and R.J. Milgram we have proved the Atiyah-Jones conjecture [32, 33]. In addition, in joint work with K. Galicki [31], we have developed a general theory for which Proposition 4.5.3 here is a special case. This is related to the work of Swann [27]. It should also be mentioned that Theorem 4.4.3 was given in a recent book by S. K. Donaldson and P. B. Knonheimer, *The Geometry of Four-Manifolds*, Oxford University Press, 1990.

2. The geometry of hyperkähler reduction

2.1 Quaternionic vector spaces. A *quaternionic structure* on a real vector space V is a pair of linear endomorphisms (I^1, I^2) that satisfy

$$(I^1)^2 = (I^2)^2 = -\operatorname{id}, \qquad I^1 I^2 + I^2 I^1 = 0.$$

When this is the case there is a third endomorphism $I^3 = I^1 I^2$ which satisfies

$$(I^3)^2 = -\,\mathrm{id} \quad \text{and} \quad I^a I^3 + I^3 I^a = 0$$

for $a = 1, 2$. Thus, for $a = 1, 2, 3$, the I^a's satisfy the algebra of the quaternions. It follows that if $z = (z_1, z_2, z_3) \in S^2$ then $I_z = z_1 I^1 + z_2 I^2 + z_3 I^3$ also satisfies $I_z^2 = -\,\mathrm{id}$. Thus, a quaternionic structure on V is equivalent to a two-sphere's worth of complex structures on V. Given a real vector space V with a quaternionic structure, we can make V a quaternionic vector space (i.e., a free right quaternionic module) by defining scalar multiplication by

$$vq = vq_0 + I^1 v q_1 + I^2 v q_2 + I^3 v q_3$$

for all $v \in V$ and $q \in \mathbb{H}$. If $(v_1, \dots, v_n)$ is a basis for V as a quaternionic vector space, then

$$(v_1, \dots, v_n, I^1 v_1, \dots, I^1 v_n, I^2 v_1, \dots, I^2 v_n, I^3 v_1, \dots, I^3 v_n)$$

is a basis as a real vector space. Clearly the dimension of such a V is necessarily divisible by 4. If V is a real vector space of dimension n, then $V \otimes_{\mathbb{R}} \mathbb{H}$ is a quaternionic vector space of real dimension $4n$.

A *hyperhermitian* inner product on a real vector space V with a quaternionic structure is a quaternionic inner product $< \cdot, \cdot >$ that satisfies

(2.1.1) $\qquad\qquad < I^a u, I^a v > = < u, v >$

for all $u, v \in V$ and for all $a = 1, 2, 3$. The real part of $< \cdot, \cdot >$ is a real positive definite inner product on V while the imaginary part of $< \cdot, \cdot >$ is symplectic structure for each $a = 1, 2, 3$.

2.2 Hyperkähler manifolds. An *almost hyperhermitian* manifold is a Riemannian manifold (M, g) whose tangent space has both a quaternionic structure and a hyperhermitian inner product at each point $x \in M$. Both these structures vary smoothly with x in such a way that the real part of the hyperhermitian inner product is the Riemannian metric g. An almost hypermitian manifold is called *hyperhermitian* if all the complex structures I_z are integrable for all $z \in S^2$, and it is called *hyperkähler* if all the complex structures are parallel with respect to the Levi-Civita connection on M. If ω denotes the imaginary part of the hyperhermitian inner product, then ω defines a hyperkähler structure if and only if it is closed. By choosing i, j, k as a basis for the imaginary quaternions $\mathbb{R}^3$, we can write

(2.2.1) $\qquad\qquad \omega = \omega^1 i + \omega^2 j + \omega^3 k.$

Hence, we have three (and thus a S^2's worth) Kähler two-forms on M, which satisfy $\omega^a(X, Y) = g(I^a X, Y)$ where X, Y are any vector fields on M. It follows from the fact that the I^a's are covariantly constant that the two-forms ω^a are closed. Furthermore, the nondegeneracy of g and I^a imply that ω^a is also nondegenerate. Thus, a hyperkähler manifold (M, g, I^a)

describes a manifold with three independent Kähler structures which generate a two sphere's worth of Kähler structures. We write (M, g, ω) to denote a hyperkähler manifold. It is well known [8] that the Ricci curvature of g vanishes on any hyperkähler manifold.

2.3 Hyperkähler reduction. Now suppose that the Lie group G acts on (M, g, ω) by isometries, and let $T_a : M \to M$ denote the action map for any $a \in G$. Then we say that G acts by *hyperkähler isometries* if, in addition to $T_a^* g = g$, we have

$$(2.3.1) \qquad T_a^* \omega = \omega$$

for all $a \in G$, or equivalently $T_a^* I^b = I^b$ for all $a \in G$ and $b = 1, 2, 3$. For computational purposes it is more convenient to consider the infinitesimal version of these conditions. Let $\mathfrak{g}$ denote the Lie algebra of G and for any $\xi \in \mathfrak{g}$ let ξ^* denote the vector field on M corresponding to ξ through the differential of the action map. Then the infinitesimal version of (2.3.1) is

$$(2.3.2) \qquad 0 = \mathscr{L}_{\xi^*} \omega = d(\iota_{\xi^*} \omega)$$

where $\mathscr{L}$ denotes the Lie derivative and ι the interior product. Thus, the one-form $\iota_{\xi^*} \omega$ which takes its values in $\mathbb{R}^3$ is closed, and, if the de Rham cohomology group $H^1(M, \mathbb{R})$ vanishes, it is exact. In this case there is an $\mathbb{R}^3$ valued function μ_ξ on M such that

$$(2.3.3) \qquad d\mu_\xi = \iota_{\xi^*} \omega.$$

Thus, we can define a map [22], $\mu : M \longrightarrow \mathfrak{g}^* \otimes \mathbb{R}^3$ called the *hyperkähler moment map* by setting $< \mu(x), \xi >= \mu_\xi(x)$. It will be convenient to write μ in terms of its components in $\mathbb{R}^3$ as done with ω in (2.2.1). We write

$$(2.3.4) \qquad \mu = \mu^1 i + \mu^2 j + \mu^3 k.$$

Notice that μ_ξ, and hence μ, is defined only up to a constant in $\mathbb{R}^3$. However, further considerations often determine this constant uniquely. Since G acts on M and on $\mathfrak{g}^*$ via the natural coadjoint action, it is natural to ask whether μ is equivariant with respect to these actions. In general, it is not, as there is an obstruction, associated to the coadjoint representation in the Lie group cohomology $H^1(G, \mathfrak{g}^*)$. However, this group often vanishes, for example, if G is semisimple. In this case, μ can be chosen to be equivariant by a suitable choice of the constant. Even in the case that μ gives a nontrivial cocycle in $H^1(G, \mathfrak{g}^*)$, there is an action on $\mathfrak{g}^*$ which makes μ equivariant. For the most part we shall deal with Ad^* equivariant moment maps in this paper; however, there is one exception when we consider the "reduced" moduli space where a nontrivial cocycle appears.

The moment map can be used to obtain new hyperkähler manifolds from old ones by a procedure known as hyperkähler reduction [22].

This is the hyperkähler analogue of the symplectic reduction of Marsden and Weinstein [19]. This new space is obtained as a quotient of the zero set of the moment map. In general, the reduction space is not a manifold. In the most general case that we deal with in this paper, the reduction space is a stratified space in the sense of Whitney. However, it is important to analyze the special case when the reduction is a manifold. This can be achieved by allowing G to act freely and properly on M. Thus, we assume for the rest of this section that G is a compact Lie group which acts freely on M by hyperkähler isometries. Then we have the following diagram:

$$
(2.3.5) \qquad
\begin{array}{ccc}
\mu^{-1}(0) & \xrightarrow{\ \iota\ } & M \\
{\scriptstyle \pi_0} \downarrow & & \\
\mu^{-1}(0)/G & &
\end{array}
$$

where the horizontal map is a Riemannian embedding and the vertical map is a Riemannian submersion. Then we have the following theorem of Hitchin, Karlhede, Lindström, and Roček:

THEOREM 2.3.6 [22]. *The quotient manifold* $\mu^{-1}(0)/G$ *has a hyperkähler structure* $(\bar{g}, \bar{\omega})$ *such that* $\bar{g}$ *is the unique Riemannian metric making* π_0 *a Riemannian submersion and* $\bar{\omega}$ *satifies* $\pi_0^*\bar{\omega} = \iota^*\omega$.

The diagram (2.3.5) can be completed to the commutative diagram

$$
(2.3.7) \qquad
\begin{array}{ccc}
\mu^{-1}(0) & \xrightarrow{\ \iota\ } & M \\
{\scriptstyle \pi_0} \downarrow & & \downarrow {\scriptstyle \pi} \\
\mu^{-1}(0)/G & \xrightarrow{\ \bar{\iota}\ } & M/G
\end{array}
$$

in such a way that the unique Riemannian metric $\check{g}$ on M/G making π a Riemannian submersion satisfies $\bar{\iota}^*\check{g} = \bar{g}$. Notice that, since G acts freely on M, the vertical maps can be considered as principal G-bundles and the pair $(\bar{\iota}, \iota)$ is a map of principal G-bundles.

2.4 The vector bundles vert and Q. There are several interesting decompositions of the tangent bundle TM arising from (2.3.7). Let $\xi \in \mathfrak{g}$ and let ξ^* be the fundamental vertical vector field on M corresponding to ξ. Let Vert_x denote the tangent space to the fibers of the Riemannian submersion $\pi : M \to M/G$ at $x \in M$. Vert_x is spanned by the fundamental vertical vectors $\xi^*(x)$ for each $x \in M$, and, since the action of G on M is free, a basis of $\mathfrak{g}$ gives a global frame which trivializes Vert. The restriction of g to Vert describes a family of metrics $g_{[x]}$ along the fibers $F_{[x]}$ which depends on $[x] \in M/G$. Since there is a one-to-one correspondence between elements in $\mathfrak{g}$ and fundamental vertical vector fields on M, there is a unique family of positive definite bilinear forms $B_{[x]}$ on $\mathfrak{g}$ such that

$$
B_{[x]}(\xi, \eta) = g_{[x]}(\xi^*, \eta^*).
$$

The restriction of the hyperkähler form ω to Vert is related to the moment map. We have

PROPOSITION 2.4.1. *The restriction of the forms ω^a to* Vert *coincides with the symplectic structure on the coadjoint orbit $\mathscr{O}_{\mu^a(x)}$, explicitly*

$$\omega(\xi^*, \eta^*) = <\mu(x), [\xi, \eta]>.$$

In particular, if $x \in \mu^{-1}(0)$ the fibers $F_{[x]}$ are isotropic submanifolds of M with respect to all of the Kähler forms. Moreover, if $\mathfrak{g}$ is semisimple the converse also holds; that is, $F_{[x]}$ is isotropic if and only if $\mu(x) = 0$.

PROOF. The formula follows from the equivariance of the moment map. Set $T_t = \exp t\xi$ and differentiate the equivariance condition $\mu(T \cdot x) = \mathrm{Ad}_T^* \mu(x)$ to get

$$< d\mu(\xi^*), \eta > = <\mu(x), [\xi, \eta]>.$$

This together with (2.3.3) gives the formula and it is a well-known result of Kostant and Souriau that the right-hand side defines a symplectic structure on $\mathscr{O}_{\mu^a(x)}$. □

REMARK 2.4.2. *Notice that, for each complex structure I^a, the moment map μ^a on M defines a map on M/G by sending $[x]$ to the coadjoint orbit $\mathscr{O}_{\mu^a(x)}$.*

Another consequence of the free action of G is that, for each $a = 1, 2, 3$, the vector fields $I^a\xi^*$ generate a trivial subbundle $\mathscr{N}^a$ of TM as ξ runs through the Lie algebra $\mathfrak{g}$. Furthermore, since the I^a's are independent, together with Vert, we get four linearly independent subbundles each of dimension dim $\mathfrak{g}$. Thus we have the following subbundles of $T(M)$:

$$(2.4.3) \qquad C^a = \mathrm{Vert} + \mathscr{N}^a \qquad \mathscr{N} = \bigoplus_{a=1}^{3} \mathscr{N}^a \qquad Q = \mathrm{Vert} + \mathscr{N}.$$

By construction, the bundles C^a have the structure of complex vector bundles, while Q has the structure of a quaternionic vector bundle.

PROPOSITION 2.4.4. *For each $a = 1, 2, 3$ the subbundle C^a is integrable. Moreover, the vector fields ξ^* and $I^a\xi^*$ tangent to the leaves generate the Lie algebra $\mathfrak{g}^{\mathbb{C}}$ of the complexification $G^{\mathbb{C}}$ of G. Hence, if the vector fields $I^a\xi^*$ are complete for fixed a and all $\xi \in \mathfrak{g}$, then they integrate to a free action of the group $G^{\mathbb{C}}$ on M. Furthermore, the metric g and Kähler forms ω^c restricted to Q satisfy*

(1) $g(\xi^*, I^a\eta^*) = <\mu^a(x), [\eta, \xi]>.$
(2) $g(I^a\xi^*, I^c\eta^*) = <\mu^b(x), [\eta, \xi]> \varepsilon^{abc}$ *for $a \neq c$.*
(3) $g(I^a\xi^*, I^a\eta^*) = g(\xi^*, \eta^*).$
(4) $\omega^c(\xi^*, I^a\eta^*) = <\mu^b(x), [\xi, \eta]> \varepsilon^{abc}$ *for $c \neq a$.*
(5) $\omega^a(\xi^*, I^a\eta^*) = g(\xi^*, \eta^*).$

(6) $\omega^c(I^a\xi^*, I^b\eta^*) = g(\xi^*, \eta^*)\varepsilon^{abc}$ *for* a, b, *and* c *all different.*

(7) $\omega^a(I^a\xi^*, I^b\eta^*) = <\mu^b(x), [\xi, \eta]>.$

Thus, restricted to the zero set of μ,

(1) *The complex bundle* C^a *is isotropic with respect to the Kähler forms* ω^b *with* $b \neq a$.

(2) $\mathcal{N}$ *is the normal bundle to the embedding* $\mu^{-1}(0) \to M$.

(3) Q *is a trivial hyperhermitian vector bundle with fiber* $\mathfrak{g} \otimes \mathbb{H}$ *and metric that is the orthogonal direct sum of four copies of the metric on* Vert.

PROOF. To show the integrability of C^a, notice that the invariance of I^a under the action of G implies

(2.4.5) $[\xi^*, I^a\eta^*] = I^a[\xi, \eta]^*,$

so it suffices to show that $[I^a\xi^*, I^a\eta^*]$ is a section of Vert. This follows from the vanishing of the Nijenhuis tensor, viz.

$$0 = \tfrac{1}{2}N(\xi^*, \eta^*) = [I^a\xi^*, I^a\eta^*] - [\xi^*, \eta^*] - I^a[\xi^*, I^a\eta^*] - I^a[I^a\xi^*, \eta^*]$$

which, using (2.4.5), gives

(2.4.6) $[I^a\xi^*, I^a\eta^*] = -[\xi^*, \eta^*].$

Equations (2.4.5) and (2.4.6) also imply that the ξ^* and $I^a\xi^*$ generate the Lie algebra $\mathfrak{g}^{\mathbb{C}}$. The formulas for g and ω are easily obtained from Proposition 2.4.1 by routine manipulations. The isotropy of C^a and the form of the metric on Q restricted to $\mu^{-1}(0)$ now follow easily from these formulas. Finally, the fact that $\mathcal{N}$ is the normal bundle on $\mu^{-1}(0)$ is standard [21]. $\square$

2.5 The horizontal subbundle $\mathcal{H}$. Let $\mathcal{H}_x$ denote the orthogonal complement to Vert_x in $T_x M$ with respect to the Riemannian metric g. Then we have the decomposition,

$$TM = \text{Vert} \oplus \mathcal{H}$$

and for each $x \in M$, $\mathcal{H}_x$ is isomorphic to the tangent space of the quotient, $T_{[x]}M/G$. Moreover, to every vector field $\bar{X}$ on M/G there corresponds a unique projectible horizontal vector field X on M. There is a similar decomposition for the Riemannian submersion π_0. Now let us decompose the tangent space to M at $x \in \mu^{-1}(0)$ into normal and tangential pieces. As seen above in Proposition 2.4.4 the normal bundle is just $\mathcal{N}$ restricted to $\mu^{-1}(0)$; that is, $\iota^*\mathcal{N}$, while the tangential summand $T_x\mu^{-1}(0)$ is given by the kernel, $\ker(L_x)$, of the differential of the moment map $L_x : T_x M \to \mathfrak{g}^* \otimes \mathbb{R}^3$. Thus, for $x \in \mu^{-1}(0)$, we have the orthogonal decomposition,

$$T_x M = \ker(L_x) \oplus \mathcal{N}_x.$$

Since $\mu^{-1}(0)$ is invariant under the G action, $\text{Vert}_x \subset \ker(L_x)$ for each $x \in \mu^{-1}(0)$ and is the vertical subspace for the Riemannian submersion π_0.

The horizontal subbundle $\hat{\mathscr{H}}$ for π_0 is determined by

$$(2.5.1) \qquad\qquad \hat{\mathscr{H}}_x = \mathscr{H}_x \cap \ker(L_x)$$

and the pullback bundle $\iota^*\mathscr{H}$ splits as $\hat{\mathscr{H}} \oplus \iota^*\mathscr{N}$. Thus, for each $x \in \mu^{-1}(0)$, we have the decomposition

$$(2.5.2) \qquad\qquad T_x M = \hat{\mathscr{H}}_x \oplus \operatorname{Vert}_x \oplus \mathscr{N}_x = \hat{\mathscr{H}}_x \oplus Q_x.$$

Since the I^a's determine a quaternionic structure on Q, and (2.5.2) is an orthogonal decomposition, the I^a's also determine a quaternionic structure on $\hat{\mathscr{H}}$. This is the essence of the proof of Theorem 2.3.6.

2.6 The second fundamental form. The second fundamental form s of the embedding $\mu^{-1}(0) \hookrightarrow M$ is a section of the bundle $\mathscr{N} \otimes \operatorname{Sym}^2(T^*\mu^{-1}(0))$. Let D denote the Levi-Civita connection on M and ∇ the induced connection on $\mu^{-1}(0)$. Let $v \in T_x\mu^{-1}(0)$ and X a vector field extending the vector $X_x \in T_b\mu^{-1}(0)$. Then Gauss' formula is $D_v X = \top(D_v X) \oplus \bot(D_v X) = \nabla_v X \oplus s(v, X)$ and we have

$$0 = D_v(L_x X) = D_v(X\mu) = d\mu(D_v X) + (D_v d\mu)(X) = L_x(D_v X) + (D_v d\mu)(X)$$

which implies, by Gauss' formula,

$$L_x(s(v, X_x)) = -(D_v d\mu)(X_x).$$

Applying the adjoint operator $L_x^\dagger$ to this equation and inverting gives

PROPOSITION 2.6.1. *For any* $x \in \mu^{-1}(0)$ *and for all* $u, v \in T_x\mu^{-1}(0)$ *we have*

$$s(u, v) = -(L_x^\dagger L_x)^{-1} L_x^\dagger (D_v d\mu)(u).$$

Consider the embedding $\mu^{-1}(0)/G \hookrightarrow M/G$. The normal bundle of this embedding, $\bar{\mathscr{N}}$, can be obtained by "pushing down" the normal bundle $\mathscr{N}$ on $\mu^{-1}(0)$. Indeed, since the G action commutes with the complex structures I^a, it follows that the normal bundle $\mathscr{N}$ is G-invariant. Thus, we can form the associated bundle $\mu^{-1}(0) \times_G \mathscr{N}$ over $\mu^{-1}(0)/G$. Furthermore, this is easily identified with the normal bundle $\bar{\mathscr{N}}$ by noticing that the fiber for both at $[x] \in \mu^{-1}(0)/G$ is spanned by the vectors $I^a\xi_x^*$ as ξ runs through $\mathfrak{g}$ and $a = 1, 2, 3$. However, the vector fields $I^a\xi^*$ are not projectible (they transform like the adjoint representation of $\mathfrak{g}$). A projectible section of $\mathscr{N}$ will be called a *basic normal field*. There is a one-to-one correspondence between the basic normal fields on $\mu^{-1}(0)$ and sections of $\bar{\mathscr{N}}$.

The second fundamental form $\bar{s}$ of the embedding $\hat{\imath}: \mathscr{M}_k \to \mathscr{V}_k$ is

$$(2.6.2) \qquad\qquad \bar{s}(X, Y) = \bot(\bar{\nabla}_X Y)$$

where X and Y are vector fields on $\mu^{-1}(0)/G$ and $\bar{\nabla}$ is the Levi-Civita connection on M/G. Hence, we have

PROPOSITION 2.6.3. *Let X and Y be vector fields on $\mu^{-1}(0)/G$, then at $x \in \mu^{-1}(0)$ we have*

$$\bar{s}_{[x]}(X, Y) = \pi_* s_x(\hat{X}, \hat{Y}) = -\pi_*(L_x^\dagger L_x)^{-1} L_x^\dagger (D_{\hat{X}} d\mu)(\hat{Y})$$

where $\hat{X}$ is the horizontal lift of the vector field X.

PROOF. Fix $x \in \mu^{-1}(0)$ and let h denote the horizontal projection of $T_x M$ onto $\mathscr{H}_x$. Then

$$\perp(\bar{\nabla}_X Y) = \perp(\pi_* h D_{\hat{X}} \hat{Y}) = \pi_* \perp(h D_{\hat{X}} \hat{Y}) = \pi_* \perp(D_{\hat{X}} \hat{Y}) = \pi_* s(\hat{X}, \hat{Y}),$$

and the last equality follows from Proposition 2.6.1. □

3. Quaternionic Lie algebras

3.1 The quaternionic adjoint map. Let us consider the "quaternionification" of a Lie algebra. For any real Lie algebra $\mathfrak{g}$, we shall refer to $\mathfrak{g} \otimes_{\mathbb{R}} \mathbb{H}$ with a certain algebraic structure to be defined below as a *quaternionic Lie algebra*. The important algebraic properties of $\mathfrak{g} \otimes \mathbb{H}$ are that of a quaternionic vector space and the multiplicative structure given by the Lie bracket on the first factor and quaternionic multiplication on the second. Indeed, given any associative algebra $\mathbb{A}$ over $\mathbb{R}$ and any real Lie algebra $\mathfrak{g}$, we can form an algebra on the real vector space $\mathfrak{g} \otimes \mathbb{A}$ by defining multiplication on simple elements of the form $\xi \otimes p$ by

$$(\xi \otimes p, \eta \otimes q) \mapsto [\xi, \eta] \otimes pq$$

and extending by linearity. When $\mathbb{A}$ is commutative, $\mathfrak{g} \otimes \mathbb{A}$ is a Lie algebra with this multiplication. We are interested in the noncommutative case when $\mathbb{A} = \mathbb{H}$ so $\mathfrak{g} \otimes \mathbb{A}$ does not have the structure of a Lie algebra as the Jacobi identities are not satisfied. Actually it will be more convienient to compose quaternionic multiplication with the quaternionic conjugation map on one factor. Thus,

DEFINITION 3.1.1. *The quaternionic adjoint map,*

$$\mathbf{ad} : \mathfrak{g} \otimes \mathbb{H} \times \mathfrak{g} \otimes \mathbb{H} \longrightarrow \mathfrak{g} \otimes \mathbb{H}$$

is defined on simple elements $x = a \otimes p, y = b \otimes q$ with $a, b \in \mathfrak{g}$ and $p, q \in \mathbb{H}$ by $\mathbf{ad}(x, y) = [a, b] \otimes \bar{p}q$ and extended to all of $\mathfrak{g} \otimes \mathbb{H} \times \mathfrak{g} \otimes \mathbb{H}$ by the sesquilinearity conditions

   (1) $\mathbf{ad}(x, y) = -\overline{\mathbf{ad}(y, x)}$.
   (2) $\mathbf{ad}(x, y + z) = \mathbf{ad}(x, y) + \mathbf{ad}(x, z)$.

We will denote $\mathfrak{g} \otimes \mathbb{H}$ with this algebraic structure by $\mathfrak{g}(\mathbb{H})$.

The following proposition is an immediate consquence of Definition 3.1.1.

PROPOSITION 3.1.2. *In terms of quaternionic components, the quaternionic adjoint map* **ad** *satisfies*

$$\mathbf{ad}(x,y)^0 = \sum_{a=0}^{3}[x^a,y^a]$$

$$\mathbf{ad}(x,y)^a = [x^0,y^a] - [x^a,y^0] + \varepsilon^{abc}[x^c,y^b]$$

for $a = 1,2,3$.

As usual by fixing $x \in \mathfrak{g}(\mathbb{H})$ we define a map $\mathbf{ad}_x : \mathfrak{g}(\mathbb{H}) \to \mathfrak{g}(\mathbb{H})$ by $\mathbf{ad}_x y = \mathbf{ad}(x,y)$. If $\mathfrak{g}$ is semisimple, then there is a nondegenerate invariant bilinear form $(\cdot,\cdot)_0$ given by a multiple of the Killing form on $\mathfrak{g}$ which satifies

$$(3.1.3) \qquad\qquad ([a,b],c)_0 = (a,[b,c])_0.$$

Thus, we can define a nondegenerate invariant bilinear form $(\cdot,\cdot)$ on $\mathfrak{g}(\mathbb{H})$ by

$$(3.1.4) \qquad\qquad (x,y) = \sum_{a=0}^{3}(x^a,y^a)_0.$$

In general, this form is not definite unless $\mathfrak{g}$ is a compact Lie algebra. For noncompact semisimple Lie algebras $\mathfrak{g}$ over $\mathbb{R}$, recall [20] that every such Lie algebra is an orthogonal symmetric Lie algebra $(\mathfrak{g},s)$, where s is an involution on $\mathfrak{g}$ whose fixed point set is a compactly embedded Lie subalgebra $\mathfrak{k}$. Then we have the decomposition of $\mathfrak{g}$ into the plus and minus eigenspaces $\mathfrak{k}$ and $\mathfrak{p}$ of s, respectively, $\mathfrak{g} = \mathfrak{k} + \mathfrak{p}$ which satisfy $[\mathfrak{k},\mathfrak{k}] \subset \mathfrak{k}$, : $[\mathfrak{k},\mathfrak{p}] \subset \mathfrak{p}$, : $[\mathfrak{p},\mathfrak{p}] \subset \mathfrak{k}$. Moreover, the bilinear form $(\cdot,\cdot)_0$ is negative definite on $\mathfrak{k}$ and positive definite on $\mathfrak{p}$, so the form defined by $-(a,sb)_0$ is positive definite on $\mathfrak{g}$. Thus, we can now define a positive definite bilinear form on $\mathfrak{g}(\mathbb{H})$ by

$$(3.1.5) \qquad\qquad (x,y)_s = -\sum_{a=0}^{3}(x^a,sy^a)_0.$$

Moreover, since $\mathfrak{g}(\mathbb{H})$ is a quaternionic vector space, it has a quaternionic structure given by I^a, $a = 1,2,3$. So we can define a hyperhermitian form $<\cdot,\cdot>$ on $\mathfrak{g}(\mathbb{H})$ whose real part is the $(\cdot,\cdot)_s$ and whose imaginary part is defined by

$$(3.1.6) \qquad\qquad \omega^a(x,y) = (I^a x,y)_s.$$

We will need the adjoint of the map $\mathbf{ad}_x$ with respect to the bilinear form (3.1.5). We have

PROPOSITION 3.1.7. *For any* $x,y \in \mathfrak{g}(\mathbb{H})$ *we have* $\mathbf{ad}_x^\dagger y = -\mathbf{ad}_{s\bar{x}}y$.

PROOF. This is a straightforward computation using (3.1.5), (3.1.3), and Proposition 3.1.2.   □

Let K and G be Lie groups with Lie algebras $\mathfrak{k}$ and $\mathfrak{g}$, respectively. The positive definite form $(\cdot, \cdot)_s$ and hence the hyperhermitian form $< \cdot, \cdot >$ is not G invariant; however, they are both K invariant. At the Lie group level we shall only be interested in the subgroup K, not in G nor the symmetric space G/K. The properties of $\mathfrak{p}(\mathbb{H}) = \mathfrak{p} \otimes \mathbb{H}$ that will concern us are the algebraic properties mentioned above, its K module structure, and its flat Riemannian structure. As mentioned previously we shall be particularly interested in the case that $\mathfrak{g}$ is an orthogonal symmetric Lie algebra. Consider the decomposition

$$\mathfrak{g}(\mathbb{H}) = \mathfrak{g} \otimes \mathbb{H} = \mathfrak{k} \otimes \mathbb{H} + \mathfrak{p} \otimes \mathbb{H},$$

where the sum is a direct sum as K modules. Let us write $\mathfrak{p}(\mathbb{H})$ for the quaternionic vector space $\mathfrak{p} \otimes \mathbb{H}$. $\mathfrak{p}(\mathbb{H})$ and $\mathfrak{g}(\mathbb{H})$ are K modules by the adjoint action, i.e.

$$(3.1.8) \qquad\qquad x \mapsto \mathrm{Ad}_{T^{-1}} x$$

for $T \in K$. We shall be interested in the real orthogonal symmetric Lie algebras of noncompact type with $\mathfrak{k} = \mathfrak{o}(k)$, namely, $gl(k, \mathbb{R})$ and $o(k, 1)$. In the former case $\mathfrak{p}$ is $\mathrm{Sym}^2(\mathbb{R}^k)$, whereas in the latter case it is just $\mathbb{R}^k$. In either of these cases $K = O(k)$. The theory can be developed to handle the more general situation where K is any compact Lie group, but for simplicity and our interest in the ADHM construction we only develop the theory for $K = O(k)$. The case $K = U(k)$ should be of interest in Donaldson's treatment [14] of the ADHM construction.

3.2 The flat hyperhermitian structure and the sesquilinear form S. Let $\mathfrak{g}$ be one of the real classical Lie algebras. Then $\mathfrak{g}$ is a subset of the algebra $M_{n,n}(\mathbb{R})$ of n by n matrices over $\mathbb{R}$, so that $\mathfrak{g}(\mathbb{H}) \subset M_{n,n}(\mathbb{H})$. On $M_{n,n}(\mathbb{H})$ there is an involution $*$ given by sending a quaternionic matrix to the transpose of the quaternionic conjugate. Thus, there is a sesquilinear map $S : M_{n,n}(\mathbb{H}) \times M_{n,n}(\mathbb{H}) \to M_{n,n}(\mathbb{H})$ defined by $S(x, y) = x^* y$. This form is selfadjoint in the sense that $S(x, y)^* = S(y, x)$. By abuse of notation we let S denote the restriction of this map to $\mathfrak{g}(\mathbb{H}) \times \mathfrak{g}(\mathbb{H})$. In the case $\mathfrak{g} = gl(k, \mathbb{R})$, $*$ is the transpose of the quaternionic conjugate and the multiplication is matrix multiplication, whereas when $\mathfrak{g} = o(k, 1)$, $*$ is just quaternionic conjugation and the multiplication is the Kronecker product. The sesquilinear form S restricted to $\mathfrak{p}(\mathbb{H})$ contains much of the crucial information needed to understand the geometry of the ADHM construction. Notice that the involution $*$ commutes with this K action; i.e., for $x \in \mathfrak{g} \otimes \mathbb{H}$, we have $(\mathrm{ad}_K x)^* = \mathrm{ad}_K x^*$. If $\mathfrak{p}(\mathbb{H})$ has quaternionic dimension k then S restricted to $\mathfrak{p}(\mathbb{H}) \times \mathfrak{p}(\mathbb{H})$ lies in $M_{k,k}(\mathbb{H})$. Using the identification of $M_{k,k}(\mathbb{H})$ with $gl(k, \mathbb{R}) \otimes \mathbb{H}$, we have for $b \in \mathfrak{p}(\mathbb{H})$ the equivariance condition

$$(3.2.1) \qquad\qquad S(\mathrm{ad}_K b', \mathrm{ad}_K b) = \mathrm{ad}_K S(b', b).$$

We want to study $\mathfrak{g}(\mathbb{H})$ and $\mathfrak{p}(\mathbb{H})$ as flat hyperkähler manifolds. For the quaternionic inner product $< \cdot \, , \cdot >$ on $\mathfrak{g}(\mathbb{H})$ we have

PROPOSITION 3.2.2. $< x \, , y > = \, \text{tr}(S(x \, , y))$.

PROOF. We may write $\text{tr}(S(x \, , y)) = \text{tr}(x^* y)$ as

$$\text{tr}[((x^0)^t - i(x^1)^t - j(x^2)^t - k(x^3)^t)(y^0 + iy^1 + jy^2 + ky^3)]$$

$$= \text{tr}[\sum_{a=0}^{3}(x^a)^t y^a + i((x^0)^t y^1 - (x^1)^t y^0 + (x^3)^t y^2 - (x^2)^t y^3) + \cdots =]$$

$$= (x \, , y)_s + i(I^1 x \, , y)_s + j(I^2 x \, , y)_s + k(I^3 x \, , y)_s.$$

where the last equality uses the relation $sx = -x^t$. □

The flat hyperhermitian form $g_0 + \omega_0$ is obtained from $< \cdot \, , \cdot >$ by identifying the vector space $\mathfrak{g}(\mathbb{H})$ with its tangent space. This gives $\mathfrak{g}(\mathbb{H})$ the structure of a flat hyperkähler manifold. By construction K is a subgroup of the hyperkähler isometry group of g_0. Thus, since K is semisimple, the Ad-action on $\mathfrak{g} \otimes \mathbb{H}$ gives rise to a K-equivariant hyperkähler moment map

$$\mu : \mathfrak{g}(\mathbb{H}) \longrightarrow \mathfrak{k}^* \otimes \mathbb{R}^3$$

defined by (2.3.3). Furthermore, the pairing of $\mathfrak{k}^*$ with $\mathfrak{k}$ can be accomplished by the trace form

(3.2.3) $$(\eta \, , \xi)_0 = - \text{tr}(\eta \xi).$$

Henceforth, we shall consider the moment map as the map $\mu : \mathfrak{g} \otimes \mathbb{H} \longrightarrow \mathfrak{k} \otimes \mathbb{R}^3$ obtained through the identification with the trace form (3.2.3).

PROPOSITION 3.2.4. On $\mathfrak{g}(\mathbb{H})$ the moment map μ is given by $2\mu^a(x) = \text{ad}^a(x^t, x)$.

PROOF. Through the identification of $\mathfrak{g}(\mathbb{H})$ with its tangent space, the fundamental vertical vector field ξ^* is identified with $[x \, , \xi]$ by (3.1.8), and the symplectic form ω_0^a is identified with $\text{tr}\, S^a$ by Proposition 3.2.2. So that the one-form $\iota_{\xi^*} \omega^a$ is identified with $\text{tr}\, S^a([x \, , \xi], \cdot)$. Thus, we compute for $a = 1 \, , 2 \, , 3$,

$$\text{tr}\, S^a([x \, , \xi], y) = \text{tr}([x \, , \xi]^* y)^a = \text{tr}([x^* \, , \xi] y)^a$$

$$= \text{tr}(([(x^0)^t, \xi] - i[(x^1)^t, \xi] - j[(x^2)^t, \xi]$$

$$-k[(x^3)^t, \xi])(y^0 + iy^1 + jy^2 + ky^3))^a$$

$$= \text{tr}([(x^0)^t, \xi] y^a - [(x^a)^t, \xi] y^0 + \varepsilon^{abc}[(x^c)^t, \xi] y^b)$$

$$= - \text{tr}\, \xi([(x^0)^t, y^a] - [(x^a)^t, y^0] + \varepsilon^{abc}[(x^c)^t, y^b])$$

$$= - \text{tr}\, \xi d\mu_x^a(y)$$

which implies that the differential of the moment map L_x is obtained by projecting the expression in parenthesis onto its antisymmetric part; i.e.,

$$L_x(y)^a = \tfrac{1}{2}([(x^0)^t, y^a] - [(x^a)^t, y^0] - [(y^a)^t, x^0] + [(y^0)^t, x^a]$$
$$+ \varepsilon^{abc}([(x^c)^t, y^b] + [(y^c)^t, x^b]))$$

Now on a flat hyperkähler manifold, the two-form ω is homogeneous of degree two. Furthermore, since the action of $K = O(k)$ is linear, the moment map μ is also homogeneous of degree two. Hence, $L_x(x)^a = 2\mu(x)$, and the result follows from (3.2.5) and the second equation of Proposition 3.1.2. □

Consider the diagonal map $\Delta : \mathfrak{g} \otimes \mathbb{H} \to \mathfrak{g} \otimes \mathbb{H} \times \mathfrak{g} \otimes \mathbb{H}$ induces a map

$$\Delta^* S : \mathfrak{g}(\mathbb{H}) \longrightarrow M_{k,k}(\mathbb{H}).$$

Let us decompose $\Delta^* S$ into its real and imaginary parts.

(3.2.6) $\Delta^* S = \Re + \Im.$

One easily sees that

PROPOSITION 3.2.7. *The following properties hold for* $\Delta^* S$:
  (1) $\Re$ *is symmetric and* $< x, x > = \operatorname{tr} \Re(x)$.
  (2) $\Im$ *is antisymmetric.*

Let us restrict the above setup to the K module $\mathfrak{p}(\mathbb{H})$. In this case we get explicit expressions for the moment map, its differential, and the adjoint of the differential map which will be important later. These expressions can be derived in a straightforward way using Propositions 3.2.4, 3.1.7, and 3.1.2. Summarizing, we have

PROPOSITION 3.2.8. *For* $b, v \in \mathfrak{p}(\mathbb{H})$ *and* $\xi \in \mathfrak{o}(k) \otimes \mathbb{R}^3$, *we have*
  (1) $2\mu = \Delta^* \mathbf{ad} = \Im$.
  (2) $\mu^a(b) = [b^0, b^a] + \frac{\varepsilon^{abc}}{2}[b^c, b^b]$.
  (3) *The differential of* μ^a *is:* $L_b(v)^a = \mathbf{ad}(b, v)^a = [b^0, v^a] - [b^a, v^0] + \varepsilon^{abc}[b^c, v^b]$.
  (4) $L_b^\dagger \xi = ([\xi^a, b^a], -[\xi^a, b^0] + \varepsilon^{abc}[\xi^b, b^c])$.
  (5) $(L_b L_b^\dagger \xi)^a = [b^0, [b^0, \xi^a]] + [b^c, [b^c, \xi^a]] + [\xi^c, [b^c, b^a]] + \varepsilon^{abc}[\xi^b, [b^0, b^c]]$.

Now the real part $\Re$ of $\Delta^* S$ can be used to define a quadratic form on the Lie algebra $\mathfrak{k}$ for each $b \in \mathfrak{p}(\mathbb{H})$. In fact we have

PROPOSITION 3.2.9. *The quadratic form defined by* $(\eta, \xi)_\Re = (\eta, \Re(b)\xi)$ *is nonnegative for all* $b \in \mathfrak{p}(\mathbb{H})$, *and positive definite on the subset where* K *acts locally freely.*

PROOF. We have $(\xi, \Re(b)\xi) = (\xi, b^* b \xi) = ((b\xi)^*, b\xi) \geq 0$, which proves the first statement. Now suppose that $(\xi, \Re(b)\xi) = 0$ for some $\xi \in \mathfrak{k}$, then $b\xi = 0$. In the case $\mathfrak{g} = \mathfrak{o}(k, 1)$, this coincides with the infinitesimal action and we are done. For $\mathfrak{g} = \mathfrak{gl}(k, \mathbb{R})$, since b is symmetric and ξ is antisymmetric, $b\xi = 0$ implies $[b, \xi] = 2b\xi = 0$. This proves the second statement. $\square$

3.3 Hyperkähler reduction on $\mathfrak{p}(\mathbb{H})$. In what follows we shall be interested in direct products of quaternionic vector spaces of the form $\mathfrak{p}(\mathbb{H})$. The structure described above then carries over additively to the product situation. Namely, on $\mathfrak{p}(\mathbb{H}) = \mathfrak{p}_1(\mathbb{H}) + \mathfrak{p}_2(\mathbb{H})$ we have quadratic forms given by $S = S_1 + S_2$ and $\mathbf{ad}_1 + \mathbf{ad}_2$. In particular, we have the moment map $\mu = \mu_1 + \mu_2$ on $\mathfrak{p}(\mathbb{H})$. Of course, this construction can be iterated any finite number of times. Now let us suppose that we have such a quaternionic vector space $\mathfrak{p}(\mathbb{H})$. Furthermore, suppose that K acts freely on some open subset $\mathfrak{p}^{\mathrm{fr}}(\mathbb{H})$, and let $\hat{\mu}$ denote the restriction of μ to $\mathfrak{p}^{\mathrm{fr}}(\mathbb{H})$. Then we can apply (1) of Proposition 3.2.8 together with the hyperkähler quotient Theorem 2.3.6 to obtain

THEOREM 3.3.1. *The quotient $\hat{\mu}^{-1}(0)/K$ is a hyperkähler manifold where $\mu = \Delta^* \mathbf{ad}$. Furthermore, the hyperhermitian form is obtain by restricting the hyperhermitian form $g_0 + \omega_0$ to the horizontal subbundle $\mathscr{H}$ over $\hat{\mu}^{-1}(0)$.*

Actually there is a more general stratified version of this theorem. However, since the stratification that is of interest to us comes from the ADHM construction we shall formulate this in that context only in the next section. We shall now see how, when applied to the spaces $\mathfrak{p}^{\mathrm{fr}}(\mathbb{H})$, the geometry of the preceeding section is all encoded in the quadratic forms S and $\mathbf{ad}$. Consider diagram 2.3.7 with $M = \mathfrak{p}^{\mathrm{fr}}(\mathbb{H})$ together with its flat hyperkähler structure. We have the following:

PROPOSITION 3.3.2. *For $M = \mathfrak{p}^{\mathrm{fr}}(\mathbb{H})$, we have (1) $\mathscr{H}_b = \ker(\mathrm{Re}(\mathbf{ad}_b))$ for $b \in \mathfrak{p}^{\mathrm{fr}}(\mathbb{H})$.*

(2) $T_b \hat{\mu}^{-1}(0) = \ker(\mathrm{Im}(\mathbf{ad}_b))$ for $b \in \hat{\mu}^{-1}(0)$.

(3) $\hat{\mathscr{H}}_b = \ker(\mathbf{ad}_b)$.

PROOF. First, (2) follows immediately from (3) of Proposition 3.2.8, and (3) follows from (1) and (2). To prove (1), we have that $v \in \mathscr{H}_b$ for any $b \in \mathfrak{p}^{\mathrm{fr}}(\mathbb{H})$ if and only if $g_0(v, \xi^*) = 0$ for all $\xi \in \mathfrak{k}$. Now let $\mathscr{E}_{ij}$ denote the matrix in $\mathfrak{o}(k) = \mathfrak{k}$ with a 1 in the ith row and jth column and a -1 in the jth row and ith column. Then the fundamental vertical vector field corresponding to $\mathscr{E}_{ij}$ is

$$(3.3.3) \qquad \mathscr{E}_{ij}^* = \sum \left(b_{ir}^a \frac{\partial}{\partial b_{jr}^a} - b_{jr}^a \frac{\partial}{\partial b_{ir}^a} \right),$$

and a straightforward calculation using (3.3.3) gives $g_0(v, \mathscr{E}_{ij}^*) = \sum [b^a, v^a]$ which is the first equation of Proposition 3.1.2. $\square$

Next consider the vertical subbundle Vert on $\mathfrak{p}^{\text{fr}}(\mathbb{H})$. Since the action of $O(k)$ on $\mathfrak{p}^{\text{fr}}(\mathbb{H}$ is free, the fundamental vertical vector fields $\mathscr{E}_{ij}^*$ give a global frame for Vert and, thus, a trivialization of Vert. Consider the flat metric restricted to Vert. This gives us a smooth family $g_{[b]}$ of metrics on the fibers of π. We have

PROPOSITION 3.3.4. Let ξ, $\eta \in o(k)$, then the family of metrics in the fiber is given by $g_{[b]}(\xi^*, \eta^*) = 4(\xi, \eta)_{\Re(b)} = 4(\xi, \Re(b)\eta)$.

PROOF. We first check that this equation is well defined: The right-hand side of the above equation depends only on the equivalence class $[b]$, since the quadratic form is $O(k)$ invariant, and Proposition 3.2.9 says that $(\cdot, \cdot)_R$ is positive definite on $\mathfrak{p}^{\text{fr}}(\mathbb{H})$. So we need only show that $g_0(\xi^*, \eta^*) = 4(\xi, \Re(b)\eta)$. Once again this is a straightforward computation performed by expanding ξ^* and η^* in terms of the basis $\mathscr{E}_{ij}^*$ and using (3.3.3). □

Since the trace form $(\cdot, \cdot)$ is $K = O(k)$ invariant, we can diagonalize $\Re(b)$ at each $b \in \mathfrak{p}^{\text{fr}}(\mathbb{H})$. Let $\lambda_i(b)$ denote the eigenvalues which are all positive. The set $\{\lambda_1(b), \ldots, \lambda_k(b)\}$ depends only on the equivalence class $[b]$; however, there is no canonical ordering depending only on $[b]$. But the elementary symmetric functions $\{\sigma_1(b), \ldots, \sigma_k(b)\}$ formed from the eigenvalues give k functions depending only on $[b]$ and thus define functions on $\mathfrak{p}(\mathbb{H})/K$. Restricting to $\hat{\mu}^{-1}(0)$, we get $4k$ horizontal vector fields $(\text{grad}\,\sigma_i, I^a\,\text{grad}\,\sigma_i)$ that are independent on a dense open set.

4. The ADHM construction

4.1 The construction. Here we briefly review the well-known construction of all gauge equivalence classes of self-dual connections on S^4 in terms of quaternionic linear algebra due to Atiyah, Drinfeld, Hitchin, and Manin. In this section we restrict ourselves to the case $G = \text{Sp}(n)$, the compact symplectic group, although using Lemma 1.19 of [13] the results obtained in §§5 and 6 hold for any compact simple Lie group. For full details we encourge the reader to consult the lecture notes of Atiyah [3]. Principal $\text{Sp}(n)$ bundles P on S^4 are classified up to isomorphism by an integer $k \in \pi_3(\text{Sp}(n))$. Let E denote the associated vector bundle corresponding to the left action of $\text{Sp}(n)$ on a quaternionic vector space. The Gauss map corresponding to the embedding of E into the trivial bundle H^{n+k} of quaternionic dimension $n + k$ gives the short exact sequence of vector bundles

$$(4.1.1) \qquad 0 \longrightarrow E \longrightarrow H^{n+k} \longrightarrow kL \longrightarrow 0$$

where kL denotes k copies of the quaternionic Hopf bundle. We wish to find an orthogonal splitting of this sequence, with respect to the flat metric on H^{n+k}; that is, we write the flat bundle H^{n+k} as the direct sum $E \oplus kL$.

The embedding of kL into H^{n+k} can then be represented by the matrix

$$(4.1.2) \qquad v(x) = \begin{pmatrix} \Lambda \\ B - xI \end{pmatrix}$$

where Λ and B are certain n by k and k by k quaternionic matrices, and $I = I_k$ denotes the k by k unit matrix. In order to guarantee that the matrix 4.1.2 represents an actual embedding of a k dimensional bundle, the following rank condition is imposed:

$$(4.1.3) \qquad \operatorname{rank} v(x) = k \quad \text{for all } x \in \mathbb{HP}(1).$$

Now connections A on E can be obtained by Gauss' formula from the flat connection on H^{n+k} and the curvature can be computed explicitly [**3**, II-3.11], viz.,

$$(4.1.4) \qquad F^A = N dx \rho^{-2} d\bar{x} N^*,$$

where N is a projection operator and

$$(4.1.5) \qquad \rho^2(x, b) = v^* v = (B^* - \bar{x}I)(B - xI) + \Lambda^* \Lambda.$$

Thus, we obtain connections $A^{(\Lambda, B)}$ on E parameterized by the quaternionic matrices (Λ, B). Hence, we get a map from the open set of matrices in $M_{n,k}(\mathbb{H}) \times M_{k,k}(\mathbb{H})$ which satisfy the rank condition (4.1.3) to the affine space $\mathscr{A}_k$ of smooth connections on E. The real orthogonal group $O(k)$ acts naturally on $M_{n+k,k}(\mathbb{H}) \simeq M_{n,k}(\mathbb{H}) \times M_{k,k}(\mathbb{H})$ by sending

$$(4.1.6) \qquad (\Lambda, B) \mapsto (\Lambda T, T^{-1} B T)$$

where $T \in O(k)$. This action clearly preserves the rank condition (4.1.3), and it can be easily checked that the connection $A^{(\Lambda, B)}$ depends only on the $O(k)$ equivalence class of (Λ, B). Let us define the space

$$(4.1.7) \qquad \mathscr{B}_k = \{(\Lambda, B) \in M_{n,k}(\mathbb{H}) \times M_{k,k}(\mathbb{H}) : \\ \text{condition } 4.1.3 \text{ is satisfied}\}/O(k).$$

Thus, we have a map from $\mathscr{B}_k$ to $\mathscr{A}_k$. Now the based gauge group $\mathscr{G}_k^b$ acts freely on $\mathscr{A}_k$ and the quotient space $\mathscr{C}_k$ is homotopy equivalent [**7**] to the four-fold loop space $\Omega^4 B \operatorname{Sp}(n)$. So composing the above map with the natural projection, we get a continous map $\alpha : \mathscr{B}_k \to \mathscr{C}_k$ sending the equivalence class $[(\Lambda, B)]$ to the based gauge equivalence class of $A^{(\Lambda, B)}$.

The condition that the connection $A^{(\Lambda, B)}$ be self-dual is precisely the condition that the k by k quaternionic matrix $\rho^2(x, b)$ be real, and this is equivalent to the following two conditions:

$$(4.1.8) \qquad \begin{array}{ll} \text{(i)} & B \text{ is symmetric} \\ \text{(ii)} & B^* B + \Lambda^* \Lambda \text{ is real.} \end{array}$$

Furthermore, the real orthogonal group $O(k)$ preserves both of these conditions. Thus, denoting by $\mathscr{M}_k$ the subspace of $\mathscr{C}_k$ consisting of based gauge equivalence classes of self-dual connections and by $\hat{\mathscr{M}}_k$ the subspace of $\mathscr{B}_k$ satisfying (4.1.8), we see that α restricts to a map $\hat{\alpha} : \hat{\mathscr{M}}_k \to \mathscr{M}_k$. It is a remarkable theorem of Atiyah, Drinfeld, Hitchin, and Manin [4] that the above construction gives **all** instantons of charge k, namely

THEOREM 4.1.9. *The map* $\hat{\alpha} : \hat{\mathscr{M}}_k \to \mathscr{M}_k$ *is a diffeomorphism.*

Hereafter, we shall refer to instanton moduli space $\mathscr{M}_k$ with the above identification being implicit.

4.2 Symmetries. In this section we present a discussion of an important transformation group on $\mathscr{B}_k$. Let $H_\infty = (SO(4) \times \mathbb{R}^*) \oslash \mathbb{R}^4$ denote the subgroup of the conformal group of the sphere S^4 which fixes the north pole, where $\oslash$ denotes semidirect product given by the natural action of $SO(4) \times \mathbb{R}^*$ on $\mathbb{R}^4$ by rotations and dilatations. The group H_∞ is just the similtitude group consisting of Euclidean transformations of $\mathbb{R}^4$ together with dilatations. This action on $\mathbb{R}^4$ induces a natural action on the space of connections $\mathscr{A}_k$ [28] which commutes with the action of the based gauge group $\mathscr{G}_k^b$. Furthermore, there is a natural action of the universal covering group of H_∞ on $M_{n+k,k}(\mathbb{H})$ given by

$$(4.2.1) \qquad\qquad T_{q,p}(\Lambda, B) = (\Lambda p, qBp)$$

for $(q, p) \in SU(2) \times SU(2)$ and by

$$(4.2.2) \qquad\qquad T_a T_\delta(\Lambda, B) = (\delta\Lambda, \delta B + aI)$$

for $\delta \in \mathbb{R}^*$, and $a \in \mathbb{H} = \mathbb{R}^4$.

Now it is easily seen that the rank condition (4.1.3), the symmetry condition and the reality condition (4.1.8) are all unaffected by the action of H_∞ given in (4.2.1) and (4.2.2) so this action leaves invariant the important subspaces $\mathscr{M}_k' \subset \mathscr{V}_k' \subset \mathscr{B}_k'$ as well as each strata $\mathscr{V}_{k,j}'$. Furthermore, since this action of the universal $\mathbb{Z}_2$ cover of H_∞ commutes with the action (4.1.6) of $O(k)$, we obtain well-defined actions of H_∞ on the various quotient spaces. Moreover, one easily verifies

PROPOSITION 4.2.3. *The map* $\alpha_k : \mathscr{B}_k \to \mathscr{C}_k$ *is equivariant with respect to the natural actions of* H_∞ *described above.*

Another subgroup of symmetries of interest is the group of constant gauge transformations. The based gauge group $\mathscr{G}_k^b$ is a normal subgroup of the full gauge group $\mathscr{G}_k$ and the factor group $\mathscr{G}_k/\mathscr{G}_k^b$ can be identified with the constant gauge transformations $Sp(n)/\mathbb{Z}_2$. Furthermore, the group $Sp(n)$ acts on $M_{n+k,k}(\mathbb{H})$ by sending (Λ, B) to $(A\Lambda, B)$ for $A \in Sp(n)$, and it is easily seen that this leaves the subspace $\mathscr{M}_k'$ invariant. Moreover, this passes to an action of $Sp(n)/\mathbb{Z}_2$ on the quotient $\mathscr{M}_k'/O(k)$ which coincides with the action of the constant gauge transformations under the identification given by Theorem 4.1.9. It is easily seen that this group commutes with the group H_∞ so that the full symmetry group acting on the moduli space $\mathcal{M}_k$ is the direct product $G = (\mathrm{Sp}(n)/\mathbb{Z}_2) \times H_\infty$.

4.3 Symmetric Lie algebras. Consider the orthogonal symmetric Lie algebras $gl(k, \mathbb{R})$ and $o(k, 1)$ in their symmetric space decomposition $\mathfrak{k} + \mathfrak{p}$. Let $\mathfrak{p}_0$ and $\mathfrak{p}_1$ denote the -1 eigenspace of s for the Lie algebras $gl(k, \mathbb{R})$ and $o(k, 1)$, respectively, and let $\mathfrak{p}_1^n$ denote n copies of $\mathfrak{p}_1$. Define $\mathfrak{p} = \mathfrak{p}_0 + \mathfrak{p}_1^n$. Then we easily see that $\mathfrak{k} = \mathfrak{o}(k)$, the Lie algebra of $O(k)$, and $\mathfrak{p}(\mathbb{H}) = (\mathrm{Sym}^2\mathbb{R}^k + \mathbb{R}^{nk}) \otimes \mathbb{H}$. Thus, $\mathfrak{p}(\mathbb{H})$ can be identified with the subset of $n + k$ by k quaternionic matrices $M_{n+k, k}(\mathbb{H})$ that satisfies condition (i) of (4.1.8). It is important to consider $\mathfrak{p}(\mathbb{H})$ as a stratified set (see [2 p. 29]) stratified by a rank condition. However, the rank condition of the ADHM construction is stronger than the usual definition of the quaternionic rank of a matrix in $M_{n+k, k}(\mathbb{H})$. Using the identification with quaternionic matrices let us write any $b \in \mathfrak{p}(\mathbb{H})$ as

$$(4.3.1) \qquad\qquad \begin{pmatrix} \Lambda \\ B \end{pmatrix}$$

where $\Lambda \in \mathbb{H}^{nk}$ and $B \in \mathrm{Sym}^2\mathbb{H}^k$. Consider the generator I of the center of the Lie algebra $gl(k, \mathbb{R})$ and let $x \in \mathbb{H}$. The element $xI \in \mathfrak{p}_0 \otimes \mathbb{H}$ gives rise to an action of the quaternionic translation group $\mathbb{R}^4$ ((4.2.2) with $\delta = 1$) on $\mathfrak{p}(\mathbb{H})$ by sending (Λ, B) to $(\Lambda, B + xI)$. Let T_x denote this action. Then the rank condition (4.1.3) can be rephrased as $\mathrm{rank}(T_{-x}b) = k$ for all $x \in \mathbb{H}$. We now make the following

DEFINITION 4.3.2. $\mathfrak{p}^{(k)}(\mathbb{H}) = \{b \in \mathfrak{p}(\mathbb{H}) : \mathrm{rank}(T_{-x}b) = k \text{ for all } x \in \mathbb{H}\}$.

We shall need

LEMMA 4.3.3. *The orthogonal group $O(k)$ acts freely on $\mathfrak{p}^{(k)}(\mathbb{H})$.*

PROOF. Assume to the contrary that $b \in \mathfrak{p}^{(k)}(\mathbb{H})$ is a fixed point for some $T \in O(k)$. We show that the rank condition must fail. Now, up to conjugation, any element $T \in O(k)$ is either a reflection or a rotation in a two-plane. In the standard representation of $O(k)$ we may assume that the reflection is given by $T(x_1, \ldots, x_n) = (-x_1, \ldots, x_n)$ and the rotation R is a rotation in the 1-2-plane. If the reflection T fixes $b = (\Lambda, B)$ then we must have

$$\Lambda = \begin{pmatrix} 0 & \Lambda' \end{pmatrix}$$

where $\Lambda' \in M_{n, k-1}(\mathbb{H})$, 0 is the zero n-vector, and

$$B = \begin{pmatrix} \beta & 0 \ldots 0 \\ 0 & \\ \vdots & B' \\ 0 & \end{pmatrix}$$

where $\beta \in \mathbb{H}$ and $B' \in \mathrm{Sym}^2 \mathbb{H}^{k-1}$. But then b loses rank at $x = \beta$.

Similarly, if R is a nontrivial rotation in the 1-2-plane that fixes $b = (\Lambda, B)$ then $\Lambda = (0 \ 0 \ \Lambda')$ and

$$B = \begin{pmatrix} aI & 0' \\ 0 & B' \end{pmatrix}$$

where $a \in \mathbb{H}$. Again b loses rank at $x = a$.   □

Let us define $\mathscr{V}_k = \mathfrak{p}^{(k)}(\mathbb{H})/O(k)$, then, since $O(k)$ is compact, we have

PROPOSITION 4.3.4. *The natural projection* $\pi : \mathfrak{p}^{(k)}(\mathbb{H}) \to \mathscr{V}_k$ *is a principal* $O(k)$ *bundle.*

The reality condition for the ADHM construction is that the k by k quaternionic matrix $\rho^2(x, b)$ given by (4.1.5) be real. By considering the translation T_x in (4.2.2) above, we can write this in a more succinct form, viz.

$$(4.3.5) \qquad \rho^2(x, b) = (T_{-x}b)^*(T_{-x}b) = \Delta^* S(T_{-x}b).$$

Decomposing this into its real and imaginary parts as in (3.2.6), we have

$$\rho^2(x, b) = \Re(T_{-x}b) + \mu(b).$$

Therefore, the reality condition of the ADHM construction is just

$$(4.3.6) \qquad \mathrm{Im}(\rho^2(x, b)) = \mu(b) = 0.$$

Notice that this constraint is *independent* of x and coincides with the vanishing of the moment map. Thus, Theorem 3.3.1 reduces in this case to Hitchin's obervation:

THEOREM 4.3.7. *The ADHM construction gives* $\mathscr{M}_k$ *the structure of a hyperkähler manifold.*

4.4 The stratified version. We define the rank of b at x, denoted by $r_b(x)$, to be the quaternionic rank of $T_{-x}b$. Note that $r_b(x)$ is less than k for at most a finite number of points $x_i \in \mathbb{H}$. Let $c_b(x) = k - r_b(x)$ denote the corank of b at x. Then $c_b(x)$ is different from 0 for at most a finite number of points x_i. Moreover, the total corank $c(b) = \sum_i c_b(x_i)$ is at most k. Define the *rank* of b by $r(b) = k - c(b)$. Then the rank condition $r(b) = k$ is precisely the rank condition of the ADHM construction for $\mathrm{Sp}(n)$ instantons given by Atiyah [3]. We now generalize Definition 4.3.2:

DEFINITION 4.4.1. $\mathfrak{p}^{(r)}(\mathbb{H}) = \{b \in \mathfrak{p}(\mathbb{H}) : r(b) = r\}$.

PROPOSITION 4.4.2. *The* rank $r(b)$ *partitions* $\mathfrak{p}(\mathbb{H})$ *into the disjoint union*

$$\coprod_{r=0}^{k} \mathfrak{p}^{(r)}(\mathbb{H}).$$

Furthermore, each stratum $\mathfrak{p}^{(r)}(\mathbb{H})$ *is a locally closed smooth submanifold of* $\mathfrak{p}(\mathbb{H})$ *of real dimension* $2r(2n + 2k - r + 1) + 4(k - r)$, *and the maximal stratum* $\mathfrak{p}^{(k)}(\mathbb{H})$ *is a dense open set in* $\mathfrak{p}(\mathbb{H})$.

PROOF. Fixing an $x \in \mathbb{H}$, this is a standard result (cf. [2, p. 29]), and the dimension count gives $2r(2n + 2k - r + 1)$. But there are precisely $k - r$ such points x over $\mathbb{H}$, so the real dimension of the submanifold is that given above. $\square$

THEOREM 4.4.3. *There is a homeomorphism*

$$\mu^{-1}(0)/O(k) \simeq \mathcal{M}_k \cup \bigcup_{l=1}^{k} \mathcal{M}_{k-l} \times SP^l(\mathbb{H})$$

where the topology of the right-hand side is that of sequential convergence with respect to the metric g.

Hereafter, we denote the right-hand side of the above equation by $\overline{\mathcal{M}}_k$. This is precisely the completion of $\mathcal{M}_k$ with respect to the hyperkähler metric g. There is a compactification of unbased moduli spaces due to Donaldson [15, 16]. It would be of interest to understand the relation between our completion and Donaldson's compactification for instantons on the four sphere.

We show that there is a set bijection. The fact that the topologies are equivalent will then follow easily. We begin with

LEMMA 4.4.4. *If* $b \in \mu^{-1}(0)$ *has rank* r *over* $\mathbb{H}$, *then* b *has rank* r *over* $\mathbb{R}$.

PROOF. Since $b \in \mu^{-1}(0) \cap \mathfrak{p}^{(r)}(\mathbb{H})$, there are $k - r = c$ points $x_1, \ldots, x_c$ of $\mathbb{H}$ such that $\det \Re(x_i, b) = 0$. We first consider the case that all the x_i's are distinct, thus having multiplicity one. In this case the matrix $\Re(x, b)$ has quaternionic rank $k - 1$ at $x = x_i$ for $i = 1, \ldots, c$ and quaternionic rank k at $x \neq x_i$. Let us fix one of the x_i's, say x_1. By translation we can take $x_1 = 0$. Then the condition that b has rank $k - 1$ over $\mathbb{H}$ at x_1 is

(4.4.5)
$$\sum b_{il} q_l = 0$$

for some quaternionic column vector $q = (q_l)$. By a permutation in $O(k)$ we can take $q_1 \neq 0$, and since the quaternionic vector q is defined only up to right multiplication by a nonzero quaternion, we can take $q_1 = -1$. Hence, we can solve for the first column b_{i1} as a quaternionic linear combination of the remaining columns; i.e., $b_{i1} = \sum_{l=2}^{k} b_{il} q_l$. Now let q denote the quaternionic vector $(0, q_2, \ldots, q_k)$, then, since $\mu(b) = 0$, we get $0 = \mu(b)_{i1} = (\text{Im}(b^* b))_{i1} = \text{Im}(\rho^2(b)q)_i = (R(b) \text{Im}(q))_i$ for $i = 1, \cdots, k$. But since R has rank $k - 1$, we must have $\text{Im } q = 0$. So the real rank of b at x_1 is $k - 1$. The same argument clearly holds at the remainding $c - 1$ points. Now suppose that x_1 has multiplicity l, then there are l linearly independent relations of the form (4.4.5) over $\mathbb{H}$, and some $k - l$ by

$k - l$ minor of $\Re(b)$ is invertible. A similar argument as above shows that l columns can be written as linear combinations over of the remaining $k - l$ columns.   □

PROOF OF 4.4.3. Suppose $b \in \mu^{-1}(0) \cap \mathfrak{p}^{(k-c)}(\mathbb{H})$, and let x_1 be one of the points where b drops rank. Again by a translation T_{x_1} we can put $x_1 = 0$. Then by Lemma 4.4.4 there is a real dependence relation between the columns of b. Thus, there is an $O(k)$ transformation which makes the first column identically zero. By translating back the first column becomes zero except for $B_{11} = x_1$. We can do this at each of the c points (including multiplicity) where b drops rank. Thus using the symmetry of B, b can be put in the form

$$(4.4.6) \qquad b = \begin{pmatrix} 0 & \cdots & 0 & \Lambda' \\ x_1 & \cdots & 0 & 0 \\ 0 & \ddots & 0 & \vdots \\ 0 & & x_c & 0 \\ 0 & 0 & 0 & B' \end{pmatrix}$$

where $b' = (\Lambda', B')$ has rank $k - c$. Furthermore, the form of b in (4.4.6) implies that $\mu(b') = 0$, so b' represents an instanton of charge $k - c$. Let Σ_c denote the symmetric group on c letters. Then by quotienting with $\Sigma_c \times O(k - c)$, (4.4.6) defines a map from $\mu^{-1}(0) \cap \mathfrak{p}^{(k-c)}(\mathbb{H})$ to $SP^c(\mathbb{H}) \times \mathscr{M}_{k-c}$ which is clearly surjective. But, since the points where b drops rank are uniquely specified, together with their multiplicity, the element $([x_1, \ldots, x_c], [b'])$ of $SP^c(\mathbb{H}) \times \mathscr{M}_{k-c}$ is the image of a unique element $[b]$ in the set of $O(k)$ equivalence classes of b. Thus we have a well-defined bijection from $\mu^{-1}(0) \cap \mathfrak{p}^{(k-c)}(\mathbb{H})/O(k)$ to $SP^c(\mathbb{H}) \times \mathscr{M}_{k-c}$.   □

REMARK 4.4.7. *The space $\mathscr{M}_k$ is contractible by the straight line homotopy obtained from the dilatation T_δ of (4.2.2). In fact, the one point union of $\mathscr{M}_k$ with $\{[0]\}$ is contractible.*

4.5 The hyperkähler structure on $\mathscr{M}_k$. In this section we discuss the geometry of the hyperhermitian form $h = g + \omega$ on $\mathscr{M}_k$. It is clear from Theorem 4.4.3 that the metric g on $\mathscr{M}_k$ is not complete, and its completion is the stratified space $\mu^{-1}(0)/O(k)$. Furthermore, each stratum $\mathscr{M}_{k-l} \times SP^l(\mathbb{H})$ has a natural hyperkähler structure off of its singular set. The hyperkähler structure on $\mathscr{M}_k$ can easily be determined from the results of §3. In what follows we let U and V be vector fields on $\mathscr{M}_k$ and let u_b and v_b denote the unique vectors in $\mathscr{H}_b$ corresponding to the evaluation U_b and V_b, respectively. Then we have

PROPOSITION 4.5.1. *The hyperhermitian form on $\mathscr{M}_k$ is given by*

$$h(U_b, V_b)_{\mathscr{M}_k} = \langle u_b, v_b \rangle = \mathrm{tr}(u_b^* v_b).$$

Let us briefly consider the special case $k = 1$ and $n = 1$. In this case there are no quadratic constraints and $O(1) = \mathbb{Z}_2$, so the metric on $\mathcal{M}_1$ is the flat hyperkähler metric on $(\mathbb{H}^2 - \mathbb{H})/\mathbb{Z}_2$, where the deleted quaternionic line is $\Lambda = 0$. This can be identified with the half-space $\mathbb{H} \times \mathbb{R}^+$ times $SO(3) = \mathbb{RP}^3$, where $\mathbb{R}^+$ denotes the positive real numbers. The more common model of the based moduli space $\mathcal{M}_1$ as $SO(3)$ times the open five ball B^5 is then obtained from our model by the Cayley transform. This gives the flat metric on the unbased moduli space $\tilde{\mathcal{M}}_1 \simeq B^5$. It is, of course, conformally equivalent to both the metric discussed by Groisser and Parker [17, 18] and the standard hyperbolic Poincaré metric on $\tilde{\mathcal{M}}_1$ [6].

Concerning symmetries we have

PROPOSITION 4.5.2. *The group* $G = (Sp(n)/\mathbb{Z}_2) \times H_\infty$ *described in* §4.2 *acts as homotheties on* $\mathcal{M}_k$, *while the subgroup* I *obtained by putting* $\delta = 1$ *in equation* (4.2.2) *acts as isometries, and the subgroup* H *obtained by putting* $\delta = 1$ *in* (4.2.2) *and* $p = 1$ *in* (4.2.1) *acts as hyperkähler isometries on* $\mathcal{M}_k$. *Furthermore,* I *is normal in* G *and is isomorphic to* $(Sp(n)/\mathbb{Z}_2) \times (SO(4) \oslash \mathbb{H})$, *whereas* H *is normal in* I *and is isomorphic to* $(Sp(n)/\mathbb{Z}_2) \times (Sp(1) \oslash \mathbb{H})$. *The factor group* I/H *is isomorphic to* $Sp(1)$ *and induces a transitive action on the two-sphere of complex structures on* $\mathcal{M}_k$.

A consequence of the last statement is that the de Rham cohomology classes of the Kähler two forms ω^a on $\mathcal{M}_k$ vanish. Furthermore, the homotheties determined by δ of (4.2.2) and this $Sp(1)$ action combine to give an $\mathbb{H}^*$ action on $\mathcal{M}_k$. Thus, $\mathcal{M}_k$ has a projectivization $\mathbb{P}\mathcal{M}_k$ and we have

PROPOSITION 4.5.3. *There is a principal fibration* $\mathbb{H}^* \to \mathcal{M}_k \to \mathbb{P}\mathcal{M}_k$ *and the quotient manifold* $\mathbb{P}\mathcal{M}_k$ *is a quaternionic Kähler manifold, whenever* $k \geq 2$.

Thus, $\mathcal{M}_k$ is an example of an associated hyperkähler bundle to the quaternionic Kähler manifold $\mathbb{P}\mathcal{M}_k$ described by Swann in his Oxford thesis [27].

Recall that Proposition 2.6.3 gives the second fundamental form of the embedding $\mu^{-1}(0)/G \to M/G$ in terms of the Hessian of the moment map. We now wish to apply this to the embedding $\mathcal{M}_k \to \mathcal{V}_k$. Since the moment map μ is a quadratic homogeneous map (cf. (2.3.3)), it is easy to see that the Hessian is precisely the quaternionic adjoint map, viz.,

(4.5.4) $$D^2\mu(u, v) = \mathbf{ad}(u, v).$$

Notice that $D^2\mu$ is a constant map independent of b. We now have

PROPOSITION 4.5.5. *The second fundamental form of the embedding* $\mathcal{M}_k \to \mathcal{V}_k$ *is given by* $\bar{s}_{[b]}(U, V) = -\pi_*(L_b^\dagger L_b)^{-1} L_b^\dagger \mathbf{ad}(u_b, v_b)$.

Now consider the Euler vector field Ψ on $\mathcal{M}_k$ which generates scale transformations given by T_δ of (4.2.2). Furthermore, Ψ_b corresponds to the vector b in the identification referred to above. It thus follows from Propositions 4.5.5 and 3.3.2 that for any $v \in \hat{\mathcal{H}}_b$ we have $s(\Psi_b, v) = 0$. Hence,

COROLLARY 4.5.6. *The radial lines generated by* Ψ *are geodesics, and any two-surface containing a radial line through each point is totally geodesic.*

Consider the ordered configuration space $\tilde{C}_k(\mathbb{H})$ of k ordered distinct points in $\mathbb{H} = \mathbb{R}^4$. The map sending $(a_1, \ldots, a_k) \in \mathbb{H}^k$ to $\Lambda = (1, \ldots, 1)$,

(4.5.7)
$$B = \begin{pmatrix} a_1 & 0 & \cdots & 0 \\ 0 & a_2 & \cdots & \vdots \\ \vdots & 0 & \ddots & 0 \\ 0 & \cdots & 0 & a_k \end{pmatrix}$$

is easily seen to be an embedding of $\tilde{C}_k(\mathbb{H})$ into $\hat{\mu}^{-1}(0)$. Moreover, the subgroup of $O(k)$ which stabilizes this subset is the symmetric group on k letters, Σ_k, and the quotient space is the unordered configuration space $C_k(\mathbb{H})$. Thus, we have the natural commutative diagram,

(4.5.8)
$$\begin{array}{ccccc} \tilde{C}_k(\mathbb{H}) & \longrightarrow & \hat{\mu}^{-1}(0) & \longrightarrow & \mathfrak{p}^{(k)}(\mathbb{H}) \\ \downarrow & & \downarrow & & \downarrow \\ C_k(\mathbb{H}) & \longrightarrow & \mathscr{M}_k & \longrightarrow & \mathscr{V}_k \end{array}$$

PROPOSITION 4.5.9. *The embedded submanifold* $C_k(\mathbb{H}) \hookrightarrow \mathscr{M}_k$ *is a totally geodesic flat hyperkähler submanifold of dimension* $4k$.

PROOF. This follows from the fact that the tangent space to $C_k(\mathbb{H})$ is spanned at each of its points by the partial derivatives $\frac{\partial}{\partial B_{ii}}$ and these correspond to the vectors in $\mathfrak{p}(\mathbb{H})$ given by Λ identically zero and B the matrix with a one in the ith diagonal and zeroes elsewhere. □

A larger submanifold of $\mathscr{M}_k$ obtained from configurations is the weighted configuration space $W_k(\mathbb{H})$ of dimension $5k + 3$. This has B given by equation (4.5.7), but now Λ has the form $(r_1 q, \ldots, r_k q)$ where q is a unit quaternion and the r's are positive real numbers. Then $W_k(\mathbb{H})$ is $W_k(\mathbb{H}) = \tilde{C}_k(\mathbb{H}) \times_{\Sigma_k} (\mathbb{R}^+)^k \times SO(3)$. Thus, the submanifold $W_k(\mathbb{H})$ is flat but not hyperkähler.

4.6 The reduced moduli space $\mathscr{M}_k^0$. In this section we study the reduced moduli space of centered instantons obtained by factoring out the "center of mass." We begin by noticing that the map

(4.6.1)
$$(\Lambda, B) \mapsto \left(\Lambda, B - \frac{\operatorname{tr} B}{k} I, \operatorname{tr} B \right)$$

defines an isomorphism of quaternionic vector spaces $\mathfrak{p}(\mathbb{H}) \simeq (\mathfrak{p}(\mathbb{H}))^0 \times \mathbb{H}$ where $(\mathfrak{p}(\mathbb{H}))^0 = \{b \in \mathfrak{p}(\mathbb{H}) : \operatorname{tr}(B) = 0\}$. Notice also by definition the rank is not effected by the map (4.6.1), so we get a similar decomposition on the various strata, viz.,

(4.6.2)
$$\mathfrak{p}^{(r)}(\mathbb{H}) \simeq (\mathfrak{p}^{(r)}(\mathbb{H}))^0 \times \mathbb{H}$$

where $(\mathfrak{p}^{(r)}(\mathbb{H}))^0 = \{b \in \mathfrak{p}^{(r)}(\mathbb{H}) : \mathrm{tr}(B) = 0\}$. Furthermore, since the $(\mathrm{tr}(B))I$ is in the center of $gl(k, \mathbb{R}) \otimes \mathbb{H}$, the moment map (2.3.3) restricts to $(\mathfrak{p}(\mathbb{H}))^0$ as well the various strata $(\mathfrak{p}^{(r)}(\mathbb{H}))^0$. Thus, we can perform the hyperkähler reduction procedure of §2 on the reduced space $(\mathfrak{p}(\mathbb{H}))^0$. For example, restricting to the maximal strata, we obtain the hyperkähler quotient $(\mathcal{M}_k^0, g^0)$ and this coincides with $(\hat{\mu}^{-1}(0) \mid (\mathfrak{p}(\mathbb{H}))^0)/O(k)$ with the restricted hyperkähler metric. We have

PROPOSITION 4.6.3. *The diffeomorphism* $\mathcal{M}_k \simeq \mathcal{M}_k^0 \times \mathbb{H}$ *induced by the map* (4.6.1) *is a hyperkähler isometry where* $\mathcal{M}_k^0 \times \mathbb{H}$ *has the product metric with* g^0 *on* $\mathcal{M}_k^0$ *and the flat hyperkähler metric on* $\mathbb{H}$.

Thus, $\mathrm{tr}(B)$ splits off a quaternionic line, and the holonomy group of $\mathcal{M}_k$ reduces to $\mathrm{Sp}(2k - 1)$. It seems likely that, as in the case of monopoles [5], the holonomy representation of the reduced moduli space $\mathcal{M}_k^0$ is irreducible. There is also an quaternionic Kähler manifold associated to the reduced moduli spaces $\mathcal{M}_k^0$ as in Proposition 4.5.3.

The reduced moduli spaces $\mathcal{M}_k^0$ can also be obtained from a hyperkähler quotient construction on $\mathcal{M}_k$. Consider the action of the translation group on $\mathcal{M}_k$ generated by T_a of (4.2.2). Let $\mathfrak{t}$ denote its Lie algebra which can be identified with $\mathbb{R}^4 = \mathbb{H}$. Since T_a preserves the hyperkähler structure on $\mathfrak{p}(\mathbb{H})$ and commutes with the action (4.1.6) of $O(k)$, this gives rise to a moment map $J : \mathcal{M}_k \to \mathfrak{t}^* \otimes \mathbb{R}^3$. A straightforward computation gives $J_0^a = \mathrm{tr}(B^a)$

$$(4.6.4) \qquad J_c^a = -\mathrm{tr}(B^0)\delta_c^a + \varepsilon^{abc}\,\mathrm{tr}(B^b)$$

where the lower index on J indicates its component in $\mathfrak{t}^* = \mathbb{R}^4$. However, since $\mathfrak{t}$ is abelian, this moment map is not equivariant, but defines a coadjoint cocycle. In any case the hyperkähler reduction procedure goes through to give a hyperkähler structure on the reduced space $\mathcal{M}_k^0 = J^{-1}(0)$ which coincides with the one discussed above.

5. Hyperkähler Morse theory and geometric invariant theory

5.1 The generalized Euler flow. In this section following a suggestion of Cliff Taubes we use the gradient of the moment map to construct a flow which gives an exponential decay for the moment map. Thus the zero set of the moment map $\mu^{-1}(0)$ is a global attractor for all of $\mathfrak{p}(\mathbb{H})$. Furthermore, it turns out that the flow equations have the form of generalized Euler equations.

The gradient of the moment map is the linear map

$$(5.1.1) \qquad L_b = D\mu_b : \mathfrak{p}(\mathbb{H}) \longrightarrow o(k) \otimes \mathbb{R}^3$$

and with the standard identification of a vector space with its tangent space,

L_b is given by the Lie bracket expression in Proposition 3.2.8. We want to construct a flow on $\mathfrak{p}(\mathbb{H})$ which provides an exponential decay for μ.

This is done as follows [28]: Put $\dot{b} = L_b^\dagger \xi$ for some $\xi \in o(k) \otimes \mathbb{R}^3$, where $L_b^\dagger$ denotes the adjoint map of L_b, the dot denotes differentiation with respect to the flow parameter t. Note that we have identified $T_b^* \mathfrak{p}(\mathbb{H})$ with $T_b \mathfrak{p}(\mathbb{H})$ by the Euclidean metric and $T_b \mathfrak{p}(\mathbb{H})$ with $\mathfrak{p}(\mathbb{H})$ itself. The Fredholm alternative states that a solution to the equation

$$(5.1.2) \qquad L_b L_b^\dagger \xi = -\mu(b)$$

exists if and only if $\mu(b)$ lies in the orthogonal complement to the kernel of $L_b L_b^\dagger$, and furthermore, the solution is unique on $(\ker L_b L_b^\dagger)^\perp$. Hence, we can invert equation (5.1.2) and the flow equations become

$$(5.1.3) \qquad \dot{b}(t) = -L_b^\dagger (L_b L_b^\dagger)^{-1} \mu(b)(t)$$

$$(5.1.4) \qquad b(0) = b_0$$

where $(L_b L_b^\dagger)^{-1} : (\ker L_b L_b^\dagger)^\perp \longrightarrow (\ker L_b L_b^\dagger)^\perp$ denotes the inverse linear transformation. That these flow equations (5.1.3) and (5.1.4) are well defined follows from Lemma 5.1.8 below. One easily sees that μ satisfies

$$(5.1.5) \qquad \mu(b(t)) = \mu(b_0)e^{-t}.$$

For convenience we introduce the following notation for the vector field which generates the flow:

$$(5.1.6) \qquad \sigma(b) = -L_b^\dagger (L_b L_b^\dagger)^{-1} \mu(b).$$

Recall that the norm on $\mathfrak{p}(\mathbb{H})$ is the trace norm described in Proposition 3.2.7. For notational convenience we write this norm as $\| \cdot \|$. A simple computation gives the norm squared

$$(5.1.7) \qquad \| \sigma(b) \|^2 = \mu(b) \cdot (L_b L_b^\dagger)^{-1} \mu(b).$$

The following lemma implies that $\mu(b)$ lies in the orthogonal complement to the kernel of $L_b L_b^\dagger$, so the construction above is valid. Let $\mathfrak{g}_b$ denote the Lie algebra of the stability subgroup of b and $\mathfrak{g}_b^\perp$ denote its orthogonal complement with respect to the trace norm in $o(k)$.

LEMMA 5.1.8. *We have the following*: (1) $\operatorname{im} L_b = \mathfrak{g}_b^\perp \otimes \mathbb{R}^3$.
(2) $\ker L_b^\dagger = \ker L_b L_b^\dagger = \mathfrak{g}_b \otimes \mathbb{R}^3$.
(3) $\mu(b) \in \mathfrak{g}_b^\perp \otimes \mathbb{R}^3$.

PROOF. (1) This is the hyperkähler version of a standard result in symplectic geometry [19, p. 184]) and is easily verified.

(2) This follows from the orthogonal decomposition $o(k) \otimes \mathbb{R}^3 = \operatorname{im} L_b \oplus \ker L_b^\dagger$ and (1). Finally,

(3) is a consequence of the homogeneity of μ, namely

$$(5.1.9) \qquad\qquad b \cdot D\mu(b) = 2\mu(b)$$

and (1). □

It is clear that the vector field σ and its norm are smooth at points $b \in \mathfrak{p}(\mathbb{H})$ where $L_b L_b^\dagger$ is invertible. However, in order to understand the flow on $\mathfrak{p}(\mathbb{H})$, a careful investigation at points where invertibility fails is necessary. The key technical point is an a priori estimate (A.2) which is given in the appendix.

LEMMA 5.1.10. *The vector field σ is locally Lipschitz on $\mathfrak{p}(\mathbb{H})$.*

PROOF. To prove that σ is Lipschitz we notice the inequalities

$$\begin{aligned}
\| \sigma(b) - \sigma(b_0) \| &\leq \| (L_b^\dagger - L_{b_0}^\dagger)(L_b L_b^\dagger)^{-1}\mu(b) \| \\
&\quad + \| L_{b_0}^\dagger [(L_b L_b^\dagger)^{-1}\mu(b) - (L_{b_0} L_{b_0}^\dagger)^{-1}\mu(b_0)] \| \\
&\leq \| L_b^\dagger - L_{b_0}^\dagger \| \| (L_b L_b^\dagger)^{-1}\mu(b) \| \\
&\quad + \| L_{b_0}^\dagger \| \| P^\perp (L_b L_b^\dagger)^{-1} P^\perp \mu(b) - P^\perp (L_{b_0} L_{b_0}^\dagger)^{-1} P^\perp \mu(b_0) \| \\
&\quad + \| L_{b_0}^\dagger \| \| P^\perp (L_b L_b^\dagger)^{-1} P^0 \mu(b) \| .
\end{aligned}$$

The first term is Lipschitz by Lemma A.2 and the fact that $L_b^\dagger$ is a linear map. The middle term is differentiable and thus locally Lipschitz. The last term is locally Lipschitz by A.2, and (2) of Lemma A.4. □

Hence, the standard existence and uniqueness theorem for systems of ordinary differential equations apply to the flow equations (5.1.3) on $\mathfrak{p}(\mathbb{H})$. Furthermore, we have

PROPOSITION 5.1.11. *The vector field σ is complete on $\mathfrak{p}(\mathbb{H})$.*

PROOF. Let t', $t \in \mathbb{R}$, then by the triangle inequality for integrals, the a priori estimate (A.2), and (5.1.5), the flow $b(t)$ satisfies the estimate

$$\| b(t') - b(t) \| \leq 2K \| \mu(b_0) \|^{1/2} | e^{-t/2} - e^{-t'/2} | .$$

It follows that the flow exists for all $t \in \mathbb{R}$. □

We now have the following:

PROPOSITION 5.1.12. *The critical set of σ is precisely the zero set of μ, and is a global attractor for all of $\mathfrak{p}(\mathbb{H})$. Furthermore, the maximal strata $\mu^{-1}(0) \cap \mathfrak{p}^{(k)}(\mathbb{H}) = \mathcal{M}_k$ is a nondegenerate critical manifold in the sense of Bott [9].*

PROOF. It follows from (5.1.7) and Lemma A.4 of the Appendix that every zero of μ is a critical point of σ. Conversely, $\sigma(b) = 0$ implies that $(L_b L_b^\dagger)^{-1}\mu(b) \in \ker L_b^\dagger$, which by 2 of Lemma 5.1.8 is the stability subalgebra $\mathfrak{g}_b \otimes \mathbb{R}^3$. But by (3) of Lemma 5.1.8 $\mu(b)$ is in its perpendicular component and $L_b L_b^\dagger$ is invertible there. This implies that $\mu(b)$ vanishes. The second statement now follows immediately from completeness Proposition 5.1.11 and (5.1.5). To prove the last statement we show that the tangent space to $\mathscr{M}_k$ is precisely the null space of the Hessian of the Morse function $\| \mu(b) \|^2$. Computing the Hessian we have

$$D^2 \| \mu(b) \|^2 (u, v) = 2D^2 \mu(b)(u, v) \cdot \mu(b) + 2L_b u \cdot L_b v.$$

Thus, for $b \in \mu^{-1}(0)$, we have $D^2 \| \mu(b) \|^2 (u, u) = 2 \| L_b u \|^2$. So by Proposition 3.3.2, u is in the nullspace of the Hessian if and only if $u \in T_b \mu^{-1}(0)$. Then since $\| \mu(b) \|^2$ is $O(k)$ invariant the statement follows by taking u to be horizontal.   □

Let us define $\xi(b) \in o(k)$ by $\xi(b) = (L_b L_b^\dagger)^{-1} \mu(b)$, then the flow equation (5.1.3) can be written in the form of the generalized Euler equations, namely

$$(5.1.13) \qquad\qquad \frac{db}{dt} = -\mathbf{ad}_b^\dagger \xi(b)(t).$$

We summarize our results as

THEOREM 5.1.14. *The generalized Euler equations* (5.1.13) *provide a strong deformation retraction of the orbit space* $\mathfrak{p}(\mathbb{H})/O(k)$ *onto* $\mathscr{M}_k$.

PROOF. We construct an $O(k)$ equivariant homotopy $H : [0, 1] \times \mathfrak{p}(\mathbb{H}) \to \mathfrak{p}(\mathbb{H})$ satisfying (1) $H(0, b) = b$.

(2) $H_1 = H(1, \cdot) : \mathfrak{p}(\mathbb{H}) \to \mu^{-1}(0)$ is a retraction.

(3) $H(s, b) = b$ for all $s \in [0, 1]$ and all $b \in \mu^{-1}(0)$.

Let $\phi_t(b)$ denote the flow of the generalized Euler equation (5.1.13). By Proposition 5.1.11, ϕ is defined for all $t \in \mathbb{R}$. Define the homotopy $H(s, b)$ by $H(s, b) = \phi_{\frac{s}{1-s}\|\mu(b)\|}(b)$, where $H(1, b) = \lim_{s \to 1} H(s, b)$. This is an $O(k)$ equivariant homotopy since the flow equation (5.1.13) and hence, the flow ϕ are manifestly $O(k)$ equivariant. Conditions 1 and 3 above are easily verified, and 2 follows from Lemma 5.1.12.   □

5.2 Invariant sets of the flow. We show that the various strata $\mathfrak{p}^{(r)}(\mathbb{H})$ and hence, the singularity set $\mathscr{S}_k$, are invariant sets.

PROPOSITION 5.2.1. *Each strata* $\mathfrak{p}^{(r)}(\mathbb{H})$, *and thus the singularity set* $\mathscr{S}_k$, *is invariant under the flow generated by* σ.

PROOF. We show that the tangency condition in Nagumo's theorem [1, p. 201] holds on each stratum. Since σ is locally Lipschitz by Lemma 5.1.10, this will show that $\mathscr{S}_k$ is + invariant. But applying the theorem to vector field $-\sigma$ gives the time reversed solution, so $\mathfrak{p}^{(r)}(\mathbb{H})$ will be invariant if we can show that the tangency condition holds.

For any $b \in \mathfrak{p}(\mathbb{H})$ define the distance to $\mathfrak{p}^{(r)}(\mathbb{H})$ by

$$d(b, \mathfrak{p}^{(r)}(\mathbb{H})) = \inf_{s \in \mathfrak{p}^{(r)}(\mathbb{H})} \| b - s \|.$$

It suffices to show that if $b \in \mathfrak{p}^{(r)}(\mathbb{H})$ then $d(b + h\sigma(b), \mathfrak{p}^{(r)}(\mathbb{H})) = 0$ for h small enough. Again by a choice of basis and the fact that an element b is in $\mathfrak{p}^{(r)}(\mathbb{H})$ if and only if its translate by T_a is in $\mathfrak{p}^{(r)}(\mathbb{H})$, we can assume that b has the form given in equation (A.1), where there are precisely N columns of zeros. We show that $\sigma(b)$ also has the form (A.1), for then by continuity we can choose h small enough so that $(b + h\sigma(b)) \in \mathfrak{p}^{(r)}(\mathbb{H})$. Let us consider the case $N = 1$ so $\alpha = 1$ as the general case is similar. Write $\sigma(b) = -L_b^\dagger \xi(b)$ where $\xi(b) = (L_b L_b^\dagger)^{-1}\mu(b)$. From Proposition 3.2.8 and the fact that b_{1j} vanishes, it suffices to consider the $1j$ component of $\xi(b)$ which is represented by $M^{-1}(b)_{1j}^{rs}\mu(b)_{rs}$. Since $\mu(b)_{1j}$ vanishes we have $2 < r < s < k$. We show that the cofactor $M(rs|1j)$ vanishes. By permuting the basis of $o(k) \otimes \mathbb{R}^3$ if necessary we can choose $r = k - 1$, $s = k$, and $j = 2$. In the first k columns the entries of the first $(k - 1)$ rows are the only nonzero entries. Thus, expanding down the first column, we see, by induction, that all minor determinants vanish. Hence, $\sigma(b)$ has rank no greater than r, so $(b + h\sigma(b)) \in \mathfrak{p}^{(r)}(\mathbb{H})$ for h small enough, and the proof is completed.   □

Finally recall from (4.6.1) that $\operatorname{tr} B$ splits off a quaternionic line from each strata. Furthermore, since $(\operatorname{tr} B)I$ is in the center of $gl(k, \mathbb{R})$, it follows from the generalized Euler equation (5.1.13) that

PROPOSITION 5.2.2. *The trace of B is an invariant of the generalized Euler flow. Thus for each r the flow on $\mathfrak{p}^{(r)}(\mathbb{H})$ restricts to a flow on $\mathfrak{p}^{(r)}(\mathbb{H})^0$ with absolute minima $\mu^{-1}(0) \cap \mathfrak{p}^{(r)}(\mathbb{H})^0$.*

Since the various strata are invariant sets for the generalized Euler flow, it seems natural to conjecture that the strata are also invariant in the limit as the flow parameter t tends to ∞.

However, such a matter can be quite delicate. Indeed, we shall give an example to show that in the degenerate strata, the flow can jump strata in the limit. Nevertheless, we conjecture that this does not happen for the generic strata. Thus,

CONJECTURE 5.2.3. *There is a strong deformation retraction of principal fiber bundles*

$$
\begin{array}{ccc}
\hat{\mu}^{-1}(0) & \xleftarrow{\; r_t \;} & \mathfrak{p}^{(k)}(\mathbb{H}) \\
\pi_0 \downarrow & & \downarrow \pi \\
\mathscr{M}_k & \xleftarrow{\; \bar{r}_t \;} & \mathscr{V}_k
\end{array}
$$

Of course, Conjecture 5.2.3 holds trivially for $k = 1$ since there are no quadratic constraints in this case. Thus, we assume $k \geq 2$ in what follows.

5.3 The flow for $k = 2$. The proof of the following theorem has already been given in [12], but we include it here for completeness.

THEOREM 5.3.1. *Conjecture 5.2.3 holds for* $k = 2$.

PROOF. We work $O(k)$ equivariantly on $\mathfrak{p}(\mathbb{H})$.

The idea is to assume that in the limit as $t \to \infty$ the generalized Euler flow (5.1.13) converges to a degenerate (ideal) instanton. Then, by carefully analyzing the flow equations, prove that it must have been degenerate for all time t. First using the translation subgroup of H_∞, we can take B to be traceless. We consider the case $Sp(1)$ only, as the argument easily generalizes to the $Sp(n)$ case. Then we have

$$\Lambda = (\alpha, \beta), \qquad B = \begin{pmatrix} a & c \\ c & -a \end{pmatrix}$$

where a, c, α, β are quaternions. For $o(2)$ the moment map has dimension one over the imaginary quaternions, so $L_b L_b^\dagger$ is multiplicative. We find from (5) of Proposition 3.2.8 $L_b L_b^\dagger = 4(|a|^2 + |c|^2) + |\alpha|^2 + |\beta|^2$. Hereafter, we denote this quantity by D. The moment map can be written as $\mu = I\mathscr{E}$ where $\mathscr{E}$ is the 2×2 antisymmetric matrix with 1 in the upper right corner and -1 in the lower left corner, and $I = \mathrm{Im}(2\bar{a}b + \bar{\alpha}\beta)$. By (5.1.4) $I = I_0 e^{-t}$ along the flow. Using the $SU(2)$ subgroup of $\hat{H}_\infty$, we can rotate I_0 so that it only has an $\mathbf{i}$ component and then by dilatation take $I_0 = \mathbf{i}$.

The flow equations, which we denote as equations (5.3.2), can now be written as

$$\dot{a}^0 = -2D^{-1}e^{-t}c^1 \qquad \dot{a}^1 = 2D^{-1}e^{-t}c^0$$
$$\dot{c}^0 = 2D^{-1}e^{-t}a^1 \qquad \dot{c}^1 = -2D^{-1}e^{-t}a^0$$
$$\dot{\alpha}^0 = -D^{-1}e^{-t}\beta^1 \qquad \dot{\alpha}^1 = D^{-1}e^{-t}\beta^0$$
$$\dot{\beta}^0 = D^{-1}e^{-t}\alpha^1 \qquad \dot{\beta}^1 = -D^{-1}e^{-t}\alpha^0$$
$$\dot{a}^3 = -2D^{-1}e^{-t}c^2 \qquad \dot{a}^2 = 2D^{-1}e^{-t}c^3$$
$$\dot{c}^3 = 2D^{-1}e^{-t}a^2 \qquad \dot{c}^2 = -2D^{-1}e^{-t}a^3$$
$$\dot{\alpha}^3 = -D^{-1}e^{-t}\beta^2 \qquad \dot{\alpha}^2 = D^{-1}e^{-t}\beta^3$$
$$\dot{\beta}^3 = D^{-1}e^{-t}\alpha^2 \qquad \dot{\beta}^2 = -D^{-1}e^{-t}\alpha^3.$$

Using the method of characteristics, we find the following constants of the motion

(5.3.3)
$$\zeta_1 = (a^0)^2 - (c^1)^2 \qquad \zeta_2 = (a^1)^2 - (c^0)^2$$
$$\zeta_3 = (a^2)^2 - (c^3)^2 \qquad \zeta_4 = (a^3)^2 - (c^2)^2$$
$$\zeta_5 = (\alpha^0)^2 - (\beta^1)^2 \qquad \zeta_6 = (\alpha^1)^2 - (\beta^0)^2$$
$$\zeta_7 = (\alpha^3)^2 - (\beta^2)^2 \qquad \zeta_8 = (\alpha^2)^2 - (\beta^3)^2$$
$$\zeta_9 = a^0 c^0 + a^1 c^1 \qquad \zeta_{10} = a^2 c^2 + a^3 c^3$$
$$\zeta_{11} = \alpha^0 \beta^0 + \alpha^1 \beta^1 \qquad \zeta_{12} = \alpha^2 \beta^2 + \alpha^3 \beta^3.$$

Now consider the limit $b_\infty = (\Lambda_\infty, B_\infty)$. Let us assume that b_∞ has rank one. (The case of rank zero is similar.) Since $\mu(b_\infty) = 0$, as in the proof of Theorem 4.4.3, there is an $O(2)$ transformation such that b_∞ takes the form

$$b = \begin{pmatrix} \alpha_\infty & 0 \\ a_\infty & 0 \\ 0 & -a_\infty \end{pmatrix}$$

so that $\beta_\infty = c_\infty = 0$. At this point on the $O(2)$ orbit the constants of the motion satisfy the conditions $\zeta_i \geq 0$ for $i = 1, \ldots, 8$ and $\zeta_i = 0$ for $i = 9, \ldots, 12$. Moreover, since the ζ_i's are independent of t, these conditions hold for all t. Now let us assume that $a_\infty \neq 0$, then some component, say a_∞^0 does not vanish. (The case where a_∞ vanishes can be treated similarly.) Notice that the constants of motion imply that if a component of a_∞ vanishes, then the same component of $a(t)$ vanishes for all t. We can now use the constants of the motion (5.3.3) together with the flow equations (5.3.2) to express all functions in terms of $a^0(t)$. We find

$$a(t) = \frac{a_\infty}{a_\infty^0} a^0(t), \qquad c(t) = \pm \frac{a_\infty}{a_\infty^0} \mathbf{i} \sqrt{(a^0(t))^2 - (a_\infty^0)^2},$$

$$\alpha(t) = \alpha_\infty \sqrt{\frac{a^0(t) + a_\infty^0}{2a_\infty^0}}, \qquad \beta(t) = \pm \alpha_\infty \mathbf{i} \sqrt{\frac{a^0(t) - a_\infty^0}{2a_\infty^0}}.$$

It is now straightforward to check that the two columns are proportional at $x = -a_\infty$ for all t. $\square$

EXAMPLE 5.3.4. *Jumping strata.*

In order to illustrate how delicate the analysis of the generalized Euler flow is, we present an example which shows that off of the generic strata the flow jumps strata in the limit as $t \to \infty$. With the same choices made as in the proof above, let us see which $b \in \mathfrak{p}(\mathbb{H})^0$ flow to the most degenerate traceless ideal instanton $b_\infty = 0$. In this case the constants of the motion (5.3.3) all vanish, and an analysis of the flow equations (5.3.2) shows that the initial b, must be of the form

$$b = \begin{pmatrix} \alpha & \pm \alpha i \\ a & \pm ai \\ \pm ai & -a \end{pmatrix}.$$

Further analysis of the flow equations (5.3.2) shows that both α and a must be given in terms of a single real function $r(t)$ of the flow parameter t as follows: $\alpha(t) = r(t)\lambda$ and $a(t) = r(t)^2 \eta$ where $\lambda, \eta \in \mathbb{H}$ and

$$r(t) = (2\sqrt{|\eta\lambda|})^{-1} \sqrt{-|\lambda|^2 + \sqrt{|\lambda|^4 + 8|\eta|^2 \, e^{-t}}}.$$

Notice that as $t \to \infty$, $r(t) \to 0$, so that b has rank one for all finite t, but gives $b_\infty = 0$ in the limit.

6. Relations to the topology of instantons

6.1 Homology calculations. Recall that, by Lemma 4.3.3, the natural $O(k)$ action on $\mathfrak{p} \otimes \mathbb{H}$ is free on the generic open subset $\mathfrak{p}^{(k)}(\mathbb{H})$ with quotient space $\mathscr{V}_k$. There is a natural inclusion

$$(6.1.1) \qquad \iota : \mathscr{M}_k \longrightarrow \mathscr{V}_k$$

induced by forgetting the reality condition (4.3.6). $\mathscr{V}_k$ can be thought of as a "linearization" of the instanton moduli space precisely because the reality condition (4.3.6) has been dropped. Thus, we can modify the fundamental commutative diagram in Atiyah and Jones, [7], to obtain

$$(6.1.2) \qquad \begin{array}{ccc} & \mathscr{V}_k(G) & \\ & \nearrow \qquad \searrow & \\ \mathscr{M}_k(G) & \longrightarrow & \Omega^4 BG \end{array}$$

Following Atiyah and Jones we have identified the based moduli space of all connections $\mathscr{C}_k$ with $\Omega^4 BG$. With this identification all maps in (6.1.2) are natural inclusions induced by forgetting some additional structure present in the domain but not the range. As mentioned in §4 we can consider the general case where G is any simple, compact Lie group by first studying the special case when $G = \mathrm{Sp}(n)$ and then applying Lemma 1.19 of [13].

Diagram (6.1.2) implies the homology constructions of [10, 13] build the same nontrivial classes in $H_*(\mathscr{V}_k(G))$ as were constructed in $H_*(\mathscr{M}_k(G))$. Of course, our Conjecture 5.2.3 implies that $\mathscr{M}_k(G)$ is a strong deformation retract of $\mathscr{V}_k(G)$ and would have as an immediate corollary that the natural forgetful map (6.1.1) would induce an isomorphism in homotopy and homology with all coefficients. Thus, Theorem 5.3.1 implies

COROLLARY 6.1.3. *For all simple compact Lie groups G and all coefficients A we have that*

$$(6.1.4) \qquad \pi_*(\mathscr{M}_2(G); A) \cong \pi_*(\mathscr{V}_2(G); A)$$

$$(6.1.5) \qquad H_*(\mathscr{M}_2(G); A) \cong H_*(\mathscr{V}_2(G); A).$$

Recently, Y. Kamiyama [23], has proven (6.1.5) for the case when $G = \mathrm{Sp}(1)$ and $A = \mathbb{Z}/2$ using methods quite different from the ones used in this paper.

6.2 Instantons and iterated loop spaces. C_n operads and C_n operad spaces play a fundamental role in the iterated n-fold loop space theory of May [26]. In studying $\mathscr{M}_k(G)$ we realized that the work of Atiyah and Jones [7] and Taubes [28] strongly suggested that the disjoint union of the instanton moduli spaces should have such a C_4 structure. However, the presence of the reality condition in $\mathscr{M}_k(G)$ forced us to consider the weaker notion of a "homotopy" C_4 operad space. While this weaker notion was sufficient to obtain many new homological results it left some questions unanswered. We shall now see that this extra structure exists in $\mathscr{V}_k(G)$. More precisely, the homology analysis of $\mathscr{M}_k(G)$ in [10, 13] depended on the construction of a homotopy C_4 structure on

$$(6.2.1) \qquad \mathscr{M}(G) = \coprod_{k=1}^{\infty} \mathscr{M}_k(G).$$

This construction, in turn, depended in a critical way on the fundamental work of Taubes [28]. The delicate part of the construction was in controlling the reality constraint throughout the entire construction. Since the $\mathscr{V}_k(G)$ spaces are, by definition, insensitive to this problem, the homotopy C_4 structure in (6.2.1) immediately extends to give

$$(6.2.2) \qquad \mathscr{V}(G) = \coprod_{k=1}^{\infty} \mathscr{V}_k(G)$$

a compatible homotopy C_4 structure. This observation was used by Kamiyama in [23].

Next we recall the recognition theorem of May.

THEOREM 6.2.3 [26]. *There exist C_n operads so that every n-fold loop space is an C_n operad space. Conversely, for any C_n operad, every connected C_n operad space has the weak homotopy type of an n-fold loop space.*

However, there are homotopy C_n spaces that do not even have classifying spaces so it does not follow from (6.2.1) or (6.2.2) that $\mathscr{M}(G)$ or $\mathscr{V}(G)$ even has a classifying space. We conclude this subsection by proving

THEOREM 6.2.4. *$\mathscr{V}(G)$ is a little cubes C_4 operad space.*

COROLLARY 6.2.5. *$\mathscr{V}(G)$ has a classifying space. Furthermore, $\Omega B\mathscr{V}(G)$ has the weak homotopy type of a four-fold loop space.*

Taubes' stability theorem [29] implies that $\Omega B\mathscr{V}(G) \simeq \Omega^4 BG \times \Omega^4 X$ for some X. Furthermore, if Conjecture 5.2.3 were true it would then follow that X would be a point. (6.2.4) and (6.2.5) suggest that the homotopy properties of $\mathscr{V}(G)$ are more regular and better behaved than those of $\mathscr{M}(G)$ (this is in line with the main geometric results of this paper). In fact it is for this reason that we were first led to examine these spaces.

PROOF OF THEOREM 6.2.4. The proof for general G follows from the proof for $G = Sp(n)$ and Lemma 1.19 of [13] so, for notational convenience we omit writing the group G explicitly in what follows. We begin with the unit disc in $\mathscr{V}_k$.

DEFINITION 6.2.6. $D(\mathfrak{p}^{(k)}(\mathbb{H})) = \{b \in \mathfrak{p}^{(k)}(\mathbb{H}) : \| b \| \leq 1\}$. *Here* $\| b \| = \operatorname{tr} b^* b = \operatorname{tr}(B^* B + \Lambda^* \Lambda)$ *is a standard trace norm (recall 3.2.7).*

Since the scale transformation $b \mapsto \delta b$ for $\delta \neq 0$ given by (4.2.2) does not effect the rank condition (4.1.3), $D(\mathfrak{p}^{(k)}(\mathbb{H}))$ is a strong deformation retract of $\mathfrak{p}^{(k)}(\mathbb{H})$. Furthermore, as scaling is $O(k)$-equivariant, the quotient space $D(\mathscr{V}_k) = D(\mathfrak{p}^{(k)}(\mathbb{H})/O(k)$ is a strong deformation retract of $\mathscr{V}_k$. Hence, it suffices to show that $D(\mathscr{V})$ is a little cubes C_4 operad space. The technical details are as follows:

Consider the unit cube I^4 embedded in $\mathbb{H}^1 = \mathbb{R}^4$ in the standard way. Then a point in $C_4(j)$ is identified with j disjoint open cubes in I^4 (with sides parallel to the axes). For each i between 1 and j let q_i denote the center (the points whose coordinates are given by the midpoints of each side) of the ith cube and let r_i denote the length of its side.

DEFINITION 6.2.7. *Let* $b_i = (\Lambda_i, \mathbb{B}_i) \in \mathscr{V}'_{k_i}$ *for* $1 \leq i \leq j$. *Then*

$$(6.2.8) \qquad \phi_\delta(c_{12\ldots j}, b_1, \ldots, b_j) = b_\delta = (\Lambda_\delta, B_\delta) \in \mathscr{V}'_k$$

where (1) $k = \sum_{i=1}^{j} k_i$.

(2) $\Lambda_\delta = \delta(r_1\Lambda_1, \ldots, r_j\Lambda_j)$.

(3) $B_\delta = \mathrm{diag}(q_1 I + \delta r_1 B_1, \ldots, q_j I + \delta r_j B_j)$, *the square block diagonal matrix with the* k_i *by* k_i *matrix* $(q_i I + \delta r_i B_i)$ *in the* ith *diagonal block.*

(4) *the* q_i's *and* r_i's *are uniquely determined from* $c_{12\ldots j} \in C_4(j)$ *in the manner described in the paragraph above.*

REMARK 6.2.9. (1) Definition 6.2.7 is compatible with the $O(k)$ action on $M_{n+k,k}(\mathbb{H})$ given by (4.1.6). In addition, the norm restriction on the elements in $\mathscr{V}'_k$, the configurations of the little cubes, and a simple extension of Lemma 5.11 of [10] combine to guarentee that ϕ_δ is well defined.

(2) ϕ_δ restricted to $\mathscr{M}$ gives a multiplication on $\mathscr{M}$ that is homotopy equivalent, although not identical to, the original multiplication given in [10].

But now setting $\phi = \phi_1$ we obtain the map we require to prove (6.2.4). This is now completely routine. We need to verify three conditions (for example see 3.4 of [10]). Conditions 2 and 3 are immediate and condition 1 is a straightforward diagram chase. Explicitly, if

(1) $(c_1, \ldots, c_m) \in C_4(m)$,
(2) $(d_{1,j_l}, \ldots, d_{j_l,j_l}) \in C_4(j_l)$ for $1 \leq l \leq m$,
(3) $(b_1, \ldots, b_j) \in \mathscr{V}^j$,

then the resulting element $(B, \Lambda) \in \mathscr{V}$ obtained by chasing diagram 3.5 of [10] in either direction is given by

(1) B is the square block diagonal matrix whose i, j, lth block is given by $\{q_i I + r_{q_i} s_{i,j} I + r_{q_i} r_{s_{i,j}} B_{i,j,l}\}$.

(2) $\Lambda = (r_{q_1} r_{s_{1,1}} \Lambda_1, \ldots, r_{q_m} r_{s_{j_l,j_l}} \Lambda_j)$. $\square$

In addition, we have

THEOREM 6.2.10. *The following diagram commutes up to homotopy:*

(6.2.11)

$$
\begin{array}{ccc}
C_4(j) \times_{\Sigma_j} (\Omega^4 BG)^j & \xrightarrow{\vartheta} & \Omega^4 BG \\
\uparrow & & \uparrow \\
C_4(j) \times_{\Sigma_j} (\mathscr{V})^j & \xrightarrow{\phi} & \mathscr{V} \\
\uparrow & & \uparrow \\
C_4(j) \times_{\Sigma_j} (\mathscr{M})^j & \xrightarrow{\phi} & \mathscr{M}
\end{array}
$$

PROOF. As ϕ_1 is homotopic to ϕ_δ we may replace ϕ by ϕ_δ where δ is sufficiently small so that ϕ_δ maps $C_4(j) \times_{\Sigma_j} (\mathscr{V})^j$ and $C_4(j) \times_{\Sigma_j} (\mathscr{M})^j$ into $\mathscr{V}$ and the Taubes tubular neighborhood of $\mathscr{M}$ respectively. This choice of δ clearly yields a diagram that is homotopy equivalent to (6.2.11). The remainder of the proof is now completely analogous to those of Theorems 6.3 and 6.10 of [**10**]. That is, there is a continuous way, which depends only on the $c_{1...j}$'s and not on the b_i's, of deforming $\phi_\delta(c_{1...j}, b_1, \ldots, b_j)$, first to a map of $c_{1...j}$ into $\mathbb{HP}(m)$ which is constant off the little cubes, and then to a map which is defined by the b_i's on those little cubes. Details are left to the reader. $\square$

REMARK 6.2.12. *Finally, notice that*

$$
\operatorname{rank}(\phi_\delta(c_{12...j}, b_1, \ldots, b_j)) = \sum_{i=1}^{j} \operatorname{rank}(b_i).
$$

This implies that the maps in Definition 6.2.6 can be defined on the various strata discussed in section four and not just on the generic strata as discussed here.

A. Appendix

We now give the a priori estimate used in §5. We begin by investigating the singularity structure of the linear map $L_b L_b^\dagger : \mathfrak{o}(k) \otimes \mathbb{R}^3 \longrightarrow \mathfrak{o}(k) \otimes \mathbb{R}^3$. Suppose that $L_b L_b^\dagger$ is not invertible at $b_0 \in \mathfrak{p}(\mathbb{H})$, then by Lemma 5.1.8 b_0 has a nontrivial stabilizer algebra $\mathfrak{g}_0 \subset \mathfrak{o}(k)$. Thus by Lemma 4.3.3 b_0 lies in the singularity set $\mathscr{S}_k'$. Let b_0 drop rank by N at $x = a \in \mathbb{H}$, then, by a change of basis if necessary, we can assume that b_0 has the form

$$
b = \begin{pmatrix} 0 & \Lambda \\ aI & 0 \\ 0 & B \end{pmatrix}
$$

where $a \in \mathbb{H}$, and I is the $N \times N$ unit matrix. Furthermore, by a computation like the one in the proof of 4.3.3, one sees that $\mathfrak{g}_{b_0} = \mathfrak{o}(N)$.

Now using the action of the translation group $\mathbb{H}$ given in (4.2.2) we can assume that b_0 drops rank at $x = 0$, so that b_0 has the form

(A.1)
$$b = \begin{pmatrix} 0 & \Lambda \\ 0 & 0 \\ 0 & B \end{pmatrix}.$$

We shall prove the following a priori estimate:

LEMMA A.2. *There is a positive constant K such that*

$$\| \sigma(b) \| \leq K \| \mu(b) \|^{1/2} .$$

PROOF.

$$\| \sigma(b) \|^2 = G(b) = \mu(b) \cdot (L_b L_b^\dagger)^{-1} \mu(b)$$
$$\leq \| \mu(b) \| \| (L_b L_b^\dagger)^{-1} \mu(b) \| \leq K^2 \| \mu(b) \|$$

where

$$K^2 = \sup_{b \in \mathfrak{p}(\mathbb{H})} \| (L_b L_b^\dagger)^{-1} \mu(b) \| .$$

It suffices to show that K^2 is finite. Since $\| (L_b L_b^\dagger)^{-1} \mu(b) \|$ is a homogeneous function of degree zero, it suffices to show that it remains bounded when $L_b L_b^\dagger$ drops rank. Now suppose that $L_b L_b^\dagger$ fails to be invertible at b_0. By lower semicontinuity there is an ε disc D_ε centered at b_0 such that dim: $\mathfrak{g}_b^\perp \geq$ dim: $\mathfrak{g}_{b_0}^\perp$. Notice that, in general, there may be various strata in D_ε. Let P^0 and $P^\perp$ denote projectors from $\mathfrak{o}(k) \otimes \mathbb{R}^3$ onto $\mathfrak{g}_{b_0} \otimes \mathbb{R}^3$ and $\mathfrak{g}_{b_0}^\perp \otimes \mathbb{R}^3$, respectively. We have

(A.3)
$$\| (L_b L_b^\dagger)^{-1} \mu(b) \|$$
$$\leq \| P^0 (L_b L_b^\dagger)^{-1} P^0 \mu(b) \| + \| P^0 (L_b L_b^\dagger)^{-1} P^\perp \mu(b) \|$$
$$+ \| P^\perp (L_b L_b^\dagger)^{-1} P^0 \mu(b) \| + \| P^\perp (L_b L_b^\dagger)^{-1} P^\perp \mu(b) \| .$$

By Lemma 5.1.8 $P^\perp (L_b L_b^\dagger)^{-1} P^\perp$ is bounded as $b \to b^0$. Thus the last term is bounded.

Now choose a basis for $\mathfrak{o}(k) \otimes \mathbb{R}^3$ as follows: First list a basis $\mathscr{E}_{\alpha\beta} \otimes e^a$ of $\mathfrak{g}_{b_0} \otimes \mathbb{R}^3$. As mentioned previously $\mathfrak{g}_{b_0} = \mathfrak{o}(l)$ for some $1 \leq l \leq k$. Furthermore, the centralizer of $\mathfrak{o}(l)$ in $\mathfrak{o}(k)$ is $\mathfrak{o}(k - l)$. Next give a basis $\mathscr{E}_{\alpha r} \otimes e^a$ for the vector space $\mathfrak{g}_{b_0}^\perp / \mathfrak{o}(k - l) \otimes \mathbb{R}^3$, and finally the basis $\mathscr{E}_{rs} \otimes e^a$ for $\mathfrak{o}(k - l) \otimes \mathbb{R}^3$. Here we have the following ranges: $1 \leq \alpha < \beta \leq l$, $l + 1 \leq r < s \leq k$, and $1 \leq a \leq 3$. We run through the Lie algebra indices first and then the $\mathbb{R}^3$ indices. Let $M(b)$ denote the symmetric matrix of $L_b L_b^\dagger$ with respect to this basis. We need estimates on the singular behavior of the inverse matrix $M(b)^{-1}$ as $b \to b_0$. In particular, the first two terms of (A.3) have a decomposition

$$(M^{-1})_{\alpha\beta}^{ij} \mu_{ij} = (M^{-1})_{\alpha\beta}^{\gamma\delta} \mu_{\gamma\delta} + (M^{-1})_{\alpha\beta}^{\gamma r} \mu_{\gamma r} + (M^{-1})_{\alpha\beta}^{rs} \mu_{rs},$$

whereas the third term consists of two terms, namely $(M^{-1})^{\alpha\beta}_{\gamma r}\mu_{\alpha\beta}$ and $(M^{-1})^{\alpha\beta}_{\gamma\delta}\mu_{\alpha\beta}$. To get estimates on M^{-1}, we first study the behavior of $M(b)$ near b_0. In (5) of Proposition 3.2.8 the first two terms are diagonal as maps from $\mathbb{R}^3$ to $\mathbb{R}^3$ and have the form $E^{kl}_{ij} = \delta_{lj}b_{ir}b_{rk} + \delta_{ik}b_{lr}b_{rj} - 2b_{ik}b_{lj}$ while the last two terms are off diagonal as maps from $\mathbb{R}^3$ to $\mathbb{R}^3$ and have the form $F^{kl}_{ij} = \varepsilon^{abc}(\delta_{lj}\mu^b_{ik} - \delta_{ik}\mu^b_{lj})$. Here, for the matrices M, E and F, the superscripts denote the columns while the subscripts denote the rows. Keeping in mind the above ranges for the indices, and that $r = \| b - b_0 \| < \varepsilon$, this gives the following estimates on the entries of M,

(A.4)
$$| M^{\alpha\beta}_{\gamma\delta} | \leq r^2, \qquad | M^{\alpha\beta}_{\gamma r} | \leq r^2, \qquad | M^{\alpha\beta}_{\alpha r} | \leq r,$$
$$| M^{\alpha\beta}_{rs} | \leq r^2, \qquad | M^{\alpha r}_{\gamma s} | \leq r, \qquad | M^{\alpha t}_{rs} | \leq r,$$

where we have suppressed the $\mathbb{R}^3$ indices, and the Greek indices are all distinct unless otherwise noted. There are no other apriori estimates on the entries of M, except those obtained from the symmetry condition on M. In particular, the entry $M^{\alpha r}_{\alpha s}$ is of order 1. For convenience let us assign weights 0, 1, or 2 to the entries of M corresponding to the power of r in the estimates (A.4) where it is understood that weight 0 means no apriori estimate. Again for notational convenience let us write $V^0 = \mathfrak{g}_{b_0} \otimes \mathbb{R}^3$ and $V^\perp = \mathfrak{g}^\perp_{b_0} \otimes \mathbb{R}^3$. Then $V \wedge V = V^0 \wedge V^0 \oplus V^0 \otimes V^\perp \oplus V^\perp \wedge V^\perp$, and writing M as a big matrix with 3 blocks by 3 blocks corresponding to this decomposition, we see that the first diagonal block has weight 2, while the diagonal entries of the middle diagonal block have weights 0, whereas the entire last diagonal block has weight 0. Thus, assuming M has maximal rank, we have the following estimate

(A.5)
$$c_0 r^{3l(l-1)} \leq \det M \leq c_1 r^{3l(l-1)}$$

for some constants c_0 and c_1.

LEMMA A.6. *There are constants* c_2, c_3, *and* c_4 *such that* (1) $| (M^{-1})^{\gamma\delta}_{\alpha\beta} | < c_2 r^{-2}$.

(2) $| (M^{-1})^{\gamma r}_{\alpha\beta} | < c_3 r^{-1}$.

(3) $| (M^{-1})^{rs}_{\alpha\beta} | < c_4 r^{3l}$.

PROOF. We need estimates on the cofactors $M(\gamma\delta \mid \alpha\beta)$, $M(\gamma r \mid \alpha\beta)$, and $M(rs \mid \alpha\beta)$. Up to a change of basis of $o(l)$ the first cofactor amounts to deleting the first row and first column. Thus, this minor has weight $3l(l-1) - 2$. So the first estimate of the lemma follows by (A.5). Likewise the second cofactor can be obtained by deleting the first row and the column corresponding to $1l + 1$. The first $3l$ elements of the last row of the first block (i.e row $1l + 1$) of this cofactor now have weight 1. Hence, the first block of this cofactor has weight $3l(l-1) - 1$. Furthermore, the middle block as well as the right lower block still have weight 0. Thus, the second estimate follows from (A.5).

To prove the last estimate we begin by again noticing that up to a change of basis which preserves the decomposition $V = V^0 + V^\perp$ it suffices to show that the weight of the minor $\det M((k-1)k \mid 12)$ obtained by deleting the first row and the last column is $3l^2$. Thus, each row is shifted up by one. In particular the $(3l(l-1)/2)$ th row now corresponds to $1l+1$, and the upper left block still has weight $3l(l-1)$. Now the diagonal elements of the middle block are not all of weight zero. There are precisely $3l$ elements of weight one on the diagonal. The first is M^{1k}_{2l+1} and the last is M^{lk}_{l+1l+2}. Again the lower right block has weight zero. This gives the last estimate and proves the lemma. $\square$

Using the symmetry of the matrix M^{-1}, Lemma A.2 now follows from Lemma A.6, the inequality (A.3) and the following easily verified inequalities:

$$\text{(A.7)} \qquad\qquad\qquad |\mu_{\alpha r}| < c_5 r.$$

$$\text{(A.8)} \qquad\qquad\qquad \| P^0 \mu(b) \| \le c_6 r^2$$

where $r = \| b - b_0 \|$ and c_5 and c_6 are constants. $\square$

REFERENCES

1. R. Abraham, J.E. Marsden, and T. Ratiu, *Manifolds, tensor analysis, and applications*, Addison-Wesley Publ. Co., 1983.

2. V.I. Arnold, S.M. Gusein-Zade, and A.N. Varchenko, *Singularities of differentiable maps*, Vol. 1, Birkhäuser, Boston, 1985.

3. M.F. Atiyah, *The geometry of Yang-Mills fields*, Ann. Scuola Norm. Sup. Pisa (1979).

4. M.F. Atiyah, V.G. Drinfeld, N.J. Hitchin and Y.I. Manin, *Construction of instantons*, Phys. Lett. A **65** (1978), 185–187.

5. M.F. Atiyah and N.J. Hitchin, *The geometry and dynamics of magnetic monopoles*, Princeton Univ. Press, Princeton, NJ, 1988.

6. M.F. Atiyah, N.J. Hitchin and I. Singer, *Self-duality in four dimensional Riemannian geometry*, Proc. Roy. Soc. London Ser. A **362** (1978), 425–461.

7. M.F. Atiyah and J.D. Jones, *Topological aspects of Yang-Mills theory*, Comm. Math. Phys. **61** (1978), 97–118.

8. A.L. Besse, *Einstein manifolds*, Springer-Verlag, New York, 1987.

9. R. Bott, *Nondegenerate critical manifolds*, Ann. of Math. (2) **60** (1954), 248–261.

10. C.P. Boyer and B.M. Mann, *Homology operations on instantons*, J. Differential Geom. **28** (1988), 423-465.

11. _____, *Instantons and homotopy*, Homotopy Theory, Lecture Notes in Math., vol. 1370, Springer-Verlag, Berlin and New York, 1989, pp. 87–102.

12. _____, *Instantons, quaternionic Lie algebras, and hyperkähler geometry*, Proc. Ann. Sem. Can. Math. Soc., Lie Theory, Differential Equations and Representation Theory (1990 pp. 119–136), Univ. de Montreal Press.

13. C.P. Boyer, B.M. Mann and D. Waggoner, *On the homology of $SU(n)$ instantons*, Trans. Amer. Math. Soc., Amer. Math. Soc. **323** (1991), 529–561.

14. S.K. Donaldson, *Instantons and geometric invariant theory*, Comm. Math. Phys. **93** (1984), 453–460.

15. _____, *Connections, cohomology and the intersection forms of 4-manifolds*, J. Differential Geom. **24** (1986), 275–341.

16. _____, *Compactification and completion of Yang-Mills moduli spaces*, Lecture Notes in Math. vol. 1410, Springer-Verlag (1990), 145–160.

17. D. Groisser and T.H. Parker, *The Riemannian geometry of the Yang-Mills moduli space*, Comm. Math. Phys. **112** (1987), 663–689.
18. _____, *The geometry of the Yang-Mills moduli space for definite manifolds*, J. Differential Geom. **29** (1989), 499–544.
19. V. Guillemin and S. Sternberg, *Symplectic techniques in physics*, Cambridge Univ. Press, 1984.
20. S. Helgason, *Differential geometry, Lie groups, and symmetric spaces*, Academic Press, NY, 1978.
21. N.J. Hitchin, *Monopoles, minimal surfaces and algebraic curves*, Press de L'Université de Montréal, 1987.
22. N.J. Hitchin, A. Karlhede, U. Lindström and M. Roček, *Hyperkähler metrics and supersymmetry*, Comm. Math. Phys. **108** (1987), 535–589.
23. Y. Kamiyama, *Topology of moduli space of certain $SU(2)$ connections over S^4 with second Chern number 1 and 2*, preprint 1990.
24. F.C. Kirwan, *Cohomology of quotients in symplectic and algebraic geometry*, Princeton Univ. Press, Princeton, NJ, 1984.
25. P.B. Kronheimer and H. Nakajima, *Yang-Mills instantons on ALE gravitational instantons*, Math. Ann. **288** (1990), 263–307.
26. J.P. May, *The geometry of iterated loop spaces*, Lecture Notes in Math., vol. 271, Springer-Verlag, 1972.
27. A.F. Swann, *Hyperkähler and quaternionic Kähler geometry*, Math. Ann. **289** (1991), 421–450.
28. C.H. Taubes, *Path-connected Yang-Mills moduli spaces*, J. Differential Geo. **19** (1984), 337–392.
29. _____, *The stable topology of self-dual moduli spaces*, J. Differential Geom. **29** (1989), 163–230.
30. A. Maciocia, *Metrics on the moduli space of instantons over Euclidean 4-sphere*, Comm. Math. Phys. **135** (1991), 467–482.
31. C.P. Boyer, K. Galicki, and B.M. Mann, *Quaternionic reduction and Einstein manifolds*, preprint, Univ. of New Mexico, 1991.
32. C.P. Boyer, J.C. Hurtubise, B.M. Mann, and R.J. Milgram, *The Atiyah-Jones conjecture*, Bull. Amer. Math. Soc. (N.S.) **26(2)** (1992), 317–321.
33. _____, *The topology of instanton monduli spaces I: The Atiyah-Jones conjecture*, Ann. of Math. (2)(to appear).

UNIVERSITY OF NEW MEXICO
E-mail address: cboyer@ramanujan.unm.edu
          mann@math.unm.edu

Proceedings of Symposia in Pure Mathematics
Volume **54** (1993), Part 2

Non-Abelian Vortices
and a New Yang-Mills-Higgs Energy

STEVEN B. BRADLOW

1. Introduction

Our aim in this paper is to explore a particular type of Yang-Mills-Higgs energy functional. The version we will consider is an adaptation of the classical Yang-Mills-Higgs functional defined for $U(1)$ bundles over $\mathbb{R}^2$ (cf. [J-T]). The minima of that functional are called vortices, and we will adopt the same terminology for the minima of our functional. The "non-Abelian" nature of our vortices is due to the fact that in the present case the vortices are associated with a unitary bundle whose rank need not be one, i.e., for which the structure group need not be Abelian. In addition, the base manifold of the bundles we consider need not be $\mathbb{R}^2$. We will retain the restriction that the dimension should be even, and will require further that the base manifold has a complex structure and admits a Kähler metric. We will make one further modification, which is to compactify the base manifold. This results in a stronger link between the topology of the bundle and the energy functional than is evident when the base is $\mathbb{R}^2$.

With these modifications, the functional we consider can be compared to the Yang-Mills energy for unitary bundles over closed Kähler manifolds. In this setting, the minima of the Yang-Mills functional are described by the Hermitian-Einstein equations (cf. [Do, U-Y]). The minimization condition for our functional can be described as the Hermitian-Einstein equation with an extra zeroth order term due to a smooth bundle section.

We begin with the definition of the functional. Let $E \to X$ be a complex rank R vector bundle over a closed Kähler manifold of complex dimension n. We assume that X has a fixed Kähler metric with associated Kähler form $\omega \in \Omega^2(X, \mathbb{C})$, and that E has a Hermitian metric denoted by H. Let $\mathscr{A}$ denote the set of complex connections on E and let $\mathscr{A}(H)$ consists of those

1991 *Mathematics Subject Classification.* Primary 53C07, 32G13.
This paper is in final form and no version of it will be submitted for publication elsewhere.

connections compatible with the metric.

In addition to connections on E, we consider also smooth sections. The space of such sections is denoted by $\Omega^0(X, E)$, and in general the space of k-forms with values in E is denoted by $\Omega^k(X, E)$. In particular, the covariant derivative of a section takes its values in $\Omega^1(X, E)$, while the curvature of a connection takes its values in $\Omega^2(X, \text{End } E)$. Here End E denotes the (general linear) endomorphism bundle associated to E.

We define a real valued "Yang-Mills-Higgs" energy on $\mathscr{A}(H) \times \Omega^0(X, E)$ by

$$(1) \qquad \mathbf{YMH}_\tau(D, \phi) = \|F_D\|_{L^2}^2 + \|D\phi\|_{L^2}^2 + \frac{1}{4} \int_X |\phi \otimes \phi^* - \tau \mathbf{I}|^2 d \text{ vol} .$$

Here $F_D = D^2$ is the curvature of the connection D and $D\phi$ is the covariant derivative of the section ϕ. The L^2 norms on $\Omega^2(X, \text{End } E)$ and $\Omega^1(X, E)$ are constructed from the inner product H on E and the inner product induced on $\Omega^2(X, \mathbb{C})$ by the Kähler metric on X. In the last term, ϕ^* is the adjoint taken with respect to the metric H. Hence $\phi \otimes \phi^* \in \Omega^0(X, E \otimes E^*) \cong \Omega^0(X, \text{End } E)$. The symbol $\mathbf{I}$ denotes the identity section in $\Omega^0(X, \text{End } E)$ and τ is a real parameter.

In the case where the rank of E is one, the last term reduces to $\frac{1}{4} \int_X (|\phi|^2 - \tau)^2 d \text{ vol}$. Apart from the variable parameter τ, the resulting functional is identical to the Yang-Mills-Higgs energy which yields the classical Abelian Vortices over $\mathbb{R}^2$. The extra parameter is required when the base manifold is taken to be *compact*. We will see that in that case topological constraints prevent the energy functional from attaining its minimum unless the parameter is in a suitable range. In the case of Abelian Vortices over (the noncompact) $\mathbb{R}^2$, such constraints do not apply. They are reflected, however, in the Vortex Number which is associated with each minimizing configuration of the functional. This is an integer and can be interpreted as a topological constraint on a bundle over a compactification of $\mathbb{R}^2$ [J-T]. As we go from $U(1)$ bundles over a noncompact manifold of complex dimension one, to $U(r)$ bundles over compact manifolds of complex dimension n, the Vortex Number is replaced by a combination of the parameter τ and the first and second Chern classes of the bundle.

2. Topological lower bound and minimization criteria

An important feature of the functional $\mathbf{YMH}_\tau$ is that it has a lower bound determined by the topology of the bundle E. This is a feature common to many of the most interesting functionals in Gauge Theory. It is true, for example of the Yang-Mills energy when defined on $SU(2)$ bundles over four-manifolds (cf. [F-U]). In that setting the lower bound emerges as a consequence of the special nature of four-dimensional geometry (i.e., the fact that two-forms decompose as self-dual and anti-self-dual forms). The minimization conditions for the Yang-Mills energy in this setting are the Self-Dual or Anti-Self-Dual equations.

In the case of our functional, it is the special nature of complex Kähler manifolds that leads to the topological lower bounds. We use the bigrading on the space of complex forms which allows one to decompose the space of k-forms into the direct sum of spaces of forms of holomorphic type (p, q), where $p + q = k$. This splitting extends to forms with values in E or End E and leads to a decomposition of a connection D and its curvature F_D. We get

$$D = D^{0,1} + D^{1,0}. \tag{2}$$

$$F_D = F^{0,2} + F^{1,1} + F^{2,0}. \tag{3}$$

We also make use of the Kähler identities (which relate $D^{0,1}$, $D^{1,0}$, and the Kähler form ω), and an identity for the norm of a two-form of type $(1, 1)$ (cf. [W, K1]). These lead to the next result.

PROPOSITION 1 [Br1]. *The functional* $YMH_\tau(D, \phi)$ *can be written as*

$$YMH_\tau(D, \phi) = \|F^{0,2}\|_{L^2}^2 + 2\|D^{0,1}\phi\|_{L^2}^2 + \left\|\sqrt{-1}(F_D, \omega) + \frac{1}{2}\phi \otimes \phi^* - \frac{\tau}{2}\mathbf{I}\right\|_{L^2}^2 \tag{4}$$
$$+ 2\pi\tau \int_X c_1(E) \wedge \frac{\omega^{n-1}}{(n-1)!} - 8\pi^2 \int_X \mathrm{ch}_2(E) \wedge \frac{\omega^{n-2}}{(n-2)!}.$$

Here $c_1(E)$ *and* $\mathrm{ch}_2(E)$ *are the first Chern class and second Chern character of* E *respectively.*

In the third term in (4), the expression (F_D, ω) is a section in $\Omega^0(X, \text{End } E)$. It is obtained via the pairing

$$(\; , \;): \Omega^2(X, \text{End } E) \times \Omega^2(X, \mathbb{C}) \to \Omega^0(X, \text{End } E)$$

induced by the Kähler metric on X. It can also be thought of as the contraction of the form part of F_D against the Kähler form ω.

COROLLARY 2. *The functional* $YMH_\tau(D, \phi)$ *attains its topological lower bound if and only if*

$$F^{0,2} = 0, \tag{5a}$$

$$D^{0,1}\phi = 0, \tag{5b}$$

$$\sqrt{-1}(F_D, \omega) + \frac{1}{2}\phi \otimes \phi^* = \frac{\tau}{2}\mathbf{I}. \tag{5c}$$

3. Interpretation of the minimizing equations

The first equation, (5a), is equivalent to the condition $(D^{0,1})^2 = 0$. The antiholomorphic part of a connection is a $\mathbb{C}$-linear first order differential operator $D^{0,1}: \Omega^0(X, E) \to \Omega^{0,1}(X, E)$. It satisfies the Leibniz formula

$\overline{\partial}_E(f\sigma) = \overline{\partial}(f)\sigma + f\overline{\partial}_E(\sigma)$ for $f \in C^\infty(X)$ and $\sigma \in \Omega^0(X, E)$. In short, $D^{0,1}$ is very much like a $\overline{\partial}$-operator for sections of E; the only property missing is the condition on the square of the operator, viz.$(D^{0,1})^2 = 0$. When this condition is satisfied one can indeed choose a holomorphic structure for E so that a section $\phi \in \Omega^0(X, E)$ is holomorphic if and only if $D^{0,1}\phi = 0$. In fact the choice of a holomorphic structure on E is equivalent to the specification of a $\overline{\partial}$-operator for sections of E, i.e. of a $\mathbb{C}$-linear first order differential operator $\overline{\partial}_E: \Omega^0(X, E) \to \Omega^{0,1}(X, E)$ which satisfies the Leibniz formula and the integrability condition $(\overline{\partial}_E)^2 = 0$ (cf. [W]).

If $D^{0,1}$ determines a holomorphic structure, then the second equation, $D^{0,1}\phi = 0$, can be interpreted as a holomorphicity condition on the section ϕ.

DEFINITION 3. Let $\mathscr{C}$ denote the set of all *holomorphic structures* on E. That is, $\mathscr{C} = \{\overline{\partial}_E: \Omega^0(X, E) \to \Omega^{0,1}(X, E)/\overline{\partial}_E$ satisfies the Leibniz Rule and the integrability condition$\}$. Define a subset of $\mathscr{C} \times \Omega^0(X, E)$, called the set of *holomorphic pairs*, by

$$\mathscr{H} = \{(\overline{\partial}_E, \phi) \in \mathscr{C} \times \Omega^0(X, E)/\overline{\partial}_E\phi = 0\}.$$

The first two equations (5a, b) can thus be interpreted as saying that $(D^{0,1}, \phi)$ determines a point in $\mathscr{H}$. It is a basic feature of the geometry of complex bundles that given any $\overline{\partial}_E \in \mathscr{C}$ and a hermitian metric H on E, there is a unique connection $D \in \mathscr{A}(H)$ such that $D^{0,1} = \overline{\partial}_E$. We will denote this connection by $D_{\overline{\partial}_E, H}$ and its curvature by $F_{\overline{\partial}_E, H}$. The connection D in equations (5) is thus fully determined by $D^{0,1}$ and the metric H on E. The third equation (5c) can thus be thought of as an equation in $(\overline{\partial}_E, \phi, H)$ and can be written

$$(6) \qquad \sqrt{-1}(F_{\overline{\partial}_E, H}, \omega) + \frac{1}{2}\phi \otimes \phi^{*H} = \frac{\tau}{2}\mathbf{I}.$$

With the metric held fixed, this equation becomes a condition on points $(\overline{\partial}_E, \phi) \in \mathscr{H}$. We will refer to (6) as the τ-**Vortex equation** and will denote the set of solutions to this equation by $\mathscr{V}_\tau(H)$. We emphasize that $\mathscr{V}_\tau(H)$ is a subset of $\mathscr{H}$. The above discussion shows, however that there is a bijective correspondence between the elements of $\mathscr{V}_\tau(H)$ and the "topological minima" of $\mathbf{YMH}_\tau$. More precisely, let

$$\mathrm{Min}_\tau(H) = \left\{(D, \phi) \in \mathscr{A}(H) \times \Omega^0(X, E)/\mathbf{YMH}_\tau(D, \phi)\right.$$

$$\left. = 2\pi\tau \int_X c_1(E) \wedge \frac{\omega^{n-1}}{(n-1)!} - 8\pi^2 \int_X \mathrm{ch}_2(E) \wedge \frac{\omega^{n-2}}{(n-2)!}\right\}.$$

PROPOSITION 4. *There is a bijective correspondence*

(7) $$\mathrm{Min}_\tau(H) \leftrightarrow \mathscr{V}_\tau(H),$$

between these two sets.

4. The τ-Vortex equation as an equation for the metric

In the definition of $\mathscr{V}_\tau(H)$ we hold the metric H fixed and interpret (6) as an equation on points in $\mathscr{H}$. If instead we hold $(\overline{\partial}_E, \phi) \in \mathscr{H}$ fixed, then the τ-Vortex equation becomes an equation for the metric H. In this interpretation we start with a holomorphic bundle with a prescribed holomorphic section and look for a hermitian metric specified by a curvature condition on the corresponding metric connection. In the absence of the section, the equation specifying the metric reduces to the Hermitian-Einstein equation, viz.,

(8) $$\sqrt{-1}(F_{\overline{\partial}_E, H}, \omega) = \mathrm{const}\ \mathbf{I}.$$

When the metric is considered fixed and the unknown in this equation is taken to be the operator $\overline{\partial}_E$, this equation is none other than the minimization criterion for the Yang-Mills energy defined on $\mathscr{A}(H)$, i.e. for $\mathbf{YM}(D) = \|F_D\|_{L^2}^2$ (cf. [**Do**]). The constant in (8) is fixed by the topology of E. By using the fact that $c_1(E) = \frac{\sqrt{-1}}{2\pi} \mathrm{Tr}(F)$, one obtains easily that the constant must be

$$\frac{2\pi}{\mathrm{Vol}(X)} \frac{1}{\mathrm{Rank}(E)} \int_X c_1(E) \wedge \frac{\omega^{n-1}}{(n-1)!}.$$

The existence of metrics satisfying the Hermitian-Einstein equation is closely related to the property of stability (in the sense of Mumford) for a holomorphic bundle. The notion of a holomorphic bundle being stable is an important concept in the theory of moduli spaces for holomorphic bundles (cf. [**N**]). In the language of sheafs of germs of holomorphic sections, stability is defined in terms of degrees and ranks. More exactly, let $(E, \overline{\partial}_E)$ be the holomorphic bundle corresponding to $\overline{\partial}_E \in \mathscr{C}$. If $F \subset E$ is any holomorphic subbundle, we can define its **degree** by

(9) $$\deg(F) = \int_X c_1(F) \wedge \frac{\omega^{n-1}}{(n-1)!}$$

where $c_1(F)$ is the first Chern class of F. The **slope** of F is then defined to be

(10) $$\mu(F) = \frac{\deg(F)}{\mathrm{Rank}(F)}.$$

Let $\mathscr{E}$ be the sheaf of germs of holomorphic sections of $(E, \overline{\partial}_E)$. The above notions of degree and slope extend to coherent analytic subsheaves of $\mathscr{E}$. The holomorphic bundle is called **semistable** if for all reflexive subsheaves $\mathscr{F} \subset \mathscr{E}$ with $\mathrm{rank}(\mathscr{F}) > 0$ we have

(11) $$\mu(\mathscr{F}) \le \mu(E).$$

The bundle is called **stable** if the inequality is strict whenever $0 < \mathrm{rank}(\mathscr{F}) < \mathrm{rank}(E)$.

The significance of solutions to the Hermitian-Einstein equation is contained in the following two theorems.

THEOREM 5 [K2, L]. *Let* $(E, \overline{\partial}_E)$ *be a holomorphic bundle over a closed Kähler manifold* X. *Suppose that* E *supports a Hermitian metric which satisfies the Hermitian-Einstein equation. Then* E *splits holomorphically as a direct sum of stable bundles, each with slope equal to* $\mu(E)$.

THEOREM 6 [U-Y]. *Let* $(E, \overline{\partial}_E)$ *be a holomorphic bundle over a closed Kähler manifold* X. *If the bundle is stable then it supports a solution to the Hermitian-Einstein equation. The metric is unique if its determinant is fixed.*

The existence of metrics satisfying the τ-vortex equation can similarly be understood in terms of a stability criterion. The relevant notion of stability now applies to a pair $(\overline{\partial}_E, \phi) \in \mathscr{H}$ and is defined as follows.

DEFINITION 7. Let $(E, \overline{\partial}_E)$ be a holomorphic bundle over a closed Kähler manifold. Let $\mathscr{E}$ be the sheaf of germs of holomorphic sections of E. Let ϕ be a prescribed nontrivial holomorphic section of E. Define

$$\mu_M = \mathrm{Sup}\{\mu(\mathscr{F})/\mathscr{F} \subset \mathscr{E} \text{ is a reflexive subsheaf with } 0 < \mathrm{rank}\,\mathscr{F}\},$$

$$\mu_m(\phi) = \mathrm{Inf}\{\mu(\mathscr{E}/\mathscr{F})/\mathscr{F} \subset \mathscr{E} \text{ is a reflexive subsheaf with } \mathrm{rank}\,\mathscr{F} < R$$
$$\text{and } \phi \in \mathscr{F}\}.$$

We say that the pair $(\overline{\partial}_E, \phi)$ is stable if

(12) $$\mu_M < \mu_m(\phi).$$

We say that the pair is τ-**stable** if it is stable and

(13) $$\mu_M < \frac{\tau \mathrm{Vol}(X)}{4\pi} < \mu_m(\phi).$$

The analog of Theorem 5 is

THEOREM 8A [Br2]. *Let* $(E, \overline{\partial}_E, \phi)$ *be as above. Suppose that for a given value of the real parameter* τ *there exists a Hermitian metric* H *on* E *such that*

$$\sqrt{-1}(F_{\overline{\partial}_E, H}, \omega) + \frac{1}{2}\phi \otimes \phi^{*_H} = \frac{\tau}{2}\mathbf{I}.$$

Then E *splits holomorphically as* $E = E_\phi \oplus E_{SS}$ *where*

(a) E_{SS}, *if nonempty, is semistable and decomposes as a direct sum of stable bundles each of slope* $\frac{\tau \mathrm{Vol}(X)}{4\pi}$,

(b) *the section* ϕ *has no component in* E_{SS}, *i.e.* $\phi \in \Omega^0(X, E_\phi)$,

(c) *on* E_ϕ, *the pair* $(\overline{\partial}_E, \phi)$ *forms a* τ-*stable pair.*

Notice that if $\frac{\tau \mathrm{Vol}(X)}{4\pi} < \mu(E)$, then no pair $(\overline{\partial}_E, \phi)$ can be τ-stable. It follows from Theorem 8 that the τ-Vortex equation cannot have solutions for such choices of τ. This restriction can also be seen simply by applying the Chern-Weil formula for $c_1(E)$ to the τ-Vortex equation. The need for the adjustable parameter τ in the definition of **YMH**$_\tau$ can be traced directly to such obstructions. The converse to Theorem 8, i.e. the analog of Theorem 6 is also true:

THEOREM 8B [Br2]. *Let* $(E, \overline{\partial}_E, \phi)$ *be as above. Suppose that* $(\overline{\partial}_E, \phi)$ *is a* τ-*stable pair. Then there is a unique smooth Hermitian metric* H *for which the* τ-*Vortex equation*

$$\sqrt{-1}(F_{\overline{\partial}_E, H}, \omega) + \frac{1}{2}\phi \otimes \phi^{*_H} = \frac{\tau}{2}\mathbf{I}$$

is satisfied.

DEFINITION 9. Let $\mathscr{V}_\tau$ and $\mathscr{S}_\tau$ be the subsets of $\mathscr{H}$ defined by

$$\mathscr{V}_\tau = \left\{ (\overline{\partial}_E, \phi) \in \mathscr{H} / \sqrt{-1}(F_{\overline{\partial}_E, H}, \omega) + \frac{1}{2}\phi \otimes \phi^{*_H} = \frac{\tau}{2}\mathbf{I} \text{ for } \textit{some} \text{ choice of } H \right\}$$

$$\mathscr{S}_\tau = \left\{ (\overline{\partial}_E, \phi) \in \mathscr{H} / (\overline{\partial}_E, \phi) \text{ is a } \tau\text{-stable pair} \right\}.$$

The above theorems can be more succinctly phrased as

THEOREM 10. (a) $\mathscr{S}_\tau \subseteq \mathscr{V}_\tau$.
  (b) *For any* $(\overline{\partial}_E, \phi)$ *in* $\mathscr{V}_\tau - \mathscr{S}_\tau$ *the bundle splits holomorphically as* $E = E_\phi \oplus E_{SS}$ *where* E_ϕ *and* E_{SS} *are as in Theorem* 8A.

Notice that the semistable summand in (b) is constrained to have slope $\frac{\tau \text{Vol}(X)}{4\pi}$. Let $M(E)$ denote the set of all slopes of subbundles, i.e.,

$$(14) \qquad M(E) = \{\mu \in \mathbb{R} / \mu \text{ is the slope of some subbundle of } E\}.$$

Clearly, if $\frac{\tau \text{Vol}(X)}{4\pi}$ is *not* an element in $M(E)$, then $\mathscr{V}_\tau - \mathscr{S}_\tau$ must be empty. When E is a bundle over an algebraic manifold, the Kähler form on X can be chosen so that all subbundles have integer degree and $M(E)$ is therefore a discrete set. It follows that Theorem 10 can be interpreted as saying that *for generic* τ, $\mathscr{S}_\tau = \mathscr{V}_\tau$.

5. Gauge invariance

Fixing a metric on the bundle E enables one to talk of *unitary* bundle automorphisms; these are precisely the automorphisms $g: E \to E$ which preserve the unitary structure imposed by the metric H. Such automorphisms form a group which we call the unitary Gauge group and denote by **G**. The group **G** acts on $\mathscr{A}(H)$ and $\Omega^0(X, E)$ by

$$(15) \qquad\qquad\qquad g(D) = g \circ D \circ g^{-1},$$

$$(16) \qquad\qquad\qquad g(\phi) = g\phi.$$

The functional $\mathbf{YMH}_\tau(D, \phi)$ is clearly invariant under such gauge transformations and this invariance is reflected in the **G**-invariance of $\text{Min}_\tau(H)$. Indeed we do not wish to distinguish between the triple (E, D, ϕ) and the gauge equivalent triple $(g(E), g(D), g\phi)$. The geometrically significant set is thus not $\text{Min}_\tau(H)$, but the set of gauge equivalence classes, $\text{Min}_\tau(H)/\mathbf{G}$.

The above prescription of the action of **G** on $\mathscr{A}(H)$ also describes an action of $\mathscr{C}$, viz. $g(\overline{\partial}_E) = g \circ \overline{\partial}_E \circ g^{-1}$. However when considering holomorphic structures on a complex bundle one does not impose a unitary constraint and the appropriate notion of equivalence is that of equivalence under any general complex linear bundle automorphism. We let $\mathbf{G_C}$ denote the group of complex linear bundle automorphisms. The G-actions extend to actions of $\mathbf{G_C}$ on $\mathscr{A}(H)$ and $\Omega^0(X, E)$. We thus have an action of $\mathbf{G_C}$ on $\mathscr{A}(H) \times \Omega^0(X, E)$. With the action thus constructed, $\mathscr{H}$ is a $\mathbf{G_C}$ invariant subset. Furthermore,

PROPOSITION 11 [**Br1**]. $\mathscr{V}_\tau$ *is* $\mathbf{G_C}$ *invariant. In fact if the τ-Vortex equation is satisfied by the triple $(\overline{\partial}_E, \phi, H)$ then it is satisfied by the triple $(g^{-1}(\overline{\partial}_E), g^{-1}\phi, Hg^*g)$ for any g in $\mathbf{G_C}$.*

REMARK. The metric $K = Hg^*g$ is defined by $K(s, t) = H(gs, gt)$ for s, t in $\Omega^0(X, E)$. Any two metrics can be related in this way by an element $g \in \mathbf{G_C}$.

The next proposition reunites the interpretation of the Vortex equation as a minimization condition for the Yang-Mills-Higgs energy $\mathbf{YMH}_\tau$ on (E, H), and its interpretation as a prescription for a special metric on $(E, \overline{\partial}_E, \phi)$.

PROPOSITION 12 [**Br1**]. *There is a bijective correspondence as sets between* $\mathrm{Min}_\tau(H)/\mathbf{G}$ *and* $\mathscr{V}_\tau/\mathbf{G}_C$.

Recall that Theorem 10 says that for generic choice of the parameter τ, $\mathscr{V}_\tau$ and $\mathscr{S}_\tau$ coincide. For such τ we thus have

$$(17) \qquad \mathrm{Min}_\tau(H)/\mathbf{G} \leftrightarrow \mathscr{V}_\tau/\mathbf{G_C} = \mathscr{S}_\tau/\mathbf{G_C}.$$

6. Results for line bundles

When the rank of E is one, the results of Theorems 8 and 9 become

THEOREM 13 [**Br1**]. *Let $(L, \overline{\partial}_L)$ be a holomorphic line bundle over a Kähler manifold X. Let $\phi \in \Omega^0(X, L)$ be a nontrivial holomorphic section. Then the condition $\frac{\tau \mathrm{Vol}(X)}{4\pi} > \deg(L, \omega)$ is necessary and sufficient for the existence of a unique hermitian metric H such that*

$$(18) \qquad \sqrt{-1}(F_{\overline{\partial}_E, H}, \omega) + \frac{1}{2}|\phi|_H^2 = \frac{\tau}{2}.$$

If X is algebraic, then the zero set $Z(\phi)$ of the section ϕ determines an effective divisor, $\mathscr{D}$, in X. The Kähler form determines a two-cycle $[\omega] \in H_2(X, \mathbb{C})$, namely the cycle Poincaré dual to the element in $H^{2n-2}(X, \mathbb{C})$ represented by $\frac{\omega^{n-1}}{(n-1)!}$. If we take the intersection pairing between $\mathscr{D}$ and $[\omega]$ we get (cf. [**G-H**]),

$$(19) \qquad \#(\mathscr{D}, [\omega]) = \deg(L, \omega).$$

Conversely, given any effective divisor on X with $\#(\mathscr{D}, [\omega]) = \deg(L, \omega)$, there is a canonically determined holomorphic line bundle $L_{\mathscr{D}}$ such that

   (a) $\deg(L_{\mathscr{D}}, \omega) = \#(\mathscr{D}, [\omega]) = \deg(L, \omega)$,

   (b) there is a canonically determined holomorphic section $\phi \in \Omega^0(X, L_{\mathscr{D}})$ with $Z(\phi) = \mathscr{D}$.

We thus find

THEOREM 14 [Br1]. *Let L be a $U(1)$ bundle over a Kähler manifold X. Suppose that X is algebraic and that the Kähler form ω, is chosen to be integral. Let $\deg(L, \omega) = N$. Then for every τ such that $\frac{\tau \operatorname{Vol}(X)}{4\pi} > N$, there is a bijective correspondence between $\operatorname{Min}_\tau(H)/\mathbf{G}$ and the set of effective divisors on X with $\#(\mathscr{D}, [\omega]) = N$.*

When X is a Riemann surface, the divisors which occur in Theorem 14 are simply sets of N (not necessarily distinct) points on X. In that case Theorem 14 corresponds exactly to the results of Taubes [T] for vortices on $\mathbb{R}^2$. This result for Abelian Vortices on compact Riemann surfaces was also obtained by Noguchi in [No].

7. The moduli space of solutions

In view of Proposition 12 we can study $\operatorname{Min}_\tau(H)/\mathbf{G}$ by studying $\mathscr{V}_\tau/\mathbf{G}_{\mathrm{C}}$ or (provided $\frac{\tau \operatorname{Vol}(X)}{4\pi}$ is *not* an element in $M(E)$)$\mathscr{S}_\tau/\mathbf{G}_{\mathrm{C}}$. At least in the case where the base manifold of E is a closed Riemann surface we can give a fairly complete description.

Suppose then that X is a closed Riemann surface of genus g. Assume further that $\frac{\tau \operatorname{Vol}(X)}{4\pi}$ is *not* contained in $M(E)$. We can examine $\mathscr{S}_\tau/\mathbf{G}_{\mathrm{C}}$ by considering it as a subset

$$(20) \qquad \mathscr{S}_\tau/\mathbf{G}_{\mathrm{C}} \subseteq \mathscr{H}/\mathbf{G}_{\mathrm{C}} \subseteq (\mathscr{C} \times \Omega^0(X, E))/\mathbf{G}_{\mathrm{C}}$$

and by looking at infinitesmal deformations of a point $(\bar{\partial}_E, \phi)$. The techniques employed are quite standard (cf. [AHS, I, K1]). Deformations of $(\bar{\partial}_E, \phi)$ in $\mathscr{H}/\mathbf{G}_{\mathrm{C}}$ can be described as elements in the first cohomology of the elliptic complex,

(C)

$$0 \to \Omega^0(X, \operatorname{End} E) \xrightarrow{d_1} \Omega^{0,1}(X, \operatorname{End} E) \oplus \Omega^0(X, E) \xrightarrow{d_2} \Omega^{0,1}(X, E) \to 0$$

where the maps d_1 and d_2 are defined by

$$(21) \qquad\qquad d_1(u) = (-\bar{\partial}_E(u), u\phi),$$

$$(22) \qquad\qquad d_2(\alpha, \eta) = \bar{\partial}_E \eta + \alpha\phi.$$

Since X is a Riemann surface, the integrability condition $(\bar{\partial}_E)^2 = 0$ is vacuus and the space $\mathscr{C}$ is an affine space modelled on $\Omega^{0,1}(X, \operatorname{End} E)$. The space $\Omega^{0,1}(X, \operatorname{End} E) \oplus \Omega^0(X, E)$ can thus be identified with the tangent space to $\mathscr{C} \times \Omega^0(X, E)$ at the point $(\overline{\partial}_E, \phi)$. The map d_2 is the differential of the map $F: \mathscr{C} \times \Omega^0(X, E) \to \Omega^{0,1}(X, E)$ given by

$$(23) \qquad\qquad F(\overline{\partial}_E, \phi) = \overline{\partial}_E \phi.$$

It follows that deformations of $(\overline{\partial}_E, \phi) \in \mathscr{C} \times \Omega^0(X, E)$ which are deformations within $\mathscr{H}$ correspond to the kernel of d_2. The map d_1 is the differential of the $\mathbf{G}_{\mathbf{C}}$ action on $\mathscr{H}$. Hence the image of this map corresponds to deformations which are tangent to a $\mathbf{G}_{\mathbf{C}}$ orbit.

It is clear therefore that if $\mathscr{H}/\mathbf{G}_{\mathbf{C}}$ is locally a complex manifold in the neighborhood of a point $[\overline{\partial}_E, \phi]$, then its tangent space is $H^1(C)$, i.e. the first cohomology of the complex (C). The question remains however as to when such a manifold structure actually exists. The answer to this question is the following:

PROPOSITION 15 [**B-D**]. *Let $(\overline{\partial}_E, \phi)$ be a point in $\mathscr{H}$. Suppose that $H^0(C)$ and $H^2(C)$ both are trivial. Then $\mathscr{H}/\mathbf{G}_{\mathbf{C}}$ is a nonsingular complex manifold in a neighborhood of $[\overline{\partial}_E, \phi]$.*

The condition $H^2(C) = 0$ is the requirement that the map d_2 be surjective. In view of the interpretation of this map as the differential of the map F, and the fact that $\mathscr{H} = F^{-1}(0)$, this condition guarantees that $\mathscr{H}$ is a manifold near the point $(\overline{\partial}_E, \phi)$. The condition $H^0(C) = 0$ rules out a kernel for d_1 and thus prevents the occurrence of singularities in the quotient $\mathscr{H}/\mathbf{G}_{\mathbf{C}}$ due to a nontrivial stabilizer of $(\overline{\partial}_E, \phi)$ in $\mathbf{G}_{\mathbf{C}}$.

The condition $H^0(C) = 0$ is the analog of the notion of *simplicity* for a holomorphic bundle. A holomorphic bundle $(E, \overline{\partial}_E)$ is called simple if the only holomorphic bundle endomorphisms are the constant multiples of the identity, i.e. if $\overline{\partial}_E(u) = 0$ implies $u = \lambda \mathbf{I}$, for $u \in \Omega^0(X, \text{End } E)$. The important properties of simple holomorphic structures on a bundle E over a Riemann surface are (cf. [**K1, N, OSS**]),

(a) the space equivalence classes of holomorphic structures is a complex manifold in a neighborhood of each simple structure,

(b) if a holomorphic structure is stable then it is simple,

(c) the moduli space of simple holomorphic structures is not necessarily Hausdorff, but its restriction to the stable points is Hausdorff.

We find similar characteristics for the points $(\overline{\partial}_E, \phi)$ which satisfy $H^0(C) = 0$.

PROPOSITION 16 [**B-D**]. (a) *If $(\overline{\partial}_E, \phi)$ is a stable pair then $H^0(C) = 0$.*

(b) *If $H^2(C) = 0$ and $(\overline{\partial}_E, \phi)$ is a stable pair, then $\mathscr{H}/\mathbf{G}_{\mathbf{C}}$ is Hausdorff in a neighborhood of $[\overline{\partial}_E, \phi]$.*

We thus call $\mathscr{H}_0 = \{(\overline{\partial}_E, \phi) \in \mathscr{H}/H^0(C) = 0\}$ the set of *simple pairs*. Proposition 15 is a statement about $\mathscr{H}_0/\mathbf{G}_{\mathbf{C}}$. The condition $H^2(C) = 0$ in that the proposition can be guaranteed in certain special cases. The condition applies, for example, when $\overline{\partial}_E$ is such that the holomorphic bundle E has vanishing first sheaf cohomology, i.e. $H^1(E) = 0$. For semistable holomorphic structures, the vanishing of $H^1(E)$ can be achieved by taking a bundle of sufficiently high degree.

PROPOSITION 17 [cf. N]. *Let E be a rank R over a Riemann surface of genus g. Suppose that $\deg(E) > R(2g - 2)$. Let $\overline{\partial}_E \in \mathscr{C}$ be a semistable holomorphic structure. Then $H^1(E) = 0$.*

An interesting special case to consider is thus the one where X is a Riemann surface of genus g, E is a bundle over X with $\deg(E) >$ $\mathrm{Rank}(E)(2g - 2)$ and where the parameter τ is chosen as follows: let

$$\mu_+ = \min\{\mu \in \mathbb{R}/\mu \text{ is the slope of a subbundle of } E \text{ and } \mu > \mu(E)\},$$

and pick τ such that

(24)
$$\mu(E) < \frac{\tau\,\mathrm{Vol}(X)}{4\pi} < \mu_+.$$

PROPOSITION 18 [B-D]. *With $E \to X$ and τ as above we have*
(a) $\mathscr{S}_\tau = \mathscr{V}_\tau$,
(b) *if $(\overline{\partial}_E, \phi) \in \mathscr{S}_\tau$, then $\overline{\partial}_E$ is a semistable holomorphic structure,*
(c) *if $(\overline{\partial}_E, \phi) \in \mathscr{S}_\tau$, then $H^0(C) = 0$ and $H^2(C) = 0$.*

It follows that

THEOREM 19 [B-D]. *With $E \to X$ and τ as above $\mathscr{S}_\tau/\mathbf{G}_{\mathbb{C}}$ is a finite dimensional Hausdorff complex manifold. Its dimension is given by*

(25)
$$\begin{aligned}
\dim_{\mathbb{C}}(\mathscr{S}_\tau/\mathbf{G}_{\mathbb{C}}) &= \dim_{\mathbb{C}} H^1(C) = \chi(E) - \chi(\mathrm{End}\ E) \\
&= \deg(E) + (g - 1)\,\mathrm{Rank}(E)(\mathrm{Rank}(E) - 1).
\end{aligned}$$

8. Moment maps and Kähler structure

By analyzing the right-hand side of the correspondence $\mathrm{Min}_\tau(H)/\mathbf{G} \leftrightarrow$ $\mathscr{S}_\tau/\mathbf{G}_{\mathbb{C}}$ we have demonstrated the existence of a *complex manifold* structure. We now examine the left-hand side, i.e. the quotient $\mathrm{Min}_\tau(H)/\mathbf{G}$. We will retain the assumptions of Propositions 18 on X, $\deg(E)$ and τ. Thus X is a Riemann surface of genus g, the degree of E is suitably large compared to its rank, and τ is generic but in the lowest permissible range. The results of this examination show that $\mathrm{Min}_\tau(H)/\mathbf{G}$ is compact and admits the structure of a symplectic manifold.

The compactness follows from the fact that the gauge transformations are restricted to be unitary, and Uhlenbeck's compactness theorem [U].

PROPOSITION 19 [B-D]. *Let $E \to X$ be a complex bundle over a closed Riemann surface of genus g. Assume that $\deg(E) > \mathrm{Rank}(E)(2g - 2)$. Fix* *a hermitian metric* H *on* E *and pick* τ *as in* (15). *The* $\text{Min}_\tau(H)/\mathbf{G}$ *is compact.*

The symplectic structure is seen by introducing a moment map. It is well known that the solutions to the Hermitian-Einstein equation can be interpreted as zeros of a moment map (cf. [A-B, Ko]), and that this interpretation leads to a natural symplectic structure on the moduli space of solutions. Solutions to the τ-Vortex equation can be similarly interpreted, and in this way $\text{Min}_\tau(H)/\mathbf{G}$ can be shown to admit a symplectic structure. The symplectic structure on $\text{Min}_\tau(H)/\mathbf{G}$ can furthermore be chosen compatible with the complex structure of $\mathscr{S}_\tau/\mathbf{G}_{\mathbb{C}}$. One thus obtains a Kähler structure.

In order to realize solutions to the τ-vortex equation as zeros of a moment map, we must treat the metric H as fixed and the pair $(\overline{\partial}_E, \phi)$ as the variables. A solution is then a zero of the map

$$(26) \qquad \Psi_\tau(\overline{\partial}_E, \phi) = (F_{\overline{\partial}_E, H}, \omega) - \frac{\sqrt{-1}}{2}\phi \otimes \phi^{*_H} + \frac{\sqrt{-1}\tau}{2}\mathbf{I}.$$

This map is defined on all of $\mathscr{C} \times \Omega^0(X, E)$, while clearly

$$(27) \qquad \text{Min}_\tau(H) = \{(\overline{\partial}_E, \phi) \in \mathscr{H}/\Psi_\tau(\overline{\partial}_E, \phi) = 0\}.$$

The map takes it values in $\Omega^0(X, \text{End } E)$, but in fact

$$(28) \qquad \Psi_\tau(\overline{\partial}_E, \phi) + \Psi_\tau(\overline{\partial}_E, \phi)^* = 0.$$

We can thus treat Ψ_τ as a map from $\mathscr{C} \times \Omega^0(X, E)$ to $\mathbf{g}$, the Lie algebra of $\mathbf{G}$. Alternatively, we can use the identification of $\mathbf{g}$ with its dual $\mathbf{g}^*$ and consider Ψ_τ as a map to $\mathbf{g}^*$. Now $\mathscr{C}$ and $\Omega^0(X, E)$ have natural hermitian inner products coming from the inner products on $\Omega^{0,1}(X, \text{End } E)$ and $\Omega^0(X, E)$. Since they are constant, these metrics are Kähler and the associated two-forms provides $\mathscr{C}$ and $\Omega^0(X, E)$ with symplectic forms. Denote these by $\Omega_\mathscr{C}$ and Ω_0 respectively. We can combine the symplectic structures on $\mathscr{C}$ and $\Omega^0(X, E)$ to give one to $\mathscr{C} \times \Omega^0(X, E)$. We take the combination $\Omega = 2\Omega_\mathscr{C} + \Omega_0$. Furthermore, the $\mathbf{G}$-action on $\mathscr{C} \times \Omega^0(X, E)$ preserves this symplectic form. It now follows directly from the definitions that

PROPOSITION 20 [B-D]. *Let* $(\mathscr{C} \times \Omega^0(X, E), \Omega)$ *denote the symplectic manifold where the symplectic form is* $\Omega = 2\Omega_\mathscr{C} + \Omega_0$. *Let the* $\mathbf{G}$-*action on* $\mathscr{C} \times \Omega^0(X, E)$ *be as in* §5. *Then* $\Psi_\tau: \mathscr{C} \times \Omega^0(X, E) \to \mathbf{g}^*$ *is a* ($\mathbf{g}$-*equivariant) moment map for the action of* $\mathbf{G}$ *on* $(\mathscr{C} \times \Omega^0(X, E), \Omega)$. *That is, for any* $u \in \mathbf{g}$ *we have*

$$(29) \qquad \Omega(X_u, Y) = d\Psi_\tau(Y)(u)$$

for all tangent vectors Y *at a point* $(\overline{\partial}_E, \phi) \in \mathscr{C} \times \Omega^0(X, E)$. *Here* X_u *is the vector field on* $\mathscr{C} \times \Omega^0(X, E)$ *generated by* u.

By verifying the hypothesis of the Marsden-Weinstein reduction theorem [M-W] one can then show

THEOREM 21 [B-D]. *Let $E \to X$ and τ be as in Proposition 18. Let H be a fixed hermitian metric on E with respect to which the functional $\mathbf{YMH}_\tau$ is defined as in (1). Let $\mathscr{C} \times \Omega^0(X, E)$ carry the symplectic structure given by $\Omega = 2\Omega_{\mathscr{C}} + \Omega_0$. Then $\mathrm{Min}_\tau(H)/\mathbf{G}$ corresponds to the symplectic reduction of $\Psi_\tau^{-1}(0)$ and is thus a symplectic manifold.*

COROLLARY 22 [B-D]. *The moduli space $\mathrm{Min}_\tau(H)/\mathbf{G} = \mathscr{S}_\tau/\mathbf{G}_{\mathbb{C}}$ has compatible complex and symplectic structures and thus admits a Kähler metric.*

9. Relation to the moduli space of stable bundles

A natural question to consider is whether the projection

$$\pi : \mathscr{C} \times \Omega^0(X, E) \to \mathscr{C}$$

leads to a map from the moduli space of stable pairs to the moduli space of stable holomorphic structures. We may consider, for example, the restriction of π to $\mathscr{S} = \bigcup \mathscr{S}_\tau$, i.e. to the set of all holomorphic pairs which are stable. This restriction is however too weak, since the image $\pi(\mathscr{S})$ contains holomorphic structures which are not even semistable (cf. [Br3]).

Suppose instead that we choose τ as small as possible (i.e. as in (24)) and look at the restriction of π to $\mathscr{S}_\tau$. We further assume that the bundle E is over a Riemann surface of genus g. By Proposition 18(b) the image of the projection map then lies in the set of semistable holomorphic structures, $\mathscr{C}_{SS}$. This is clearly also true of the restriction of π to $\Psi_\tau^{-1}(0)$. The projection is manifestly a $\mathbf{G}$-equivariant map and we thus get a map

(30) $$\tilde{\pi} : \Psi_\tau^{-1}(0)/\mathbf{G} \to \mathscr{C}_{SS}/\mathbf{G}.$$

In addition, it is known that there is a surjective map from $\mathscr{C}_{SS}/\mathbf{G}$ to $\mathscr{M}$, the Seshadri compactification of the moduli space of stable holomorphic structures. This map is constructed in [D] and is shown to be continuous. By composing with the map $\tilde{\pi}$ we thus get a continuous map from $\Psi_\tau^{-1}(0)/\mathbf{G} = \mathscr{S}_\tau/\mathbf{G}_{\mathbb{C}}$ onto $\mathscr{M}$. Call this map p.

It is known that for genus $g \geq 2$, $\mathscr{M}$ is an irreducible projective variety [S]. We also know that $\mathscr{S}_\tau/\mathbf{G}_{\mathbb{C}}$ is a Kähler manifold. One may suspect therefore that the map p is a holomorphic map. This is in fact so.

THEOREM 23 [B-D]. *With $E \to X$ and τ as above, the map $p : \mathscr{S}_\tau/\mathbf{G}_{\mathbb{C}} \to \mathscr{M}$ is holomorphic.*

REFERENCES

[A-B] M. Atiyah, and R. Bott, *The Yang-Mills equations over a Riemann surface*, Philos. Trans. Roy. Soc. London **308** (1982), 523–615.

[A-H-S] M. Atiyah, N. Hitchin, and I. Singer, *Self duality in four dimensional Riemannian geometry*, Proc. Roy Soc. London Ser. A **362**, (1978), 425–461.

[Br1] S. Bradlow, *Vortices in holomorphic line bundles over closed Kähler manifolds*, Comm. Math. Phys. **135** (1990), 1–17.

[Br2] ——, *Special metrics and stability for holomorphic bundles with global sections*, J. Differential Geom. **33** (1991), 169–213.

[Br3] ——, *φ-stability in Rank two bundles over Riemann surfaces*, MSRI Preprint series, December, 1989.

[B-D] S. B. Bradlow and G. D. Daskalopoulos, *Moduli of stable paris for holomorphic bundles over Riemann surfaces*, I. J. M. **2** (1991), 477–513.

[D] G. Daskalopoulos, *The topology of the space of stable bundles over a compact Riemann surface*, Thesis, Univ. of Chicago, 1989.

[Do] S. K. Donaldson, *Anti-self-dual Yang-Mills connections over complex algebraic surfaces and stable bundles*, Proc. London Math. Soc. (3) **50** (1985), 1–26.

[F-U] D. Freed and K. Uhlenbeck, *Instantons and four-manifolds*, Springer, NY, 1984.

[G-H] P. Griffiths and J. Harris, *Principles of algebraic geometry*, John Wiley & Sons, NY, 1978.

[I] M. Itoh, *Geometry of Anti-self-dual connections and the Kuranishi map*, J. Math. Soc. Japan **40** (1988), 9–33.

[J-T] A. Jaffe and C. Taubes, *Vortices and monopoles*, Birkhäuser, Boston, 1980.

[K1] S. Kobayashi, *Differential geometry of complex vector bundles*, Princeton Univ. Press, Princeton, NJ, 1987.

[K2] ——, *Curvature and stability of vector bundles*, Proc. Japan Acad. Ser. A Math. Sci. A4 **58** (1982), 158–162.

[L] M. Lubke, *Stability of Einstein-Hermitian vector bundles*, Manuscripta Math. **42** (1983), 245–257.

[M-W] J. Marsden and A. Weinstein, *Reduction of symplectic manifolds with symmetry*, Rep. Math. Phys. **5** (1974), 121–130.

[N] P. E. Newstead, *An introduction to moduli problems and orbit spaces*, Tata Institute for Fundamental Research, Bombay.

[No] M. Noguchi, *Yang-Mills-Higgs theory on a compact riemann surface*, J. Math. Phys. **28** (1987)

[O-S-S] C. Okonek, M. Schneider, and H. Spindler, *Vector bundles on complex projective space*, Progr. in Math. 3, Birkhäuser, Boston, 1980.

[S] C. S. Seshadri, *Space of unitary vector bundles on a compact Riemann surface*, Ann. of Math. (2) **85** (1967), 303–336.

[T] C. Taubes, *Arbitrary N-vortex solutions to the first order Ginsburg-Landau equations*, Comm. Math. Phys. **72** (1980), 277–292.

[U] K. Uhlenbeck, *Connections with L^p bounds on curvature*, Comm. Math. Phys. **83** (1982), 31–42.

[U-Y] K. Uhlenbeck and S. T. Yau, *On the existence of Hermitian-Yang-Mills connections in stable vector bundles*, Comm. Pure Appl. Math. **39** (1986), 257–293.

[W] R. O. Wells, *Differential analysis on complex manifolds*, Springer-Verlag, NY, 1980.

UNIVERSITY OF CALIFORNIA, SAN DIEGO
E-mail address: sbradlow@ucsd.edu

Proceedings of Symposia in Pure Mathematics
Volume **54** (1993), Part 2

What Are the Best Almost-Complex Structures on the 6-Sphere?

EUGENIO CALABI AND HERMAN GLUCK

ABSTRACT. The most familiar almost-complex structures on the 6-sphere S^6 are defined in terms of Cayley multiplication. We observe here, using a twistor type transformation and the method of calibrated geometries, that these structures are "best" in the sense that they have the smallest volume when pictured as cross-sections of the appropriate bundle $E(S^6) = O(8)/U(4)$ over S^6.

To carry this out, we use the Triality Theorem for SO(8) to construct an explicit isometry of each component of $E(S^6)$ with the real Grassmann manifold $G_2 R^8$ of oriented 2-planes in 8-space, which matches these smallest cross-sections with the subgrassmannians $G_1 R^7$. These in turn are known from an earlier paper to be volume-minimizing in their homology class because they are calibrated by the Euler 6-form on $G_2 R^8 \cong G_6 R^8$.

A *complex structure* on a real vector space V is a linear map $J : V \to V$ such that $J^2 = -I$. An *almost-complex structure* on a smooth manifold M is a smoothly varying choice of complex structures on the tangent spaces to M at each of its points. If a complex structure on a real inner product space V preserves the inner product, then we call it a *hermitian structure*. Likewise, if an almost-complex structure on a Riemannian manifold M preserves the inner product on each tangent space, then we call it an *almost-hermitian structure*.

If (M^{2n}, g) is a Riemannian manifold, let $E(M)$ denote the set of all pairs (x, J_x), where x is a point of M and J_x is a hermitian structure on the tangent space to M at x. $E(M)$ is a bundle over M with fibre the set of all hermitian structures on R^{2n}. An almost-hermitian structure $\mathscr{J}$ on M is a smooth cross-section of this bundle, and its image is a smooth submanifold of $E(M)$. We define the *volume* of the almost-hermitian structure to be the volume of this submanifold, measured in the natural Riemannian

1991 *Mathematics Subject Classification.* Primary 53C15.
This paper is in final form and no version of it will be submitted for publication elsewhere.

metric induced on $E(M)$ by (M, g) and its Levi-Civita connection. By *best almost-hermitian structures* on M, we mean those of minimum volume.

Let S^n denote the unit n-sphere in R^{n+1}. Both S^2 and S^6 admit almost-complex structures; see below for details. In 1947, Hopf [**Ho 1**] and Ehresmann [**Eh 1**] each showed that S^4 does not admit an almost-complex structure. In the same year, Kirchhoff [**Ki**] showed that if S^n admits an almost-complex structure, then S^{n+1} is parallelizable; see also [**St, p. 216**]. The next year, Hopf [**Ho 2**] showed that S^8 does not admit an almost-complex structure. In 1950, Ehresmann [**Eh 2**] pointed out that Kirchhoff's result implies that S^{4n} does not admit an almost-complex structure. Finally, in 1953, Borel and Serre [**BS**] showed that S^2 and S^6 are the only spheres which do admit almost-complex structures.

$E(S^2)$ consists simply of two copies of the base S^2, corresponding to its two possible orientations, so there are exactly two almost-hermitian structures on S^2. By contrast, $E(S^6)$ is a 12-dimensional manifold, with two disjoint copies of the complex projective space CP^3 as fibres.

There are many almost-hermitian structures on S^6: the most familiar example $\mathscr{J}$ is obtained by thinking of R^7 as the purely imaginary Cayley numbers, and of S^6 as those of unit length. For each $x \in S^6$, let J_x denote, say, right Cayley multiplication by x, acting on the tangent space to S^6 at x. Note that this tangent space consists of all Cayley numbers orthogonal to both 1 and x, and hence is taken to itself by right multiplication by x. Since multiplication by a unit Cayley number is an orthogonal transformation, each J_x is a hermitian structure.

Simple modifications of this example are obtained by using O(7) to alter the orthogonal identification of S^6 with the unit imaginary Cayley numbers. We refer to all of these almost-hermitian structures on S^6 as *Cayley structures*, and will prove

THEOREM. *The almost-hermitian structures of minimum volume on S^6 are the Cayley structures, and no others.*

In 1951, Echmann and Fröhlicher [**EF**] showed that these Cayley structures on S^6 do not come from complex structures. Indeed, the question as to whether or not S^6 admits a complex structure seems still to be undecided.

We thank Lyndon Woodward of the University of Durham for suggesting this problem to us, and the National Science Foundation for its support.

1. Cayley structures on S^6

We observe in this section that the space $CaS(S^6)$ of Cayley structures on S^6 is homeomorphic to the union of two disjoint copies of the 7-dimensional real projective space RP^7.

Let $\mathscr{J}$ be our original Cayley structure on S^6, in which $J_x = R_x =$ right Cayley multiplication by $x \in S^6$. Write $\mathscr{J} = \{(x, R_x) : x \in S^6\}$. The other

Cayley structures can be written as

$$\mathscr{J}^A = \{(A^{-1}(x), A^{-1}R_x A) : x \in S^6\},$$

where $A \in O(7)$, viewed as the subgroup of $O(8)$ fixing the Cayley number 1. It is easy to see that $\mathscr{J}^A = \mathscr{J}$ precisely when A is an automorphism of the Cayley algebra. Hence $\mathrm{CaS}(S^6)$ is a copy of $G_2 \backslash O(7)$, where G_2 is the 14-dimensional exceptional Lie group of automorphisms of the Cayley algebra. Each component of $\mathrm{CaS}(S^6)$ is a copy of $G_2 \backslash SO(7)$.

To help see the homeomorphism $G_2 \backslash SO(7) \cong RP^7$, we will exhibit $G_2 \backslash SO(7)$ as a bundle over S^4 with fibre RP^3.

Let $B \in SO(7)$, and interpret B as acting on the Cayley numbers with $B(1) = 1$. Let S^4 denote the 4-sphere of unit Cayley numbers orthogonal to 1, i and j. Define a map $p : SO(7) \to S^4$ by

$$p(B) = B^{-1}(B(i)B(j)).$$

Since $B(1) = 1$, we note that $B(i)$ and $B(j)$ are orthogonal purely imaginary Cayley numbers, and hence their product $B(i)B(j)$ is a purely imaginary Cayley number orthogonal to both of them. Hence $p(B) = B^{-1}(B(i)B(j))$ is a purely imaginary Cayley number orthogonal to i and j, and therefore an element of the 4-sphere defined above.

Now suppose $B = CD$, where $C \in G_2$ is an automorphism of the Cayley algebra. Then

$$\begin{aligned}
p(CD) &= (CD)^{-1}(CD(i)CD(j)) \\
&= D^{-1}C^{-1}(C(D(i)D(j))) = D^{-1}(D(i)D(j)) \\
&= p(D).
\end{aligned}$$

Hence p is constant on right cosets $G_2 D$, and therefore descends to a map $p : G_2 \backslash SO(7) \to S^4$.

This map p is a fibration, and we want to examine the fibre $p^{-1}(k)$. That means we are looking at those $B \in SO(7)$ which satisfy $B(k) = B(i)B(j)$. That is, the restriction of B to the subalgebra of quaternions is an isomorphism into the Cayley algebra. Let ε be a unit Cayley number orthogonal to the quaternions. There is a unique automorphism C of the Cayley algebra which agrees with B on the $1, i, j, k, \varepsilon$-plane. Again writing $B = CD$, we see that $D = C^{-1}B$ is the identity on this 5-plane. Let D' denote the restriction of D to the orthogonal 3-plane. The correspondence $G_2 B \leftrightarrow D'$ matches the fibre $p^{-1}(k)$ with the $SO(3)$ of orientation-preserving isometries of this 3-plane. Hence the fibre of p is homeomorphic to a real projective space RP^3.

Thus each component $G_2 \backslash SO(7)$ of the space of Cayley structures on S^6 is a bundle over S^4 with fibre RP^3. The space RP^7 also appears as a bundle over S^4 with fibre RP^3: just take the Hopf fibration of S^7 by great

3-spheres and divide out by the antipodal map. One checks easily that these two bundles are the same.

Indeed, the correspondence between $G_2 \backslash \mathrm{SO}(7)$ and $RP(7)$ can be confirmed at the end of the paper, when we have seen an explicit isometry between $E^+(S^6)$ and the Grassmann manifold $G_2 R^8$, matching the individual right-handed Cayley structures on S^6 with the individual subgrassmannians $G_1 R^7$. These subgrassmannians consist of all oriented 2-planes through the origin in R^8 which contain a fixed line. There is an RP^7's worth of such lines, hence an RP^7's worth of such subgrassmannians, and hence an RP^7's worth of right-handed Cayley structures.

2. Complex structures and Grassmann manifolds

Let x be a point of S^6 and J_x a hermitian structure on the tangent space $T_x S^6$. We translate $T_x S^6$ with its hermitian structure to the parallel 6-plane through the origin, and then extend J_x to a hermitian structure on $R^8 \cong \mathrm{Ca}$ by defining $J_x(1) = x$ and $J_x(x) = -1$. Vice versa, if we are given a hermitian structure J on $R^8 \cong \mathrm{Ca}$, let $x = J(1)$ and let J_x be the restriction of J to the orthogonal complement of the 2-plane spanned by 1 and x, that is (in effect), to $T_x S^6$. In this way, we have a one-to-one correspondence between points (x, J_x) of $E(S^6)$ and hermitian structures J on $R^8 \cong \mathrm{Ca}$. See [St, pp. 215–216]. The set of these hermitian structures on R^8 is just the homogeneous space $\mathrm{O}(8)/\mathrm{U}(4)$; hence

$$E(S^6) = \mathrm{O}(8)/\mathrm{U}(4).$$

This space has two isometric components; we focus on one of them:

$$E^+(S^6) = \mathrm{SO}(8)/\mathrm{U}(4).$$

In this section we describe a specific isometry between the above space and the Grassmann manifold

$$G_2 R^8 = \mathrm{SO}(8)/(\mathrm{SO}(2) \times \mathrm{SO}(6))$$

of oriented 2-planes through the origin in 8-space. In his 1950 paper [Eh 2], Ehresmann remarked that $\mathrm{SO}(8)/\mathrm{U}(4)$ is the complex hyperquadric in CP^7, which in turn is well known to be a copy of $G_2 R^8$. The isometry we give here is based on the

TRIALITY THEOREM FOR $\mathrm{SO}(8)$. *Consider any triple of elements A, B and C in $\mathrm{SO}(8)$ such that*

$$A(mu) = B(m)C(u), \quad \text{for all } m, u \in \mathrm{Ca},$$

where Cayley multiplication is used on both sides of the equation. If any one of these three orthogonal transformations is preassigned, then the other two exist and are unique up to changing sign for both of them.

We will write $(A, B, C) \in \mathrm{Tr}$ to indicate that this triality relation holds.

For proofs and further discussion of this theorem, the reader is referred to Cartan [**Cr,** pp. 370 and 373] as well as to [**Fr,** pp. 14–16] and [**GWZ,** pp. 192–194].

Suppose now that B is a hermitian structure on R^8. Thus $B^2 = -I$. Let A and C be the elements of SO(8) promised by the Triality Theorem. Then

$$A^2(mu) = A(A(mu)) = A(B(m)C(u))$$
$$= B^2(m)C^2(u) = -mC^2(u).$$

By the uniqueness part of the Triality Theorem, we either have $A^2 = I$ and $C^2 = -I$ or else $A^2 = -I$ and $C^2 = I$. That is, either A is an involution and C is an hermitian structure, or vice versa. Which of these two alternatives occurs depends on whether the hermitian structure B is "right-handed" or "left-handed."

Let J be a hermitian structure on R^{2n}. Fix an orientation for R^{2n}, and suppose that $x_1, \ldots, x_n$ are chosen there so that the vectors

$$x_1, J(x_1), x_2, J(x_2), \ldots, x_n, J(x_n)$$

form an orthonormal basis. If this ordered basis agrees with the orientation, we say that the hermitian structure J is *right-handed*; otherwise, we say it is *left-handed*. This definition is independent of the choice of elements $x_1, \ldots, x_n$.

Next we see how the handedness of the hermitian structure B determines which of A and C is the involution and which the hermitian structure.

Let x be a purely imaginary Cayley number of unit length. Let R_x and L_x denote right and left multiplication by x. Both are hermitian structures on R^8. Choose the orientation of R^8 so that R_x is right-handed. Then L_x will be left-handed.

LEMMA. $(R_x, R_x, -L_xR_x) \in \mathrm{Tr}$.

PROOF. To verify this, we must show that

$$(mu)x = -(mx)(xux).$$

To do this, we use the Moufang identity [**Mo**]

$$z(xyx) = ((zx)y)x,$$

one of three identities representing vestiges of associativity which hold for the Cayley numbers. Note that no parentheses appear in the expression xyx on the left-hand side of this identity, since associativity holds whenever only two letters are involved; in particular, the maps R_x and L_x commute with one another. In this Moufang identity, put $x = x$, $y = u$ and $z = mx$, getting (since $x^2 = -1$)

$$(mx)(xux) = ((mxx)u)x = -(mu)x,$$

as desired.

Denoting this "charter member" of Tr by (A, B, C), we observe that

$$C^2 = (-L_x R_x)(-L_x R_x) = (L_x)^2 (R_x)^2$$
$$= (-I)(-I) = I,$$

since R_x and L_x commute. Thus C is an involution. And clearly $A = R_x$ is a hermitian structure.

So in this particular example, in which the hermitian structure B is right-handed, it is C which turns out to be the involution and A the hermitian structure. We see next that the same alternative occurs for every right-handed hermitian structure B, which one naturally expects by continuity.

To do this, first notice the "multiplicative structure" of the triality relation. That is, suppose (A, B, C) and (A', B', C') satisfy the relation. Then so do (AA', BB', CC') and (A^{-1}, B^{-1}, C^{-1}). This is easily checked, and reflects the fact that the correspondences between orthogonal transformations in a triality triple can be expressed as automorphisms of either Spin(8) or PSO(8), though not, incidentally, of SO(8). Here, Spin(8) denotes the simply connected double cover of SO(8), while PSO(8) denotes the quotient of SO(8) by its center $\{I, -I\}$.

Suppose now that B is a right-handed hermitian structure on R^8. Since any two hermitian structures of the same handedness on R^8 are conjugate within SO(8), there is an element $G \in$ SO(8) such that $B = GR_x G^{-1}$. Suppose $(F, G, H) \in$ Tr. By the lemma, $(R_x, R_x, -L_x R_x) \in$ Tr. By the multiplicative structure of triality,

$$(A, B, C) = (FR_x F^{-1}, GR_x G^{-1}, H(-L_x R_x)H^{-1}) \in \text{Tr}.$$

Hence C is the involution and A the hermitian structure, as claimed.

By contrast, if B is a left-handed hermitian structure on R^8 and $(A, B, C) \in$ Tr, then A is the involution and C the hermitian structure.

Return once more to the triple

$$A = R_x, \quad B = R_x, \quad C = -L_x R_x,$$

and focus on the involution C. Notice that

$$C(1) = -L_x R_x(1) = -x1x = 1,$$

and

$$C(x) = -L_x R_x(x) = -xxx = x,$$

since $x^2 = -1$. On the other hand, if y is orthogonal to both 1 and x, then

$$C(y) = -L_x R_x(y) = -xyx = xxy = -y,$$

since orthogonal pure imaginaries anticommute. Thus C restricts to the identity on the 2-plane P_x spanned by 1 and x, but to minus the identity on the orthogonal 6-plane. So to the hermitian structure $B = R_x$ on R^8, we intend to associate the 2-plane P_x.

Now suppose that B is any right-handed hermitian structure on R^8. As before, we have $B = GR_x G^{-1}$ for some $G \in SO(8)$. Let $(F, G, H) \in \text{Tr}$. Then the involution $C = H(-L_x R_x)H^{-1}$ restricts to the identity on the 2-plane $H(P_x)$, and to minus the identity on the orthogonal 6-plane. So to the hermitian structure B we associate the 2-plane $H(P_x)$, and denote it by P_B. By uniqueness of H up to sign, this 2-plane is well-determined by B.

If the hermitian structure B on R^8 is right-handed, then so is $-B$. If, in the triality relation, B is replaced by $-B$, then exactly one of A and C will change sign. Hence the 2-plane $P_{-B} = P_B$. By uniqueness in the Triality Theorem, one can only have $P_B = P_{B'}$, if $B' = B$ or $-B$.

Thus we get a one-to-one correspondence between pairs $(B, -B)$ of hermitian structures on R^8 and unoriented 2-planes $P_B = P_{-B}$ in R^8. Passing to the simply connected double covers, we get a one-to-one correspondence between hermitian structures B and oriented 2-planes, still denoted by P_B. That is, we have a bijection between $E^+(S^6) = SO(8)/U(4)$ on the one hand, and the Grassmannian $G_2 R^8 = SO(8)/(SO(2) \times SO(6))$ of oriented 2-planes in 8-space on the other.

It is easy to see that this bijection is an isometry between these two symmetric spaces, when each is given the Riemannian submersion metric induced from a common bi-invariant metric on $SO(8)$. The triality "correspondence" in $SO(8)$, which takes $B = R_x$ to $C = -L_x R_x$, descends to an automorphism

$$T : PSO(8) \rightarrow PSO(8),$$

which is an isometry by the uniqueness up to scale of bi-invariant metrics on simple Lie groups. Noting that $U(4)$ and $SO(2) \times SO(6)$ are the centralizers in $SO(8)$ of B and C, respectively, one checks easily that T takes $P(U(4))$ to $P(SO(2) \times SO(6))$. Thus T descends to an isometry taking

$$E^+(S^6) = PSO(8)/P\,U(4)$$

to

$$G_2 R^8 = PSO(8)/P(SO(2) \times SO(6)).$$

Now suppose the right-handed almost-hermitian structure $\mathscr{J}$ on S^6 is our primitive Cayley structure: $\mathscr{J} = \{(x, R_x) : x \in S^6\}$. The image of J in $G_2 R^8$ under the above isometry then consists of all the oriented 2-planes spanned by 1 and x, for $x \in S^6$. But this is a subgrassmannian of the type $G_1 R^7$, and as mentioned in the introduction, is known to be calibrated by the Euler 6-form on $G_2 R^8$, and is hence volume-minimizing in its homology class. The reader is referred to [HL] or [GMZ] for the definition of "calibrate," and to [GMZ] for the proof of this assertion.

Since all the other right-handed Cayley structures are images of this primitive one under the isometric action of $SO(8)$, they too are volume-minimizing in this homology class, and correspond to all the other realizations of $G_1 R^7$ inside $G_2 R^8$.

It is easy to see that all right-handed almost-hermitian structures on S^6 lie in the same homology class in $E^+(S^6)$. In fact, as Ehresmann [Eh 2] pointed out, $\pi_6 E^+(S^6)$ is infinite cyclic, and hence all the cross-sections of this bundle are homotopic.

Likewise, all the left-handed Cayley structures are volume-minimizing in their homology class, in the other component of $E(S^6)$.

Since it was proved in [GMZ] that the subgrassmannians $G_1 R^7$ are the unique volume minimizers in their homology class in $G_2 R^8$, and since the Cayley structures correspond to these subgrassmannians, it follows that the Cayley structures are the unique volume minimizers in their two homology classes, completing the proof of the theorem.

REFERENCES

[BS] A. Borel and J.-P. Serre, *Groupes de Lie et puissances réduites de Steenrod*, Amer. J. Math. **75** (1953), 409–448.

[Ca] E. Calabi, *Construction and properties of some 6-dimensional almost complex manifolds*, Trans. Amer. Math. Soc. **87** (1958), 407–438.

[Cr] E. Cartan, *Le principe de dualité et la théorie des groupes simples et semi-simples*, Bull. Sci. Math. **49** (1925), 361–374.

[Ec] B. Eckmann, *Complex-analytic manifolds*, Proc. Internat. Congr. Math., Cambridge, Mass., 1950, pp. 420–427.

[EF] B. Eckmann and A. Fröhlicher, *Sur L'integrabilite des structures presque-complexes*, C. R. Acad. Sci. Paris **232** (1951), 2284–2286.

[Eh 1] C. Ehresmann, *Sur la théorie des espaces fibrés*, Topologie Algebrique—Paris 1949, Edition CNRS, Paris, 1949, pp. 3–15.

[Eh 2] ____, *Sur les variétés presque complexes*, Proc. Internat. Congr. Math., Cambridge, Mass., 1950, pp. 412–419.

[Fr] H. Freudenthal, *Oktaven, Ausnahmegruppen und Octaven geometrie*, Utrecht, 1951.

[GMZ] H. Gluck, F. Morgan and W. Ziller, *Calibrated geometries in Grassmann manifolds*, Comment. Math. Helv. **64** (1989), 256–268.

[GWZ] H. Gluck, F. Warner, and W. Ziller, *The geometry of the Hopf fibrations*, Enseign. Math. **32** (1986), 173–198.

[HL] R. Harvey and H. B. Lawson, Jr., *Calibrated geometries*, Acta Math. **147** (1981), 47–157.

[Ho 1] H. Hopf, *Sur les champs d'éléments de surface dans les variétés à 4 dimensions*, Topologie Algebrique—Paris 1947, Editions CNRS, Paris, 1949, pp. 55–59.

[Ho 2] ____, *Zur Topologie der komplexen Mannigfaltigkeiten*, Studies and Essays Presented to R. Courant on his 60th Birthday, Interscience 1948, pp. 167–185.

[Ki] A. Kirchhoff, *Sur l'existence de certains champs tensoriels sur les spheres*, C. R. Acad. Sci. Paris **225** (1947), 1258–1260.

[Mo] R. Moufang, *Zur Struktur von Alternativkörpern*, Math. Ann. **110** (1935), 416–430.

[St] N. Steenrod, *The topology of fibre bundles*, Princeton Univ. Press (1951).

UNIVERSITY OF PENNSYLVANIA

Proceedings of Symposia in Pure Mathematics
Volume **54** (1993), Part 2

The Failure of Complex and Symplectic Manifolds to be Kählerian

LUIS A. CORDERO, MARISA FERNÁNDEZ, AND ALFRED GRAY

1. Introduction

There are many well-known topological properties of compact Kähler manifolds (see conditions (K1)–(K5) below). Much less is known about compact symplectic manifolds or about compact complex manifolds. The purpose of this paper is to show how Massey Products and the Frölicher spectral sequence can be used to measure the failure of manifolds in each of these two classes to be Kählerian.

2. Symplectic manifolds

Let M be a differentiable manifold. Then M is called *symplectic* provided there is a globally defined closed two-form F on M that is nondegenerate at each point. It is easy to see that any symplectic manifold must be orientable and have even dimension $2n$.

One way that symplectic manifolds arise is via Kähler metrics. To explain this notion first let us recall the notion of *almost complex manifold*. This is a differentiable manifold equipped with a tensor field J of type $(1, 1)$ such that $J^2 = -I$. Like symplectic manifolds any almost complex manifold must be orientable and even dimensional. In fact any symplectic form gives rise to an almost complex structure.

Now assume that in addition to being almost complex suppose that M is a pseudo-Riemannian manifold, that is a differentiable manifold equipped with an everywhere nondegenerate symmetric bilinear form $\langle \, , \, \rangle$. Suppose also that $\langle \, , \, \rangle$ and J are compatible in the sense that $\langle JX, JY \rangle = \langle X, Y \rangle$ for all vector fields X and Y on M. The formula $F(X, Y) = \langle JX, Y \rangle$ defines a nondegenerate two-form on M. We say that $\langle \, , \, \rangle$ is a *Kähler*

1991 *Mathematics Subject Classification*. Primary 53B35, 32C10, 53C15, 55F20, 55H99.
This paper is in final form, and no version will be submitted for publication elsewhere.

metric provided that F is closed and J is integrable. A *Kähler manifold* is a manifold equipped with a positive definite Kähler metric.

There are well-known topological conditions for a compact manifold to admit a positive definite Kähler metric.

THEOREM 1. *If M is a $2n$-dimensional compact manifold that carries a positive definite Kähler metric, then the following conditions must be satisfied:*
(K1) *the Betti numbers $b_{2i}(M)$ are nonzero for $1 \leq i \leq n$;*
(K2) *the Betti numbers $b_{2i-1}(M)$ are even;*
(K3) *$b_i(M) \geq b_{i-2}(M)$ for $1 \leq i \leq n$;*
(K4) *the strong Lefschetz theorem holds for M, namely,*

$$[F]^{n-i} : H^i(M) \longrightarrow H^{2n-i}(M)$$

is an isomorphism for $0 \leq i \leq n$,

(K5) *the minimal model of M is formal (so in particular all Massey products of M vanish).*

Conditions (K1)–(K4) are due to Weil (See [**Weil**]). Condition (K5) is more recent (see [**DGMS**]).

QUESTION. When is a compact symplectic manifold a Kähler manifold? This question has two senses. One can ask if a given symplectic form arises from a positive definite Kähler metric, or we can ask if a symplectic manifold carries a positive definite Kähler metric that may or may not have anything to do with the given symplectic form.

In this paper we shall be concerned answering the question negatively in the second sense. To do so means that topological reasons must be given for a compact symplectic manifold not to carry a positive definite Kähler metric. Note that any compact symplectic manifold M satisfies (K1), so one way is to prove that M fails to satisfy at least one of the conditions (K2)–(K5).

For many years it was not known if there existed compact symplectic manifolds that possessed no positive definite Kähler metric. The first such example, although sometimes attributed to Thurston [**Th**], was actually found by Kodaira [**Kod**] some years earlier.

3. Massey products

Massey products are a very powerful tool for distinguishing Kähler manifolds from those compact symplectic manifolds which carry no positive definite Kähler metric. Let us recall the definition of Massey products; we shall give the definition in terms of differential forms on manifolds, because that is usually all the generality that is needed in the study of symplectic manifolds. Let M be a compact differentiable manifold and suppose that there are cohomology classes $[\alpha] \in H^p(M)$, $[\beta] \in H^q(M)$ and $[\gamma] \in H^r(M)$ (represented by differential forms α, β and γ) such that $[\alpha][\beta] = [\beta][\gamma] = 0$. Then

$$\alpha \wedge \beta = df \quad \text{and} \quad \beta \wedge \gamma = dg$$

for some choice of differential forms f and g. Let $\mu = \alpha \wedge g + (-1)^{p-1} f \wedge \gamma$; then μ is a closed form of degree $p + q + r - 1$. The cohomology class $[\mu]$ depends on the choice of f and g, but the class

$$[[\mu]] \in \frac{H^{p+q+r-1}(M)}{[\alpha] \cdot H^{q+r-1}(M) \oplus [\gamma] \cdot H^{p+q-1}(M)}$$

is independent of the choice of f and g. It is called the *Massey product* of $[\alpha]$, $[\beta]$ and $[\gamma]$, and we write $\langle [\alpha], [\beta], [\gamma] \rangle$ for $[[\mu]]$.

Although Massey products have been well known for some time in algebraic topology, it is only recently that their importance in differential geometry has been recognized. In particular, the main theorem of [DGMS], which is condition (K5), is of fundamental importance. It turns out the complex parallelizable manifolds (described below) have Betti numbers that satisfy (K1), (K2) and (K3), so that from the point of view of Betti numbers they are indistinguishable from Kähler manifolds. Thus to show that a complex parallelizable manifold, not a torus, carries no positive definite Kähler metric we must use (K4) or (K5).

4. Complex manifolds and the Frölicher spectral sequence

A complex manifold is a different generalization of a Kähler manifold. Just as properties (K1)–(K5) may fail for compact symplectic manifolds, they may also fail for compact complex manifolds.

For complex manifolds in addition to ordinary de Rham cohomology one has Dolbeault cohomology. This arises as follows.

Let M be a complex manifold of complex dimension n. Recall that a differential form of degree (r, s) on M is a complex differential form φ which can be written in local complex coordinates $(z_1, \ldots, z_n)$ as

$$\varphi = \sum a_{i_1 \cdots i_r j_1 \cdots j_s} dz_{i_1} \wedge \cdots \wedge dz_{i_r} \wedge d\bar{z}_{j_1} \wedge \cdots \wedge d\bar{z}_{j_s}.$$

This definition is independent of the choice of local coordinates. Denote by $\Lambda^{r,s}(M)$ the space of differential forms of degree (r, s) on M. Then the space $\Lambda^p(M)$ of all differential forms can be decomposed as

$$\Lambda^p(M) \otimes \mathbb{C} = \sum_{r+s=p} \Lambda^{r,s}(M).$$

For an almost complex manifold it turns out that

$$d(\Lambda^{p,q}(M)) \subseteq \Lambda^{p-1,q+2}(M) \oplus \Lambda^{p,q+1}(M) \oplus \Lambda^{p+1,q}(M) \oplus \Lambda^{p+2,q-1}(M),$$

and for a complex manifold this inclusion can be improved to $d(\Lambda^{p,q}(M)) \subseteq \Lambda^{p+1,q}(M) \oplus \Lambda^{p,q+1}(M)$. Thus for a complex manifold we get a decomposition of d as $d = \partial + \bar{\partial}$ where

$$\partial(\Lambda^{p,q}(M)) \subseteq \Lambda^{p+1,q}(M) \quad \text{and} \quad \bar{\partial}(\Lambda^{p,q}(M)) \subseteq \Lambda^{p,q+1}(M).$$

Since $\overline{\partial}^2 = 0$, it can be used to define the *Dolbeault groups*

$$H^{p,q}(M) \cong \frac{\{\alpha \in \Lambda^{p,q}(M) | \overline{\partial}\alpha = 0\}}{\overline{\partial}(\Lambda^{p,q-1}(M))}.$$

Write $h^{p,q}(M) = \dim H^{p,q}(M)$.

For Kähler manifolds we have the following result. Let $H^s(M)$ denote the sth de Rham group of M.

THEOREM 2. *Let M be a compact complex manifold with a positive definite Kähler metric. Then*

(1) $$H^s(M) \otimes \mathbb{C} \cong \sum_{p+q=s} H^{p,q}(M)$$

for $1 \le s \le 2n$. Also,

(2) $$H^{p,q}(M) \cong H^{q,p}(M)$$

for all p and q.

For a proof of this well-known theorem due to Weil [**Weil**] see, for example, [**Wells**].

COROLLARY 3. *Let M be a compact complex manifold with a positive definite Kähler metric. Then a form α on M which satisfies $\overline{\partial}\alpha = 0$ also satisfies $d\alpha = 0$.*

For a general compact complex manifold M Theorem 1 and Corollary 3 may be false. Moreover, there is a much more complicated set of groups, $\{E_r^{p,q}(M)\}$ called the *Frölicher spectral sequence* associated to the complex manifold M. Here

$$E_1^{p,q}(M) = H^{p,q}(M).$$

The general definition of $E_r^{p,q}(M)$ for $r \ge 2$, although standard, is quite complicated and inappropriate to give here. (For details see [**CFG1, CFG2**].) Instead we describe a special case. It is best summarized by saying that for a compact complex manifold

some differential forms are more exact than others.

It turns out that

$$E_r^{p,0}(M) = \frac{\{\alpha \in \Lambda^{p,0}(M) | d\alpha = 0\}}{\{\alpha \in \Lambda^{p,0}(M) | \alpha = d(\beta_{p-1,0} + \beta_{p-2,1} + \cdots + \beta_{p-r+1,r-2})\}}.$$

The space $E_r^{p,0}(M)$ can be considered as a refinement of $E_1^{p,0}(M)$. It is clear that for $r \ge 2$ we have a sequence of epimorphisms

$$E_2^{p,0}(M) \twoheadrightarrow E_3^{p,0}(M) \twoheadrightarrow \cdots \twoheadrightarrow E_r^{p,0}(M) \twoheadrightarrow \cdots.$$

But are these groups really different? (Frölicher [**Fr**] showed that for a manifold M with a positive definite Kähler metric one has $E_r^{p,q}(M) \cong E_1^{p,q}(M)$ for all p, q, r.)

5. Nilmanifolds

Let G be a real nilpotent Lie group and let $\{\omega_1, \ldots, \omega_n\}$ be a basis for the left invariant one-forms on G. Then we may write

$$d\omega_i = \sum_{j<k<i} A_{ijk} \omega_j \wedge \omega_k,$$

where the A_{ijk}'s are constants. If the A_{ijk}'s are rational then there is a discrete subgroup Γ such that the space of left cosets $\Gamma \backslash G$ is compact (see [**Ma**]). The space $\Gamma \backslash G$ is called a *compact nilmanifold*.

A compact nilmanifold $\Gamma \backslash G$ is a parallelizable manifold of a special type. The point is that the parallelization is defined by a basis $\{\omega_1, \ldots, \omega_n\}$ of the one-forms on M and with respect to this basis the exterior derivative has constant coefficients. This is the reason that the computation of the cohomology of a compact nilmanifold is straightforward.

Consider, for example, the simplest nonabelian nilpotent Lie group, the Heisenberg group H. It consists of the matrices

$$\left\{ \begin{pmatrix} 1 & a & c \\ 0 & 1 & b \\ 0 & 0 & 1 \end{pmatrix} \middle| a, b, c \in \mathbb{R} \right\}.$$

There are many discrete subgroups Γ such that $\Gamma \backslash H$ is compact, but the simplest is

$$\Gamma = \left\{ \begin{pmatrix} 1 & a & c \\ 0 & 1 & b \\ 0 & 0 & 1 \end{pmatrix} \middle| a, b, c \in \mathbb{Z} \right\}.$$

Let $\pi: H \to \Gamma \backslash H$ be the projection. Let us compute the cohomology of $\Gamma \backslash H$. First let x, y, z be the globally defined coordinate functions on H, that is,

$$x \left(\begin{pmatrix} 1 & a & c \\ 0 & 1 & b \\ 0 & 0 & 1 \end{pmatrix} \right) = a, \qquad y \left(\begin{pmatrix} 1 & a & c \\ 0 & 1 & b \\ 0 & 0 & 1 \end{pmatrix} \right) = b,$$

$$z \left(\begin{pmatrix} 1 & a & c \\ 0 & 1 & b \\ 0 & 0 & 1 \end{pmatrix} \right) = c.$$

Denote by $L_A: H \to H$ left multiplication by

$$A = \begin{pmatrix} 1 & a & c \\ 0 & 1 & b \\ 0 & 0 & 1 \end{pmatrix};$$

then it is easy to check that

$$x \circ L_A = x + a, \qquad y \circ L_A = y + a, \qquad z \circ L_A = z + c + ay,$$

so that

$$L_A^*(dx) = dx, \qquad L_A^*(dy) = dy, \qquad L_A^*(dz) = dz + a\,dy.$$

Thus $\{dx, dy, dz - xdy\}$ are a basis for the left invariant one-forms on H. In particular, they are preserved by Γ, and so they descend to one-forms on the nilmanifold $\Gamma\backslash H$. Otherwise said, there are one-forms α, β, γ on $\Gamma\backslash H$ such that

$$(3) \qquad \pi^*(\alpha) = dx, \qquad \pi^*(\beta) = dy, \qquad \pi^*(\gamma) = dz - xdy.$$

From (3) it is easy to see that

$$(4) \qquad\qquad d\alpha = d\beta = 0 \quad \text{and} \quad d\gamma = -\alpha \wedge \beta.$$

Furthermore,

$$d(\alpha \wedge \beta) = d(\alpha \wedge \gamma) = d(\beta \wedge \gamma) = 0.$$

It follows from (4) that the real cohomology groups are

$$H^0(\Gamma\backslash H, \mathbb{R}) = \{[1]\},$$

$$H^1(\Gamma\backslash H, \mathbb{R}) = \{[\alpha], [\beta]\}, \qquad H^2(\Gamma\backslash H, \mathbb{R}) = \{[\alpha \wedge \gamma], [\beta \wedge \gamma]\},$$
$$H^3(\Gamma\backslash H, \mathbb{R}) = \{[\alpha \wedge \beta \wedge \gamma]\}.$$

In 1954 Nomizu [**Nomizu**], generalizing a result of Matsushima [**Mat**], proved

THEOREM 4. *Let G be a connected nilpotent Lie group with discrete subgroup Γ such that the space of left cosets $\Gamma\backslash G$ is compact. Then there is an isomorphism of cohomology groups*

$$(5) \qquad\qquad H^*(\mathfrak{g}) \cong H^*(\Gamma\backslash G),$$

*where $H^*(\mathfrak{g})$ denotes the Chevalley-Eilenberg cohomology of the Lie algebra $\mathfrak{g}$ of G (see [**CE**]) and $H^*(\Gamma\backslash G)$ denotes the de Rham cohomology of $\Gamma\backslash G$.*

Nomizu's proof goes along the following lines. First he shows that the nilmanifold $M = \Gamma\backslash G$ is a principal T^k-bundle over a lower dimensional nilmanifold M' for some k-dimensional torus T^k. The Serre spectral sequence is used to express the de Rham cohomology of M in terms of that of T^k and M'. Since T^k is a compact Lie group acting on M, we may consider the T^k-invariant cohomology of M. Again the Serre spectral sequence can be used to express the T^k-invariant cohomology of M in terms of the invariant cohomology of T^k and the cohomology of M'. Since Theorem 4 is clear in the case that G is abelian, the general case follows by induction.

Hattori [**Hattori**] has proved a generalization of Theorem 4 for certain solvable Lie groups that have compact quotients.

6. Massey products on nilmanifolds

THEOREM 5. *Any compact two-step nilmanifold has nonzero Massey products.*

PROOF. Let $M = G/\Gamma$ where G is a nilpotent Lie group and Γ is a discrete subgroup, and let $\mathfrak{g}$ be the Lie algebra of G. We choose a basis $\alpha_1, \ldots, \alpha_n$ of the dual space $\mathfrak{g}^*$ such that $\alpha_1, \ldots, \alpha_p$ is a basis for the one-forms that vanish on $[\mathfrak{g}, \mathfrak{g}]$ and

$$d\alpha_i = \sum_{i \leq j < k < i} \alpha_j \wedge \alpha_k.$$

Suppose that $d\alpha_{p+1}$ has rank s. Then $\alpha_1, \ldots, \alpha_p$ can be rechosen so that

$$d\alpha_{p+1} = \alpha_1 \wedge \alpha_2 + \cdots + \alpha_{2s-1} \wedge \alpha_{2s}.$$

Then

$$\langle [\alpha_1 \wedge \alpha_3 \wedge \cdots \wedge \alpha_{2s-1}], [\alpha_{2s}], [\alpha_{2s}] \rangle$$

is a nonzero Massey product. To prove this we first note that

$$\alpha_1 \wedge \alpha_3 \wedge \cdots \wedge \alpha_{2s-1} \wedge \alpha_{2s} = (-1)^{s-1} d(\alpha_1 \wedge \alpha_3 \wedge \cdots \wedge \alpha_{2s-3} \wedge \alpha_{p+1})$$

so that

$$[\alpha_1 \wedge \alpha_3 \wedge \cdots \wedge \alpha_{2s-1}] \cdot [\alpha_{2s}] = 0.$$

Thus the Massey product is represented by

$$\alpha_1 \wedge \alpha_3 \wedge \cdots \wedge \alpha_{2s-3} \wedge \alpha_{2s} \wedge \alpha_{p+1}.$$

The cohomology class $[\alpha_1 \wedge \alpha_3 \wedge \cdots \wedge \alpha_{2s-3} \wedge \alpha_{2s} \wedge \alpha_{p+1}]$ is nonzero because $\mathfrak{g}$ is two-step nilpotent. Not only is $[\alpha_1 \wedge \alpha_3 \wedge \cdots \wedge \alpha_{2s-3} \wedge \alpha_{2s} \wedge \alpha_{p+1}]$ a nonzero cohomology class, it is also nonzero in the quotient

$$\frac{H^{s+1}(M)}{[\alpha_1 \wedge \alpha_3 \wedge \cdots \wedge \alpha_{2s-1}] \cdot H^1(M) \oplus [\alpha_{2s}] \cdot H^s(M)}.$$

7. Complex and symplectic structures on nilmanifolds

Since a nilmanifold is parallelizable, it has many almost complex structures, provided it is even-dimensional. There is no guarantee, however, that a nilmanifold have complex or symplectic structures.

In fact [FGG] we gave the first example of a compact symplectic manifold M that carries no complex structure. More examples have recently been found by Yamato [Yam]. The proof that the manifold M (a three-step compact four-dimensional nilmanifold) carried no complex structure used Kodaira's classification of compact complex surfaces (see [BPV]). Therefore, it would be difficult to extend the proof to higher dimensions.

On the other hand, there are many interesting compact nilmanifolds with complex structures. Let G be a real Lie group. Instead of describing the

Lie algebra $\mathfrak{g}$ of G in terms of its bracket we use the exterior differential on the dual space $\mathfrak{g}^*$. The two are equivalent because of the formula $d\alpha(X, Y) = -\alpha([X, Y])$, where α is any one-form on $\mathfrak{g} \otimes \mathbb{C}$ and $X, Y \in \mathfrak{g} \otimes \mathbb{C}$. Now suppose that G has an almost complex structure J. We choose a $\mathbb{C}$-basis $\{\omega_1, \ldots, \omega_n\}$ for the complex forms on $\mathfrak{g} \otimes \mathbb{C}$; then $\{\omega_1, \ldots, \omega_n, \overline{\omega}_1, \ldots, \overline{\omega}_n\}$ is a real basis of $\mathfrak{g}^*$. The structure equations (equivalent to the bracket) have the form

$$(6) \qquad d\omega_i = \sum_{j<k} A_{ijk}\omega_j \wedge \omega_k + \sum_{j,k} B_{ijk}\omega_j \wedge \overline{\omega}_k + \sum_{j<k} C_{ijk}\overline{\omega}_j \wedge \overline{\omega}_k,$$

for $1 \leq i \leq n$. In general such an almost complex structure on G will not be integrable because of the presence of $(0, 2)$-forms on the right-hand side of (6). However, there is a simple criterion for a Lie group to have a left invariant complex structure

THEOREM 6. *Let G be a Lie group with a left invariant complex structure. Then the structure equations for the Lie algebra $\mathfrak{g}$ of G have the form*

$$(7) \qquad d\omega_i = \sum_{j<k} A_{ijk}\omega_j \wedge \omega_k + \sum_{j,k} B_{ijk}\omega_j \wedge \overline{\omega}_k, \qquad (1 \leq i \leq n).$$

Conversely, the structure equations (7) define a Lie group G with left invariant complex structure. Hence quotients of G have complex structures.

PROOF. Defining an almost complex structure J on G is equivalent to saying that $\{\omega_1, \ldots, \omega_n\}$ form a basis for the $(1, 0)$-forms on G. Since on a complex manifold the differential of a $(1, 0)$-form must be the sum of forms of types $(2, 0)$ and $(1, 1)$, we see that a necessary condition for an almost complex structure J be integrable is that all the coefficients C_{ijk} in (6) vanish. Moreover, this condition is also sufficient. (See for example [KN, volume 2, Theorem 2.8].)

8. Complex parallelizable manifolds

Compact quotients of complex nilpotent Lie groups provide the best known examples of complex structures on nilmanifolds. Let us compute a portion of the Frölicher spectral sequence for the Iwasawa manifold I_3. Explicitly $I_3 = \Gamma \backslash G$ where G is the group of complex matrices of the form

$$(8) \qquad \begin{pmatrix} 1 & z_1 & z_3 \\ 0 & 1 & z_2 \\ 0 & 0 & 1 \end{pmatrix}$$

and Γ is the subgroup of G consisting of those matrices whose entries are Gaussian integers. The computation of the cohomology of I_3 is similar to that of the Heisenberg manifold $\Gamma \backslash G$ considered in §4. A basis for the left invariant one-forms of type $(1, 0)$ is $\{dz_1, dz_2, dz_3 - z_1 dz_2\}$. These differential forms descend to $(1, 0)$-forms $\omega_1, \omega_2, \omega_3$ on the quotient I_3 such that

$$(9) \qquad d\omega_1 = d\omega_2 = 0 \quad \text{and} \quad d\omega_3 = -\omega_1 \wedge \omega_2.$$

Hence $H^1(I_3, \mathbb{C}) = \{\omega_1, \omega_2, \overline{\omega}_1, \overline{\omega}_2\}$, and the other cohomology groups can be computed in a similar fashion. Thus the first, second and third Betti numbers of I_3 are

$$b_1(I_3) = 4, \qquad b_2(I_3) = 8, \qquad b_3(I_3) = 10,$$

and the others are determined by Poincaré duality. Thus, just like a compact Kähler manifold, the Iwasawa manifold satisfies conditions (K1), (K2) and (K3).

Let us determine part of the spectral sequence for I_3. Note that $\overline{\partial}\omega_3 = 0$ but ω_3 is not $\overline{\partial}$-exact. It follows that

$$E_1^{1,0}(I_3) \cong \{\omega_1, \omega_2, \omega_3\},$$

$$E_2^{1,0}(I_3) \cong \{\omega_1, \omega_2\} \cong E_\infty^{1,0}(I_3).$$

For more details on the Iwasawa manifold see [**FG1**].

It follows from Corollary 3 that the natural complex structure on I_3 has no compatible positive definite Kähler metric. However, a stronger assertion is true. Since the Iwasawa manifold has nonzero Massey products, it follows from the Main Theorem of [**DGMS**] that no complex structure on I_3 has a compatible positive definite Kähler metric. The Iwasawa manifold I_3 is an example of a complex parallelizable manifold. This notion was introduced by Wang [**Wang**]. In analogy with the real case, a complex manifold M of complex dimension n is called complex *parallelizable* provided there exist n holomorphic 1-forms linearly independent at each point. Wang proved that, in contrast with the real case, complex parallelizable manifolds are quite special:

THEOREM 7. *Any compact complex parallelizable manifold is the quotient of a complex Lie group by a discrete subgroup.*

(Here one must be careful about the definition of complex Lie group. We require that both left multiplication and right multiplication are holomorphic. As we shall see shortly there are Lie groups that are left holomorphic but not right holomorphic.)

Any compact complex parallelizable nilmanifold satisfies (K1), (K2) and (K3); this is a consequence of the following theorem of Sakane [**Sa**]:

THEOREM 8. *If M is a compact complex parallelizable nilmanifold then the following relations between the $h^{p,q}(M)$ and the Betti numbers $b_i(M)$ hold*

$$b_{2i+1}(M) = 2\{h^{0,2i+1}(M) + h^{0,2i}(M) + h^{0,1}(M) + \cdots + h^{0,i+1}(M)h^{0,i}(M)\},$$

$$b_{2i}(M) = 2\{h^{0,2i}(M) + h^{0,2i-1}(M)h^{0,1}(M) + \cdots + h^{0,i+1}(M)h^{0,i-1}(M)\}$$
$$+ (h^{0,i}(M))^2.$$

The Frölicher spectral sequence is also especially simple for complex parallelizable nilmanifolds. Instead of collapsing at E_1 (the case for Kähler manifolds), Sakane [**Sa**] proved

THEOREM 9. *The Frölicher spectral sequence of a compact parallelizable manifolds collapses at* E_2.

9. Real parallelizable complex nilmanifolds

Of course any complex parallelizable manifold is real parallelizable. But it can happen that a complex manifold is real parallelizable but not complex parallelizable. Using holomorphic forms instead of holomorphic vector fields, it is easy to see that complex parallelizable manifolds are not the only interesting real parallelizable complex manifolds. Let $\{\omega_1, \ldots, \omega_n\}$ be the linearly independent holomorphic one-forms that define a complex parallelization of a complex manifold M. Then

(10)
$$d\omega_i = \sum_{j<k} A_{ijk}\omega_j \wedge \omega_k \qquad (1 \le i \le n).$$

Notice that (10) is simpler than (7). In other words, for a real parallelizable complex manifold the $d\omega_i$'s can have $(1, 1)$ components, but this is prohibited for a complex parallelizable manifold.

The Kodaira-Thurston manifold is the lowest dimensional example of a real parallelizable complex manifold that is not complex parallelizable. It is not the compact quotient of a complex nilpotent Lie group. Let G be the nilpotent group of complex matrices of the form

$$\begin{pmatrix} 1 & \bar{z} & w \\ 0 & 1 & z \\ 0 & 0 & 1 \end{pmatrix},$$

and let Γ be the subgroup of G whose entries are Gaussian integers.

It is interesting that the left translations on G are holomorphic, but not the right translations. This follows from the equations

$$z \circ L_A = z + a, \qquad z \circ R_A = z + a,$$
$$w \circ L_A = w + c + \bar{a}z, \qquad w \circ R_A = w + c + a\bar{z},$$

where L_A and R_A denote left and right multiplication by the matrix

$$A = \begin{pmatrix} 1 & \bar{a} & c \\ 0 & 1 & a \\ 0 & 0 & 1 \end{pmatrix}.$$

The Kodaira-Thurston manifold can be realized as the quotient manifold $KT = \Gamma\backslash G$; then KT is a compact complex manifold, because $\{z, w\}$ is a system of complex coordinates (global for G and local for KT).

In fact, KT has an indefinite Kähler metric (see [FGG]). We have described only one of the complex structures on KT; we call it the natural complex structure. It is easy to see that dz and $dw - \bar{z}dz$ and their conjugates form a basis for the complex one-forms on G. The dual vector fields are the left invariant vector fields

$$\frac{\partial}{\partial z} + \bar{z}\frac{\partial}{\partial w}, \qquad \frac{\partial}{\partial w}$$

and their conjugates. All of these objects descend to the quotient space. If $\pi: G \longrightarrow KT$ denotes the projection, then we may write

$$\pi^*(\varphi) = dz, \qquad \pi^*(\psi) = dw - \overline{z}dz,$$

and

$$\pi_* \left(\frac{\partial}{\partial z} + \overline{z}\frac{\partial}{\partial w} \right) = Z, \qquad \pi_* \left(\frac{\partial}{\partial w} \right) = W.$$

Then $d\varphi = 0$ and $d\psi - \varphi \wedge \overline{\varphi}$. The only nonzero bracket is $[Z, \overline{Z}] = \overline{W} - W$. Note that although W is holomorphic, Z is not.

It turns out that not only is the Kodaira-Thurston manifold complex but it also has many interesting symplectic forms. Moreover, each symplectic form gives rise to an indefinite Kähler metric. The following theorem is proved in [AFGM1]:

THEOREM 10. *The manifold KT with its natural left invariant complex structure has a real four-parameter family of left invariant indefinite Kähler metrics. Furthermore, any left invariant indefinite Kähler metric on KT is flat.*

10. Indefinite Kähler metrics

Let M be an almost complex manifold and $\langle \, , \, \rangle$ an indefinite almost Hermitian metric on M. Denote by ∇ the Riemannian connection of M, and let $\mathfrak{X}_{\mathbb{C}}(M)$ be the complex vector fields on M. Here is the precise definition of an indefinite Kähler manifold.

DEFINITION. We say that $\langle \, , \, \rangle$ is an indefinite Kähler metric on M provided

$$\nabla_X JY - J\nabla_X Y = 0$$

for all $X, Y \in \mathfrak{X}_{\mathbb{C}}(M)$. In this case M is said to be an indefinite Kähler manifold.

Next we recall some equivalent definitions of an indefinite Kähler manifold.

LEMMA 11. *Let M be an almost Hermitian manifold. Then the following two conditions are equivalent:*

(i) *M is an indefinite Kähler manifold;*

(ii) *M is always a complex manifold and the Kähler form F of M is closed.*

PROOF. Let M be any almost Hermitian manifold. For $X, Y \in \mathfrak{X}_{\mathbb{C}}(M)$ let us write

$$\nabla_X(J)(Y) = \nabla_X JY - J\nabla_X Y.$$

Denote by S the Nijenhuis tensor of J, that is,

$$S(X, Y) = [X, Y] + J[JX, Y] + J[X, JY] - [JX, JY]$$

for $X, Y \in \mathfrak{X}_C(M)$. Then following formulas can be checked:

(11)
$$
\begin{aligned}
S(X, Y) = & -\nabla_X(J)(JY) + \nabla_{JY}(J)(X) \\
& -\nabla_{JX}(J)(Y) + \nabla_Y(J)(JX),
\end{aligned}
$$

(12)
$$
\begin{aligned}
dF(X, Y, Z) = & \nabla_X(F)(Y, Z) + \nabla_Z(F)(X, Y) \\
& + \nabla_Y(F)(Z, X)
\end{aligned}
$$

for $X, Y, Z \in \mathfrak{X}_C(M)$. Hence, the vanishing of ∇J implies the vanishing of S and dF. Thus, an indefinite Kähler manifold is automatically both complex and symplectic, that is (i) implies (ii).

That the converse is also true follows from the formula

(13)
$$
\begin{aligned}
2\nabla_X(F)(Y, Z) = & dF(X, Y, Z) - dF(X, JY, JZ) \\
& + \langle X, S(Y, JZ) \rangle
\end{aligned}
$$

(For formulas (11)–(13) see [**Gr**].) Hence (ii) implies (i).

Thus, the class $\mathscr{IK}$ of indefinite Kähler manifolds is intermediate between the class $\mathscr{K}$ of manifolds with positive definite Kähler metrics and the class $\mathscr{S}$ of all symplectic manifolds:

(14)
$$
\mathscr{K} \subsetneqq \mathscr{IK} \subsetneqq \mathscr{S}.
$$

At the same time $\mathscr{IK}$ is intermediate between $\mathscr{K}$ and the class $\mathscr{C}$ of all complex manifolds

(15)
$$
\mathscr{K} \subsetneqq \mathscr{IK} \subsetneqq \mathscr{C}.
$$

The proper inclusions in (14) and (15) can be taken in a very strong sense. In [**FGG**] it is shown that there are compact symplectic manifolds that possess no integrable almost complex structure whatsoever. Similar statements apply to the other inclusions.

Just as with positive definite Kähler metrics the connection, curvature, and Ricci curvature of an indefinite Kähler metric simplify greatly when they are evaluated on holomorphic vector fields. On the other hand, in contrast to the positive definite case there are many easy to find examples of compact manifolds with Ricci flat indefinite Kähler metrics. The following theorem is proved in [**AFGM2**]:

THEOREM 12. *On a Kodaira-Thurston manifold KT let J be any left invariant complex structure. Then there is a four-parameter family of left invariant indefinite Kähler metrics on KT compatible with J. Furthermore, any such metric is flat.*

In [**AFGM1**] it is shown that there is an indefinite Kähler metric on the Iwasawa manifold I_3 (with respect to a nonstandard complex structure) for which the curvature is Ricci flat but not flat.

11. Compact symplectic manifolds with free circle actions

Compact nilmanifolds are not the only compact symplectic manifolds that carry no positive definite Kähler metrics. The following theorem is proved in [**FGM**].

THEOREM 13. *Let U be a compact manifold and assume Ω_U is a symplectic form on U. Suppose there is a symplectic diffeomorphism $\varphi\colon U \to U$ such that the induced cohomology map $\varphi^*\colon H^1(U, \mathbb{Z}) \to H^1(U, \mathbb{Z})$ has an element $\mathbf{b} \in H^1(U, \mathbb{Z})$ with $\varphi^*\mathbf{b} = \mathbf{b}$. Then there is a principal circle bundle $\pi\colon E \to U_\varphi$ (where U_φ denotes the mapping torus of φ) such that Ω_U gives rise to a symplectic form Ω_E on E. If $\mathbf{b}$ is nontrivial in $H^1(U, \mathbb{R})$, then E carries no positive definite Kähler metric.*

Another series of examples has recently been found by Yamato [**Yam**].

THEOREM 14. *Let Σ_g be a compact Riemann surface of genus g. Suppose that a compact symplectic manifold N admits a homomorphism $\rho\colon \pi_1(N) \to \mathrm{Diff}(\Sigma_g)$ such that the image of ρ is generated by certain Dehn twists. Denote by $\tilde{N}$ the universal covering space of N and define a $\pi_1(N)$-action on $\tilde{N} \times \Sigma_g$ by*

$$\Phi_h(\tilde{x}, z) = (\sigma(h)(\tilde{x}), \rho(h)(z)) \quad \text{for } h \in \pi_1(N),$$

where $\sigma(h)$ is the covering transformation corresponding to $h \in \pi_1(N)$. Then the quotient space M of $\tilde{N} \times \Sigma_g$ by the $\pi_1(N)$-action is a symplectic manifold. M also has nonzero Massey products, so M carries no positive definite Kähler metric.

12. Are there obstructions to the existence of a complex structure?

We have already noted that (K1)–(K5) are obstructions to the existence of a positive definite Kähler metric on a compact manifold M. There are also well-known obstructions for the existence of an almost complex structure on M; for example, the odd Stiefel-Whitney classes of M must vanish. But are there obstructions, different from these, for the existence of a complex structure on a compact manifold M?

The only known examples of compact almost complex manifolds that carry no complex structure occur in dimension four. The first examples were given by Van de Ven [**VdV**]. Also, there is the compact nilmanifold of [**FGG**] mentioned above, as well as examples of [**Yau, Br, Yam**]. In all cases the proof that a four-dimensional almost complex manifold carries no complex structure ultimately depends on Kodaira's classification of surfaces.

The most famous example of an almost complex manifold for which it is not known if a complex structure exists is of course the sphere S^6. But no other examples are known.

The situation with compact nilmanifolds is as follows. The least twisted compact nilmanifold is a torus, which of course has a positive definite Kähler metric. As we have observed above slightly twisted compact nilmanifolds can have complex structures, and some of these carry indefinite Kähler metrics. It appears, however, that there are very twisted compact nilmanifolds that carry no complex structure. One such example is described in [CFG3].

In [FG2] we showed that there is a compact four-dimensional solvmanifold that has no complex structure in spite of the fact conditions (K1)–(K5) are satisifed. Furthermore the manifold is parallelizable, so that Van de Ven's obstructions are insufficient to prove the nonexistence of a complex structure on it. Thus implicit in Kodaira's classification are further obstructions to the existence of a complex structure. It would be useful to have explicit descriptions of these obstructions; possibly they could be generalized to higher dimensions.

The example is as follows. Let $G(k)$ be the connected solvable Lie group of dimension three consisting of matrices of the form

$$a = \begin{pmatrix} e^{kz} & 0 & 0 & x \\ 0 & e^{-kz} & 0 & y \\ 0 & 0 & 1 & z \\ 0 & 0 & 0 & 1 \end{pmatrix}$$

where $x, y, z \in \mathbb{R}$ and k is a real number such that $e^k + e^{-k}$ is an integer different from two. Then a global system of coordinates $\{x, y, z\}$ for $G(k)$ is given by $x(a) = x$, $y(a) = y$, $z(a) = z$, and a standard calculation shows that a basis for the right invariant one-forms on $G(k)$ consists of

$$\{dx - kxdz, dy + kydz, dz\}.$$

Let $\Gamma(k)$ be a discrete subgroup of $G(k)$ such that the quotient space $M^3(k) = G(k)/\Gamma(k)$ is compact. (Such a subgroup $\Gamma(k)$ always exists; see for example [AGH].) Hence the forms $dx - kxdz$, $dy + kydz$, dz descend to one-forms α, β, γ on $M^3(k)$.

Now let us consider the product $M^4(k) = M^3(k) \times S^1$. Then there are one-forms $\{\alpha, \beta, \gamma, \eta\}$ on $M^4(k)$ such that

$$d\alpha = -k\alpha \wedge \gamma, \qquad d\beta = k\beta \wedge \gamma, \qquad d\gamma = d\eta = 0,$$

and such that at each point of $M^4(k)$, $\{\alpha, \beta, \gamma, \eta\}$ is a basis for the one-forms on $M^4(k)$. Moreover, it is easy to use de Rham's theorem to compute the real cohomology of $M^4(k)$,

$$H^0(M^4(k), \mathbb{R}) = \{1\},$$
$$H^1(M^4(k), \mathbb{R}) = \{[\gamma], [\eta]\},$$
$$H^2(M^4(k), \mathbb{R}) = \{[\alpha \wedge \beta], [\gamma \wedge \eta]\},$$
$$H^3(M^4(k), \mathbb{R}) = \{[\alpha \wedge \beta \wedge \gamma], [\alpha \wedge \beta \wedge \eta]\},$$
$$H^4(M^4(k), \mathbb{R}) = \{[\alpha \wedge \beta \wedge \gamma \wedge \eta]\}.$$

Thus

$$b_1(M^4(k)) = b_2(M^4(k)) = b_3(M^4(k)) = 2.$$

Hence $M^4(k)$ satisfies conditions (K1)–(K3); a calculation shows that it satisfies condition (K4) also.

THEOREM 15. *The minimal model of $M^4(k)$ is formal.*

PROOF. Let $(\Lambda^* M^4(k), d)$ be the exterior algebra of differential forms on $M^4(k)$; although $(\Lambda^* M^4(k), d)$ is free it is not the minimal model of $M^4(k)$ because it does not satisfy the nilpotency condition described in [DGMS].

In fact the minimal model $(\mathcal{M}, d_{\mathcal{M}})$ of $(\Lambda^* M^4(k), d)$ is as follows. Let $\mathcal{M}$ be the free algebra with generators $\mathbf{a}_1, \mathbf{a}_2, \mathbf{b}, \mathbf{c}$, where $\mathbf{a}_1$ and $\mathbf{a}_2$ have degree one, $\mathbf{b}$ has degree two, and $\mathbf{c}$ has degree three. The differential $d_{\mathcal{M}}$ is given as follows

$$d_{\mathcal{M}}(\mathbf{a}_1) = d_{\mathcal{M}}(\mathbf{a}_2) = d_{\mathcal{M}}(\mathbf{b}) = 0, \qquad d_{\mathcal{M}}(\mathbf{c}) = \mathbf{b}^2.$$

Let $\Phi: (\mathcal{M}, d_{\mathcal{M}}) \longrightarrow (\Lambda^*(M^4(k)), d)$ and $\Psi: (\mathcal{M}, d_{\mathcal{M}}) \mapsto (H^*(\mathcal{M}), d = 0)$ be the homomorphisms of differential algebras defined by

$$\Phi(\mathbf{a}_1) = \gamma, \qquad \Phi(\mathbf{a}_2) = \eta, \qquad \Phi(\mathbf{b}) = \alpha \wedge \beta, \qquad \Phi(\mathbf{c}) = 0,$$

and

$$\Psi(\mathbf{a}_1) = [\mathbf{a}_1], \qquad \Psi(\mathbf{a}_2) = [\mathbf{a}_2], \qquad \Psi(\mathbf{b}) = [\mathbf{b}], \qquad \Psi(\mathbf{c}) = 0.$$

It is easy to show that Φ and Ψ are homomorphisms between differential algebras each of which induces the identity in cohomology. Thus by definition the minimal model of $M^4(k)$ is $(\mathcal{M}, d_{\mathcal{M}})$ and it is formal.

THEOREM 16. *The manifold $M^4(k)$ has a symplectic structure but carries no complex structure (compatible with the symplectic structure or otherwise).*

PROOF. Let F be the two-form on $M^4(k)$ defined by $F = \alpha \wedge \beta + \gamma \wedge \eta$. A simple calculation shows that F is closed and has maximal rank so that F is a symplectic form on $M^4(k)$.

To show that $M^4(k)$ has no complex structure, suppose the contrary. Since $M^4(k)$ is parallelizable, its Euler characteristic and first Pontryagin number vanish; thus the Chern numbers $c_1^2(M^4(k))$ and $c_2(M^4(k))$ also vanish. Since $b_1(M^4(k)) = 2$, Theorem 3 of [Kod] implies that the geometric genus p_g of $M^4(k)$ vanishes. Then Kodaira's classification of complex surfaces (more precisely Theorem 10 of [Kod]) implies that $M^4(k)$ is algebraic.

Moreover, for $M^4(k)$ we can make use of Yau's classification of compact complex surfaces with topologically trivial tangent bundle. Since $b_1(M^4(k)) = 2 \neq 1$, it follows from Yau's classification [Yau, p. 53] that $M^4(k)$ is a ruled surface of genus one, the complex torus or an elliptic surface. But these cases have already been excluded. Therefore, $M^4(k)$ has no complex structure.

More generally, we have

THEOREM 17. *Let M be a four-dimensional parallelizable compact manifold, and suppose that $b_1(M) = 2$. Then M carries no complex structure.*

REFERENCES

[AFGM1] L. C. de Andrés, M. Fernández, A. Gray, and J. J. Mencía, *Compact manifolds with indefinite Kähler metrics*, Proceedings of the Sixth International Colloquium on Differential Geometry, Universidade de Santiago de Compostela, September, 1988, pp. 25–50.

[AFGM2] ____, *Moduli spaces of complex structures on compact four dimensional nilmanifolds*, Boll. Un. Mat. Ital. (to appear).

[AGH] L. Auslander, L. Green, and F. Hahn, *Flows on homogeneous spaces*, Ann. of Math. Studies, no. 53, Princeton Univ. Press, Princeton, NJ, 1963.

[BPV] W. Barth, C. Peters, and A. Van de Ven, *Compact complex surfaces*, Ergebnisse der Mathematik (3) vol. 4, Springer-Verlag, Berlin, Heidelberg, 1984.

[BG] C. Benson and C. Gordon, *Kähler and symplectic structures on nilmanifolds*, Topology **27** (1988), 513–518.

[Br] N. Brotherthon, *Some parallelizable four manifolds not admitting a complex structure*, Bull. London Math. Soc. **10** (1978), 303–304.

[CE] C. Chevalley and S. Eilenberg, *Cohomology theory of Lie groups and Lie algebras*, Trans. Amer. Math. Soc. **63** (1948), 85–124.

[CFG1] L. A. Cordero, M. Fernández, and A. Gray, *La suite spectrale de Frölicher et les nilvariétés complexes compactes*, C. R. Acad. Sci. Paris **305** (1987), 753–756.

[CFG2] ____, *The Frölicher spectral sequence for compact nilmanifolds*, Illinois Math. J. **35** (1991), 56–67.

[CFG3] ____, *Lie groups with no left invariant complex structures*, Portugal Math. **47** (1990), 183–190.

[DGMS] P. Deligne, P. Griffiths, J. Morgan, and D. Sullivan, *Real homotopy theory of Kähler manifolds*, Invent. Math. **29** (1975), 245–274.

[FGG] M. Fernández, M. Gotay, and A. Gray, *Four-dimensional parallelizable symplectic and complex manifolds*, Proc. Amer. Math. Soc. **103** (1988), 1209–1212.

[FG1] M. Fernández and A. Gray, *The Iwasawa manifold*, Differential Geometry, Peñíscola, 1985, Lecture Notes in Math., vol. 1209, Springer-Verlag, 1986, 157–159.

[FG2] ____, *Compact symplectic solvmanifolds not admitting complex structures*, Geometria Dedicata **34** (1990), 295–299.

[FGM] M. Fernández, A. Gray, and J. Morgan, *Compact symplectic manifolds with free circle actions*, **38** (1991), 271–283.

[Fr] A. Frölicher, *Relations between the cohomology groups of Dolbeault and topological invariants*, Proc. Nat. Acad. Sci. U.S.A. **41** (1955), 641–644.

[Gr] A. Gray, *Minimal varieties and almost Hermitian submanifolds*, Michigan Math. J. **12** (1965), 273–287.

[GH] P. Griffiths and J. Harris, *Principles of algebraic geometry*, John Wiley and Sons, NY, 1978.

[Hattori] A. Hattori, *Spectral sequence in the de Rham cohomology of fiber bundles*, J. Fac. Sci. Univ. Tokyo Sect. **18** (1960), 289–331.

[KN] S. Kobayashi and K. Nomizu, *Foundations of differential geometry*, vol. 2, John Wiley and Sons, NY, 1969.

[Kod] K. Kodaira, *On the structure of complex analytic surfaces*. I, Amer. J. Math. **86** (1964), 751–798.

[Ma] А. И. Мал′цемии, Об одном классе однородных пространств, Известиа Академии Наук СССР.Серия Математическая 13 (1949), 9–32. English translation A. I. Mal′čev, *A class of homogeneous spaces*, Amer. Math. Soc. Transl. **39** (1951).

[Mat] Y. Matsushima, *On the discrete subgroups and homogeneous spaces of nilpotent Lie groups*, Nagoya Math. J. **2** (1951), 95–110.

[Nomizu] K. Nomizu, *On the cohomology of compact homogeneous spaces of nilpotent Lie groups*, Ann. of Math. (2) **59** (1954), 531–538.

[Pi] H. Pittie, *The nondegeneration of the Hodge-de Rham spectral sequence*, Bull. Amer. Math. Soc. N.S. **20** (1989), 19–22.

[Sa] Y. Sakane, *On compact parallelizable solvmanifolds*, Osaka J. Math. **13** (1976), 187–212.

[Th] W. P. Thurston, *Some examples of symplectic manifolds*, Proc. Amer. Math. Soc. **55** (1976), 467–468.

[VdV] A. Van de Ven, *On the Chern numbers of certain complex and almost complex manifolds*, Proc. Nat. Acad. Sci. U.S.A. **55** (1966), 1624–1627.

[Wang] H. C. Wang, *Complex parallelisable manifolds*, Proc. Amer. Math. Soc. **5** (1954), 771–776.

[Weil] A. Weil, *Introduction à l''etude des variétés kählériennes*, Actualités Sci. Indust. no. 1267, Hermann, Paris, 1958.

[Wells] R. Wells, *Differential analysis on complex manifolds*, (2nd ed.), Springer-Verlag, 1980.

[Yam] K. Yamato, *Examples of non-Kähler symplectic manifolds*, Osaka J. Math. **27** (1990), 431–439.

[Yau] S. T. Yau, *Parallelizable manifolds without complex structure*, Topology **15** (1976), 51–53.

UNIVERSIDAD DE SANTIAGO DE COMPOSTELA, SPAIN

UNIVERSIDAD DEL PAÍS VASCO, SPAIN

UNIVERSITY OF MARYLAND

Proceedings of Symposia in Pure Mathematics
Volume 54 (1993), Part 2

Nonabelian Hodge Theory

KEVIN CORLETTE

0. Introduction

Questions about finite-dimensional representations of the fundamental group of a Kähler manifold arise in response to a number of fundamental problems in the subject. Among these are problems concerning the characterization of manifolds which are uniformized by bounded symmetric domains, problems related to the study of stable holomorphic vector bundles, and problems coming from the study of variations in the Hodge-theoretic data associated with families of Kähler manifolds. In recent years, the moduli space parametrizing isomorphism classes of representations has itself become the subject of a Hodge-theoretic approach which embeds many of these issues in a common theoretical framework. This framework is nowadays called nonabelian Hodge theory, and it is the subject of this survey. Another recent overview of the subject is found in [29].

In approximate terms, the fundamental object of interest in nonabelian Hodge theory is a nonlinear analogue of the Hodge structure on the cohomology groups $H^k(M, \mathbb{R})$ of a compact Kähler manifold M. One replaces the ordinary cohomology groups by the nonabelian cohomology set $H^1(M, \mathrm{GL}(n, \mathbb{R}))$, which is naturally interpreted as the set of isomorphism classes of real rank n representations of $\pi_1(M)$, and tries to find an analogue of the "decomposition into (p, q)-types" in this setting. What is required is, first, a nonlinear analogue of the notion of a harmonic representative of a cohomology class and, second, a nonlinear version of the well-known equality

$$\Delta = \Delta' + \Delta'' = 2\Delta'',$$

where Δ is the de Rham Laplacian and Δ', Δ'' are the Laplacians associated to ∂, $\overline{\partial}$, respectively. The first requirement is fulfilled by the harmonic map equation in the case of nonabelian "de Rham" theory, and a modification

1991 *Mathematics Subject Classification*. Primary 58A14, 32J25.

This paper is in final form and no version of it will be submitted for publication elsewhere.

of the Hermitian-Yang-Mills equation in the case of nonabelian "Dolbeault" cohomology. The appropriate analogue of the "equality of Laplacians" then turns out to be a Bochner formula for harmonic maps which first emerged in work of Siu, and was further developed by Sampson, Carlson-Toledo, and others. One of the main tasks of this paper will be the description of a framework in which the analogy with ordinary Hodge theory can be easily seen.

There are a number of sources for the material to be described here. Among the most important is a paper [16] of Hitchin in which many of the main features of nonabelian Hodge theory were laid out in the case of projective algebraic curves and representations of their fundamental groups in SL(2, ℂ). Independently, growing out of work of Eells and Sampson and the work of Siu and Sampson mentioned above, nonlinear versions of the Hodge-de Rham theorem were worked out in [5] and [9] (as well as [10], which was written in direct response to the concerns of [16]). Meanwhile, the nonabelian analogue of Dolbeault theory was worked out in a succession of generalizations by Narasimham-Seshadri, Donaldson, Uhlenbeck-Yau, and Simpson. A visionary synthesis and extension of these developments into a full-fledged nonabelian Hodge theory was presented by Simpson in [27]; this paper also made contact with related work [2] by Carlson and Toledo on the classification of harmonic mappings from Kähler manifolds to locally symmetric spaces and work [11] by Goldman and Millson on the infinitesimal Hodge theory of moduli spaces of representations of fundamental groups.

My talk in Los Angeles was devoted in part to nonabelian Hodge theory in an L^2 setting and relationships with intersection cohomology. The idea was to give nonabelian analogues of theorems of Zucker [31], Cattani-Kaplan-Schmid [3] and Kashiwara-Kawai [20]. Considerations of time have forced me to leave this aspect of the subject untouched; I expect to deal with it on another occasion.

1. Abelian Hodge theory

Let V be a real vector space. A (real) Hodge structure of weight k on V is a decomposition

$$V \otimes \mathbb{C} = \bigoplus_{p+q=k} V^{p,q}$$

of the complexification of V into subspaces such that $\overline{V^{p,q}} = V^{q,p}$, where the bar represents the natural conjugation map $V \otimes \mathbb{C} \to V \otimes \mathbb{C}$. A polarization of the Hodge structure on V is a nondegenerate bilinear pairing $\langle \, , \, \rangle : V \otimes V \to \mathbb{R}$ which is symmetric if k is even, antisymmetric otherwise, and whose complex bilinear extension satisfies:

1. $\langle v_1, v_2 \rangle = 0$ whenever $v_1 \in V^{p,q}$, $v_2 \in V^{p',q'}$, and $p \neq q'$;
2. if $v = \sum_{p+q=k} v^{p,q} \in V$, then $\sum_{p+q=k} i^{p-q} \langle v^{p,q}, v^{q,p} \rangle > 0$ when $v \neq 0$.

Consider the multiplicative group $\mathbb{C}^\times$ of nonzero complex numbers; this is the set of complex points of the algebraic group $\mathbb{G}_m$. Restricting scalars to $\mathbb{R}$, we get an algebraic group $\mathbb{G} = \text{Res}_{\mathbb{C}/\mathbb{R}} \mathbb{G}_m$ whose set of real points is isomorphic to $\mathbb{C}^\times$. It is an observation of Deligne's [7] that the category of real Hodge structures is equivalent to the category of real representations of $\mathbb{G}$; the correspondence goes as follows. If V carries a real Hodge structure, then there is an induced action of $\mathbb{G}(\mathbb{C}) = \mathbb{C}^\times \times \mathbb{C}^\times$ on $V \otimes \mathbb{C}$ by

$$(z, w)v = z^p w^q v$$

if $v \in V^{p,q}$. The set of real points of $\mathbb{G}$ is $\{(z, \overline{z}) | z \in \mathbb{C}^\times\}$, and one checks easily that $\mathbb{G}(\mathbb{R})$ preserves the real structure on $V \otimes \mathbb{C}$. Conversely, any complex representation of $\mathbb{G}$ decomposes into a direct sum of representations of type (p, q), and these will be related in the appropriate way by complex conjugation if the representation is defined over $\mathbb{R}$.

The Weil automorphism C of a Hodge structure is the automorphism induced by $i \in \mathbb{G}(\mathbb{R})$. One can define a polarization to be a bilinear form $\langle \ , \ \rangle$ on V which is invariant under the maximal compact subgroup of G such that $\langle \cdot, C \cdot \rangle$ is symmetric and positive-definite.

Now suppose that M is a compact Kähler manifold with Kähler form ω. Let $\mathscr{A}^k(M)$ be the set of k-forms on M, and $\mathscr{A}^{p,q}(M)$ the set of forms of type (p, q). The de Rham cohomology groups of M are defined to be the cohomology groups of the complex

$$0 \to \mathscr{A}^1(M) \xrightarrow{d} \mathscr{A}^2(M) \xrightarrow{d} \cdots \xrightarrow{d} \mathscr{A}^{2n}(M) \to 0;$$

these are denoted by $H^k(M, \mathbb{R})$. The space of harmonic forms is defined to be

$$\mathscr{H}^k = \{\theta \in \mathscr{A}^k(M) | \Delta\theta = 0\},$$

where $\Delta = dd^* + d^*d$. The Hodge theorem implies that each class in $H^k(M, \mathbb{R})$ has a unique harmonic representative, so there is a canonical isomorphism

$$H^k(M, \mathbb{R}) \xrightarrow{\sim} \mathscr{H}^k.$$

We can define in a similar way the Dolbeault complex

$$0 \to \mathscr{A}^{p,0}(M) \xrightarrow{\overline{\partial}} \mathscr{A}^{p,1}(M) \xrightarrow{\overline{\partial}} \cdots \xrightarrow{\overline{\partial}} \mathscr{A}^{p,n}(M) \to 0$$

for each p. The cohomology groups are the Dolbeault cohomology groups $H^q(M, \Omega^p)$, where Ω^p is the sheaf of holomorphic p-forms on M. Define spaces of Dolbeault harmonic forms by

$$\mathscr{H}^{p,q} = \{\theta \in \mathscr{A}^{p,q}(M) | \Delta''\theta = 0\},$$

where $\Delta'' = \overline{\partial}\overline{\partial}^* + \overline{\partial}^*\overline{\partial}$. There is again a canonical isomorphism $H^q(M, \Omega^p) \to \mathscr{H}^{p,q}$. It is a fundamental consequence of the equality $\Delta = 2\Delta''$ that elements of $\mathscr{H}^{p,q}$ are in $\mathscr{H}^{p+q} \otimes \mathbb{C}$ while, conversely, (p, q) components of elements of $\mathscr{H}^k$ are in $\mathscr{H}^{p,q}$. As a result, we obtain the following.

THEOREM 1.1. *The de Rham cohomology group* $H^k(M, \mathbb{R})$ *of a compact Kähler manifold carries of Hodge structure of weight* k.

The Kähler form determines a bilinear form on $H^k(M, \mathbb{R})$: for classes $[\theta]$, $[\eta] \in H^k(M, \mathbb{R})$ represented by forms θ, $\eta \in \mathscr{A}^k(M)$, define

$$\langle [\theta], [\eta] \rangle = (-1)^{k(k-1)/2} \int_M \theta \wedge \eta \wedge \omega^{n-k}.$$

This is symmetric or antisymmetric depending on the parity of k; in order to obtain a polarization, we have to restrict to an appropriate Hodge substructure of $H^k(M, \mathbb{R})$. Define an operator $L: \mathscr{A}^k(M) \to \mathscr{A}^{k+2}(M)$ by

$$L\eta = \omega \wedge \eta;$$

this operation commutes with the Laplacian, so induces a map $\mathscr{H}^k \to \mathscr{H}^{k+2}$. The space of primitive harmonic forms is given by

$$\mathscr{P}^k = \ker(L^{n-k+1}: \mathscr{H}^k \to \mathscr{H}^{2n-k+2}).$$

This is a Hodge substructure of $\mathscr{H}^k$, and $\langle \ , \ \rangle$ induces a polarization of $\mathscr{P}^k$.

The case which is most apt for our guiding metaphor is that of $H^1(M, \mathbb{R})$. Notice that in this case all classes are primitive, that the polarization is a symplectic form, and that the subspaces $H^{0,1}$ and $H^{1,0}$ are Lagrangian subspaces with respect to the induced complex symplectic structure on $H^1(M, \mathbb{C})$.

It is worth considering at this point the manner in which this data varies in families of Kähler manifolds. Suppose $\pi: X \to S$ is a holomorphic map without critical points between two Kähler manifolds; we assume that the fibers X_s, $s \in S$, are compact. The cohomology groups $H^k(X_s, \mathbb{R})$ are fibers of a flat vector bundle $E = R^k \pi_* \mathbb{R}$ over S, a higher direct image of the constant sheaf $\mathbb{R}$ on X. We denote by ∇ the flat connection on E. It is interesting to consider the manner in which the various constituents of the Hodge structures on the fibers vary with s.

Note first that the Kähler form ω_s on X_s is the restriction of the Kähler form ω on X, so the class $[\omega_s]$ is a covariant constant section of $R^2 \pi_* \mathbb{R}$. Hence the pairing $\langle \ , \ \rangle: E \otimes E \to \mathbb{R}$ is flat, and the vector bundle of primitive subspaces $\mathscr{P}E$ is a flat summand of E. The Hodge subbundles $E^{p,q}$ of $E_\mathbb{C} = E \otimes \mathbb{C}$ are in general only smooth subbundles. However, the direct sums

$$F^r = \bigoplus_{p \geq r} E^{p,q}$$

turn out to be holomorphic subbundles of $E_\mathbb{C}$ with respect to the holomorphic structure induced by the $(0, 1)$-component ∇' of ∇; this may be seen as follows. Denote by $\mathscr{O}_S$ the sheaf of holomorphic functions on S, and by $\pi^* \mathscr{O}_S$ the sheaf of fiberwise constant holomorphic functions on X. Let

$\Omega^p_{X/S}$ be the sheaf of relative holomorphic p-forms on X, i.e., the sheaf $\Omega^p_X/\pi^*\Omega^p_S$. Here,

$$\pi^*\Omega^p_S = \pi^\cdot\Omega^p_S \otimes_{\pi^\cdot\mathcal{O}_S} \mathcal{O}_X,$$

where $\pi^\cdot\Omega^p_S$ is the sheaf of pullbacks of holomorphic p-forms on S. The holomorphic de Rham theorem says that $H^k(X_s, \mathbb{C})$ is isomorphic to the hypercohomology $\mathbb{H}^k(X_s, \Omega^*_{X_s})$ of the holomorphic de Rham complex; the filtration of the holomorphic de Rham complex obtained by omitting the first $p - 1$ terms induces a filtration F on its hypercohomology, hence on $H^k(X_s, \mathbb{C})$. The equality of Laplacians in this context may be interpreted as the statement that the associated spectral sequence degenerates, so that

$$F^r H^k(X_s, \mathbb{C}) = \bigoplus_{p \geq r} H^q(X_s, \Omega^q_{X_s}).$$

This argument may be applied to the relative holomorphic de Rham complex, and one finds that the bundles F^r are holomorphic subbundles of $E_{\mathbb{C}}$.

A important property of the connection ∇ is that it satisfies *Griffiths' transversality*. Let $\mathcal{F}^r$ be the sheaf of holomorphic sections of F^r and ∇' the $(1, 0)$-component of ∇. Griffiths' transversality means that

$$\nabla'\mathcal{F}^r \subset \mathcal{F}^{r-1} \otimes \Omega^1_S.$$

Griffiths incorporated all of this data in the notion of a variation of Hodge structure.

DEFINITION 1.2. A (real) variation of Hodge structure of weight k over S is a flat real vector bundle (E, ∇) together with a smooth direct sum decomposition

$$E_{\mathbb{C}} = E \otimes \mathbb{C} = \bigoplus_{p+q=k} E^{p,q}$$

satisfying the following conditions:

1. $F^r = \bigoplus_{p \geq r} E^{p,q}$ is a holomorphic subbundle of $E_{\mathbb{C}}$ relative to the holomorphic structure induced by ∇'';

2. $\nabla'\mathcal{F}^r \subset \mathcal{F}^{r-1} \otimes \Omega^1_S$, where $\mathcal{F}^r$ is the sheaf of holomorphic sections of F^r.

3. $\overline{E^{p,q}} = E^{q,p}$.

A polarization of a variation of Hodge structure is a flat bilinear pairing $E \otimes E \to \mathbb{R}$ which induces a polarization of each fiber.

A fundamental result of Griffiths [7] is the following:

THEOREM 1.3. *If $\pi: X \to S$ is a proper holomorphic map of Kähler manifolds, then $\mathcal{P}E = \ker(L: R^k\pi_*\mathbb{R} \to R^{k+2}\pi_*\mathbb{R})$ is a polarized variation of Hodge structure.*

2. Nonabelian de Rham theory

Now we take up the problem of Hodge structures for nonabelian cohomology. If M is a topological space and G is a topological group, then

$H^1(M, G)$ is the set of isomorphism classes of principal G-bundles over M whose transition functions are continuous functions into G. In particular, if G has the discrete topology, $H^1(M, G)$ is just the set of isomorphism classes of homomorphisms $\pi_1(M) \to G$. Consider the special case where M is a manifold and G is a reductive real algebraic group (taken with the classical topology; when we wish to use the discrete topology, we will write G^d). The set $H^1(M, G^d)$ then parametrizes local systems on M with structure group G. As such, we can use the following de Rham model for $H^1(M, G^d)$. Let $\mathscr{C}$ be the union of infinite-dimensional affine spaces of connections on smooth principal G-bundles over M. $H^1(M, G^d)$ is then the quotient of the subset of flat connections in $\mathscr{C}$ by the relation "connection-preserving isomorphism of G-bundles." More briefly, it is the set of flat connections modulo gauge equivalence.

The main question now is: if M is a compact Riemannian manifold, what is the appropriate notion of a harmonic representative for a class in $H^1(M, G^d)$? We propose the following answer. Let K be a maximal compact subgroup of G (in the classical topology); we refer to a smooth reduction of the structure group of a principal G-bundle P to K as a "metric," by analogy with the case of $\mathrm{GL}(n)$. A metric on P may be interpreted as a smooth section h of the associated bundle

$$Y = P \times_G (G/K).$$

If P is endowed with a flat connection, then h may be interpreted locally on M as a map into the Riemannian symmetric space G/K. It thus makes sense to require that h be a harmonic map: this means that

$$D^* dh = 0,$$

where dh is the differential of h as a map from an open set in M to G/K, and D is the pullback of the Levi-Civita connection from G/K. Our proposal is then that a harmonic flat bundle is a triple (P, ∇, h), where P is a principal G-bundle, ∇ is a flat connection on P, and h is a harmonic section of the associated bundle Y.

Philosophically, this definition might be justified by the fact that a choice of metric is necessary in order to do Hodge theory in the ordinary sense for the de Rham complex with coefficients in any vector bundle associated to P. More concretely, we might recall that harmonic 1-forms on M may be interpreted locally as harmonic maps from M into the real line, and the natural generalization of this is to replace the real line by the Riemannian symmetric space associated to G. In any case, we will see that this choice makes possible the development of a reasonable analogue of Hodge theory in the nonabelian setting. The first step is the following analogue of the Hodge theorem.

THEOREM 2.1. *Let* (P, ∇) *be a principal G-bundle with a flat connection.* (P, ∇) *admits a harmonic metric if and only if the Zariski closure of the holonomy group of* ∇ *is a reductive subgroup of* G.

This result is of course motivated by the well-known existence result of Eells and Sampson for harmonic maps between compact manifolds. This version was proved independently by Diederich-Ohsawa [9], Donaldson [10], and the author [5]. For extensions to the case of twisted harmonic maps into nonsymmetric manifolds, see Jost-Yau [19], Labourie [21], and [4].

From this, we see that harmonic representatives exist only for certain classes in $H^1(M, G^d)$; the restriction of our attention to this subset is in fact reasonable from another point of view. If one attempts to construct an algebraic variety parametrizing isomorphism classes of flat G-bundles, one immediately runs into the basic problem confronting quotient constructions in algebraic geometry: orbits need not be closed, so the quotient spaces need not be Hausdorff. Geometric invariant theory was created in order to circumvent this difficulty. If Z is an algebraic variety with an action of an algebraic group G, then there is a distinguished class Z^{ss} of points (the semistable ones) together with an equivalence relation which approximates the relation of lying in the same orbit. The quotient of Z^{ss} by this equivalence relation is an algebraic variety, and is generally as close an approximation to Z/G as an algebraic variety can come. In our case, Z is the variety which parametrizes all homomorphisms from $\pi_1(M)$ to G, and G acts on Z by conjugation. The quotient construction of geometric invariant theory in this case gives precisely the set of isomorphism classes of representations with reductive Zariski closures. We will denote this moduli space by $X(M, G)$.

Let us now set our sights on the analogue of the "equality of Laplacians" in ordinary Hodge theory. In the case of a harmonic 1-form θ, this says that any function f such that $df = \theta$ locally is pluriharmonic, that is, is harmonic upon restriction to any complex curve in M. That the same is true in the nonabelian context is a consequence of a Bochner formula for harmonic maps from Kähler manifolds into manifolds with sufficiently nonpositive curvature. This formula first arose in work of Siu [30], and was refined by Sampson [25], Carlson-Toledo [2], and others. To make the appropriate statement in the setting of interest here, we require some notation. Suppose (P, ∇, h) is a harmonic flat bundle. Let $\mathrm{ad}(P)$ be the vector bundle associated to P by the adjoint representation of G on its Lie algebra $\mathfrak{g}$. The K-reduction h determines a splitting of $\mathrm{ad}(P)$ into direct summands $\mathrm{ad}_{\mathfrak{k}}(P)$ and $\mathrm{ad}_{\mathfrak{p}}(P)$ corresponding to the splitting $\mathfrak{g} = \mathfrak{k} \oplus \mathfrak{p}$ of $\mathfrak{g}$ into the Lie algebra of K and a K-invariant complement. There is a corresponding splitting

$$\nabla = D + \theta,$$

where D is a connection preserving the K-reduction and θ is a one-form with values in $\mathrm{ad}_{\mathfrak{p}}(P)$. As previously, we denote the $(1, 0)$-components of

D and θ by D' and θ', and we indicate the $(0, 1)$-components by D'' and θ''. The Bochner formula is then the following.

THEOREM 2.2. *Suppose that* M *is a compact Kähler manifold, and* (P, ∇, h) *is a harmonic flat bundle over* M. *The following hold*:
1. $D'' \circ D'' = 0$;
2. $D''\theta' = 0$;
3. $[\theta', \theta'] = 0$.

The proof of this result is based on the identity

$$\overline{\partial}\partial|\theta'|^2 = |D''\theta|^2 + |[\theta', \theta']|^2;$$

the left side integrates to zero over M, while the right side is nonnegative. It is worth remarking that similar formulas are known for harmonic maps emanating from manifolds with other special holonomy groups; see [4].

Hitchin, in the case where M is a Riemann surface, and Simpson, in the general case, observed that this Bochner formula permits the definition of a natural circle action on $X(M, G)$. For any triple (P, ∇, h) and any complex number z of norm one, define

$$\nabla_z = D + z\theta' + \overline{z}\theta''.$$

THEOREM 2.3. *If* (P, ∇, h) *is a harmonic flat bundle, then* (P, ∇_z, h) *is a harmonic flat bundle for any* z *with norm one.*

PROOF. The condition that ∇_z is flat is that $\nabla_z^2 = 0$. We find

$$\nabla_z^2 = D^2 + zD\theta' + \overline{z}D\theta'' + z^2[\theta', \theta'] + 2[\theta', \theta''] + \overline{z}^2[\theta'', \theta''].$$

However, $D^2 + 2[\theta', \theta'']$ is the $\mathfrak{k}$-component of the curvature of ∇, so it vanishes. $D''\theta''$ and $D'\theta'$ are also components of the curvature of ∇, and thus vanish. The remaining terms vanish by the Bochner formula. Similarly, the harmonic equation

$$D^*(z\theta' + \overline{z}\theta'') = 0$$

continues to be satisfied, since

$$D^*\theta' = \Lambda D''\theta' = 0,$$
$$D^*\theta'' = -\Lambda D'\theta'' = 0,$$

where Λ is the adjoint of L. Q.E.D.

This is the compact part of the $\mathbb{C}^\times$-action on $X(M, G)$; we must wait until we have discussed nonabelian Dolbeault theory before describing the full action. Nevertheless, interesting applications are available at this stage.

We begin the discussion of these applications by determining the significance of fixed points for the $U(1)$-action. The following line of thought is due to Simpson, although certain elements are found in [16]. An equivalence class $[(P, \nabla, h)]$ of harmonic flat bundles is fixed by the $U(1)$-action if (P, ∇_z, h) is gauge equivalent to (P, ∇, h) for every $z \in U(1)$. Thus, for each z, there must be some metric-preserving gauge transformation g_z of P such that $g_z^* \nabla = \nabla_z$. In particular, g_z must preserve D, so also the holomorphic structure induced by D'' on the associated $G_{\mathbb{C}}$-bundle $P_{\mathbb{C}}$. We fix some z having infinite order in $U(1)$ and an associated g_z.

Now suppose that (V, ρ) is a finite-dimensional real representation of G; the associated vector bundle and its complexification will be denoted by E and $E_{\mathbb{C}}$. The action of g_z on the fibers of $E_{\mathbb{C}}$ induces a decomposition into eigenspaces on which g_z acts by z^p; because g_z is holomorphic and unitary, these eigenspaces vary smoothly over M, so we get a decomposition $E_{\mathbb{C}} = \bigoplus E^{-p,p}$ into smooth subbundles (N.B.: p need not be an integer.) These are holomorphic subbundles with respect to the holomorphic structure determined by D'', since g_z commutes with D. Furthermore, $g_z^* \theta'' = \overline{z} \theta''$, so θ'' is a $(0, 1)$-form with values in $\bigoplus_p \text{Hom}(E^{-p,p}, E^{1-p,p-1})$; by an analogous argument θ' takes values in $\bigoplus_p \text{Hom}(E^{-p,p}, E^{-p-1,p+1})$. These interact with the real structure by

$$\overline{E^{-p,p}} = E^{p,-p}.$$

Define

$$F^r = \bigoplus_{-p \geq r} E^{-p,p}.$$

The operator $\nabla'' = D'' + \theta''$ preserves each F^r, while $\nabla' = D' + \theta'$ satisfies Griffiths' transversality:

$$\nabla' \mathscr{F}^r \subset \Omega^1 \otimes \mathscr{F}^{r-1}.$$

Let ε be a gauge automorphism such that $\varepsilon^* \nabla = \nabla_{-1}$. Since $h(\cdot, \cdot)$ is parallel with respect to D and θ takes values in selfadjoint endomorphisms, we find that $h(\cdot, \varepsilon \cdot)$ is an indefinite Hermitian form parallel with respect to ∇. Hence, the complex bilinear form

$$\langle \cdot, \cdot \rangle = h(\cdot, \overline{\varepsilon \cdot})$$

is also parallel. This adds up to the following observation, originally due to Simpson.

THEOREM 2.4. *The fixed points of the action of $U(1)$ on $X(M, G)$ are the monodromy representations arising from variations of Hodge structure over M.*

Suppose that M is a smooth projective variety of dimension three or more. The Lefschetz theorem says that the fundamental group of any smooth hyperplane section H of M maps isomorphically onto that of M. This isomorphism induces an isomorphism $X(H, G) \to X(M, G)$ of algebraic varieties. The $U(1)$-action is functorial, so fixed points in $X(H, G)$ correspond to fixed points in $X(M, G)$.

COROLLARY 2.5. *Let H be a smooth hyperplane section of a smooth complex projective variety M of dimension at least three. Any polarized variation* *of Hodge structure over* H *extends to a polarized variation of Hodge structure over* M.

This statement was given by Simpson [27] for complex variations of Hodge structure. Certain families of algebraic varieties can be reconstructed from associated variations of Hodge structure; pre-eminent among these are families of abelian varieties. The previous corollary implies

COROLLARY 2.6. *Let* M *and* H *be as above. Any family of abelian varieties over* H *extends to a family of abelian varieties over* M.

Examples of such families are provided by compact Shimura varieties.

Another consequence of this characterization of variations of Hodge structure is the following.

COROLLARY 2.7. *Suppose the equivalence class of* $\rho\colon \pi_1(M) \to G$ *in* $X(M, G)$ *is isolated. Then* ρ *arises from a polarized variation of Hodge structure.*

This follows from the fact that the $U(1)$-action is continuous, so fixes any isolated point. Simpson derived as a consequence of this certain restrictions on fundamental groups of compact Kähler manifolds. To appreciate this result, let us recall first a conjecture of Carlson and Toledo.

CONJECTURE 2.8. *Suppose* G *is a simple Lie group whose associated symmetric space is not Hermitian, and* Γ *is a lattice in* G. Γ *is not the fundamental group of any compact Kähler manifold.*

There are by now a number of known restrictions on fundamental groups of compact Kähler manifolds, almost all depending on Hodge theory in some more or less traditionals sense; see Goldman-Millson [11], Gromov [12], Johnson-Rees [18] and Morgan [24] for examples. This conjecture represents one of the first attacks on a class of groups which does not yield to Hodge theory in this sense.

Carlson and Toledo were able to prove their conjecture in the case of the group $G = \mathrm{SO}(n, 1)$, $n \geq 3$. The Siu-Sampson formula implies that the rank of a harmonic map from a compact Kähler manifold to a real hyperbolic manifold can have rank at most two. The suggests, and Carlson and Toledo were able to prove, that such a map must actually factor through a holomorphic map to a Riemann surface or a map to a closed geodesic. Neither sort of map can induce an isomorphism onto a lattice in $\mathrm{SO}(n, 1)$ when $n > 2$.

Simpson's approach to this problem is based on the following observation.

PROPOSITION 2.9. *Suppose that* $\rho\colon \pi_1(M) \to G$ *is a homomorphism whose image is dense (in the real Zariski topology on* G), *and* G *is semisimple. If the equivalence class of* ρ *is fixed by the action of* $U(1)$, *then* G *has a compact Cartan subgroup.*

PROOF. Let (E, ∇) be the flat vector bundle associated to ρ by the adjoint representation of G; as indicated above, it underlies a polarized variation of Hodge structure. We denote the polarization of $E_{\mathbb{C}}$ by $\langle \cdot, \cdot \rangle$. The complex Zariski closure $G_{\mathbb{C}}$ of the holonomy group of ∇ at $m \in M$ preserves $\langle \cdot, \cdot \rangle$. There is a natural complex conjugation map $\sigma \colon E_{\mathbb{C}} \to E_{\mathbb{C}}$; the real Zariski closure of the holonomy is the subgroup of $G_{\mathbb{C}}$ preserving σ. There is an element $\varepsilon \in G$ whose square induces the identity on the fiber of $E_{\mathbb{C}}$ over m. Letting ε act on elements of $G_{\mathbb{C}}$ by conjugation, we see that the group fixed by $\varepsilon\sigma$ is the subgroup of $G_{\mathbb{C}}$ preserving $\langle \cdot, \varepsilon\sigma \cdot \rangle$, which is a positive-definite Hermitian form. Thus, $G_{\mathbb{C}}^{\varepsilon\sigma}$ is a compact real form of $G_{\mathbb{C}}$. This group contains ε, so contains a Cartan subgroup containing ε. This is also contained in $G_{\mathbb{C}}^{\sigma} = G$, so G has a compact Cartan subgroup.   Q.E.D.

We will refer to a semisimple group with the property in the conclusion of this result as a group of Hodge type. Examples of groups which are not of Hodge type include complex semisimple groups, $SO(2n + 1, 1)$, and $SL(n, \mathbb{R})$, $n \geq 3$. If we combine the previous two results with the Weil rigidity theorem (and its successors) for lattices in semisimple groups, we obtain the following result of Simpson's.

THEOREM 2.10. *Suppose* Γ *is a cocompact irreducible lattice in any semisimple Lie group which is not of Hodge type, or is an irreducible lattice in any semisimple group which is not of Hodge type and is not locally isomorphic to* $SL(2, \mathbb{C})$. *Then* Γ *is not isomorphic to the fundamental group of any compact Kähler manifold.*

It should be noted that the theorem of Carlson and Toledo is not subsumed by this result; the groups $SO(2n, 1)$ are of Hodge type. The group F_4^{-20}, which is also of Hodge type, has recently been treated by Carlson and Hernandez [1]. There are also results in the converse direction, stating that monodromy representations of certain variations of Hodge structure must be rigid; see, e.g., [6].

The existence of harmonic metrics also has consequences for the Hodge theory of de Rham cohomology with coefficients. If (E, ∇) is a flat vector bundle with a parallel positive-definite metric, then the arguments of the previous section extend immediately to show that there is an equality

$$\Delta = \Delta' + \Delta'' = 2\Delta''$$

between the Laplacians associated to ∇, ∇' and ∇'', respectively. This shows that the de Rham cohomology groups $H^k(M, E)$ can be identified with Dolbeault cohomology groups having coefficients in appropriate holomorphic vector bundles. This no longer remains true if ∇ does not preserve a positive-definite metric; however, some years ago Deligne [31] showed that a slight modification of this scheme works if (E, ∇) comes from a polarized variation of Hodge structure. His observation was that one should replace the operator ∇' by

$$\mathscr{D}' = D' + \theta'',$$

and ∇'' by

$$\mathscr{D}'' = D'' + \theta'.$$

Taking Δ', Δ'' to be the Laplacians associated to $\mathscr{D}', \mathscr{D}''$ rather than ∇', ∇'', we regain the desired equality of Laplacians. This idea in fact works in the general context of harmonic flat bundles, leading to a purely holomorphic description of the de Rham groups $H^k(M, E)$ whenever E is a reductive flat complex vector bundle. The cohomology groups of the following complex are canonically isomorphic to the de Rham groups:

$$0 \to \mathscr{A}^0(M, E) \xrightarrow{\mathscr{D}''} \mathscr{A}^1(M, E) \xrightarrow{\mathscr{D}''} \cdots \xrightarrow{\mathscr{D}''} \mathscr{A}^{2n}(M, E) \to 0.$$

One consequence of this fact, due to Goldman and Millson [11], is that, if E is associated to a principal bundle by the adjoint representation of the structure group, then the differential graded Lie algebra $\mathscr{A}^*(M, E)$ is "formal," i.e. is quasi-isomorphic to its cohomology algebra. This is a generalization of the formality of the de Rham complex of a compact Kähler manifold, proved by Deligne, Griffiths, Morgan and Sullivan [8]. Goldman and Millson used this fact to study the local structure of the $X(M, G)$.

3. Nonabelian Dolbeault theory

We now take up the problem of describing nonabelian Dolbeault cohomology. The Bochner formula of Siu and Sampson suggests the correct notion.

DEFINITION 3.1. Let M be a compact Kähler manifold, let $K_{\mathbb{C}}$ be the complexification of the group K from the previous section, and $\mathfrak{p}_{\mathbb{C}}$ the complexification of the K-invariant complement of the Lie algebra of K inside the Lie algebra of G. A Higgs bundle over M is a pair $(P_{\mathbb{C}}', \theta')$, where $P_{\mathbb{C}}'$ is a holomorphic principal $K_{\mathbb{C}}$-bundle over M, and θ' is a holomorphic $(1, 0)$-form on M with values in the vector bundle associated to $P_{\mathbb{C}}'$ by the representation of $K_{\mathbb{C}}$ on $\mathfrak{p}_{\mathbb{C}}$.

This definition was given in the case of curves by Hitchin (the last condition is vacuous in this case); the general version was given by Simpson. As in the case of flat bundles, there is a notion of harmonic metric for Higgs bundles. Suppose h is a metric on $P_{\mathbb{C}}'$: this means a smooth reduction of the structure group from $K_{\mathbb{C}}$ to K. Let D'' be the $\bar{\partial}$-operator which is determined by the holomorphic structure on $P_{\mathbb{C}}'$ or any associated vector bundle. It is widely known that there is a unique connection D on $P_{\mathbb{C}}'$ for which h is parallel and whose $(0, 1)$-component is D''. Similarly, h determines a $(0, 1)$-form with values in the vector bundle associated to $P_{\mathbb{C}}'$ with fiber $\mathfrak{p}_{\mathbb{C}}$ by taking the adjoint of θ'. Adding these together gives a connection

$$\nabla_h = D + \theta' + \theta''$$

on the principal $G_{\mathbb{C}}$-bundle $P_{\mathbb{C}}$ associated to $P'_{\mathbb{C}}$. We will denote the curvature 2-form of ∇_h by F_h; it can be easily checked that this is a $(1, 1)$-form with values in the Lie algebra of K.

It will simplify the discussion if we assume henceforth that the group G is $GL(n, \mathbb{R})$. We denote by E the vector bundle associated to $P'_{\mathbb{C}}$ by the standard representation of $K_{\mathbb{C}}$ on $\mathbb{C}^n$; it is a holomorphic vector bundle with a holomorphic nondegenerate symmetric bilinear pairing. θ' is a holomorphic 1-form taking values in symmetric endomorphisms of E. The pair (E, θ') determines the Higgs bundle. A "metric" in this case is a real structure on E together with an orthogonal structure on the resulting real vector bundle. In this case, the appropriate harmonicity condition for a metric is

$$\Lambda F_h = 0;$$

recall that Λ is the adjoint of exterior multiplication by the Kähler form. This equation is a generalization of the Hermitian-Einstein (or Hermitian-Yang-Mills) equation, and was studied by Hitchin, in the case of Riemann surfaces, and generally by Simpson. As in the case of harmonic metrics for flat vector bundles, harmonic metrics can be found only on certain types of Higgs bundles.

DEFINITION 3.2. A Higgs bundle (of the sort under discussion) is stable if every torsion-free coherent subsheaf $\mathscr{E}'$ of the sheaf $\mathscr{E}$ of holomorphic sections of E which is mapped to $\Omega^1 \otimes \mathscr{E}'$ by θ' satisfies the condition

$$\int_M c_1(\mathscr{E}') \wedge \omega^{n-2} < 0,$$

where $c_1(\mathscr{E}')$ is the first Chern class of $\mathscr{E}'$.

This is a direct imitation of the notion of a stable vector bundle in algebraic geometry, and the introduction of the concept has the same beneficial effects as in the case of harmonic metrics on flat bundles. Namely, we have the following existence result from [16, 26].

THEOREM 3.3. *A Higgs bundle* (E, θ') *admits a Hermitian-Einstein metric if and only if it is a direct sum of stable Higgs bundles, i.e.,* $E = \bigoplus E_i$, *where each* E_i *is a holomorphic subbundle on which the symmetric form restricts to a nondegenerate one, the* E_i's *are mutually orthogonal,* θ' *maps* E_i *to* $\Omega^1 \otimes E_i$, *and each* $(E_i, \theta'|_{E_i})$ *is stable.*

We are really interested in constructing flat connections, so observe that the condition $\mathrm{ch}_2(E) = 0$ implies that

$$\int_M \mathrm{Tr}\, F^2 \wedge \omega^{n-2} = \int_M |F|^2 \omega^n = 0,$$

provided $\Lambda F = 0$. Here, $\mathrm{ch}_2(E)$ is the component in degree four of the Chern character of E.

COROLLARY 3.4. *There is a canonical bijection between irreducible local systems of real vector spaces over a compact Kähler manifold and stable Higgs bundles* (E, θ'), *where* E *is a holomorphic bundle with a nondegenerate holomorphic symmetric inner product and* $\mathrm{ch}_2(E) = 0$, *and* θ' *is a holomorphic* $(1, 0)$-*form with values in symmetric endomorphisms of* E *satisfying* $[\theta', \theta'] = 0$.

This result is the nonabelian analogue of the equivalence between de Rham and Dolbeault notions of harmonicity, and gives two distinct avenues of approach to $X(M, \mathrm{GL}(n, \mathbb{R}))$. A similar statement can be given for any other linear reductive group G.

One of the first applications of this equivalence was Hitchin's description of the moduli spaces $X(S, \mathrm{PSL}(2, \mathbb{R}))$, where S is a compact Riemann surface. This moduli space is partitioned into a finite union of subsets by the Euler class of the topological bundle associated to a representation; it is well known that, for a flat $\mathrm{PSL}(2, \mathbb{R})$-bundle over a surface, the Euler class satisfies $|\chi(P)| \leq 2g - 2$, where g is the genus of the surface. To give a Higgs bundle description of a flat bundle with Euler class χ, we need first a $\mathrm{PSO}(2, \mathbb{C})$-bundle P with Euler class χ. The representation of $K_{\mathbb{C}} = \mathbb{C}^{\times}$ on $\mathfrak{p}_{\mathbb{C}}$ is in this case a direct sum of two irreducible one-dimensional subspaces V and V^{-1}, on which $z \in \mathbb{C}^{\times}$ acts by multiplication by z and z^{-1}, respectively. Hence, θ' is a holomorphic 1-form with values in $L \oplus L^{-1}$, where L is a holomorphic line bundle with Euler class χ, and L^{-1} is its dual. For the sake of convenience, we assume that the degree of L is nonnegative, i.e., that $\chi \geq 0$. We denote the component of θ' with values in L by b, and the component with values in L^{-1} by c.

Since we are dealing with $\mathrm{PSL}(2, \mathbb{R})$ and not its double cover $\mathrm{SL}(2, \mathbb{R})$, we use the adjoint representation rather than the standard two-dimensional representation of $\mathrm{SL}(2, \mathbb{R})$. Define $E_{\mathbb{C}}$ to be $L \oplus \mathscr{O} \oplus L^{-1}$, where $\mathscr{O}$ is the trivial line bundle over S. This bundle has a nondegenerate symmetric pairing, since it is the adjoint bundle of the holomorphic $\mathrm{PSL}(2, \mathbb{C})$-bundle associated to P. Furthermore, θ' takes values in symmetric endomorphisms of $E_{\mathbb{C}}$, using the identification of $E_{\mathbb{C}}$ with a subbundle of $\mathrm{End}(E_{\mathbb{C}})$ via the natural fiberwise Lie bracket on $E_{\mathbb{C}}$. Hence, $(E_{\mathbb{C}}, \theta')$ defines a Higgs bundle.

If L is positive, then $(E_{\mathbb{C}}, \theta')$ is a stable Higgs bundle if and only if L and $L \oplus \mathscr{O}$ are not Higgs subbundles. Hence, the necessary and sufficient condition for stability in this case is that c be nonzero. We have $c \in H^0(S, \Omega^1 \otimes L^{-1}) - \{0\}$; c and L are determined (up to scale, in the case of c) by the divisor of c. The latter is an element in the $(2g - k - 2)$nd symmetric power $\Sigma^{2g-k-2}S$ of S, where k is the degree of L. The set of pairs (L, c) where L has degree k is a $\mathbb{C}^{\times}$-bundle over $\Sigma^{2g-k-2}S$. The Higgs bundle structure is determined by the choice of b, which is an element of $H^0(S, \Omega^1 \otimes L)$. The latter is a vector space of dimension $g + k - 1$, so the space of triples (L, b, c) is parametrized by a complex vector bundle of rank $g + k - 1$ over a $\mathbb{C}^\times$-bundle over $\Sigma^{2g-k-2}S$. However, there is a gauge equivalence between the Higgs bundles defined by (L, b, c) and $(L, \lambda b, \lambda^{-1} c)$, so the set of isomorphism classes of stable Higgs bundles arising from flat $\mathrm{PSL}(2, \mathbb{R})$-bundles with Euler class $k > 0$ is parametrized by a complex vector bundle over $\Sigma^{2g-k-2}S$. By the discussion above, this is equivalent to the set of isomorphism classes of Zariski dense representations giving rise to bundles with Euler class k. Furthermore, it is easy to see that any representation without Zariski dense image must have Euler class zero. Hence, we arrive at the following result of Hitchin's.

THEOREM 3.5. *If S is a surface of genus at least 2, then the subset of $X(S, \mathrm{PSL}(2, \mathbb{R}))$ corrsponding to representations of Euler class $k > 0$ is homeomorphic to a complex vector bundle of rank $g + k - 1$ over $\Sigma^{2g-k+2}S$.*

Notice that the case $k = 2g - 2$ gives a proof of the classical result that Teichmüller space is homeomorphic to $\mathbb{C}^{3g-3}$. (Hitchin [14] has found that a similar component exists when $\mathrm{PSL}(2, \mathbb{R})$ is replaced by a split real form of a simple complex group.) In fact, the existence of a solution to the generalized Hermitian-Yang-Mills equation in this case gives a proof of the uniformization theorem for compact Riemann surfaces of genus at least two: the resulting flat connection and the associated mapping $\widetilde{S} \to \mathrm{PSL}(2, \mathbb{R})/\mathrm{PSO}(2)$ turn out to give the hyperbolic metric in the appropriate conformal class. Simpson [26] gave a much more general uniformization result which characterizes varieties whose universal covering is a bounded symmetric domain; these results overlap with uniformization results coming from Yau's solution of the Calabi conjecture.

We now return to the main thread of our discussion. The Higgs bundle description of $X(M, G)$ allows us to extend the $\mathrm{U}(1)$-action of the previous section to an action of $\mathbb{C}^\times$. This is now quite simple: $z \in \mathbb{C}^\times$ sends an equivalence class of Higgs bundles represented by $(P'_\mathbb{C}, \theta')$ to the class represented by $(P'_\mathbb{C}, z\theta')$. It is easy to see that this action preserves the property of stability, so there is an induced action of $\mathbb{C}^\times$ on $X(M, G)$. This is the nonabelian Hodge structure at which we have been aiming. One of the important properties [16, 28] of this action is that limits as z approaches zero exist, at least when M is projective.

THEOREM 3.6. *Suppose M is a smooth complex projective variety. For any $\rho \in X(M, G)$, $\lim_{z\to 0} z\rho$ exists in $X(M, G)$. In other words, any homomorphism $\pi_1(M) \to G$ can be deformed to the monodromy representation of a polarized variation of Hodge structure.*

Simpson refers to this phenomenon as the "ubiquity of variations of Hodge structure."

4. Polarizations

We take up finally the question of what a polarization should be in the nonabelian setting. To first approximation, the position we take is that it is a naturally occurring hyperkähler metric on $X(M, G_{\mathbb{C}})$.

The deformations of a flat G-bundle P are measured to first order by elements of $H^1(M, \text{ad}(P))$. Given any $\eta \in H^1(M, \text{ad}(P))$, there is a sequence of obstructions in $H^2(M, \text{ad}(P))$ to extending a first-order deformation represented by η to an actual deformation. The first of these obstructions is the cup product $[\eta, \eta] \in H^2(M, \text{ad}(P))$; it is a theorem of Goldman and Millson [11] that, when M is Kähler, the vanishing of this first obstruction implies the vanishing of all further obstructions, so η can be extended to a deformation if and only if $[\eta, \eta] = 0$. If $[\eta, \eta] = 0$ for all $\eta \in H^1(M, \text{ad}(P))$ and P has no nontrivial connection-preserving automorphisms, then $X(M, G)$ is smooth in a neighborhood of the point representing P, and its tangent space there is canonically isomorphic to $H^1(M, \text{ad}(P))$. For simplicity, consider the case where G is semisimple. Using the Killing form $B(\cdot, \cdot)$ on the Lie algebra of G (which we may take to be complex-valued if G is a complex Lie group), we get an antisymmetric pairing Ω: $\wedge^2 H^1(M, \text{ad}(P)) \to \mathbb{R}$ (or $\mathbb{C}$) defined by

$$\Omega(\eta, \xi) = \int_M B(\eta, \xi) \wedge \omega^{n-1}.$$

The Hard Lefschetz theorem together with Poincaré duality implies that this pairing is nondegenerate, so Ω is a (possibly complex) symplectic form on $H^1(M, \text{ad}(P))$.

Let $\widehat{X}(M, G)$ be the open subset of $X(M, G)$ consisting of representations with Zariski dense image for which the cup product pairing on H^1 is trivial. Ω defines a nondegenerate 2-form on $\widehat{X}(M, G)$. We concentrate our attention on some component Y of $\widehat{X}(M, G)$. Ω is in fact a descendent of a closed 2-form defined on the affine space $\mathscr{C}$ of all connections on the smooth principal bundle underlying the flat bundles in Y. The tangent space to $\mathscr{C}$ at any point is isomorphic to $\mathscr{A}^1(M, \text{ad}(P))$, the space of 1-forms with values in $\text{ad}(P)$. As in the case of $H^1(M, \text{ad}(P))$, we can define a nondegenerate antisymmetric pairing $\widetilde{\Omega}$ on $\mathscr{A}^1(M, \text{ad}(P))$ by the formula

$$\widetilde{\Omega}(\eta, \xi) = \int_M B(\eta, \xi) \wedge \omega^{n-1}.$$

This form is independent of the point in $\mathscr{C}$ at which we calculate it, so it is a closed 2-form. Furthermore, it is gauge-invariant, and its restriction to the subvariety of flat connections is the pullback of the form Ω under the natural projection map. Hence, Ω is closed, so is a symplectic form on $\widehat{X}(M, G)$. This is meant to remind the reader of the symplectic pairing on $H^1(M, \mathbb{R})$ (or $H^1(M, \mathbb{C})$) discussed in the first section.

If we fix a metric on P, there is an L^2-metric on $\mathscr{A}^1(M, \mathrm{ad}(P))$, and this induces a Riemannian structure on $\mathscr{C}$ which is invariant under metric-preserving gauge automorphisms. Restricting to the subvariety of flat connections for which the fixed metric is harmonic, we obtain a Riemannian metric which descends to a metric on $\widehat{X}(M, G)$. One can check that the form Ω is parallel with respect to this metric. There are two cases to consider. If G is not a complex group, then Ω is an ordinary symplectic form, and the Riemannian metric is a Kähler metric for which Ω is the Kähler form. If we complexify G, then Ω is a nondegenerate form of type $(2, 0)$ on $\widehat{X}(M, G_{\mathbb{C}})$, and the Riemannian metric is a hyperkähler metric with respect to which Ω is parallel. (Recall that a hyperkähler metric on a manifold of real dimension $4n$ is a metric with holonomy group contained in $\mathrm{Sp}(n)$, the compact real form of the symplectic group.)

The existence of metrics with these special properties was first observed by Hitchin [15, 16] in the case of curves. He recognized further that the symplectic structure on $X(M, G_{\mathbb{C}})$ was natural from another point of view: $X(M, G_{\mathbb{C}})$ is a partial compactification of the cotangent bundle of $X(M, G^c)$, where G^c is the compact real form of $G_{\mathbb{C}}$. Of course, Hitchin's discussion only treated the case where M is a curve, but one can easily extend to the general case. For simplicity, we consider the subset $\widehat{X}(M, G^c)$ consisting of equivalence classes of homomorphisms with Zariski dense image contained in $\widehat{X}(M, G_{\mathbb{C}})$. The tangent space to $\widehat{X}(M, G^c)$ at ρ is the Dolbeault group $H^1(M, \mathrm{ad}(P_\rho) \otimes \mathbb{C})$, where P_ρ is the principal G^c-bundle associated to ρ. Let K be the canonical bundle of M. Using Serre duality, the cotangent space of $\widehat{X}(M, G^c)$ at ρ is canonically isomorphic to $H^{n-1}(M, K \otimes_{\mathbb{C}} \mathrm{ad}(P_\rho))$. On the other hand, using the Hard Lefschetz theorem, we see that the map

$$H^0(M, \Omega^1 \otimes_{\mathbb{C}} \mathrm{ad}(P_\rho)) \to H^{n-1}(M, K \otimes_{\mathbb{C}} \mathrm{ad}(P_\rho))$$

given by wedge product with ω^{n-1} is an isomorphism. Hence, the set of pairs $\{(\rho, \theta') | \rho \in \widehat{X}(M, G^c), \ \theta' \in H^0(M, \Omega^1 \otimes_{\mathbb{C}} \mathrm{ad}(P_\rho))\}$ is isomorphic to the cotangent bundle of $\widehat{X}(M, G^c)$. On the other hand, each element of $\widehat{X}(M, G^c)$ gives rise to a stable holomorphic $G_{\mathbb{C}}$-bundle, so these pairs all give rise to stable Higgs bundles. Hence, there is an open subset of $X(M, G_{\mathbb{C}})$ which is canonically identified with the cotangent bundle of $\widehat{X}(M, G^c)$. It is not difficult to check that the restriction of the symplectic form Ω to this subset is the canonical symplectic form on the cotangent bundle, up to normalization.

It is a remarkable fact, observed by Hitchin, that, when $M = S$ is a Riemann surface of genus greater than one and $G_{\mathbb{C}}$ is one of the classical simple Lie groups, $\widehat{X}(M, G_{\mathbb{C}})$ underlies an algebraic completely integrable system. This means the following. There is a natural algebraic structure on $\widehat{X}(S, G_{\mathbb{C}})$ (one of many!) specified by its identification with a moduli space of Higgs bundles. Suppose that the complex dimension of $\widehat{X}(M, G_{\mathbb{C}})$ is $2N$; it must be even because $\widehat{X}(M, G_{\mathbb{C}})$ is a complex symplectic manifold. A completely integrable system is a collection of N algebraic functions $f_1, \ldots, f_N$ on $\widehat{X}(S, G_{\mathbb{C}})$ which Poisson commute and whose differentials are generically linearly independent on $\widehat{X}(S, G_{\mathbb{C}})$. Hitchin products these functions in the following manner. Let $p_1, \ldots, p_k$ be a minimal set of homogeneous generators for the algebra of invariant polynomials on the Lie algebra of $G_{\mathbb{C}}$. Since S is one-dimensional, we have $\Omega^1 = K$, so each p_i defines a map

$$H^0(S, \mathrm{ad}(P)) \to H^0(S, K^{d_i}),$$

where P is a holomorphic $G_{\mathbb{C}}$-bundle and d_i is the degree of p_i. Hence, we obtain an algebraic mapping

$$p: \widehat{X}(S, G_{\mathbb{C}}) \to \bigoplus_{i=1}^{k} H^0(S, K^{d_i}).$$

The basic facts about this mapping are contained in the following result of Hitchin's [17].

THEOREM 4.1. *The mapping p is surjective. If g_1, g_2 are linear functions on*

$$\bigoplus_{i} H^0(S, K^{d_i}),$$

then the functions $f_1 = g_1 \circ p$, $f_2 = g_2 \circ p$ Poisson commute on $\widehat{X}(S, G)$. Furthermore, $\bigoplus_i H^0(S, K^{d_i})$ is a vector space of dimension N, so the functions which arise in this manner give rise to an algebraic completely integrable system.

The fiber of p over 0 is the moduli space of stable $G_{\mathbb{C}}$ bundles over S. On the other hand, the generic fiber of p is an (open subset of) an abelian variety, which is either the Jacobian or a Prym variety of some Riemann surface covering S. An idea of why covers of S arise and why line bundles on them are relevant may be given as follows (cf. [28]). The main observation is that a Higgs bundle, given as a holomorphic vector bundle E on S together with a holomorphic 1-form $\theta' \in H^0(S, K \otimes \mathrm{End}(E))$, determines a sheaf $\widehat{\mathscr{E}}$ on the cotangent bundle T^*S of S. To describe $\widehat{\mathscr{E}}$, we must recall what the structure sheaf $\mathscr{O}_{T^*S}$ of T^*S looks like. On the inverse image V in T^*S of any proper open subset U of S, $\mathscr{O}_{T^*S}(V)$ is a polynomial algebra over $\mathscr{O}_S(U)$ generated by an element q, corresponding to some nonvanishing vector field X on U. E determines a module $\mathscr{E}(U)$ over each $\mathscr{O}_S(U)$. The Higgs bundle structure gives a canonical way of extending $\mathscr{E}(U)$ to an $\mathscr{O}_{T^*S}(V)$-module: let $q \in \mathscr{O}_{T^*S}(V)$ act on elements of $\mathscr{E}(U)$ by the endomorphism $\theta'(X)$. The resulting family of $\mathscr{O}_{T^*S}(V)$-modules determines the sheaf $\widehat{\mathscr{E}}$. Conversely, one can reconstruct the Higgs bundle from the sheaf $\widehat{\mathscr{E}}$. The support of $\widehat{\mathscr{E}}$ is the subvariety

$$\det(\lambda - \theta')^{-1}(0) \subset T^*S,$$

where λ is a point in T^*S, regarded as taking values in constant endomorphisms of E, and det is the natural determinant map

$$T^*S \otimes \operatorname{End}(E) \to (T^*S)^r,$$

in case E has rank r. The set on which $\widehat{\mathscr{E}}$ is supported is always one-dimensional, and surjects onto S under the natural projection from T^*S to S. In the generic case, it is a covering of degree r (the generic $r \times r$ matrix has r distinct eigenvalues), and is a smooth Riemann surface. In this event, $\widehat{\mathscr{E}}$ comes from a line bundle over this covering.

These ideas have intriguing points of contact with Witten's approach to the Jones polynomial (see [13]) and the Langlands' conjecture for groups over function fields [23, 24].

References

1. J. Carlson and L. Hernandez, *Harmonic mappings of Kähler manifolds to exceptional hyperbolic spaces*, Preprint.
2. J. Carlson and D. Toledo, *Harmonic mappings of Kähler manifolds to locally symmetric spaces*, Inst. Hautes Études Sci. Publ. Math. **69** (1989), 173–201.
3. E. Cattani, A. Kaplan, and W. Schmid, L^2 *and intersection cohomologies for a polarizable variation of Hodge structure*, Invent. Math. **87** (1987), 217–252.
4. K. Corlette, *Archimedean superrigidity and hyperbolic geometry*, Ann. of Math. **135** (1992), 165–182.
5. ____, *Flat G-bundles with canonical metrics*, J. Differential Geom. **28** (1988), 361–382.
6. ____, *Rigid representations of Kählerian fundamental groups*, J. Differential Geom. **33** (1991), 239–252.
7. P. Deligne, *Travaux de Griffiths*, Séminaire Bourbaki no. 376, 1969/70.
8. P. Deligne, P. Griffiths, J. Morgan, and D. Sullivan, *Real homotopy theory of Kähler manifolds*, Invent. Math. **29** (1975), 245–274.
9. K. Diederich and T. Ohsawa, *Harmonic mappings and disc bundles over compact Kähler manifolds*, Publ. Res. Inst. Math. Sci. **21** (1985), 819–833.
10. S. K. Donaldson, *Twisted harmonic maps and the self-duality equations*, Proc. London Math. Soc. **55** (1987), 127–131.
11. W. Goldman and J. Millson, *Deformation theory of representations of fundamental groups of compact Kähler manifolds*, Inst. Hautes Études Sci. Publ. Math. **67** (1988), 43–96.
12. M. Gromov, *Sur le groupe fondamental d'une variété kählérienne*, C. R. Acad. Sci. Paris Sér. I Math. **308** (1989), 67–70.
13. N. Hitchin, *Flat connections and geometric quantization*, Comm. Math. Phys. **131** (1990), 347–380.
14. ____, *Lie groups and Teichmüller space*, Preprint.
15. ____, *Metrics on moduli spaces*, The Lefschetz Centennial Conference, Part I (D. Sundararaman, ed.), Contemp. Math., no. 58, Amer. Math. Soc., Providence, RI, 1986, pp. 157–178.
16. ____, *The self-duality equations of a Riemann surface*, Proc. London Math. Soc. **55** (1987), 59–126.
17. ____, *Stable bundles and integrable systems*, Duke Math. J. **54** (1987), 91–114.
18. F. Johnson and E. Rees, *On the fundamental group of a complex algebraic manifold*, Bull. London Math. Soc. **19** (1987), 463–466.
19. J. Jost and S.-T. Yau, *Harmonic maps and group representations*, Preprint.

20. M. Kashiwara and T. Kawai, *The Poincaré lemma for a variation of polarized Hodge structure*, Proc. Japan Acad. Ser. A Math. Sci. **61** (1985), 164–167.
21. F. Labourie, *Existence d'applications harmoniques tordues à valeurs dans les variétés à courbure négative*, Proc. AMS **111** (1991), 877–882.
22. G. Laumon, *Correspondance de Langlands géométrique pour les corps de fonctions*, Duke Math. J. **54** (1987), 309–359.
23. G. Laumon, *Un analogue global du cône nilpotent*, Duke Math. J. **57** (1988), 647–671.
24. J. Morgan, *The algebraic topology of smooth algebraic varieties*, Inst. Hautes Études Sci. Publ. Math. **48** (1978), 137–204.
25. J. Sampson, *Applications of harmonic maps to Kähler geometry*, Contemp. Math. no. 49, Amer. Math. Soc., Providence, RI, 1986, pp. 125–133.
26. C. Simpson, *Constructing variations of Hodge structure using Yang-Mills theory and applications to uniformization*, J. Amer. Math. Soc. **1** (1988), 867–918.
27. ____, *Higgs bundles and local systems*, Preprint.
28. ____, *Moduli of representations of the fundamental group of a smooth projective variety*, Preprint.
29. ____, *Nonabelian Hodge theory*, talk at Kyoto ICM (1990).
30. Y.-T. Siu, *The complex-analyticity of harmonic maps and the strong rigidity of compact Kähler manifolds*, Ann. of Math. (2) **112** (1980), 73–112.
31. S. Zucker, *Hodge theory with degenerating coefficients*: L_2 *cohomology in the Poincaré metric*, Ann. of Math. (2) **109** (1979), 415–476.

UNIVERSITY OF CHICAGO

Proceedings of Symposia in Pure Mathematics
Volume **54** (1993), Part 2

Geometric Invariants and Their Adiabatic Limits

XIANZHE DAI

1. Introduction

Analytic interpretation of a topological invariant often leads to deeper understanding of the invariant. A classical example is the Atiyah-Singer index theorem. Such analytic interpretation usually involves the spectrum of a (natural) elliptic operator, as do many geometric invariants (e.g., the eta invariant and the Ray-Singer analytic torsion). When more than just 0-spectrum is involved, the study of these geometric invariants becomes analytically difficult. Recent development shows that the method of adiabatic limit provides a useful and often powerful tool.

The adiabatic limit refers to the geometric process of blowing up a metric in some directions while leaving it fixed in the others. Equivalently this is the same as shrinking the metric in the others since geometric invariants often have simple scaling properties. The situation appears naturally when one considers an open manifold, e.g., a locally symmetric space of finite volume. In fact, the work of Atiyah-Donnelly-Singer [**ADS**] has a natural interpretation in this context, namely, as relating the adiabatic limit of η-invarint (spectral invariant) with the value of L-function (number-theoretic quantity). In this respect the recent works of Müller [**Mü1, Mü2**] and Stern [**S**] are also closely related.

The original motivation, however, comes from Witten [**W**] who relates the adiabatic limit of η-invariant with the global anomaly, i.e., the holonomy of determinant line bundle. Witten's work has been given full mathematical treatment by Bismut-Freed [**BF**] and Cheeger [**C2**] and is extended by Bismut-Cheeger to the more general case [**BC2**]. Using this generalization, Bismut-Cheeger [**BC6**] were able to extend the work of [**ADS**].

On the analytic side, the study of adiabatic limit also presents a very interesting problem, i.e., a degenerating elliptic problem. By employing a general

1991 *Mathematics Subject Classification.* Primary 58Gxx.
Supported in part by the National Science Foundation.
The detailed version of this paper will be submitted for publication elsewhere.

theory developed by Melrose et al., Mazzeo and Melrose [MM] constructed a pseudodifferential calculus well adapted to the degeneracy. Very recently, Epstein and Melrose [EM] applied successfully the adiabatic limit to solve an outstanding conjecture on the Toeplitz correspondence of Boutet de Monvel and Guillemin.

A very interesting variant of adiabatic limit applies although not restricted to manifold with boundary; cf. [Si], see also [DW]. Here the boundary of a manifold is being "pushed" (so to speak) to the infinity. In the end, one is often able to separate the contribution from the interior of the manifold with that of the boundary. The study of this kind of adiabatic limit is intimately related to the problem of the behavior of a geometric invariant under cutting and pasting. It is also related to Taubes's program of computing the Donaldson invariants.

In this paper, we will discuss our work on the adiabatic limit of η-invariant and our joint work with Melrose on analytic torsion. The rest of the paper is organized as follows:

§2. η-invariant and analytic torsion,
§3. adiabatic limit,
§4. ideas of proof,
§5. Leray spectral sequence.

2. η-invariant and analytic torsion

To an odd-dimensional closed oriented Riemannian manifold M^n, one can associate two very interesting and closely related geometric invariants, the η invariant of Atiyah-Patodi-Singer [APS] and the analytic torsion of Ray-Singer [RS]. Let $\rho: \pi_1(M) \to U(k)$ be a unitary representation. It gives rise to a hermitian flat bundle $\xi = \widetilde{M} \times_\rho C^k$. The operator

$$A = \omega(d + \delta): \Lambda^* M \otimes \xi \to \Lambda^* M \otimes \xi,$$

where $\omega|_{\Lambda^p M \otimes \xi} = i^{(n+1)/2+p(p-1)} *$, is called the signature operator on the odd-dimensional manifold M. One has $A^2 = \Delta$, the (twisted) Hodge Laplacian. The η-function is defined as

$$\eta_A(s) = \frac{1}{\Gamma(s + \frac{1}{2})} \int_0^{+\infty} t^{(s-1)/2} \mathrm{tr}(Ae^{-tA^2})\, dt$$

$$= \sum_{\lambda \in \mathrm{spec}\, A,\ \lambda \neq 0} \frac{\mathrm{sign}\,\lambda}{|\lambda|^s}.$$

This is well defined for $\mathrm{Re}\, s \gg 0$, and it admits a meromorphic continuation to C with $s = 0$ a regular point. The η-invariant of A is defined to be

$$\eta(A) = \eta_A(0).$$

Thus formally $\eta(A) = \#\{\text{positive eigenvalues}\} - \#\{\text{negative eigenvalues}\}$ measures the spectral asymmetry.

To introduce the analytic torsion, one needs the so-called number operator N,

$$N|_{\Lambda^p M \otimes \xi} = p,$$

i.e., multiplying each (homogeneous) differential form by its degree. Let tr_s denote the supertrace associated with the natural $\mathbf{Z}_2$-grading. Define the supersymmetric zeta function

$$\zeta_T(s) = \frac{1}{\Gamma(s)} \int_0^{+\infty} t^{s-1} \text{tr}_s(Ne^{-t\underline{\Delta}})\,dt,$$

where $\underline{\Delta} = \Delta|_{(\ker \Delta)^\perp}$. Once again this is well defined for $\text{Re}\,s \gg 0$ and admits a meromorphic continuation to C with $s = 0$ a regular point. The analytic torsion $T_\rho(M)$ is defined so that

$$\ln T_\rho(M) = \zeta_T'(0).$$

Formally $T_\rho(M)$ is a ratio of determinants:

$$T_\rho(M) = \frac{(\det \Delta_1)(\det \Delta_3)^3 \cdots}{(\det \Delta_2)^2 (\det \Delta_4)^4 \cdots},$$

where $\Delta_p = \Delta|_{\Lambda^p M \otimes \xi}$.

REMARK. The analytic torsion defined here is the square of the original one. This applies also to the others discussed below.

The significance of η invariant lies in the fact that it gives the boundary correction term to the index theory on manifold with boundary.

THEOREM 2.1 [APS]. *Let N^{n+1} be a compact oriented manifold with $\partial N = M$ and $\rho: \pi_1(N) \to U(k)$ a representation. Denote by $\text{sign}_\rho(N)$ the signature of N in the local coefficient system ρ. Then*

$$\text{sign}_\rho(N) = \int_N \mathscr{L} - \frac{1}{2}\eta(A).$$

Here $\mathscr{L}$ is the (modified) Hirzebruch polynomial, evaluated via the Chern-Weil homomorphism, with the curvature tensor coming from any Riemannian metric on N which is of product type near the boundary M. M has the induced metric.

For analytic torsion, the main interest is that it is the analytic counterpart of the Reidemeister torsion $\tau_\rho(M)$. In fact, this is the original motivation of Ray-Singer for constructing the analytic torsion. Ray-Singer conjectured that their analytic torsion should be equal to the Reidemeister torsion, which is subsequently proven by Cheeger [C1] and Müller [Mü3] independently.

THEOREM 2.2 [C1, Mü3]. *The analytic torsion equals the Reidemeister torsion:*

$$T_\rho(M) = \tau_\rho(M).$$

We remark that when ρ is acyclic, i.e., $H^*(M, \rho) = 0$, $\tau_\rho(M)$ is a simple homotopy invariant, hence a topological invariant.

The analytic torsion has been used by D. Fried to study the question of "counting closed orbits" in dynamical system [F1]. Recently, Quillen adopted Ray-Singer's construction to give a metric on the determinant line bundle of a family of elliptic operators. Here the analytic torsion appears as the weighting factor. Quillen's idea had a dramatic impact; see the work of Bismut-Freed [BF], Bismut-Gillet-Soulé [BGS] and Bismut-Lebeau [BL].

We conclude this section by making a comparison of the two invariants concerning the small time asymptotic expansions of the heat kernels involved.

THEOREM 2.3 [BF]. *For small time $t > 0$, one has*

$$\text{tr}(Ae^{-tA^2}) = O(t^{1/2}).$$

Consequently,

$$\eta(A) = \frac{1}{\sqrt{\pi}} \int_0^\infty t^{-1/2} \text{tr}(Ae^{-tA^2})\, dt.$$

Thus for η-invariant, the small time asymptotic expansion of the heat kernel involved is very well behaved: there is no singular term. And we have a very nice heat kernel representation for η-invariant. For analytic torsion, things are not so nice.

THEOREM 2.4 [DM]. *For small time $t > 0$, we have a pointwise asymptotic expansion*

$$(2.1) \qquad\qquad \text{tr}_s(Ne^{-t\Delta}) = a_{-1/2}t^{-1/2} + O(t^{1/2}),$$

where $a_{-1/2}$ is the top Lipschitz-Killing curvature:

$$a_{-1/2} = c(n) \sum_{k=1}^{n} (-1)^k \text{Pf}(\Omega_k) \wedge \omega_k.$$

Here Ω_k denotes the matrix obtained by deleting the kth row and column of the curvature matrix in a local orthonormal basis for which $\{\omega_k\}$ is a dual basis and $\text{Pf}(\Omega_k)$ denotes the Pfaffian of Ω_k. Hence for any $\delta > 0$,

$$\ln T_\rho(M) = -2\delta^{-1/2} \int_M a_{-1/2} - (c + \ln\delta)\chi_2(M, \rho)$$

$$+ \int_0^\delta t^{-1}(\text{tr}_s(Ne^{-t\Delta}) - a_{-1/2}t^{-1/2})\, dt$$

$$+ \int_\delta^\infty t^{-1}\text{tr}_s(Ne^{-t\Delta})\, dt,$$

where $\chi_2(M, \rho) = \sum_{p=1}^{n}(-1)^p pb_p$, $b_p = \dim H^p(M, \rho)$ and c is the Euler constant.

The singular term in the small time asymptotic expansion for analytic torsion presents new difficulty in our study of its adiabatic limit.

REMARK. After we proved (2.1) Prof. Gilkey pointed out that essentially the same result was proven in [GS]. We would like to thank him for the reference and also for kindly communicating to us his own proof by his invariance theory. Our proof extends to the more general situation where partial number operator is allowed; see §4. This generalization is in fact very important in our study of the adiabatic limit of analytic torsion.

3. Adiabatic limit

The first nontrivial case to consider is when M^n has a fibration structure; i.e., M is fibered smoothly over a closed manifold B with typical fiber a closed manifold Y:

$$Y \to M \xrightarrow{\pi} B.$$

If $g = \pi^* g_B + g_Y$ is a submersion metric, then

$$g_x = x^{-2} \pi^* g_B + g_Y$$

is a family of metrics obtained by blowing up along the base direction (when $x \downarrow 0$).

In [W] Witten considers an odd-dimensional manifold M that fibers over the circle S^1. There is on the even-dimensional fibers a family of elliptic operators A_Y^+, namely the usual signature operators. It defines a line bundle over S^1, called the determinant line bundle:

$$\det A_Y^+ = (\Lambda^{\text{top}} \ker A_Y^-)^* \otimes (\Lambda^{\text{top}} \ker A_Y^+).$$

Physicists are interested in the curvature and the holonomy of this line bundle (the local and global anomaly). The local family index theorem [B] yields the curvature. Witten gave a heuristic argument leading to

THEOREM 3.1 [BF, C2]. *If* hol$(\det A_Y^+)$ *denotes the holonomy of the line bundle* $\det A_Y^+$ *and* A_x *denotes the signature operator on* M *equipped with the blowing up metric* g_x, *then*

$$\lim_{x \to 0} \eta(A_x) = \text{hol}(\det A_Y^+).$$

This has been extended to the case of higher-dimensional base space by Bismut-Cheeger [BC2]. The study of adiabatic limit in this case to a higher-dimensional analog of the η-invariant, the $\tilde{\eta}$ form. This is a canonically constructured differential form on the base B. Its definition involves Bismut's Levi-Civita superconnection [B], which we recall briefly.

The metric g determines a splitting

$$TM = T^{\text{V}}M \oplus T^{\text{H}}M,$$

with $T^{\text{V}}M$ the vertical bundle. The projections of the Levi-Civita connection ∇^L onto $T^{\text{V}}M$ and $T^{\text{H}}M$ define connections on these bundles. Let ∇ denote the direct sum connection of the two connections thus obtained. Also let $m(\cdot)$ be the mean curvature operator of the fibers and $T(\cdot, \cdot)$ the curvature operator of the fibration; namely, if $\{e_i\}$ is a local orthonormal basis of $T^V M$ and $\{f_\alpha\}$ is a local orthonormal basis of TB, then

$$m(f_\alpha^H) = \sum \langle \nabla_{e_i}^L e_i, f_\alpha^H \rangle,$$
$$T(f_\alpha^H, f_\beta^H) = -[f_\alpha^H, f_\beta^H]^V,$$

where f_α^H denotes the horizontal lift of f_α and $(\ )^V$ indicates taking the vertical component. Bismut's Levi-Civita superconnection is

$$B = f_\alpha^H \wedge \left(\nabla_{f_\alpha^H} - \frac{1}{2} m(f_\alpha^H) \right) + A_Y - \frac{1}{8} \langle T(f_\alpha^H, f_\beta^H), e_i \rangle f_\alpha^H \wedge f_\beta^H \wedge c(e_i)$$
$$\stackrel{\text{def}}{=} \widetilde{\nabla}^u + A_Y - \frac{c(T)}{4},$$

where $A_Y = c(e_i)\nabla_{e_i}$ is the family of signature operators along the fibers and $c(e_i) = e_i \wedge - i_{e_i}$ the Clifford multiplication.

The rescaled Bismut superconnection is

$$(3.2) \qquad\qquad B_t = \widetilde{\nabla}^u + t^{1/2} A_Y - \frac{c(T)}{4t^{1/2}}.$$

The basic property here is that the curvature B_t^2 is a smooth family of fiberwise differential operators with the leading part the fiberwise Laplacian. Hence its heat kernel $e^{-B_t^2}$ is a smooth family of fiberwise smoothing operators.

The $\hat{\eta}$ form is defined as

$$\hat{\eta} = \begin{cases} \frac{1}{2\sqrt{\pi}} \int_0^\infty t^{-1/2} \text{tr}_s \left[\left(A_Y + \frac{c(T)}{4t} \right) e^{-B_t^2} \right] dt & \text{if } \dim Y \text{ even}, \\ \frac{1}{2\sqrt{\pi}} \int_0^\infty t^{-1/2} \text{tr}^{\text{even}} \left[\left(A_Y + \frac{c(T)}{4t} \right) e^{-B_t^2} \right] dt & \text{if } \dim Y \text{ odd}. \end{cases}$$

Here tr_s denotes the fiberwise supertrace and tr^{even} indicates taking the even-form part of the fiberwise trace. Thus $\hat{\eta}$ is an odd (even) differential form on the base B when $\dim Y$ is even (odd). The act that $\hat{\eta}$ is well defined depends on some remarkable cancellation properties of Bismut's superconnection. For example

$$\text{tr}_s \left[\left(A_Y + \frac{c(T)}{4t} \right) e^{-B_t^2} \right] = O(t^{1/2}) \quad \text{as } t \to 0,$$

and

$$\text{tr}_s \left[\left(A_Y + \frac{c(T)}{4t} \right) e^{-B_t^2} \right] = O(t^{-1}) \quad \text{as } t \to \infty.$$

We refer to [**BC2, BGV, D**] for detail.

We renormalize $\hat{\eta}$ by multiplying a p-form by $(\frac{1}{2\pi i})^{[(p+1)/2]}$. Denote the resulting form by $\tilde{\eta}$. Note that the zero-form part of $\tilde{\eta}$ is exactly the η-invariant of A_Y; and the 1-form part of $\tilde{\eta}$ is Witten's covariant anomaly (i.e., the connection 1-form of $\det A_Y^+$). More importantly, this canonical differential form plays exactly the role of η-invariant in Bismut-Cheeger's

Families Index Theorem for manifolds with boundary [**BC3, BC4**]. Also in a recent work J. Lott studied a noncommutative analog of $\tilde{\eta}$ form [**L**].

THEOREM 3.2 [**BC2**]. *Assume that* $H^*(Y, \rho|_Y) = 0$. *Then*

$$\lim_{x \to 0} \eta(A_x) = 2 \int_B \mathscr{L}\left(\frac{R^B}{2\pi}\right) \wedge \tilde{\eta}.$$

The most general case is treated in [**D**]. In particular, we have

THEOREM 3.3 [**D**]. *Without the acyclicity,*

$$(3.3) \qquad \lim_{x \to 0} \eta(A_x) = 2 \int_B \mathscr{L}\left(\frac{R^B}{2\pi}\right) \wedge \tilde{\eta} + \eta(A_B \otimes \mathscr{H}^*(Y, \rho|_Y)) + \sum_{r \geq 2} \sigma_r,$$

where σ_r' *s are topological invariants computable from the Leray spectral sequence; see* §5.

Recently, in collaboration with R. Melrose, we are able to prove an analog of the above for Ray-Singer's analytic torsion. Let (E_r, d_r) denote the E_r-term of the Leray spectral sequence and τ_r its torsion (see §5). For a Z-graded vector space $V = \bigoplus_i V^i$, define $\chi_2(V) = \sum(-1)^i i \dim V^i$.

THEOREM 3.4 [**DM**]. *For small* $x > 0$,

$$\ln T_x(M, \rho) = -2\ln x \left[\sum_{r \geq 2} r(\chi_2(E_r) - \chi_2(E_{r+1})) - \chi(Y)\chi_2(B, \rho)\right]$$

$$+ \chi(B)\ln T(Y, \rho) + \ln T(B, \mathscr{H}^*(Y, \rho)) + \sum_{r \geq 2} \ln \tau_r + O(x^\alpha),$$

where α *is some positive number and* $T(B, \mathscr{H}^*(Y, \rho))$ *denotes the analytic torsion of* B *associated to the representation of* $\pi_1(B)$ *on* $H^*(Y, \rho)$.

REMARK. Under certain acyclicity conditions this formula reduces to some purely topological formulae of D. Fried concerning the Reidemeister torsion [**F1**]. Recently Prof. Fried informed us that he could prove a similar formula for the Reidemeister torsion [**F2**].

The author is very grateful to Jeff Cheeger for several discussions that occurred before the preparation of our work [**DM**] concerning the torsion of Leray spectral sequence.

4. Ideas of proof

The starting point for the proof of Theorem 3.3 is the work of Bismut-Cheeger.

THEOREM 4.1 [**BC2**]. *There exists a large positive constant* K *such that for small* x,

$$\mathrm{tr}(A_x e^{-tA_x^2}) = 2\sqrt{\pi} \int_B \mathscr{L}\left(\frac{R^B}{2\pi}\right) \wedge \tilde{\eta} + O(x(1 + t^K)).$$

Note that the error term grows with large t. In the acyclic case, Bismut-Cheeger showed that there is a uniform positive lower bound on the first eigenvalues of A_x^2. Hence the large time contribution is shown to be negligible. In the general case, however, there are infinitely many eigenvalues of A_x^2 that approach zero. This is in fact the essential difficulty here. Thus our first result concerns the behavior of the eigenvalues of A_x as $x \downarrow 0$.

THEOREM 4.2 [D]. *For $x > 0$ the eigenvalues of A_x depend analytically on x. Thus there are (countably many) analytic functions λ_x such that $\mathrm{spec}(A_x) = \{\lambda_x\}$ for all $x > 0$. Moreover,*

(1) *there exists a positive constant λ_0 such that for small x, either*
$$|\lambda_x| \geq \lambda_0 > 0,$$
or λ_x has a complete asymptotic expansion at $x = 0$,

(4.4) $$\lambda_x \sim \lambda_1 x + \lambda_2 x^2 + \cdots ;$$

(2) *there is a constant C (> 0) such that if (4.4) holds and $\lambda_1 \neq 0$, then*
$$\lambda_x = \lambda_1 x + x^2 C(x)\lambda_1^2,$$
with $|C(x)| \leq C$ uniformly bounded;

(3) *the number of eigenvalues that decay at least like x^2 (i.e., $\lambda_x \sim \lambda_2 x^2 + \cdots$) is finite and equal to the dimension of the E_2-term of the Leray spectral sequence of the fibration. Moreover the number of eigenvalues that decay at least like x^r $(r \geq 2)$ is equal to the dimension of the E_r-term of the Leray spectral sequence. Furthermore the leading coefficients are determined by the Leray spectral sequence (see §5 for a precise statement).*

The proof of this result made essential use of the Melrose theory on the degenerate elliptic problems [M1, M2] and the perturbation theory. In particular the construction of Mazzeo-Melrose [MM] is very important in showing that the resolvent family of $\frac{1}{4}A_x$ is a smooth family of L^2-operators down to $x = 0$. We would like to emphasize that the third part of the above theorem should be viewed as a Hodge type theorem relating the spectrum of the (rescaled) Laplacian to the global topology of the fibration.

Applying Theorem 4.2 and the finite propagation speed technique of [CGT], we obtain a corresponding result about the heat kernel in the adiabatic limit.

THEOREM 4.3 [D]. *There are positive constants C, C' and K sufficiently large such that*
$$\left| \mathrm{tr}' \left(\frac{1}{x} A_x e^{-t(A_x/x)^2} \right) - \mathrm{tr}(A_0 e^{-tA_0})^2 \right| \leq \frac{Cx}{t^K} e^{-C't},$$
where tr' indicates omitting the contributions from the eigenvalues that decay at least quadratically. And $A_0 = A_B \otimes \mathscr{H}^*(Y, \rho)$.

In order to combine Theorem 4.1 and Theorem 4.3 we need a result on the uniform asymptotic expansion of $\mathrm{tr}(\frac{1}{x}A_x e^{-t(A_x/x)^2})$. This is done by constructing a suitable parametrix. We remark that a stronger result can be obtained via Melrose's blow-up technique.

Theorem 4.3 also indicates that we should deal separately with the eigenvalues that decay at least quadratically in the large time analysis. Indeed by Theorem 4.2 they are finite in number and can be dealt with by a simple Mellin transform. They give rise to the Leray spectral sequence contribution in the adiabatic limit formula; see §5.

In fact the same analysis leads to the corresponding term in the case of analytic torsion. For the other two terms, however, new difficulties arise from the "incomplete" cancellation as presented in Theorem 2.4. Fortunately we find that Melrose's blow-up technique enables us to combine cancellation results and estimates to obtain optimum results on uniform asymptotic expansions. We believe that this is very interesting and applicable to other situations.

Let us now briefly indicate the ideas behind the proof of Theorem 3.4. Motivated by the case of product fibration we split the supersymmetric zeta function into a "vertical" part and a "horizontal" part together with the contribution from the eigenvalues that decay at least quadratically,

$$\zeta_{T,x}(s) = \zeta_{1,x}(s) + x^{-2s}\zeta_{2,x}(s) + \zeta_{3,x}(s),$$

where

$$\zeta_{1,x}(s) = \frac{1}{\Gamma(s)} \int_0^{+\infty} t^{s-1}\mathrm{tr}_s(N_Y e^{-t\underline{\Delta}_x})\,dt,$$

$$\zeta_{2,x}(s) = \frac{1}{\Gamma(s)} \int_0^{+\infty} t^{s-1}\mathrm{tr}_s(N_B e^{-(t/x^2)\underline{\Delta}_x})\,dt,$$

Here $\underline{\Delta}_x$ denotes the Laplacian Δ_x restricted to the orthocomplement of the eigenspaces associated to the very small eigenvalues. And N_Y, N_B are the partial number operators referred to previously, which count only the vertical and horizontal degrees respectively.

PROPOSITION 4.4. *There exists* $\alpha > 0$ *such that*

$$\zeta'_{1,x}(0) = \chi(B)\zeta'_{T,Y}(0) + O(x^\alpha),$$

where $\zeta_{T,Y}(s) = \frac{1}{\Gamma(s)} \int_0^{+\infty} t^{s-1}\mathrm{tr}_s(N_Y e^{-t\Delta_Y})\,dt$.

PROPOSITION 4.5. *There exists* $\alpha > 0$ *such that*

$$\zeta'_{2,x}(0) = \zeta'_{T,B}(0) + O(x^\alpha),$$

where $\zeta_{B,Y}(s) = \frac{1}{\Gamma(s)} \int_0^{+\infty} t^{s-1}\mathrm{tr}_s(N_B e^{-t\Delta_0^2})\,dt$.

Here in the proofs of the two propositions essential use has been made of Melrose's blow-up technique. It furnishes us with some uniform behavior of the "adiabatic heat kernel" on a blow-up space. However, what we need is uniform behavior on the original (blow-down) space. To pass from the blow-up space to the original space certain cancellation results (i.e., vanishing at the front face) have to be shown. Details will appear in [**DM**].

5. Leray spectral sequence

Consider the fibration

$$Y \to M^n \xrightarrow{\pi} B$$

and the local coefficient system $\rho: \pi_1(M) \to U(k)$. Let $(E_r(\rho), d_r)$ be the associated Leray spectral sequence. Thus

$$E_2(\rho) = H^*(B, \mathscr{H}^*(Y, \rho)),$$
$$E_\infty(\rho) = H^*(M, \rho),$$
$$H(E_r(\rho), d_r) = E_{r+1}(\rho).$$

If $\rho: \pi_1(M) \to U(1)$ is the trivial representation, we will denote $E_r(\rho)$ just by E_r.

Each $E_r(\rho)$ comes with a natural Z-grading and a graded multiplicative structure:

$$E_r(\rho) = \bigoplus_{k=0}^{n} E_r^k(\rho),$$

and

$$E_r^k(\rho) \otimes E_r^l(\rho) \to E_r^{k+l}.$$

When both the vertical bundle and the base are oriented, the orientations give rise to an isomorphism:

$$E_2^n \cong C,$$

with a canonical basis $\xi_2 \in E_2^n$. By a construction of [**CHS**], this yields a canonical basis $\xi_r \in E_r^n \cong C$ for each $r \geq 2$. Using these bases, we can define for each $r \geq 2$ a pairing

$$E_r^k(\rho) \otimes E_r^{n-k-1}(\rho) \to C,$$
$$\varphi \otimes \psi \to \langle \varphi \cdot d_r \psi, \xi_r \rangle.$$

One readily checks that its restriction to $E_r^{(n-1)/2}(\rho)$ is symmetric when $n = 4m - 1$, skew-symmetric when $n = 4m - 3$. Denote by σ_r the signature of this restriction (if skew-symmetric take the corresponding symmetric one). Then the topological invariant σ appearing in the adiabatic limit formula for η-invariant (Theorem 3.3) is

$$\sigma = \sum_{r \geq 2} \sigma_r.$$

Since the Leray spectral sequence degenerates at finite steps, this is a finite sum.

To see how this invariant comes in, we need the following more precise statement of part (3) of Theorem 4.2.

THEOREM 5.1. *Let* $E_r =$ *the limit of space spanned by* λ_x-*eigenforms associated to eigenvalues* λ_x *such that* λ_x *is* $O(x^r)$ $(r \geq 2)$ *in the adiabatic limit. Then* $(E_r, x^{-r}d)$ *forms a spectral sequence which is isomorphic to the Leray spectral sequence of the fibration. Moreover, the* $*$ *map induced by the metric* g_x *gives rise to the duality map.*

It follows from this theorem that the leading coefficients of the very small eigenvalues are determined by the differentials d_r together with the duality operator $*_r$ defined by

$$E_r^k(\rho) \otimes E_r^{n-k}(\rho) \to C,$$

$$\varphi \otimes \psi \to \langle \varphi \cdot \psi, \xi_r \rangle.$$

Namely, if $\lambda_x \sim \lambda_r x^r + \cdots$ then λ_r is an eigenvalue of $*_r d_r + d_r *_r$, and every such eigenvalue appears as a leading coefficient.

To define the torsion τ_r of $E_r(\rho)$, let

$$\Delta^{(r)} = (*_r d_r + d_r *_r)^2,$$

and $\Delta_k^{(r)} = \Delta^{(r)}|_{E_r^k(\rho)}$. Then

$$\tau_r = \frac{(\det' \Delta_1^{(r)})(\det' \Delta_3^{(r)})^3 \cdots}{(\det \Delta_2^{(r)})^2 (\det \Delta_4^{(r)})^4 \cdots},$$

where $\det' \Delta_k^{(r)} = \det' \Delta_k^{(r)} = \det(\Delta_k^{(r)}|_{[\ker \Delta_k^{(r)}]^\perp})$. Note that τ_r depends only on the choice of the orientations on the vertical bundle $T^V M$ and the base B.

REFERENCES

[ADS] M. F. Atiyah, H. Donnelly, and I. M. Singer, *Eta invariants, signature defect of cusps and values of L-functions*, Ann. of Math. (2) **118** (1983), 131–177.

[APS] M. F. Atiyah, V. K. Patodi, and I. M. Singer, *Spectral asymmetry and Riemannian geometry*, I, Math. Proc. Cambridge Philos. Soc. **77** (1975), 43–69.

[BGV] N. Berline, E. Getzler, and M. Vergne, *Heat kernels and Dirac operators*, manuscript, 1990.

[B] J.-M. Bismut, *The Atiyah-Singer index theorem for families of Dirac operators: two heat equation proofs*, Invent. Math. **83** (1986), 91–151.

[BC1] J.-M. Bismut and J. Cheegar, *Invariants eta et indice des familles pour varietes a bord*, C. R. Acad. Sci. Paris Sér. I Math. **305** (1987), 127–130.

[BC2] ——, *η-invariants and their adiabatic limits*, J. Amer. Math. Soc. **2** (1989), 33–70.

[BC3] ——, *Families index for manifolds with boundary, superconnections and cones*. I, J. Funct. Anal. **89** (1990), 313–363.

[BC4] ——, *Families index for manifolds with boundary, superconnections and cones*. II, J. Funct. Anal. **90** (1990), 306–354.

[BC5] ——, *Remarks on the index theorem for families of Dirac operators on manifolds with boundary*, Differential Geometry (Lawson & Tenenblat, ed.) Longman (1991).

[BC6] ——, *Transgressed Euler classes of* SL(2n, Z) *vector bundles, adiabatic limits of eta invariants and special values of L-functions*, Preprint (1991).

[BF] J.-M. Bismut and D. Freed, *The analysis of elliptic families*. I, II, Comm. Math. Phys. **106** (1986), 159–176; **107** (1986), 103–163.

[BGS] J.-M. Bismut, H. Gillet, and C. Soule, *Analytic torsion and holomorphic determinants*. I, II, III, Comm. Math. Phys. **115** (1988), 49–78; 79–126, 301–351.

[BL] J.-M. Bismut and G. Lebeau, *Immersions complexes et metriques de Qullen*, C. R. Acad. Sci. Paris. Sér. I Math. **309** (1989) 487–491.

[C1] J. Cheeger, *Analytic torsion and the heat equation*, Ann. of Math. (2) **109** (1979), 259–322.

[C2] ____, *η-invariants, the adiabatic approximation and conical singularities*, J. Differential Geom. **26** (1987), 175–221.

[CGT] J. Cheeger, M. Gromov, and M. Taylor, *Finite propagation speed, kernel estimates for functions of the Laplace operator, and the geometry of complete Riemannian manifolds*, J. Differential Geom. **17** (1982), 15–53.

[CHS] S. S. Chern, F. Hirzebruch, and J.-P. Serre, *On the index of a fibered manifold*, Proc. Amer. Math. Soc. **8** (1957), 587–596.

[D] X. Dai, *Adiabatic limits, the non-multiplicativity of signature and Leray spectral sequence*, J. Amer. Math. Soc. **4** (1991), 265–321.

[DM] X. Dai, R. B. Melrose, *Adiabatic limit of the Ray-Singer analytic torsion*, in preparation.

[DW] R. Douglas and K. Wojciechowski, *The adiabatic limit of the η-invariants. I. The odd-dimensional Atiyah-Patodi-Singer problem*, Preprint.

[EM] C. L. Epstein and R. B. Melrose, *Shrinking tubes and the $\bar{\partial}$-Neumann problem*, Preprint (1990).

[F1] D. Fried, *Lefschtz formulas for flows*, Contemp. Math., no. 58, Amer. Math. Soc., Providence, RI, 1987, pp. 19–69.

[F2] ____, *Torsion for non-unitary representation*, in preparation.

[GS] P. Günther and R. Schimming, *Curvature and spectrum of compact Riemannian manifolds*, J. Differential Geom. **12** (1977), 599–618.

[L] J. Lott, *Superconnections and noncommutative de Rham homology*, Preprint (1990).

[MM] R. Mazzeo and R. Melrose, *The adiabatic limit, Hodge cohomology and Leray's spectral sequence for a fibration*, J. Differential Geom. **31** (1990), 185–213.

[M1] R. Melrose, *Pseudodifferential operators on manifolds with corners*, preprint.

[M2] ____, *Analysis on manifolds with corners*, M.I.T. Notes (1988).

[Mü1] W. Müller, *Signature defects of cusps of Hilbert modular varieties and values of L-series at s = 1*, J. Differential Geom. **20** (1984), 55–119.

[Mü2] ____, *Manifolds with cusps of rank one*, Lecture Notes in Math., vol. 1244, Springer-Verlag, 1987.

[Mü3] ____, *Analytic torsion and R torsion of Riemannian manifolds*, Adv. in Math. **28** (1978), 233–305.

[RS] D. B. Ray and I. M. Singer, *R-torsion and the Laplacian on Riemannian manifolds*, Adv. in Math. **7** (1971), 145–210.

[Si] I. M. Singer, *The η-invariant and the index*, Mathematical Aspects of String Theory (S. T. Yau, ed.), World Scientific, Singapore, 1987.

[S] M. Stern, *L^2-index theorems on locally symmetric spaces*, Invent. Math. **96** (1989), 231–282.

[W] E. Witten, *Global gravitational anomalies*, Comm. Math. Phys. **100** (1985), 197–229.

MASSACHUSETTS INSTITUTE OF TECHNOLOGY

Proceedings of Symposia in Pure Mathematics
Volume **54** (1993), Part 2

Geometry of Elementary Particles

ANDRZEJ DERDZINSKI

Introduction

This is a presentation of the *standard model of elementary particles*, i.e.,
their generally accepted theory, discussed here on the *classical* level (without
field quantization). Our basic reference is the book [**5**], for which the present
paper forms a self-contained summary. The author also intends to provide in
[**6, 7**] a more detailed treatment of particle geometry, along with its quantized
version.

Although this material is covered in numerous physics textbooks (such
as [**2–4, 9, 14, 15, 18, 20–23**]), the approach adopted here may be particularly suitable for a mathematician reader due to its consistently geometric
language. Presentations of more general topics are also available in the mathematical literature; see, for instance, [**1, 8, 10, 12, 13, 16, 17, 19**].

The theory outlined below does *not*, by itself, provide a usable model of
the real world, even though it accounts well for many qualitative effects.
Its inadequacy as a source of quantitative predictions is mainly due to its
classical (macroscopic) character, as opposed to the predominantly quantum
(small-scale) properties displayed by elementary particles in nature. However, instead of discarding the classical model for this reason, one uses field
quantization to transform it into a *quantum theory*, which supplies the required "microscopic" corrections.

Cross-references in the text are indicated by arrows $(\rightarrow)$.

1. Particles and interactions in nature

1.0. **Experimental data** show that there are over 200 species (kinds) of
subatomic particles, such as the *electron* e, *proton* p, *neutron* n, *photon*

1991 *Mathematics Subject Classification*. Primary 53C80; Secondary 81T13, 53B50, 81V05,
81V10, 81V15, 81V22.
   Key words and phrases. Standard model, electroweak model, quark model, gauge theory.
   Supported in part by NSF Grant DMS-8601282.
   This paper is in final form and no version of it will be submitted for publication elsewhere.

γ, *electronic neutrino* ν_e, *neutral pion* π^0, etc. ($\rightarrow$ 1.5, 1.6). Particles interact in four basic ways, known as the *strong, electromagnetic, weak* and *gravitational* interactions (forces), ordered here by decreasing *strength* of the interaction, i.e., probability of its occurrence in the given circumstances. We will not discuss gravitational forces, negligibly weak on the microscopic level; see, however, 2.0.

1.1. Particle invariants assign to particle species elements of various Abelian semigroups. Here belong, for instance, the *mass* $m \in \mathbb{R}_+ = [0, \infty)$, *electric charge* $Q \in \mathbb{Z}$ ($\rightarrow$ 1.2), *average lifetime* $\tau \in (0, \infty]$, as well as *spin* $s \in \frac{1}{2}\mathbb{Z}_+ = \{0, \frac{1}{2}, 1, \ldots\}$ and *parity* $\varepsilon \in \mathbb{Z}_2 = \{1, -1\}$. The spin measures the capacity of the particle to carry internal angular momentum (as if it were rotating about its axis), while the parity of the particle describes symmetry properties of its configurations with respect to space reflections ($\rightarrow$ 3.3, 3.5). Particles with $s \in \mathbb{Z}$ (resp., $s \notin \mathbb{Z}$) are known as *bosons* (resp., *fermions*).

1.2. Quantization of the electric charge. According to experimental evidence (Millikan, 1909) and theoretical arguments ($\rightarrow$ 5.1(iv)), the electric charge always comes in integral multiples of the electron charge q (by convention, $q < 0$). Charges of individual particle species may be indicated by superscripts such as $^-, ^0, ^+, ^{++}$, so that $e = e^-$, $p = p^+$, $n = n^0$, $\gamma = \gamma^0$, $\nu_e = \nu_e^0$, etc. ($\rightarrow$ 1.0).

1.3. Interaction carriers are the few particle species that serve as agents mediating interactions. These are: the photon γ ($\rightarrow$ 5.1(ii)) for electromagnetism, the *weak bosons* W^+, W^-, Z^0 for the weak interaction ($\rightarrow$ 5.3(ii) (b), (c), eight kinds of *gluons* ($\rightarrow$ 5.2(ii)) for the strong force and, probably, *gravitons* for gravity. Except for gravitons with $s = 2$, their spins ($\rightarrow$ 1.1) all equal 1.

1.4. Matter particles, i.e., those particle species which are not interaction carriers, may in turn be classified into *hadrons* and *leptons*, depending on whether they can or cannot participate in the strong interaction.

1.5. Leptons consist of 12 known species, all of spin $\frac{1}{2}$ ($\rightarrow$ 1.1): the electron e, *muon* μ and *tauon* τ (with positive masses and electric charge -1), the electrically neutral, massless *neutrinos* ν_e, ν_μ, ν_τ, as well as their antiparticles ($\rightarrow$ 3.2(i)): $e^+, \mu^+, \tau^+, \overline{\nu_e}, \overline{\nu_\mu}, \overline{\nu_\tau}$.

1.6. Hadrons which are fermions (resp., bosons, $\rightarrow$ 1.1) are called *baryons* (resp., *mesons*), with about 100 known species of either. For instance, π^0 is a meson. Baryons can further be divided into the disjoint classes of *baryons proper* (such as p, n), and their antiparticles ($\rightarrow$ 3.2(i)), the *antibaryons*.

1.7. Classification of elementary particles, as outlined above:

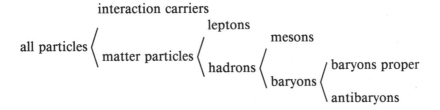

2. Bundles over the spacetime

2.0. **Spacetime.** We work with a fixed *spacetime* $(\mathcal{M}, g)$, which is a four-manifold $\mathcal{M}$ endowed with a pseudo-Riemannian metric g of signature $-+++$ and with a *time orientation*, i.e., a continuous choice $x \mapsto C_x^+$ of one ("future") component of the timelike cone $C_x = \{v \in T_x\mathcal{M} : g(v, v) < 0\}$ at each $x \in \mathcal{M}$. Although one often assumes that $\mathcal{M}$ is an affine space and g is translation invariant (the *Minkowski spacetime*), nonflat spacetimes are also used, as models of gravity (in general relativity).

2.1. **Vector bundles over the spacetime** $(\mathcal{M}, g)$, used below, include the product line bundle $1 = \mathcal{M} \times \mathbb{C}$, the tangent bundle $T = T\mathcal{M}$, its dual T^*, the bundle $\Lambda^4 T^*$ of volume forms (pseudoscalars) on $\mathcal{M}$ and, for each $k \in \mathbb{Z}_+$, the bundle $S_0^k T^*$ whose fiber over $x \in \mathcal{M}$ consists of all real k-linear symmetric forms on $T_x\mathcal{M}$ with g-contraction zero (i.e., of all pseudo-spherical harmonics in $T_x\mathcal{M}$).

2.2. **Weyl spinors.** Whenever necessary, the spacetime $(\mathcal{M}, g)$ will be assumed to be an orientable spin manifold. Denoting L, R the orientations of $\mathcal{M}$, we may then choose fixed *Weyl spinor bundles* σ_L, σ_R over $(\mathcal{M}, g)$, with $\sigma_R = \overline{\sigma_L}$ ($\to$ 3.2). They are complex vector bundles of fiber dimension 2, obtained from a common spin structure over $(\mathcal{M}, g)$ via two mutually conjugate nontrivial representations of $\mathrm{Spin}(3, 1) = SL(2, \mathbb{C})$ in $\mathbb{C}^2$. The (Levi-Civita) spinor connection $\overset{\circ}{\nabla}$ in σ_L and σ_R then gives rise to the Dirac operator $\mathscr{D}$ sending sections of σ_L to those of σ_R and vice versa.

2.3. **Dirac spinors.** If $(\mathcal{M}, g)$ is orientable and spin, we choose a fixed *Dirac spinor bundle* over $(\mathcal{M}, g)$ to be the direct sum $\sigma = \sigma_L + \sigma_R$ of the Weyl spinor bundles selected as in 2.2, with *unordered* summands, so that no orientation of $\mathcal{M}$ is distinguished. The operators $\overset{\circ}{\nabla}$ and $\mathscr{D}$ now are also defined in σ.

3. Models of free matter particles

3.0. **Particle models.** Each particle species is represented by ("lives in") a specific fiber bundle η, endowed with some additional geometric structure, over the spacetime manifold $\mathcal{M}$ ($\to$ 2.0). The sections ψ of η are to be thought of as "semiclassical states" of the particle, evolving with time in a manner described by suitable *field equations* (which form a part of the geometry of η).

3.1. Models of matter particles ($\to$ 1.4) are *vector* bundles and their field equations are *linear*. For those matter particles which are *free*, i.e., not subject to interactions, the choices of bundles and equations are quite specific ($\to$ 3.3 – 3.5). On the other hand, interaction carriers are represented by a special type of *affine* bundles ($\to$ 4.0) which, for many "practical" purposes, may also be regarded as vector bundles ($\to$ 4.4, 5.3(ii), (iii)).

3.2. Physical meaning of vector bundle operations. Equalities between vector or affine bundles, such as $(\eta^*)^* = \eta$, stand for natural (functorial) isomorphisms, the category in question being usually clear from the context. For a complex vector bundle η over $\mathcal{M}$, let $\overline{\eta}$ be its *conjugate bundle*, with each fiber $\overline{\eta}_x$, $x \in \mathcal{M}$, consisting of all antilinear maps $\eta_x^* \to \mathbb{C}$. Thus, we have $\overline{\eta} = \eta^*$ whenever the geometry of η involves a fixed Hermitian fiber metric (which may even be indefinite), as is the case for all models of free matter particles except neutrinos ($\to$ 3.4).

(i) For a particle species represented by a complex vector bundle η ($\to$ 3.0, 3.1), the conjugate $\overline{\eta}$, along with the corresponding "conjugate geometry," may be expected to host another, related particle species, called the *antiparticle* of the original one. The resulting *antiparticle formation* (denoted $^-$), leaves most of the relevant particle invariants ($\to$ 1.1) either completely unchanged (e.g., ε for bosons, m, τ, s), or just changes their signs (as in the case of ε for fermions and Q ; $\to$ 3.5, 5.1(iii)). Antiparticles also make sense for interaction carriers ($\to$ 4.4, 5.3(ii) (b), (c), and, in fact, turn out to exist for all particles known in nature. However, some species (referred to as *strictly neutral*) coincide with their antiparticles and then it is natural to represent them by *real* rather than complex bundles. For instance, $\overline{\pi^0} = \pi^0$, $\overline{\gamma} = \gamma$, $\overline{Z^0} = Z^0$, while $(W^+)^- = W^-$, $\overline{e} = e^+ \neq e$, $\overline{\nu_e} \neq \nu_e$, $\overline{p} \neq p$, $\overline{n} \neq n$ (notation of 1.0, 1.2, 1.3).

(ii) Given k particle species living in vector bundles η_j, the direct sum $\eta = \eta_1 + \cdots + \eta_k$ stands for their *common generalization*, which is *not* a particle species in the usual sense. (For instance, the *nucleon*, generalizing protons and neutrons, does not have a well-defined electric charge.) Conversely, if a bundle η describing some particles happens to be *naturally reducible*, i.e., admit a direct sum decomposition that is natural (functorial), these particles should be regarded as forming several distinct species represented by the summands, for which η provides a common generalization.

(iii) Putting together k particles of (not necessarily different) species that live in vector bundles η_j, one obtains a *composite object* whose evolving states ($\to$ 3.0) are sections of the tensor-product bundle $\eta_1 \cdots \eta_k$. Such objects include subatomic particles (hadrons, $\to$ 5.2), as well as nuclei, atoms, or even molecules. However, $\eta_1 \cdots \eta_k$ is often naturally reducible ($\rightarrow$ (ii)), and then it stands for *several* particle species, represented by its summands. Using the projection onto any summand η, one may thus characterize the formation of such a *composite particle* living in η as a natural (functorial) surjective bundle morphism $\eta_1 \cdots \eta_k \rightarrow \eta$.

3.3. Matter bosons of spin $k \in \mathbb{Z}_+$ and parity $(-1)^k$ (resp., $(-1)^{k-1}$, $\rightarrow$ 1.1) live in the bundle $\eta = S_0^k T^*$ (resp., in the tensor product

$$\eta = (S_0^k T^*)\Lambda^4 T^*),$$

if they are strictly neutral, and in its complexification $\eta^{\mathbb{C}}$ otherwise ($\rightarrow$ 2.1, 3.2(i)). Their field equations ($\rightarrow$ 3.0) consist of the *Klein-Gordon equation*

$$\tag{1} \square \psi = (mc/\hbar)^2 \psi$$

and, if $k \geq 1$, also of the *divergence condition*

$$\tag{2} \operatorname{div} \psi = 0$$

imposed on sections ψ of η or $\eta^{\mathbb{C}}$. Here $\square = \operatorname{Trace}_g \circ \overset{\circ}{\nabla}{}^2$ is the d'Alembertian (wave operator) of the Levi-Civita connection $\overset{\circ}{\nabla}$ in η and $\operatorname{div} \psi$ stands for the obvious g-contraction of $\overset{\circ}{\nabla}\psi$, while $\hbar$, m and c are, respectively, Planck's constant divided by 2π, the mass of the particle in question, and the speed of light. In particular, states of a particle with mass m, spin 0 and parity $+1$ are real or complex valued functions ψ on the spacetime $(\mathscr{M}, g)$ satisfying (1) with the pseudo-Riemannian Laplacian $\square$ of $(\mathscr{M}, g)$. As another example, sections ψ of $T^* = T^*\mathscr{M}$, with (1) and (2), describe the states of a strictly neutral particle with mass m, spin 1 and parity -1.

3.4. Neutrinos of all species ν_e, ν_μ, ν_τ ($\rightarrow$ 1.5) are represented by a fixed Weyl spinor bundle σ_L ($\rightarrow$ 2.2), while the corresponding antineutrinos $\overline{\nu_e}, \overline{\nu_\mu}, \overline{\nu_\tau}$ live in $\sigma_R = \overline{\sigma_L}$. In all cases, the field equations consist of *Weyl's equation*

$$\tag{3} \mathscr{D}\psi = 0$$

for sections ψ of σ_L or σ_R, $\mathscr{D}$ being the Dirac operator. As the choice of bundles indicates, each (anti)neutrino species *distinguishes an orientation of space*. This is manifested through the phenomenon known as *parity violation*, predicted by Lee and Yang and discovered by Wu in 1956.

3.5. Fermions other than neutrinos, with spin $k + \frac{1}{2}$, $k \in \mathbb{Z}_+$ and parity $(-1)^k$ (resp., $(-1)^{k-1}$, $\rightarrow$ 1.1) live in the subbundle η of the tensor product $(S_0^k T^*)\sigma$ obtained by requiring that the Clifford product involving T^* and

σ be zero (resp., in its conjugate $\overline{\eta}$), where σ is a fixed Dirac spinor bundle over $(\mathcal{M}, g)$ ($\rightarrow$ 2.3, 2.1). The field equations consist of *Dirac's equation*

$$(4) \qquad\qquad (\mathcal{D} + mc/\hbar)\psi = 0$$

and, if $k \geq 1$, also of the divergence condition (2), imposed on sections ψ of η or $\overline{\eta}$, with $m, c, \hbar$ as in 3.3. In particular, particles of spin $\frac{1}{2}$ such as the electron e, proton p, neutron n live in σ, while their antiparticles (the *positron* e^+, *antiproton* $\overline{p}$, *antineutron* $\overline{n}$) are represented by $\overline{\sigma}$, and each is governed by the Dirac equation (4) with the appropriate mass m. (It is convenient here to distinguish σ from $\overline{\sigma}$, even though they are naturally isomorphic. Also, in contrast with hadrons, lepton parities are not well defined, i.e., cannot be determined by experiment, and so the choice of σ rather than $\overline{\sigma}$ for e is just a matter of convention.)

4. The Yang-Mills description of interactions

4.0. Interaction bundles. In the formalism of Yang and Mills (1954), a given interaction is described by an *interaction bundle* δ, which is a real or complex vector bundle, of some fiber dimension N, over the spacetime manifold $\mathcal{M}$ ($\rightarrow$ 2.0). Moreover, δ is endowed with a fixed geometry, consisting mainly of a G-structure, i.e., a reduction P of the full principal frame bundle of δ to a (usually compact) subgroup G of $GL(N, \mathbb{F})$, where $\mathbb{F}$ is $\mathbb{R}$ or $\mathbb{C}$. (In most cases we will replace P by an equivalent tensorial object in δ, such as a Hermitian fiber metric when $G = U(N) \subset GL(N, \mathbb{C})$.) Suppressing P from the notation, we denote $\mathscr{C}(\delta)$ the affine bundle over $\mathcal{M}$, the C^∞ sections of which coincide with the connections ∇ in δ, compatible with P. The translation-space bundle (i.e., the associated vector bundle) of $\mathscr{C}(\delta)$ then is the tensor product $\mathfrak{g}(\delta)T^*$, where $T^* = T^*\mathcal{M}$ and $\mathfrak{g}(\delta)$ stands for the G-adjoint bundle of Lie algebras, corresponding to P. The interaction-carrier particles now live in $\mathscr{C}(\delta)$, that is, their states are just G-connections ∇ in δ, evolving so as to obey the *Yang-Mills equation*

$$(5) \qquad\qquad \operatorname{div} R^\nabla = 0,$$

where the curvature R^∇ of ∇ is regarded as a $\mathfrak{g}(\delta)$-valued 2-form on $\mathcal{M}$.

A natural question to ask now is what interaction bundles and in particular which structure groups G correspond in this way to the known physical interactions ($\rightarrow$ 1.0). The answer ($\rightarrow$ 5.1 – 5.3) is currently believed to be $G = U(1)$ for electromagnetism, $G = U(2)$ for the unified *electroweak* (electromagnetic plus weak) interaction, and $G = SU(3)$ for the strong force. On the other hand, the weak interaction, on its own, is not of the Yang-Mills type ($\rightarrow$ 5.3(iii)).

4.1. Interactions involving matter. Let η be the *free-particle bundle* (or, *generic-particle bundle*, $\rightarrow$ 4.4) of the given matter-particle species, i.e., the vector bundle over $\mathcal{M}$ where the particle lives when it is considered free ($\rightarrow$

3.1). Subject to an interaction described by the interaction bundle δ, this particle is represented by the *interacting-particle bundle* α obtained from δ and η via a specific natural (functorial) construction, a basic example of which is the tensor product $\alpha = \delta\eta$. (As $\overline{\delta\eta} = \overline{\delta}\overline{\eta}$, the role of δ for the corresponding antiparticle species must be played by $\overline{\delta}$, $\rightarrow$ 3.2(i).) To account for the interaction, the field equations now are to be imposed on pairs (ψ, ∇) consisting of sections ψ of α, and ∇ of $\mathscr{C}(\delta)$, so that they govern simultaneous evolution of matter particles and interaction carriers, and have the "coupled" form

$$\mathscr{F}(\psi, \overset{\circ}{\nabla} \otimes \nabla) = 0, \qquad \operatorname{div} R^{\nabla} = \mathscr{J}(\psi),$$

where

(i) $\overset{\circ}{\nabla}$ stands for the canonical (Levi-Civita) connection in η, and $\overset{\circ}{\nabla}\otimes\nabla$ for the connection in α naturally induced by $\overset{\circ}{\nabla}$ and ∇,

(ii) $\mathscr{F}$ is such that the free field equations ((1) and (2), or (3), or (4) and (2)) can be written as $\mathscr{F}(\psi, \overset{\circ}{\nabla}) = 0$ for sections ψ of η, and so $\mathscr{F}$ must be sufficiently general to also make sense in α,

(iii) $\mathscr{J}$ is a differential operator (to be determined in each case from natural considerations), sending sections of α onto $\mathrm{g}(\delta)$-valued one-forms on $\mathscr{M}$.

4.2. **Lack of naturality.** Genuine (observable) physical objects can move in space and "age" with time. The bundles η representing particles should therefore be *natural* in the sense that isometries between open subsets of $(\mathscr{M}, g)$ have (single or multiple-valued) functorial lifts to bundle maps in η. It is so, in fact, for the models of free matter particles ($\rightarrow$ 3.3 – 3.5), but not, in general, for interaction bundles δ, or the bundles $\mathscr{C}(\delta)$ and α as in 4.0, 4.1, where interaction carriers and interacting matter particles live. To achieve some sort of naturality in the latter cases, one uses additional procedures (formation of bound states or breaking of symmetry), as described below.

4.3. **Bound states.** The interacting-particle bundles $\alpha_1, \ldots, \alpha_k$ obtained as in 4.1 using some $\eta_1, \ldots, \eta_k$ and a fixed δ, may sometimes admit natural surjective morphisms of their tensor product $\alpha_1 \cdots \alpha_k$ onto a *natural* bundle η which is the free-particle bundle of a matter particle. The resulting composite particle ($\rightarrow$ 3.2(iii)), living in η, then may be called a *bound state* of the original k particles, as it is held together by the given interaction (force), yet, being free, does not exert comparably strong forces of this type on its environment. To obtain such bound states, one only needs to eliminate δ-related factors present in the α_j using natural multilinear bundle maps (examples: $\rightarrow$ 5.1(i), 5.2(i)).

4.4. **Symmetry breaking** means enriching the original geometry of a given interaction bundle δ ($\rightarrow$ 4.0), mainly by choosing a reduction of its

G-structure to some proper subgroup H of G. Very often this procedure is just formal and lacks direct physical meaning. When that is the case, H is usually assumed trivial, and the resulting choice of a trivialization for δ and the G-structure leads to the identification $\mathscr{C}(\delta) = T^* + \cdots + T^*$. Thus, the interaction carriers, living in $\mathscr{C}(\delta)$ ($\to$ 4.0) may to some extent be regarded as forming $\dim G$ separate species of "matter-like," strictly neutral particles with spin 1 and parity -1 ($\to$ 3.2(ii), 3.3). Similarly, under such a trivialization, interacting-particle bundles α ($\to$ 4.1) become direct sums of natural bundles (e.g., if $\alpha = \delta\eta$, one obtains $\alpha = \eta + \cdots + \eta$ with N summands, N being the fibre dimension of δ). Consequently, for $N > 1$, a single free-particle bundle η may lead to *several* "observed" matter-particle species, which justifies referring to η as the *generic-particle bundle*. However, the decompositions in question depend on the trivialization used and so have no "absolute" physical significance. For instance, a state of a *single* species for one trivialization will usually correspond to a *mixture* of species for another.

4.5. Spontaneous symmetry breaking takes place when the reduction from G to H in 4.4 is actually present in nature. This may only happen if the interaction is sufficiently *weak*. (Solidifying of fluids at *low temperatures* is a useful analogy.) The reason why it is then worthwhile to keep G (instead of just H) in the picture is the purely accidental manner in which the specific reduction is selected, so that the possible ways the symmetry *could* have become broken still enjoy full G-symmetry (example: $\to$ 5.3(i)).

5. The standard model

5.0. The standard model of elementary particles is a system of theories that, in view of experimental evidence, as well as its internal coherence, is generally accepted as a correct description of the microworld. Besides the field quantization and renormalization procedures, not discussed here, its principal ingredients are the *electroweak* and *quark-gluon models* (both based on the Yang-Mills formalism, $\to$ 4.0, 4.1), the main ideas of which are outlined below.

5.1. Electromagnetism (Weyl, 1929). The *electromagnetism bundle* (i.e., the electromagnetic interaction bundle, $\to$ 4.0), corresponding to the electron charge q, is a complex line bundle λ with a Hermitian fiber metric $\langle\,,\,\rangle$ (a $U(1)$-structure) over the spacetime $(\mathscr{M}, g)$. For a matter-particle species of charge kq, $k \in \mathbb{Z}$ ($\to$ 1.2), represented by the free-particle bundle η, the interacting-particle bundle ($\to$ 4.1) is the tensor product $\alpha = \lambda^k \eta$ (where $\lambda^{-1} = \bar{\lambda}$, $\to$ 3.2).

(i) Since $\lambda^{k_1} \cdots \lambda^{k_n} = \lambda^{k_1 + \cdots + k_n}$ due to the natural isomorphism induced by $\langle\,,\,\rangle : \lambda\bar{\lambda} \to \lambda^0 = 1 = \mathscr{M} \times \mathbb{C}$, electromagnetic bound states, arising from mutual "cancellation" of the λ factors ($\to$ 4.3), can only be obtained when the charges of the constituent particles add up to zero (so that the system they form is electrically neutral). This is, e.g., the case for nonionized atoms, but not for nuclei (which would fly apart due to electric repulsion, were it not for the strong force).

(ii) Electromagnetic forces are strong enough for their $U(1)$ symmetry not to be broken spontaneously ($\rightarrow$ 4.5). From formal symmetry breaking ($\rightarrow$ 4.4) we obtain $\mathscr{C}(\lambda) = T^*$ and $\alpha = \lambda^k \eta = \eta$ (with η, α as above), so that there is a single interaction-carrier particle species (the photon γ, living in $\mathscr{C}(\lambda)$), which is strictly neutral, with spin 1 and parity -1 ($\rightarrow$ 3.3), while matter particles subject to the electromagnetic interaction lead to the same picture as the free ones.

(iii) As $\overline{\lambda^k \eta} = \lambda^{-k} \overline{\eta}$, the electric charge becomes reversed under antiparticle formation ($\rightarrow$ 3.2(i)).

(iv) It is the electric charge quantization ($\rightarrow$ 1.2) that enables a *single* bundle λ to account for electromagnetic properties of *all* matter particles (which in turn is the first step toward a unified description of particle interactions).

5.2. The quark model (Gell-Mann, Zweig, 1964) presumes that all hadrons are bound states (composites) of peculiar particles called *quarks*, which come in 6 *flavors* (species) u, d, s, c, b, t, and of their antiparticles, the *antiquarks* $\overline{u}, \ldots, \overline{t}$. The complicated strong forces involving hadrons then may be viewed as residual effects of the much stronger (and simpler) interactions of (anti)quarks, just as some interatomic forces (of electromagnetic origin) are caused by uneven distribution of electric charge in each (neutral) atom.

All quarks (resp., antiquarks) have spin $\frac{1}{2}$ and parity $+1$ (resp., -1), so that, by 3.5, the free-particle bundle ($\rightarrow$ 4.1) for each flavor is a fixed Dirac spinor bundle σ over the spacetime $(\mathscr{M}, g)$ (resp., its conjugate $\overline{\sigma}$). The *strong-interaction bundle* ($\rightarrow$ 4.0) is a complex vector bundle ρ of fiber dimension 3 with a Hermitian fiber metric $\langle , \rangle$ and a fixed section Ω of $\Lambda^3 \rho^*$ which are *compatible* in the sense that $|\Omega| = 1$, i.e., $\Omega(\xi_1, \xi_2, \xi_3) = 1$ for some orthonormal basis ξ_1, ξ_2, ξ_3 of each fiber ρ_x, $x \in \mathscr{M}$. Obviously, the pair consisting of $\langle , \rangle$ and Ω is nothing else than an $SU(3)$-structure in ρ ($\rightarrow$ 4.0). The interacting-particle bundle ($\rightarrow$ 4.1) of each quark (resp., antiquark) flavor is the tensor product $\rho\sigma$ (resp., $\overline{\rho}\overline{\sigma}$).

(i) Bound states of quarks are obtained as in 4.3 by "naturally cancelling" ρ and $\overline{\rho}$ in the $\rho\sigma$, $\overline{\rho}\overline{\sigma}$ factors, which, essentially, can only be done using one of the bundle morphisms $\langle , \rangle, \Omega, \overline{\Omega}$ of the tensor products $\rho\overline{\rho}, \rho^3, \overline{\rho}^3$ onto $1 = \mathscr{M} \times \mathbb{C}$. The resulting composite particles are quark-antiquark pairs, three-quark systems, or three-antiquark systems, and may be easily identified with mesons, baryons, and, respectively, antibaryons ($\rightarrow$ 1.6).

(ii) The interquark forces are far too strong to allow spontaneous breaking of symmetry ($\rightarrow$ 4.5). Under formal symmetry breaking ($\rightarrow$ 4.4), one may regard *gluons*, i.e., the strong-interaction carriers, living in $\mathscr{C}(\rho)$, as forming $\dim SU(3) = 8$ species, while each (anti)quark flavor, when subject to the strong interaction, appears to come in three versions (*colors*), with all reservations stated in 4.4.

(iii) Quarks do not seem to exist freely (outside of hadrons) in nature, which is probably due to the *extreme strength* of their mutual interaction.

(iv) Another strange property of quarks is that their electric charges are *fractional* multiples of the electron charge q ($-2q/3$ for u, c, t and $q/3$ for d, s, b). This does not contradict the charge quantization ($\rightarrow$ 1.2), both in view of (iii) and since q may be replaced by $q/3$. In fact, values of the electric charge (and, similarly, of all other particle invariants additive under the composite-particle formation), thus assigned to (anti)quarks, lead to remarkable agreement with the observed spectrum of hadrons.

5.3. The electroweak model (Glashow, Salam, Weinberg, 1961-1967) treats electromagnetism and the weak force as manifestations of a single interaction, ascribing the observed differences between them to spontaneous symmetry breaking.

The *electroweak-interaction bundle* ($\rightarrow$ 4.0) is a complex vector bundle ι of fiber dimension 2 with a Hermitian fiber metric (a $U(2)$-structure; see, however, (iv)), inducing a fiber norm denoted $|\ |$, over the spacetime $(\mathcal{M}, g)$. The geometry of ι also involves a *natural fiber metric* $(,)$ in the affine bundle $\mathscr{C}(\iota)$, i.e., a fiber metric in its translation-space bundle $\mathfrak{g}(\iota)T^*$ ($\rightarrow$ 4.0), obtained by combining g in $T^* = T^*\mathcal{M}$ with a fiber metric in $\mathfrak{g}(\iota)$ (also denoted $(,)$) that comes from an Ad-invariant inner product in the Lie algebra $\mathfrak{g} = \mathfrak{u}(2)$. Since $(X, Y) = (a - b)(\operatorname{Trace} X)(\operatorname{Trace} Y) - 2a\operatorname{Trace}(XY)$ for $X, Y \in \mathfrak{g}(\iota_x) \subset \operatorname{End} \iota_x$, $x \in \mathcal{M}$, and some constants $a, b > 0$, $(,)$ is determined up to a factor by its *Weinberg angle* $\theta \in (0, \pi/2)$ with $\tan^2\theta = a/b$.

The generic (free) particle bundle ($\rightarrow$ 4.1) is a fixed Dirac spinor bundle $\sigma = \sigma_L + \sigma_R$ ($\rightarrow$ 2.3), and it represents the *electron generation* (e, ν_e), i.e., the electron and its neutrino (with their antiparticles $e^+, \overline{\nu}_e$ accounted for implicitly). The same approach applies to either of the remaining *lepton generations* $(\mu, \nu_\mu), (\tau, \nu_\tau)$ ($\rightarrow$ 1.5). For the interacting-particle bundle ($\rightarrow$ 4.1), we choose $\alpha = \iota\sigma_L + (\Lambda^2\iota)\sigma_R$.

(i) Spontaneous symmetry breaking ($\rightarrow$ 4.5) in the electroweak model consists in selecting a section ϕ of ι with a constant length $|\phi| > 0$, which amounts to a structure-group reduction from $U(2)$ to $U(1)$. The line subbundle $\lambda = \phi^\perp$ of ι, with the corresponding fiber metric, then is interpreted as the electromagnetism bundle ($\rightarrow$ 5.1). Thus, $\iota = 1 + \lambda$ with $1 = \mathcal{M} \times \mathbb{C} = \operatorname{Span} \phi$, while $\Lambda^2\iota = \lambda$ under the isometric bundle isomorphism $\lambda_x \ni \xi \mapsto \xi \wedge \phi/|\phi| \in \Lambda^2\iota_x$. Hence $\alpha = (1 + \lambda)\sigma_L + \lambda\sigma_R$, i.e., the interacting-particle bundle $\alpha = \sigma_L + \lambda\sigma$ now stands for two particle species represented by σ_L and $\lambda\sigma$, which are obviously identified with ν_e and e, carrying their correct electric charges ($\rightarrow$ 3.2(ii), 5.1, 3.4, 3.5, 1.5).

(ii) Extending connections from $\lambda = \phi^\perp$ to ι so as to make ϕ parallel, we obtain an injective morphism $\mathscr{C}(\lambda) \to \mathscr{C}(\iota)$ of affine bundles. Also, isometric embeddings of the *real* vector bundles λ, $\mathscr{M} \times \mathbb{R}$ into $\mathfrak{g}(\iota)$ with the fiber metric $(\,,\,)$ can be defined by $\xi \mapsto \frac{1}{2}a^{-1/2}|\phi|^{-1}X$ and $(x,r) \mapsto \frac{1}{2}ira^{-1/2}(\sec\theta)Y$, where, at each $x \in \mathscr{M}$, $X, Y \in \mathrm{End}\,\iota_x$ satisfy $X\phi = |\phi|^2\xi$, $X(z\xi) = -z|\xi|^2\phi$, $z \in \mathbb{C}$ (thus, X depends on $\xi \in \lambda_x$), and $Y = \mathrm{Id}$ on $\mathrm{Span}\,\phi$, $Y = -(\cos 2\theta)\,\mathrm{Id}$ on $\phi^\perp$, with a and the Weinberg angle θ selected before. Orthogonality of the images of these morphisms now establishes a $(\,,\,)$-orthogonal direct-sum decomposition $\mathscr{C}(\iota) = \mathscr{C}(\lambda) + \lambda T^* + T^*$, which is unique and natural in a suitable category. Thus, by 3.2(ii), the electroweak-interaction carriers, living in $\mathscr{C}(\iota)$, form the following particle species, corresponding to the summands $\mathscr{C}(\lambda), \lambda T^*, T^*$:

(a) The photon γ, represented by $\mathscr{C}(\lambda)$ ($\to$ 5.1(ii)).

(b) The charged weak boson W^-, living in λT^* (and hence carrying the electron charge, $\to$ 5.1). Its antiparticle W^+ ($\to$ 3.2(i)) is another weak-interaction carrier, implicit in our discussion. Since $\lambda T^* = \lambda \otimes_\mathbb{R} T^* = \lambda \otimes_\mathbb{C} (T^*)^\mathbb{C}$, the free-particle bundle of $W^\pm$ (cf. 5.1) is the complexification $(T^*)^\mathbb{C}$ of $T^* = T^*\mathscr{M}$, which makes either of $W^\pm$ a matter particle of spin 1 ($\to$ 3.3).

(c) The neutral weak boson Z^0, living in T^* and hence strictly neutral, of spin 1 ($\to$ 3.3).

(iii) The masses m of most particle species are positive. Exceptions, with $m = 0$, are only possible when the field equations cannot, for formal reasons, contain a nonzero *mass term* (as in (1), (4)), which in fact is the case for neutrinos, satisfying (3), and interaction carriers (with unbroken symmetry), governed by (5). Thus, whether the given particle is massive or massless depends only on its free-particle bundle. (Specifically, $m = 0$ for affine bundles and σ_L, σ_R, while $m > 0$ for the remaining vector bundles in 3.3, 3.5.) Since spontaneous symmetry breaking establishes the "massive" bundles $(T^*)^\mathbb{C}$, T^* as models of the W and Z bosons, their masses must be positive, in contrast with the photon living in $\mathscr{C}(\lambda)$. This is consistent with experimental evidence such as the short range for the weak force, as opposed to the electromagnetic and (interquark) strong interactions, which are long-range. Without the electroweak unification, a description of the weak interaction based on the Yang-Mills formalism ($\to$ 4.0) would not be possible precisely because of its short-range character, i.e., massiveness of its carriers.

(iv) Physicists usually choose the structure group of the electroweak theory to be the (twofold) covering group $U(1) \times SU(2)$ of $U(2)$ rather than $U(2)$ itself. However, taken at the face value, this would amount to imposing unnecessary additional conditions on the geometry and physics of the model.

(v) A dynamical approach to the electroweak model (not presented here) also involves a device known as the *Higgs boson*, which may be just a formal mechanism, but could as well turn out to be a (still undiscovered) matter particle with some unusual properties.

6. Grand unifications

6.0. Grand unified theories are attempts to go beyond the standard model by describing both strong and electroweak forces in terms of a single interaction subject to spontaneous symmetry breaking, just as the electroweak model does for the weak force and electromagnetism ($\rightarrow$ 5.3). Examples: $\rightarrow$ 6.3, 6.5.

6.1. The physical status of grand unifications is unclear, since they all use large structure groups G ($\rightarrow$ 6.2), and hence predict the existence of additional interaction carriers (note that dim G is, basically, the number of carrier species, $\rightarrow$ 4.4). The new kind of extremely weak interactions mediated by such particles would, in particular, cause the protons to decay spontaneously (at a very slow rate). However, a decade of intensive searches brought no evidence of proton-decay processes in nature.

6.2. The most obvious direct-sum unification of all microworld forces, using the interaction bundle $\rho + \iota$ with ρ, ι as in 5.2, 5.3, and with the corresponding $SU(3) \times U(2)$-structure, is not acceptable precisely because of its reducibility, which violates physical and geometric simplicity requirements.

6.3. The $SU(5)$ grand unified theory (Georgi and Glashow, 1974) uses an interaction bundle ($\rightarrow$ 4.0) which is a complex vector bundle β of fiber dimension 5 over the spacetime manifold $\mathcal{M}$, while its geometry is an $SU(5)$-structure, i.e., consists of a Hermitian fiber metric $\langle \, , \, \rangle$ in β and a section Ω of $\Lambda^5 \beta^*$ compatible with $\langle \, , \, \rangle$ as in 5.2.

We restrict our consideration to just one, e.g., the first, of the *basic-fermion generations* $(e, \nu_e, u, d), (\mu, \nu_\mu, c, s), (\tau, \nu_\tau, t, b)$, which the model describes separately. (These generations, ordered by increasing masses, are naturally distinguished by their common pattern $(-1, 0, 2/3, -1/3)$ of electric charges, $\rightarrow$ 1.5, 5.2(iv).)

The generic (free) particle bundle ($\rightarrow$ 4.1) of the whole generation is a fixed Dirac spinor bundle $\sigma = \sigma_L + \sigma_R$ ($\rightarrow$ 2.3). As in 5.3, we let $\alpha = \beta\sigma_L + (\Lambda^2\beta)\sigma_R$ be the interacting-particle bundle. The spontaneous symmetry breaking ($\rightarrow$ 4.5) in the $SU(5)$ model involves three steps:

(a) Choosing a subbundle $\iota \subset \beta$ of complex fiber dimension 2 which, with the Hermitian fiber metric inherited from $\langle \, , \, \rangle$, is interpreted as the electroweak-interaction bundle, while the natural fiber metric $(\, , \,)$ in $\mathscr{E}(\iota)$ ($\rightarrow$ 5.3) is induced by one in $\mathscr{E}(\beta)$. Since the latter is unique, up to a factor, in view of simplicity of $SU(5)$, the $SU(5)$ theory predicts a specific value of the Weinberg angle θ, such that $\tan^2\theta = 3/5$, in some contrast with the experimental bounds $0.29 \leq \tan^2\theta \leq 0.32$. (There are plausible explanations of this discrepancy.)

(b) Selecting a complex line bundle χ with a Hermitian fiber metric and a fixed isometric isomorphism $\chi^3 = \Lambda^2\iota$ (which amounts to a structure-group reduction in a suitable larger bundle). As Ω provides the identification $1 = \Lambda^5\beta = (\Lambda^2\iota)\Lambda^3\iota^\perp = \chi^3\Lambda^3\iota^\perp$, the complex vector bundle $\rho = \overline{\chi\iota^\perp}$ of fiber dimension 3 over $\mathscr{M}$ then satisfies $\Lambda^3\rho = 1$, i.e., carries a natural $SU(3)$-structure. Thus, ρ may be regarded as the strong-interaction bundle ($\rightarrow$ 5.2).

(c) Breaking the symmetry in ι exactly as in 5.3(i), i.e., by choosing a section ϕ of constant positive length and thinking of $\lambda = \phi^\perp \subset \iota$ as the electromagnetism bundle.

Since $\lambda = \Lambda^2\iota$ ($\rightarrow$ 5.3(i)), we may write $\chi = \lambda^{1/3}$ and $\beta = \lambda^{-1/3}\overline{\rho} + \iota$. This leads to the decomposition

$$\alpha = \sigma_L + \lambda\sigma + \lambda^{-1/3}\overline{\rho}\sigma + \left[\overline{\lambda^{-2/3}\rho\sigma_L} + \lambda^{-2/3}\rho\sigma_R\right],$$

the four summands of which represent, as in 3.2(ii), the first-generation fermions along with their correct electric charges and strong-interaction properties ($\rightarrow$ 5.1, 5.2). In fact, σ_L, $\lambda\sigma$ and $\lambda^{-1/3}\overline{\rho}\sigma$ clearly correspond to ν_e, e and the $\overline{d}$ antiquark ($\rightarrow$ 3.4, 3.5, 5.2), while the fourth summand only differs from the correct model $\lambda^{-2/3}\rho\sigma = \lambda^{-2/3}\rho\sigma_L + \lambda^{-2/3}\rho\sigma_R$ of the interacting u quark ($\rightarrow$ 5.2(iv)) by having one of its own summands replaced by the conjugate (such complications cannot be avoided, for dimensional reasons).

One also has (see [5], formula (7.18)) the natural orthogonal decomposition $\mathscr{E}(\beta) = \mathscr{E}(\rho) + \mathscr{E}(\lambda) + \lambda T^* + T^* + \lambda^{4/3}\rho T^* + \lambda^{1/3}\rho T^*$, which leads, besides the known interaction carriers (gluons in $\mathscr{E}(\rho)$, the photon in $\mathscr{E}(\lambda)$, W^+, W^- in λT^*, Z^0 in T^*, $\rightarrow$ 5.2(ii), 5.3(ii)), also to new ones, represented by the last two summands. The latter, hypothetical particles, denoted $X^{-4/3}$, $Y^{-1/3}$ (and called, along with their antiparticles $X^{4/3}$, $Y^{1/3}$, the X and Y bosons), carry fractional electric charges ($\rightarrow$ 1.2, 5.1) and, if they really exist, should cause proton decay ($\rightarrow$ 6.1).

6.4. **Some spinor bundles.** For each positive integer k, a $\mathrm{Spin}(4k + 2)$-structure in a complex vector bundle ζ of fiber dimension 4^k over $\mathscr{M}$ can be described as a Hermitian fiber metric $\langle\,,\,\rangle$ in ζ along with a real vector subbundle $\kappa \subset \mathrm{End}_{\mathbb{R}}\zeta$ of fiber dimension $4k+2$ such that, for each $x \in \mathscr{M}$, $X \in \kappa_x$ and $\xi, \xi' \in \zeta_x$, $X : \zeta_x \rightarrow \zeta_x$ is antilinear, while X^2 is a multiple of Id and $\langle\xi, X\xi'\rangle = \langle\xi', X\xi\rangle$. A natural construction of an example is based on using a complex vector bundle β of fiber dimension $2k+1$ with an $SU(2k + 1)$-structure, i.e., a Hermitian fiber metric (also denoted $\langle\,,\,\rangle$) and a compatible $\Omega \in \Lambda^{2k+1}\beta^*$ ($\rightarrow$ 5.2). The induced Hermitian fiber metric $\langle\,,\,\rangle$ in the exterior-algebra bundle $\Lambda\beta^* = \Lambda^{\mathrm{even}}\beta^* + \Lambda^{\mathrm{odd}}\beta^*$ then is also sesquilinear in the bundle $\Lambda^{\mathrm{even}}\beta^* + \overline{\Lambda^{\mathrm{odd}}\beta^*} = \Lambda^{\mathrm{even}}\overline{\beta} + \Lambda^{\mathrm{odd}}\beta$, i.e., in $\Lambda\beta^*$ endowed with the new complex structure for which the multiplication by

i equals the old $(-1)^j i \cdot \mathrm{Id}$ on $\Lambda^j \beta^*$. The Hodge star $*$ determined by $\langle \, , \rangle$ and Ω, although antilinear in $\Lambda \beta^*$, is a linear endomorphism of $\Lambda^{\mathrm{even}}\overline{\beta} + \Lambda^{\mathrm{odd}}\beta$. We now define a complex vector bundle $\zeta = \zeta[\beta]$ with a Spin$(4k+2)$-structure by $\zeta[\beta] = \mathrm{Ker}\,(* - \mathrm{Id}) \subset \Lambda^{\mathrm{even}}\overline{\beta} + \Lambda^{\mathrm{odd}}\beta$, with the Hermitian fiber metric obtained by restricting $\langle \, , \rangle$ and with $\kappa \subset \mathrm{End}_{\mathbb{R}}(\zeta[\beta])$ given as the image of the real bundle morphism $\beta^* \to \mathrm{End}_{\mathbb{R}}(\Lambda\beta^*)$ sending ψ to $e_\psi + i_\psi$, where $e_\psi(\omega) = \psi \wedge \omega$ and

$$i_\psi(\psi_1 \wedge \cdots \wedge \psi_k) = \sum_{j=1}^{k} (-1)^{j-1} \langle \psi_j, \, \psi \rangle \psi_1 \wedge \cdots \wedge \widehat{\psi_j} \wedge \cdots \wedge \psi_k.$$

Under the projection $\frac{1}{2}(*+\mathrm{Id}) : \Lambda^{\mathrm{even}}\overline{\beta} + \Lambda^{\mathrm{odd}}\beta \to \zeta$, ζ becomes isomorphic to $1 + \beta + \Lambda^2\overline{\beta} + \Lambda^3\beta + \cdots$ (summation up to $\Lambda^k \beta$ or $\Lambda^k\overline{\beta}$). In particular, β is isometrically embedded in $\zeta[\beta]$.

6.5. The Spin(10) grand unification (Fritzsch and Minkowski, 1975), usually called the $SO(10)$ theory, uses an interaction bundle ($\to$ 4.0) which is a complex vector bundle ζ of fiber dimension 16 with a Spin(10)-structure, as in 6.4 with $k = 2$. The generic-particle bundle for a whole basic-fermion generation (e, ν_e, u, d) ($\to$ 6.3) is, this time, a fixed Weyl spinor bundle σ_L ($\to$ 2.2), and one chooses the interacting-particle bundle ($\to$ 4.1) to be $\alpha = \zeta \sigma_L$.

The first step of spontaneous symmetry breaking consists in selecting a bundle β with an $SU(5)$-structure along with a structure-preserving isomorphism $\zeta = \zeta[\beta]$ ($\to$ 6.4, with $k = 2$), so that $\beta \subset \zeta$ and $\zeta = 1 + \beta + \Lambda^2\overline{\beta}$, where $1 = \mathscr{M} \times \mathbb{C}$. Regarding β as the interaction bundle of the $SU(5)$ theory, one then proceeds with the further steps of symmetry breaking ($\to$ 6.3(a), (b), (c)). Since now $\alpha = (1 + \beta + \Lambda^2\overline{\beta})\sigma_L = \sigma_L + \beta\sigma_L + \overline{(\Lambda^2\beta)\sigma_R}$, with the last two summands resembling the choice of α in the $SU(5)$ model ($\to$ 6.3), we obtain, after 6.3(a), (b), (c), $\alpha = (\sigma_L + \sigma_L) + \lambda\sigma_L + \overline{\lambda}\sigma_L + \lambda^{-1/3}\overline{\rho}\sigma_L + \lambda^{1/3}\rho\sigma_L + \lambda^{-2/3}\rho\sigma_L + \lambda^{2/3}\overline{\rho}\sigma_L$. Because $\sigma = \sigma_L + \sigma_R$ with $\sigma_R = \overline{\sigma_L}$, this decomposition differs from the expression $\sigma + \lambda\sigma + \lambda^{1/3}\rho\sigma + \lambda^{-2/3}\rho\sigma$ by containing, instead of some summands, their conjugates. Ignoring such discrepancies (as in 6.3), we interpret the latter four summands ($\to$ 3.2(ii)) as models of a "modified" neutrino ν_e ($\to$ 6.6), the electron e, and the u, d quarks, along with their correct electromagnetic and strong-interaction properties ($\to$ 3.5, 5.1, 5.2).

As $\dim \mathrm{Spin}(10) = 45$, the Spin(10) theory predicts even more interaction-carrier species than the $SU(5)$ model (cf. 6.1). See [11] for details.

6.6. Massive neutrinos. Neutrinos might conceivably have very small, positive masses (the experimental evidence makes this appear rather unlikely, but is still inconclusive). Then they would be described by Dirac's equation (4) in the "chiral" Dirac spinor bundle $\sigma = \sigma_L + \sigma_R$ the summands of which are *ordered*, in contrast with 2.3, to account for parity violation ($\to$ 3.4).

References

1. E.Binz, J. Śniatycki and H. Fischer, *Geometry of classical fields*, North-Holland Mathematics Studies, no. 154, North-Holland, Amsterdam, 1988.
2. N. D. Birrell and P. C. W. Davies, *Quantum fields in curved space*, Cambridge Univ. Press, Cambridge, 1983.
3. N. N. Bogolyubov [N. N. Bogolubov], A. A. Logunov, A. I. Oksak and I. T. Todorov, *General principles of quantum field theory*, Mathematical Physics and Applied Mathematics, no. 10, Kluwer, Dordrecht, 1990.
4. N. N. Bogolyubov [N. N. Bogoliubov] and D. V. Shirkov, *Quantum fields*, Benjamin/Cummings, Reading, Mass., 1983.
5. A. Derdzinski, *Geometry of the standard model of elementary particles*, Texts and Monographs in Physics, Springer-Verlag, Berlin-Heidelberg-New York, 1992.
6. _____, *Quantum electrodynamics: A mathematical introduction*, (in preparation).
7. _____, *Notes on particle physics*, (in preparation).
8. G. G. Emch, *Mathematical and conceptual foundations of 20th-century physics*, North-Holland Mathematics Studies, no. 100, North-Holland, Amsterdam, 1984.
9. B. Felsager, *Geometry, particles and fields*, Odense Univ. Press, Odense, 1983.
10. F. G. Friedlander, *The wave equation on a curved space-time*, Cambridge Univ. Press, Cambridge, 1975.
11. H. Fritzsch and P. Minkowski, *Unified interactions of leptons and hadrons*, Ann. Phys. **93** (1975), 193-266.
12. R. Hermann, *Vector bundles in mathematical physics*. I, II, Benjamin, New York, 1970.
13. _____, *Interdisciplinary mathematics* I–XXIV, Math Sci Press, Brookline, Mass., 1973-1988.
14. K. Huang, *Quarks, leptons and gauge fields*, World Scientific, Singapore, 1982.
15. J. Leite Lopes, *Gauge field theories*, Pergamon Press, Oxford, 1981.
16. Y. I. Manin, *Mathematics and physics*, Progress in Physics, no. 3, Birkhäuser, Boston, 1981.
17. _____, *Gauge field theory and complex geometry*, Springer-Verlag, Berlin, 1988.
18. R. N. Mohapatra, *Unification and supersymmetry*, Contemporary Physics, Springer-Verlag, New York, 1986.
19. R. Penrose and W. Rindler, *Spinors and space-time*. I, II, Cambridge Univ. Press, Cambridge, 1984-1986.
20. P. Ramond, *Field theory*, Benjamin/Cummings, Reading, Mass., 1981.
21. L. H. Ryder, *Elementary particles and symmetries*, Documents on Modern Physics, Gordon and Breach, New York, 1986.
22. _____, *Quantum field theory*, Cambridge Univ. Press, Cambridge, 1985.
23. A. Sudbery, *Quantum mechanics and the particles of nature: An outline for mathematicians*, Cambridge Univ. Press, Cambridge, 1986.

Ohio State University
E-mail address: andrzej@mps.ohio-state.edu

Proceedings of Symposia in Pure Mathematics
Volume **54** (1993), Part 2

Existence of Connections
with Prescribed Yang-Mills Currents

DENNIS DeTURCK, HUBERT GOLDSCHMIDT,
AND JANET TALVACCHIA

1. Yang-Mills currents

Let (M, g) be a Riemannian manifold of dimension $n \geq 3$. By $\bigotimes^m T^*$, $\bigwedge^k T^*$ and $S^l T^*$, we shall mean the mth tensor power, the kth exterior product and the lth symmetric product of the cotangent bundle T^* of M, respectively. If E is a vector bundle over M, we denote by $\mathscr{E}$ the sheaf of sections of E over M. If s is a section of E over a neighborhood of $x \in M$, then $j_k(s)(x)$ is the k-jet of s at x; if $j_{k-1}(s)(x) = 0$, we identify $j_k(s)(x)$ with an element of $S^k T^* \otimes E$.

Let G be a Lie group, whose Lie algebra we denote by $\mathfrak{g}$, and let E be a vector bundle over M with structure group G. The action of G on E induces for each $x \in X$ a representation $\rho_x : \mathfrak{g} \to \operatorname{End}(E_x)$ on the fiber E_x of E. We henceforth assume that these representations are faithful and consider the sub-bundle $\mathfrak{g}(E)$ of $\operatorname{End}(E) = E^* \otimes E$, whose fiber at x is $\rho_x(\mathfrak{g})$. If U is an open subset of M, endowed with a coordinate system $(x^1, \ldots, x^n)$, over which E is trivial, a connection on E can be viewed as a $\mathfrak{g}$-valued one-form

$$
(1) \qquad A = \sum_{i=1}^{n} dx^i \otimes A_i,
$$

where the A_i are $\mathfrak{g}$-valued functions on U. This connection enables us to define the covariant derivative of a section of E and a covariant

1991 *Mathematics Subject Classification.* Primary 35N10, 53B15, 53C07, 81T13.

The first author was supported in part by NSF Grant DMS 90-01707, the Pew Science Program and the Institute for Advanced Study, the second author was supported in part by NSF Grant DMS 87-04209, and the third author was supported in part by the Pew Science Program.

This paper is in final form and no version will be submitted for publication elsewhere.

differentiation operator

$$d^A : \bigwedge\nolimits^j \mathcal{T}^* \otimes \mathcal{E} \to \bigwedge\nolimits^{j+1} \mathcal{T}^* \otimes \mathcal{E}$$

acting on E-valued differential forms. Since $\mathfrak{g}(E)$ acts on itself via the adjoint action, the connection A on E induces in a natural way a connection on $\mathfrak{g}(E)$, and we also obtain a covariant differentiation operator d^A on $\bigwedge^j \mathcal{T}^* \otimes \mathfrak{g}(\mathcal{E})$. As is well known, the curvature F^A of A is an E-valued two-form which measures the extent to which the operators d^A fail to give rise to a complex, i.e., $d^A d^A \omega = F^A \cdot \omega$, for any E-valued form ω. Over U, viewing F^A as a $\mathfrak{g}$-valued two-form, we have

$$F^A = dA + \frac{1}{2}[A, A] = \frac{1}{2} \sum_{i,j=1}^n dx^i \wedge dx^j \otimes F_{ij},$$

where

(2) $$F_{ij} = \frac{\partial A_j}{\partial x^i} - \frac{\partial A_i}{\partial x^j} + [A_i, A_j].$$

In gauge theory, the Yang-Mills equation arises from the variational problem of minimizing the functional $\int_M |F^A|^2$, where the norm is taken with respect to the metric g and a G-invariant inner product on $\mathfrak{g}(E)$. The Euler-Lagrange equation for this problem turns out to be $\delta^A F^A = 0$, where δ^A is the formal L^2-adjoint of d^A. By analogy with Maxwell's equations, we define the $\mathfrak{g}(E)$-valued one-form $J^A = \delta^A F^A$ to be the *Yang-Mills current* of the connection A. The subject of this paper is the following:

PROBLEM. Given a $\mathfrak{g}(E)$-valued one-form J on M, is there a connection A on E such that $J^A = J$?

To solve this problem evidently requires us to prove the existence of a solution of a system of (nonlinear) second-order partial differential equations. If J is not identically zero, we will refer to this system as the inhomogeneous Yang-Mills equation. Note that, unlike the (homogeneous) Yang-Mills equation, our system is *not* gauge invariant, but only gauge-equivariant. Indeed, for any gauge transformation ϕ (i.e., a G-valued change of basis in the bundle E) and any solution A of $J^A = 0$, it is true that $\phi^* A$ is another solution. Our situation is different: if A is a solution of $J^A = J$, then $B = \phi^* A$ will not solve the same equation, but rather the equation $J^B = \phi^* J$. One of the effects of this modified invariance property is that the local problem for the inhomogeneous Yang-Mills equation is already hard (and interesting). Therefore, we will concentrate on the problem of local existence; more particularly, we will assume that all the objects under consideration (g, J, etc.) are real-analytic with respect to the local coordinates $(x^1, \ldots, x^n)$, and seek a convergent power series solution A of $J^A = J$ about some point of U. Our principal result is:

THEOREM 1. *If the stucture group G of the bundle E over the real-analytic Riemannian manifold (M, g) is semisimple, then for a generic analytic $\mathfrak{g}(E)$-*

valued one-form J *given in a neighborhood of* $x \in M$, *there exists a real-analytic connection* A *on* E *such that* $\delta^A F^A = J$ *in a neighborhood of* x.

The genericity conditions on J will be made somewhat more precise below. Genericity in our sense is only a sufficient condition for local solvability; in fact, we shall also give examples to show that there may be no local existence results for certain J, which do not satisfy our conditions.

Our strategy consists in showing the existence of a formal power series solution of finite order which is *strongly prolongable*. Consider a system of differential equations of order k on $T^* \otimes \mathfrak{g}$. For $l \geq 0$ and $x \in M$, a formal solution of order $k + l$ of this system at x is a $k + l$-jet of a section of $T^* \otimes \mathfrak{g}$ at x, which satisfies these equations to order l at x; let $R_{k+l,x}$ be the set of all such formal solutions of order $k + l$ at x. A jet $p \in R_{k+l,x}$ is called strongly prolongable if, for any element q of $R_{k+l+m,x}$ which agrees with p up to order $k + l$, there is a jet $q' \in R_{k+l+m+1,x}$ which agrees with q up to order $k+l+m$. According to a result of Malgrange [7], if the system admits a strongly prolongable formal solution of finite order at x, the system also has a convergent power series solution at x. In other words, having solved for derivatives of the solution up to a certain order, we must be able to solve for the next-order derivatives to guarantee the existence of a convergent series solution.

The aspect of the Yang-Mills problem, which is especially interesting from the differential systems point of view, is that the strong prolongability criterion can not be applied directly to the inhomogeneous Yang-Mills equation. In attempting to prolong formal solutions of our equation, we are led to discover a series of identities which act as integrability conditions for A and J. First the Bianchi identity appears (as in similar problems, see [2]), and then, if G is semisimple, a sequence of new and unexpected identities is encountered. The number and nature of these latter identities depend in an essential way on the nature of the structure group G. In attempting to satisfy these identities, we are forced to impose genericity assumptions on J.

It should also be pointed out that we do not rely on the general theorem on the existence of formal solutions of [4] and that we prove existence directly.

2. First obstructions

We begin our consideration of the Yang-Mills equation $J^A = J$ by computing the symbol of the equation and of its prolongations. For simplicity of exposition, we restrict our attention to the case where $M = \mathbb{R}^n$, with its standard coordinate system $(x^1, \ldots, x^n)$, endowed with the standard Euclidean metric g, and where E is a trivial vector bundle (the general case is treated in [3]); we denote by $\mathfrak{g}$ the trivial vector bundle $\mathfrak{g}(E) = M \times \mathfrak{g}$. For $1 \leq i_1, \ldots, i_k \leq n$, we write

$$\partial_{i_1 \cdots i_k} = \frac{\partial^k}{\partial x^{i_1} \cdots \partial x^{i_k}},$$

and, if V is a vector space and u is an element of $\bigotimes^k T_x^* \otimes V$, with $x \in M$ (or a section of $\bigotimes^k T^* \otimes V$ over M), we set

$$u_{i_1 \cdots i_k} = u\left(\frac{\partial}{\partial x^{i_1}}, \ldots, \frac{\partial}{\partial x^{i_k}}\right).$$

The section J is written as $J = \sum_{i=1}^n dx^i \otimes J_i$, where the J_i are $\mathfrak{g}$-valued functions on M. Under these assumptions, if A is the connection given by (1), we have

$$\delta^A F^A = -\sum_{i,j=1}^n dx^j \otimes \left(\frac{\partial F_{ij}}{\partial x^i} + [A_i, F_{ij}]\right),$$

where F_{ij} is given by (2). It follows that the Yang-Mills equation is quasi-linear and that the symbol $\sigma : S^2 T^* \otimes T^* \otimes \mathfrak{g} \to T^* \otimes \mathfrak{g}$ of the corresponding Yang-Mills operator is given by

$$(\sigma(u))_j = \sum_{i=1}^n (u_{iji} - u_{iij}),$$

for $u \in S^2 T^* \otimes T^* \otimes \mathfrak{g}$. If the kth prolongation

$$\sigma_k : S^{k+2} T^* \otimes T^* \otimes \mathfrak{g} \to S^k T^* \otimes T^* \otimes \mathfrak{g}$$

of σ (see [1, Chapter IX]) were surjective for all $k \geq 0$, we could apply Malgrange's theorem directly to obtain a convergent power series solution. However, although σ is surjective, if

$$\mathrm{Tr} : S^k T^* \otimes T^* \otimes \mathfrak{g} \to S^{k-1} T^* \otimes \mathfrak{g}$$

is the trace mapping defined by

$$(\mathrm{Tr}\, v)_{j_1 \cdots j_{k-1}} = \sum_{j=1}^n v_{jjj_1 \cdots j_{k-1}},$$

for all $v \in S^k T^* \otimes T^* \otimes \mathfrak{g}$, we have the following:

LEMMA 1. *The sequences*

$$S^{k+2} T^* \otimes T^* \otimes \mathfrak{g} \xrightarrow{\sigma_k} S^k T^* \otimes T^* \otimes \mathfrak{g} \xrightarrow{\mathrm{Tr}} S^{k-1} T^* \otimes \mathfrak{g} \to 0$$

are exact, for $k \geq 1$.

As a consequence of this lemma, we see that a formal solution $j_{k+1}(A)(x)$ of order $k+1$ at $x \in M$ of the inhomogeneous Yang-Mills equation, which satisfies $j_{k-1}(\delta^A F^A - J)(x) = 0$, can be extended to a formal solution of order $k+2$ if and only if

(3) $$\mathrm{Tr}\, j_k(\delta^A F^A - J)(x) = 0.$$

This reasoning means that there should be an integrability condition which guarantees that our solutions of order $k+1$ can be prolonged to solutions of higher order. It is obtained by considering the case $k = 1$, and arises from the "Bianchi identity" for the current, namely $\delta^A(\delta^A F^A) = 0$. If we apply the operator δ^A to our equation and take this identity into account, we see that a solution A of our system must also satisfy

$$(4) \qquad \delta^A J = \sum_{i=1}^{n} \left(-\frac{\partial J_i}{\partial x_i} + [J_i, A_i] \right) = 0.$$

Note that this equation is actually of order zero (i.e., algebraic) in A. If a formal solution $j_{k+1}(A)(x)$ of order $k + 1$ of the Yang-Mills equation satisfies the equation (4) to order $k - 1$, then the relation (3) holds and, by Lemma 1, we can modify the derivatives of A of order $k + 2$ in order to satisfy the equation $\delta^A J^A = J$ to kth order at x.

We are thus led to consider the (overdetermined) system consisting of the two equations

$$(5) \qquad \delta^A F^A - J = 0, \qquad j_2(\delta^A J) = 0,$$

which are both of order 2.

An easy consequence of the Bianchi identity is the construction of examples of sections J that can not be Yang-Mills currents in a neighborhood of a point $x_0 \in M$. Namely, if J is a section of $T^* \otimes \mathfrak{g}$ satisfying $J(x_0) = 0$ and $\sum_{i=1}^{n}(\partial J_i/\partial x^i)(x_0) \neq 0$, then one sees that (4) cannot possibly hold at x_0. Thus there is no connection A whose Yang-Mills current is equal to this section J on any neighborhood of x_0.

3. Second obstructions

Our first task in considering the combined system (5) is to compute its symbol. Since (4) is linear in A, the symbol $\tau : T^* \otimes \mathfrak{g} \to \mathfrak{g}$ of the operator corresponding to equation (4) is given by $\tau(A) = \sum_{i=1}^{n} [J_i, A_i]$, for $A \in T^* \otimes \mathfrak{g}$. The kth prolongation

$$\tau_k : S^k T^* \otimes T^* \otimes \mathfrak{g} \to S^k T^* \otimes \mathfrak{g}$$

of τ is equal to $\mathrm{id} \otimes \tau$, and the kth prolongation of the symbol of the combined system (5) is

$$\sigma_k \oplus \tau_{k+2} : S^{k+2} T^* \otimes T^* \otimes \mathfrak{g} \to (S^k T^* \otimes T^* \otimes \mathfrak{g}) \oplus (S^{k+2} T^* \otimes \mathfrak{g}).$$

Since σ_k is not surjective for $k \geq 1$, we can not expect this mapping to be surjective. However, according to our discussion in the previous section, we conclude that in order to apply Malgrange's strong prolongability theorem it would be sufficient to know that the prolongations τ_{k+2} of τ are surjective when restricted to the kernel of σ_k.

From the definition of τ, we see that the morphism τ_k at $x \in M$ is a linear mapping whose coefficients depend on the $J_i(x)$ and whose variables belong to the Lie algebra $\mathfrak{g}$. To understand τ_k properly, it is necessary to digress briefly and discuss some aspects of linear algebra over $\mathfrak{g}$.

Suppose that G is connected and that $\mathfrak{g}$ is a semisimple Lie algebra of rank r. This last assumption implies that the dimension of the kernel of $\operatorname{ad} C : \mathfrak{g} \to \mathfrak{g}$ is at least r, for all $C \in \mathfrak{g}$; in fact, the principal elements $C \in \mathfrak{g}$ (i.e., for which the dimension of the kernel of $\operatorname{ad} C$ is equal to r) form a nonempty Zariski-open subset of $\mathfrak{g}$. For example, the Lie algebras $\mathfrak{sl}(r+1, \mathbb{R})$, $\mathfrak{so}(2r)$ are of rank r. We therefore do not expect even to be able to solve the simplest linear equation

$$(6) \qquad\qquad [C, X] = D,$$

for $X \in \mathfrak{g}$, where C, D are given elements of $\mathfrak{g}$. To give a necessary (and often sufficient) condition for solvability, we consider the algebra $\mathscr{I}(\mathfrak{g})$ of all invariant polynomials on $\mathfrak{g}$. These are the polynomial functions p on $\mathfrak{g}$ which satisfy $p(X) = p(\operatorname{Ad} g \cdot X)$, for all $X \in \mathfrak{g}$ and $g \in G$. Examples of such polynomials are given by $q_k(X) = \operatorname{Tr}(\operatorname{ad} X)^k$, for $X \in \mathfrak{g}$, $k \geq 2$; the polarization of q_2 is the Killing form of $\mathfrak{g}$. Chevalley's theorem (see [8]) asserts that the algebra $\mathscr{I}(\mathfrak{g})$ is generated by 1 and by precisely r algebraically independent homogeneous polynomials. We fix such a set $\{p_1, \ldots, p_r\}$ of generating elements of $\mathscr{I}(\mathfrak{g})$.

We view the polarization of a homogeneous polynomial $p \in \mathscr{I}(\mathfrak{g})$ of degree d as an element of the dth symmetric power of $\mathfrak{g}^*$. The invariance condition on p implies that

$$(7) \quad \begin{aligned} &p([Y_1, X], Y_2, \ldots, Y_d) \\ &\quad + p(Y_1, [Y_2, X], \ldots, Y_d) + \cdots + p(Y_1, Y_2, \ldots, [Y_d, X]) = 0, \end{aligned}$$

for all $X, Y_1, \ldots, Y_d \in \mathfrak{g}$. Clearly, if we take $Y_1 = \cdots = Y_d = C$ in the above equality, we obtain

$$p(C, \ldots, C, [C, X]) = -(d-1)p([C, C], C, \ldots, C, X) = 0.$$

Thus the element $D \in \mathfrak{g}$ must satisfy the condition $p_j(C, \ldots, C, D) = 0$, for all $1 \leq j \leq r$, in order to be able to solve the equation (6). If C is a principal element of $\mathfrak{g}$, a theorem of Kostant [6] asserts that this condition is in fact both necessary and sufficient for the solvability of (6). We now understand the simplest linear system over $\mathfrak{g}$, namely one equation in one unknown.

To give a flavor for the kind of reasoning employed in the proofs of exactness for the sequences given below, we consider one more case, namely that of one equation in several unknowns. We wish to solve the equation

$$(8) \qquad\qquad [C_1, X_1] + [C_2, X_2] + \cdots + [C_n, X_n] = D,$$

where $C_1, \ldots, C_n, D$ are given elements of $\mathfrak{g}$, for $X_1, \ldots, X_n \in \mathfrak{g}$; we suppose that C_1, C_2 are principal elements of $\mathfrak{g}$. Let $X_3, \ldots, X_n$ be arbitrary elements of $\mathfrak{g}$ and set $E = D - \sum_{k=3}^{n}[C_k, X_k]$. The equation (8) is now equivalent to $[C_1, X_1] + [C_2, X_2] = E$. In order to solve our original problem, it suffices to find an element $Y \in \mathfrak{g}$ such that

$$p_j(C_1, \ldots, C_1, Y) = p_j(C_1, \ldots, C_1, E), \qquad p_j(C_2, \ldots, C_2, Y) = 0,$$

for all $1 \leq j \leq r$. Indeed, by the result of the previous paragraph, if we have an element $Y \in \mathfrak{g}$ satisfying these relations, we may successively solve the equations $[C_2, X_2] = Y$ for $X_2 \in \mathfrak{g}$, and $[C_1, X_1] = E - Y$ for $X_1 \in \mathfrak{g}$. A sufficient condition for the existence of Y is that the $2r$ linear functionals

$$X \mapsto p_j(C_1, \ldots, C_1, X), \qquad X \mapsto p_j(C_2, \ldots, C_2, X),$$

where $X \in \mathfrak{g}$, $1 \leq j \leq r$, belonging to $\mathfrak{g}^*$ be linearly independent. A result of Raïs asserts that the Zariski-open subset U of $\mathfrak{g} \times \mathfrak{g}$ consisting of those elements (C_1, C_2) which satisfy this condition is nonempty (see [3]). It relies on the construction of principal S-triples for complex semisimple Lie algebras due to Kostant [5]. In fact, according to Kostant [6], if $(C_1, C_2) \in U$, both C_1 and C_2 are principal elements of $\mathfrak{g}$. Thus the system (8) is solvable for all $D \in \mathfrak{g}$, provided the pair (C_1, C_2) is generic.

For our analysis of the prolongations of the symbol of the system (5), we need another immediate consequence of the invariance equation (7) (see [3, Proposition 2.1]):

LEMMA 2. *Suppose that p is a homogeneous polynomial of $\mathscr{I}(\mathfrak{g})$ of degree d. If $u \in S^d T^* \otimes \mathfrak{g}$, then we have*

$$\sum_{i_1, \ldots, i_d = 1}^{n} p(J_{i_1}, \ldots, J_{i_{d-1}}, [J_{i_d}, u_{i_1 \ldots i_d}]) = 0.$$

If p is a homogeneous polynomial of $\mathscr{I}(\mathfrak{g})$ of degree d, for $k > d$ we define a morphism of vector bundles $\psi_p : S^k T^* \otimes \mathfrak{g} \to S^{k-d-1} T^*$ by

$$\psi_p(w)(\partial_{j_1}, \ldots, \partial_{j_{k-d-1}}) = \sum_{i_1, \ldots, i_{d-1}, j=1}^{n} p(J_{i_1}, \ldots, J_{i_{d-1}}, w_{i_1 \ldots i_{d-1} j j j_1 \ldots j_{k-d-1}}),$$

for $w \in S^k T^* \otimes \mathfrak{g}$. For $k \geq 2$, it is easily seen that $\psi_p \cdot \tau_k(u) = 0$, for all $u \in \operatorname{Ker} \sigma_{k-2}$; we thus obtain the complex

$$(9) \qquad \operatorname{Ker} \sigma_{k-2} \xrightarrow{\tau_k} S^k T^* \otimes \mathfrak{g} \xrightarrow{\Psi} \bigoplus_{1 \leq j \leq r} S^{k-d_j-1} T^*,$$

where $\Psi = \bigoplus_{1 \leq j \leq r} \psi_{p_j}$ and d_j is the degree of p_j. According to Theorem 2.2 of [3], the sequences (9) are exact for all $k \geq 2$, if J is generic in the sense that $(J_1, \ldots, J_n)$ takes its values in a certain nonempty Zariski-open subset $\mathscr{O}_1$ of $\mathfrak{g}^n$. Consequently, under the assumption that J is generic, we see that a jet $j_{k+1}(A)(x)$ of order $k+1$ of a connection A at $x \in M$, satisfying $j_{k-1}(\delta^A F^A - J)(x) = 0$ and $j_k(\delta^A J)(x) = 0$, can be modified by an element of $S^{k+1} T_x^* \otimes T_x^* \otimes \mathfrak{g}$ to become a formal solution of (5) of order $k+1$ at x if and only if

$$(10) \qquad \Psi j_{k+1}(\delta^A J)(x) = 0.$$

If p is an element of our generating set for $\mathscr{I}(\mathfrak{g})$ of degree k, then (10) includes the relation

$$(11) \qquad \sum_{i_1,\ldots,i_{k-1}=1}^{n} p(J_{i_1},\ldots,J_{i_{k-1}},\partial_{i_1\ldots i_{k-1}}\Delta\delta^A J)(x) = 0,$$

where $\Delta = \sum_{i=1}^{n}\partial_{ii}$ is the Laplacian.

We now use the equation $j_{k-1}(\delta^A F^A - J)(x) = 0$ and Lemma 2 to rewrite the condition (11). In fact, if the connection A is given by (1) and $\delta^A F^A = -\sum_{j=1}^{n} dx^j \otimes \theta_j$, then we have

$$(12) \qquad \begin{aligned} &\sum_{i_1,\ldots,i_{k-1}=1}^{n} p(J_{i_1},\ldots,J_{i_{k-1}},\partial_{i_1\ldots i_{k-1}}\Delta\delta^A J) = -\Phi_p(A) \\ &+ \sum_{i_1,\ldots,i_{k-1},i=1}^{n} p(J_{i_1},\ldots,J_{i_{k-1}},[J_i,\partial_{i_1\ldots i_{k-1}}(\theta_i + J_i)]), \end{aligned}$$

where $\Phi_p(A)$ is an expression which only involves derivatives of A up to order k. In fact, we have

$$\Phi_p(A) = \sum_{i_1,\ldots,i_{k-1}=1}^{n} p(J_{i_1},\ldots,J_{i_{k-1}},\phi_p(A)_{i_1\ldots i_{k-1}}) + \rho_p(J,A),$$

where

$$\begin{aligned} \phi_p(A)_{i_1\ldots i_{k-1}} = \sum_{i,l=1}^{n} &\Big([J_i,[\partial_{li_1\ldots i_{k-1}}A_l,A_i] + 2[A_l,\partial_{li_1\ldots i_{k-1}}A_i]] \\ &- 2[\partial_l J_i,\partial_{li_1\ldots i_{k-1}}A_i] - \sum_{q=1}^{k-1}[\partial_{i_q}J_i,\partial_{lli_1\ldots\widehat{i_q}\ldots i_{k-1}}A_i]\Big) \end{aligned}$$

and $\rho_p(J,A)$ is an expression which only involves derivatives of A up to order $k-1$. Clearly a solution A of (5) also satisfies the equation $\Phi_p(A) = 0$ of order k. From the equality (12), under the above assumptions on $j_{k+1}(A)(x)$, we deduce that the relation (11) is equivalent to the condition $\Phi_p(A)(x) = 0$ imposed on $j_k(A)(x)$. More precisely, under the assumption that J is generic, we see that, for $1 \le j \le r$, the new identity $\Phi_{p_j}(A) = 0$ must be satisfied to the appropriate order $k - d_j$ at x so that we can modify $j_{k+1}(A)(x)$ to become a solution of (5) of order $k+1$ at x.

We now adjoin these r scalar-valued identities to our system (5). We set $d = \sup(d_j)$ and analyze the system

$$(13) \qquad \begin{aligned} &j_{d-2}(\delta^A F^A - J) = 0, \\ &j_{d-d_j}(\Phi_{p_j}(A)) = 0, \qquad j_d(\delta^A J) = 0, \end{aligned}$$

with $1 \le j \le r$, of order d for the connection A over M by the methods used above. The details are carried out in [3]. If p is an element of our generating set of $\mathscr{I}(\mathfrak{g})$ of degree k and $B = \sum_{j=1}^{n} dx^j \otimes B_j$ is an element of $T_x^* \otimes \mathfrak{g}$, with $x \in M$, the symbol

$$\sigma(\Phi_p)_B : S^k T_x^* \otimes T_x^* \otimes \mathfrak{g} \to \mathbb{R}$$

of the operator Φ_p at B is given by

$$\sigma(\Phi_p)_B(u) = \sum_{i_1, \ldots, i_{k-1}, i, l=1}^{n} p(J_{i_1}, \ldots, J_{i_{k-1}}, v_{i_1 \cdots i_{k-1} il}),$$

for $u \in S^k T_x^* \otimes T_x^* \otimes \mathfrak{g}$, where

$$v_{i_1 \cdots i_{k-1} il} = [J_i(x), [u_{l i_1 \cdots i_{k-1} l}, B_i] + 2[B_l, u_{l i_1 \cdots i_{k-1} i}]]$$

$$- 2[(\partial_l J_i)(x), u_{l i_1 \cdots i_{k-1} i}] - \sum_{q=1}^{k-1} [(\partial_{i_q} J_i)(x), u_{l l i_1 \cdots \widehat{i_q} \cdots i_{k-1} i}].$$

We denote by $\sigma((\Phi_p)_B)_{+l} : S^{k+l} T_x^* \otimes T_x^* \otimes \mathfrak{g} \to S^l T_x^*$ the lth prolongation of $\sigma(\Phi_p)_B$ and consider the morphism

$$\chi_{k,B} = \bigoplus_{1 \le j \le r} (\sigma(\Phi_{p_j})_B)_{+(k-d_j)} : S^k T_x^* \otimes T_x^* \otimes \mathfrak{g} \to \bigoplus_{1 \le j \le r} S^{k-d_j} T_x^*.$$

In [3], a certain nonempty Zariski-open subset $\mathscr{O}_2$ (depending only on $\mathfrak{g}$ and on our choice of generators of $\mathscr{I}(\mathfrak{g})$) of $\mathfrak{g}^m$, with $m = n(n+1)$, is constructed which has the following property: Whenever $((J_i, \partial J_i / \partial x^l)(x))_{1 \le i, l \le n}$ belongs to $\mathscr{O}_2$, with $x \in M$, then $(J_1(x), \ldots, J_n(x))$ is an element of $\mathscr{O}_1$ and we can find an element $B \in T_x^* \otimes \mathfrak{g}$ which satisfies the equation

$$\sum_{i=1}^{n} [J_i(x), B_i] = \sum_{i=1}^{n} \frac{\partial J_i}{\partial x^i}(x),$$

and for which the sequences

$$\mathrm{Ker}\, \sigma_{k-2,x} \xrightarrow{\tau_k \oplus \chi_{k,B}} (S^k T_x^* \otimes \mathfrak{g}) \oplus \bigoplus_{1 \le j \le r} S^{k-d_j} T_x^* \xrightarrow{\Psi} \bigoplus_{1 \le j \le r} S^{k-d_j-1} T_x^* \to 0$$

are exact, for all $k \ge 2$. Using this property of $\mathscr{O}_2$, Lemma 1 and the exactness of the sequences (9), if $(J_i, \partial J_i / \partial x^l)(x)$ belongs to $\mathscr{O}_2$, in [3] we prove the existence of strongly prolongable formal solutions of the system (13) at x. From Malgrange's theorem we then deduce the existence of a real-analytic connection A on E satisfying $\delta^A F^A = J$ on a neighborhood of x.

The argument given in [3] also proves that the equation (13) is formally integrable in the sense of [4] (see also [1, Chapter IX]).

THEOREM 2. *If $(J_i, \partial J_i / \partial x^l)$ takes its values in $\mathscr{O}_2$ on an open subset U of M, then the equation (13) is formally integrable on U and its symbol is involutive.*

Theorem 2 holds only in the case of flat metrics and does not require the real-analyticity of E. Theorem 1 can be seen to be valid for arbitrary metrics by considering directly the system (13) at $x \in M$ in terms of a system of normal coordinates at x.

REFERENCES

1. R. Bryant, S. S. Chern, R. Gardner, H. Goldschmidt, and P. Griffiths, *Exterior differential systems*, Math. Sci. Res. Inst. Publ., vol. 18, Springer-Verlag, New York, Berlin, Heidelberg, 1991.

2. D. DeTurck, *Existence of metrics with prescribed Ricci curvature: local theory*, Invent. Math. **65** (1981), 179–207.

3. D. DeTurck, H. Goldschmidt, and J. Talvacchia, *Connections with prescribed curvature and Yang-Mills currents: The semi-simple case*, Ann. Sci. École Norm. Sup. (4) **24** (1991), 57–112.

4. H. Goldschmidt, *Integrability criteria for systems of non-linear partial differential equations*, J. Differential Geom. **1** (1967), 269–307.

5. B. Kostant, *The principal three-dimensional subgroup and the Betti numbers of a complex simple Lie group*, Amer. J. Math. **81** (1959), 973–1032.

6. ———, *Lie group representations on polynomial rings*, Amer. J. Math. **85** (1963), 327–404.

7. B. Malgrange, *Équations de Lie*. II, J. Differential Geom. **7** (1972), 117–141.

8. V. S. Varadarajan, *Lie groups, Lie algebras and their representations*, Graduate Texts in Math., vol. 102, Springer-Verlag, New York, Berlin, Heidelberg, 1984.

UNIVERSITY OF PENNSYLVANIA
E-mail address: deturck@math.upenn.edu

COLUMBIA UNIVERSITY
E-mail address: hg@shire.math.columbia.edu

SWARTHMORE COLLEGE
E-mail address: talvacchia@campus.swarthmore.edu

Proceedings of Symposia in Pure Mathematics
Volume **54** (1993), Part 2

Vector Bundles over Homogeneous Spaces and Complete, Locally Symmetric Spaces

LANCE D. DRAGER AND ROBERT L. FOOTE

ABSTRACT. Let M be a reductive homogeneous space of the Lie group G with Lie algebra $\mathfrak{g}$. Let $\rho : G \to GL(V)$ be a finite-dimensional representation, and suppose $\mathscr{V}^1 \oplus \mathscr{V}^2 \to M$ is an equivariant, Whitney sum decomposition of the trivial bundle $M \times V \to M$. The reductivity gives rise to the well-known, canonical connection on the principal bundle $G \to M$, which leads to a covariant derivative ∇ on $M \times V \to M$. We use the Whitney sum decomposition to define a second covariant derivative $\widehat{\nabla}$ on $M \times V \to M$, and we determine necessary and sufficient conditions on ρ to have $\nabla = \widehat{\nabla}$.

We apply this to the Whitney sum decomposition $\mathfrak{m} \oplus \mathfrak{h} \to M$ of the trivial bundle $M \times \mathfrak{g} \to M$. Here the fiber $\mathfrak{h}_p$ of $\mathfrak{h}$ over $p \in M$ is the Lie algebra of the isotropy group H_p of p, and the fiber $\mathfrak{m}_p$ of $\mathfrak{m}$ is an appropriate $\mathrm{Ad}(H_p)$-invariant complement in $\mathfrak{g}$.

THEOREM. *M is locally symmetric if and only if $\nabla = \widehat{\nabla}$.*

In earlier work [**DFM**] we investigated the relationships between curves in Grassmannians, sub-Grassmannians, and linear control systems. We were led to study the Whitney sum decomposition $U \oplus U^\perp \to G_k(\mathbb{R}^n)$ of the trivial bundle $G_k(\mathbb{R}^n) \times \mathbb{R}^n \to G_k(\mathbb{R}^n)$. ($U^\perp \to G_k(\mathbb{R}^n)$ is the co-universal bundle, defined below.) We used the projections onto the summands to define a covariant derivative $\widehat{\nabla}$ on this bundle. Since $G_k(\mathbb{R}^n)$ is a reductive homogeneous space of $O(\mathbb{R}^n)$, the reductive structure also induces a covariant derivative ∇. We found that the two derivatives are equal. In [**DFMW**] we studied the corresponding situation for $\Sigma_k(\mathbb{R}^n)$, a generalized Grassmannian (defined below), and found again that the two derivatives are equal.

In this paper we describe the general context in which these two derivatives

---

1991 *Mathematics Subject Classification.* Primary 53C30.

Research of the second author supported in part by the Byron K. Trippet research fund at Wabash College.

The final version of this paper has been submitted for publication elsewhere.

are both defined and how they are related. We use this to characterize the complete, locally symmetric spaces among the reductive homogeneous spaces. Most proofs are omitted. Complete details will appear elsewhere.

Whitney sum decompositions of trivial bundles

Let M be a manifold and let V be a finite-dimensional vector space. Suppose that $V = V_p^1 \oplus V_p^2$ is a smooth decomposition of V parameterized by $p \in M$. By smooth we mean that the maps $p \mapsto V_p^i$, $i = 1, 2$, into appropriate Grassmannians are smooth. Then the trivial bundle $M \times V \to M$ splits into the Whitney sum $\mathscr{V}^1 \oplus \mathscr{V}^2 \to M$, where the fiber of $\mathscr{V}^i \to M$ over p is V_p^i, $i = 1, 2$. We will call $\{V_p^i\}$ a smooth family of subspaces of V parameterized by M. The notation $\mathscr{V}^i$ will also be used to the denote the family.

A familiar example is when $M \subset \mathbb{R}^n$. For each $p \in M$ we have $\mathbb{R}^n = T_pM \oplus \nu_pM$, where T_pM and ν_pM are, respectively, the tangent and normal spaces of M at p. We have the decomposition $TM \oplus \nu M = M \times \mathbb{R}^n \to M$.

For a second example, fix $1 \le k \le n - 1$ and let $G_k(\mathbb{R}^n)$ be the Grassmannian of k-dimensional subspaces of $\mathbb{R}^n$. For $S \in G_k(\mathbb{R}^n)$ let $S^\perp$ be its orthogonal complement. Let $U \to G_k(\mathbb{R}^n)$ be the universal bundle and $U^\perp \to G_k(\mathbb{R}^n)$ the co-universal bundle, in which the fibers over S are, respectively, S and $S^\perp$. Then $U \oplus U^\perp = G_k(\mathbb{R}^n) \times \mathbb{R}^n \to G_k(\mathbb{R}^n)$.

Our third example generalizes the second, and will play a central role in the study of the general case. Fix $1 \le k \le n - 1$, $n = \dim V$, and let

$$\Sigma_k(V) = \{(V^1, V^2) \in G_k(V) \times G_{n-k}(V) \mid V^1 \oplus V^2 = V\}.$$

We call $\Sigma_k(V)$ a splitting space. There are two "universal" bundles U^1, $U^2 \to \Sigma_k(V)$; the fiber of $U^i \to \Sigma_k(V)$ over (V^1, V^2) is V^i, $i = 1, 2$. We have $U^1 \oplus U^2 = \Sigma_k(V) \times V \to \Sigma_k(V)$. The splitting space enters into the study of the general smooth decomposition $V = V_p^1 \oplus V_p^2$ via the map $f: M \to \Sigma_k(V)$ defined by $f(p) = (V_p^1, V_p^2)$, where k is the fiber dimension of $\mathscr{V}^1 \to M$. It is clear that the Whitney sum $\mathscr{V}^1 \oplus \mathscr{V}^2 \to M$ is the pull-back of $U^1 \oplus U^2 \to \Sigma_k(V)$ via f.

For the balance of the paper we will assume that M is a reductive homogeneous space of the Lie group G. Many of our constructions do not require this much structure, but this is the setting common to all of them. Our main references for homogeneous spaces are [H] and [KN].

Let $\rho: G \to GL(V)$ be a representation. There is an obvious action of G on the trivial bundle $M \times V \to M$ determined by ρ, namely $g(p, v) = (gp, \rho(g)v)$. We will assume the decomposition $V = V_p^1 \oplus V_p^2$ is invariant under this action, that is, $\rho(g)V_p^i = V_{gp}^i$ for $p \in M$, $g \in G$, $i = 1, 2$. We will say that the families $\mathscr{V}^i$ are $\rho(G)$-invariant. Another way to say this is that G acts on the Whitney sum $\mathscr{V}^1 \oplus \mathscr{V}^2 \to M$. The bundles $U \oplus U^\perp \to G_k(\mathbb{R}^n)$ and $U^1 \oplus U^2 \to \Sigma_k(V)$ are of this type. The groups are $O(\mathbb{R}^n)$ and $GL(V)$, respectively, and the representations are the identity in both cases.

Our main example involves the reductive structure itself. Let $\mathfrak{g}$ be the Lie algebra of G. For each $p \in M$ let H_p be the isotropy of p in G, and let $\mathfrak{h}_p$ be its Lie algebra. The family $\mathfrak{h} = \{\mathfrak{h}_p\}$ of subspaces of $\mathfrak{g}$ is $\mathrm{Ad}(G)$-invariant, where $\mathrm{Ad}: G \to GL(\mathfrak{g})$ is the adjoint representation. Reductivity means that there is a second $\mathrm{Ad}(G)$-invariant family $\mathfrak{m} = \{\mathfrak{m}_p\}$ of subspaces of $\mathfrak{g}$ such that $\mathfrak{m}_p \oplus \mathfrak{h}_p = \mathfrak{g}$ for all $p \in M$. For each $p \in M$, $\mathfrak{m}_p$ is $\mathrm{Ad}(H_p)$-invariant. The whole family can be recovered from an $\mathrm{Ad}(H_p)$-invariant $\mathfrak{m}_p$ for a single p (the usual definition of reductivity) by setting $\mathfrak{m}_{gp} = \mathrm{Ad}(g)\mathfrak{m}_p$ for $g \in G$. We have the bundles $\mathfrak{m} \to M$ and $\mathfrak{h} \to M$ with fibers $\mathfrak{m}_p$ and $\mathfrak{h}_p$ over p, respectively, and the decomposition $\mathfrak{m} \oplus \mathfrak{h} = M \times \mathfrak{g} \to M$.

Given a reductive structure $\mathfrak{m}$, the bundles $\mathfrak{m} \to M$ and $TM \to M$ are equivalent. For each $p \in M$ we have the map $\pi_p: G \to M$ given by $\pi_p(g) = gp$. The equivalence is induced by the isomorphisms $\pi_{p*}: \mathfrak{m}_p \to T_pM$, $p \in M$. We will use this equivalence to identify $\mathfrak{m}$ and TM. Then TM is a subbundle of $M \times \mathfrak{g}$, and the action of G on TM is given by $(g, X) \mapsto \mathrm{Ad}(g)X$.

Covariant derivatives

We next define two covariant derivatives, ∇ and $\widehat{\nabla}$, on the bundles $\mathscr{V}^1 \oplus \mathscr{V}^2 = M \times V \to M$ and $\mathscr{V}^i \to M$, $i = 1, 2$.

The reductive structure $\mathfrak{m}$ induces the "canonical connection" on the principal bundle $G \xrightarrow{\pi_p} M$ in a standard way. This connection in turn induces a covariant derivative on every vector bundle induced by a representation of H_p (see [KN]). The bundles $M \times V \to M$ and $\mathscr{V}^1, \mathscr{V}^2 \to M$ are induced by the restricted representations $\rho: H_p \to GL(V)$ and $\rho: H_p \to GL(V_p^i)$, $i = 1, 2$, respectively. (Note that the $\rho(G)$-invariance of the family $\mathscr{V}^i = \{V_p^i\}$ implies the $\rho(H_p)$-invariance of the subspace V_p^i.) We denote the induced covariant derivative by ∇.

The second covariant derivative depends on the Whitney sum decomposition $\mathscr{V}^1 \oplus \mathscr{V}^2 = M \times V \to M$ for its definition. Let v be a smooth section of $M \times V \to M$, which we think of as a function $v: M \to V$. (We will take all vector bundle sections to be smooth.) For $X \in T_pM$, let $D_X v = Xv$, thought of as an element in the fiber over p. (Literally, $D_X v = (p, Xv)$, but we will drop the first component when the fiber is clear.) We will think of D as the trivial covariant derivative on $M \times V \to M$. Let $P^i: \mathscr{V}^1 \oplus \mathscr{V}^2 = M \times V \to \mathscr{V}^i$, $i = 1, 2$, be the vector bundle projections and define $\widehat{\nabla}_X v = P^1 D_X(P^1 v) + P^2 D_X(P^2 v)$. This is analogous to the standard way of inducing a covariant derivative on $TM \to M$ from the ambient derivative D when M is a submanifold of $\mathbb{R}^n$.

The subbundles $\mathscr{V}^1, \mathscr{V}^2 \to M$ of $M \times V \to M$ are parallel for both ∇ and $\hat{\nabla}$. In a sense, the two derivatives both measure the "rotation" of the decomposition $V = V_p^1 \oplus V_p^2$, $p \in M$. In many familiar situations they are equal, but in general they are different.

In order to study $\hat{\nabla}$ on $\mathscr{V}^1 \oplus \mathscr{V}^2 \to M$, we need to understand the tangent bundle of $\Sigma_k(V)$ and the nature of $\hat{\nabla}$ on $U^1 \oplus U^2 \to \Sigma_k(V)$, which we do next in some detail.

The tangent bundle of $\Sigma_k(V)$

The group $GL(V)$ acts on $\Sigma_k(V)$ by $A(V^1, V^2) = (AV^1, AV^2)$. This action is clearly transitive, and so $\Sigma_k(V)$ is a homogeneous space of $GL(V)$. The isotropy $H_{(V^1, V^2)}$ of $(V^1, V^2) \in \Sigma_k(V)$ is the subgroup of $GL(V)$ of all A such that $AV^1 = V^1$ and $AV^2 = V^2$. Its Lie algebra $\hat{\mathfrak{h}}_{(V^1, V^2)}$ is the subalgebra of $\mathfrak{gl}(V)$ of all X such that $XV^1 \subset V^1$ and $XV^2 \subset V^2$. Let $\hat{\mathfrak{m}}_{(V^1, V^2)}$ be the subspace of $\mathfrak{gl}(V)$ of all X such that $XV^1 \subset V^2$ and $XV^2 \subset V^1$, that is, X "switches" the subspaces V^1 and V^2. Clearly $\mathfrak{gl}(V) = \hat{\mathfrak{m}}_{(V^1, V^2)} \oplus \hat{\mathfrak{h}}_{(V^1, V^2)}$. It is easily verified that the family $\hat{\mathfrak{m}} = \{\hat{\mathfrak{m}}_{(V^1, V^2)}\}$ is invariant under the adjoint action of $GL(V)$ on $\mathfrak{gl}(V)$, and so $\hat{\mathfrak{m}}$ is a reductive structure.

As with any reductive structure, the spaces $T_{(V^1, V^2)}\Sigma_k(V)$ and $\hat{\mathfrak{m}}_{(V^1, V^2)}$ are isomorphic. Before identifying them, we would like to make this isomorphism a bit more concrete.

PROPOSITION 1. *Let $X \in T_{(V^1, V^2)}\Sigma_k(V)$, and let L_X be the corresponding element of $\hat{\mathfrak{m}}_{(V^1, V^2)}$. Then L_X is the difference tensor between $\hat{\nabla}_X$ and D_X. More precisely, if v is a section of $U^1 \oplus U^2 = \Sigma_k(V) \times V \to \Sigma_k(V)$, then $\hat{\nabla}_X v = D_X v - L_X v$.*

PROOF. Let $\gamma(t) = \exp(tL_X)(V^1, V^2)$. Then $\gamma'(0) = X$. Let $v_0 \in V^1$, and consider $v(t) = \exp(tL_X)v_0$ as a section of $U^1 \to \Sigma_k(V)$ along γ. Then $\hat{\nabla}_X v(0) = P^1 v'(0) = P^1 L_X v_0 = 0$, since $L_X v_0 \in V^2$. Since $D_X v(0) = v'(0) = L_X v_0$, we have $\hat{\nabla}_X v(0) = D_X v(0) - L_X v_0$. This last formula is easily seen to hold if we start with $v_0 \in V^2$ as well. Since the difference between D_X and ∇_X is a tensor, $\hat{\nabla}_X v = D_X v - L_X v$ holds for all sections v.

For the remainder of the paper we will identify $T_{(V^1, V^2)}\Sigma_k(V)$ and $\hat{\mathfrak{m}}_{(V^1, V^2)}$, writing X for L_X. The formula in Proposition 1 then becomes $\hat{\nabla}_X v = D_X v - Xv$. Note that Xv does *not* denote a derivative here. Rather it denotes the image of v under the linear map $X : V \to V$.

It is interesting to note that there is a semi-Riemannian structure on $\Sigma_k(V)$. For $X, Y \in T_{(V^1, V^2)}\Sigma_k(V) = \hat{\mathfrak{m}}_{(V^1, V^2)}$, let $\langle X, Y \rangle = \operatorname{tr} XY$. This is

$GL(V)$-invariant and is easily seen to be non-degenerate, although not definite, and so it is a pseudo-metric. The canonical connection is the Rieman-
nian connection for $\langle\,\cdot\,,\,\cdot\,\rangle$. It is worth noting that $\Sigma_k(V)$ is also a reductive
homogeneous space of the semi-simple group $SL(V)$.

Comparison of ∇ and $\widehat{\nabla}$

We return to the study of ∇ and $\widehat{\nabla}$ on the bundle $\mathscr{V}^1 \oplus \mathscr{V}^2 = M \times V \to$
M. Recall that the families $\mathscr{V}^i = \{V_p^i\}$, $i = 1, 2$, are $\rho(G)$-invariant for
the representation $\rho\colon G \to GL(V)$. In addition, we have the map $f\colon M \to$
$\Sigma_k(V)$, where k is the fiber dimension of $\mathscr{V}^1$, defined by $f(p) = (V_p^1, V_p^2)$,
and $\mathscr{V}^1 \oplus \mathscr{V}^2 = f^*(U^1 \oplus U^2)$.

PROPOSITION 2. *The derivative $\widehat{\nabla}$ on $\mathscr{V}^1 \oplus \mathscr{V}^2 \to M$ is the pull-back of $\widehat{\nabla}$
on $U^1 \oplus U^2 \to \Sigma_k(V)$. If v is a section of $\mathscr{V}^1 \oplus \mathscr{V}^2 \to M$ and $X \in T_pM$,
then*

$$\widehat{\nabla}_X v = D_X v - f_*(X)v \quad and \quad \nabla_X v = D_X v - \rho_*(X)v.$$

*Furthermore, $f_*(X) = \rho_*(X)_{\widehat{\mathfrak{m}}_{f(p)}}$, and so the difference tensor between ∇ and
$\widehat{\nabla}$ on $\mathscr{V}^1 \oplus \mathscr{V}^2 \to M$ is given by $\widehat{\nabla}_X v - \nabla_X v = \rho_*(X)_{\widehat{\mathfrak{h}}_{f(p)}} v$.*

These formulas for $\widehat{\nabla}$ and ∇ make sense as a result of our identifications of
$T_{f(p)}\Sigma_k(V)$ with $\widehat{\mathfrak{m}}_{f(p)}$ and T_pM with $\mathfrak{m}_p$: $f_*(X)$ is a linear map $V \to V$
that switches V_p^1 and V_p^2, and X is in the domain of ρ_*. The subscripts
$\widehat{\mathfrak{m}}_{f(p)}$ and $\widehat{\mathfrak{h}}_{f(p)}$ refer to components relative to the decomposition $\mathfrak{gl}(V) =$
$\widehat{\mathfrak{m}}_{f(p)} \oplus \widehat{\mathfrak{h}}_{f(p)}$.

As a special case, consider ∇ and $\widehat{\nabla}$ on $U^1 \oplus U^2 = \Sigma_k(V) \times V \to \Sigma_k(V)$.
Computing directly or using the proposition (the representation $GL(V) \to$
$GL(V)$ is the identity), one finds that $\nabla_X v = D_X v - Xv$. This is the same
as the formula for $\widehat{\nabla}$ given in Proposition 1, and so $\nabla = \widehat{\nabla}$ on $U^1 \oplus U^2 \to$
$\Sigma_k(V)$. It is also easily seen that $\nabla = \widehat{\nabla}$ on $U \oplus U^\perp \to G_k(\mathbb{R}^n)$. In general,
we have the following theorem. Its proof follows easily from Proposition 2.

THEOREM 1. *The following statements are equivalent*: (1) $\nabla = \widehat{\nabla}$,
(2) $\rho_*(X) = f_*(X)$ for all $X \in TM = \mathfrak{m}$,
(3) $\rho_*(X)V_p^1 \subset V_p^2$ and $\rho_*(X)V_p^2 \subset V_p^1$ for all $p \in M$ and $X \in \mathfrak{m}_p$,
(4) $\rho_*(\mathfrak{m}_p) \subset \widehat{\mathfrak{m}}_{f(p)}$ for all $p \in M$, and
(5) $\rho_*(X)_{\widehat{\mathfrak{h}}_{f(p)}} = 0$ for all $p \in M$ and $X \in \mathfrak{m}_p$.

The conditions $\rho_*(X)V_p^1 \subset V_p^2$ and $\rho_*(X)V_p^2 \subset V_p^1$ of (3) are easily seen to
be equivalent, respectively, to the vanishing of the difference tensor $\rho_*(X)_{\widehat{\mathfrak{h}}_{f(p)}}$
on V_p^1 and V_p^2.

As a consequence of Proposition 2 and Theorem 1 we can give interpre-
tations of $\widehat{\nabla}$ and ∇. Both derivatives are sensitive to the "rotation" of the

decomposition $V = V_p^1 \oplus V_p^2$ as $p \in M$ moves. The infinitesimal rotation is measured geometrically by $f_*(X)$ thought of as a vector tangent to $\Sigma_k(V)$. It affects $\widehat{\nabla}$ and ∇ by the action of $f_*(X) = \rho_*(X)_{\widehat{\mathfrak{m}}_{f(p)}}$ thought of as a linear map. In addition, ∇ is sensitive to any "internal rotations" of V_p^1 and V_p^2 caused by ρ, since $\rho_*(X)$ can have a component in $\widehat{\mathfrak{h}}_{f(p)}$. The two derivatives are equal when ρ causes no internal rotations of the summands.

Characterization of complete, locally symmetric spaces

We now apply the general theory of the previous sections to our main example. Recall the setting. We have the $\mathrm{Ad}(G)$-invariant families $\mathfrak{h} = \{\mathfrak{h}_p\}$ and $\mathfrak{m} = \{\mathfrak{m}_p\}$ of subspaces of $\mathfrak{g}$, where $\mathrm{Ad}: G \to GL(\mathfrak{g})$ denotes the adjoint representation. These give rise to the bundles $\mathfrak{h} \to M$ and $\mathfrak{m} \to M$ and the Whitney sum decomposition $\mathfrak{m} \oplus \mathfrak{h} = M \times \mathfrak{g} \to M$. The splitting space involved is $\Sigma_n(\mathfrak{g})$, where $n = \dim M$. We have $f: M \to \Sigma_n(\mathfrak{g})$ given by $f(p) = (\mathfrak{m}_p, \mathfrak{h}_p)$.

Our main result is roughly that $\nabla = \widehat{\nabla}$ if and only if M is locally symmetric. Recall that a locally symmetric space is a manifold M with a covariant derivative on $TM \to M$ such that the local geodesic symmetry at each point (the map that reverses geodesics) is affine (see [KN, H]). To be precise, we need to specify which derivative we are using for the sufficiency part of the claim. As it turns out, it can be either. (Note that ∇ and $\widehat{\nabla}$ make sense as derivatives on $TM \to M$ since $TM = \mathfrak{m} \to M$ is a parallel subbundle of $\mathfrak{m} \oplus \mathfrak{h} \to M$ for both derivatives.)

THEOREM 2. *The following statements are equivalent*: (1) $\nabla = \widehat{\nabla}$ *on* $\mathfrak{m} \oplus \mathfrak{h} \to M$,

(2) (M, ∇) *is locally symmetric*,

(3) $(M, \widehat{\nabla})$ *is locally symmetric, and*

(4) $[\mathfrak{m}, \mathfrak{m}] \subset \mathfrak{h}$.

The condition $[\mathfrak{m}, \mathfrak{m}] \subset \mathfrak{h}$ means, of course, $[\mathfrak{m}_p, \mathfrak{m}_p] \subset \mathfrak{h}_p$ for all $p \in M$. This latter condition is $\mathrm{Ad}(G)$-invariant, so if it holds for one p, it holds for all. This is a familiar property of symmetric spaces, and so its connection with local symmetry is not too surprising.

SKETCH. The proof is an application of Theorem 1 and Proposition 2. A space is locally symmetric if and only if its torsion tensor vanishes and its curvature tensor is parallel [KN]. The curvature for ∇ is always ∇-parallel since it is G-invariant [KN]. (The distinguishing feature here is that ∇ is the derivative induced by the reductive structure.) Thus (M, ∇) is locally symmetric if and only if its torsion vanishes.

Let Δ be the difference tensor between the derivatives: $\Delta_X Y = \widehat{\nabla}_X Y - \nabla_X Y$, where $X \in T_p M = \mathfrak{m}_p$ and Y is a section of $\mathfrak{m} \oplus \mathfrak{h} \to M$. A simple computation using Proposition 2 shows that $\Delta_X Y = 0$ when $Y \in \mathfrak{h}_p$ and $\Delta_X Y = -\mathrm{Tor}(X, Y) = \widehat{\mathrm{Tor}}(X, Y)$ when $Y \in T_p M = \mathfrak{m}_p$, where Tor and

$\widehat{\mathrm{Tor}}$ are the torsion tensors. It is then clear that (1), (2), and (3) are each equivalent to the simultaneous vanishing of Δ, Tor, and $\widehat{\mathrm{Tor}}$.

The equivalence of (1) and (4) is an application of Theorem 1. The difference tensor already vanishes on $\mathfrak{h}_p$, as noted above. It turns out that it vanishes on $\mathfrak{m}_p$ if and only if $[\mathfrak{m}_p, \mathfrak{m}_p] \subset \mathfrak{h}_p$.

We close with the following remark.

REMARK. Consider ∇ and $\widehat{\nabla}$ on $\mathfrak{m} \oplus \mathfrak{h} \to M$ when $\nabla \neq \widehat{\nabla}$. Since $\nabla = \widehat{\nabla}$ on $\mathfrak{h}$, the difference between the derivatives appears when they are applied to sections of $\mathfrak{m} = TM \to M$. Thus, $\widehat{\nabla}$ is a canonical, G-invariant, covariant derivative on $TM \to M$, different from the standard one, which does not seem to have been previously studied. The difference tensor on TM is $\Delta_X Y = \mathrm{Ad}_*(X)_{\widehat{\mathfrak{h}}_{f(p)}} Y = [X, Y]_{\mathfrak{m}_p}$, which is antisymmetric, and so ∇ and $\widehat{\nabla}$ have the same geodesics. Since $\widehat{\mathrm{Tor}} = -\mathrm{Tor}$, the torsion-free derivative with the same geodesics is $\frac{1}{2}(\nabla + \widehat{\nabla})$.

REFERENCES

[DFM]   L.D. Drager, R.L. Foote, and C.F. Martin, *Controllability of linear systems,* Differential Geometry of Curves in Grassmannians, and Riccati Equations, Lecture Notes, Dept. of Math., Texas Tech Univ., 1986.

[DFMW] L.D. Drager, R.L. Foote, C.F. Martin, and J. Wolper, *Controllability of linear systems,* Differential Geometry of Curves in Grassmannians and Generalized Grassmannians, and Riccati Equations, Acta Appl. Math. **16** (1989), 281–317.

[H]   S. Helgason, *Differential geometry, Lie groups, and symmetric spaces,* Academic Press, New York, 1978.

[KN]   S. Kobayashi and K. Nomizu, *Foundations of differential geometry,* 2 Vols., Wiley-Inter-science, New York, 1963.

TEXAS TECH UNIVERSITY
E-mail address: drager@ttmath.ttu.edu

WABASH COLLEGE
E-mail address: footer@wabash.bitnet

Proceedings of Symposia in Pure Mathematics
Volume **54** (1993), Part 2

Margulis Space-Times

TODD A. DRUMM

0. M is a complete affinely flat manifold

0.1. Complete affinely flat manifolds can be viewed as $M = \mathbb{R}^n/\Gamma$, where Γ is a subgroup of the affine group in dimension n, $\mathrm{Aff}(\mathbb{R}^n)$, which acts properly discontinuously on $\mathbb{R}^n$. Note that $\pi_1(M) \cong \Gamma$.

What is the nature of $\pi_1(M)$?

0.2. For instance, if M is Riemannian flat, i.e., Γ is restricted to the group of isometries of $\mathbb{R}^n$, then a classical theorem of Bieberbach's [**Bi**] states that $\pi_1(M)$ is torsion free, finitely generated, and virtually abelian. Further, M can be chosen to be compact.

Does Bieberbach's theorem, or a close relative, apply for all affinely flat manifolds?

CONJECTURE. (Milnor [**Mi**]). If M is a complete affinely flat manifold then $\pi_1(M)$ is virtually polycyclic.

0.3. Auslander's [**A**] weaker conjecture, where M is assumed to be compact, is known to be true for $\dim M \le 5$ [**K, FG, T**]. However, Milnor's conjecture is known to be false for $\dim M = 3$ and it is this gap which will be the subject of this paper.

Margulis was the first to show noncompact complete affinely flat manifolds with free fundamental group exist [**Ma1, Ma2**]. The shape of these manifolds and the moduli space of these manifolds will be discussed later [**DG, D1, D2**].

1. $\pi_1(M)$ is virtually polycyclic or ...

1.1. G is a polycyclic group if there exists a normal series $G \supset G_1 \supset G_2 \supset \cdots \supset G_n = \{1\}$ such that each factor G_i/G_{i+1} is cyclic. G is virtually polycyclic if there exists a subgroup of finite index which is polycyclic.

1991 *Mathematics Subject Classification.* Primary 51-XX, 52-XX.

This paper is in final form and no version will be submitted for publication elsewhere.

THEOREM (Milnor [Mi]). *If G is torsion free and virtually polycyclic then there exists a complete affinely flat manifold M such that $\pi_1(M) \cong G$.*

Are these all possible $\pi_1(M)$? (Conjecture 0.2)

1.2. Define $l\colon \mathrm{Aff}(\mathbb{R}^3) \to \mathrm{GL}(n, \mathbb{R})$ as the usual projection from an affine automorphism to its linear part.

THEOREM (Fried and Goldman [FG]). *If Γ acts freely and properly discontinuously on three-space then Γ is virtually polycyclic if*
(a) $\mathbb{R}^3/\Gamma$ *is compact,*
(b) $l(\Gamma)$ *is not conjugate to a subgroup of* $\mathrm{SO}(2,1)$.

This establishes Auslander's conjecture for $\dim M = 3$, and demonstrates that any noncompact complete affinely flat manifold with free fundamental group must be Lorentzian flat.

1.3. For Lorentzian flat manifolds Mess [Me] has shown that $l(\Gamma)$ is either virtually polycyclic or conjugate to a cocompact finite index subgroup of $\mathrm{SO}(2,1)$. Γ is either polycyclic or free of more than one generator and these two cases are mutually exclusive.

2. ... $\pi_1(M)$ is free

2.1. Let $\mathbf{E} = \mathbb{R}^{2,1}$ be the three-dimensional vector space with Lorentzian inner product $\mathbb{B}(x, y) = x_1 y_1 + x_2 y_2 - x_3 y_3$. Complete affinely flat manifolds M are "Margulis space-times" if $\pi_1(M)$ is free of rank ≥ 2. They are isomorphic to $\mathbf{E}/\Gamma$ where $\Gamma \subset \mathrm{SO}(2,1) \ltimes \mathbf{E}$ is free and acts properly discontinuously on $\mathbf{E}$.

Which Γ's, if any, act properly discontinuously on $\mathbf{E}$?

2.2. Consider hyperbolic elements of $g \in \mathrm{SO}(2,1)$, i.e., elements with three real distinct eigenvalues $\lambda(g) < 1 < \lambda(g)^{-1}$ and their corresponding eigenvectors $x^-(g)$, $x^0(g)$, $x^+(g)$. Eigenvectors $x^{\pm}(g)$ are chosen so that their Euclidean norm is one and their third coordinate is positive, and $x^0(g)$ is chosen so that $\mathbb{B}(x^0(g), x^0(g)) = 1$ and $\{x^0(g), x^-(g), x^+(g)\}$ is a right-handed basis for $\mathbb{R}^{2,1}$.

For every $h \in \mathrm{SO}(2,1) \ltimes \mathbb{R}^{2,1}$ such that $l(h) = g$ is hyperbolic and h acts freely on $\mathbb{R}^{2,1}$ there exists a unique invariant line A_h parallel to $x^0(g)$. Choose $x \in A_h$ and define $\alpha(h) = \mathbb{B}(h(x) - x, x^0(g))$. $\alpha(h)$ is the "Lorentzian length" of the simple closed geodesic in $\mathbb{R}^{2,1}/\langle h \rangle$. If $\alpha(h) = 0$ then h has a fixed point. The minimum Euclidean distance between $h(x)$ and x can be approximated by $\alpha(h)$.

2.3. Note that $\alpha(h^n) = |n|\alpha(h)$, and in particular $\alpha(h^{-1}) = \alpha(h)$. For affine maps h_1 and h_2 whose eigenvectors are distinct and $\lambda(h_i) \ll 1$, Margulis's observed that $\alpha(h_1 h_2) \approx \alpha(h_1) + \alpha(h_2)$.

THEOREM (Margulis [**Ma1, Ma2**]). *For hyperbolic* h_1, $h_2 \in \mathrm{SO}(2, 1) \rtimes \mathbf{E}$
if $\alpha(h_1)$ *and* $\alpha(h_2)$ *are of different signs then the group generated by* g *and*
h *does not act properly discontinuously on* $\mathbb{R}^{2,1}$.

There are several results which give necessary conditions on the linear part
of a free subgroup which acts properly discontinuously on three-space. This
is the only nontrivial necessary condition on the translational part.

2.4. THEOREM (Margulis [**Ma1, Ma2**]). *For hyperbolic* $g_i \in \mathrm{SO}(2, 1)$
whose eigenvectors are distinct, and where the $\lambda(g_i)$ *are sufficiently small,*
and the $\alpha(g_i)$ *are of the same sign, there exist* Γ *which acts properly discon-*
tinuously on $\mathbb{R}^{2,1}$ *and* $l(\Gamma) = \langle g_1, g_2, \ldots, g_n \rangle$.

The above theorem was proved by demonstrating that $\{\gamma \in \Gamma \mid |\alpha(\gamma)| < k\}$
is finite for any $k > 0$. This implies $\{\gamma \in \Gamma \mid \gamma(K) \cap K \neq \varnothing\}$ is finite for any
compact K and Γ acts properly discontinuously on $\mathbb{R}^{2,1}$.

2.5. Let C' be the projectivization of the cone $C = \{x \in \mathbb{R}^{2,1} \mid \mathbb{B}(x, x) = 0\}$. $\mathscr{G}$ is a Schottky group in $\mathrm{SO}(2, 1)$ if there are generators $g_1, g_2, \ldots,$
g_n and mutually distinct closed neighborhoods $A_i^{\pm\prime}$ of C' such that $g_i'(A_i^{-\prime})$
$= \mathrm{closure}(C' - A_i^{+\prime})$, where g_i' is the induced action of g_i on C'. $\mathscr{G}$ is a
Schottky group if and only if it is free and purely hyperbolic.

THEOREM (Drumm and Goldman [**DG**]). *If* $\mathscr{G}$ *is a Schottky group in*
$\mathrm{SO}(2, 1)$ *then there exists an affine group* Γ *such that* $l(\Gamma) = \mathscr{G}$ *and* Γ
acts properly discontinuously on $\mathbb{R}^{2,1}$.

The proof, as for Theorem 2.4, depended on using $\alpha(\gamma)$ to show that
$\{\gamma \in \Gamma \mid \gamma(K) \cap K \neq \varnothing\}$ is finite for every compact K. However, a lower
bound estimate for $\alpha(\gamma)$ was constructed slightly differently. γ was viewed
as a reduced word in the free group on n generators. The amount that the
translational part of each factor of γ contributed to $\alpha(\gamma)$ was then bounded
from below.

3. The shape of M if $\pi_1(M)$ is free

3.1. What do Margulis space-times look like?
Given $\mathscr{G}$ a Schottky group in $\mathrm{SO}(2, 1)$ let g_i be a generator and $A_i^\pm$ be
the part of C which is projected onto the $A_i^{\pm\prime}$ defined in 2.5. A fundamental
domain for the action of g_i on C-{eigendirections} is the closure of the
complement of $A_i^- \cup A_i^+$.

3.2. A fundamental domain for the action of g_i on C-{half-planes tangent
to eigendirections} is bounded by a pair of "wedges" W_i^- and W_i^+. The
wedges are composed of two half-planes tangent to C and a pair of infinite
triangles bounded by C. Furthermore, $W_i^- \cap C = A_i^-$, $W_i^+ \cap C = A_i^+$, and
$g_i(W_i^-) = W_i^+$.

If $h_i(x) = g_i(x) + v_i$ then

$$h_i(W_i^- - \tfrac{1}{2}g_i^{-1}(v_i)) = (g_i(W_i^-) - \tfrac{1}{2}v_i) + v_i = W_i^+ + \tfrac{1}{2}v_i$$

and translational parts of affine generators correspond to the translation of both wedges. A fundamental domain for the action of h_i on **E** is the closed region bounded by the translated wedges if the wedges to not intersect.

3.3. An alternate proof of Theorem 2.5 is a consequence of

THEOREM [D1]. *If $\mathscr{G}$ is a Schottky group in* SO(2, 1) *of n generators then there exists an affine group Γ such that $l(\Gamma) = \mathscr{G}$ and there exist n pairs of distinct translated wedges which bound a fundamental domain of the action of Γ on* **E**.

Let X be the closed region bounded by the translated wedges. Certainly different $x, y \in$ (interior of X) are not equivalent under the action of Γ. The difficulty lies in showing that for every $y \in$ **E** there exist an $x \in X$ and $\gamma \in \Gamma$ such that $\gamma(y) = x$.

3.4. THEOREM [D1]. *The topological type of the $\mathbb{R}^{2,1}/\Gamma$ for Γ described in Theorem* 3.4 *is that of a solid handlebody of genus n*.

3.5. The construction of the wedges can be expanded to include parabolic elements.

THEOREM [D2]. *There exist affine Γ acting properly discontinuously on* **E** *where $l(\Gamma)$ contains parabolic elements*.

3.6. A generalized Schottky group is one in which only interiors of the $A_i^\pm$ need be distinct. The intersection of two untranslated wedges may be a half-plane. The wedges may still be translated so that they separate. Γ is a free and discrete group if and only if it is a generalized Schottky group [D2, P].

THEOREM [D2]. *There exists a free affine group Γ which acts properly discontinuously on* **E** *such that $l(\Gamma) = \mathscr{G}$ if and only if $\mathscr{G}$ is a free discrete subgroup of* SO(2, 1).

3.7. The reader is referred to [Be] for the definitions of a Fuchsian group of the first and second kind. Schottky groups are Fuchsian of the second kind, but generalized Schottky groups are Fuchsian of the first and second kind.

4. And some unanswered questions

4.1. Does there exist a convex fundamental domain?

4.2. For given wedges, which translations separate them? In the rank 2 case, the "separating translations" $\mathscr{T}$ lie in $\mathbb{R}^{2,1} \times \mathbb{R}^{2,1}$. [D1] defines convex $A, B \subset \mathbb{R}^{2,1}$ where $A \times B \subset \mathscr{T}$ for the Schottky group case. The generalized Schottky case leads one to guess that $\mathscr{T}$ is not convex.

4.4. What is the moduli space? It seems likely that Γ gives rise to a Margulis space-time if and only if separated wedges may be constructed. This would imply that all Margulis space-times have the topological type of a solid handlebody. The difficulty in proving this seems to be in making some type of canonical choice of wedges.

REFERENCES

[A] L. Auslander, *The structure of compact locally affine manifolds*, Topology **3** (1964), 131–139.

[Be] A. F. Beardon, *The geometry of discrete groups*, Springer-Verlag, New York, 1983.

[Bi] L. Bieberbach, *Über die Bewegungsgruppen der Euklidischen Räume*, Math. Ann. **70** (1911), 297–336; **72** (1912), 400–412.

[D1] T. A. Drumm, *Fundamental polyhedra for Margulis space-times* (to appear Mich. Math. J.).

[D2] ____, *Linear holonomy of Margulis space-times* (in preparation).

[DG] T. A. Drumm and W. Goldman, *Complete flat Lorentz 3-manifolds with free fundamental group*, Internat. J. Math. **1** (1990), 149–161.

[FG] D. Fried and W. Goldman, *Three-dimensional affine crystallographic groups*, Adv. Math. **47** (1983), 1–49.

[K] N. H. Kuiper, *Sur les surfaces localement affines*, Géométrie Differentielle. Colloq. Internat. du Centre National de la Recherche Scientifique Strasbourg, 1953, pp. 78–87.

[Ma1] G. A. Margulis, *Free properly discontinuous groups of affine transformations*, Dokl. Akad. Nauk SSSR **272** (1983), 937–940.

[Ma2] ____, *Complete affine locally flat manifolds with free fundamental group*, J. Soviet Math. **134** (1987), 129–134.

[Me] G. Mess, *Flat Lorentz spacetimes* (to appear).

[Mi] J. W. Milnor, *On fundamental groups of complete affinely flat manifolds*, Adv. in Math. **25** (1977), 178–187.

[P] J. Parker, *2-generator Möbius subgroups*, Doctoral dissertation, 1990.

[T] Tomanov, *The virtual solvability of the fundamental group of a generalized Lorentz space form*, J. Differential Geom. **32** (2) (1990), 539–549.

BRANDEIS UNIVERSITY

Proceedings of Symposia in Pure Mathematics
Volume **54** (1993), Part 2

Incomplete Flat Homogeneous Geometries

D. DUNCAN AND E. IHRIG

ABSTRACT. Recent results concerning the classification of incomplete flat homogeneous pseudo-Riemannian manifolds are discussed.

1. Introduction

Starting with Wolf's book [W2], a large amount of information has been compiled over the last few decades concerning complete spaces of constant curvature. However, relatively little work has been done with incomplete spaces. A notable exception to this has been the study of incomplete flat spaces.

Fried, Goldman and Hirsch have investigated compact affine manifolds (compact manifolds with a flat connection) with the ultimate objective of resolving the conjecture that a compact affine manifold is complete if and only if it admits a nonzero covariant constant volume form. They have shown, among other things, that this conjecture is correct in the case in which the holonomy group is nilpotent [FGH2] or in the case in which the holonomy group is solvable with solvable rank less than the dimension of the manifold [GH]. [FGH1, Y] are related references.

In another direction, Vinberg [V1, V2] initiated an extensive study of a special kind of affine manifold; namely, homogeneous convex open subsets of affine space. Vinberg was able to establish a one-to-one correspondence between such domains and certain algebras, called Vinberg algebras (see [V1]). For more information on this topic see [S2] and the references therein.

Our interest has been in attempting to give a classification of flat homogeneous pseudo-Riemannian spaces (connected manifolds with a flat, not necessarily positive definite, metric which has a transitive group of isometries). When the metric is positive definite, a classical result says that the metric must be complete, and thus its universal cover must be Euclidean space. Wolf has provided a complete classification in this case (see [W1] and

---

1991 *Mathematics Subject Classification.* Primary 53-XX; Secondary 58-XX.
This paper is in final form and no version will be submitted for publication elsewhere.

also [W2]). A natural next step is to consider spacetimes; that is, connected manifolds with a pseudo-metric of signature $-, +, +, \ldots, +$. Minkowski space ($\mathbb{R}^n$ with the metric $-(dx_1)^2 + \sum_{i=2}^{n}(dx_i)^2$) is the only simply connected complete homogeneous flat spacetime. We have shown that any flat homogeneous spacetime must either have Minkowski space, or the subset $\{(x_1, \ldots, x_n)/x_1 > x_2\}$ of Minkowski space, as its universal cover. This enabled us to complete the classification of flat homogeneous spacetimes in [DI1]. Classification of homogeneous flat pseudo-metrics of other signatures has thus far avoided conquest.

In this article we will indicate how the problem of analyzing flat homogeneous pseudo-Riemannian manifolds relates to the ideas found in the work on compact affine manifolds and homogeneous affine domains. We will also announce some partial results engendered by these ideas. We give brief sketches of proof; the details will appear elsewhere.

2. The development map and algebraic groups

The main thing that makes the study of incomplete homogeneous flat pseudo-metrics tractable is the developing map. This enables one to translate the brunt of the problem to that of finding all homogeneous open subsets of $\mathbb{R}^n_s$. $\mathbb{R}^n_s$ denotes $\mathbb{R}^n$ with the metric

$$-\sum_{i=1}^{s}(dx_i)^2 + \sum_{i=s+1}^{n}(dx_i)^2.$$

The developing map is defined in the following way. Let M be a connected homogeneous flat pseudo-Riemannian manifold. Let $\widetilde{M}$ denote its universal cover, which is also a connected homogeneous flat pseudo-Riemannian manifold. The development map, denoted dev, is given by

$$\mathrm{dev}: \widetilde{M} \to \mathbb{R}^n_s$$

where $\mathrm{dev}(m) = (f_1(m), \ldots, f_n(m))$ and df_i are mutually orthogonal covariant constant one-forms each of whose length is either 1 or -1. dev is an isometric immersion from $\widetilde{M}$ onto an open subset of $\mathbb{R}^n_s$. Moreover $U = \mathrm{dev}(\widetilde{M})$ is also homogeneous. The crucial fact is that dev is a covering projection (see [GH]). This enables one to first seek all homogeneous domains U in $\mathbb{R}^n_s$, then construct their universal covers $\widetilde{M}$, and finally find all possible groups of deck transformations Δ so that $\widetilde{M}/\Delta$ will still have a transitive isometry group. This condition is equivalent to finding all Δ whose centralizer in the group of isometries of $\widetilde{M}$ has an open orbit (see [W3]).

The second useful ingredient is that the symmetry group of an open homogeneous domain in $\mathbb{R}^n_s$ has finite index in an algebraic group. The idea, due to Vinberg (see [V2]), is that if a group, G, has an open orbit, then the identity component of the normalizer of G, $N(G)_0$, must leave that orbit invariant. Thus $N(G)_0 \subset G \subset N(G)$. But the normalizer of a connected

Lie subgroup of a linear algebraic group is algebraic since x is in this group if and only if $x \mathfrak{g} x^{-1} = \mathfrak{g}$ where $\mathfrak{g}$ is the Lie algebra of G. This can in turn be expressed as polynomial conditions on x using a basis for $\mathfrak{g}$.

These two ingredients can be used to produce the following interesting result which applies to arbitrary signature.

2.1 THEOREM (Goldman and Hirsch [GH]). *Let M be a compact homogeneous flat pseudo-Riemannian manifold. Then M is complete.*

The proof of this result centers on the algebraic closure $\mathscr{A}$ of $\text{dev}(\Delta)$ where Δ is the group of deck transformations of $\widetilde{M}$. Using the radiance obstruction, they are able to show that $\mathscr{A}$ must have all of $\mathbb{R}^n_s$ as an orbit whenever M is a compact flat pseudo-Riemannian manifold. Now when M is homogeneous, $\text{dev}(\Delta)$ is a subgroup of the identity component G of the isometry group of $\text{dev}(\widetilde{M})$. Thus $\mathscr{A}$ is contained in $\mathscr{A}(G)$, the algebraic closure of G, which means $\mathscr{A}(G)$ acts transitively on $\mathbb{R}^n_s$. But G has finite index in $\mathscr{A}(G)$, so G itself acts transitively on $\mathbb{R}^n_s$. This says $\text{dev}(\widetilde{M}) = \mathbb{R}^n_s$. dev is a covering projection, and hence an isometry since $\mathbb{R}^n_s$ is simply connected.

A similar idea can be used to show the following result.

2.2 THEOREM (see [DI2]). *Let M be a flat pseudo-Riemannian manifold upon which some nilpotent subgroup of the isometry group acts transitively. Then M is complete.*

The proof of this proceeds as follows. Again, using the developing map, translate the situation into one in which U has a connected transitive nilpotent group of symmetries G. Let $\mathscr{A}$ be the Zariski connected component of the identity of the Zariski closure of G. $\mathscr{A}$ has an open orbit. Using an idea of Wolf [W3], we can show that any element of the center of $\mathscr{A}$ is unipotent. Thus $\mathscr{A}$ itself must be unipotent. Rosenlicht (see [R]) has shown that all the orbits of $\mathscr{A}$ must be Zariski closed. Hence $\mathscr{A}$ has an open orbit which is closed, and hence $\mathscr{A}$ acts transitively on $\mathbb{R}^n_s$. Again G has finite index in $\mathscr{A}$, so the rest of the proof proceeds as above.

Result 2.2 is analogous to the result in [FGH2] in which it was shown that compact affine manifolds with nilpotent holonomy are complete if they have a parallel volume. Also note that Theorem 2.1 follows from this result and the idea of Wolf mentioned above which can be used to show that the holonomy of a flat homogeneous pseudo-metric space is always unipotent.

Again, using similar ideas, one can show results such as the following:

2.3 THEOREM. *Let M be a homogeneous flat pseudo-Riemannian manifold. If M has nontrivial holonomy, then M must have at least one complete geodesic.*

This result is an exact analogy of [FGH2] since the holonomy is automatically unipotent. The result of the theorem is weaker because the lack of a compactness condition allows the holonomy group to be small.

Another interesting completeness result is one by Helmstetter (see [H]) which says that if M is a flat pseudo-Riemannian space with a simply transitive unimodular group of isometries then M is complete.

The general impression left by the results of this section may be that there will probably be few incomplete homogeneous flat pseudo-Riemannian manifolds. The examples given in the next section will serve to illustrate that this is far from the truth.

3. Incomplete spaces

In this section we will give some examples of incomplete flat homogeneous pseudo-Riemannian manifolds. We do this by showing how to construct such a manifold out of a homogeneous open subset of affine space. We then give some of the many examples of these domains that are known.

3.1 CONSTRUCTION. Let V be an open subset of $\mathbb{R}^s$. Let G be a subgroup of the group of affine transformations on $\mathbb{R}^s$ which has V as an orbit. if $g \in G$ and $x \in \mathbb{R}^s$, then $g(x) = \lambda(g)x + \tau(g)$ where $\lambda(g) \in Gl(s)$ and $\tau(g) \in \mathbb{R}^s$. Let

$$M = V \times \mathbb{R}^s \times \mathbb{R}^{n-2s} \subseteq \mathbb{R}^s \times \mathbb{R}^s \times \mathbb{R}^{n-2s}.$$

Define pseudo-metric g_{ij} on M by

$$(g_{ij}) = \begin{bmatrix} 0 & I_s & 0 \\ I_s & 0 & 0 \\ 0 & 0 & I_m \end{bmatrix}.$$

Here I_k indicates the $k \times k$ identity matrix and $m = n - s$.

If $g \in G$, define $\rho(g)$, an affine transformation on $\mathbb{R}^n$, by

$$\rho(g)(x) = \begin{bmatrix} \lambda(g) & 0 & 0 \\ 0 & \lambda(g^{-1})^t & 0 \\ 0 & 0 & I \end{bmatrix} x + \begin{bmatrix} \tau(g) \\ 0 \\ 0 \end{bmatrix}$$

for all $x \in \mathbb{R}^n$. $\rho(g)$ gives rise to an isometry on M for each $g \in G$. Also translations by vectors of the form $[0, v, w]^t$ give rise to isometries on M as well. Together, these isometries generate M as a single orbit. Thus M is a homogeneous flat pseudo-Riemannian manifold.

We now give two basic examples of homogeneous affine domains. These domains, as well as any products of them, may be used in 3.1.

3.2 EXAMPLE. (a) Let $\mathscr{k} = \mathbb{R}$, $\mathbb{C}$, or $\mathbb{H}$. Let $n \geq m$.

$$V_m(\mathscr{k}^n) = \{X : X \in \mathrm{Hom}_{\mathscr{k}}(\mathscr{k}^m, \mathscr{k}^n), \ker(X) = 0\}.$$

$V_m(\mathscr{k}^n)$ is an open subset in $\mathscr{k}^{nm}$ and is called a Stiefel domain. $Gl(n, \mathscr{k})$ acts transitively on $V_m(\mathscr{k}^n)$ by

$$\rho: Gl(n, \mathscr{k}) \times V_m(\mathscr{k}^n) \to V_m(\mathscr{k}^n), \qquad \rho(g, X) = gX.$$

(b) Let J be any nondegenerate inner product on $\mathbb{R}^{k-1}$. Define

$$Q_\pm(J) = \{(x, a)^t : x \in \mathbb{R}^{k-1},\ a \in \mathbb{R} \text{ and } \pm J(x, x) < \pm a\}.$$

Let

$$G = \left\{ \begin{bmatrix} xI & 0 \\ v^t & x^2 \end{bmatrix} : x \in \mathbb{R} - \{0\},\ v \in \mathbb{R}^{k-1} \right\}$$

where v^t satisfies $v^t w = J(w, v)$ for all w. Define $\rho(g)$ by

$$\rho(g)x = gx + \begin{bmatrix} v/2x \\ J(v, v)/4x^2 \end{bmatrix}.$$

G acts transitively on $Q_\pm(J')$.

If we take G in 3.1 to be $Gl(1, \mathbb{R})$, then the symmetry group described in 3.1 has an abelian commutator. This shows that 2.2 cannot be generalized to even two-step solvable groups. The examples generated by 3.2(b) show that the group of isometries need not be generated by its linear elements together with its pure translations. Also these examples show that, even in the simply transitive case, the isometry group need not be solvable (contrast with [A, S1]).

However there is one common aspect to all these examples. If V is the subgroup of the isometry group consisting of pure translations, then $V^\perp \subset V$ when V is identified with a subspace of $\mathbb{R}^n_s$. We say that M is translationally isotropic if the isometry group of $\mathrm{dev}(\widetilde{M})$ satisfies this condition. We have shown that if M is translationally isotropic, then $\mathrm{dev}(\widetilde{M})$ is one of the domains constructed in 3.1 (see [DI2]). Thus if every flat homogeneous pseudo-Riemannian space is translationally isotropic, then the problem of finding all of these manifolds reduces to the difficult problem of finding all homogeneous domains in $\mathbb{R}^s$. However, when s is small, a complete classification is possible so that certain other signatures can be completely resolved.

This means that the next major step to a fuller understanding of flat homogeneous pseudo-Riemannian spaces is to determine whether it is possible to have one which is not translationally isotropic.

REFERENCES

[A]   L. Auslander, *Simply transitive groups of affine motions*, American J. Math. **99** (1977), 809–826.

[DI1] D. Duncan and E. Ihrig, *Homogeneous spacetimes of zero curvature*, Proc. Amer. Math. Soc. **107** (1989), 785–795.

[DI2] ——, *Flat pseudo-Riemannian manifolds with a nilpotent transitive group of isometries* (to appear).

[FGH1] D. Fried, W. Goldman, and M. Hirsch, *Affine manifolds and solvable groups*, Bull. Amer. Math. Soc. (N.S.) **3** (1980), 1045–1047.

[FGH2] ——, *Affine manifolds with nilpotent holonomy*, Comment. Math. Helv. **56** (1981), 487–523.

[GH] W. Goldman and M. Hirsch. *Affine manifolds and orbits of algebraic groups*, Trans. Amer. Math. Soc. **295** (1986), 175–198.

[H]   J. Helmstetter, *Algèbres symmétriques à Gauche*, C. R. Acad. Sci. Paris **272** (1971), 1088–1091.

[R] M. Rosenlicht, *On quotient varieties and the affine embedding of certain homogeneous spaces*, Trans. Amer. Math. Soc. **101** (1961), 211–233.

[S1] J. Scheuneman, *Translations in certain groups of affine motions*, Proc. Amer. Math. Soc. **47** (1975), 223–228.

[S2] S. Shimizu, *A remark on homogeneous convex domains*, Nagoya Math. J. **105** (1987), 1–7.

[V1] E. Vinberg, *The theory of convex homogeneous cones*, Trans. Moscow Math. Soc. **12** (1963), 340–403.

[V2] ____, *The structure of the group of automorphisms of a homogeneous convex cone*, Trans. Moscow Math. Soc. **13** (1965), 63–93.

[W1] J. Wolf, *Homogeneous manifolds of zero curvature*, Trans. Amer. Math. Soc. **104** (1962), 462–469.

[W2] ____, *Homogeneous manifolds of zero curvature*, Comment. Math. Helv. **39** (1964), 21–64.

[W3] ____, *Spaces of constant curvature*, McGraw-Hill, New York, 1967.

[Y] K. Yagi, *On compact homogeneous affine manifolds*, Osaka J. Math. **7** (1970), 457–475.

CALIFORNIA STATE UNIVERSITY

ARIZONA STATE UNIVERSITY

Proceedings of Symposia in Pure Mathematics
Volume **54** (1993), Part 2

Geodesic and Causal Behavior of Gravitational Plane Waves: Astigmatic Conjugacy

PAUL E. EHRLICH AND GERARD G. EMCH

A key ingredient for singularity theory in General Relativity is the imposition of physically motivated curvature conditions such as the strong energy condition and the generic condition which imply that every complete non-spacelike geodesic in the given space–time (M, g) contains a pair of *conjugate points*. In this manner, "geodesic focusing in all directions" is partly responsible for the development of nonspacelike geodesic incompleteness. A separate issue that comes up in this context is *geodesic connectivity*; i.e., the property that for every $P, Q \in M$, there exists a geodesic from P to Q.

In gravitational plane waves, a class of geodesically complete, hence nonsingular, space–times is encountered which satisfy the strong energy condition (but not the null generic condition) and which exhibit a weaker kind of geodesic focusing called *astigmatic conjugacy*. Certain points P in M have the property that a two-dimensional subset of points lying in a null hyperplane $\Pi_{u_1}^3$ in the "future" of P are all conjugate to P with multiplicity one. As this conjugacy arises, geodesic connectivity from P *fails* to hold; namely, a subset of points in $\Pi_{u_1}^3$ open in the relative topology, fails to be joined to P by any geodesic. Yet $\exp_P(T_P M)$ is dense in M, unlike the case of the universal cover of anti de Sitter space–time. Astigmatic conjugacy is a genuinely non–Riemannian phenomena, since the Hopf–Rinow Theorem for complete Riemannian manifolds includes as a consequence that geodesic completeness implies geodesic connectivity.

In the sense that the gravitational plane waves are exact solutions of Einstein's equations, they provide natural examples of space–times which *fail* to be globally hyperbolic (cf. Penrose [14]) while still satisfying weaker causality conditions including *strong causality*. An important feature of the causality

1991 *Mathematics Subject Classification*. Primary 53C50, 83C35.
Research supported by NSF Grant DMS-8802672.
This paper is in final form and no version of it will be submitted for publication elsewhere.

and geodesic geometry of these models is that if Q is in the chronological future of P, but prior to the development of astigmatic conjugacy at $\Pi^3_{u_1}$, not only does there exist a geodesic segment from P to Q, but up to reparametrization, there is a unique geodesic segment from P to Q, which is both maximal and timelike. Thus "causality" coincides with *geodesic causality* prior to astigmatic conjugacy. Our proofs exploit systematically the behavior of the isometry group acting on geodesics emanating from P.

The space–times (M^4, g) referred to as *plane gravitational waves* are $M^4 = R^4$ (which, for our purposes, it is convenient to write as $M^4 = \{(\xi; v, u)|\xi \in R^2; v \in R, u \in R\}$) and $g = \eta + \langle \xi, h(u)\xi\rangle du^2$ where $\eta = d\xi^2 + 2 du\, dv$ is the usual Minkowski metric on R^4 with $d\xi^2 = dy^2 + dz^2$ denoting the usual Euclidian metric on R^2; (u, v) denoting the null coordinates in Minkowski space $u = \frac{1}{\sqrt{2}}(t - x)$ and $v = -\frac{1}{\sqrt{2}}(t + x)$; and $h : u \in R \mapsto h(u) \in M_0(2, R)$ with $M_0(2, R)$ denoting the two-by-two real matrices that are hermitian and of trace zero (the latter condition is equivalent to $G = 0$, i.e., to Ric $= 0$); the corresponding "plane gravitational wave" is said to be *polarized* if $h(u)$ can be simultaneously diagonalized (for all $u \in R$) by an orthogonal matrix U independent of u; in this case we shall choose the y–axis and the z–axis in R^2 to be the principal directions of $h(u)$, and thus write $h_{ij}(u) = (-1)^{i+1}\delta_{ij}f(u)$ where i and j each runs over the indices 1 (for y) and 2 (for z).

The basic geometric properties of these space–times have been known for some time, cf. [4, 7, 11, 13, 14, 15, 17]; we briefly recall those properties which we shall need. All plane gravitational wave space–times are *geodesically complete* and under very mild conditions on $h(u)$ (namely, $h \neq 0$ and $(\det h)' \neq 0$), admit a five dimensional Lie algebra of Killing vector fields spanned by

$$(1) \qquad X_v = \partial_v \quad \text{and} \quad X_\phi = \langle \phi, \nabla\rangle - \langle \dot{\phi}, \xi\rangle\partial_v$$

where ϕ run over the four dimensional space of solutions of the ODE $\ddot{\phi} = h\phi$. (Incidentally, this very same equation occurs also in the explicit description of geodesics, as well as in the Jacobi equation.)

Note that X_v is the only parallel vector field of this geometry; that each of the hyperplanes $\Pi^3_u = R^2 \times R \times \{u\}$ is stable under the Lie group G^5 generated by these Killing vector fields; that G^5 acts transitively on each Π^3_u; and that the metric $g_u = d\xi^2$, induced on $\Pi^3_u \subset M^4$, is degenerate of degree 1. Each (Π^3_u, g_u) is therefore a *doubly Galilean space–time* [9] (i.e., there are two Galilean time dimensions, namely $\xi \in R^2$, and one Galilean space dimension, namely $v \in R$). These remarks have three consequences that are important in the context of our study, namely: (i) the null hyperplanes (Π^3_u, g_u) are totally geodesic and flat; (ii) the null geodesics $\{(\xi; v, u)|v \in R\}$ with ξ and u fixed, are null lines that do not satisfy the generic condition, nor admit any conjugate or null cut points; all other geodesics in a Π_u^3 are spacelike; moreoever, it can be seen that u is weakly increasing along any smooth future–directed causal curve not lying in some Π_u^3; thus plane gravitational waves *are causal* ; and (iii) the symmetry group G^5 of M^4 is isomorphic to the group obtained by eliminating the rotations in the two– dimensional plane $R^2 \times \{v\} \times \{u\}$ from the six–dimensional group of Galilean transformations acting on Π_u^3.

A deeper study of the causal properties of (M^4, g) is predicated on the knowledge of the geodesic geometry, especially on the properties of the "light cones" $\{K^{\pm}(P)|P \in M^4\}$, where $K^+(P)$ (resp. $K^-(P)$) denotes the family of future- (resp. past-) directed, null geodesics issuing from any point $P = (\xi; v, u) \in M^4$, cf. [14]. Since G^5 acts transitively on $\Pi_{u_0}^3$, and the image through G^5 of any future- (resp. past-) directed null geodesic is again a future- (resp. past-) directed null geodesic, it is sufficient to describe $K^{\pm}(P_0)$ with $P_0 = (0; 0, u_0)$,

$$(2) \qquad K^+(P_0) = G^2(P_0)\left[\gamma_\nu^+\right] \cup \gamma_\pi^+$$

where γ_ν is the null geodesic $\{(0; 0, u)|u \in R\}$; $\gamma_\nu^+ = \{\gamma_\nu(u)|u \geq u_0\}$; γ_π is the only null geodesic issuing from P_0 in the null hyperplane $\Pi_{u_0}^3$, namely $\{(0; -v, u_0)|v \in R\}$; $\gamma_\pi^+ = \{\gamma_\pi(v)|v \geq 0\}$; and $G^2(P_0)$ is the isotropy group of P_0; this group is generated by the Killing vector fields X_ϕ [see equation (1)] where now ϕ runs only over the two dimensional space of solutions of the initial value ODE $\ddot{\phi} = h\phi \ with \ \phi(u_0) = 0$.

The level surfaces $K^+(P_0) \cap \Pi_u^3$ are paraboloids. However, these surfaces can become degenerate at an astigmatic conjugate pair: the light cone issuing from P_0 intersects the hyperplane $\Pi_{u_1}^3$, not along a two dimensional surface, but along a planar parabola $N(P_0)$.

Here with u_0 and $u_1 \neq u_0$ in R, we say that (u_0, u_1) is an *astigmatic conjugate pair* with respect to the hermitian $n \times n$ matrix–valued function h of u, if for every pair (ϕ, ψ) of R^n valued functions of $u \in R$, which are linearly independent solutions of the initial value ODE $\ddot{\chi} = h\chi$ with $\chi(u_0) = 0$, the vectors $\phi(u_1)$ and $\psi(u_1)$ are linearly dependent and at least one of them, say $\phi(u_1)$, is $\neq 0$. We further say that our differential system is *astigmatic conjugate (resp. disconjugate) in an interval* (a, b) if it admits (resp. does not admit) an astigmatic conjugate pair in that interval.

This is the essence of the phenomena discovered in [14]; one of the problems we have investigated is to make explicit some implicit assumption usually made to the effect that the gravitational wave is "sufficiently weak and tame" to avoid there being "too many" u_n $(n = 1, 2, ...)$ for which this phenomena occurs. We established the following preliminary result, suggested by an analogy with geometrical optics, and proved by analyzing the phase portrait of the appropriately reparametrized Riccati equation.

LEMMA 1. *In the polarized case, suppose that:* (1) $f(u) = f(-u) \geq 0 \ \forall u \in R$,

(2) $u \dot{f}(u) \leq 0 \ \forall u \in R$,

(3) $0 < \int_{-\infty}^{+\infty} du \sqrt{f(u)} \leq \pi$.

Then there exists a unique $u_f > 0$ *such that* $u_0 \leq -u_f$ *implies that there exists a unique* u_1 *for which* (u_0, u_1) *is an astigmatic conjugate pair, and then* $u_1 \geq u_f$; *the symmetric statement holds for* $u_0 \geq u_f$; *for* $u_0 \in (-u_f, u_f)$ *there does not exist any* $u_1 \in R$ *such that* (u_0, u_1) *is an astigmatic conjugate pair, i.e., h is astigmatic disconjugate on* $(-\infty, u_f)$ *and on* $(-u_f, \infty)$, *but it is astigmatic conjugate on* R.

In fact, we proved this result under more general circumstances; in particular, the symmetry condition (1) can be relaxed, but some form of the monotonicity condition (2) is essential, and so is condition (3).

Let (u_0, u_1) be an astigmatic conjugate pair for h and $P_1 = (0; 0, u_1) \in N(P_0)$. Then there is a one–dimensional family Γ of null geodesics issuing from P_0 that focus at P_1: it is obtained from γ_ν by the action of the one–parameter subgroup of symmetries for which both P_0 and P_1 are fixed points. The null geodesic half–line $\gamma_\pi^- = \{\gamma_\pi(v) | v \in R_-\}$ is a limiting curve for Γ. Hence $\gamma_\pi^- \in \overline{J^+(P_0)}$. This implies immediately that (M^4, g) is *not globally hyperbolic* [14]. Moreover, to prove now the new result that (M^4, g) is *not causally simple*, it is sufficient to prove that $\gamma_\pi^- \notin J^+(P_0)$. This follows as a consequence of the following result stated for simplicity in the polarized case, but which we have obtained for an arbitrary gravitational plane wave space–time in the presence of astigmatic conjugacy.

THEOREM 2. *Let* (u_0, u_1) *be an astigmatic conjugate pair satisfying the conclusion of Lemma 1;* $P_0 = (0; 0, u_0)$; $P_1 = (0; 0, u_1)$; $\mathscr{S}$ *be the strip* $\{S | u(S) \in (u_0, u_1)\}$; $\mathscr{R}$ *be the set* $\{R \in \Pi_{u_1}^3 | z(R) \neq 0\}$ *and* $\mathscr{L}$ *be the set* $\{L \in \Pi_{u_1}^3 | z(L) = 0\}$,

Then

(1) *for every* $S \in \mathscr{S}$, *there is a unique geodesic from* P_0 *to* S, *and* $\exp_{P_0}$ *is a diffeomorphism onto* $\mathscr{S}$,

(2) *for every* $R \in \mathscr{R}$, *there is no geodesic from* P_0 *to* R,

(3) *for every* $L \in \mathscr{L}$, *L is conjugate to* P_0 *along a one–parameter family of geodesics, and all these geodesics have the same tangent length,*

(4) $\mathscr{L}$ *is the conjugate locus of* P_0,

(5) *the future null conjugate locus of* P_0 *in* $\mathscr{L}$ *is the planar parabola* $N(P_0)$ *described above,*

(6) *the future causal cut locus of* P_0 *in* $\Pi_{u_1}^3$ *is the first future causal conjugate locus of* P_0,

(7) *for every* $S \in \mathscr{S} \cup \mathscr{L}$ *with* $P_0 \ll S$, *there exists a maximal timelike geodesic from* P_0 *to* S,

(8) *for every* $S \in \mathscr{S} \cup \mathscr{L}$ *with* $S \in J^+(P_0) \backslash I^+(P_0)$, *there exists a maximal null geodesic from* P_0 *to* S.

Note that these results can be translated from $P_0 = (0; 0, u_0)$ to any $Q_0 = (\xi_0; v_0, u_0) \in \Pi_{u_0}^3$ by the flow generated by an appropriately chosen Killing vector field; in particular the first future null conjugate locus of Q_0 is a planar parabola $N(Q_0) \subset \Pi_{u_1}^3$ parallel to $N(P_0)$.

COROLLARY 3. (M^4, g) *is not causally simple, but it is strongly causal.*

Hence, in particular, (M^4, g) is distinguishing, and the Alexandrov topology, induced by the open sets $I^+(P) \cap I^-(Q)$, is Hausdorff and coincides with the natural manifold topology of M^4. The physical relevance of this result is highlighted by the fact that the "diamonds" $I^+(P) \cap I^-(Q)$ are primary ingredients in the Araki-Haag-Kastler algebraic approach to quantum field theories (cf. e.g. [8, Chapter 4] or [9]; and compare with [10]).

Also, we have observed that the astigmatically conjugate gravitational plane waves provide a new class of nonlobally hyperbolic space–times which are *causally disconnected* in the sense of [1, 3]. In view of Theorem 11.41 of [1, p. 390], it is precisely the failure of the null generic condition to be satisfied along the null geodesic lines in the hyperplanes Π_u^3 which put the gravitational plane wave space–times just outside the purview of the classical singularity theory machinery in General Relativity.

The maximality of nonspacelike geodesic join from P_0 up to and including the first astigmatic conjugate hyperplane $\Pi_{u_1}^3$ may be obtained geometrically in a general context as follows. Consider a space–time (M, g) which admits a smooth function $f : M \to R$ with ∇f everywhere nonzero, past directed null and put $N_\lambda = f^{-1}(\lambda)$. (In the gravitational plane wave case, $f = u$ and $N_\lambda = \Pi_\lambda^3$.)

We have established the following result.

PROPOSITION 4. *Fix* $p \in N_\lambda$ *and for* $\mu > \lambda$, *let* $S := f^{-1}((\lambda, \mu))$. *Suppose that* $\exp_p$ *is a diffeomorphism onto* $S \cap I^+(p)$, *i.e., for each* m *in* $S \cap I^+(p)$, *there is (up to reparametrization) a unique geodesic from* p *to* m.
Then

    (i) *there are no null or spacelike geodesics from* p *to any point of* $S \cap I^+(p)$,

    (ii) $q \in S \cap I^+(p)$ *iff there exists a unique maximal timelike geodesic from* p *to* q *in* S.

The concept of a (smooth) *global time function*, i.e., a smooth function $f : M \to R$ such that ∇f is everywhere past directed timelike, has proven useful to causality theory in General Relativity [11, 17]: if a space-time (M, g) admits a global time function, then (M, g) is stably causal. A global time function is strictly increasing along all future causal curves.

Our study of the gravitational plane wave space-times points to the fact that a weaker concept can also be useful. If ∇f is allowed to be null as well as timelike, then the property that f is strictly increasing along "most" future causal curves still gives some control on the causal behavior of the space-time. This is evinced by Theorem 2 and Corollary 3. For gravitational plane waves, the isotimic surfaces $N_\lambda = f^{-1}(\lambda)$ are the hyperplanes Π_u^3, each of which is a *quasi-Cauchy surface* in the sense that a large class of causal curves exists that intersect every leaf of the foliation

$$(3) \qquad\qquad M^4 = \bigcup_{u \in R} \Pi_u^3$$

exactly once. This class includes all inextendible time-like curves γ with $g(\dot\gamma, \dot\gamma)$ bounded away from zero; and thus, in particular, all inextendible time-like geodesics (for an application of the latter property to the construction of "test" quantum fields on (M^4, g), cf. [10]); this class also contains all inextendible null geodesics that are not entirely confined to a single Π_u^3 (i.e., through any point $P \in M^4$, all inextendible null geodesics through P, except one, namely $\gamma_\pi = \{ (\xi(P), v, u(P)) \mid v \in R \}$.

For general purposes however, the condition that ∇f be everywhere nonzero, past directed and causal needs some refinement which we provide below. Indeed, the cylinder $M = S^1 \times R$ with metric $g = d\theta \times dt$ contains smooth closed null geodesics; therefore this (M, g) fails to be causal, although it is chronological and admits a smooth function, namely $f = -t$, whose gradient is everywhere past directed null and globally parallel.

In order to build on the positive aspects manifested in plane gravitational waves, while ruling out the above counterexample, we were led to introduce the following concept.

DEFINITION 5. A smooth function $f : M \to R$ is said to be a quasi-time function if it satisfies the following two conditions:

(a) ∇f is everywhere (nonzero) past directed and causal;

(b) every null geodesic segment β, for which $f \circ \beta$ is constant, is injective.

We proved that a space-time which admits a quasi-time function is causal as well as chronological. Moreover, in light of Proposition 4, the behavior exhibited by the cut and conjugate loci in Theorem 2 can be regarded as stemming from the existence of a quasi-time function which is well adapted (in the sense of Proposition 4) to the exponential map. We thus hope that the results we obtained in this study will also be of use elsewhere, in particular for further discusssions of the singularities that develop in space—times involving colliding gravitational waves (cf. [5, 12, 18, 19]).

REFERENCES

1. J.K. Beem and P.E. Ehrlich, *Global lorentzian geometry*, Dekker, New York, 1981.
2. _____, *The space–time cut locus*, Gen. Rel. Grav. **11** (1979), 89–103.

3. ____, *Constructing maximal geodesics in strongly causal space–times*, Math. Proc. Cambridge Philos Soc. **90** (1981), 183–190.

4. H. Bondi and F. Pirani, *Gravitational waves in general relativity* XIII: *caustic properties of plane waves*, Proc. Roy. Soc. London A **421** (1989), 395–410.

5. S. Chandrasekhar and B. Xanthopoulos, *A new type of singularity created by colliding gravitational waves*, Proc. Roy. Soc. London A **408** (1986), 175–208.

6. C. Chicone and P.E. Ehrlich, *Line integration of Ricci curvature and conjugate points in Lorentzian and Riemannian manifolds*, Manuscripta Math. **31** (1980), 297–316.

7. J. Ehlers and W. Kundt, *Exact solutions of the gravitational field equations*, Gravitation, (L. Witten, ed.), Wiley, Chichester, (1962), pp. 49–101.

8. G.G. Emch, *Algebraic methods in statistical mechanics and quantum field theory*, Wiley-Interscience, New York, 1972.

9. ____, *Mathematical and conceptual foundations of 20th century physics*, North- Holland, Amsterdam, 1984.

10. G. Gibbons, *Quantized fields propagating in plane-wave space-times*, Commun. Math. Phys. **45** (1975), 191–202.

11. S.W. Hawking and G.F.R. Ellis, *The large scale structure of space–time*, Cambridge Univ. Press, Cambridge, 1973.

12. K. Khan and R. Penrose, *Scattering of two impulsive gravitational plane waves*, Nature **229** (1971), 185–186.

13. C.W. Misner, K.S. Thorne, and J.A. Wheeler, *Gravitation*, Freeman, San Francisco, 1973.

14. R. Penrose, *A remarkable property of plane waves in general relativity*, Rev. Modern Phys. **37** (1965), 215–220.

15. ____, *Structure of space–time, Battelle Rencontres 1967*, Lectures in Mathematics and Physics, C. DeWitt and J. Wheeler, eds; Benjamin, New York (1968).

16. ____, *Techniques of topology in relativity*, SIAM Regional Conference Series in Applied Mathematics 7 (1972).

17. R.K. Sachs and H. Wu, *General relativity for mathematicians*, Springer-Verlag, New York, 1977.

18. P. Szekeres, *Colliding plane gravitational waves*, J. Math. Phys. **13** (1972), 286–294.

19. A.H. Taub, *On the collision of planar impulsive gravitational waves; Collision of impulsive gravitational waves followed by dust clouds*, J. Math. Phys. **29** (1988), 690–695; 2622–2627.

20. F. Tipler, *General relativity and conjugate ordinary differential equations*, J. Differential Equations **30** (1978), 165–174.

UNIVERSITY OF FLORIDA

Proceedings of Symposia in Pure Mathematics
Volume **54** (1993), Part 2

Curvature of Singular Spaces via the Normal Cycle

JOSEPH H. G. FU

Introduction

There is a tradition in differential geometry of studying the curvature of certain types of singular spaces. This goes back (at least) to the famous formula of Steiner for the volume of a tubular neighborhood of a convex set in Euclidean space [**Sa**, p. 220]; it continues through Blaschke's ideas of integral geometry [**B1**], through Federer's studies [**Fe1**] of sets with positive reach and Banchoff's PL curvature theory [**Ba**]; and among contemporary work it includes that of Cheeger, Müller and Schrader on the Regge calculus [**CMS**]. In the early 1980s, a simple but extremely fundamental observation was made independently by two German mathematicians, M. Zähle and the late P. Wintgen, which has served to bring the entire subject into sharper focus.

The idea of Wintgen and Zähle is to associate to certain subsets X of euclidean space a current $N(X)$, called the *normal cycle* of X, which yields in a natural way the generalized curvatures of X. This procedure is well illustrated in the simple case of a compact convex region K in the euclidean plane $\mathbb{E}^2$. We begin by recalling the classical viewpoint on this situation. Steiner's formula states in this case that

$$\text{area}(K_r) = \text{area}(K) + r\,\text{Perimeter}(K) + r^2\pi,$$

where $K_r := \{x \in \mathbb{E}^2 : \text{dist}(x, K) \leq r\}$ (see Figure 1 on next page). This formula may be localized as follows: for $x \in \mathbb{E}^2$, let $\xi_K(x) \in K$ be the unique point $p \in K$ such that $|x - p| = \text{dist}(x, K)$. Then for any Borel set $U \subset K$ we have

$$\text{area}(K_r \cap \xi_K^{-1}(U)) = \Phi_2^K(U) + r\Phi_1^K(U) + \pi r^2 \Phi_0^K(U)$$

for some $\Phi_0^K(U) \geq 0$ and where $\Phi_1^K(U) = \text{length}(\text{bdry}\,K \cap U)$, $\Phi_2(U) =$

1991 *Mathematics Subject Classification.* Primary 53C65; Secondary 32C30.
This paper is in final form and no version of it will be submitted for publication elsewhere.

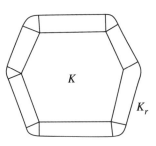

FIGURE 1

area$(K \cap U)$. The coefficients of this polynomial are in fact *measures*, which coincide with integrals of curvature if K has smooth boundary.

Now put $N(K) := \{(x, v) \in \mathbb{E}^2 \times S^1 : x \in \text{bdry} K,\ v \cdot (y - x) \leq 0$ for every $y \in K\}$. Thus the fiber of $N(K)$ over x is simply the cone dual to the tangent cone to K at x. Then $N(K)$ is naturally bilipschitz-homeomorphic to bdry K_r, $r > 0$, via the map $(x, v) \mapsto x + rv$. As such, $N(K)$ is a lipschitz submanifold of $S\mathbb{E}^2 = \mathbb{E}^2 \times S^1$ with a natural orientation. Now let $\alpha, \kappa_0, \kappa_1$ be the one-forms on $S\mathbb{E}^2$ given by

$$\alpha(x, v)(\sigma, \tau) = v \cdot \sigma$$
$$\kappa_0(x, v)(\sigma, \tau) = v^{\perp} \cdot \tau,$$
$$\kappa_1(x, v)(\sigma, \tau) = v^{\perp} \cdot \sigma,$$

where $(\sigma, \tau) \in T_x\mathbb{E}^2 \times T_v S^1 \subset T_x\mathbb{E}^2 \times T_v\mathbb{E}^2$. Then $\int_{N(K)} \varphi \alpha = 0$ for any function $\varphi : \mathbb{E}^2 \times S^1 \to \mathbb{R}$ and

$$\int_{N(K) \cap p^{-1}(U)} \kappa_1 = \Phi_1^K(U),$$

$$\int_{N(K) \cap p^{-1}(U)} \kappa_0 = 2\pi \Phi_0^K(U)$$

for any Borel set $U \subset K$, where p is the projection $\mathbb{E}^2 \times S^1 \to \mathbb{E}^2$. That is: *as a current, $N(K)$ annihilates α, and its contractions against κ_0, κ_1 give the measures Φ_0, Φ_1.*

In dimension m, one obtains in a similar way a sequence of measures $\Phi_0^K, \ldots, \Phi_m^K$ from some distinguished forms $\kappa_0, \ldots, \kappa_{m-1}$ of degree $m-1$ in $S\mathbb{E}^m$. From the observation above it is a simple matter to define the curvature measures Φ_i^X for a finite union of convex sets, if we recall the basic principle (Gauss-Bonnet theorem) that the total curvature of a set should equal its Euler characteristic. Since

$$\chi(K \cup L) = \chi(K) + \chi(L) - \chi(K \cap L)$$

we are led to put for convex $K, L \subset \mathbb{E}^2$,

$$N(K \cup L) = N(K) + N(L) - N(K \cap L)$$

as a sum of *currents*, and to extend to general finite unions X by induction. Then Φ_i^X is defined by contraction against κ_i, followed by pushing down into the base space $\mathbb{E}^m$. There is of course some question of whether this description of $N(X)$ depends on the chosen decomposition into convex sets, but it turns out that it does not. The current $N(X)$ has (current-theoretic) boundary zero, hence may be called the *normal cycle* of X.

Wintgen and Zähle used this method to study sets X which, like the finite unions of convex sets above, may be decomposed into simple pieces: riemannian simplices in Wintgen's case [**Wi, CMS2**], or sets with positive reach in Zähle's [**Zä2**]. Here we announce the results and some of the methods of a program to study the normal cycles of sets X defined by *analytic* equations and maps, which generally do not admit such decompositions. The key point is to characterize directly the normal cycle of a singular subspace, without using the method outlined above. The normal cycle of an "analytic" subset may then be constructed in a straightforward way. We then show that certain invariants constructed from the normal cycle have meanings analogous to those of classical curvatures.

1. A uniqueness theorem for the normal cycle

We state first a general uniqueness theorem for integral currents. Let M be an oriented manifold of dimension m, and let T^*M be its cotangent bundle with projection $\pi: T^*M \to M$ and natural symplectic two-form ω. An integral current T of dimension m and living in T^*M is said to be *lagrangian* iff $T \llcorner \omega = 0$ (here $\llcorner$ is contraction; cf. [**Fe2** 4.1.7]). Thus if $V \subset T^*M$ is an m-dimensional oriented submanifold and $T =$ integration over M, then T is lagrangian iff V is a lagrangian submanifold.

1.1 THEOREM [**Fu2**]. *Suppose that $T \in \mathbf{I}_m(T^*M)$ is a closed lagrangian integral current which is locally vertically bounded, i.e. if $K \subset M$ is compact, then so is $\pi^{-1}(K) \cap \operatorname{spt} T$. Let β be a volume form on M. If $T \llcorner \pi^* \beta = 0$, then $T = 0$.*

For our purposes this theorem has two main consequences.

1.2 COROLLARY. *Given a locally lipschitzian function $f: M \to \mathbb{R}$, there is at most one closed, lagrangian, locally vertically bounded integral current T such that for any smooth $\varphi: T^*M \to \mathbb{R}$,*

$$T(\varphi(x, y)\pi^* \beta) = \int_M \varphi(x, df(x))\beta.$$

If T exists, we denote it by $[df]$. This current exists whenever f is smooth, and is given by integration over $\operatorname{graph}(df)$.

To prepare the second corollary, let S^*M denote the cotangent ray space at M, i.e. $S^*M = T^*M\text{-(zero-section)}/\sim$, where $\xi \sim \eta$ iff $\xi = t\eta$ for some $t > 0$. Then S^*M carries a natural contact structure and admits a global contact one-form α. An integral current $T \in \mathbf{I}_{m-1}(S^*M)$ is called *legendrian* iff $T \llcorner \alpha = 0$.

Specializing to the case $M = \mathbb{E}^m$, let $X \subset \mathbb{E}^m$ be compact, and for $(p, v) \in \mathbb{E}^m \times S^{m-1}$ put

$$\iota_X(p, -v) = \lim_{r \downarrow 0} \lim_{\varepsilon \to 0} [\chi(B(p, r) \cap X \cap \{x : x \cdot v \le x \cdot p + h\})]|_{h=-\varepsilon}^{h=+\varepsilon}.$$

If X is a smooth submanifold this expression becomes quite simple. Given $v \in S^{m-1}$, let $h_v(x) := x \cdot v$ be the height function in the direction v. Then

$$\iota_X(p, v) = \begin{cases} (-1)^\lambda & \text{if } p \text{ is a critical point of } h_v|X \text{ of Morse index } \lambda, \\ 0 & \text{otherwise.} \end{cases}$$

It is well known that the Gauss curvature K_X of X may be expressed in terms of integrals of ι_X over S^{m-1}; namely,

$$(1a) \qquad \int_{X \cap U} K_X(p) \, dp = \omega_{m-1}^{-1} \int_{S^{m-1}} \left(\sum_{p \in U} \iota_X(p, v) \right) dv,$$

where ω_{m-1} is the volume of S^{m-1}. In fact this was the method used by Banchoff [**Ba**] to study the PL case. The next corollary states that this expression for the curvature essentially determines the normal cycle of X. It is most conveniently stated in terms of the tangent sphere bundle $S\mathbb{E}^m \cong \mathbb{E}^m \times S^{m-1} \cong S^*\mathbb{E}^m$. Let κ_0 be the pull-back to $S\mathbb{E}^m$ of the volume form of S^{m-1}.

1.3 COROLLARY. *Let* $X \subset \mathbb{E}^m$ *be compact. There is at most one compactly supported, closed, legendrian integral current* $T \in \mathbf{I}_{m-1}(S\mathbb{E}^m)$ *such that, for any smooth* $\varphi : S\mathbb{E}^m \to \mathbb{R}$,

$$T(\varphi \cdot \kappa_0) = \int_{S^{m-1}} \sum_p \iota_X(p, v) \varphi(p, v) \, dv.$$

If this T exists, we call it the *normal cycle* of X and denote it by $N(X)$.

2. The kinematic formula

The "principal kinematic formula" has occupied a place of honor in the study of generalized curvature ever since its introduction by Blaschke [**B1**] (cf. [**Ch2, Fe1, CMS2**]). To state the formula, let $G = SO(m) \ltimes \mathbb{R}^m$ be the group of orientation-preserving euclidean motions of $\mathbb{E}^m$, with bi-invariant measure $d\gamma$, normalized appropriately. The formula states that there are constants c_{ij} such that if X and Y are "geometrically nice" compact bodies in $\mathbb{E}^m$ then

$$(2a) \qquad \int_G \chi(X \cap \gamma Y) \, d\gamma = \sum_{i+j=m} c_{ij} \Phi_i(X) \Phi_j(Y),$$

where we have put $\Phi_i(Z) := \Phi_i^Z(Z)$.

The right way to approach this formula is use the Gauss-Bonnet theorem to rewrite the left-hand side of (2a) as $\int_G \Phi_0(X \cap \gamma Y) \, d\gamma$, which may then be considered a double integral. If X and Y are convex, then $X \cap \gamma Y$ is also, and this last integral makes sense; part of the beauty of Federer's approach [**Fe1**] to the subject is that if X and Y have positive reach, then so does $X \cap \gamma Y$ for a.e. $\gamma \in G$. Thus he was able to handle the convex and the smooth cases under one heading. This formulation suggests generalizations where Φ_0 is replaced on the left-hand side by $\Phi_1, \Phi_2, \dots$; these were first formulated and proved by Federer.

Using the normal cycle, the integral in (2a) may be written as

$$\int_G N(X \cap \gamma Y)(\kappa_0) \, d\gamma.$$

It turns out that (2a) admits a conceptually very simple proof [**Fu4, Ro-Zä**] based on the idea that the normal cycles $N(X \cap \gamma Y)$ may be computed from $N(X)$ and $N(Y)$ in a purely formal way. Federer had already observed that, if X and Y have positive reach, then a general normal vector to $X \cap \gamma Y$ at $p \in \mathrm{bdry}\, X \cap \mathrm{bdry}\, \gamma Y$ may be written as $sv + t\gamma_* w$ for some $s, t \geq 0$, where v is a normal vector to X at p and w is normal to Y at $q = \gamma^{-1} p$ (see Figure 2).

In other words, every normal to every set $X \cap \gamma Y$, $\gamma \in G$, corresponds to some pair of normals v, w as above; conversely, given such unit normals, we obtain an arc of unit normals to $X \cap \gamma Y$, namely the arc of the unit sphere connecting v and $\gamma_* w$. There is a left coset of $SO(m)$ in G taking q to p; as γ varies over this coset, v and $\gamma_* w$ produce arcs of normals as above. To model this situation, we take $v = w = e_m$, and consider the set $C \subset SO(m) \times S^{m-1}$ given by

$C := \{(\alpha, \theta): \theta$ *lies on a minimizing geodesic connecting* e_m *to* $\alpha e_m\}$.

(See Figure 3 on next page.) Then C carries an orientation making it into an integral current that we call **connect**.

To prove (2a), we consider the bundle $\mathcal{B}: S\mathbb{E}^m \times G \to \mathbb{E}^m \times \mathbb{E}^m$, with fiber $SO(m) \times S^{m-1}$, given by the projection $(\xi, \gamma) \mapsto (\pi\xi, \gamma^{-1}\pi\xi)$, and pull it back to a bundle $\overline{\mathcal{B}}$ via the natural projection $S\mathbb{E}^m \times S\mathbb{E}^m \to \mathbb{E}^m \times \mathbb{E}^m$. By the invariance properties of **connect** we may define a product current

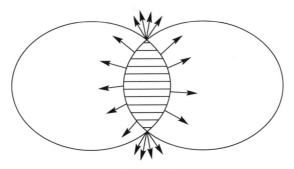

FIGURE 2

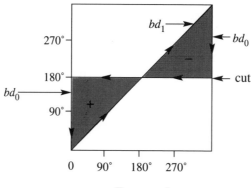

FIGURE 3

$\Omega := N(X) \times N(Y) \times_{\overline{\mathscr{B}}}$ **connect** living in the total space E of $\overline{\mathscr{B}}$; then, for nice sets $X, Y \subset \mathbb{E}^m$, the currents $N(X \cap \gamma Y)$ are obtained from Ω by projection and slicing. The kinematic formula (2a) follows quickly.

2.2 REMARK. The current **connect** is not closed. In fact its boundary decomposes naturally as the sum of three closed summands, of which the most interesting is called **cut** and is supported on the set $\{\alpha \in SO(m): \alpha e_m = -e_m\} \times S^{m-1}$. Since **cut** enjoys the same invariance properties as **connect**, it is natural to expect that another "kinematic formula" should result if we consider instead the current $N(X) \times N(Y) \times_{\overline{\mathscr{B}}}$ **cut**. This is indeed the case: in particular, if we take $Y = N(\text{half-space})$ then the resulting formula is precisely the integral geometric formula (1a) for the Gauss curvature of X [**Fu7**, 3.3.1.2].

3. Construction of the (co-)normal cycle in the subanalytic case

In a general manifold M, it is of course more natural in the absence of a riemannian structure to seek a current $N^*(X)$ living in S^*M. To construct it, we generalize the following obvious procedure for *smooth* submanifolds V. Let $f: M \to [0, \infty)$ be a nice function with $f^{-1}(0) = V$; for example, one might choose $f(x) := \text{dist}(x, V)$. Let N_r be the manifold of outward-pointing covectors to $f^{-1}[0, r]$, and take $N(V) = \lim_{r \to 0} N_r$ in an appropriate space of currents.

The subanalytic category [**Ha, Hi, Bi-Mi**] is a particularly convenient setting in which to seek a generalization, due to

3.1 PROPOSITION [**Fu2**]. *Let M be a real analytic manifold, and $f: M \to \mathbb{R}$ a lipschitzian subanalytic function. Then the current $T =: [df]$ of Corollary 1.2 exists.*

Using this proposition, let $\nu: T^*M\text{-(zero-section)} \to S^*M$ be the natural "normalization" map. Given a subanalytic set $X \subset M$, let $f: M \to [0, \infty)$ be any lipschitzian subanalytic function with $f^{-1}(0) = X$; for technical reasons we assume also that X is compact and f proper. We define N_f as a limit of slices of $[df]$ by $\pi^* f$, in the sense of [**Fe2**, §4.3]; namely,

$$N_f := \lim_{r \to 0} \nu_\#\langle [df], \pi^* f, r\rangle.$$

It is clear that N_f is a closed legendrian integral current. We then have the key

3.2 THEOREM [**Fu7**]. *The current N_f depends only on X, and not on the choice of f.*

SKETCH OF PROOF. Using analytic local coordinates we may assume that X is a compact subanalytic subset of $\mathbb{E}^m$. We want to invoke the uniqueness Theorem 1.3 to show that any such N_f is $N(X)$. It is easy to prove by approximation that $N_f(\kappa_0) = \omega_{m-1}\chi(X)$, i.e., that the Gauss-Bonnet theorem holds if the curvature of X is measured using N_f in place of the normal cycle of X. For $(v, t) \in S^{m-1} \times \mathbb{R}$, let $h_{v,t} : \mathbb{E}^m \to [0, \infty)$ be the function $h_{v,t}(x) := \max\{0, v \cdot x - t\}$. Thus,

(3a)
$$N_{f+h_{v,t}}(\kappa_0) = \omega_{m-1}\chi(X \cap H_{v,t})$$

where $H_{v,t}$ is the half-space $h_{v,t}^{-1}(0)$.

Meanwhile, one may show that $N_{f+h_{v,t}}$ is obtained by projection and slicing from $N_f \times N(\text{half-space}) \times_{\overline{\mathscr{B}}}$ **connect** as in the previous section; that is, $N_{f+h_{v,t}} = \mathscr{F}(N_f, v, t)$, a formal construction from N_f. Recalling the relation between **connect** and its boundary component **cut**, we use Remark 2.2 to show that (3a) implies the hypothesis of Theorem 1.3.

In view of Theorem 3.2 we may denote the current N_f above by $N^*(X)$. Looking at the Proof of 3.2 more closely, we obtain also the general formula

3.3 COROLLARY. $N^*(X \cup Y) = N^*(X) + N^*(Y) - N^*(X \cap Y)$ *for any closed subanalytic sets $X, Y \subset M$.*

4. Intrinsic invariants from the normal cycle

Although the normal cycle of a subanalytic set X obviously depends on the embedding $X \subset M$, nevertheless it is possible to obtain *intrinsic* information from it. This situation is analogous to the genesis of riemannian geometry.

Assume first that M is a *complex* manifold of complex dimension m and that X is a complex analytic subvariety. In this case it turns out that $N^*(X)$ is again a holomorphic object, and it is perhaps more natural to think of its projectivized version $\mathbb{P}N^*(X)$, a current of dimension $2m - 2$ living inside the projectivized cotangent bundle $\mathbb{P}^*(M)$ of M. By analogy with the classical relation between the Chern cohomology classes and the curvature of a smooth variety, one expects that the characteristic homology classes of a singular variety X (cf. [**MacP**]) should be computable from $\mathbb{P}N^*(X)$. In fact this is true. Let $\pi : \mathbb{P}^* M \to M$ be the natural projection.

4.1 THEOREM [**Fu5**]. *There are closed differential forms* $\gamma_0, \ldots, \gamma_{m-1}$, $\deg \gamma_i = 2(m - i - 1)$, *such that if* X *is any compact complex analytic subvariety of* M *then the currents* $\pi_\#(\mathbb{P}N^*(X) \llcorner \gamma_i)$ *represent the Chern-MacPherson characteristic homology classes* $\hat{c}_i(X)$ *of* X.

SKETCH OF PROOF. In fact we may show directly that these classes satisfy the required axioms (cf. [**MacP**]), thereby providing a new proof of MacPherson's theorem on the existence of these classes. The key point is to show that, if $f: X \to Y$ is analytic morphism between proper subvarieties $X \subset M$ and $Y \subset N$, then

(4a)
$$f_* \hat{c}_i(X) = \sum_j n_j \hat{c}_i(V_j),$$

where the closed subvarieties $V_j \subset Y$ and integers n_j are determined by the relation

(4b)
$$\sum_j n_j 1_{V_j}(q) = \chi(f^{-1}(q)), \qquad q \in Y.$$

To see this we consider the graph $\Gamma \subset M \times N$ of f and its projectivized conormal cycle $\mathbb{P}N^*(\Gamma) \in I_{2m+2n-2}(\mathbb{P}^*(M \times N))$. If we blow up the space $\mathbb{P}^*(M \times N)$ over the subspace $(\mathbb{P}^*M \times N) \cup (M \times \mathbb{P}^*N)$, we obtain a proper transform $\mathbb{P}\widetilde{N}^*(\Gamma) \in I_{2m+2n-2}(\widetilde{\mathbb{P}}^*(M \times N))$. We then consider the contractions

$$\mathbb{P}\widetilde{N}^*(\Gamma) \llcorner \left(B \wedge \sum_{i+j=k} \gamma_i \wedge g_j \right),$$

where the γ_i are as above and the g_j are the analogous forms in $\mathbb{P}^*N$, and where B is a certain singular one-form. The boundary of this current is essentially the difference of currents living in the exceptional divisors over $\mathbb{P}^*M \times N$ and $M \times \mathbb{P}^*N$, which project into $\mathbb{P}^*M$ and $\mathbb{P}^*N$ to give $\mathbb{P}N^*(X) \llcorner \gamma_k$ and $\sum n_j \mathbb{P}N^*(V_j) \llcorner g_k$ respectively, for some n_j, V_j. As those are homologous over the graph Γ, this means that $f_*[\pi_\#(\mathbb{P}N^*(X) \llcorner \gamma_k)] = [\pi_\# \sum n_j \mathbb{P}N^*(V_j) \llcorner g_k]$.

It remains to check that (4b) holds. This is accomplished by leaving the complex category: we let $S \subset M \times N$ be the inverse image of a small ball around q, and consider the conormal cycle of the semianalytic set $\Gamma \cap S$. We form the contraction $N^*(\Gamma \cap S) \llcorner \Pi \wedge P$, where Π and P are the transgression forms for M and N respectively; taking its boundary and applying the Gauss-Bonnet theorem to one side, (4b) follows.

A similar argument yields an interesting result in the real case. Let M^m be a C^ω manifold with a C^∞ riemannian metric. Let $\Pi_M \in \bigwedge^{m-1}(SM)$ be the geodesic curvature form of Chern [**Ch1**], and let $\Omega_M \in \bigwedge^m(M)$ be the Gauss-Bonnet-Chern form. Then we may form the *Gauss curvature measure* Φ_0^X of a closed subanalytic set $X \subset M$ by taking

$$\Phi_0^X = \pi_\#(N(X) \llcorner \Pi_M) + [X] \llcorner \Omega_M.$$

4.2 THEOREM [Fu7]. *Let N^n be another such manifold, and $Y \subset N$ a closed subanalytic set. Let $f : X \to Y$ be a subanalytic homeomorphism preserving the lengths of curves. Then $f_\# \Phi_0^X = \Phi_0^Y$.*

SKETCH OF PROOF. We must show that $\Phi_0^X(U) = \Phi_0^Y(f(U))$ for any open set $U \subset X$. To this end we approximate the composition $M \times N \to M \xrightarrow{1_U} \mathbb{R}$ by smooth functions φ, and seek instead to show that

$$(4c) \qquad \Phi_0^X(\varphi(x, f(x))) = \Phi_0^Y(\varphi(f^{-1}(y), y))$$

for all smooth $\varphi : M \times N \to \mathbb{R}$. Assuming for simplicity that X and Y have positive codimension in M and N respectively, we consider again the graph $\Gamma \subset M \times N$ of f, and its normal cycle $N(\Gamma) \in \mathbf{I}_{m+n-1}(S(M \times N))$.

Blowing up $S(M \times N)$ over $(SM \times N) \cup (M \times SN)$, we obtain a proper transform $\widetilde{N}(\Gamma) \in \mathbf{I}_{m+n-1}(\widetilde{S}(M \times N))$ and consider the contraction $\widetilde{N}(\Gamma) \llcorner \varphi \Pi_M \wedge \Pi_N$. This is a one-dimensional current with boundary

$$\partial \widetilde{N}(\Gamma) \llcorner \varphi \Pi_M \wedge \Pi_N \pm \widetilde{N}(\Gamma) \llcorner d\varphi \wedge \Pi_M \wedge \Pi_N.$$

The first term turns out to be a difference of 0-currents (i.e. measures) living over $SM \times N$ and $M \times SN$, with total measures given by the two sides of (4c). The last term turns out to have total measure zero, and (4c) follows.

REFERENCES

[A-W] C. B. Allendoerfer and A. Weil, *The Gauss-Bonnet theorem for Riemannian polyhedra*, Trans. Amer. Math. Soc. **53** (1943), 101–129.

[Ba] T. Banchoff, *Critical points and curvature for embedded polyhedra*, J. Differential Geom. **1** (1967), 245–256.

[Ba2] ____, *Critical points and curvature for embedded polyhedra. II*, J. Differential Geom. Proc., Special Year, Maryland, Progr. Math., vol. 32, Birkhäuser, Boston, 1983, 34–55.

[Ba3] ____, *Critical points and curvature for embedded polyhedral manifolds*, Amer. Math. Monthly **77** (1970), 475–485.

[Bi-Mi] E. Bierstone and P. Milman, *Semianalytic and subanalytic sets*, Publ. Math. Inst. Hautes Études Sci. **67** (1988), 5–42.

[Bl] W. Blaschke, *Vorlesungen über Integralgeometrie*, (3rd ed.), Deutsch. Verlag, Wiss., Berlin, 1955.

[Bo-Fe] T. Bonnesen and W. Fenchel, *Theorie der konvexen Körper*, Springer, Berlin 1934.

[B-K] U. Brehm and W. Kühnel, *Smooth approximation of polyhedral surfaces regarding curvatures*, Geom. Dedicata **12** (1982), 435–461.

[Br] I. A. Brin, *Gauss-Bonnet theorems for polyhedra*, Uspekhi Mat. Nauk **3** (1948), 226–227. (Russian)

[Bu] L. Budach, *Lipschitz-Killing curvatures of angular partially ordered sets*, Adv. Math. **78** (1989), 140–167.

[CMS1] J. Cheeger, W. Müller and R. Schrader, *On the curvature of piecewise flat spaces*, Comm. Math. Phys. **92** (1984), 405–454.

[CMS2] ____, *Kinematic and tube formulas for piecewise linear spaces*, Indiana Univ. Math. J. **35** (1986), 737–754.

[CMS3] ____, *Lattice gravity, or riemannian structure on piecewise linear spaces*, Unified Theories of Elementary Particles (Heisenberg Symposium, 1981) (P. Breitenlohner and H. P. Durr, eds.), Lecture Notes in Physics, Springer-Verlag, New York, 1982.

[Ch1] S. S. Chern, *On the curvatura integra in a Riemannian manifold*, Ann. of Math. (2) **46** (1945), 674–684.

[Ch2] ____, *On the kinematic formula in the euclidean space of n dimensions*, Amer. J. Math. **74** (1952), 227–236.

[Ch3] ____, *On the kinematic formula in integral geometry*, J. Math. Mech. **16** (1966), 101–118.

[Ch4] ____, *Differential geometry and integral geometry*, Proc. Internat. Congr. Math., Edinburgh, 1958, Cambridge Univ. Press, New York, 1960, pp. 441–449.

[Ch-La] S.S. Chern and R. K. Lashof, *On the total curvature of immersed manifolds*, Amer. J. Math. **79** (1957), 306–318.

[Fe1] H. Federer, *Curvature measures*, Trans. Amer. Math. Soc. **93** (1959), 418–491.

[Fe2] ____, *Geometric measure theory*, Springer-Verlag, New York, 1969.

[F1] F. Flaherty, *Curvature measures for piecewise linear manifolds*, Bull. Amer. Math. Soc. **79** (1973), 100–102.

[Fu1] J. H. G. Fu, *Curvature measures and generalized Morse theory*, J. Differential Geom. **30** (1989), 619–642.

[Fu2] ____, *Monge-Ampère functions. I*, Indiana Univ. Math. J. **38** (1989), 745–771.

[Fu3] ____, *Monge-Ampère functions. II*, Indiana Univ. Math. J. **38** (1989), 773–789.

[Fu4] ____, *Kinematic formulas in integral geometry*, Indiana Univ. Math. J. **39** (1990) 1115–1154.

[Fu5] ____, *Curvature measures and Chern classes of singular varieties* (to appear).

[Fu6] ____, *On Verdier's specialization formula for Chern classes* Math. Annalen **291** (1991), 247–251.

[Fu7] ____, *Curvature measures of subanalytic sets* to appear in Amer. J. Math.

[Gr] P. Griffiths, *Complex differential and integral geometry and curvature integrals associated to singularities of complex analytic varieties*, Duke Math. J. **45** (1978), 427–512.

[Ha1] R. M. Hardt, *Topological properties of subanalytic sets*, Trans. Amer. Math. Soc. **211** (1975), 57–70.

[Ha2] ____, *Some analytic bounds for subanalytic sets*, Differential Geometric Control Theory, Birkhaüser, Boston, 1983.

[Had] H. Hadwiger, *Vorlesungen über Inhalt, Oberfläche and Isoperimetrie*, Springer-Verlag, Berlin, 1957.

[Hi] H. Hironaka, *Subanalytic sets*, Number Theory, Algebraic Geometry and Commutative Algebra, Kinokuniya, Tokyo, 1973, pp. 453–493.

[Kp] N. Kuiper, *Morse relations for curvature and tightness*, Proc. Liverpool, Singularities Symp. II, Lecture Notes in Math., vol. 209, Springer-Verlag, New York, 1971.

[Ku] W. Kühnel, *Total absolute curvature of polyhedral manifolds with boundary in E^n*, Geom. Dedicata 8 (1979), 1–12.

[La] J. Lafontaine, *Measures de courbure des variétés lisses et des polyedres*, Astérisque **145–146** (1987), 241–256.

[Lg] R. Langevin, *Courbure et singularités complexes*, Comm. Math. Helv. **54** (1979), 6–16.

[Ni] A. Nijenhuis, *On Chern's kinematic formula in integral geometry*, J. Differential Geom. **9** (1974), 475–482.

[Nö] G. Nöbeling, *Über die Hauptformel der ebenen Kinematik von L. A. Santalo und W. Blaschke*, Math. Ann. **120** (1949), 585–614; 615–633.

[Oh] D. Ohmann, *Eine Verallgemeinerung der Steinersche Formel*, Math. Ann. **129** (1955), 209–212.

[Po] G. Pólya, *An elementary analogue to the Gauss-Bonnet theorem*, Amer. Math. Monthly **61** (1954), 601–603.

[Ro-Zä] W. Rother and M. Zähle, *A short proof of the principal kinematic formula and extensions*, Trans. Amer. Math. Soc. **321** (1990), 547–558.

[Re] T. Regge, *General relativity without coordinates*, Nuovo Cimento **19** (1961), 551–571.

[Sa] L. A. Santaló, *Integral geometry and geometric probability*, Addison-Wesley, Reading, MA, 1976.

[Sch1] R. Schneider, *Kinematische Berührmasse für konvexe Körper*, Abh. Math. Sem. Univ. Hamburg **44** (1975), 101–134.

[Sch2] ____, *Curvature measures of convex bodies*, Ann. Math. Pura ed Appl. **116** (1978), 101–134.

[Sch3] ____, *Kritische Punkte and Krümmung für die Mengen des Konvexrings*, L'Enseignement Math. **23** (1977), 1–6.

[Sh] T. Shifrin, *The kinematic formula in complex integral geometry*, Trans. Amer. Math. Soc. **264** (1981), 255–293.

[Sh2] ____, *Curvature integrals and Chern classes of singular varieties*, Contemp. Math. **63** (1987), 279–298.

[St] J. Steiner, *Über parallel flächen*, J. Ber. Preuss. Akad. Wiss. **114–118** (1840); or Gesammelte Werke, vol. II, Chelsea, New York, 1971, pp. 171–177.

[Sto1] D. Stone, *Sectional curvature in piecewise linear manifolds*, Bull. Amer. Math. Soc. **79** (1973), 1060–1063.

[Sto2] ____, *Geodesics in piecewise linear manifolds*, Trans. Amer. Math. Soc. **215** (1976), 1–44.

[VA] E. Vidal Abascal, *A generalization of Steiner's formula*, Bull. Amer. Math. Soc. **53** (1947), 841–844.

[Wa] R. Walter, *A generalized Allendoerfer-Weil formula and an inequality of the Cohn-Vossen type*, J. Differential Geom. **10** (1975), 167–180.

[Wi] P. Wintgen, *Normal cycle and integral curvature for polyhedra in Riemannian manifolds*, Differential Geometry (Gy. Soos and J. Szenthe, eds.), North-Holland, Amsterdam, 1982.

[Wu] T. J. Wu, *Integralgeometrie 26 Über die kinematische Hauptformel*, Math. Z. **43** (1937), 212–227.

[Zä1] M. Zähle, *Integral and current representations of Federer's curvature measures*, Arch. Math. **46** (1986), 557–567.

[Zä2] ____, *Curvatures and currents for unions of sets with positive reach*, Geom. Dedicata **23** (1987), 155–171.

UNIVERSITY OF GEORGIA

Proceedings of Symposia in Pure Mathematics
Volume **54** (1993), Part 2

The Nonintegrable Phase Factor and Gauge Theory

RONALD O. FULP

ABSTRACT. We obtain a rigorous proof of a result of A. M. Polyakov which uses so-called "variational derivatives" to show that gauge theory may be regarded as a theory of differential forms on loop space. Our proof is obtained by introducing a Fréchet manifold structure on the path space $\mathscr{P}^*P$ of a principal fiber bundle P in such a way that $\mathscr{P}^*P \to (\mathscr{P}^*P)/\mathrm{Diff}([0, 1])$ is a principal fiber bundle with structure group the Fréchet Lie group $\mathrm{Diff}([0, 1])$. We then show that if ω is a connection on P then the mapping $\mathscr{I} : \mathscr{P}^*P \to G$ defined by $\mathscr{I}(\gamma) = P \exp(\int_\gamma \omega)$ for $\gamma \in \mathscr{P}^*P$ is smooth and

$$(d_\gamma \mathscr{I})(\delta) = \mathscr{I}(\gamma) \left\{ [\omega(\delta(1)) - \omega(\delta(0))] \right.$$
$$\left. -P \exp \int_0^1 \mathrm{Ad}(V(s)^{-1})\Omega(\dot{\gamma}(s), \delta(s)) \, ds \right\}$$

for an appropriately defined mapping V from I to G. If Θ is the Maurer-Cartan form on G, $\mathscr{I}^*(\Theta)$ defines a flat connection on any trivial G-bundle over $\mathscr{P}^*P$ and Yang-Mills equations may be expressed in terms of $\mathscr{I}^*(\Theta)$. Thus gauge theory on P is a theory of flat connections on $\mathscr{P}^*P$.

Introduction

A. M. Polyakov [7] has given a heuristic proof of the fact that gauge theory can be reformulated as a theory of "differential forms" on the loop space of a manifold M. Specifically he considers the function ψ on the loop space LM of M defined by $\psi(C) = P \exp \int_C A_\mu dx^\mu$ where $A_\mu dx^\mu$ is a gauge field on M and the integral is the path-ordered exponential of $A_\mu dx^\mu$ on the loop C. He shows that the variational derivative $F_\mu(s, C) = \frac{\delta\psi(C)}{\delta x_\mu(s)}$ is given by

$$F_\mu(s, C) = \dot{x}_\nu P \exp \left(\int_o^s A_\mu \dot{x}_\mu \, dt \right) F_{\mu\nu}(x(s)) P \exp \left(-\int_o^s A_\mu \dot{x}_\mu \, dt \right)$$

1991 *Mathematics Subject Classification.* Primary 58D15, 53C05; Secondary 53C80.
This paper is in final form and no version of it will be submitted for publication elsewhere.

and that

$$\frac{\delta F_\mu(s, C)}{\delta x_\nu(s')} - \frac{\delta F_\nu(s, C)}{\delta x_\mu(s')} + [F_\mu(s, C), F_\nu(s', C)] = 0.$$

He then obtains a variational derivative form of the Yang-Mills equation. Thus gauge theory becomes a theory of flat connections on loop space.

More recently L. Gross [4] has developed a framework relative to which many of Polyakov's results may be formulated and proven. His paper is a significant contribution and in many ways goes beyond the Polyakov program. The author of the present work was unaware of the work of Gross prior to the symposium and independently developed another framework in which Polyakov's results may be proven. There are reasonably significant differences in this paper and that of Gross although some overlap is inevitable. We have chosen to formulate our results on rather general principal fiber bundles and have given the path space a Fréchet manifold structure. In [4] Gross endowed the path space with a Banach manifold structure and, for the most part, developed his formalism on trivial bundles over $\mathbb{R}^n$. Much of his work is valid in more general contexts however, and the Banach manifold structure permits him to prove a number of deep results which are presently not available to us. On the other hand the group of path reparameterizations acts smoothly on our path space and the quotient space relative to this action has an induced Fréchet manifold structure. There are obvious benefits as our results descend to the Fréchet manifold of "parameter independent" paths. Since the group of reparameterizations does *not* act smoothly on the path spaces of Gross, his techniques do not provide a differentiable structure on his corresponding quotient space so that it is not meaningful in his context to develop geometrical ideas on his space of "parameter independent" paths. In any case we take the view that each of these approaches provide insight into Polyakov's ideas and that the present paper provides us with a framework for future development of the geometry of the space of "parameter-independent" path space.

The next few paragraphs provide an overview of the paper and a summary of the results. Let $P \to M$ be a principal fiber bundle with structure group G and let $\mathscr{P}^*P$ denote the set of all paths in P. Thus $\gamma \in \mathscr{P}^*P$ iff γ is a smooth mapping from the interval $I = [0, 1]$ into P such that $\dot{\gamma}(s) \neq 0$ for all $s \in I$. It is known that $C^\infty(I, P)$ is a Fréchet manifold. We modify results of this type to show that $\mathscr{P}^*P$ may be given the structure of a Fréchet manifold in such a way that it has the following properties

(1) the mapping from $\mathscr{P}^*P \times \text{Diff}(I) \to \mathscr{P}^*P$ defined by $(\gamma, \varphi) \mapsto \gamma \circ \varphi$ for $(\gamma, \varphi) \in \mathscr{P}^*P \times \text{Diff}(I)$ is a smooth right action of the Fréchet Lie group $\text{Diff}(I)$ on $\mathscr{P}^*P$, and

(2) $\mathscr{P}^*P/\text{Diff}(I)$ is a Fréchet manifold.

Using these techniques we show that $\mathscr{P}^*P \to \mathscr{P}^*P/\text{Diff}(I)$ is in fact a principal fiber bundle. This structure permits us to work both with "parameterized paths" and with "parameter independent" paths in the spirit of Polyakov's work.

If ω is a connection on P we define the phase integral $\mathscr{I} : \mathscr{P}^*P \to G$ by $\mathscr{I}(\gamma) = P \exp \int_\gamma \omega$ for $\gamma \in \mathscr{P}^*P$. To avoid difficulties with the path-ordered exponential we assume that G is a subgroup of $Gl(n, \mathbb{R})$ (although most such difficulties can be resolved using techniques of [3, 5]). We show that $\mathscr{I}$ is a smooth mapping and that its differential $d\mathscr{I}$ is given by

$$(d_\gamma\mathscr{I})(\delta) = \mathscr{I}(\gamma) \left\{ [\omega(\delta(1)) - \omega(\delta(0))] - \int_o^1 \mathrm{Ad}(V(s)^{-1})\Omega(\dot{\gamma}(s), \delta(s))\, ds \right\}$$

where $\Omega = D\omega$ is the curvature of ω, $\gamma \in \mathscr{P}^*P$, $\delta \in T_\gamma(\mathscr{P}^*P)$, and where $V : I \to G$ is defined by $V(s) = P\exp(\int_s^1 \omega(\dot{\gamma}(t))\, dt)$ for $s \in I$. If Θ is the Maurer-Cartan form of the Lie group G and $\alpha = \mathscr{I}^*(\Theta)$ then clearly α is a smooth g-valued one-form on $\mathscr{P}^*P$ given by the formula

$$\alpha_\gamma(\delta) = \omega(\delta(1)) - \omega(\delta(0)) - \int_0^1 \mathrm{Ad}(V(s)^{-1})\Omega(\dot{\gamma}(s), \delta(s))\, ds.$$

Moreover α defines a flat connection on any trivial G-bundle over $\mathscr{P}^*P$. If α is restricted to the submanifold LP of loops on P then we may formally recover Yang-Mills equations in terms of $\alpha|LP$. More precisely we derive an identity

$$D(*\alpha)_\gamma(\delta_1, \delta_2) \equiv \mathrm{hor}(\hat{d}(*\alpha))_\gamma(\delta_1, \delta_2)$$

$$= \int_0^1 R_{V(s)}^*(D(*\Omega))(\dot{\gamma}(t), \delta_1(t), \delta_2(t))\, dt$$

where $\gamma \in LP$, and $\delta_1, \delta_2 \in T_\gamma(LP)$.

The operators $*$, $\hat{d}$, and hor are formal operators defined only on those differential forms on LP which some authors [1] call loop-forms. They are defined near the end of the next section and are utilized in the last few paragraphs of the paper.

Differentiable structure and path forms on path space

To say that γ is a *path* in a manifold M means that γ *is a smooth mapping* from an interval $[0, r] \subseteq \mathbb{R}$, $r > 0$, into M. The set of all paths in M is denoted $\mathscr{P}M$. Let $I = [0, 1]$ and let $\mathscr{D}$ be the set of all smooth mappings $\varphi : I \to I$ such that $\varphi'(x) > 0$ for each $x \in I$. The multiplicative group of positive real numbers is denoted by $\mathbb{R}^+$ and the group $\mathscr{D} \times \mathbb{R}^+$ is denoted by $\mathscr{G}$. Define a right action of $\mathscr{G}$ on $\mathscr{P}M$ by $[\gamma \cdot (\varphi, s)](t) = \gamma\left(r\varphi\left(\frac{t}{rs}\right)\right)$ where $\gamma \in \mathscr{P}M$, $(\varphi, s) \in \mathscr{G}$, $r \in \mathbb{R}^+$ such that $\mathrm{dom}\, \gamma = [0, r]$, and $0 \le t \le rs$. The space of orbits $\mathscr{P}M/\mathscr{G}$ is identified as the space of "parameter independent" paths in M. Note that if γ is a constant path in M and $(\varphi, s) \in \mathscr{G}$ then $\gamma \cdot (\varphi, s)$ is also constant and we regard the orbits determined by such γ as being trivial. On the other hand if we define $\mathscr{P}^*M$

to be the set of all $\gamma \in \mathscr{P}M$ such that $\dot{\gamma}(t) \neq 0$ for each $t \in \mathrm{dom}\,(\gamma)$ then the action of $\mathscr{G}$ on $\mathscr{P}M$ leaves $\mathscr{P}^*M$ invariant and it is easy to show the action of $\mathscr{G}$ on $\mathscr{P}^*M$ is a free action (use the fact that each $\gamma \in \mathscr{P}^*M$ is locally invertible). It follows that each of the orbits of the action of $\mathscr{G}$ on $\mathscr{P}^*M$ is in one-to-one correspondence with $\mathscr{G}$.

It is our aim to show in this section that the set $\mathscr{P}^*M$ of parameterized paths in M is a principal fiber bundle over the set $\mathscr{P}^*M/\mathscr{G}$ of parameter independent paths in M. To accomplish this we must provide a manifold structure for $\mathscr{P}^*M$ such that $\mathscr{P}^*M/\mathscr{G}$ also acquires a compatible manifold structure. Thus we prove the following theorem.

THEOREM 1. *There is a differentiable structure on $\mathscr{P}^*M$ such that*

(1) *$\mathscr{P}^*M$ is a Fréchet manifold,*
(2) *the action of $\mathscr{G}$ on $\mathscr{P}^*M$ is smooth,*
(3) *The quotient space $\mathscr{P}^*M/\mathscr{G}$ acquires a differentiable structure relative to which it becomes a Fréchet manifold,*
(4) *The projection $\mathscr{P}^*M \to \mathscr{P}^*M/\mathscr{G}$ defines a principal fiber bundle structure on $\mathscr{P}^*M$ with structure group $\mathscr{G}$.*

We first show that it is sufficient to work with paths having domain the unit interval $I = [0, 1]$. Let $\mathscr{P}_1^*M$ denote the set of all $\gamma \in \mathscr{P}^*M$ such that $\mathrm{dom}\,\gamma = I$. Define a mapping $\psi : \mathscr{P}^*M \to \mathscr{P}_1^*M \times \mathbb{R}^+$ by $\psi(\gamma) = (\gamma_r, r)$ where $\mathrm{dom}\,\gamma = [0, r]$ and where $\gamma_r \in \mathscr{P}_1^*M$ is defined by $\gamma_r(t) = \gamma(rt)$ for each $t \in I$. Clearly ψ is a bijection and there is an induced action of $\mathscr{G}$ on $\mathscr{P}_1^*M \times \mathbb{R}^+$ given by $(\gamma_r, r) \cdot (\varphi, s) = (\gamma_r \circ \varphi, rs)$. If we identify $\mathscr{P}^*M$ with $\mathscr{P}_1^*M \times \mathbb{R}^+$ via the bijection ψ then we see that $\mathscr{P}^*M \to \mathscr{P}^*M/\mathscr{G}$ may be given a differential structure which satisfies the theorem provided $\mathscr{P}_1^*M \to \mathscr{P}_1^*M/\mathscr{D}$ has such a structure since utilizing the product structure on $\mathscr{P}_1^*M \times \mathbb{R}^+$ leads to $\mathscr{P}^*M/\mathscr{G} = \dfrac{\mathscr{P}_1^*M \times \mathbb{R}_+}{\mathscr{D} \times \mathbb{R}^+} = \dfrac{\mathscr{P}_1^*M}{\mathscr{D}}$.

Thus we focus attention on $\mathscr{P}_1^*M$ and the action of $\mathscr{D}$ on $\mathscr{P}_1^*M$ given by $\gamma \cdot \varphi = \gamma \circ \varphi$ for $\gamma \in \mathscr{P}_1^*M$ and $\varphi \in \mathscr{D}$.

In order to construct a manifold structure on $\mathscr{P}_1^*M$ we find it convenient to utilize tubular neighborhoods of paths in $\mathscr{P}_1^*M$; thus we choose an arbitrary but fixed positive definite Riemannian metric g on M. Since $\dot{\gamma}(t) \neq 0$ for each $t \in I$ and $\gamma \in \mathscr{P}_1^*M$ we know that for a given γ there exists $\delta > 0$ such that $\gamma|J$ is injective for every subinterval J of I such that $l(J) \leq \delta$ ($l(J)$ denotes the length of J). Thus $\gamma|J$ has a tubular neighborhood which lies in M which we denote by $N(\gamma|J)$.

More explicitly let $T_\gamma^\perp M = \{(t, w) | t \in I, w \in T_{\gamma(t)}M, w \in \dot{\gamma}(t)^\perp\}$. Then $T_\gamma^\perp M \to I$ is a vector bundle and for each subinterval $J \subseteq I$ with $l(J) \leq \delta$ we define $U_\varepsilon(\gamma|J)$ by

$$U_\varepsilon(\gamma|J) = \{(t, w) \in T_\gamma^\perp M | \ t \in J, \ g_{\gamma(t)}(w, w) < \varepsilon^2\}$$

and $N_\varepsilon(\gamma|J)$ by $N_\varepsilon(\gamma|J) = \{\exp_{\gamma(t)}(w) | (t, w) \in U_\varepsilon(\gamma|J)\}$. Clearly $N_\varepsilon(\gamma|J)$

is well defined if we choose $\varepsilon > 0$ sufficiently small and it is also clear that there exists $\varepsilon > 0$ such that $N_\varepsilon(\gamma|J)$ is well-defined for every subinterval $J \subseteq I$ such that $l(J) \leq \delta$. Moreover $U_\varepsilon(\gamma|J) \to J$ and $N_\varepsilon(\gamma|J) \to \gamma(J)$ are fiber bundles and the mapping $(t, w) \to \exp_{\gamma(t)}(w)$ is a diffeomorphism from $U_\varepsilon(\gamma|J)$ onto $N_\varepsilon(\gamma|J)$, which carries the fiber $U_\varepsilon(\gamma|J)_t$ onto the fiber $N_\varepsilon(\gamma|J)_{\gamma(t)}$ for $t \in J$.

Let $\tilde{N}_\varepsilon(\gamma) = \{(t, p)|t \in I$ and $p \in N_\varepsilon(\gamma|J)$ for some $J \subseteq I$ such that $l(J) \leq \delta\}$ and let $\tilde{\pi}_\gamma : \tilde{N}_\varepsilon(\gamma) \to I$ be the projection defined by $\tilde{\pi}_\gamma(t, p) = t$ for $(t, p) \in \tilde{N}_\varepsilon(\gamma)$. Then $\tilde{N}_\varepsilon(\gamma) \to I$ is a fiber bundle over I and for each $J \subseteq I, l(J) \leq \delta$, there is a fiber bundle isomorphism $N_\varepsilon(\gamma|J) \to \tilde{N}_\varepsilon(\gamma)|J$ which identifies the fiber of $N_\varepsilon(\gamma|J)$ over $\gamma(t)$ with the fiber of $\tilde{N}_\varepsilon(\gamma|J)$ over t for each $t \in J$. Let $\mathrm{Exp}_\gamma : U_\varepsilon(\gamma) \to \tilde{N}_\varepsilon(\gamma)$ be defined by $\mathrm{Exp}_\gamma(t, w) = (t, \exp_{\gamma(t)}(w))$ for $(t, w) \in U_\varepsilon(\gamma)$. Then Exp_γ is a fiber preserving diffeomorphism from the fiber bundle $\pi_\gamma : U_\varepsilon(\gamma) \to I$ onto the fiber bundle $\tilde{\pi}_\gamma : \tilde{N}_\varepsilon(\gamma) \to I$. Observe that for $t_1 \neq t_2$ in I such that $\gamma(t_1) = \gamma(t_2)$ the fibers $\tilde{N}_\varepsilon(\gamma)_{t_1}$ and $\tilde{N}_\varepsilon(\gamma)_{t_2}$ have no points in common although the subsets $N_\varepsilon(\gamma)_{t_1}$ and $N_\varepsilon(\gamma)_{t_2}$ intersect at $\gamma(t_1) = \gamma(t_2) \in M$.

If μ is any path in M we adopt the convention that $\tilde{\mu}$ will denote the mapping from I into $I \times M$ defined by $\tilde{\mu}(t) = (t, \mu(t))$ for $t \in I$. Observe that if μ is a path in $N_\varepsilon(\gamma) \subseteq M$, $\tilde{\mu}$ may or may not be a section of $\tilde{N}_\varepsilon(\gamma) \overset{\tilde{\pi}_\gamma}{\to} I$. We denote sections of $\tilde{N}_\varepsilon(\gamma)$ by $\Gamma(\tilde{N}_\varepsilon(\gamma))$ and we denote the set of all paths $\mu \in \mathscr{P}_1^* M$ which lie in $N_\varepsilon(\gamma)$ and "lift" to sections $\tilde{\mu}$ of $\tilde{N}_\varepsilon(\gamma)$ by $\Gamma(N_\varepsilon(\gamma))$. Thus $\Gamma(N_\varepsilon(\gamma))$ denotes paths in the "tubular neighborhood" of γ which lift to a section of the "normal bundle neighborhood" $\tilde{N}_\varepsilon(\gamma)$ of γ.

One might expect that sets of the type $\Gamma(N_\varepsilon(\gamma))$ would provide chart domains in $\mathscr{P}_1^* M$ about various $\gamma \in \mathscr{P}_1^* M$ since they define and are defined by an open subsets $\Gamma(U_\varepsilon(\gamma))$ of a Fréchet space $\Gamma(T_\gamma^\perp M)$ but if one were to define a structure based on these spaces one would find that the mapping from $\mathscr{P}_1^* M \times \mathscr{D}$ to $\mathscr{P}_1^* M$ defined by $(\mu, \varphi) \mapsto \mu \circ \varphi$ not only fails to be smooth but in fact is not continuous. The problem is that $\Gamma(N_\varepsilon(\gamma)) \cap \Gamma(N_\varepsilon(\gamma \circ \varphi)) = \phi$ for $\varphi \in \mathscr{D} - \{\mathrm{id}\}$. This suggests enlarging the chart domains to include all the sets $\Gamma(N_\varepsilon(\gamma \circ \varphi))$ along with $\Gamma(N_\varepsilon(\gamma))$. Let

$$\mathscr{S}(N_\varepsilon(\gamma)) = \bigcup\{\Gamma(N_\varepsilon(\gamma \circ \varphi))|\varphi \in \mathscr{D}\}.$$

LEMMA 1. *The family of sets* $\{\mathscr{S}(N_\varepsilon(\gamma))|\gamma \in \mathscr{P}_1^* M\}$ *is a family of chart domains for a differentiable structure on* $\mathscr{P}_1^* M$. *For each* $\gamma \in \mathscr{P}_1^* M$, *the mapping* $\Phi_\gamma : \mathscr{S}(N_\varepsilon(\gamma)) \to \Gamma(U_\varepsilon(\gamma)) \times \mathscr{D}$ *defined by* $\Phi_\gamma(\mu) = (\mathrm{Exp}_\gamma^{-1} \circ \widetilde{\mu \circ \varphi^{-1}}, \varphi)$ *for* $\varphi \in \mathscr{D}$ *and* $\mu \in \Gamma(N_\varepsilon(\gamma \circ \varphi))$ *is a chart in the structure on* $\mathscr{P}_1^* M$ *provided that* $\Gamma(U_\varepsilon(\gamma)) \times \mathscr{D}$ *is identified as an open subset of an appropriate Fréchet space (see the Remark below).*

REMARK. The Fréchet space which contains the image of the chart Φ_γ is the direct product of two Fréchet spaces $\Gamma(T_\gamma^\perp M)$ and $C_{0,1}^\infty(I)$ which are defined as follows. The space $C_{0,1}^\infty(I)$ is defined to be the set of all $f \in C^\infty(I, \mathbb{R})$ such that $f(0) = 0 = f(1)$. Clearly $C_{0,1}^\infty(I)$ is a closed subspace of the Fréchet space $C^\infty(I, \mathbb{R})$ and thus is itself a Fréchet space. Moreover it is clear that if $U = \{f \in C_{0,1}^\infty(I) |\ f'(x) < 1 \text{ for all } x \in I\}$ then U is open in $C_{0,1}^\infty(I)$ and the mapping from $\mathscr{D}$ onto U defined by $\varphi \to \mathrm{id}_I - \varphi$ is a bijection. Thus we identify $\mathscr{D}$ as an open subset of the Fréchet space $C_{0,1}^\infty(I)$.

To see that $\Gamma(T_\gamma^\perp M)$ is a Fréchet space choose an orthonormal basis of $(T_\gamma^\perp M)_0$ and parallel transport it along γ to obtain an orthonormal basis $\{l_i(t)\}$ of $(T_\gamma^\perp M)_t$ for each $t \in I$. For each section σ of $T_\gamma^\perp M \to I$ we have that $\sigma(t) = \sum_i \sigma^i(t) l_i(t)$ for each $t \in I$. Thus σ may be identified with the n-tuple $(\sigma_1, \sigma_2, \ldots, \sigma_n)$ which in turn may be identified as an element of $C^\infty(I, \mathbb{R}^n)$. Since $C^\infty(I, \mathbb{R}^n)$ is a well-known Fréchet space [6] so is $\Gamma(T_\gamma^\perp M)$.

We proceed to prove the lemma. Let $\mathscr{S}_i = \mathscr{S}(N_{\varepsilon_i}(\gamma_i))$, $i = 1, 2$ be two intersecting chart domains. If $\mu_0 \in \mathscr{S}_1 \cap \mathscr{S}_2$ then clearly there exists $\varphi_1, \varphi_2 \in \mathscr{D}$ such that $\mu_0 \in \Gamma(N_{\varepsilon_i}(\gamma_i \circ \varphi_i))$ for $i = 1, 2$. We eventually show that $\Phi_{\gamma_1}(\mu_0)$ is in an open subset of $\Gamma(U_{\varepsilon_1}(\gamma_1)) \times \mathscr{D}$ which is contained in the domain of $\Phi_{\gamma_2} \circ \Phi_{\gamma_1}^{-1}$ and that $\Phi_{\gamma_2} \circ \Phi_{\gamma_1}^{-1}$ maps this open set smoothly onto an open subset of $\Gamma(U_{\varepsilon_2}(\gamma_2)) \times \mathscr{D}$ containing $\Phi_{\gamma_2}(\mu_0)$.

Since $\mu_0 \in \Gamma(N_{\varepsilon_1}(\gamma_1 \circ \varphi_1)) \cap \Gamma(N_{\varepsilon_2}(\gamma_2 \circ \varphi_2))$ it is clear that

$$\tilde{\mu}_0(I) \subseteq \tilde{N}_{\varepsilon_1}(\gamma_1 \circ \varphi_1) \cap \tilde{N}_{\varepsilon_2}(\gamma_2 \circ \varphi_2).$$

Moreover if $\psi_1, \psi_2 \in \mathscr{D}$, $\psi_1 \neq \psi_2$, we have that

$$\Gamma(N_{\varepsilon_1}(\gamma \circ \psi_1)) \cap \Gamma(N_{\varepsilon_2}(\gamma \circ \psi_2)) = \phi$$

for arbitrary γ and ε. Thus

$$\mathscr{S}_1 \cap \mathscr{S}_2 = \bigcup_{\varphi \in \mathscr{D}} [\Gamma(N_{\varepsilon_1}(\gamma_1 \circ \varphi_1 \circ \varphi)) \cap \Gamma(N_{\varepsilon_2}(\gamma_2 \circ \varphi_2 \circ \varphi))].$$

Obviously $\widetilde{N}_{\varepsilon_1}(\gamma_1 \circ \varphi_1 \circ \varphi) \cap \widetilde{N}_{\varepsilon_2}(\gamma_2 \circ \varphi_2 \circ \varphi)$ is an open fiber bundle neighborhood of $(\widetilde{\mu_0 \circ \varphi})(I)$ in both bundles $\widetilde{N}_{\varepsilon_i}(\gamma_i \circ \varphi_i \circ \varphi)$, $i = 1, 2$ for each $\varphi \in \mathscr{D}$. Thus the set of sections of $\widetilde{N}_{\varepsilon_1}(\gamma_1 \circ \varphi_1 \circ \varphi) \cap \widetilde{N}_{\varepsilon_2}(\gamma_2 \circ \varphi_2 \circ \varphi)$ is open in both

$$\Gamma(\widetilde{N}_{\varepsilon_1}(\gamma_1 \circ \varphi_1 \circ \varphi)) \quad \text{and} \quad in \Gamma(\widetilde{N}_{\varepsilon_2}(\gamma_2 \circ \varphi_2 \circ \varphi)).$$

Let $U_{\varepsilon_1 \varepsilon_2}(\gamma_1 \circ \varphi_1) = \mathrm{Exp}_{\gamma_1 \circ \varphi_1}^{-1}(\widetilde{N}_{\varepsilon_1}(\gamma_1 \circ \varphi_1) \cap \widetilde{N}_{\varepsilon_2}(\gamma_2 \circ \varphi_2))$ and $U_{\varepsilon_1 \varepsilon_2}(\gamma_2 \circ \varphi_2) = \mathrm{Exp}_{\gamma_2 \circ \varphi_2}^{-1}(\widetilde{N}_{\varepsilon_1}(\gamma_1 \circ \varphi_1) \cap \widetilde{N}_{\varepsilon_2}(\gamma_2 \circ \varphi_2))$. These are open in $U_{\varepsilon_1}(\gamma_1 \circ \varphi_1) \subseteq T_{\gamma_1 \circ \varphi_1}^\perp M$ and in $U_{\varepsilon_2}(\gamma_2 \circ \varphi_2) \subseteq T_{\gamma_2 \circ \varphi_2}^\perp M$, respectively. Since the image of Φ_{γ_i} is in

$\Gamma(U_{\varepsilon_i}(\gamma_i)) \times \mathscr{D}$ we consider the mapping $\tilde{\varphi}_i : U_{\varepsilon_i}(\gamma_i) \to U_{\varepsilon_i}(\gamma_i \circ \varphi_i)$ defined by $\tilde{\varphi}_i(t, w) = (\varphi_i^{-1}(t), w)$ for $(t, w) \in U_{\varepsilon_i}(\gamma_i)$, $i = 1, 2$. Let $U_{\varepsilon_1 \varepsilon_2}(\gamma_i) = \tilde{\varphi}_i^{-1}(U_{\varepsilon_1 \varepsilon_2}(\gamma_i \circ \varphi_i))$ for i=1,2. We have a sequence of mappings of bundles

$$U_{\varepsilon_1 \varepsilon_2}(\gamma_1) \xrightarrow{\tilde{\phi}_1} U_{\varepsilon_1 \varepsilon_2}(\gamma_1 \circ \varphi_1) \xrightarrow{\tilde{\phi}_{12}} U_{\varepsilon_1 \varepsilon_2}(\gamma_2 \circ \varphi_2) \xrightarrow{\tilde{\phi}_2^{-1}} U_{\varepsilon_1 \varepsilon_2}(\gamma_2)$$
$$\downarrow \qquad\qquad \downarrow \qquad\qquad\qquad \downarrow \qquad\qquad\qquad \downarrow$$
$$I \xrightarrow{\varphi_1^{-1}} I \qquad = \qquad I \xrightarrow{\varphi_2} I$$

where $\varphi_{12} = (\text{Exp}_{\gamma_2 \circ \varphi_2}^{-1} \circ \text{Exp}_{\gamma_1 \circ \varphi_1}) | U_{\varepsilon_1 \varepsilon_2}(\gamma_1 \circ \varphi_1)$. These mappings are diffeomorphisms and are bundle morphisms. There are induced mappings $\Gamma(U_{\varepsilon_1 \varepsilon_2}(\gamma_1 \circ \varphi_1)) \xrightarrow{\hat{\phi}_{12}} \Gamma(U_{\varepsilon_1 \varepsilon_2}(\gamma_2 \circ \varphi_2))$, $\Gamma(U_{\varepsilon_1 \varepsilon_1}(\gamma_i)) \xrightarrow{\hat{\phi}_i} \Gamma(U_{\varepsilon_1 \varepsilon_2}(\gamma_i \circ \varphi_i))$ defined by $\hat{\phi}_{12}(\tilde{\nu}) = \varphi_{12} \circ \tilde{\nu}$ and by $\hat{\phi}_i(\tilde{\lambda}) = \tilde{\phi}_i \circ \tilde{\lambda} \circ \varphi_i$ for $\tilde{\nu} \in \Gamma(U_{\varepsilon_1 \varepsilon_2}(\gamma_1 \circ \varphi_1))$, $\tilde{\lambda} \in \Gamma(U_{\varepsilon_1 \varepsilon_2}(\gamma_i))$. It is not difficult to show that the mappings $\hat{\phi}_1$, $\hat{\phi}_2$, $\hat{\phi}_{12}$ are smooth using the techniques and results of Hamilton [6].

Let $\varphi_{12}^* = \hat{\phi}_{12} \times \text{id}_{\mathscr{D}}$ and $\varphi_i^* = \hat{\phi}_i \times \text{id}_{\mathscr{D}}$ for $i = 1, 2$. A tedious but straightforward computation shows that

$$(\Phi_{\gamma_2} \circ \Phi_{\gamma_1}^{-1}) | [\Gamma(U_{\varepsilon_1 \varepsilon_2}(\gamma_1)) \times \mathscr{D}] = ((\varphi_2^{-1})^* \circ \varphi_{12}^* \circ \varphi_1^*) | [\Gamma(U_{\varepsilon_1 \varepsilon_2}(\gamma_1)) \times \mathscr{D}].$$

We leave the verification of this fact to the reader. It follows that $\Phi_{\gamma_2} \circ \Phi_{\gamma_1}^{-1}$ agrees with the mapping $(\varphi_2^{-1})^* \circ \varphi_{12}^* \circ \varphi_1^*$ on the open set $\Gamma(\tilde{U}_{\varepsilon_1 \varepsilon_2}(\gamma_1)) \times D$ and thus is smooth. It is clear that $\Phi_{\gamma_2} \circ \Phi_{\gamma_1}^{-1}$ is smooth at each $\tilde{\mu}_o$ in its domain and by a symmetrical argument we see that $\Phi_{\gamma_1} \circ \Phi_{\gamma_2}^{-1}$ is also smooth. The lemma follows.

It should now be clear that statement (1) of Theorem 1 now follows. Statement (2) of Theorem 1 is a consequence of our next lemma.

LEMMA 2. *Let σ be the mapping from $\mathscr{P}_1^* M \times \mathscr{D}$ into $\mathscr{P}_1^* M$ defined by $\sigma(\mu, \varphi) = \mu \circ \varphi$ for $(\mu, \varphi) \in \mathscr{P}_1^* M \times \mathscr{D}$. Then σ is a smooth right action of $\mathscr{D}$ on $\mathscr{P}_1^* M$.*

PROOF. We know that σ is smooth in a neighborhood of an arbitrary element γ of $\mathscr{P}_1^* M$. Let $\mathscr{S}(N_\varepsilon(\gamma)) \xrightarrow{\Phi_\gamma} \Gamma(U_\varepsilon(\gamma)) \times \mathscr{D}$ be a chart at γ and observe that for an arbitrary $\mu \in \mathscr{S}(N_\varepsilon(\gamma))$, $\mu \in \Gamma(N_\varepsilon(\gamma \circ \varphi))$ for some $\varphi \in \mathscr{D}$. If $\mu = \lambda \circ \varphi$ for $\lambda \in \Gamma(N_\varepsilon(\gamma))$ then $\Phi_\gamma(\sigma(\mu, \psi)) = \Phi_\gamma(\lambda \circ \varphi \circ \psi) = (\text{Exp}_\gamma^{-1} \circ \tilde{\lambda}, \varphi \circ \psi)$. If $\tilde{\sigma}$ is the mapping from $[\Gamma(U_\varepsilon(\gamma)) \times \mathscr{D}] \times \mathscr{D}$ into $\Gamma(U_\varepsilon(\gamma)) \times D$ defined by $\tilde{\sigma}((\nu, \varphi), \psi) = (\nu, \varphi \circ \psi)$ then $\tilde{\sigma}$ is smooth and the diagram

$$\mathscr{S}(N_\varepsilon(\gamma)) \times \mathscr{D} \xrightarrow{\quad\sigma\quad} \mathscr{S}(N_\varepsilon(\gamma))$$
$$\Phi_\gamma \times \text{id}_{\mathscr{D}} \downarrow \qquad\qquad\qquad \downarrow \Phi_\gamma$$
$$[\Gamma(U_\varepsilon(\gamma)) \times \mathscr{D}] \times \mathscr{D} \xrightarrow{\quad\tilde{\sigma}\quad} \Gamma(U_\varepsilon(\gamma)) \times \mathscr{D}$$

is commutative. The lemma follows.

Statement (2) of Theorem 1 now follows. We now discuss how one obtains a differentiable structure on the quotient space $\mathscr{P}_1^* M/\mathscr{D}$. Observe first that the chart domains $\mathscr{S}(N_\varepsilon(\gamma))$, $\gamma \in \mathscr{P}_1^* M$ were constructed in such a manner that each of them is invariant under the action of $\mathscr{D}$. Thus if $\mathscr{P}_1^* M/\mathscr{D}$ denotes the orbit space of $\mathscr{P}_1^* M$ under the action of $\mathscr{D}$ then $\mathscr{S}(N_\varepsilon(\gamma))/\mathscr{D}$ is meaningful and may be regarded as a subset of $\mathscr{P}_1^* M/\mathscr{D}$. Moreover one has an obvious action of $\mathscr{D}$ on $\Gamma(U_\varepsilon(\gamma)) \times \mathscr{D}$ defined by $(\mu, \varphi) \cdot \psi = (\mu, \varphi \circ \psi)$ for $(\mu, \varphi) \in \Gamma(U_\varepsilon(\gamma)) \times \mathscr{D}$, $\psi \in \mathscr{D}$. If we denote elements of an orbit space under the action of $\mathscr{D}$ by $\chi_\mathscr{D}$ where χ is in the space on which $\mathscr{D}$ acts then it is easy to show that there is a well-defined mapping $\tilde{\Phi}_\gamma$ from $\mathscr{S}(N_\varepsilon(\gamma))/\mathscr{D}$ to $[\Gamma(U_\varepsilon(\gamma)) \times \mathscr{D}]/\mathscr{D}$ given by $\tilde{\Phi}_\gamma(\nu_\mathscr{D}) = \Phi_\gamma(\nu)_\mathscr{D}$ for $\nu \in \mathscr{S}(N_\varepsilon(\gamma))$. There is also a well-defined mapping

$$\nu_\gamma : [\Gamma(U_\varepsilon(\gamma)) \times \mathscr{D}]/\mathscr{D} \to \Gamma(U_\varepsilon(\gamma))$$

given by $\nu_\gamma((\lambda, \psi)_\mathscr{D}) = \lambda$ for $(\lambda, \psi) \in \Gamma(U_\varepsilon(\gamma)) \times \mathscr{D}$.

LEMMA 3. *The set of pairs* $(\mathscr{S}(N_\varepsilon(\gamma))/\mathscr{D}, \nu_\gamma \circ \tilde{\Phi}_\gamma)$ *is a differentiable structure on the orbit space* $\mathscr{P}_1^* M/\mathscr{D}$.

PROOF. First observe that if $\gamma_1, \gamma_2 \in \mathscr{P}_1^* M$ and $\mathscr{S}_i = \mathscr{S}(N_{\varepsilon_i}(\gamma_i))$ for $i = 1, 2$ then $(\mathscr{S}_1/\mathscr{D}) \cap (\mathscr{S}_2/\mathscr{D}) = (\mathscr{S}_1 \cap \mathscr{S}_2)/\mathscr{D}$. If $\mathscr{S}_1 \cap \mathscr{S}_2$ is nonempty then we may utilize the same notation as that in the proof of Lemma 1 as well as the arguments given there to see that Φ_{γ_i} maps $\mathscr{S}_1 \cap \mathscr{S}_2$ onto $\Gamma(\tilde{U}_{\varepsilon_1 \varepsilon_2}(\gamma_i)) \times \mathscr{D}$ for an appropriate bundle neighborhood $U_{\varepsilon_1 \varepsilon_2}(\gamma_i)$ of the zero section of the normal bundle to γ_i in TM. It follows that we will have the desired structure on $\mathscr{P}_1^* M/\mathscr{D}$ provided we show the two charts $(\mathscr{S}_i, \nu_{\gamma_i} \circ \tilde{\Phi}_{\gamma_i})$ are compatible and this will follow iff the mapping $(\nu_{\gamma_2} \circ \tilde{\Phi}_{\gamma_2}) \circ (\nu_{\gamma_1} \circ \tilde{\Phi}_{\gamma_1})^{-1}$ is a smooth function from $\Gamma(U_{\varepsilon_1 \varepsilon_2}(\gamma_1))$ onto $\Gamma(U_{\varepsilon_1 \varepsilon_2}(\gamma_2))$. Observe first that for $(\tilde{\nu}, \psi) \in \Gamma(\tilde{N}_\varepsilon(\gamma)) \times \mathscr{D}$, $\Phi_\gamma^{-1}(\tilde{\nu}, \psi) = \exp_{\gamma \circ \psi}(\nu \circ \psi)$ and consequently

$$\Phi_\gamma^{-1}((\tilde{\nu}, \psi) \cdot \varphi) = \Phi_\gamma^{-1}(\tilde{\nu}, \psi) \cdot \varphi$$

for each $\varphi \in \mathscr{D}$. Thus we define $\widetilde{\Phi_\gamma^{-1}}$ by $\widetilde{\Phi_\gamma^{-1}}((\tilde{\nu}, \psi)_\mathscr{D}) = \Phi_\gamma^{-1}(\tilde{\nu}, \psi)_\mathscr{D}$ and we have that $(\tilde{\Phi}_\gamma)^{-1} = \widetilde{\Phi_\gamma^{-1}}$. Utilizing the notation of the proof of Lemma 1 we recall that $\Phi_{\gamma_2} \circ \Phi_{\gamma_1}^{-1} = (\varphi_2^{-1})^* \circ \varphi_{12}^* \circ \varphi_1^*$ where $\Gamma(N_{\varepsilon_1}(\gamma_1 \circ \varphi_1)) \cap \Gamma(N_{\varepsilon_2}(\gamma_2 \circ \varphi_2)) \neq \varnothing$. Thus, for $\lambda \in \Gamma(U_{\varepsilon_1 \varepsilon_2}(\gamma_1))$ we have

$$[(\nu_{\gamma_2} \circ \tilde{\Phi}_{\gamma_2}) \circ (\nu_{\gamma_1} \circ \tilde{\Phi}_{\gamma_1})^{-1}](\lambda) = (\nu_{\gamma_2} \circ \widetilde{\Phi_{\gamma_2} \circ \Phi_{\gamma_1}^{-1}})((\lambda, \mathrm{id}_I)_\mathscr{D})$$

$$= \nu_{\gamma_2}(((\Phi_{\gamma_2} \circ \Phi_{\gamma_1}^{-1})(\lambda, \mathrm{id}_I))_\mathscr{D})$$

$$= \nu_{\gamma_2}(((\varphi_2^{-1})^* \circ \varphi_{12}^* \circ \varphi_1^*)(\lambda, \mathrm{id}_I)_\mathscr{D})$$

$$= (\hat{\varphi}_2^{-1} \circ \hat{\varphi}_{12} \circ \hat{\varphi}_1)(\lambda).$$

Thus $(\nu_{\gamma_2} \circ \tilde{\Phi}_{\gamma_2}) \circ (\nu_{\gamma_1} \circ \tilde{\Phi}_{\gamma_1})^{-1}$ is smooth and the lemma follows. Statement (3) of Theorem 1 is a consequence.

If $\tau : \mathscr{P}_1^* M \to \mathscr{P}_1^* M / \mathscr{D}$ is the canonical projection defined by $\tau(\mu) = \mu_{\mathscr{D}}$ for $\mu \in \mathscr{P}_1^* M$ then it is clear that τ is a smooth mapping since we may choose $\nu_\gamma \circ \tilde{\Phi}_\gamma \circ \tau \circ \Phi_\gamma^{-1}$ as its local coordinate representative at $\gamma \in \mathscr{P}_1^* M$ and since $(\nu_\gamma \circ \tilde{\Phi}_\gamma \circ \tau \circ \Phi_\gamma^{-1})(\nu, \varphi) = \nu_\gamma(\tilde{\Phi}_\gamma((\Phi_\gamma^{-1}(\nu, \varphi))_{\mathscr{D}})) = \nu$ for arbitrary $(\nu, \varphi) \in \Gamma(U_\varepsilon(\gamma)) \times \mathscr{D}$. The theorem now follows trivially.

One consequence of Lemma 3 is useful in many contexts.

COROLLARY. *The manifold* $(P_1^* M)/\mathscr{D}$ *is a quotient manifold of* $\mathscr{P}_1^* M$ *in the sense that every smooth mapping* f *from* $\mathscr{P}_1^* M$ *into a Fréchet manifold* N *which is constant on the fibers of* $\tau : \mathscr{P}_1^* M \to (\mathscr{P}_1^* M)/\mathscr{D}$ *admits a factorization* $f = g \circ \tau$ *for some smooth mapping* g *from* $\mathscr{P}_1^* M / \mathscr{D}$ *into* N.

The proof follows trivially from Lemma 3 and is omitted. This corollary provides one with a proof technique. One first proves a proposition using parameterized paths and then shows that the property being proven is reparametrization invariant. The proposition then follows for parameter independent paths according to the corollary.

In the next section of the paper we have occasion to utilize certain differential forms on the path space $\mathscr{P}_1^* M$. Since $\mathscr{P}_1^* M$ is a Fréchet manifold one naturally knows what it means to have a smooth vector field or a smooth one-form on $\mathscr{P}_1^* M$ as they are simply smooth sections of bundles of the type $T(\mathscr{P}_1^* M) \to \mathscr{P}_1^* M$ or $T^*(\mathscr{P}_1^* M) \to \mathscr{P}_1^* M$, respectively. There are special differential forms on $\mathscr{P}_1^* M$ which arise from corresponding forms on M as follows. Given a differential $(k+1)$-form $\tilde{\alpha}$ on M let α be defined on $\mathscr{P}_1^* M$ by $\alpha_\gamma(\nu_1, \ldots, \nu_k) = \int_0^1 \tilde{\alpha}_{\gamma(t)}(\dot{\gamma}(t), \nu_1(t), \ldots, \nu_k(t)) \, dt$. We claim that α is a (smooth) k-form on $\mathscr{P}_1^* M$. These forms will be called *path-forms* on $\mathscr{P}_1^* M$ and have been studied by many authors on a formal basis (see [1]).

Preliminary to showing path-forms are smooth we need a lemma. The proof of the lemma requires that we understand how to decompose tangent vectors at points of $\mathscr{P}_1^* M$ into a "normal component" and a "tangential component."

Let $\gamma \in \mathscr{P}_1^* M$ and let Φ_γ be a standard chart with chart domain $\mathscr{S}(N_\varepsilon(\gamma))$ in the differential structure of $\mathscr{P}_1^* M$. For $\mu \in \mathscr{S}(N_\varepsilon(\gamma))$ and $\delta_\mu \in T_\mu \mathscr{P}_1^* M$ we know that there is a curve $t \mapsto \mu_t$, $-\varepsilon < t < \varepsilon$, such that $\delta_\mu = \frac{d}{dt}(\mu_t)|_{t=0}$, $\mu_0 = \mu$. Since $\mu_t \in \mathscr{S}$ we have that $\mu_t = \lambda_t \circ \varphi_t$ where $\lambda_t \in \Gamma(N_\varepsilon(\gamma))$ and $\varphi_t \in \mathscr{D}$ for each t. Thus

$$\frac{d}{dt}(\mu_t)\bigg|_{t=0} = \frac{d}{dt}(\lambda_t)\bigg|_{t=0} \circ \varphi_0 + (d\lambda_0)\left(\frac{d}{dt}(\varphi_t)\bigg|_{t=0}\right).$$

It follows that

$$\delta_\mu = \delta_\mu^N \circ \varphi_0 + \delta_\mu^T \dot{\lambda}_0$$

where $\delta_\mu^N = \frac{d}{dt}(\lambda_t)|_{t=0}$ is an "infinitesimal deformation" of μ in a direction normal to γ and where δ_μ^T represents a "shift" tangential to μ. Using the fact that Φ_γ trivializes $\mathscr{S}(N_\varepsilon(\gamma))$ it is easy to show that the mapping π_N which projects (μ, δ_μ) to $(\mu \circ \varphi_0^{-1}, \delta_\mu^N)$ is a smooth mapping from $T(\mathscr{S}(N_\varepsilon(\gamma)))$ onto $T(\Gamma(N_\varepsilon(\gamma)))$ (note that $\Gamma(N_\varepsilon(\gamma))$ is a submanifold of $\mathscr{S}(N_\varepsilon(\gamma))$ since $\Phi_\gamma(\Gamma(N_\varepsilon(\gamma))) = \Gamma(\mathscr{U}_\varepsilon(\gamma)) \times \{\mathrm{id}\})$.

LEMMA 4. *Let $\tilde\alpha$ be a smooth $(k+1)$-form on a manifold M and let δ_1, $\delta_2, \ldots, \delta_k$ be smooth vector fields on $\mathscr{P}_1^* M$. Then the function A from $\mathscr{P}_1^* M$ to $\mathscr{C}^\infty(I, \mathbb{R})$ defined by $A(\mu)(t) = \tilde\alpha_{\mu(t)}(\dot\mu(t), \delta_1(\mu)(t), \ldots, \delta_k(\mu)(t))$ for $\mu \in \mathscr{P}_1^* M$, $t \in I$, is smooth.*

PROOF. We prove the lemma by expressing $\tilde\alpha$ in terms of "geodesic coordinates" on an appropriate subset of M. Actually these coordinates, which were defined in the remark following Lemma 1, are not really meaningful on a subset of M but rather are defined on fiber bundle neighborhoods $\tilde N_\varepsilon(\gamma)$ locally identifiable with subsets of M. Thus A must be reformulated in terms of forms defined on $\tilde N_\varepsilon(\gamma)$ for $\gamma \in \mathscr{P}_1^* M$. It follows that we need to rewrite the arguments $\mu, \dot\mu, \delta_1(\mu), \ldots, \delta_k(\mu)$ of $\tilde\alpha$ all in terms of curves and vector fields along curves of $\tilde N_\varepsilon(\gamma)$. Moreover the modified fields must vary smoothly as functions of $\mu \in (\mathscr{S}(N_\varepsilon(\gamma))$.

Begin by observing that for each $\mu \in \mathscr{S}(N_\varepsilon(\gamma))$, $\mu \in \Gamma(N_\varepsilon(\gamma \circ \varphi))$ for some unique $\varphi \in \mathscr{D}$. Thus for each $\mu \in \mathscr{S}$ there is a unique $\varphi_\mu \in \mathscr{D}$ such that the curve $\mu \circ \varphi_\mu^{-1}$ is in $\Gamma(N_\varepsilon(\gamma))$. It is clear that the mapping $\tilde\varphi$: $\mathscr{S} \to \mathscr{D}$ defined by $\tilde\varphi(\mu) = \varphi_\mu$, $\mu \in \mathscr{S}$, is smooth since $(\tilde\varphi \circ \Phi_\gamma^{-1})(\nu, \varphi) = \tilde\varphi(\exp_{\gamma \circ \varphi}(\nu \circ \varphi)) = \varphi$ for $(\nu, \varphi) \in \Gamma(U_\varepsilon(\gamma)) \times \mathscr{D}$. It follows from this fact that the mapping $\hat\varphi : \mathscr{S} \to \Gamma(N_\varepsilon(\gamma))$ defined by $\hat\varphi(\mu) = \mu \circ \varphi_\mu^{-1}$, $\mu \in \mathscr{S}$, is smooth. Indeed $\hat\varphi$ is the composite of smooth mappings $\mu \mapsto (\mu, \varphi_\mu^{-1}) \mapsto \mu \circ \varphi_\mu^{-1}$ and its image lies in the subset $\Gamma(N_\varepsilon(\gamma))$ of $\mathscr{S}$. Moreover $\Gamma(N_\varepsilon(\gamma))$ is actually a submanifold of $\mathscr{S}$ since $\Phi_\gamma(\Gamma(N_\varepsilon(\gamma))) = \Gamma(U_\varepsilon(\gamma))$. Thus $\hat\varphi$ is smooth and consequently so is $E_\gamma \circ \Phi_\gamma \circ \hat\varphi$ where $E_\gamma : \Gamma(U_\varepsilon(\gamma)) \to \Gamma(\tilde N_\varepsilon(\gamma))$ is defined by $E_\gamma(\nu) = \mathrm{Exp}_\gamma \circ \nu$ for $\nu \in \Gamma(U_\varepsilon(\gamma))$. Thus the mapping from $\mathscr{S}$ to $\Gamma(\tilde N_\varepsilon(\gamma))$ defined by $\mu \mapsto \widetilde{\mu \circ \varphi}_\mu^{-1} = (E_\gamma \circ \Phi_\gamma \circ \hat\varphi)(\mu)$ is smooth. It is this mapping which provides us with a transition from $\mathscr{S}$ to $\Gamma(\tilde N_\varepsilon(\gamma))$ on which there is defined a global set of coordinates. We must now transform the other arguments of $\tilde\alpha$ in the definition of A to $\Gamma(\tilde N_\varepsilon(\gamma))$.

Next we rewrite the mapping $\mu \to \dot\mu$ in terms of curves in $\tilde N_\varepsilon(\gamma)$. First recall that the mapping from $C^\infty(I, \mathbb{R}^n)$ to $\mathscr{C}^\infty(I, T\mathbb{R}^n)$ defined by $f \mapsto (f, \dot f)$ is smooth. Since $T_\gamma^\perp M \to I$ is a vector bundle the set of sections of $T_\gamma^\perp M \to I$ may be identified with $\mathscr{C}^\infty(I, \mathbb{R}^n)$ for some n and in fact this is how we obtained our Fréchet structure on $\Gamma(T_\gamma^\perp M)$. Since

$U = U_\varepsilon(\gamma)$ is open in $T_\gamma^\perp M$ it is easily seen that the mapping $\hat{\rho}$ from $\Gamma(U)$ to $C^\infty(I, TU)$ defined by $\hat{\rho} : f \mapsto (f, \dot{f})$, $f \in \Gamma(U)$, is smooth (one may use the global coordinates of $T_\gamma^\perp M$ to establish this fact). We use this result to show that the mapping ρ from $\Gamma(\tilde{N}_\varepsilon(\gamma))$ to $\mathscr{C}^\infty(I, T\tilde{N})$ defined by $\rho : g \mapsto (g, \dot{g})$, $g \in \Gamma\tilde{N}$, is smooth. To see this note that ρ admits a factorization $\Gamma(\tilde{N}_\varepsilon(\gamma)) \xrightarrow{E_\gamma^{-1}} \Gamma(U_\varepsilon(\gamma)) \xrightarrow{\hat{\rho}} \mathscr{C}^\infty(I, TU) \xrightarrow{dE_\gamma} \mathscr{C}^\infty(I, T\tilde{N})$ where the notation dE_γ is a slight abuse of notation. It is clear that each of these factors is smooth and thus so is ρ once we clarify what we mean by dE_γ. One reason it is clear that E_γ is smooth is that the mapping from $\mathscr{C}^\infty(I, U)$ to $\mathscr{C}^\infty(I, \tilde{N})$ defined by $f \mapsto \operatorname{Exp}_\gamma \circ f$ for $f \in \mathscr{C}^\infty(I, U)$ is smooth ($\operatorname{Exp}_\gamma$ is a diffeomorphism). Since E_γ is the restriction of this mapping to the submanifold ΓU, E_γ itself is smooth. On the other hand $d(\operatorname{Exp}_\gamma) : TU \to T\tilde{N}$ is also a diffeomorphism and consequently the mapping from $\mathscr{C}^\infty(I, TU)$ to $\mathscr{C}^\infty(I, T\tilde{N})$ defined by $\tilde{f} \mapsto d(\operatorname{Exp}_\gamma) \circ \tilde{f} \in \mathscr{C}^\infty(I, TU)$, is smooth. It is this mapping we have denoted by dE_γ. The reader should observe that in fact $dE_\gamma(\hat{\rho}(E_\gamma^{-1}(g))) = (g, \dot{g})$ for $g \in \Gamma\tilde{N}$. It now follows that the mapping from $\mathscr{S}$ to $\mathscr{C}^\infty(I, T(\tilde{N}_\varepsilon(\gamma)))$ defined by $\mu \mapsto (\widetilde{\mu \circ \varphi_\mu^{-1}}, \widetilde{\dot{\mu} \circ \varphi_\mu^{-1}})$ for $\mu \in \mathscr{S}$ is smooth. Indeed it admits a factorization in terms of the mappings we have already shown to be smooth

$$\mu \xrightarrow{\hat{\rho}} \mu \circ \varphi_\mu^{-1} \xrightarrow{E_\gamma \circ \Phi_\gamma} \widetilde{\mu \circ \varphi_\mu^{-1}} \xrightarrow{\rho} (\widetilde{\mu \circ \varphi_\mu^{-1}}, \widetilde{\dot{\mu} \circ \varphi_\mu^{-1}}).$$

Before proceeding with similar arguments we show how these facts will be utilized. Observe that for $\mu \in \mathscr{S}$, $t \in I$,

$$\tilde{\alpha}_{\mu(t)}(\dot{\mu}(t), \delta_1(\mu)(t), \ldots, \delta_k(\mu)(t)) = \varphi_\mu'(t) \tilde{\alpha}_{(\mu \circ \varphi_\mu^{-1})(\bar{t})}((\mu \circ \varphi_\mu^{-1})'(\bar{t}),$$

$$(*) \qquad\qquad (\delta_1(\mu) \circ \varphi_\mu^{-1})(\bar{t}), \ldots, (\delta_k(\mu) \circ \varphi_\mu^{-1})(\bar{t}))$$

where $\bar{t} = \varphi_\mu(t)$. The curve $\mu \circ \varphi_\mu^{-1}$ is a curve in $N_\varepsilon(\gamma)$ which defines a curve in $\tilde{N}_\varepsilon(\gamma)$ which we have been denoting by $\widetilde{\mu \circ \varphi_\mu^{-1}}$. Locally $\tilde{N}_\varepsilon(\gamma)$ and $N(\gamma)$ are identified and under this local identification $\widetilde{\mu \circ \varphi_\mu^{-1}}$ is just $\mu \circ \varphi_\mu^{-1}$. Moreover $(\mu \circ \varphi_\mu^{-1})^{\cdot}$ is tangent to $\mu \circ \varphi_\mu^{-1}$ at each point so $\widetilde{\dot{\mu} \circ \varphi_\mu^{-1}}$ is the corresponding tangent vector to $\widetilde{\mu \circ \varphi_\mu^{-1}}$ at each point of $\widetilde{\mu \circ \varphi_\mu^{-1}}$. The form $\tilde{\alpha}$ may also be locally identified as a form on $\tilde{N}_\varepsilon(\gamma)$. If we call $\hat{\alpha}$ the form on $\tilde{N}_\varepsilon(\gamma)$ corresponding to $\tilde{\alpha}$ on $N_\varepsilon(\gamma)$ and if we can identify the vector fields along $\widetilde{\mu \circ \varphi_\mu^{-1}}$ which correspond to the fields $\delta_i(\mu) \circ \varphi_\mu^{-1}$ defined at points of $\mu \circ \varphi_\mu^{-1}$ then it will be possible to rewrite $A(\mu)$ in terms of data on $\tilde{N}_\varepsilon(\gamma)$ where we can use our global coordinates. Actually this is quite easy. The mapping which carries $\mu \circ \varphi_\mu^{-1}$ to $\widetilde{\mu \circ \varphi_\mu^{-1}}$ is simply $E_\gamma \circ [\Phi_\gamma | \Gamma(N_\varepsilon(\gamma))]$

and its differential $d(E_\gamma \circ [\Phi_\gamma | \Gamma(N_\varepsilon(\gamma))])$ carries $(\mu \circ \varphi_\mu^{-1}, (\mu \circ \varphi_\mu^{-1})^\cdot)$ to $(\widetilde{\mu \circ \varphi}_\mu^{-1}, \widetilde{\mu \circ \varphi}_\mu^{-1})$. Thus $E_\gamma \circ [\Phi_\gamma | \Gamma(N_\varepsilon(\gamma))]$ is the mapping which identifies the points of $\Gamma(N_\varepsilon(\gamma))$ with those of $\Gamma(\tilde{N}_\varepsilon(\gamma))$ as well as any vector fields and forms which we might want to transform. Now $\delta_i(\mu) \circ \varphi_\mu^{-1}$ is a vector field along $\mu \circ \varphi_\mu^{-1}$ but it is not generally in $T(\Gamma(N_\varepsilon(\gamma)))$. On the other hand

$$(\delta_i(\mu) \circ \varphi_\mu^{-1})^N + (\delta_i(\mu) \circ \varphi_\mu^{-1})^T (\mu \circ \varphi_\mu^{-1})^\cdot$$

is the decomposition of $\delta_i(\mu) \circ \varphi_\mu^{-1}$ into its normal and tangential components along $\mu \circ \varphi_\mu^{-1}$. If we write $\hat{\delta}_i(\mu) = (\delta_i(\mu) \circ \varphi_\mu^{-1})^N = \pi_N(\delta_i(\mu) \circ \varphi_\mu^{-1})$ then the fact that $(\mu \circ \varphi_\mu^{-1})^\cdot(\bar{t})$ occurs in the first argument of $\tilde{\alpha}$ in equation $(*)$ assures that the tangential component of $\delta_i(\mu) \circ \varphi_\mu^{-1}$ vanishes in $(*)$ so that for $\bar{t} = \varphi_\mu(t)$,

$$A(\mu)(t) = \varphi_\mu'(t)\tilde{\alpha}_{(\mu \circ \varphi_\mu^{-1})(\bar{t})}((\mu \circ \varphi_\mu^{-1})^\cdot(\bar{t}), \tilde{\delta}_1(\mu)(\bar{t}), \ldots, \tilde{\delta}_k(\mu)(\bar{t})).$$

If we utilize the local identification of $N_\varepsilon(\gamma)$ and $\tilde{N}_\varepsilon(\gamma)$ defined above we have

$$A(\mu)(t) = \varphi_\mu'(t)\hat{\alpha}_{\hat{\mu}(\bar{t})}(\hat{\mu}^\cdot(\bar{t}), \hat{\delta}_1(\mu)(\bar{t}), \ldots, \hat{\delta}_k(\mu)(\bar{t}))$$

where $\hat{\mu} = \widetilde{\mu \circ \varphi}_\mu^{-1}$ and $\hat{\delta}_i(\mu) = d(E_\gamma \circ \Phi_\gamma)(\tilde{\delta}_i(\mu))$ for each $\mu \in \mathscr{S}$. If $y^0, y^1, \ldots, y^n$ are global coordinates on $\tilde{N}_\varepsilon(\gamma)$, then for $\bar{t} = \varphi_\mu(t)$,

$$A(\mu)(t) = \varphi_\mu'(t)\hat{\alpha}_{i_1 \cdots i_{k+1}}(\hat{\mu}(\bar{t}))dy^{i_1}(\hat{\mu}^\cdot(\bar{t}))dy^{i_2}(\hat{\delta}_1(\mu)(\bar{t}))dy^{i_{k+1}}(\hat{\delta}_k(\mu)(\bar{t})).$$

Thus $A(\mu)$ is a finite sum of terms each of which is a finite product of elements of $C^\infty(I, \mathbb{R})$ and if we can show that each factor of each term is smooth as a function from $\mathscr{S}$ into $C^\infty(I, \mathbb{R})$ then the lemma will follow. The functions we examine are all from $\mathscr{S}$ to $C^\infty(I, \mathbb{R})$ and are defined by

$$\mu \mapsto \varphi_\mu',$$

$$\mu \mapsto \hat{\alpha}_{i_1 \cdots i_{k+1}} \circ \hat{\mu} \circ \varphi_\mu, \qquad \mu \mapsto dy^{i_1} \circ \hat{\mu}^\cdot \circ \varphi_\mu,$$

$$\mu \mapsto dy^{i_{j+1}} \circ \hat{\delta}_j(\mu) \circ \varphi_\mu \quad \text{for } \mu \in \mathscr{S} \quad \text{(no sum)}.$$

It is a tedious but easy exericise to show that each of the mappings is smooth. The details are left to the reader. The lemma follows.

THEOREM 2. *Let $\tilde{\alpha}$ be a smooth $(k+1)$-form on a manifold M and, for each $\gamma \in \mathscr{P}_1^* M$, let $\alpha_\gamma : T_\gamma \mathscr{P}_1^* M \times \cdots \times T_\gamma \mathscr{P}_1^* M \to \mathbb{R}$ be defined by $\alpha_\gamma(\delta_1, \ldots, \delta_k) = \int_0^1 \tilde{\alpha}_{\gamma(t)}(\dot{\gamma}(t), \delta_1(t), \ldots, \delta_k(t))\,dt$. Then α is a differential form on the manifold $\mathscr{P}_1^* M$.*

PROOF. We need only show that if $\delta_1, \delta_2, \ldots, \delta_k$ are smooth vector fields on $\mathscr{P}_1^* M$ then the function from $\mathscr{P}_1^* M$ to $\mathbb{R}$ defined by

$$\mu \mapsto \alpha_\mu(\delta_1(\mu), \ldots, \delta_k(\mu))$$

is smooth. Lemma 4 implies that for $\gamma \in \mathscr{P}_1^* M$ the mapping $A(\mu)$ from $\mathscr{S}(N_\varepsilon(\gamma))$ to $C^\infty(I, \mathbb{R})$ defined by

$$A(\mu)(t) = \tilde{\alpha}_{\mu(t)}(\delta_1(\mu)(t), \ldots, \delta_k(\mu)(t))$$

for $\mu \in \mathscr{S}$, $t \in I$, is smooth. But for $\mu \in \mathscr{S}$,

$$\alpha_\mu(\delta_1(\mu), \ldots, \delta_k(\mu)) = \int_0^1 A(\mu)(t)\, dt$$

and it is well known [6] that the mapping from $C^\infty(I, \mathbb{R})$ to $\mathbb{R}$ defined by $f \mapsto \int_0^1 f(t)\, dt$ is smooth. It follows that $\mu \mapsto \alpha_\mu(\delta_1(\mu), \ldots, \delta_k(\mu))$ is smooth on the open subset $\mathscr{S}$ of $\mathscr{P}_1^* M$. Thus α is a differential form on $\mathscr{P}_1^* M$.

Many of the usual transformations which act on differential forms defined on a manifold M extend to path-forms on $\mathscr{P}_1^* M$. For example if $\tilde{\alpha}$ is a $(k + 1)$-form on M and

$$\alpha_\gamma(\delta_1, \ldots, \delta_k) = \int_0^1 \tilde{\alpha}_{\gamma(t)}(\dot{\gamma}(t), \delta_1(t), \ldots, \delta_k(t))\, dt$$

then we define $\hat{d}\alpha$ to be the path-form given by

$$(\hat{d}\alpha)_\gamma(\delta_1, \ldots, \delta_{k+1}) = -\int_0^1 (d\tilde{\alpha})_{\gamma(t)}(\dot{\gamma}(t), \delta_1(t), \ldots, \delta_{k+1}(t))\, dt.$$

It is clear that $\hat{d}\alpha$ is a new differential form on $\mathscr{P}_1^* M$ since it is a path-form. We show below that although $\hat{d}$ does not coincide with exterior differentiation on $\mathscr{P}_1^* M$, it is the case that for one-forms α on loop space $\hat{d}\alpha = d\alpha$.

Other operations may be extended to path-forms. For example, if M has a metric then we can define a duality operator on path-forms without explicitly having a metric on $\mathscr{P}_1^* M$. One simply defines $*\alpha$ by

$$(*\alpha)_\gamma(\delta_1, \ldots, \delta_l) = \int_0^1 (*\tilde{\alpha})_{\gamma(t)}(\dot{\gamma}(t), \delta_1(t), \ldots, \delta_l(t))\, dt.$$

Clearly $*\alpha$ is a differential form on $\mathscr{P}_1^* M$ as it is a path-form.

In the next section we will want to examine Yang-Mills equations on a principal bundle P in terms of the path space $\mathscr{P}_1^* P$. If we have a metric on M then $*$ on M induces a Hodge operator $\bar{*}$ on equivariant horizontal forms on P (see [2]). Thus if $\tilde{\alpha}$ is an equivariant horizontal form on P we define $(\bar{*}\alpha)_\gamma(\delta_1, \ldots, \delta_l) = \int_0^1 (\bar{*}\alpha)_{\gamma(t)}(\dot{\gamma}(t), \delta_1(t), \ldots, \delta_l(t))\, dt$. Similarly, if α is any path-form on $\mathscr{P}_1^* P$ we define

$$(\mathrm{hor}\,\alpha)_\gamma(\delta_1, \ldots, \delta_k) = \int_0^1 (\mathrm{hor}\,\tilde{\alpha})_{\gamma(t)}(\dot{\gamma}(t), \delta_1(t), \ldots, \delta_k(t))\, dt.$$

These operations on forms on $\mathscr{P}_1^* P$ will be used in the next section.

In the last few paragraphs of the paper we derive a formula which is a loop space form of Yang-Mills equation. The derivation of this formula involves the operator d on one-forms on $\mathscr{P}_1^* M$. In the next few paragraphs we show that if α is a one-form on loop space LM then $\hat{d}\alpha = d\alpha$ where d is the usual exterior derivative of α on the Fréchet manifold LM. Let $\tilde{\alpha}$ be a two-form on a manifold M and let α be its corresponding path form,

$$\alpha_\gamma(\delta) = \int_0^1 \tilde{\alpha}_{\gamma(t)}(\dot{\gamma}(t),\, \delta(t))\, dt.$$

If $\hat{\delta}_1$, $\hat{\delta}_2$ are vector fields on $\mathscr{P}_1^* M$ the usual exterior derivative of a one-form α satisfies the condition

$$(d\alpha)(\hat{\delta}_1,\, \hat{\delta}_2) = \hat{\delta}_1(\alpha(\hat{\delta}_2)) - \hat{\delta}_2(\alpha(\hat{\delta}_1)) - \alpha([\hat{\delta}_1,\, \hat{\delta}_2])$$

We derive a formula for $d\alpha$ which will allow easy comparison with $\hat{d}\alpha$. To do this let $\gamma \in \mathscr{P}_1^* M$ and let δ_1, $\delta_2 \in T_\gamma(\mathscr{P}_1^* M)$. For $i = 1, 2$, δ_i is a vector field along γ and, for each $t \in I$, $\delta_i(t) = \delta_i^a(t)(\frac{\partial}{\partial y^a}\big|_{\gamma(t)})$ where y^0, y^1, ..., y^n are geodesic coordinates on $\tilde{N}_\varepsilon(\gamma)$ and where we have used the local identification of $\tilde{N}_\varepsilon(\gamma)$ used extensively above. Define vector fields $\tilde{\delta}_1$, $\tilde{\delta}_2$ on $\tilde{N}_\varepsilon(\gamma)$ by $\tilde{\delta}_i(p) = \delta_i^a(\tilde{\pi}_\gamma(p))(\frac{\partial}{\partial y^a}\big|_p)$ (recall that $\tilde{\pi}_\gamma : \tilde{N}_\varepsilon(\gamma) \to I$ is the bundle projection). Clearly, for each $\mu \in \mathscr{S}(N_\varepsilon(\gamma))$, $\tilde{\delta}_i(\mu) : t \mapsto \tilde{\delta}_i(\mu(t))$ defines a vector field along μ which may be identified with an element of $T_\mu(\mathscr{P}_1^* M)$. (Elements of $T_\mu(\mathscr{P}_1^* M)$ are vector fields along μ such that the "tangential components" vanish at $t = 0, 1$. Since the tangential component of $\tilde{\delta}_i \circ \mu$ is $\delta^T(\mu \circ \varphi^{-1})\dot{}$ for some $\varphi \in \mathscr{D}$ where $\delta^T \in T_\varphi \mathscr{D}$ and since $\delta^T(0) = \delta^T(1) = 0$, $\tilde{\delta}_i \circ \mu$ satisfies the requisite condition.) We have

$$d_\gamma\alpha(\delta_1,\, \delta_2) = d_\gamma\alpha(\hat{\delta}_1(\gamma),\, \hat{\delta}_2(\gamma))$$
$$= (\hat{\delta}_1)_\gamma(\alpha(\hat{\delta}_2)) - (\hat{\delta}_2)_\gamma(\alpha(\hat{\delta}_1)) - \alpha_\gamma([\hat{\delta}_1,\, \hat{\delta}_2]).$$

To compute the latter let $s \mapsto \psi_s^i$ denote the flow of the vector field $\tilde{\delta}_i$ on $\tilde{N}_\varepsilon(\gamma)$. If we define $\mu_s^i = \psi_s^i \circ \gamma$ then $s \mapsto \mu_s^i$ is a curve in $\mathscr{P}_1^* M$ such that $\mu_0^i = \gamma$ and $\frac{d}{ds}(\mu_s^i)\big|_{s=0} = \frac{d}{ds}(\psi_s^i \circ \gamma)\big|_{s=0} = \tilde{\delta}_i \circ \gamma = \hat{\delta}_i(\gamma)$. Thus

$$(\hat{\delta}_1)_\gamma(\alpha(\hat{\delta}_2)) = d_\gamma(\alpha(\hat{\delta}_2))(\hat{\delta}_1(\gamma)) = d_\gamma(\alpha(\hat{\delta}_2))\left(\frac{d}{ds}(\mu_s^1)\big|_{s=0}\right)$$
$$= \frac{d}{ds}[\alpha(\hat{\delta}_2)(\mu_s^1)]\big|_{s=0} = \frac{d}{ds}[\alpha_{\mu_s^1}(\hat{\delta}_2)]\big|_{s=0}$$
$$= \frac{d}{ds}\left[\int_0^1 \tilde{\alpha}(\mu_s^1(t))\left(\frac{d}{dt}(\mu_s^1(t)),\, \hat{\delta}_2(\mu_s^1)(t)\right) dt\right]\Bigg|_{s=0}.$$

We will derive a formula for the latter expression in terms of Lie derivatives.

To do this observe that if $\psi_s = \psi_s^1$ then

$$\tilde{\alpha}(\mu_s^1(t))\left(\frac{d}{dt}(\mu_s^1(t)),\, \hat{\delta}_2(\mu_s^1)(t)\right)$$

$$= \tilde{\alpha}(\psi_s(\gamma(t)))\left(\frac{d}{dt}(\psi_s(\gamma(t))),\, \tilde{\delta}_2(\psi_s(\gamma(t)))\right)$$

$$= (\psi_s^*\tilde{\alpha})(\gamma(t))(\dot{\gamma}(t),\, d\psi_s^{-1}(\tilde{\delta}_2(\psi_s(\gamma(t)))))$$

and

$$\frac{d}{ds}\left[\tilde{\alpha}(\mu_s^1(t))\left(\frac{d}{dt}(\mu_s^1(t)),\, \tilde{\delta}_2(\mu_s^1(t))\right)\right]\Bigg|_{s=0}$$

$$= \frac{d}{ds}(\psi_s^*\tilde{\alpha})(\dot{\gamma}(t),\, d\psi_s^{-1}(\tilde{\delta}_2(\psi_s(\gamma(t)))))\Bigg|_{s=0}$$

$$= \frac{d}{ds}[(\psi_s^*\tilde{\alpha})]\Bigg|_{s=0}(\dot{\gamma}(t),\, \tilde{\delta}_2(\gamma(t))) + \frac{d}{ds}[\tilde{\alpha}(\dot{\gamma}(t),\, d\psi_s^{-1}(\tilde{\delta}_2(\psi_s(\gamma(t)))))]\Bigg|_{s=0}$$

$$= (L_{\tilde{\delta}_1}\tilde{\alpha})_{\gamma(t)}(\dot{\gamma}(t),\, \tilde{\delta}_2(\gamma(t))) + \tilde{\alpha}_{\gamma(t)}\left(\dot{\gamma}(t),\, \frac{d}{ds}(d\psi_s^{-1}(\tilde{\delta}_2(\psi_s(\gamma(t)))))\Bigg|_{s=0}\right)$$

$$= (L_{\tilde{\delta}_1}\tilde{\alpha})_{\gamma(t)}(\dot{\gamma}(t),\, \tilde{\delta}_2(\gamma(t))) + \tilde{\alpha}_{\gamma(t)}(\dot{\gamma}(t),\, [\tilde{\delta}_1,\, \tilde{\delta}_2](\gamma(t))).$$

It follows that

$$(\hat{\delta}_1)_\gamma(\alpha(\hat{\delta}_2)) = \int_0^1 [(L_{\tilde{\delta}_1}\tilde{\alpha})_{\gamma(t)}(\dot{\gamma}(t),\, \tilde{\delta}_2(\gamma(t))) + \tilde{\alpha}(\dot{\gamma}(t),\, [\tilde{\delta}_1,\, \tilde{\delta}_2](\gamma(t)))]\, dt.$$

Thus we have the following identity

$$d_\gamma\alpha(\delta_1,\, \delta_2) = \int_0^1 [(L_{\tilde{\delta}_1}\tilde{\alpha})(\dot{\gamma}(t),\, \delta_2(t)) - (L_{\tilde{\delta}_2}\tilde{\alpha})(\dot{\gamma}(t),\, \delta_1(t))]\, dt$$

$$+ 2\int_0^1 \tilde{\alpha}(\gamma(t))(\dot{\gamma}(t),\, [\tilde{\delta}_1,\, \tilde{\delta}_2](\gamma(t)))\, dt - \alpha_\gamma([\hat{\delta}_1,\, \hat{\delta}_2]).$$

In order to complete the argument it is useful to know that for $\mu \in \mathscr{P}_1^*M$, $[\hat{\delta}_1, \hat{\delta}_2](\mu) = [\tilde{\delta}_1, \tilde{\delta}_2] \circ \mu$ for $\mu \in \mathscr{P}_1^*M$. We show that this result follows from consideration of the local representatives of the vector fields involved. Indeed given charts $\Phi_\gamma : \mathscr{S} \to E$, $y : \tilde{N}_\varepsilon(\gamma) \to \mathbb{R}^{n+1}$ where $E = C^\infty(I, \mathbb{R}^{n+1})$, we may write

$$\hat{\delta}_i(\mu) = (\mu,\, \xi_i(\mu)), \qquad [\hat{\delta}_1,\, \hat{\delta}_2](\mu) = (\mu,\, \xi_{12}(\mu))$$

$$\tilde{\delta}_i(p) = (p,\, \tilde{\xi}_i(p)), \qquad [\tilde{\delta}_1,\, \tilde{\delta}_2](p) = (p,\, \tilde{\xi}_{12}(p))$$

for $\mu \in \mathscr{S}$, $p \in \tilde{N}_\varepsilon(\gamma)$, $i = 1, 2$. Here we have abused notation by identifying the various vector fields with their local representatives. The functions ξ_1, ξ_2, ξ_{12} are smooth functions from $\Phi_\gamma(\mathscr{S}) \approx \mathscr{S}$ to E and $\tilde{\xi}_1, \tilde{\xi}_2, \tilde{\xi}_{12}$ are smooth functions from $y(\tilde{N}_\varepsilon(\gamma))$ to $\mathbb{R}^{n+1}$. It follows that

$$\xi_{12}(\mu) = d_\mu\xi_2(\xi_1(\mu)) - d_\mu\xi_1(\xi_2(\mu))$$

$$\tilde{\xi}_{12}(p) = d_p\tilde{\xi}_2(\tilde{\xi}_1(p)) - d_p\tilde{\xi}_1(\tilde{\xi}_2(p))$$

for $\mu \in \mathscr{S}$, $p \in \tilde{N}_\varepsilon(\gamma)$. Moreover $d\xi_2(\xi_1(\mu))(t) = d_{\mu(t)}\tilde{\xi}_2(\tilde{\xi}_1(\mu(t)))$, for $t \in I$, with a similar formula for $d\xi_1(\xi_2(\mu))$. Thus $\xi_{12}(\mu) = \tilde{\xi}_{12} \circ \mu$ for $\mu \in \mathscr{S}$ and $[\hat{\delta}_1, \hat{\delta}_2](\mu) = [\tilde{\delta}_1, \tilde{\delta}_2] \circ \mu$ as asserted. With this fact in hand it now follows that

$$\alpha_\gamma([\hat{\delta}_1, \hat{\delta}_2]) = \int_0^1 \tilde{\alpha}_{\gamma(t)}(\dot{\gamma}(t), [\hat{\delta}_1, \hat{\delta}]_\gamma(t))\, dt$$

$$= \int_0^1 \tilde{\alpha}_{\gamma(t)}(\dot{\gamma}(t), [\tilde{\delta}_1, \tilde{\delta}_2](\gamma(t)))\, dt.$$

Thus

$$d_\gamma\alpha(\delta_1, \delta_2) = \int_0^1 [(L_{\hat{\delta}_1}\tilde{\alpha})(\dot{\gamma}(t), \delta_2(t)) - (L_{\hat{\delta}_2}\tilde{\alpha})(\dot{\gamma}(t), \delta_1(t))]\, dt$$

$$+ \int_0^1 \tilde{\alpha}(\gamma(t))(\dot{\gamma}(t), [\tilde{\delta}_1, \tilde{\delta}_2](\gamma(t)))\, dt$$

$$= \int_0^1 (-i_{\tilde{\delta}_2}L_{\tilde{\delta}_1} + i_{\tilde{\delta}_1}L_{\tilde{\delta}_2})(\tilde{\alpha})_{\gamma(t)}(\dot{\gamma}(t))\, dt$$

$$+ \int_0^1 \tilde{\alpha}(\gamma(t))(\dot{\gamma}(t), [\tilde{\delta}_1, \tilde{\delta}_2](\gamma(t)))\, dt.$$

Using the identity $i_{[\tilde{\delta}_2, \tilde{\delta}_1]}\tilde{\alpha} = L_{\tilde{\delta}_2}(i_{\tilde{\delta}_1}\tilde{\alpha}) - i_{\tilde{\delta}_1}(L_{\tilde{\delta}_2}\tilde{\alpha})$ we have that

$$-i_{\tilde{\delta}_2}L_{\tilde{\delta}_1} + i_{\tilde{\delta}_1}L_{\tilde{\delta}_2} = -d\tilde{\alpha}(\tilde{\delta}_1, \tilde{\delta}_2, \cdot) + \tilde{\alpha}([\tilde{\delta}_1, \tilde{\delta}_2], \cdot) + d(\tilde{\alpha}(\tilde{\delta}_1, \tilde{\delta}_2))(\cdot).$$

It follows that

$$d_\gamma\alpha(\delta_1, \delta_2) = \int_0^1 \{-d\tilde{\alpha}(\tilde{\delta}_1, \tilde{\delta}_2, \dot{\gamma}) + \tilde{\alpha}([\tilde{\delta}_1, \tilde{\delta}_2], \dot{\gamma})$$

$$+ d(\tilde{\alpha}(\tilde{\delta}_1, \tilde{\delta}_2))(\dot{\gamma}) + \tilde{\alpha}(\dot{\gamma}, [\tilde{\delta}_1, \tilde{\delta}_2])\}\, dt$$

$$= -\int_0^1 d\tilde{\alpha}(\dot{\gamma}, \tilde{\delta}_1, \tilde{\delta}_2)\, dt + \int_0^1 d(\tilde{\alpha}(\tilde{\delta}_1, \tilde{\delta}_2))(\dot{\gamma})\, dt$$

$$= \tilde{\alpha}_{\gamma(1)}(\delta_1(1), \delta_2(1)) - \tilde{\alpha}_{\gamma(0)}(\delta_1(0), \delta_2(0)) + \hat{d}_\gamma\alpha(\delta_1, \delta_2).$$

These remarks have shown that if α is the one-form defined by

$$\alpha_\gamma(\delta) = \int_0^1 \tilde{\alpha}_{\gamma(t)}(\dot{\gamma}(t), \delta(t))\, dt$$

then

$$d_\gamma\alpha(\delta_1, \delta_2) = \tilde{\alpha}_{\gamma(1)}(\delta_1(1), \delta_2(1)) - \tilde{\alpha}_{\gamma(0)}(\delta_1(0), \delta_2(0))$$

$$- \int_0^1 (d\tilde{\alpha})_{\gamma(t)}(\dot{\gamma}(t), \delta_1(t), \delta_2(t))\, dt$$

Phase factors and gauge theory

In this section we consider the mapping $\mathscr{I}$ defined on the path space $\mathscr{P}_1^* P$ of a principal fiber bundle P by $\mathscr{I}(\gamma) = P \exp \int_0^1 \omega(\dot{\gamma}(t))\, dt$ for $\gamma \in \mathscr{P}_1^* P$.

The integral is the path ordered exponential or product integral discussed in detail in [3, 5] and ω denotes a connection on P.

The mapping $\mathscr{I}$ has played an important role in physics beginning, perhaps, with Dirac's early discussions of magnetic monopoles. It was essential in the Wu-Yang nonintegrable phase factor approach to gauge theory [8]. More recently Polyakov [7] gave a heuristic proof of a remarkable result which recasts gauge theory in terms of the mapping $\mathscr{I}$.

In this section we show that $\mathscr{I}$ is smooth and that the pull-back relative to $\mathscr{I}$ of the Maurer-Cartan form on the structure group G of P defines a flat connection on the trivial G-bundle over $\mathscr{P}_1^* P$. In some sense one can reformulate nontrivial gauge theories on P in terms of flat connections on $\mathscr{P}_1^* P$ since the Yang-Mills equations also have an analogue on $\mathscr{P}_1^* P$.

Polyakov's version of these facts inspired the present work but his techniques involve the variational derivative, a concept more closely related to distributional derivatives than to classical derivatives. Our proof utilizes derivatives in a more "classical sense" although we are forced to develop the ideas on infinite dimensional manifolds.

THEOREM 3. *Let G be a Lie subgroup of $Gl(n, \mathbb{R})$ for some n and (P, M, π, G) a principal fiber bundle with structure group G. Let ω be any connection on P and $\mathscr{I}$ the mapping from $\mathscr{P}_1^* P$ to G defined by*

$$\mathscr{I}(\gamma) = P \exp \int_0^1 \omega(\dot{\gamma}(t)) \, dt$$

for $\gamma \in \mathscr{P}_1^* P$. Then $\mathscr{I}$ is a smooth mapping and

$$d_\gamma \mathscr{I}(\delta) = \mathscr{I}(\gamma)[\omega(\delta(1)) - \omega(\delta(0))] + \left[\int_0^1 Ad(U(s))\Omega(\delta(s), \dot{\gamma}(s)) \, ds \right] \mathscr{I}(\gamma)$$

where $U(s) = P \exp \int_0^s \omega(\dot{\gamma}(u)) du$, for $s \in I$.

PROOF. To show that $\mathscr{I}$ is smooth we proceed as follows. Let $\gamma \in \mathscr{P}_1^* P$ and choose a chart $(\mathscr{S}(N_\varepsilon(\gamma)), \Phi_\gamma)$ centered at γ. We will show that $\mathscr{I}$ is smooth on $\mathscr{S}(N_\varepsilon(\gamma))$ by showing that $\mathscr{I}$ can factored as the composite of two mappings Θ and ε which will be shown to be smooth. These mappings are defined as follows:

$$\Theta : \mathscr{S}(N_\varepsilon(\gamma)) \to C^\infty(I, \mathfrak{g}) \qquad \Theta(\mu) = \omega \circ \mu$$

$$\varepsilon : C^\infty(I, \mathfrak{g}) \to G \qquad \varepsilon(\beta) = P \exp \int_0^1 \beta(t) \, dt$$

for $\mu \in \mathscr{S}(N_\varepsilon(\gamma))$, $\beta \in C^\infty(I, \mathfrak{g})$ (here $\mathfrak{g}$ is the Lie algebra of G).

To show that Θ is a smooth at γ we show that the local coordinate representative of Θ, $\Theta \circ \Phi_\gamma^{-1} \circ \psi_\gamma^{-1}$, is smooth. For (u, φ) in an appropriate open subset $\mathscr{O}$ of $C^\infty(I, \mathbb{R}^n) \times \mathscr{D}$ we have

$$(\Theta \circ \Phi_\gamma^{-1} \circ \psi_\gamma^{-1})(u, \varphi) = \Theta(\Phi_\gamma^{-1}(\widetilde{u^i l_i}, \varphi))$$
$$= \Theta(\exp_{\gamma \circ \varphi}((u^i l_i) \circ \varphi))$$
$$= \omega \circ \frac{d}{dt}(\exp_{\gamma \circ \varphi}((u^i l_i) \circ \varphi)).$$

(Here the $\{l_i\}$ are defined as in the Remark following the statement of Theorem 1 and the mapping $\psi_\gamma : \Gamma(U_\varepsilon(\gamma)) \times \mathscr{D} \to C^\infty(I, \mathbb{R}^n) \times \mathscr{D}$ is defined in such a way that $\psi_\gamma^{-1}(u^1, \ldots, u^n, \varphi) = \widetilde{u^i l_i} \circ \varphi$.) To show that the latter is smooth it is convenient to introduce coordinates on $N_\varepsilon(\gamma)$ but since there is no clear way to obtain global coordinates on $N_\varepsilon(\gamma)$ we transform to $\tilde{N}_\varepsilon(\gamma)$ where we can define a global coordinate system. This transformation requires two steps. We must show that ω induces a mapping from $T(\tilde{N}_\varepsilon(\gamma))$ to $\underline{g}$ and that the curve $\mu(t) = \exp_{\gamma(t)}(u^i(t) l_i(t))$ in $N_\varepsilon(\gamma)$ naturally defines a curve $\tilde{\mu}$ in $\tilde{N}_\varepsilon(\gamma)$ such that $\tilde{\omega} \circ (\frac{d\tilde{\mu}}{dt}) = \omega \circ \frac{d\mu}{dt}$. Recall that if $J \subseteq I$ is a subinterval whose length is not too large we have that $\tilde{N}_\varepsilon(\gamma)|J \cong N_\varepsilon(\gamma|J)$. Thus for each such $J \subseteq I$ we have a bundle isomorphism $\phi_J : \tilde{N}_\varepsilon(\gamma)|J \to N_\varepsilon(\gamma|J)$ and moreover for $J_1 \subseteq I$, $J_2 \subseteq I$ such that $J_1 \cap J_2 \neq \phi$ we have that $\phi_{J_1}|[\tilde{N}_\varepsilon(\gamma)|(J_1 \cap J_2)] = \phi_{J_2}|[\tilde{N}_\varepsilon(\gamma)|(J_1 \cap J_2)]$. Since the restriction of ω to any one of the bundles $N_\varepsilon(\gamma|J)$ is a $\underline{g}$-valued one-form on $N_\varepsilon(\gamma|J)$, clearly its pullback $\phi_J^* \omega$ is a one-form on $\tilde{N}_\varepsilon(\gamma)|J$. Since $\phi_{J_1}^* \omega = \phi_{J_2}^* \omega$ for J_1, J_2 such that $J_1 \cap J_2 \neq \phi$ we have a well-defined $\underline{g}$-valued one form $\tilde{\omega}$ on all of $\tilde{N}_\varepsilon(\gamma)$.

Since $\tilde{N}_\varepsilon(\gamma)$ and $N_\varepsilon(\gamma)$ are locally identical we have been distinguishing between the two by writing the points of $\tilde{N}_\varepsilon(\gamma)_t$ as ordered pairs (t, p) where $p \in N(\gamma)_t$ for $t \in I$. Thus if μ is a curve in $N\varepsilon(\gamma)$ and $\tilde{\mu} = \phi_J^{-1} \circ \mu$ is its pullback to $\tilde{N}_\varepsilon(\gamma)$ we have been writing $\tilde{\mu}(s) = (s, \mu(s))$ for each s in the domain of μ. Thus we have

$$\tilde{\omega}\left(\frac{d}{dt}(t, \mu(t))\right) = \tilde{\omega}\left(\frac{d}{dt}(\phi_J^{-1} \circ \mu)(t)\right)$$
$$= \omega(d\phi_J\left(\frac{d}{dt}(\phi_J^{-1} \circ \mu)(t)\right)$$
$$= \omega\left(\frac{d\mu}{dt}(t)\right)$$

as required.

It follows that if $(u, \varphi) \in \mathscr{O} \subseteq C^\infty(I, \mathbb{R}^n) \times \mathscr{D}$ then

$$(\Theta \circ \Phi_\gamma^{-1} \circ \psi_\gamma^{-1})(u, \varphi) = \tilde{\omega}\left(\frac{d}{dt}(\phi_J^{-1} \circ (\exp_{\gamma \circ \varphi}((u^i l_i) \circ \varphi)))\right)$$

$$= \tilde{\omega}\left(\frac{d}{dt}(\phi_J^{-1} \circ \exp_\gamma(u^i l_i) \circ \varphi)\right)$$

$$= \tilde{\omega}\left(\left(\frac{d}{dt}(\phi_J^{-1} \circ \exp_\gamma(u^i l_i)) \circ \varphi\right)\varphi'\right)$$

$$= \varphi' \tilde{\omega}\left(\frac{d}{dt}(\phi_J^{-1} \circ \exp_\gamma(u^i l_i)) \circ \varphi\right).$$

Utilizing the convention above

$$\frac{d}{dt}(\phi_J^{-1} \circ \exp_\gamma(u^i l_i))(t) = \frac{d}{dt}(t, \exp_{\gamma(t)}(u^i(t) l_i(t)))$$

$$= \frac{d}{dt}((\mathrm{Exp}_\gamma \circ \widetilde{u^i l_i})(t)).$$

Now $t \mapsto (\mathrm{Exp}_\gamma \circ \widetilde{u^i l_i})(t)$ is a curve in $\tilde{N}_\varepsilon(\gamma)$ on which we have defined global coordinates above. These coordinates were defined as follows. Choose an orthonormal frame of $(T_\gamma^\perp P)_0$ and parallelly propagate it along γ so that one obtains an orthornormal frame $l_1(t), \ldots, l_n(t)$ of $(T_\gamma^\perp P)_t$ as above. For each $w \in (T_\gamma^\perp P)_t$ define $x^0(w) = t$ and define $x^i(w)$, $1 \le i \le n$, by requiring that $w = x^i(w) l_i(t)$. Clearly $(x^0, x^1, \ldots, x^n)$ is a global chart on the normal bundle $T_\gamma^\perp P$ and thus on $U_\varepsilon(\gamma) \subseteq T_\gamma^\perp P$. Now define $y^\mu = x^\mu \circ \mathrm{Exp}_\gamma^{-1}$. It follows that $(y^0, y^1, \ldots y^n)$ is a global coordinate system on $\tilde{N}_\varepsilon(\gamma)$. Thus if $\tilde{\mu} = \mathrm{Exp}_\gamma \circ \widetilde{u^i l_i}$ then

$$\left(\left[\frac{d}{dt}(\mathrm{Exp}_\gamma \circ \widetilde{u^i l_i})\right] \circ \varphi\right)(t) = \frac{d}{d\bar{t}}[\tilde{\mu}(\bar{t})]\bigg|_{\bar{t}=\varphi(t)}$$

$$= \frac{d}{d\bar{t}}(y^\nu \circ \tilde{\mu})(\bar{t})\left(\frac{\partial}{\partial y^\nu}\bigg|_{\tilde{\mu}(\bar{t})}\right)\bigg|_{\bar{t}=\varphi(t)}$$

$$= \frac{d}{d\bar{t}}(x^\nu \circ \widetilde{u^i l_i})(\bar{t})\left(\frac{\partial}{\partial y^\nu}\bigg|_{\tilde{\mu}(t)}\right)\bigg|_{\bar{t}=\varphi(t)}$$

$$= \frac{du^\nu}{d\bar{t}}(\varphi(t))\left(\frac{\partial}{\partial y^\nu}\bigg|_{\tilde{\mu}(\varphi(t))}\right)$$

and

$$(\Theta \circ \Phi_\gamma^{-1} \circ \psi_\gamma^{-1})(u, \varphi)(t) = \varphi'(t)\tilde{\omega}\left(\frac{du^\nu}{dt}(\varphi(t))\left(\frac{\partial}{\partial y^\nu}\bigg|_{\tilde{\mu}(\varphi(t))}\right)\right)$$

If $A_\nu : \tilde{N}_\varepsilon(\gamma) \to \mathfrak{g}$ is defined by $A_\nu(t, w) = \tilde{\omega}(\frac{\partial}{\partial y^\nu}|_{(t,w)})$ for each $(t, w) \in \tilde{N}_\varepsilon(\gamma)$ then

$$(\Theta \circ \Phi_\gamma^{-1} \circ \psi_\gamma^{-1})(u, \varphi) = \varphi'\left(\frac{du^\nu}{dt} \circ \varphi\right)(A_\nu \circ \mathrm{Exp}_\gamma \circ \widetilde{u^i l_i} \circ \varphi)$$

for each pair $(u, \varphi) \in C^\infty(I, \mathbb{R}^n) \times \mathscr{D}$. Arguments similar to the ones above and to the arguments in [6] show that $\Theta \circ \Phi_\gamma^{-1} \circ \psi_\gamma^{-1}$ is indeed smooth.

We have yet to show that the mapping $\varepsilon : C^\infty(I, \mathfrak{g}) \to G$ defined by $\varepsilon(\beta) = P \exp \int_0^1 \beta(t) \, dt$ is smooth. Observe that ε is continuous due to estimates of the type found on p. 33 of [3]. We first examine the directional derivatives of ε. Let $\lambda \mapsto \beta_\lambda$, $-\delta_0 < \lambda < \delta_0$, be a smooth curve in $C^\infty(I, \mathfrak{g})$. It is well known that the mapping $\lambda \mapsto P \exp \int_0^1 \beta_\lambda(t) \, dt$ is differentiable and that

$$\frac{d}{d\lambda}[\varepsilon(\beta_\lambda)] = \int_0^1 \left[P \exp \int_0^s \beta_\lambda(t) \, dt \right] \frac{d\beta_\lambda}{d\lambda}(s) \left[P \exp \int_s^1 \beta_\lambda(t) \, dt \right] d s.$$

This is shown on p. 35 of [3] but the reader should note that the time ordering convention in [3] is opposite that used in Polyakov [7] and the present paper. If we define $U(s, \alpha) = P \exp \int_0^s \alpha(t) \, dt$ for each $\alpha \in C^\infty(I, \mathfrak{g})$ then we have

$$\frac{d}{d\lambda}[\varepsilon(\beta_\lambda)] = \left[\int_0^1 \mathrm{Ad}(U(s, \beta_\lambda)) \left[\frac{d}{d\lambda}(\beta_\lambda)(s) \right] ds \right] \varepsilon(\beta_\lambda).$$

Since $\lambda \mapsto \beta_\lambda$ is a smooth curve in $C^\infty(I, \mathfrak{g})$, $\frac{d}{d\lambda}(\beta_\lambda)|_{\lambda=0}$ is a tangent vector in $T_{\beta_0}(C^\infty(I, \mathfrak{g}))$. Let $\delta(\beta_0) = \frac{d}{d\lambda}(\beta_\lambda)|_{\lambda=0}$. Then we see that the directional derivative is

$$(D_{\delta(\beta_0)}\varepsilon)(\beta_0) = \left[\int_0^1 \mathrm{Ad}(U(s, \beta_0))(\delta(\beta_0)(s)) \, ds \right] \varepsilon(\beta_0).$$

Since $C^\infty(I, \mathfrak{g})$ is actually a Fréchet *vector space* (g may be identified with $\mathbb{R}^d$ for some d) $T_{\beta_0}(C^\infty(I, \mathfrak{g}))$ may be identified with $C^\infty(I, \mathfrak{g})$ itself so that we regard $\delta(\beta_0) \in C^\infty(I, \mathfrak{g})$. In the special case that $\beta_\lambda = \beta_0 + \lambda\delta$ for some $\delta \in C^\infty(I, \mathfrak{g})$ we see that

$$(D\varepsilon)(\beta_0, \delta) = \left[\int_0^1 \mathrm{Ad}(U(s, \beta_0))\delta(s) \, ds \right] \varepsilon(\beta_0).$$

It is easy to show that $D\varepsilon$ is continuous as a mapping from $C^\infty(I, \mathfrak{g}) \times C^\infty(I, \mathfrak{g})$ into $G \subseteq gl(n, \mathbb{R})$ using the fact that the topology of $C^\infty(I, \mathfrak{g})$ is the topology of uniform convergence of β_0 and its derivatives and using estimates such as those on p. 33 of [3]. Thus ε is a class C^1 mapping. To see that ε is of class C^2 proceed similarly. We have

$$(D^2\varepsilon)(\beta_0, \delta_1, \delta_2)$$

$$= \lim_{\lambda \to 0} \left(\frac{1}{\lambda}\right)\{D\varepsilon(\beta_0 + \lambda\delta_2, \delta_1) - D\varepsilon(\beta_0, \delta_1)\}$$

$$= \lim_{\lambda \to 0} \left(\frac{1}{\lambda}\right)\left[\int_0^1 \text{Ad}(U(s, \beta_0 + \lambda\delta_2))\delta_1(s)ds(\varepsilon(\beta_0 + \lambda\delta_2) - \varepsilon(\beta_0))\right]$$

$$+ \lim_{\lambda \to 0} \left(\frac{1}{\lambda}\right)\left[\int_0^1 \text{Ad}[U(s, \beta_0 + \lambda\delta_2) - U(s, \beta_0)]\delta_1(s)ds\right]\varepsilon(\beta_0)$$

$$= \int_0^1 \text{Ad}(\tilde{U}(\beta_0)(s))\delta_1(s)ds\, D\varepsilon(\beta_0, \delta_2)$$

$$+ \int_0^1 \text{Ad}(D\tilde{U}(\beta_0, \delta_2)(s))\delta_1(s)ds\, \varepsilon(\beta_0)$$

where $\tilde{U} : C^\infty(I, \mathfrak{g}) \to C^\infty(I, Gl(n, \mathbb{R}))$ is defined by $\tilde{U}(\beta)(s) = U(s, \beta) = P\exp\int_0^s \beta(t)\,dt$. It is easily shown that $D^2\varepsilon$ is continuous on $C^\infty(I, \mathfrak{g})^3$ and thus that ε is of class C^2. One may proceed inductively and in general we obtain that ε is of class C^m for each m with

$$(D^m\varepsilon)(\beta_0, \delta_1, \ldots, \delta_m)$$

$$= \left(\int_0^1 \text{Ad}(\tilde{U}(\beta_0)(s))\delta_1(s)ds\right)(D^{m-1}\varepsilon)(\beta_0, \delta_2, \ldots, \delta_m)$$

$$+ \sum_{k=1}^m \sum_{2 \le i_1 < \cdots < i_k \le m} \int_0^1 \text{Ad}((D^k\tilde{U})(\beta_0, \delta_{i_1} \cdots \delta_{i_k})(s))\delta_1(s)\,ds(D^{m-k-1}\varepsilon)$$

where $D^{m-k-1}\varepsilon$ is evaluated at the $(m - k - 1)$-tuple obtained by deleting $\delta_1, \delta_{i_1}, \ldots, \delta_{i_k}$ from the m-tuple $(\delta_1, \delta_2, \ldots, \delta_m)$.

It follows that both Θ and ε are smooth and thus so is $\mathcal{J}$. To conclude the proof of Theorem 2 we have only to find $d_\gamma\mathcal{J}$ for $\gamma \in \mathcal{P}_1^*P$. To accomplish this we first need a lemma.

LEMMA 5. *Let* $t \mapsto \gamma_t$, $-\delta_0 < t < \delta_0$, *be a smooth curve in* $\mathcal{P}_1^*P$ *and define a function* $\tilde{\gamma} : I \times (-\delta_0, \delta_0) \to P$ *by* $\tilde{\gamma}(s, t) = \gamma_t(s)$ *for* $(s, t) \in I \times (-\delta_0, \delta_0)$. *If* ω *is a connection on* P *then*

$$\frac{\partial}{\partial t}\left[\omega\left(\frac{\partial\tilde{\gamma}}{\partial s}\right)\right] - \frac{\partial}{\partial s}\left[\omega\left(\frac{\partial\tilde{\gamma}}{\partial t}\right)\right] = d\omega\left(\frac{\partial\tilde{\gamma}}{\partial t}, \frac{\partial\tilde{\gamma}}{\partial s}\right).$$

PROOF. Write $\omega = A_\mu dx^\mu$ in a local chart. Then

$$d\omega = (\partial_\nu A_\mu)(dx^\nu \wedge dx^\mu)$$

and

$$\frac{\partial}{\partial t}\left[\omega\left(\frac{\partial\tilde\gamma}{\partial s}\right)\right] - \frac{\partial}{\partial s}\left[\omega\left(\frac{\partial\tilde\gamma}{\partial t}\right)\right]$$

$$= \frac{\partial}{\partial t}\left[A_\mu\left(\frac{\partial(x^\mu\circ\tilde\gamma)}{\partial s}\right)\right]$$

$$- \frac{\partial}{\partial s}\left[A_\mu\left(\frac{\partial(x^\mu\circ\tilde\gamma)}{\partial t}\right)\right]$$

$$= (\partial_\nu A_\mu)\left(\frac{\partial(x^\nu\circ\tilde\gamma)}{\partial t}\frac{\partial(x^\mu\circ\tilde\gamma)}{\partial s}\right)$$

$$- \partial_\nu A_\mu\left(\frac{\partial(x^\nu\circ\tilde\gamma)}{\partial s}\frac{\partial(x^\mu\circ\tilde\gamma)}{\partial t}\right)$$

$$= d\omega\left(\frac{\partial\tilde\gamma}{\partial t}, \frac{\partial\tilde\gamma}{\partial s}\right).$$

To derive a formula for $d_{\gamma_0}\mathscr{I}$, let $t\mapsto\gamma_t$, $-\delta_0 < t < \delta_0$, be a curve in $\mathscr{P}_1^* P$ through γ_0 and let $\delta = \frac{d\gamma_t}{dt}\big|_{t=0}$ be an arbitrary tangent vector to $\mathscr{P}_1^* P$ at γ_0. We have that

$$d_{\gamma_0}\mathscr{I}(\delta) = \frac{d}{dt}[\mathscr{I}(\gamma_t)]\bigg|_{t=0}$$

$$= \frac{d}{dt}P\exp\left[\int_0^1\omega\left(\frac{d\gamma_t}{ds}\right)ds\right]\bigg|_{t=0}$$

To evaluate this integral introduce functions $\tilde\gamma : I\times(-\delta_0, \delta_0)\to P$ and $U : I\times(-\delta_0, \delta_0)\to Gl(n, \mathbb{R})$ defined by $\tilde\gamma(s, t) = \gamma_t(s)$ and by $U(s, t) = P\exp\int_0^s\omega(\frac{\partial\tilde\gamma}{\partial s}(u, t))du$ for $(s, t)\in I\times(-\delta_0, \delta_0)$. We have

$$\frac{d}{dt}[U(1, t)] = \int_0^1\left[P\exp\int_0^s\omega\left(\frac{\partial\tilde\gamma}{\partial s}(u, t)\right)du\right]$$

$$\times\frac{\partial}{\partial t}\left[\omega\left(\frac{\partial\tilde\gamma}{\partial s}(s, t)\right)\right]\left[P\exp\int_s^1\omega\left(\frac{\partial\tilde\gamma}{\partial s}(u, t)\right)du\right]ds$$

$$= \int_0^1\text{Ad}(U(s, t))\frac{\partial}{\partial t}\left[\omega\left(\frac{\partial\tilde\gamma}{\partial s}(s, t)\right)\right]ds\, U(1, t)$$

$$= \int_0^1\text{Ad}(U(s, t))\left\{\frac{\partial}{\partial s}\left[\omega\left(\frac{\partial\tilde\gamma}{\partial t}(s, t)\right)\right]\right.$$

$$\left. + d\omega\left(\frac{\partial\tilde\gamma}{\partial t}, \frac{\partial\tilde\gamma}{\partial s}\right)\right\}ds\, U(1, t)$$

$$= \int_0^1\text{Ad}(U(s, t))\frac{\partial}{\partial s}\left[\omega\left(\frac{\partial\tilde\gamma}{\partial t}\right)\right]ds\, U(1, t)$$

$$+ \int_0^1\text{Ad}(U(s, t))d\omega\left(\frac{\partial\tilde\gamma}{\partial t}, \frac{\partial\tilde\gamma}{\partial s}\right)ds\, U(1, t).$$

On the other hand since $\frac{\partial}{\partial s}(U(s,t)) = U(s,t)\omega(\frac{\partial \tilde{\gamma}}{\partial s}(s,t))$ (by [3])

$$\frac{\partial}{\partial s}\left[\mathrm{Ad}(U(s,t))\left(\omega\left(\frac{\partial \tilde{\gamma}}{\partial t}\right)\right)\right]$$

$$= \frac{\partial}{\partial s}[U(s,t)]\omega\left(\frac{\partial \tilde{\gamma}}{\partial t}\right)U(s,t)^{-1} + U(s,t)\frac{\partial}{\partial s}\left[\omega\left(\frac{\partial \tilde{\gamma}}{\partial t}\right)\right]U(s,t)^{-1}$$

$$+ U(s,t)\omega\left(\frac{\partial \tilde{\gamma}}{\partial t}\right)\left[-U(s,t)^{-1}\frac{\partial}{\partial s}(U(s,t))U(s,t)^{-1}\right]$$

$$= \mathrm{Ad}(U(s,t))\left(\omega\left(\frac{\partial \tilde{\gamma}}{\partial s}\right)\omega\left(\frac{\partial \tilde{\gamma}}{\partial t}\right)\right) + \mathrm{Ad}(U(s,t))\frac{\partial}{\partial s}\left[\omega\left(\frac{\partial \tilde{\gamma}}{\partial t}\right)\right]$$

$$- \mathrm{Ad}(U(s,t))\left(\omega\left(\frac{\partial \tilde{\gamma}}{\partial t}\right)\omega\left(\frac{\partial \tilde{\gamma}}{\partial s}\right)\right)$$

$$= \mathrm{Ad}(U(s,t))\left(\frac{\partial}{\partial s}\left[\omega\left(\frac{\partial \tilde{\gamma}}{\partial t}\right)\right] + \left[\omega\left(\frac{\partial \tilde{\gamma}}{\partial s}\right), \omega\left(\frac{\partial \tilde{\gamma}}{\partial t}\right)\right]\right).$$

It follows that if $\Omega = D\omega$ then

$$\frac{d}{dt}[U(1,t)] = \left\{\int_0^1 \left\{\frac{\partial}{\partial s}\left[\mathrm{Ad}(U(s,t))\left(\omega\left(\frac{\partial \tilde{\gamma}}{\partial t}\right)\right)\right]\right.\right.$$

$$\left.- \left[\mathrm{Ad}(U(s,t))\left(\left[\omega\left(\frac{\partial \tilde{\gamma}}{\partial s}\right), \omega\left(\frac{\partial \tilde{\gamma}}{\partial t}\right)\right]\right)\right]\right\}ds$$

$$\left.+ \int_0^1 \mathrm{Ad}(U(s,t))\left(d\omega\left(\frac{\partial \tilde{\gamma}}{\partial t}, \frac{\partial \tilde{\gamma}}{\partial s}\right)\right)ds\right\}U(1,t)$$

$$= \left(\int_0^1 \mathrm{Ad}(U(s,t))\left(D\omega\left(\frac{\partial \tilde{\gamma}}{\partial t}, \frac{\partial \tilde{\gamma}}{\partial s}\right)\right)ds\right)U(1,t)$$

$$+ \mathrm{Ad}(U(1,t))\left[\omega\left(\frac{\partial \tilde{\gamma}}{\partial t}(1,t)\right) - \omega\left(\frac{\partial \tilde{\gamma}}{\partial t}(0,t)\right)\right]U(1,t)$$

and

$$\frac{d}{dt}[U(1,t)]\bigg|_{t=0} = \mathrm{Ad}U(1,0)[\omega(\delta(1)) - \omega(\delta(0))]U(1,0)$$

$$+ \int_0^1 \mathrm{Ad}(U(s,1))\Omega(\delta(s), \dot{\gamma}_o(s))ds\, U(1,0)$$

$$= \mathrm{Ad}(\mathscr{I}(\gamma_o))[\omega(\delta(1)) - \omega(\delta(0))]\mathscr{I}(\gamma_o)$$

$$+ \left[\int_0^1 \mathrm{Ad}(U(s,0))\Omega(\delta(s), \dot{\gamma}_o(s))ds\right]\mathscr{I}(\gamma_o).$$

Thus

$$d_{\gamma_o}\mathscr{I}(\delta) = \mathscr{I}(\gamma_o)[\omega(\delta(1)) - \omega(\delta(0))]$$

$$+ \left(\int_0^1 \mathrm{Ad}(U(s,0))\Omega(\delta(s), \dot{\gamma}_o(s))ds\right)\mathscr{I}(\gamma_o)$$

and Theorem 3 follows.

It was shown in Theorem 3 that

$$d_\gamma \mathscr{F}(\delta) = \mathscr{F}(\gamma)[\omega(\delta(1)) - \omega(\delta(0))] + \int_0^1 \mathrm{Ad}(U(s))\Omega(\delta(s), \dot{\gamma}(s))ds \, \mathscr{F}(\gamma)$$

where

$$U(s) = P \exp \int_0^s \omega(\dot{\gamma}(u))du.$$

If we define $V(s)$ by $V(s) = P \exp \int_s^1 \omega(\dot{\gamma}(u))du$ then $U(s)V(s) = \mathscr{F}(\gamma)$ for each $s \in I$ and

$$d_\gamma \mathscr{F}(\delta) = \mathscr{F}(\gamma)\left[\omega(\delta(1)) - \omega(\delta(0)) + \int_0^1 \mathrm{Ad}(V(s)^{-1})\Omega(\delta(s), \dot{\gamma}(s))\, ds\right].$$

If Θ is the Maurer-Cartan form on G it follows that for $\gamma \in \mathscr{P}_1^* P$, $\delta \in T_\gamma(\mathscr{P}_1^* P)$,

$$\mathscr{F}^*(\Theta)_\gamma(\delta) = \omega(\delta(1)) - \omega(\delta(0)) - \int_0^1 (R_{V(s)}^* \Omega)(\dot{\gamma}(s), \delta(s))ds.$$

If $\alpha = \mathscr{F}^*(\Theta)$ then clearly α is a $\mathfrak{g}$valued one-form on $\mathscr{P}_1^* P$ and defines a *flat connection* on any trivial G-bundle over $\mathscr{P}_1^* P$.

Recall that Ω is a horizontal form on P and thus so is $R_{V(s)}^* \Omega$. Moreover if M has a metric then the Hodge duality operator $*$ on M lifts to equivariant horizontal forms on P (see [2]) and if $\bar{*}$ is this duality it is easy to show that $\bar{*}(R_{V(s)}^* \Omega) = R_{V(s)}^*(\bar{*}\Omega)$. Moreover $d(\bar{*}(R_{V(s)}^* \Omega)) = R_{V(s)}^*(d(\bar{*}\Omega))$. Finally hor$(dR_{V(s)}(X)) = dR_{V(s)}(\mathrm{hor}\, X)$ for each vector X which is tangent to P at some point of P and this fact implies that hor$(d(\bar{*}(R_{V(s)}^* \Omega))) = R_{V(s)}^*(D(\bar{*}\Omega))$. Using these properties along with the definitions of how $\bar{*}$, $\hat{d}$, and hor act on path-forms given in the last section we see that if we restrict α to the submanifold of loops then

$$\mathrm{hor}(\hat{d}(\bar{*}\alpha))_\gamma(\delta) = \int_0^1 R_{V(s)}^*(D(\bar{*}\Omega))(\dot{\gamma}(s), \delta(s))\, ds.$$

This identity provides us with the analogue of the Yang-Mills operator defined on loop space. Observe that the operators hor, $\hat{d}$, $\bar{*}$ do not derive from metrics or connection on $\mathscr{P}_1^* P$ (indeed α is flat!) but rather from the structure of P. Nonetheless $\mathscr{P}_1^* P$ does acquire extra structure if only on the set of path-forms on P.

Final remark

Notice that the set of all path-forms on $\mathscr{P}_1^* P$ is invariant under the action $\mathscr{D}$ and consequently induce smooth forms on $\mathscr{P}_1^* P/\mathscr{D}$. Moreover the operators hor, $\hat{d}$, $\bar{*}$ leave the set of path-forms invariant. It follows that there is a distinguished class of differential forms on $LP/\mathscr{D}$, namely those which descend from loop-forms on LP, on which the operators hor$/\mathscr{D}$, $\hat{d}/\mathscr{D}$, $\bar{*}/\mathscr{D}$

are well defined. Each connection ω on P induces a mapping $\mathscr{I} : \mathscr{P}_1^* P \to \underline{g}$ which descends to a mapping $\tilde{\mathscr{I}} : \mathscr{P}_1^* P / \mathscr{D} \to \underline{g}$. If $\tilde{\alpha} = \tilde{I}^*(\Theta)$ then $\widetilde{\mathrm{hor}}(\tilde{\tilde{d}}(\widetilde{\bar{*}}(\tilde{\alpha})))$ represents the analogue of the "Yang-Mills operator" in this setting. The author intends to investigate this operator more fully in future work possibly recasting gauge theory into this mold. Moreover it would be useful to find some intermediate formulation in which the best parts of this work could draw on the deeper results obtained by Gross in [4].

REFERENCES

1. R. Coquereaux and K. Pilch, *String structures on loop bundles*, Comm. Math. Phys. 120, 353, 1989.
2. D. Bleecker, *Gauge theory and variational principles*, Addison-Wesley Publishing Company, Reading, Mass., 1981.
3. J. D. Dollard and C. N. Friedman, *Product integration*, Addison-Wesley Publishing Company, 1979.
4. L. Gross, *A Poincaré Lemma for connection forms*, J. Funct. Anal. **63** 1, (1985).
5. J. F. Hamilton, *Muliplicative Riemann integration and logarithmic differentiation in Lie algebras and Lie groups*, Dissertation Indiana Univ., 1973.
6. R. S. Hamilton, *The Inverse Function Theorem of Nash and Moser*, Bull. Amer. Math. Soc. (N.S.) **7** (1982),
7. A. M. Polyakov, *Gauge fields as rings of glue*, Nuclear Phys. B **164** 171, (1979).
8. C. N. Yang and T. T. Wu, *Concept of nonintegrable phase factors and global formulation of gauge fields*, Phys. Rev. D **12** 3845, (1975).

NORTH CAROLINA STATE UNIVERSITY

Proceedings of Symposia in Pure Mathematics
Volume **54** (1993), Part 2

The Lorentzian Version
of the Cheeger-Gromoll Splitting Theorem
and its Application to General Relativity

GREGORY J. GALLOWAY

1. Introduction

In his well-known problem section appearing in the early 1980s, S.-T. Yau
[**Y**] posed the problem of establishing a Lorentzian analogue of the Cheeger-
Gromoll splitting theorem of Riemannian geometry. This problem has now
been completely resolved (cf. [**B+, E, G2, N**]) giving rise to what we shall
refer to as the Lorentzian splitting theorem. The purpose of this paper is to
discuss this theorem and its relevance to certain problems in general relativity
concerning the rigidity of the classical Hawking-Penrose singularity theorems.

We begin by establishing some terminology and notation. A Lorentzian
manifold M is a smooth manifold equipped with a metric $\langle\, , \rangle$ of Lorentzian
signature $(-+ \cdots +)$. M is *time orientable* if and only if the assignment
of a past and future null cone at each point of M can be carried out in a
continuous manner over M. For convenience, we will always assume M
is time orientable, in which case we refer to M as a spacetime. Thus, in
a spacetime M there is a global notion of past and future. A timelike (respectively, null, causal) curve is a curve $t \mapsto \gamma(t)$ whose tangent field γ' is
timelike, $\langle \gamma', \gamma' \rangle < 0$ (resp. null, $\langle \gamma', \gamma' \rangle = 0$, causal, $\langle \gamma', \gamma' \rangle \leq 0$). In
Lorentzian geometry we are primarily concerned with the geometry of causal
curves. The length of a causal curve $t \mapsto \gamma(t)$, $a \leq t \leq b$, is defined as
$\int_a^b \sqrt{-\langle \gamma', \gamma' \rangle}\, dt$. We will have occasion to make use of the following standard causal notation. For any subset $S \subset M$, $I^+(S)$, the timelike future
of S, is the set of all points that can be reached from S by future directed
timelike curves. $J^+(S)$, the causal future of S, is the set of all points that

1991 *Mathematics Subject Classification.* Primary 53C50; Secondary 83C75.
Partially supported by National Science Foundation Grant DMS-9006678.
This paper is in final form and no version of it will be submitted for publication elsewhere.

can be reached from S by future directed causal curves. $I^-(S)$ and $J^-(S)$, the timelike past and causal past of S, respectively, are defined time-dually.

Just as in the Riemannian case, a Lorentzian manifold carries a unique Levi-Civita connection ∇. Hence, geodesics in spacetime are defined in the usual manner: $t \mapsto \gamma(t)$ is a geodesic provided $\nabla_{\gamma'}\gamma' = 0$. The theory of timelike geodesics in Lorentzian geometry is, in many respects, analogous to the theory of geodesics in Riemannian geometry. There are some differences, however. For one thing, timelike geodesics locally *maximize* arc length. For another, the naive analogue of the Hopf-Rinow theorem does not hold in the Lorentzian case. For instance, geodesic completeness does not guarantee that timelike related points can be joined by a maximal timelike geodesic. The standard condition in Lorentzian geometry that guarantees this is *global hyperbolicity*. A spacetime M is globally hyperbolic if and only if it is strongly causal (i.e., there are no closed or "almost closed" causal curves in M) and the causal intervals $J^+(p) \cap J^-(q)$, $p, q \in M$, are compact. Because it insures the causal geodesic convexity of spacetime, global hyperbolicity frequently plays the role of geodesic completeness in Lorentzian geometry.

2. The Lorentzian splitting theorem

In [Y] Yau posed the following problem: show that a spacetime M which is timelike geodesically complete, obeys the *strong energy condition*, $\mathrm{Ric}(X, X) \geq 0$, for all timelike vectors X, and contains a complete timelike line splits isometrically into a product $(R \times V, -dt^2 \oplus h)$, where (V, h) is a complete Riemannian manifold. A timelike line is an inextendible timelike geodesic such that each segment gives a maximum for the length among all causal curves joining its endpoints. Yau's formulation of the problem parallels very closely the statement of the Cheeger-Gromoll splitting theorem. A natural alternative to this version of the problem is obtained by replacing the assumption of timelike geodesic completeness by global hyperbolicity. The Lorentzian splitting problem (both versions) has been completely resolved in a series of papers by Beem, Ehrlich, Markvorsen and Galloway [B+], Eschenburg [E], Galloway [G2], and Newman [N]. We state below the Lorentzian splitting theorem.

THEOREM 1. *Suppose M is a spacetime which satisfies the following*:

(1) *M either is globally hyperbolic or timelike geodesically complete*;

(2) *M obeys the strong energy condition; i.e., $\mathrm{Ric}(X, X) \geq 0$, for all timelike vectors X*;

(3) *M contains a complete timelike line.*

Then M splits isometrically into a product $(R \times V, -dt^2 \oplus h)$, where (V, h) is a complete Riemannian manifold.

It is easy to understand why the proof in the Riemannian case fails to carry over to the Lorentzian case. Whereas the Riemannian proof is based on the ellipticity of the Riemannian Laplacian, the spacetime Laplacian =

the trace of the Hessian = the d'Alembertian is *hyperbolic*. In particular, the concept of subharmonicity as used in the Riemannian proof is not directly applicable. There are other more technical difficulties involved in carrying over the Riemannian proof, as well.

Beem et al. [B+] settled the Lorentzian splitting problem in the sectional curvature case, i.e., by imposing the more stringent curvature condition that all timelike sectional curvatures (sectional curvatures on timelike planes) be non*positive*. (Because of the Lorentzian signature, the Ricci curvature of a unit timelike vector is *minus* the sum of timelike sectional curvatures.) They also assume global hyperbolicity but not timelike geodesic completeness. Subsequent to this work, Eschenburg [E] made a major advance on the problem, settling it in the Ricci curvature case, but assuming both timelike geodesic completeness and global hyperbolicity. Although both of these conditions are quite natural, the work of Beem et al. suggests that there may be some redundancy in assuming both. In [G2], we proved the splitting theorem assuming global hyperbolicity and then Newman [N] proved it assuming timelike geodesic completeness, thus finishing off the problem.

As in the Riemannian case, the proof involves an analysis of the Busemann functions $b^{\pm}$ associated to the given line γ,

$$b^+(x) = \lim_{r \to \infty} r - d(x, \gamma(r)),$$

$$b^-(x) = \lim_{r \to \infty} r - d(\gamma(-r), x),$$

where d is the Lorentzian distance function: if $q \in J^+(p)$, $d(p, q)$ is the sup of the lengths of all future directed causal curves from p to q; if $q \notin J^+(p)$, $d(p, q) = 0$. Although the definition of the Lorentzian Busemann function is formally the same as in the Riemannian case, one immediately encounters certain technical difficulties in working with it. For instance, the Lorentzian Busemann function is not in general continuous. In the Riemannian case, the continuity of the Busemann function follows from the triangle inequality. The Lorentzian distance function, however, does not obey the triangle inequality; in fact, it obeys a reverse triangle inequality. In [B+], the sectional curvature condition is used to establish the continuity of $b^{\pm}$ on $I(\gamma) = I^+(\gamma) \cap I^-(\gamma)$. In the Ricci curvature case, Eschenburg was able to establish, without any curvature assumptions, the continuity of $b^{\pm}$ near the line γ. Thus, in the Ricci curvature case, one works near the line to obtain a local splitting and then globalizes.

Using the reverse triangle inequality one obtains the inequality

$$b^+ + b^- \geq 0$$

on $I(\gamma)$, with equality holding along γ. As in the Riemannian case (where the reverse of the above inequality holds), the aim is to establish equality off the line. Recall, in the Riemannian case, this is accomplished by establishing the subharmonicity of the Busemann function under the assumption of nonnegative Ricci curvature.

The key idea in the Lorentzian proof (due to Eschenburg) is to consider $b^{\pm}$ restricted to certain spacelike hypersurfaces Σ. A spacelike hypersurface is a smooth codimension one submanifold of M with everywhere timelike normal. In the context of general relativity, a spacelike hypersurface represents the spatial universe, or some portion of it, "at a given instant of time." Of mathematical importance is the fact that the induced metric along a spacelike hypersurface Σ is Riemannian (positive definite), and so the induced Laplacian $\triangle_{\Sigma}$ is elliptic. In this way elliptic theory re-enters the picture. In [E], Eschenburg restricts b^- to the level sets $b^+ = \text{const.}$ (or more precisely, to certain smooth spacelike hypersurfaces Σ approximating $b^+ = \text{const.}$; the level sets of b^+ are not in general smooth).

In [G2], where we prove the splitting theorem without the completeness assumption, we develop a method which involves restricting $b^{\pm}$ to a *maximal* (i.e., mean curvature zero) spacelike hypersurface. This method allows one to simplify certain aspects of Eschenburg's proof and to use the Busemann functions in a more intrinsic, geometric manner. Using the ellipticity of the induced Laplacian, we show that, provided the strong energy condition holds, $b^{\pm}$ obeys a certain maximum principle: if $b^{\pm}|_{\Sigma}$ attains an interior minimum, where Σ is a maximal spacelike hypersurface situated near the line γ, then $b^{\pm}|_{\Sigma}$ is constant. This result implies that the level sets of $b^{\pm}$ are mean convex in the following sense: if Σ is a maximal spacelike hypersurface with its edge on the level set $b^{\pm} = \text{const.}$ then Σ lies to one side of the level set, $\Sigma \subseteq \{b^{\pm} \geq \text{const.}\}$. This result is then used to show that the level sets $b^{\pm} = 0$ coincide along a smooth maximal spacelike hypersurface Σ defined near and passing through $\gamma(0)$. From each point of Σ one now constructs past and future asymptotes to γ. It follows easily that these asymptotes actually form a timelike line orthogonal to Σ. Exponentiating normally off of Σ then gives a splitting of a neighborhood of γ. (Of course, without the completeness assumption, there are some difficulties in showing that the normal exponential map is defined on all of $R \times \Sigma$.) Extending the splitting to all of M is not difficult. The implementation of this method requires an appropriate existence result for maximal spacelike hypersurfaces. For this, we were able to take advantage of work of Bartnik [B2] concerning the existence and regularity of solutions to the prescribed mean curvature equation with rough boundary data.

We conclude our discussion of the proof of the Lorentzian splitting theorem with some comments about Newman's work. In [N] Newman proves the Lorentzian splitting theorem assuming timelike geodesic completeness but not global hyperbolicity. The global hyperbolicity assumption is used primarily to guarantee the existence of maximal geodesic segments joining timelike related points. Using the timelike geodesic completeness assumption, Newman shows that each point p sufficiently close to the given line γ and each point q sufficiently far in the future of p on γ can be joined by a maximal timelike geodesic segment. Newman uses a limit curve argument, which we briefly describe, to accomplish this. Since γ is a line, one has $d(p, q) < \infty$. Let $\{\sigma_n\}$ be a sequence of causal curves from p to q, the lengths of which converge to $d(p, q)$, and let σ be a limit curve of this sequence. Newman shows, by using timelike geodesic completeness, that σ either is the desired maximal segment or a null ray (half-line) from p. Using Busemann function estimates of Eschenburg, which Newman shows are still valid even without global hyperbolicity, he is able to eliminate the latter possibility.

In the next section we discuss the application of the Lorentzian splitting theorem to some open problems in general relativity.

3. Rigidity in singularity theory

The physical motivation for establishing a Lorentzian analogue of the Cheeger-Gromoll splitting theorem was to study the rigidity of the classical Hawking-Penrose singularity theory of general relativity. The singularity theorems establish, under generic circumstances, the existence in spacetime of incomplete timelike or null geodesics. Such incompleteness indicates that spacetime comes to an end either in the past or the future. In specific models past incompleteness is typically associated with a "big bang" beginning to the universe, and future incompleteness is typically associated with a "big crunch" (the time dual of a big bang) or, of a more local nature, gravitational collapse to a black hole. All the singularity theorems require *energy conditions*, such as the strong energy condition assumed in the Lorentzian splitting theorem. In addition to imposing conditions involving weak curvature inequalities, the standard singularity theorems also impose conditions, like the so-called *generic condition*, which require certain curvature objects to obey a strict inequality (sufficiently often). From the standpoint of rigidity, one would like to show that, even if the strict curvature inequalities are not required to hold, spacetime will still be singular (incomplete) except under very special circumstances. As an illustration, we consider the following basic singularity theorem.

A SINGULARITY THEOREM. *If M is a spacetime which*

(1) *contains a compact Cauchy surface S,*
(2) *satisfies the strong energy condition, $\mathrm{Ric}(X, X) \geq 0$ for all timelike vectors X, and*
(3) *satisfies the generic condition,*

then M is timelike or null geodesically incomplete.

We make some comments about the hypotheses. A Cauchy surface is a special type of spacelike hypersurface. By definition, it is a spacelike hypersurface in which each inextendible causal curve intersects it exactly once. It is a standard fact of causal theory that a spacetime is globally hyperbolic if and only if it contains a Cauchy surface. Moreover, it is known that any compact spacelike hypersurface in a globally hyperbolic spacetime is necessarily Cauchy. Thus, assumption (1) is equivalent to the assumption that M

is globally hyperbolic and contains a compact spacelike hypersurface S. In view of the compactness condition, this singularity theorem applies to spatially closed (i.e., finite universe) spacetimes. In physical terms, the strong energy condition means that gravity is attractive. The *generic condition* asserts that for each inextendible timelike and null geodesic there is a point at which a certain curvature object (constructed from the Riemann curvature tensor and the tangent to the curve) is nonzero. For timelike geodesics, the generic condition is equivalent to the existence of a nonzero sectional curvature (with plane section containing a tangent direction) at some point of each inextendible timelike geodesic. Physically speaking, the generic condition means that each freely falling observer and zero rest-mass particle encounters at some moment in its history a nonzero tidal acceleration (i.e., inhomogeneity in the gravitational field).

It will be useful for later discussion to briefly discuss the proof of the singularity theorem. The proof is analogous to the proof of the result in Riemannian geometry that a complete Riemannian manifold with positive Ricci curvature has at most one end. The key step in the proof is the construction of a timelike (or null) line γ. The construction is as follows. Take a sequence of points $\{q_n\}$ in the future of S which extends to future infinity, and a sequence $\{p_n\}$ in the past of S which extends to past infinity, such that $p_n \in I^-(q_n)$ for all n. Since M is globally hyperbolic there exists a maximal timelike geodesic segment γ_n from p_n to q_n. Because S is Cauchy, each γ_n meets S. Hence, since S is compact, the γ_n's accumulate to an inextendible causal curve γ. The maximality of the γ_n's implies that γ is a timelike, or possibly null, line; γ may be null because, roughly speaking, the γ_n's may be turning null as $n \to \infty$. The line γ must be incomplete. Indeed, if it were complete then the curvature assumptions would imply that γ has a pair of conjugate points, which would contradict its maximality.

The singularity theorem considered above is false without the generic condition. The flat spatially closed spacetime cylinder $(R \times S^1, -dt^2 \oplus d\theta^2)$ is a simple counterexample. However, this counterexample is quite special: it is a geometric product. From the standpoint of rigidity, one would like to show that the singularity theorem considered above, without the generic condition, can fail only under very special circumstances. One is led to consider the following conjecture (stated in [**B3**]).

CONJECTURE 1. *Let M be a spacetime which*

(1) *contains a compact Cauchy surface, and*

(2) *obeys the strong energy condition,* $\mathrm{Ric}(X, X) \geq 0$ *for all timelike vectors X.*

Then either M is timelike geodesically incomplete, or else M splits isometrically into the product $(R \times V, -dt^2 \oplus h)$, where (V, h) is a compact Riemannian manifold.

Conjecture 1, if proved, would provide an example of a rigid singular-

ity theorem. Forms of this conjecture were considered in the past by Avez
[A] and Geroch [Ge]. Proofs of restricted versions of the conjecture were
obtained in the 1980s. The conclusion of Conjecture 1 can be stated in the
form: if M is timelike geodesically complete then M splits. Relying on earlier work of Bartnik [B1], we [G1] gave a proof of the conjecture under the
additional completeness type assumption that M has no "observer horizons."
A spacetime M has no observer horizons provided for every inextendible
timelike curve σ, $I^-(\sigma) = I^+(\sigma) = M$. More recently, Bartnik [B3] was
able to weaken this assumption somewhat. He gives a proof of the conjecture under the additional assumption that there exists a point $p \in M$ such
that $M \setminus I^-(p) \cup I^+(p)$ is compact. (For spacetimes with compact Cauchy
surface, the no observer horizon condition is equivalent to the requirement
that Bartnik's condition hold at every point; Bartnik's condition is closely related to a condition previously considered by Geroch [Ge].) These two results
were obtained by establishing, under suitable circumstances, the existence of
a compact maximal (mean curvature zero) global spacelike hypersurface in
spacetime (thereby fulfilling the approaches first considered by Avez [A] and
Geroch [Ge]). It is not difficult to show that Conjecture 1 holds for space-
times which contain a compact maximal spacelike hypersurface (cf. [B3, G3]).
Thus, one approach to Conjecture 1 is to establish the existence of a compact maximal hypersurface under sufficiently general circumstances, that is,
without imposing any additional conditions such as those used in our result
and the result of Bartnik.

The Lorentzian splitting theorem suggests another approach to proving
Conjecture 1. (This is the approach originally advocated by Yau.) The idea
is to use the hypotheses of the conjecture, together with the assumption of
timelike geodesic completeness, to establish the existence of a timelike line.
The Lorentzian splitting theorem would then imply that spacetime splits. In
fact, the Lorentzian splitting theorem already implies the results discussed
in the previous paragraph since each of the additional conditions discussed
there is enough to guarantee the existence of a timelike line.

A proof of the following conjecture, taken in conjunction with the Lorentzian splitting theorem, would provide a proof of Conjecture 1.

CONJECTURE 2. *Suppose M is a spacetime which satisfies the following*:

(1) *M contains a compact Cauchy surface*;
(2) *M obeys the strong energy condition,* $\mathrm{Ric}(X, X) \geq 0$ *for all timelike
    vectors X*;
(3) *M is timelike geodesically complete.*
Then M contains a timelike line.

As we have seen, even without any curvature or completeness assumptions, there is a standard procedure for constructing a *timelike or null* line in
a spacetime with compact Cauchy surface. The problem, however, is to guarantee the existence of a *timelike* line. One might hope to be able to modify

the construction of the line discussed in the proof of the singularity theorem (by, for example, choosing the sequences going off to future and past infinity in some special way) to guarantee that the resulting line is timelike. Some recent work of ours with Paul Ehrlich [EhG] shows that, in the absence of any curvature conditions, this approach is futile. In [EhG] we construct an example of a timelike and null geodesically complete spacetime with compact Cauchy surface which does not contain any timelike lines. Our example does not obey the strong energy condition, but shows that Conjecture 2 is false if this condition is omitted. On the other hand, we also showed that if one strengthens the curvature condition to require that all timelike sectional curvatures be nonpositive then the conjecture holds. The proof makes use of triangle comparison techniques which, unfortunately, do not extend to the Ricci curvature case.

More recently, we have developed with Eschenburg [EsG] certain Busemann function techniques for constructing timelike lines. In fact, we have been able to obtain new sufficient conditions for the existence of timelike lines by this approach. Although the result obtained is still short of what is needed to prove Conjecture 2, the approach used is promising, since various aspects of Busemann functions are controlled by Ricci curvature.

We conclude the paper with a brief description of this recent work with Eschenburg. Let M be a spacetime which contains a compact acausal spacelike hypersurface S. (A set is acausal if no two of its points can be joined by a causal curve; spacelike hypersurfaces are always locally acausal, but may not be acausal in the large.) We do not require the more stringent condition that S be Cauchy. An S-ray is a future inextendible timelike geodesic γ emanating from a point in S such that γ realizes the Lorentzian distance to S from each of its points. Every compact acausal spacelike hypersurface S admits an S-ray. Eschenburg and I have proved the following: Let M be a timelike geodesically complete spacetime which contains a compact acausal spacelike hypersurface S. If there is an S-ray γ such that $I^-(\gamma) \supset S$ then M contains a timelike line. This result, taken in conjunction with the Lorentzian splitting theorem, yields the following theorem.

THEOREM 2. *Let M be a spacetime which*

(1) *contains a compact acausal spacelike hypersurface S, and*
(2) *obeys the strong energy condition.*

If M is timelike geodesically complete and contains an S-ray γ such that $I^-(\gamma) \supset S$ then M splits as in Conjecture 1.

Theorem 2 generalizes the main theorem in [G1] discussed earlier, and provides evidence for the more general conjecture that a spacetime which contains a compact acausal spacelike hypersurface and obeys the strong energy condition either is timelike geodesically incomplete or splits. Some additional evidence supporting this more general conjecture is provided in [G3].

References

[A]   A. Avez, *Essais de geometrie riemanniene hyperbolique globale. 1 Applications a la relativite generale*, Ann. Inst. Fourier (Grenoble) **13** (1963), 105–190.

[B1]   R. Bartnik, *Existence of maximal surfaces in asymptotically flat spacetimes*, Comm. Math. Phys. **94** (1984), 155–175.

[B2]   _____, *Regularity of variational maximal surfaces*, Acta Math. **161** (1988), 145–181.

[B3]   _____, *Remarks on cosmological spacetimes and constant mean curvature surfaces*, Comm. Math. Phys. **117** (1988), 615–624.

[B+]   J. K. Beem, P. E. Ehrlich, S. Markvorsen, and G. J. Galloway, *Decomposition theorems for Lorentzian manifolds with nonpositive curvature*, J. Differential Geom. **22** (1985), 29–42.

[EhG]   P. Ehrlich and G. J. Galloway, *Timelike lines*, Classical Quantum Gravity **7** (1990), 297–307.

[E]   J.-H. Eschenburg, *The splitting theorem for space-times with strong energy condition*, J. Differential Geom. **27** (1988), 477–491.

[EsG]   J.-H. Eschenburg and G. J. Galloway, Preprint.

[G1]   G. J. Galloway, *Splitting theorems for spatially closed space-times*, Comm. Math. Phys. **96** (1984), 423–429.

[G2]   _____, *The Lorentzian splitting theorem without completeness assumption*, J. Differential Geom. **29** (1989), 373–387.

[G3]   _____, *Some connections between global hyperbolicity and geodesic completeness*, J. Geom. Phys. **6** (1989), 127–141.

[Ge]   R. Geroch, *Singularities in closed universes*, Phys. Rev. Lett. **17** (1966), 445–447.

[N]   R. P. A. C. Newman, *A proof of the splitting conjecture of S.T. Yau*, J. Differential Geom. **31** (1990), 163–184.

[Y]   S.-T. Yau, *Problem section*, Ann. of Math. Stud., vol. 102 (S.-T. Yau, editor), Princeton Univ. Press, Princeton, N.J., 1982, pp. 669–706.

University of Miami

Proceedings of Symposia in Pure Mathematics
Volume 54 (1993), Part 2

Weyl Structures on Self-Dual Conformal Manifolds

PAUL GAUDUCHON

0. Introduction

Let M be an oriented four-manifold equipped with a (positive definite) conformal structure c.

Assume that c is self-dual, i.e., that the Weyl tensor W of c satisfies

$$(1) \qquad W = *W,$$

where $*$ is the Hodge star operator determined on two-forms by c and the orientation. Then, to the conformal manifold (M, c) is canonically attached a three-dimensional complex manifold Z, the so-called (projective, half) twistor space of (M, c), fibered over M.

The fibers of the projection of Z onto M are projective lines endowed with a canonical Kähler structure, namely the Fubini-Study metric of (constant) sectional curvature $1/2$.

In addition, the twistor space comes equipped with a real structure σ, whose restriction to each fiber coincides with the antipodal map.

Conversely, the conformal structure c is entirely recovered by the complex structure of Z, the geometry of (projective) lines and the real structure σ (cf. [AHS]).

EXAMPLE (flat case). Let (V, j) be a complex four-dimensional vector space (the twistor vector space), equipped with a quaternionic structure j ($=$ an antilinear automorphism of V of square -1).

Then, the standard conformal sphere S^4 is realized as the quaternionic projective line $\mathbf{H}P(V, j)$, while the twistor space Z is the complex projective space $\mathbf{C}P(V)$.

The twistor projection associates to any complex line l in V the

1991 *Mathematics Subject Classification.* Primary 53C21; Secondary 53C05, 53C07, 53C55, 53C56.

A detailed version of the main part of this paper appeared in the Annali della Scuola Normale Superiore di Pisa.

quaternionic line generated by l, i.e. the j-invariant (complex) two-plane generated by l and $j \cdot l$.

When M is compact, a celebrated theorem of N. Hitchin asserts the following equivalence

$$Z \text{ is Kählerian } \Leftrightarrow Z \text{ is projective}$$

$$\Leftrightarrow (M, c) \text{ is isomorphic to}$$

$$(S^4, c_0) \text{ or } (CP^2, c_0),$$

where c_0 denotes the standard conformal structure on the four-sphere S^4 or the complex projective plane CP^2, [**H2**]. (Cf. also [**FK**].) Then, F. Campana proved the following series of implications (cf. [**C**],

$$Z \text{ is bimeromorphic to } \Leftrightarrow Z \text{ is a Moïshezon variety}$$

$$\text{a Kähler manifold   (bimeromorphic to a projective one)}$$

$$\Rightarrow M \text{ is simply-connected}$$

$$\Rightarrow M \text{ is homeomorphic to } S^4 \text{ or a}$$

$$\text{connected sum } kCP^2 \text{ of } k$$

$$\text{copies of } CP^2\text{'s}.$$

Explicit examples of self-dual conformal structures have been constructed on connected sums kCP^2, first by Y. S. Poon, for $k = 2$ [**P**], then, for any k, by C. Lebrun [**LB**] (cf. also [**KK**]).

On the other hand, M. Ville proved that the algebraic dimension of Z is equal to zero (more precisely, Z does not admit any nontrivial divisor) whenever c is of negative type (= the conformal class c contains a metric with negative scalar curvature) and the first Betti number $b_1(M)$ of M is zero (cf. [**V**]). The restriction $b_1(M) = 0$ actually can be removed to get the

THEOREM 1. *Let (M, c) be a compact, oriented, self-dual conformal manifold of negative type.*

Then, the associated twistor space Z does not contain any nontrivial divisor.

Any complex hypersurface in Z is generically transverse to the fibers of the projection of Z onto M, hence may be considered as a singular, multivalues complex structure on M, adapted to the conformal structure c and the *reversed* orientation (see the definition of Z given below).

For instance, the standard conformal four-sphere S^4 admits such a "complex structure" (one-valued, singular at a point), associated to any complex projective plane CP^2 contained in $Z(S^4) = CP^3$.

Theorem 1 implies the nonexistence of singular, multivalued complex structure, a fortiori of honest complex structure, adapted to a (self-dual) conformal structure c and the opposite orientation, whenever M is compact and c is of negative type.

This applies, in particular, to any compact quotient of the hyperbolic four-space H^4, or the complex hyperbolic two-space CH^2 (the latter oriented by its natural complex structure).

Besides the argument contained in [V], the proof of Theorem 1 relies on a precise description of the set of (real) holomorphic structures on the vertical tangent bundle over Z, cf. §II, and a general Vanishing theorem given in §III. In both of them, the prominent role is played by (self-dual) Weyl structures on (M, c).

1. Weyl structures

In this section, (M, c) denotes a conformal manifold of any dimension n. A *Weyl structure* on (M, c) is a linear connection on M (= acting on the sections of the tangent bundle TM) which

    (i) is symmetric (= without torsion), and

    (ii) preserves the conformal structure.

Condition (ii) means that the Weyl structure D is induced by a $CO(n)$-equivariant connection on the $CO(n)$-principal bundle QM of c-orthonormal frames, or, introducing the *real line bundle L of scalars of weight* 1 (associated to the representation $A \to |\det A|^{1/n}$ of the linear group), that the conformal structure c, considered as a section of $S^2 T^* M \otimes L^2$, cf. [H1], is D-parallel (where $S^2 T^* M$ denotes the vector bundle of symmetric bilinear forms on M).

Equivalently, for any metric g in the conformal class c, there exists a real one-form α_g on M such that

$$(2) \qquad\qquad Dg = -2\alpha_g \otimes g.$$

Any Weyl structure D induces a linear connection D^L on the line bundle L, and it can be observed that *the correspondence*

$$(3) \qquad\qquad D \to D^L$$

is an isomorphism of affine spaces from the (affine) *space of Weyl structures on* (M, c) *and the* (affine) *space of* (all) *linear connections on* L.

NOTE. In H. Weyl's original setting. cf. [W], a linear connection on L is regarded as a *metrical relationship*, while a linear connection on M is called an *affine relationship*. Then, the isomorphism (3) is equivalent to Postulates I and II in [W, §18], respectively.

I. *The nature of space imposes no restriction on the metrical relationship.*

II. *The affine relationship is uniquely determined by the metrical relationship.*

It follows from (3) that the space of Weyl structures on (M, c), as an affine space, is modelled on the vector space of real one-forms on M.

To be precise, any two Weyl structures D_1 and D_2 on (M, c) are related by

$$(4) \qquad\qquad D_2 = D_1 + \tilde{\alpha},$$

where α is a real one-form on M, regarded as an ad QM-valued one-form by the formula

$$(5) \qquad \tilde{\alpha}(X)Y = \alpha(X)Y + \alpha(Y)X - g(X, Y)\alpha^*$$

(for any metric g in c, where α^* denotes the g-dual of α).

In particular, for any Weyl structure D and any metric g in the conformal class c, we have

$$(6) \qquad D = D^g + \tilde{\alpha}_g,$$

where D^g denotes the Levi-Civita connection of g (viewed as a Weyl structure) and α_g is the one-form determined by (2).

Hence, the theory of Weyl structures, as developed by H. Weyl in [**W**], appears as a "gauge theory," corresponding to the group $\mathbf{R}^+$ of positive real numbers, where two Weyl structures are "gauge-equivalent" whenever their difference is an exact one-form $d\log f$.

When M is oriented and four-dimensional, the "instantons" of this gauge theory are $\pm$-*self-dual Weyl structures*, such that the curvature ρ^D of the induced connection D^L is a $\pm$-self-dual (real) two-form.

If, in addition, M is compact, the "instantons" are exactly the *closed* Weyl structures, that is Weyl structures inducing a flat connection D^L on L.

Equivalently, closed Weyl structures are locally the Levi-Civita connections of (local) metrics in c.

The moduli space of closed Weyl structures on (M, c) is naturally identified with the de Rham vector space $H^1(M, \mathbf{R})$, where the origin represents the set of all (global) Levi-Civita connections of (global) metrics in c (for which the induced connection D^L is trivial).

A pair (D, g) formed by a Weyl structure D and a metric g in c is said to be a *distinguished pair*, if the one-form α_g determined by (2) or (6) is co-closed with respect to g.

In particular, for any g in c, the pair (D^g, g) is distinguished. More generally, we have the following (cf. [**G1**]),

THEOREM 2. *Let (M, c) be any compact, oriented, conformal manifold, of dimension n greater than* 2.

For any Weyl structure D on (M, c), there exists a metric g in the conformal class, unique up to a constant factor, such that the pair (D, g) is distinguished.

EXAMPLE (cf. [**G1**]). Any almost-complex structure J on M, c-orthogonal, induces a Weyl structure on (M, c).

Then, the metric g given by Theorem 2 is called the standard metric of the almost-hermitian conformal structure (c, J).

These metrics are characterized by the following condition

$$(7) \qquad dd^c\Omega^{m-1} = 0,$$

where Ω denotes the associated Kähler form $(m = n/2)$.

It follows from (7) that, for any compact complex manifold Z, equipped with a standard hermitian metric g, and any holomorphic line bundle F on Z, the *degree* $\deg_g F$ of F w.r. to g, defined by

$$(8) \qquad \deg_g F = \int_M \gamma \wedge \Omega^{m-1},$$

where γ is the Chern form determined by any (fibered) hermitian structure h on F, is independent of h, i.e. depends upon the holomorphic structure of F and the metric g only.

It can be shown to be equal to the (algebraic) volume, w.r. to g, of the divisor of any (nontrivial) meromorphic section of F.

In particular, we have the following implication:

$$(9) \qquad \deg_g F < 0 \Rightarrow H^0(Z, F^p) = \{0\}, \qquad \forall p > 0,$$

where $H^0(Z, F^p)$ denotes the space of holomorphic sections of the pth (complex) tensor power F^p of F.

2. Twistor spaces

In this section, (M, c) is an oriented, conformal manifold of dimension 4. Let AM denote the vector bundle of skew-symmetric endomorphisms of TM, with respect to the conformal structure c, and consider the orthogonal decomposition

$$(10) \qquad AM = A^+ M \oplus A^- M$$

of AM into its $\pm$-self-dual components (w.r. to the Hodge operator $*$ acting on AM).

The vector bundle AM admits a natural Euclidean structure, given by

$$(11) \qquad \langle a, \ell \rangle = -\tfrac{1}{2} \operatorname{trace}(a \circ \ell).$$

Any element a of $A^+ M$ or $A^- M$ satisfies the relation

$$(12) \qquad a^2 = -\tfrac{1}{2}|a|^2 \cdot I,$$

where I denotes the identity of TM.

It follows from (12) that the sphere bundle of radius $\sqrt{2}$ in $A^- M$ coincides with the bundle of complex structures on M, compatible with c ($= c$-orthogonal) and the opposite orientation ($=$ the orientation induced by the complex structure J on the tangent space $T_x M$ is the reversed orientation).

This fiber bundle is called the *twistor bundle*, denoted by Z, of the oriented conformal manifold (M, c).

Let π denote the natural projection from Z to M, Θ the *vertical tangent bundle* on Z.

For any point J of $Z_x = \pi^{-1}(x)$, the inclusion of Z in $A^- M$ induces a natural identification

$$(13) \qquad \Theta_J = \{ a \in A_x^- M \,|\, J \circ a = a \circ J \},$$

where $\Theta_J = T_J(Z_x)$ is the fiber of Θ at J.

It follows that Θ admits a canonical structure of complex vector bundle, of rank one, determined by

$$(14) \qquad \mathcal{J}a = J \circ a,$$

compatible with the Euclidean structure induced by the imbedding of Θ_J in $A^- M$, so that Θ comes equipped with a canonical (fibered) hermitian structure.

For each fiber Z_x, the hermitian structure of Θ can be viewed as an almost-hermitian structure, which is easily shown to be Kähler, isomorphic to the Fubini-Study Kähler structure (of constant sectional curvature $1/2$).

Choose any Weyl structure D on (M, c), and consider the associated horizontal distribution $\mathcal{H}^D$ on Z.

Let us denote by v^D the corresponding vertical projection from TZ to Θ, so that any vector U on Z, at point J, is represented by the pair

$$(15) \qquad U = (v^D(U), X),$$

where $X = \pi_*(U)$ is the projection of U in $T_x M$.

We thus obtain an almost-complex structure $\mathcal{J}$ on Z, defined by

$$(16) \qquad \mathcal{J}U = (J \circ v^D(U), JX),$$

whose restriction to Θ coincides with the natural complex structure already considered on Θ.

It is well known (cf. [AHS]), that

(i) $\mathcal{J}$ is independent of the chosen Weyl structure D, and

(ii) $\mathcal{J}$ is integrable if and only if the conformal structure c is self-dual.

(The assertion (i) is well known when D is the Levi-Civita connection of a metric in c; the—easy—proof extends readily to general Weyl structures. The assertion (ii) is easily checked by computing the Nijenhuis tensor of $\mathcal{J}$, which is essentially identical to the anti-self-dual part of the Weyl tensor W.)

NOTE. Twistors have been introduced by R. Penrose "to provide an alternative framework for physics," based on complex geometry (cf. [P].

The construction, following Penrose's ideas, of a twistor space over any self-dual, four-dimensional Riemannian manifold is fully developed in [AHS], in connection with Yang-Mills fields. See, also, [H1, S, DV]. The exposition above is close to [DV].

3. Real holomorphic structures on the vertical tangent bundle

From now on, we make the assumption that the conformal structure c is self-dual, so that Z is a three-dimensional complex manifold.

In addition, the (antiholomorphic) involution σ defined by

$$(17) \qquad\qquad\qquad \sigma(J) = -J,$$

provides Z with a *real structure*, whose restriction to each fiber Z_x coincides with the antipodal map (with respect to the natural Kähler metric).

The vertical tangent bundle Θ, viewed as complex line bundle on Z, has no preferred holomorphic structure.

In particular, Θ is never a holomorphic subbundle of the (holomorphic) tangent bundle TZ of Z (cf. [G2]).

On the other hand, due to the existence of a canonical hermitian structure on the fibers of Θ, the datum of a holomorphic structure on Θ, via the associated Cauchy-Riemann operator $\bar{\partial}$, is equivalent to the datum of a *Chern connection* on Θ, i.e. a hermitian connection ∇ whose curvature R^{∇} is $\mathcal{J}$-invariant, of which $\bar{\partial}$ is the $(0, 1)$-part.

A holomorphic structure on Θ is said *natural* if its restriction to each fiber Z_x coincides with the canonical holomorphic structure of the restriction $\Theta_{|Z_x} = TZ_x$, induced by the complex structure of Z_x.

Equivalently, the corresponding Chern connection restricts to the Levi-Civita connection $\nabla^{(x)}$ of Z_x, w.r. to its natural Kähler structure. Since, for each x, the holomorphic structure of TZ_x is unique, up to isomorphism, any holomorphic structure on Θ is isomorphic to a natural one. A holomorphic structure on Θ is said *real* if the corresponding Chern connection is σ-invariant.

We observe that, for each fiber Z_x, the Levi-Civita connection $\nabla^{(x)}$ is real in this sense, that is σ-invariant.

Given any Weyl structure D on (M, c), we obtain a hermitian connection ∇ on Θ in the following manner.

For any section ξ of Θ, viewed as a vertical vector field on Z, any J in Z_x, any U in the tangent space $T_J Z$, we put (see (15))

$$(18) \qquad\qquad\qquad \nabla_U \xi = \nabla^{(x)}_{v^D(U)} \xi + [\tilde{X}, \xi]_J$$

where $\tilde{X}$ denotes the horizontal lift, with respect to $\mathcal{H}^D$, of any vector field on M extending the projection X of U in $T_x M$, $[\tilde{X}, \xi]_J$ the value at J of the bracket of the vector fields $\tilde{X}$ and ξ. It is immediately checked that the latter depends upon X, hence U, only, and that (18) defines a R-linear connection on Θ.

Moreover, since the Kähler structure of the fibers Z_x is entirely determined by the conformal structure c, it is preserved by the flow of the D-horizontal vector field X for any Weyl structure D. It follows that ∇ is a hermitian connection as required.

We observe that the restriction of D to each fiber Z_x coincides, by the very definition, with the Levi-Civita connection $\nabla^{(x)}$.

We then have the following two assertions (see [G2] for the proof).

PROPOSITION 1. *The hermitian connection* ∇ *is a Chern connection, i.e. induces a* (*natural*) *holomorphic structure on* Θ, *if and only if the Weyl connection* D *is self-dual* (*closed, if* M *is compact*).

In this case, two Weyl structures are equivalent if and only if the corresponding holomorphic structures on Θ *are isomorphic* (= *conjugate by an automorphism of the complex line bundle* Θ).

REMARK 1. It can be checked that the Chern connections corresponding to (gauge) equivalent Weyl structures are *not* conjugate (= gauge equivalent).

PROPOSITION 2. *The holomorphic structure on* Θ *induced by a self-dual Weyl structure, via* (18), *is real.*

Conversely, any real holomorphic structure on Θ *is isomorphic to one induced by a* (*self-dual*) *Weyl structure.*

COROLLARY. *If* (M, c) *is a compact, self-dual conformal manifold, the moduli space of real holomorphic structures on* Θ *is canonically identified, via* (18), *to the moduli space of closed Weyl structures, that is to the de Rham vector space* $H^1(M, \mathbf{R})$.

REMARK 2. The correspondence

$$(19) \qquad\qquad\qquad\qquad D \to \bar{\partial}$$

which, to any self-dual Weyl structures on (M, c), associates, via (18), a (natural, real) holomorphic structure on the vertical tangent bundle Θ over the twistor space Z, identifying the moduli space of self-dual Weyl structures to the moduli space of real holomorphic structures on Θ, may be regarded as an "*Atiyah-Ward*" *correspondence* (cf. [A]) attached to the Weyl gauge theory of group $\mathbf{R}^+$.

Through this correspondence, the class of (global) Levi-Civita connections on (M, c) determines a distinguished (real) isomorphism class of holomorphic structures on Θ, inducing a natural identification between the set of isomorphism classes of holomorphic structures on Θ with the neutral component $\mathrm{Pic}^0(Z)$ of the Picard group of Z. We then infer, via the correspondence (19), a canonical identification

$$(20) \qquad\qquad\qquad \mathrm{Pic}_R^0(Z) = H_+^1(M, \mathbf{R})$$

where $\mathrm{Pic}_R^0(Z)$ denotes the real (= σ-invariant) part of $\mathrm{Pic}^0(Z)$, while $H_+^1(M, \mathbf{R})$ is the quotient by exact one-forms of the vector space of real one-forms whose exterior differential is self-dual.

When M is compact, the identification (20) reduces to

$$(21) \qquad\qquad\qquad \mathrm{Pic}_R^0(Z) = H^1(M, \mathbf{R}).$$

See also Remark 3 below.

4. A Vanishing Theorem

For any pair (D, g) formed by a Weyl connection D and a metric g in the conformal class c, we consider the $\mathscr{J}$-hermitian metric on Z obtained by lifting g on the D-horizontal subbundle $\mathscr{H}^D$.

In addition, it is required that g^D coincide with the natural metric on the vertical part Θ, and that the two subbundles $\mathscr{H}^D$ and Θ be g^D-orthogonal.

We then have (see **[G2]** for the proof).

PROPOSITION 3. *The hermitian metric g^D on the twistor space Z is standard if and only if the pair (D, g) is distinguished.*

Assume now that M is compact, and choose any metric g in the conformal class c.

For any element ω of $H^1(M, \mathbf{R})$, viewed as the moduli space of closed Weyl structures, consider the (closed) Weyl structure D in ω defined by

$$(22) \qquad\qquad D = D^g + \widetilde{\theta},$$

where θ is the g-harmonic representative of ω, cf. (6) for the notations. Then, the pair (D, g) is distinguished, and the corresponding hermitian metric g^D is standard, by Proposition 3.

On the other hand, the metric g induces a euclidean norm on the vector space $H^1(M, \mathbf{R})$, defined by

$$(23) \qquad\qquad |\omega|^2_g = \int_M |\theta|^2_g \cdot v_g,$$

where v_g denotes the volume-form of g and $|\theta|^2_g$ the square-norm of θ w.r. to g.

PROPOSITION 4. *Consider any element ω of $H^1(M, R)$ satisfying the inequality*

$$|\omega|^2_g > \frac{1}{6} \int_M \mathrm{Scal}_g \cdot v_g,$$

where Scal_g denotes the scalar curvature of g.

Let D be the (closed) Weyl structure, belonging to ω, defined by (22). Then, the degree of Θ with respect to the (standard) hermitian metric g^D and the (real) holomorphic structure $\overline{\partial}$ determined by D can be made negative by rescalling g by a constant factor.

COROLLARY (Vanishing theorem). *We have the following implication*

$$(24) \qquad |\omega|^2 > \frac{1}{6} \int_M \mathrm{Scal}_g \cdot v_g \Rightarrow H^0_{\overline{\partial}}(Z, \Theta^p) = \{0\}, \qquad \forall p > 0,$$

where $H^0_{\overline{\partial}}(Z, \Theta^p)$ denotes the space of holomorphic sections of Θ^p w.r. to any holomorphic structure $\overline{\partial}$ belonging to the isomorphism class determined by ω.

In particular, if c contains a metric with negative total scalar curvature, we have

$$H_{\overline{\partial}}(Z, \Theta^p) = \{0\}, \qquad \forall p > 0,$$

for any (real) holomorphic structure on Θ.

The proof of Proposition 4 relies on the explicit computation of the curvature R^∇ of the Chern connection ∇.

For that, we consider a slightly more general situation where $\pi\colon Z \to M$ is any fibration over a manifold M, D a connection for π, determined by a horizontal distribution $\mathscr{H}^D$ on Z, $\nabla^{(\cdot)} = \{x \to \nabla^{(x)}\}$ a (smooth) *connection field* on the fibers of π, associating to each x in M a linear connection $\nabla^{(x)}$ on the fiber Z_x.

Then, formula (18) still determines a linear connection ∇ on the vertical tangent bundle Θ over Z.

From (18), follows easily the following expression for the curvature R^∇ of ∇.

Let J be any point on any fiber Z_x, ξ any element in the fiber Θ_J, a, ℓ any vertical vectors in $\Theta_J \subset T_J Z$, $\widetilde{X}$, $\widetilde{Y}$ any D-horizontal vectors in $\mathscr{H}_J^D \subset T_J Z$, of projection X, Y in $T_x M$. Then, we have

(25a) $$R^\nabla_{a,\ell}\xi = R^{(x)}_{a,\ell}\xi,$$

(25b) $$R^\nabla_{a,\widetilde{X}}\xi = (D_X\nabla^{(\cdot)})_a\xi$$

(25c) $$R^\nabla_{\widetilde{X},\widetilde{Y}}\xi = \nabla^{(x)}_\xi(R^D_{X,Y}) + T^{(x)}(R^D_{X,Y}J, \xi),$$

where $R^{(x)}$ and $T^{(x)}$ denote respectively the curvature and the torsion of the connection $\nabla^{(x)}$, $D_X\nabla^{(\cdot)}$ the covariant derivative along X of the connection field $\nabla^{(\cdot)}$ w.r. to D, and where $R^D_{X,Y}$, the curvature of D evaluated at X, Y, is regarded as a vector field on the fiber Z_x, defined by

(26) $$R^D_{X,Y}J = v^D([\widetilde{X}, \widetilde{Y}]_J),$$

still denoting by $\widetilde{X}$, $\widetilde{Y}$ any horizontal extension of $\widetilde{X}$, $\widetilde{Y}$ in the neighborhood of J.

In the case of interest here, where π is the twistorial projection, the connection field $\nabla^{(\cdot)}$ is D-parallel since D preserves the Kähler structure of the fibers, so that *the mixed terms $R^\nabla_{a,\widetilde{X}}$ vanish*.

In addition, the torsion $T^{(x)}$ vanishes for each x in M, and (26) reduces to

(27) $$R^D_{X,Y}J = [R^D(X \wedge Y), J],$$

where, in the r.h.s., R^D denotes the curvature of the *linear* connection D, viewed as a two-form on M with values in the vector bundle $\operatorname{ad} QM = AM \oplus \mathbf{R}\cdot I$.

On the other hand, the Chern form γ^D of Θ, attached to the canonical hermitian structure of Θ and the holomorphic structure determined by the Weyl structure D, is given by

(28) $$\gamma^D = 1/2\pi i \cdot R^\nabla.$$

Then, the determination of the degree of Θ (with respect to g^D) and Proposition 4 follow easily from the computation above of R^∇. See [G2] for details. $\square$

Theorem 1 can be deduced from the Vanishing theorem above in the following way. According to [V], any nontrivial divisor on Z induces a (nontrivial) σ-invariant divisor, corresponding to a (nontrivial) holomorphic section of some (positive) tensor power Θ^p of Θ, for some (real) holomorphic structure $\bar{\partial}$.

By the corollary above, such a section does not exist in case when c contains a metric with negative total scalar curvature. $\square$

REMARK 3. Let $\bar{\partial}$ be the (natural, real) holomorphic structure on Θ determined by some (self-dual) structure D, ∇ the corresponding Chern connection, cf. (18)–(19). Let $d^\beta = d + \beta \cdot i$ be any self-dual hermitian connection on the product line bundle $M \times \mathbf{C}$ over M, where d denotes the trivial connection and β a real one-form on M, with self-dual exterior differential $d\beta$.

Then, d^β determines by pull-back a hermitian connection on the product bundle $Z \times \mathbf{C}$, whose curvature $\pi^* d\beta \otimes \mathscr{J}$ is $\mathscr{J}$-invariant.

Indeed, for any J in Z_x, any U, V in $T_J Z$, of projections X, Y in $T_x M$, we have $(\pi^* d\beta)_J(\mathscr{J}X, \mathscr{J}Y) = d\beta_x(JX, JY)$, and each self-dual two-form at x is J-invariant for any J in Z_x.

This is a particular case of the well-known Atiyah-Ward correspondence corresponding to the group S^1 of unit complex numbers (to be compared with the Atiyah-Ward like correspondence (18)–(19) corresponding to the group $\mathbf{R}^+$).

We thus get a new holomorphic structure on $Z \times \mathbf{C}$, and, tensoring by $(\Theta, \bar{\partial})$, a new holomorphic structure on Θ, of which the associated Chern connection ∇^β is equal to

$$(29) \qquad\qquad \nabla^\beta = \nabla + \pi^* \beta \otimes \mathscr{J}.$$

This new holomorphic structure is still natural, but not real except for $\beta = 0$, since the image of ∇^β by σ is clearly $\nabla^{-\beta}$.

It can be shown that each holomorphic structure on Θ is isomorphic to one of this form, cf. [G2].

In particular, (21) can be completed by the following canonical identification for the whole $\mathrm{Pic}^0(Z)$ (see, also, [H1, Theorem 4.1]):

$$(30) \qquad \mathrm{Pic}^0(Z) = H^1(M, \mathbf{R}) \times (H^1(M, \mathbf{R})/H^1(M, \mathbf{Z})),$$

where the toral part $H^1(M, \mathbf{R})/H^1(M, \mathbf{Z})$ is identified, by (29), to the moduli space of flat hermitian connections on $M \times \mathbf{C}$.

Proposition 4 still holds for any holomorphic structure $\bar{\partial}$ whose real part belongs to ω, and, similarly, its corollary holds for any holomorphic structure.

REMARK 4. Proposition 1 may be completed by observing that, for any self-dual Weyl structure D, the corresponding space of real holomorphic sections of Θ is naturally isomorphic to the kernel of the *Penrose operator* (or *twistor operator*) P^D associated to D, acting on sections of the vector bundle $A^- M$ as follows

(31) $P^d a = $ the trace-free part of $D a$.

This *Penrose correspondence*, extended here to any (self-dual) Weyl structure, admits, for Θ, the especially simple expression:

$$a \in \operatorname{Ker} P^D \leftrightarrow \{J \in Z \to a \circ J - J \circ a \in \Theta_J\} \in H^0_{\bar\partial}(Z, \Theta).$$

We can then recover the Vanishing theorem above by extending to Weyl structures—which are not metrical connections in general—the well-known techniques of Weitzenböck formulae relative to the Penrose operator, cf. [G2] for details.

REFERENCES

[A] M. F. Atiyah, *Geometry of Yang-Mills fields*, Lezioni Fermiane, Pisa, 1979.

[AHS] M. F. Atiyah, N. J. Hitchin, and I. M. Singer, *Self-duality in four-dimensional Riemannian geometry*, Proc. Roy Soc. London Ser. A Vol. 362 (1978), 425–461.

[C] F. Campana, *On twistor spaces of the class C* , J. Diff. Geom. 33 (1991), 541–549.

[DV] M. Dubois-Violette, *Structures complexes au-dessus des variétés*, Applications, Séminaire de Mathématiques de l'ENS, 1981.

[FK] T. Friedrich and H. Kurke, *Compact four-dimensional self-dual Einstein manifolds with positive scalar curvature*, Math. Nachr. 106 (1982), 271–299.

[G1] P. Gauduchon, *La 1-forme de torsion d'une variété hermitienne compacte*, Math. Ann. 267 (1984), 495–518.

[G2] ——, *Structures de Weyl et théorèmes d'annulation sur une variété conforme autoduale*, Ann. Scuola Norm. Sup. Pisa, Serie IV, Vol. XVIII, Fasc. 4 (1991), 563–629.

[H1] N. J. Hitchin, *Linear field equations on self-dual spaces*, Proc. Roy Soc. London Ser. A 370 (1980), 173–191.

[H2] ——, *Kählerian twistor spaces*, Proc. Roy Soc. London Ser. A 43 (1981), 133–150.

[KK] B. Kreussler and H. Kurke, *Twistor spaces over the connected sum of 3 projective planes*, Comp. Math. 82 (1992), 25–55.

[LB] C. LeBrun, *Explicit self-dual metrics on $CP_2 \sharp \cdots \sharp CP_2$* , Preprint, 1989.

[P] Y. S. Poon, *Compact self-dual manifolds with positive scalar curvature*, J. Differential Geom. 24 (1986), 97–132.

[S] S. Salamon, *Topics in four-dimensional Riemannian geometry*, Geometry Seminar "Luigi Bianchi", Lecture Notes in Math., vol. 1022, Springer-Verlag, 1983.

[V] M. Ville, *Twistor examples of algebraic dimension zero threefolds*, Inv. Math. 10 (1991), 537–546.

[W] H. Weyl, *Space-time-matter*, Dover, 1952 (Translation of Raum-Zeit-Materie 1921).

CENTRE NATIONAL DE LA RECHERCHE SCIENTIFIQUE, PARIS

Proceedings of Symposia in Pure Mathematics
Volume **54** (1993), Part 2

Chiral Anomalies and Dirac Families
in Riemannian Foliations

JAMES F. GLAZEBROOK AND FRANZ W. KAMBER

ABSTRACT. We describe a family of differential operators parametrized by the transversal vector potentials of a Riemannian foliation relative to the Clifford algebra of the foliation. This is a family of nonelliptic geometric Dirac operators that has a well-defined index in the K-theory of the parameter space as a result of elliptic-like regularity properties. For a foliation of even codimension, chiral anomalies are seen as cohomological elements relative to the gauge group of a foliated principal bundle. We show that a uniform description is obtained by maps from the truncated Weil algebra to the cohomology of the gauge group.

1. Introduction

Let M be an oriented compact Riemannian manifold of dimension m and $(M, \mathscr{F}, g_M)$ an oriented Riemannian foliation of M of codimension q with bundle-like metric g_M such that the induced metric on the normal bundle Q of the foliation is holonomy invariant (see, e.g., [**18, 21**]).

We recall the exact sequence

$$(1.1) \qquad 0 \to L(\mathscr{F}) \to TM \xrightarrow{\pi} Q \to 0 \,,$$

where $L(\mathscr{F})$ denotes the tangent bundle along the leaves of $\mathscr{F}$. Let E be a complex Hermitian *foliated* bundle on M with foliation lifted from that on M, whose structure is a Clifford module over $\mathrm{Cl}_\mathbb{C}(Q)$, the *transversal Clifford algebra* of $\mathscr{F}$ and let $\mathrm{cl} = \cdot : C^\infty(Q \otimes E) \to C^\infty(E)$ denote Clifford multiplication. Taking $\hat\pi$ to denote the projection

$$(1.2) \qquad \hat\pi : C^\infty(T^* M \otimes E) \to C^\infty(Q^* \otimes E) \cong C^\infty(Q \otimes E) \,,$$

1991 *Mathematics Subject Classification.* Primary 58G30; Secondary 53C12.

Key words and phrases. Basic connection, characteristic classes, chiral anomaly, family of transversal Dirac operators, gauge transformations, Riemannian foliation.

Work supported in part by a grant from the North Atlantic Treaty Organization.

This paper is in final form and no version will be submitted for publication elsewhere.

we define the *transversal Dirac operator* $\mathbb{D}'_{tr}$ as a generalized Dirac operator in the sense of [17] by

$$(1.3) \qquad\qquad \mathbb{D}'_{tr} = \text{cl} \circ \hat{\pi} \circ \nabla^E,$$

where $\nabla = \nabla^E$ denotes covariant differentiation with respect to the metric connection on E. If $\{E_\alpha\}_{\alpha=1,\ldots,q}$ is taken to be a local orthonormal projectable frame in Q, then

$$(1.4) \qquad\qquad \mathbb{D}'_{tr} = \sum_\alpha E_\alpha \cdot \nabla_{E_\alpha}.$$

In [7] it was shown that $\mathbb{D}_{tr} = \mathbb{D}'_{tr} - \frac{1}{2}\kappa\cdot$ is a *symmetric transversally elliptic differential operator*, with symbol σ satisfying $\sigma(x,\zeta) = \zeta\cdot$ for $\zeta \in Q^*_x$ and $\sigma(x,\zeta) = 0$ for $\zeta \in L^*_x$ where $\kappa \in C^\infty(Q^*)$ denotes the mean-curvature form of $(M, \mathscr{F}, g_M)$. We define the subspace $\Gamma_b(E)$ of *basic* or *holonomy invariant* sections of E by

$$(1.5) \qquad \Gamma_b(E) = \{s \in C^\infty(E) : \nabla_X s = 0, X \in C^\infty(L(\mathscr{F}))\}.$$

Let $\mathbb{D}_b = \mathbb{D}_{tr}|\Gamma_b(E)$; we refer to this operator $\mathbb{D}_b$ as the *basic Dirac operator*, mapping smooth sections of $\Gamma_b(E)$ into itself. In [7] certain vanishing theorems and index formulae were established for the operators $\mathbb{D}_{tr}$, $\mathbb{D}_b$ and their powers. Indeed these operators are essentially selfadjoint on $L^2(\Gamma_b(E))$ with $\mathbb{D}_b$ and $\mathbb{D}_b^2$ each possessing a discrete spectrum. The special case of the basic Laplace operator was studied in [20]. As a consequence, $\mathbb{D}_b^2$ satisfies full elliptic regularity and $\mathbb{D}_b^2 v = \mathbb{D}_b s \in \Gamma_b(E)$, implies that $v \in \Gamma_b(E)$. Even though the operator is not an elliptic operator on M (for $q < m$), the above considerations lead to the following elliptic-like regularity [7]:

$$(1.6) \qquad \text{If } \mathbb{D}_b u = v \text{ with } v \in \Gamma_b(E), u \in H_s(\Gamma_b(E)), \text{ then } u \in \Gamma_b(E),$$

where $H_s(\Gamma_b(E))$ denotes the Sobolev space completion. We see that such a restriction of an operator to obtain a discrete spectrum is analogous to imposing boundary conditions for an elliptic operator on a manifold with boundary.

When q is *even*, we have the following decomposition respecting the half-spin representations, $E = E^+ \oplus E^-$. In this case, the operator splits as

$$\mathbb{D}_b = \begin{bmatrix} 0 & \mathbb{D}_b^- \\ \mathbb{D}_b^+ & 0 \end{bmatrix}$$

with

$$(1.7) \qquad \mathbb{D}_b^+ : \Gamma_b(E^+) \to \Gamma_b(E^-), \qquad \mathbb{D}_b^- : \Gamma_b(E^-) \to \Gamma_b(E^+)$$

and $\mathbb{D}_b^- = (\mathbb{D}_b^+)^*$. In which case, the elliptic-like regularity takes the form

$$(1.8) \qquad \text{If } \mathbb{D}_b^+ u = v \text{ with } v \in \Gamma_b(E^-), u \in H_s(\Gamma_b(E^+)), \text{ then } u \in \Gamma_b(E^+).$$

The essential properties of the operator $\mathbb{D}_b$ may be summarized in the following:

THEOREM 1.1.

(i) *The basic heat-operator* $e^{-t\mathbb{D}_b^2}$ *is a smoothing operator of trace class and kernel* $K_{b,t} \in \Gamma_b(E^* \boxtimes E, \mathscr{F} \times \mathscr{F})$, *for* $t \geq 0$;

(ii) $\left(e^{t\mathbb{D}_b^2}\right) u \to \Pi(u)$ *in the* C^∞-*topology for* $t \uparrow \infty$ *and* $u \in L^2(\Gamma_b(E))$, *where* $\Pi : L^2\left(\Gamma_b(E)\right) \to \ker(\mathbb{D}_b^2)$ *is the orthogonal projection*;

(iii) $\mathrm{Tr}_s\left(e^{t\mathbb{D}_b^2}\right)$ *is independent of* $t > 0$, *where* Tr_s *denotes supertrace*;

(iv) *the index of* $\mathbb{D}_b^+$ *is given by the following formula*:

$$(1.9) \qquad Ind\left(\mathbb{D}_b^+\right) \equiv \ker\left(\mathbb{D}_b^+\right) - \ker\left(\mathbb{D}_b^-\right) = \lim_{t \downarrow 0} Tr_s\left(e^{-t\mathbb{D}_b^2}\right).$$

We remark that the above considerations along with (1.8) show that $\mathbb{D}_b^+$ has closed range in $L^2\left(\Gamma_b(E)\right)$ and hence the operators are Fredholm in the unbounded sense. However, boundedness may be obtained by appropriately altering norms in the Sobolev spaces.

Our basic Dirac operators are of interest since a heat expansion does exist and one expects index theorems to appear, but the spectral analysis is that of the leaf space which is highly singular in nature. Here, attention is paid to the families of these operators where the parametrization is by orbit classes under the gauge group (of the foliation) of basic vector potentials on E (cf. [3]). In fact, we regard E as associated to a foliated G–bundle P (see below) and proceed to draw upon the theory of characteristic classes of foliations, as developed in [18], to study maps from the truncated Weil algebra to forms on the gauge group. In the absolute case, this formulism can be applied to the study of various classes of anomalies arising in gauge theory (as in e.g. [3, 5, 24, 25]). In the broader setting of Riemannian foliations, we find that the structure of the foliation here plays a crucial role. Indeed, the analogous cocycles in the gauge group arising via the topological families index as described in [3], are obtained in our setting precisely when the foliation has mean curvature zero (i.e., $\mathscr{F}$ is *harmonic*).

It is a pleasure to acknowledge the hospitality and support of the Institute of Mathematics at Aarhus University and to thank Johan Dupont and Ib Madsen for helpful comments and suggestions. The first named author also wishes to thank J. Stasheff for enjoyable discussions.

2. Indexing a family of basic Dirac operators

Henceforth, we assume that Q is endowed with a $\mathrm{Spin}(q)$ structure. Thus we have the exact sequence of principal bundles

$$\mathbb{Z}_2 \to F_{\mathrm{Spin}}(Q) \to F_{\mathrm{SO}}(Q).$$

We take $E = S \otimes V$, where $S = F_{\mathrm{Spin}}(Q) \times_{\mathrm{Spin}(q)} \triangle_S$ denotes the spin bundle associated to Q, $\triangle_S$ the spin representation and V a complex coefficient bundle. We view V as the vector bundle associated to a *foliated* principal G-bundle $P \xrightarrow{\pi} M$, where G is a connected, compact Lie group. Here, the G-action permutes the leaves of the foliation $\tilde{\mathscr{F}}$ on E and the differential π_* pointwise maps $L(\tilde{\mathscr{F}})$ to $L(\mathscr{F})$ isomorphically. For instance, V is often associated to a $\mathrm{SO}(q)$-reduction of the principal frame bundle $F(Q)$.

Let $\rho : G \to \mathrm{U}(r)$ be a representation of G on $\mathbb{C}^r$ and let $V = P \times_\rho \mathbb{C}^r$ be the resulting complex vector bundle endowed with a Hermitian structure. We call the resulting bundle E as above a *foliated twisted spin bundle*. In [14] we defined a restricted gauge group $\mathscr{G}(\tilde{\mathscr{F}})$, preserving the G-invariant foliation $\tilde{\mathscr{F}}$ on the principal bundle P. $\mathscr{G}(\tilde{\mathscr{F}})$ acts freely on the (convex) subset $\mathscr{A}_b$ of *basic* connections in $\mathscr{A}$, the set of all connections (vector potentials) in P. Recall that $A \in \mathscr{A}_b$ is *basic* if for $\hat{X} \in C^\infty(L(\tilde{\mathscr{F}}))$

$$(2.1) \qquad i_{\hat{X}} A = 0 \text{ (i.e. } A \text{ is adapted)} \quad \text{and} \quad L_{\hat{X}} A = 0 .$$

For a choice of connection $A \in \mathscr{A}_b$, we have then a basic Dirac operator $(\mathbb{D}_b)_A$ depending on A, satisfying the *covariance* condition [3, 14]

$$(2.2) \qquad (\mathbb{D}_b)_{\phi \cdot A} = \hat{\phi}^{-1} (\mathbb{D}_b)_A \hat{\phi} ,$$

where $\hat{\phi}$ is the lift of $\phi \in \mathscr{G}(\tilde{\mathscr{F}})$ to E. We remark that all metrics in question remain fixed throughout. We proceed by defining a principal bundle

$$(2.3) \qquad \left(\mathscr{A}_b \times P, \mathscr{G}(\tilde{\mathscr{F}}), \mathscr{A}_b \times_{\mathscr{G}(\tilde{\mathscr{F}})} P = \tilde{P} \right) ,$$

where $\mathscr{G}(\tilde{\mathscr{F}})$ acts on $\mathscr{A}_b \times P$ by $(A, u) \to (\phi \cdot A, \phi^{-1}(u))$. Now with our choice of $\mathscr{G}(\tilde{\mathscr{F}})$ which commutes with G, $\tilde{P}$ is itself a foliated principal G-bundle

$$(2.4) \qquad G \to \tilde{P} \to \mathscr{A}_b / \mathscr{G}(\tilde{\mathscr{F}}) \times M .$$

We now establish some notation. We elect to denote $\mathscr{A}_b / \mathscr{G}(\tilde{\mathscr{F}})$ by Y and $\mathscr{A}_b / \mathscr{G}(\tilde{\mathscr{F}}) \times M$ by Z. With regards to the representation ρ, we define an "auxiliary" vector bundle

$$(2.5) \qquad \mathscr{E}^0 = \tilde{P} \times_\rho \mathbb{C}^r$$

over Z with projections

$$(2.6) \qquad M \xleftarrow{\bar{\pi}} Z \xrightarrow{\bar{p}} Y .$$

Now taking the complex spin bundle S associated to Q as a vector bundle over M, we define

$$(2.7) \qquad \mathscr{E} = \bar{\pi}^* S \otimes \mathscr{E}^0$$

and extend the family $\{(\mathbb{D}_b)_A\}$ to $\mathscr{E}$ as a family of operators $\{(\tilde{\mathbb{D}}_b)_A\}$ on $\Gamma_b(\mathscr{E})$. For q even, we similarly extend the family $\{(\mathbb{D}^\pm{}_b)_A\}$ to $\mathscr{E}^\pm$ as a family of operators

$$\{(\tilde{\mathbb{D}}^\pm{}_b)_A\} : \Gamma_b\left(\mathscr{E}^\pm\right) \to \Gamma_b\left(\mathscr{E}^\mp\right)$$

over Z.

For our purposes it will also be appropriate to view the family $\{(\tilde{\mathbb{D}}^{\pm}{}_{b})_{A}\}$ as an operator $\tilde{\mathbb{D}}_{b}^{\pm}$, acting along the fibers of the Hilbert bundle

$$(2.8) \qquad \mathscr{H}^{\pm} = \mathscr{A}_{b} \times_{\mathscr{G}(\tilde{\mathscr{F}})} L^{2}\left(\Gamma_{b}\left(E^{\pm}\right)\right) \to Y.$$

We now proceed to describe how canonical connections are endowed on $\mathscr{A}_{b} \to Y$ and $\hat{P} \to Z$ (cf. [3]). On $\mathscr{A}_{b}$, there is a canonical connection ω given by the horizontal spaces $\mathscr{H}_{A}$ at $A \in \mathscr{A}_{b}$ respecting the scalar product on $T(\mathscr{A}_{b})$. In order to work out the explicit details, we recall that

$$(2.9) \qquad \mathrm{Lie}(\mathscr{G}(\tilde{\mathscr{F}})) \cong \Omega_{b}^{0}(\tilde{\mathscr{F}}, \mathfrak{g})_{G} \cong \Omega_{b}^{0}(\mathscr{F}, \mathrm{Ad}(P)) \ ,$$

$$(2.10) \qquad T_{A}(\mathscr{A}_{b}) \cong \Omega_{b}^{1}(\tilde{\mathscr{F}}, \mathfrak{g})_{G} \cong \Omega_{b}^{1}(\mathscr{F}, \mathrm{Ad}(P)) \ .$$

These formulas follow from the (foliated) Atiyah sequence of P,

$$(2.11) \qquad 0 \to \mathrm{Ad}(P) = P \times_{\mathrm{Ad}} \mathfrak{g} \overset{A}{\hookrightarrow} TP/G \overset{\pi_{*}}{\to} TM \to 0 \ ,$$

or equivalently, from the sequence

$$0 \to P \times \mathfrak{g} \to TP \to \pi^{*}TM \to 0,$$
$$(u, \xi) \to \xi_{u}^{*}$$

of G-bundles.

Thus it makes sense to look at the *basic* DeRham complex:

$$\Omega_{b}^{0}(\mathscr{F}, \mathrm{Ad}(P)) \xrightarrow{\ d_{A}\ } \Omega_{b}^{1}(\mathscr{F}, \mathrm{Ad}(P)) \xrightarrow{\ d_{A}\ } \Omega_{b}^{2}(\mathscr{F}, \mathrm{Ad}(P))$$

$$\uparrow{\scriptstyle\cong} \qquad\qquad\qquad \uparrow{\scriptstyle\cong}$$

$$\mathrm{Lie}(\mathscr{G}(\tilde{\mathscr{F}})) \qquad\longrightarrow\qquad T(\mathscr{A}_{b})$$

The metric g_{M} on M and a bi-invariant metric g_{0} on $\mathfrak{g} = \mathrm{Lie}(G)$ define canonically a bundle-like G-invariant metric g_{P} on P, relative to $A \in \mathscr{A}_{b}$, so that TP splits orthogonally via $H_{A} = \ker A$ and we have the *isometry* $\pi_{*}: H_{A} \to \pi^{*}TM$. This induces a canonical metric $\langle\,,\,\rangle$ on $\Omega_{b}(\mathscr{F}, \mathrm{Ad}(P))$ which by construction is also $\mathscr{G}(\tilde{\mathscr{F}})$-*invariant*.

We denote by d_{A}, δ_{A}, respectively, the exterior differential with respect to $A \in \mathscr{A}_{b}$ and the corresponding adjoint codifferential. We list the following facts (compare also [13]):

1. For $X \in \mathrm{Lie}(\mathscr{G}(\tilde{\mathscr{F}}))$, $d_{A}X = X_{A}^{*}$ is the 'vertical' vector field in $\mathscr{A}_{b} \to Y$.

2. $\mathscr{H}_{A} = \{X_{A}^{*}\}^{\perp}$ is the orthogonal complement of the vertical tangent space to $\mathscr{G}(\tilde{\mathscr{F}})$-orbits at A. From [6] we conclude that there is an orthogonal decomposition

$$\Omega_{b}^{1}(\mathscr{F}, \mathrm{Ad}(P)) = \mathrm{im}\, d_{A} \oplus \ker \delta_{A} \ .$$

This implies the *"background gauge"* condition

$$(2.12) \qquad \mathscr{H}_{A} = \ker \delta_{A}$$

since $\langle X, \delta_A B \rangle = \langle d_A X, B \rangle = \langle X_A^*, B \rangle$, $\forall X \in \mathrm{Lie}(\mathscr{G}(\tilde{\mathscr{F}}))$.

3. For $X \in \mathrm{Lie}(\mathscr{G}(\tilde{\mathscr{F}}))$, $F_A = dA + \frac{1}{2}[A, A] \in \Omega_b^2(\mathscr{F}, \mathrm{Ad}(P))$ is the curvature of A. Hence

$$\langle d_A^2 X, F_A \rangle = \langle [F_A, X], F_A \rangle = \frac{1}{2} \left\{ \langle [F_A, X], F_A \rangle + \langle F_A, [F_A, X] \rangle \right\} = 0$$

(since $\mathscr{G}(\tilde{\mathscr{F}})$ is isometric on $\Omega_b^1(M, \mathrm{Ad}(P))$). We then deduce that $\langle X, \delta_A^2 F_A \rangle = \langle d_A X, \delta_A F_A \rangle = \langle X_A^*, \delta_A F_A \rangle = 0$ and hence $\delta_A^2 F_A = 0$ or $\delta_A F_A \in \ker(\delta_A) = \mathscr{H}_A$. This shows that the divergence $\delta_A F_A$ is always in the background gauge.

4. In order to compute the connection ω and its curvature Ω explicitly, we observe that we must have

(i) $\omega(X_A^*) = X$ at $T_A(\mathscr{A}_b)$, $\ker \omega_A = \mathscr{H}_A$;
(ii) $R_\phi^* \omega = \mathrm{Ad}(\phi)^{-1} \circ \omega$, $\phi \in \mathscr{G}(\tilde{\mathscr{F}})$.

The second condition is automatically fulfilled due to the isometric $\mathscr{G}(\tilde{\mathscr{F}})$-action. The first condition says that

(2.13) $\omega(d_A X) = X$, $\forall X \in \mathrm{Lie}(\mathscr{G}(\tilde{\mathscr{F}}))$,

that is, $\omega \circ d_A = \mathrm{id}$.

Now in view of fact 2 above, we have

$$\ker \omega_A = \mathscr{H}_A = \ker \delta_A$$

and

$$\delta_A(X_A^*) = \delta_A d_A(X) = \Delta_A(X),$$

where

$$\Delta_A = \delta_A \circ d_A$$

is the Laplacian on $\Omega_b^0(\mathscr{F}, \mathrm{Ad}(P))$. Further, as $d_A : X \to X_A^*$ is *injective* ($\mathscr{G}(\tilde{\mathscr{F}})$ acts effectively), Δ_A is an invertible selfadjoint (transversally) elliptic operator on $\Omega_b^0(\mathscr{F}, \mathrm{Ad}(P))$ with compact selfadjoint Green's operator $G_A = \Delta_A^{-1}$. It follows that ω must be given by

(2.14) $\omega(B) = G_A(\delta_A B)$, $B \in T_A(\mathscr{A}_b)$.

5. We now consider the curvature Ω in $\mathscr{A}_b \to Y$. For vector fields $B, B' \in C^\infty(T(\mathscr{A}_b))$, which are horizontal, i.e., $B_A, B_A' \in \mathscr{H}_A = \ker \omega_A$, it follows now by direct calculation that

$$\Omega(B, B')_A = B\omega(B') - B'\omega(B) - \omega[B, B']_A + [\omega(B), \omega(B')]$$
$$= -\omega[B, B']_A$$

and by (2.14)

$$\Omega(B, B')_A = -G_A \circ \delta_A[B, B']_A.$$

To express the vertical component of the Lie bracket of vector fields $[B, B']_A$ on $\mathscr{A}_b$ in terms of the punctual Lie bracket $[\,,\,]_{\mathfrak{g}}$ in $\mathfrak{g}$, we use the formula

$$\langle [X, B_A]_{\mathfrak{g}}, B'_A \rangle = \langle X_A^*, [B, B']_A \rangle$$

for $X \in \mathrm{Lie}(\mathscr{G}(\tilde{\mathscr{F}}))$ and horizontal vector fields $B, B' \in C^\infty(T(\mathscr{A}_b))$. Defining the map $b_b : \Omega_b^0(\mathscr{F}, \mathrm{Ad}(P)) \to \Omega_b^1(\mathscr{F}, \mathrm{Ad}(P))$ at A by $b_b(X) = [B_a, X]_{\mathfrak{g}}$, we have for X, B, B' as above:

$$\begin{aligned}
\langle X, \delta_A[B, B']_A \rangle &= \langle d_A X, [B, B']_A \rangle = \langle X_A^*, [B, B']_A \rangle \\
&= \langle [X, B_A]_{\mathfrak{g}}, B'_A \rangle = -\langle b_b(X), B'_a \rangle \\
&= -\langle X, b_b^*(B'_a) \rangle.
\end{aligned}$$

As this is valid for all $X \in \mathrm{Lie}(\mathscr{G}(\tilde{\mathscr{F}}))$, we have

$$\delta_A[B, B']_A = -b_b^*(B'_A) \ .$$

This depends only on B_A, B'_A and thus we obtain the final formula (cf. [3]):

$$(2.15) \qquad \Omega(B, B')_A = G_A \circ b_b^*(B'), \quad B, B' \in \mathscr{H}_A \ .$$

We now turn to the nature of the connection $\tilde{\omega}$ and its curvature $\tilde{\Omega}$ on $\tilde{P}$ (recalling (2.3)). We intend showing that $\tilde{\omega}$ is in fact a basic connection on $\tilde{P}$. First, we view the diagram

$$(2.16) \qquad \begin{array}{ccc}
\tilde{P} = \mathscr{A}_b \times_{\mathscr{G}(\tilde{\mathscr{F}})} P & \longrightarrow & Y \times M = Z \\
\big\uparrow {\scriptstyle \mathrm{id} \times \mu} & & \big\uparrow {\scriptstyle \tilde{\pi} \times \pi} \\
\mathscr{A}_b \times_{\mathscr{G}(\tilde{\mathscr{F}})} (\mathscr{G}(\tilde{\mathscr{F}}) \times P) & \xrightarrow{\ \cong\ } & \mathscr{A}_b \times P
\end{array}$$

where μ denotes the evaluation map. Now at (A, u)

$$(2.17) \qquad \begin{aligned}
T_{(A, u)}(\mathscr{A}_b \times P) &= \Omega_b^1(\mathscr{F}, \mathrm{Ad}(P)) \oplus T_u P \\
&= \Omega_b^1(\mathscr{F}, \mathrm{Ad}(P)) \oplus (\mathfrak{g} \oplus \ker(A)_u).
\end{aligned}$$

Thus, using the previous constructions, we have at $[A, u] \in \tilde{P}$ the orthogonal direct sum decomposition

$$(2.18) \qquad T_{[A, u]}(\mathscr{A}_b \times_{\mathscr{G}(\tilde{\mathscr{F}})} P) \cong \ker \delta_A \oplus (\mathfrak{g} \oplus (\ker A)_u) \cong \mathscr{H}_A \oplus (\mathfrak{g} \oplus (H_A)_u).$$

The connection $\tilde{\omega}$ on $\tilde{P}$ is now given by projecting orthogonally along $\mathscr{H}_A$, $(H_A)_u$ at $[A, u]$, i.e.,

$$(2.19) \qquad \tilde{\omega}(B, Y_u)_{[A, u]} = A(Y_u) \ ,$$

where B is in the background gauge at A ($\delta_A B = 0$) and $Y_u \in T_u P$. The connections ω, $\tilde{\omega}$ are related as follows:

$$(2.20) \qquad \tilde{\omega}(B, Y_u)_{[A, u]} = \omega_A(B) + A(Y_u) \ ,$$

where $B \in T_A \mathscr{A}_b$, $Y_u \in T_u P$.

The curvature $\tilde{\Omega}$ of $\tilde{\omega}$ has a decomposition of components into types respecting the product $\mathscr{A}_b \times P$

$$(2.21) \qquad \tilde{\Omega} = \tilde{\Omega}^{(0,2)} + \tilde{\Omega}^{(1,1)} + \tilde{\Omega}^{(2,0)} \ ,$$

where at $[A, u]$

(i) $\tilde{\Omega}^{(0,2)}_{[A,u]} = (F_A)_{\pi(u)}$;

(ii) $\tilde{\Omega}^{(1,1)}_{[A,u]}(B, Y_u) = B(Y_u)$;

(iii) $\tilde{\Omega}^{(2,0)}_{[A,u]}(B, B') = G_A \circ b_b^*(B')$.

Clearly we have $i_{\tilde{X}}\tilde{\omega} = 0$ for $\tilde{X} \in C^\infty(L(\tilde{\mathscr{F}}))$. Applying $i_{\tilde{X}}$ to each of the above components of $\tilde{\Omega}$, it is straightforward to deduce that $L_{\tilde{X}}\tilde{\omega} = i_{\tilde{X}}\tilde{\Omega} = 0$.

Hence we have established

PROPOSITION 2.1. *The connection $\tilde{\omega}$ on $\check{P}$ given by (2.19) is a basic connection with respect to the foliation $\tilde{\mathscr{F}}$ on $\check{P}$ induced from P.*

In the following section we will investigate the characteristic classes of the G-bundle $\check{P}$ and compute their integrated suspensions in terms of the evaluation map.

3. Characteristic classes and suspension of the gauge group

Henceforth we take $\mathscr{F}$ to be a transversally oriented foliation of *even* codimension $q = 2k$ and denote by $\chi_{\mathscr{F}}$ the characteristic form of $\mathscr{F}$ of degree $(m - q)$. We shall also assume that $\mathscr{F}$ is minimal, that is, $\kappa = 0$. Then the following operator

$$(3.1) \qquad z = \int_M \wedge \chi_{\mathscr{F}}$$

induces an isomorphism $H^q(\Omega_b(\mathscr{F})) \cong \mathbb{R}$ [19, 20]).

Let $W = W(G)$ be the Weil algebra of G and $FW = F^{2(k+1)}W(G)$ the filtration ideal defined by symmetric polynomials of degree $> 2k$ and define the *truncated Weil algebra* by $W_k = W/FW$. As W is contractible, we have a canonical isomorphism

$$(3.2) \qquad \delta : H^{\bullet-1}(W_k) \xrightarrow{\cong} H^\bullet(FW) \ .$$

We observe that the canonical inclusion $j : FI(G) \to FW$ induces a map $j_* : FI(G)^\bullet \to H^\bullet(FW) \cong H^{\bullet-1}(W_k)$ with kernel given by the *decomposable* elements in $FI(G)$, namely [10, 18]

$$I(G)^+ \cdot FI(G) \subset FI(G) \ .$$

Since we have shown that the connection $\tilde{\omega}$, and hence the curvature $\tilde{\Omega}$ on $\check{P}$ are $\mathscr{F}$-basic, there exists a homomorphism of differential algebras

$$(3.3) \qquad k(\tilde{\omega}) : W \to \Omega_b(\check{P}, \tilde{\mathscr{F}}) \ .$$

On G–basic forms, this induces the Chern-Weil homomorphism

$$(3.4) \qquad h(\tilde{\omega}) : I(G) \to \Omega_b(\tilde{P}, \tilde{\mathscr{F}})_G \cong \Omega_b(Z, \mathscr{F}) .$$

Note that the codimension of $\tilde{\mathscr{F}}$ on $\tilde{P}$ is infinite and the powers of $\tilde{\Omega}$ in (2.21) do not, in general, satisfy a vanishing condition.

The operator z in (3.1) defines a map

$$(3.5) \qquad z : \Omega_b^{\bullet+2k}(Z, \mathscr{F}) \to \Omega^{\bullet}(Y)$$

as follows. For $w \in \Omega_b^{2k+j}(Y \times M, \mathscr{F})$ an explicit description can be given by using the bigrading on $\Omega_b^{\bullet+2k}(Y \times M, \mathscr{F})$ to write

$$w = \sum_{j \leq l \leq 2k+j} w_{l, 2k+j-l}$$

and we denote by $(\partial \pm d)$ the total differential on $\Omega_b(Y \times M, \mathscr{F})$. We define then

$$(3.6) \qquad z(w) = \int_M w_{j, 2k} \wedge \chi_{\mathscr{F}} .$$

LEMMA 3.1. *The map z defines a cycle satisfying*

$$(3.7) \qquad \partial z = z(\partial \pm d) .$$

PROOF. Rummler's formula [19, 20] states that modulo $\mathscr{F}$-trivial forms,

$$(3.8) \qquad d\chi_{\mathscr{F}} + \kappa \wedge \chi_{\mathscr{F}} \equiv 0 .$$

Since we have assumed $\kappa = 0$, Green's theorem yields

$$\begin{aligned} \partial z(w) &= \partial \int_M w_{j, 2k} \wedge \chi_{\mathscr{F}} \\ &= \int_M (\partial w_{j, 2k} \pm dw_{j+1, 2k-1}) \wedge \chi_{\mathscr{F}} \pm \int_M w_{j+1, 2k-1} \wedge d\chi_{\mathscr{F}} \\ &= \int_M ((\partial \pm d)w)_{j+1, 2k} \wedge \chi_{\mathscr{F}} = z((\partial \pm d)w) . \quad \square \end{aligned}$$

Thus we obtain a composite chain map of degree $-2k$

$$(3.9) \qquad k = z \circ h(\tilde{\omega}) : FI(G) \to \Omega(Y)$$

and we denote by $k_* = z_* \circ h_*$ the induced map into the DeRham cohomology $H_{DR}(Y)$. Recalling the associated vector bundle $\mathscr{E}^{\circ}$ in (2.5), we observe that

$$(3.10) \qquad \int_M ch(\mathscr{E}^{\circ}) \wedge \chi_{\mathscr{F}} = k_*(\rho^* ch) ,$$

where ch in $I(U(r))$ is the Chern character.

The presence of the Green's operator term in the curvature $\tilde{\Omega}$ means that the integrated characteristic forms of $\tilde{P}$ are not local on $Y = B\mathscr{G}(\tilde{\mathscr{F}})$ (cf. [3]). In order to obtain local forms on $\mathscr{G}(\tilde{\mathscr{F}})$, we will determine the suspension $\Sigma\mathscr{G}(\tilde{\mathscr{F}}) \to B\mathscr{G}(\tilde{\mathscr{F}})$ by the following construction suggested by

J. L. Dupont. In particular, this will relate the characteristic classes discussed in [5] using the difference construction, with the construction in [3].

Consider the automorphism of foliated G-bundles $\bar{\mu} : \mathscr{G}(\tilde{\mathscr{F}}) \times P \to \mathscr{G}(\tilde{\mathscr{F}}) \times P$ given by $\bar{\mu} = (\mathrm{pr}_1, \mu)$ where μ denotes the evaluation map. Instead of suspending a single gauge transformation in $\mathscr{G}(\tilde{\mathscr{F}})$, as was the case in [14, 15], we suspend all of $\mathscr{G}(\tilde{\mathscr{F}})$ to obtain a G-bundle

$$(3.11) \qquad \bar{P} = \mathbb{R} \times_{\mathbb{Z}} \mathscr{G}(\tilde{\mathscr{F}}) \times P \to \mathbb{R} \times_{\mathbb{Z}} (\mathscr{G}(\tilde{\mathscr{F}}) \times M) = S^1 \times \mathscr{G}(\tilde{\mathscr{F}}) \times M$$

where $\mathbb{Z}$ denotes $\mathbb{Z}(\bar{\mu})$. The point here is that $\bar{P}$ is the pullback of $\check{P}$ by a map $f = s \times \mathrm{id}$ where $s : S^1 \times \mathscr{G}(\tilde{\mathscr{F}}) \to Y = B\mathscr{G}(\tilde{\mathscr{F}})$ realizes the suspension map. Explicitly, for any $A \in \mathscr{A}_b$, we define for $t \in [0, 1]$

$$(3.12) \qquad A_{\phi, t} = (1 - t)A + t\phi^* A$$

and $A_{\phi, t+n} = (\phi^n)^* A_{\phi, t}$, for $n \in \mathbb{Z}$. A routine calculation shows that the map

$$(3.13) \qquad \bar{f} : \mathbb{R} \times \mathscr{G}(\tilde{\mathscr{F}}) \times P \to \mathscr{A}_b \times P,$$

defined by $\bar{f}(t, \phi, u) = (A_{\phi, t}, u)$, induces a bundle map $\hat{f} : \bar{P} \to \check{P}$ of foliated G-bundles, covering $f = s \times \mathrm{id}$, where $s(t, \phi) = [A_{\phi, t}]$. In the following, we view A also as a connection on $\mathscr{G}(\tilde{\mathscr{F}}) \times P$ and on $\bar{P}$, pulled back via projection.

Taking $\bar{\omega} = \hat{f}^* \tilde{\omega}$, we observe that $\bar{\omega}_t = (\bar{\mu}^* A)_t$ on $\mathscr{G}(\tilde{\mathscr{F}}) \times P$ and $\bar{\omega}_t(\frac{\partial}{\partial t}) = 0$. Furthermore, $\bar{\mu}^* A$ is given at $[t, \phi, u]$ by

$$(3.14) \qquad \bar{\mu}^* A(X, Y) = (\phi^* A)_t(Y_u) + i_{X_0^*}(\phi^* A)_t$$

where $Y_u \in T_u P$, $X = L_{\phi *} X_0$ and $X_0 \in \mathrm{Lie}(\mathscr{G}(\tilde{\mathscr{F}}))$ (cf. [5]). The curvature $\bar{\Omega}$ of $\bar{\omega}$ is given by

$$F_{(\bar{\mu}^* A)_t} + (\bar{\mu}^* A - A) \wedge dt \,.$$

We observe that $(F_A)^{k+1} = 0$ and $(F_{\bar{\mu}^* A})^{k+1} = (\bar{\mu}^* F_A)^{k+1} = 0$, whereas this is generally not the case for arbitrary values of t.

So for $\Phi \in FI(G)$, the difference construction [18, §5]), applied to $(\bar{\mu}^* A, A)$ yields a well-defined chain map

$$(3.15) \qquad \Delta(\bar{\mu}^* A, A) : W_k \to \Omega_b(\mathscr{G}(\tilde{\mathscr{F}}) \times M, \mathscr{F}) \cong \Omega_b(\mathscr{G}(\tilde{\mathscr{F}}) \times P, \mathscr{F})_G$$

which, via the map j, is given on indecomposable elements $\Phi \in FI(G)$ by

$$(3.16) \qquad \begin{aligned} \Delta(\bar{\mu}^* A, A)(j\Phi) &= \int_{S^1} h(\bar{\omega})(\Phi) \\ &= \int_{S^1} \Phi(\bar{\mu}^* A - A, F_{(\bar{\mu}^* A)_t}, \ldots, F_{(\bar{\mu}^* A)_t}). \end{aligned}$$

Observing that the cycle z is also defined on $\Omega_b(\mathscr{G}(\tilde{\mathscr{F}}) \times M, \mathscr{F})$, we can now formulate the main result of this section.

THEOREM 3.2. *For* $\Phi \in FI(G)^{2k+j+1}$ *indecomposable and* j *odd, the form*

$$(3.17) \qquad z\Delta(\bar{\mu}^* A, A)(j\Phi)$$

in $\Omega^j(\mathscr{G}(\tilde{\mathscr{F}}))$ *represents the suspension of* $zh(\tilde{\omega})\Phi \in \Omega^{j+1}(Y)$. *In particular, the following diagram is commutative:*

$$(3.18) \qquad \begin{array}{ccc} FI(G)^{2k+j+1} & \xrightarrow{k_*} & H_{\mathrm{DR}}^{j+1}(Y) \\ \Big\downarrow{\delta^{-1} j_*} & & \Big\downarrow{\sigma} \\ H^{2k+j}(W_k) & \xrightarrow{z_* \Delta_*} & H_{\mathrm{DR}}^j(\mathscr{G}(\tilde{\mathscr{F}})) \end{array}$$

PROOF. This follows from the following calculation:

$$z\Delta(\bar{\mu}^* A, A)(j\Phi) = \int_M \left[\int_{S^1} k(\omega)(\Phi) \right] \wedge \chi_{\mathscr{F}} = \int_M \left[\int_{S^1} f^* h(\tilde{\omega})(\Phi) \right] \wedge \chi_{\mathscr{F}}$$

$$= \int_{S^1} \left[\int_M f^* h(\tilde{\omega})(\Phi) \wedge \chi_{\mathscr{F}} \right] = \int_{S^1} s^* \left[\int_M h(\tilde{\omega})(\Phi) \wedge \chi_{\mathscr{F}} \right]$$

$$= \int_{S^1} s^* z(h(\tilde{\omega})(\Phi)) ,$$

and $\int_{S^1} s^*$ realizes the suspension $H^\bullet(B\mathscr{G}(\tilde{\mathscr{F}})) \to H^\bullet(\Sigma\mathscr{G}(\tilde{\mathscr{F}}))$. $\quad\square$

To complete the section, we state several conclusions. First, Theorem 3.2 gives explicit formulas for the suspension of the integrated characteristic classes of $\check{P}$ (respectively $\mathscr{E}^0$) by local expressions on $\mathscr{G}(\tilde{\mathscr{F}})$, in terms of the secondary characteristic classes of foliations. Indeed, it should be seen that the above construction of secondary foliation classes, lends new insight even for the foliation by points, where $q = 2k = \dim M$. In this particular case, one recovers the formulas for the chiral anomalies described in the literature (see e.g., [3, 5, 25]).

Our second remark is that whereas the secondary characteristic forms constructed above generally depend on $A \in \mathscr{A}_b$, diagram (3.18) in Theorem 3.2 shows that the corresponding cohomology classes are indeed independent of A.

These constructions can be related to generalized Cheeger-Chern-Simons classes [8, 9, 10] and will be reported upon in detail elsewhere.

4. The basic analytic families index and maps to Fredholm operators

Following [2], the elliptic-like regularity of the $\tilde{\mathbb{D}}_b^+$ family leads to the well-defined element

$$(4.1) \qquad \left[\ker \tilde{\mathbb{D}}_b^+ \right] - \left[Y \times \mathbb{C}^k \right]$$

in the K-theory of Y, for some k (see [1, 2, 14]). Here, the difference in (4.1) accounts for the jumping in the dimension of the fibers. The expression in (4.1) is the *basic analytic families index* of the $\tilde{\mathbb{D}}_b^+$ family and is denoted by $\mathrm{Ind}(\tilde{\mathbb{D}}_b^+)$. In our case Y is connected and by evaluation at $[A] \in Y$, the index

$$(4.2) \qquad \mathrm{Ind}\left(\tilde{\mathbb{D}}_b^+\right)_A \in \mathbb{Z}$$

is independent of $A \in \mathscr{A}_b$.

The construction of the analytic families index in [14] (cf. also [11]) yields a unique homotopy class of maps $\Psi_{(P,\tilde{\mathscr{F}})}^+ = [\Psi^+]$,

$$(4.3) \qquad \Psi^+ : Y = \mathscr{A}_b/\mathscr{G}(\tilde{\mathscr{F}}) \to BU \times \mathbb{Z}$$

mapping Y to the fixed component of index $c = \mathrm{Ind}((\mathbb{D}_b^+)_A)$ (see [1, Appendix 1]).

A mapping in the class $\Psi_{(P,\tilde{\mathscr{F}})}^+$ may be realized as follows. As GL is contractible, the Hilbert bundle $\mathscr{H}^\pm$ over Y has a trivialization (unique up to homotopy), given by an equivariant mapping

$$(4.4) \qquad s^\pm : \mathscr{A}_b \to \mathrm{GL}^\pm = \mathrm{GL}\left(L^2\left(\Gamma_b\left(E^\pm\right)\right)\right)$$

satisfying

$$(4.5) \qquad s^\pm(\phi \cdot A) = \left(\hat{\phi}^\pm\right)^{-1} \circ s^\pm(A).$$

The covariance condition (2.2) along with (4.5) implies that the mapping

$$\Psi^+(A) = s^-(A)^{-1} \circ \left(\mathbb{D}_b^+\right)_A \circ s^+(A)$$

is constant on $\mathscr{G}(\tilde{\mathscr{F}})$-orbits and thus defines

$$(4.6) \quad \Psi^+ : Y = \mathscr{A}_b/\mathscr{G}(\tilde{\mathscr{F}}) \to \mathrm{Fred}\left(H_1\left(\Gamma_b\left(E^+\right)\right), L^2\left(\Gamma_b\left(E^-\right)\right)\right)_c \cong \mathbf{F}_c$$

which evidently realizes the class $\Psi_{(P,\tilde{\mathscr{F}})}^+$.

Passing to loop spaces [22], we observe that $\mathscr{A}_b/\mathscr{G}(\tilde{\mathscr{F}}) \simeq B\mathscr{G}(\tilde{\mathscr{F}})$ and $G \simeq \Omega BG$ for any topological group. Thus we obtain a homotopy class $\Omega\Psi_{(P,\tilde{\mathscr{F}})}^+ = [\Omega\Psi^+]$ of maps

$$(4.7) \qquad \Omega\Psi^+ : \mathscr{G}(\tilde{\mathscr{F}}) \simeq \Omega B\mathscr{G}(\tilde{\mathscr{F}}) = \Omega Y \to \Omega\mathbf{F}_c \simeq \Omega B\,\mathrm{GL}_{cpt} \simeq \mathrm{GL}_{cpt},$$

(cf. [11]) which, as in [25], may be realized along a fixed orbit for which $\left(\mathbb{D}_b^+\right)_A$ is invertible by the mapping

$$\Omega\Psi^+(\phi) = \left(\mathbb{D}_b^+\right)_A^{-1} \circ \left(\mathbb{D}_b^+\right)_{\phi \cdot A}$$

$$= \left(\mathbb{D}_b^+\right)_A^{-1} \circ \left(\hat{\phi}^-\right)^{-1}\left[\left(\mathbb{D}_b^+\right)_A, \hat{\phi}\right] + I$$

$$(4.8) \qquad = \left(\mathbb{D}_b^+\right)_A^{-1} \circ \Xi_{A,\phi}^+ + I,$$

where the bundle map $\Xi_{A,\phi}^+$ is a basic Hermitian endomorphism given by

$$(4.9) \qquad \Xi_{A,\phi}^+ = (\mathbb{D}_b^+)_{\phi \cdot A} - (\mathbb{D}_b^+)_A.$$

THEOREM 4.1 [14]. *There exist well-defined homotopy classes of maps* $\Psi^+_{(P,\tilde{\mathscr{F}})}$ *from* $B\mathscr{G}(\tilde{\mathscr{F}})$ *to* $BU \simeq \mathbf{F}_c$ *and* $\Omega\Psi^+_{P,(\tilde{\mathscr{F}})}$ *from* $\mathscr{G}(\tilde{\mathscr{F}})$ *to* $U \simeq GL_{cpt}$, *realized by* (4.4), (4.9) *such that the induced homomorphisms*

$$(4.10) \qquad \Psi^{+*}_{(P,\tilde{\mathscr{F}})} : H^\bullet(BU, \mathbb{Z}) \to H^\bullet\left(B\mathscr{G}(\tilde{\mathscr{F}}), \mathbb{Z}\right)$$

and

$$(4.11) \qquad \Omega\Psi^{+*}_{(P,\tilde{\mathscr{F}})} : H^\bullet(U, \mathbb{Z}) \to H^\bullet\left(\mathscr{G}(\tilde{\mathscr{F}}), \mathbb{Z}\right)$$

are related by transgression in the respective universal bundles. We have moreover

$$(4.12) \qquad \Psi^{+*}_{(P,\tilde{\mathscr{F}})}\left(\tilde{c}_j\right) = c_j\left(\mathrm{Ind}\left(\tilde{\mathbb{D}}^+_b\right)\right),$$

where $\tilde{c}_j$ *denotes the universal Chern class.*

Our aim is to interpret the characteristic class

$$\Psi^{+*}_{(P,\tilde{\mathscr{F}})}\left(\tilde{c}_1\right) \in H^2\left(B\mathscr{G}(\tilde{\mathscr{F}}), \mathbb{Z}\right),$$

resp. its suspension in $H^1\left(\mathscr{G}(\tilde{\mathscr{F}}), \mathbb{Z}\right)$ via the determinant map on $\mathscr{G}(\tilde{\mathscr{F}})$, i.e. as an obstruction to defining a gauge invariant determinant for the family $\{(\tilde{\mathbb{D}}^+_b)_A\}$. In analogy with the absolute case, one might therefore call these classes $\mathscr{F}$-*relative chiral anomalies.*

The case $j = 1$ in (4.12) can be dealt with as follows. The open sets $U^{(\alpha)} \subset Y$, defined by $U^{(\alpha)} = \{[A] \in Y : \alpha \notin \mathrm{spec}(\tilde{\mathbb{D}}^-_b \tilde{\mathbb{D}}^+_b)_A\}$, determine an open covering $\{U^{(\alpha)}\}$ of Y. Let $\mathscr{H}^{(\alpha)}_\pm$ denote the sum of eigenspaces of eigenvalues less than α over $U^{(\alpha)}$. The fibers of $\mathscr{H}^{(\alpha)}_\pm$ are finite dimensional and locally of constant rank. Hence $\mathscr{H}^{(\alpha)}_\pm$ forms a vector bundle over $U^{(\alpha)}$ and we obtain the exact sequence

$$(4.13) \qquad 0 \to \ker\tilde{\mathbb{D}}^-_b\tilde{\mathbb{D}}^+_b \to \mathscr{H}^{(\alpha)}_+ \xrightarrow{\tilde{\mathbb{D}}^+_b} \mathscr{H}^{(\alpha)}_- \to \ker\tilde{\mathbb{D}}^+_b\tilde{\mathbb{D}}^-_b \to 0.$$

A line bundle $\mathscr{L}^{(\alpha)}$ is defined on $U^{(\alpha)}$ by

$$(4.14) \qquad \mathscr{L}^{(\alpha)} = \left(\det\mathscr{H}^{(\alpha)}_+\right)^* \otimes \left(\det\mathscr{H}^{(\alpha)}_-\right).$$

In [16], we showed that the $\mathscr{L}^{(\alpha)}$ may be pieced together to give a globally defined line bundle $\mathscr{L} \to Y$ (cf. also [4, 12, 23]) and moreover

$$(4.15) \qquad c_1\left(\mathrm{Ind}(\tilde{\mathbb{D}}^+_b)\right) = c_1(\mathscr{L}) \in H^2(Y, \mathbb{Z}).$$

In general, it is not possible to define a gauge-invariant determinant for our operators $(\tilde{\mathbb{D}}^+_b)_A$, apart from the case where G is abelian. We shall discuss the cohomological obstructions in the next section.

5. Local and topological obstructions

For fixed A we choose an operator $(\mathbb{D}_b)_A$ satisfying $\ker(\mathbb{D}^-_b)_A = 0$ and identify $\Gamma_b(E^-)$ with a subspace of finite codimension of $\Gamma_b(E^+)$ via the fixed operator $\left(\mathbb{D}_b^-\right)_A$. Following [25], we defined in [14, 16] the operator

(5.1) $$T_\phi = \left(\mathbb{D}_b^-\right)_A \circ \left(\mathbb{D}_b^+\right)_{\phi \cdot A} : \Gamma_b(E^+) \to \Gamma_b(E^+)$$

whose ζ-function regularization is *dependent on the codimension* q of the foliation [7]. The assignment $\phi \to \det T_\phi$ defines a smooth, complex-valued function on $\mathscr{G}(\tilde{\mathscr{F}})$ and we define

(5.2) $$\mathrm{DET} : \mathscr{G}(\tilde{\mathscr{F}}) \to \mathbb{C}^*$$

by $\mathrm{DET}(\phi) = \det T_\phi$. An element of $H^1(\mathscr{G}(\tilde{\mathscr{F}}), \mathbb{Z})$ is now obtained by pulling back the generator $(2\pi i)^{-1} dz/z$ of $H^1(\mathbb{C}^*, \mathbb{Z})$ by the DET map. The image of the latter in $H^1(\mathscr{G}(\tilde{\mathscr{F}}), \mathbb{R})$ can be represented by the differential form

(5.3) $$\sigma = (2\pi i)^{-1} d_{\mathscr{G}} \mathrm{DET}/\mathrm{DET}.$$

Let $f \in \mathrm{Lie}\,(\mathscr{G}(\tilde{\mathscr{F}}))$; i.e. f is an *infinitesimal* gauge transformation. Then the closed form σ_A can be expressed for $\left(\mathbb{D}_b^+\right)_A$ invertible, by

(5.4) $$\sigma_A(f) = \mathrm{Tr}\left\{\left(T_\phi\right)^{-s} \left(\mathbb{D}_b^+\right)_{\phi \cdot A}^{-1} \left[\left(\mathbb{D}_b^+\right)_{\phi \cdot A}, \hat{f}\right]\right\}\Big|_{s=0}$$
$$= \mathrm{Tr}\left\{\left(\mathbb{D}_b^+\right)_{\phi \cdot A}^{-1} \delta \left(\mathbb{D}_b^+\right)_{\phi \cdot A} / \delta f\right\}.$$

On the other hand, for any $A \in \mathscr{A}_b$ and $\left(\mathbb{D}_b^-\right)_A$ nonsingular, we define

(5.5) $$\tilde{\sigma}_A(f) = \frac{d}{dt} \det\left\{\left(\mathbb{D}_b^-\right)_A \left(\mathbb{D}_b^+\right)_{e^{tf_A}}\right\} \Big/ \det\left\{\left(\mathbb{D}_b^-\right)_A \left(\mathbb{D}_b^+\right)_{e^{tf_A}}\right\}_{t=0}.$$

The above form in (5.5) is not closed, but is invariant and the restriction of $\tilde{\sigma}_A$ to an orbit and hence to $\mathscr{G}(\tilde{\mathscr{F}})$ agrees with $\sigma_A(f)$ (closed) at the identity and gives a 1-form τ on $\mathscr{G}(\tilde{\mathscr{F}})$, representing an element of $H^1(\mathscr{G}(\tilde{\mathscr{F}}), \mathbb{R})$. It can be shown that

(5.6) $$\tau = (2\pi i)^{-1} d\left(\det T_\phi\right) / \det T_\phi + dg = \sigma + dg,$$

that is,

$$[\tau] = [\sigma] \in H^1\left(\mathscr{G}(\tilde{\mathscr{F}}), \mathbb{R}\right).$$

Thus we see that the cohomology class

$$\left[\mathrm{DET}^*\left((2\pi i)^{-1} dz/z\right)\right],$$

coinciding with the class $[\tau]$ defined above, is the obstruction to the existence of a gauge-invariant determinant for the familiy $\{\left(\tilde{\mathbb{D}}^+{}_b\right)_A\}$. It transgresses to the class

(5.7) $$c_1\left(\mathrm{Ind}\left(\tilde{\mathbb{D}}_b^+\right)\right) = \Psi_{(P, \mathscr{F})}^{+*}(\tilde{c}_1) \in H^2\left(B\mathscr{G}(\tilde{\mathscr{F}}), \mathbb{Z}\right).$$

It is possible to express the cohomology class $[\sigma]$ in (5.3), that is, the suspension of the class

$$c_1 \left(\text{Ind} \left(\tilde{\mathbb{D}}_b^+ \right) \right) = c_1(\mathscr{L}) \in H^2(Y, \mathbb{Z})$$

in (4.15) by the foliation invariants in Theorem 3.2. To this end, we pass to the associated bundle $\mathscr{E}^0$ via the representation $\rho : G \to U(r)$ and consider the composite cohomology map of degree $-2k$

$$(5.8) \qquad H(W(U(r))_k) \xrightarrow{\rho^*} H(W(G)_k) \xrightarrow{z_* \Delta_*} H\left(\mathscr{G}(\tilde{\mathscr{F}}), \mathbb{R} \right)$$

defined in (3.17). Now the cohomology of $W(U(r))_k$ is well known (cf. [18, §5]). In degree $2k+1$ it is generated by the classes $y_i \otimes c_J$, where c_J are monomials of degree $\leq 2k$ in the Chern polynomials $c_j \in I(U(r))$ and y_i is the suspension of c_i with $\deg(c_i c_J) > 2k$. In particular, it contains the classes of Godbillon-Vey type, where $i = 1$, and the classes $y_i \otimes 1$ for c_i of degree $> 2k$, if $r > k$. We have then the following result.

THEOREM 5.1. *The cohomology class*

$$[\sigma] \in H^1\left(\mathscr{G}(\tilde{\mathscr{F}}), \mathbb{R} \right)$$

is contained in the image of $H^{2k+1}(W(U(r))_k)$, that is, the linear space generated by the classes $z_* \Delta_*(\rho^*(y_i \otimes c_J))$.

In view of (3.10) and Theorem 3.2 it is very unlikely that such a result is valid for the suspensions of the classes $c_j(\text{Ind}(\tilde{\mathbb{D}}_b^+))$ for $j > 1$; in fact, the spin-structure of the normal bundle Q will have to intervene via the $\hat{A}$-genus. Thus it might be appropriate to view the above result as a very special case of a transversal index theorem for families in the setting of Riemannian foliations.

In concluding, we remark that the case of q *odd* can be directly related to the notion of spectral flow and this is taken up in [14] and [15].

REFERENCES

1. M. F. Atiyah, *K-Theory*, Benjamin, New York 1972.
2. M. F. Atiyah and I. M. Singer, *The index of elliptic operators IV*, Ann. Math. **93** (1971), 119–138.
3. M. F. Atiyah and I. M. Singer, *Dirac operators coupled to vector potentials*, Proc. Natl. Acad. Sci. USA, **81** (1984), 2597–2600.
4. J. H. Bismut, H. Gillet, and C. Soulé, *Analytic torsion and holomorphic determinant line bundles III*, Commun. Math. Phys. **115** (1988), 301–351.
5. L. Bonora, P. Cotta-Ramusino, M. Rinaldi, and J. Stasheff, *The evaluation map in field theory, sigma models and strings I*: Commun. Math. Phys. **112** (1987), 237–282; *II*: Commun. Math. Phys. **114** (1988), 381–437.
6. J.-P. Bourguignon, D. Ebin, and J. E. Marsden, *Sur le noyau des opérateurs pseudo-différentiels à symbole surjectif et non injectif*, C. R. Acad. Sci. Paris, Série A, **282** (1976), 867–870.
7. J. Brüning and F. W. Kamber, *On the spectrum and index of transversal Dirac operators associated to Riemannian foliations*, to appear.

8. S. S. Chern and J. Simons, *Characteristic forms and geometric invariants*, Ann. Math. **99** (1974), 48–69.

9. J. Cheeger and J. Simons, *Differential characters and geometric invariants*, Lect. Notes in Math. **1167**, 50–89, Springer Verlag, Berlin-Heidelberg-New York 1985.

10. J. L. Dupont and F. W. Kamber, *On a generalization of Cheeger-Chern-Simons classes*, Illinois J. Math. **34** (1990), 221–255.

11. D. S. Freed, *An index theorem for families of Fredholm operators parameterized by a group*, Topology **27**, No. 3, (1988) p. 279–300.

12. ____, *On determinant line bundles*, Mathematical Aspects of String Theory, 189–238, World Scientific Publ., Singapore 1987.

13. D. S. Freed and K. K. Uhlenbeck, *Instantons on four-manifolds*, M.S.R.I. Publ. **1**, Springer Verlag, Berlin-Heidelberg-New York 1984.

14. J. F. Glazebrook and F. W. Kamber, *Transversal Dirac families in Riemannian foliations*, Commun. Math. Phys., **140**, (1991), 217–240.

15. ____, *On spectral flow of transversal Dirac operators and a theorem of Vafa-Witten*, Ann. Global Anal. Geom. **9** (1991), 27–35.

16. ____, *Determinant line bundles for Hermitian foliations and a generalized Quillen metric*, AMS Symp. Pure Math., **52**, Part 2, (1991), 225–232.

17. M. Gromov and H. B. Lawson, *Positive scalar curvature and the Dirac operator on complete Riemannian manifolds*, Publ. Math. I.H.E.S. **58** (1983), 83–196.

18. F. W. Kamber and Ph. Tondeur, *Foliated bundles and characteristic classes*, Lect. Notes in Math. **493**, Springer Verlag, Berlin-Heidelberg-New York 1975.

19. ____, *Foliations and metrics*, Progr. in Math. **32**, 103–152, Birkhäuser Verlag, Boston 1983.

20. F. W. Kamber and Ph. Tondeur, *DeRham-Hodge theory for Riemannian foliations*, Math. Ann. **277** (1987), 415–431.

21. P. Molino, *Riemannian foliations*, Progr. in Math. **73**, Birkhäuser Verlag, Boston 1988.

22. A. Pressley and G. Segal, *Loop groups and their representations*, Oxford Univ. Press, Oxford (1986).

23. D. Quillen, *Determinants of Cauchy-Riemann operators on a Riemann surface*, Funk. Anal. i ego Prilozhenya **19** (1985), 37–41.

24. R. Schmid, *The geometry of BRS transformations*, Illinois J. Math. **34** (1990), 87–97.

25. I.M. Singer, *Families of Dirac operators with applications to physics*, Astérisque (hors série) (1985), 323–340.

EASTERN ILLINOIS UNIVERSITY
E-mail address: cfjfg@ux1.cts.eiu.edu

UNIVERSITY OF ILLINOIS AT URBANA-CHAMPAIGN
E-mail address: kamber@symcom.math.uiuc.edu

Proceedings of Symposia in Pure Mathematics
Volume **54** (1993), Part 2

What is the Shape of Space in a Spacetime?

STEVEN G. HARRIS

1. Introduction and background

A spacetime, mathematically speaking, is a manifold with a Lorentz metric: a nondegenerate pseudo-Riemannian metric of signature $+ \cdots + -$; the canonical example is Minkowski n-space ($\mathbb{L}^n$): $\mathbb{R}^n$ with metric $(dx^1)^2 + \cdots + (dx^{n-1})^2 - (dt)^2$. Physically, a spacetime represents a melding together of the separate Newtonian notions of space and time, and general relativity is a method of formulating large-scale physics in such a model; $\mathbb{L}^4$ represents simple empty spacetime with no forces, matter, or energy, and special relativity is its physics. Although it is the union of space and time which provides the proper perspective and context in which to do experimentally verifiable physics, we can none the less find it profitable—for understanding both our universe and the potential models we build for it—to divorce the two notions again and ask questions about space separately.

Of course, we do not end up with Newton's Euclidean 3-space again. Rather, the objects of discussion in this paper are spacelike hypersurfaces of spacetimes, broadly defined. Depending on context, these could be initial data surfaces for the evolution problem in relativity (the Cauchy problem), surfaces of simultaneity for some field of observers or cosmic time-function, or possible representations of the shape of space.

A vector X in a spacetime V is called *timelike* if $\langle X, X \rangle < 0$, *null* (or *lightlike*) if $\langle X, X \rangle = 0$ (but $X \neq 0$), and *spacelike* if $\langle X, X \rangle > 0$ (or $X = 0$); the null vectors at a point form a double cone (the *light cone*) with the timelike vectors in the two components of the cone's interior, the spacelike vectors in the exterior. A *causal* vector is timelike or null. A subspace of the tangent space is timelike, null, or spacelike, according as the induced metric is respectively Lorentzian, degenerate, or Riemannian. The perpendicular

1991 *Mathematics Subject Classification.* Primary 53C50.
This paper is in final form and no version will be submitted for publication elsewhere.

space to a vector is spacelike, null, or timelike, according as the vector is timelike, null, or spacelike.

A curve in the spacetime is called timelike, null, or causal if its velocity vector always has the corresponding character; two points are called *causally related* (respectively, *chronologically related*) if there is a causal (respectively, timelike) curve connecting them. A *time orientation* is a global choice of one component of timelike vectors in each tangent space as the future-pointing ones; this allows us to speak of one point causally (or chronologically) preceding another. The fundamental physical assumption is that the entirety of events in the future, present, and past of our universe can be represented as a four-dimensional spacetime with the world-history of each particle (i.e., the set of its events throughout time) being represented as a causal curve: a null curve for a particle with zero rest-mass (such as a photon), a timelike one for a massive particle. Since all causal influences are assumed mediated by particles of one sort or another, this implies that one event can cause an influence on a second only if the first causally precedes the second. Aspects of a spacetime depending only on the notions of timelike, null, and spacelike are collectively called the *causal structure*; this structure depends solely on the conformal class of the metric. Actual geometry enters the picture by assuming that the world-history of each particle is a geodesic (if the particle is not subject to extraneous forces, such as a rocket engine) and that the Ricci curvature of the metric reflects the structure of matter and energy present at each point (the Einstein field equations).

A submanifold of a spacetime is called timelike, null, or spacelike if its tangent space can be so characterized at each point. Thus, for instance, if a spacetime is foliated by a field of observers—timelike curves—which is sufficiently coherent that the set of vectors perpendicular to the observers forms an involutive distribution, then an integral submanifold of that distribution will be a spacelike hypersurface—a *restspace* for that field of observers, a set of events which, infinitesimally, is seen as simultaneous by each observer: a representation of "all of space at one instant."

Spacelike hypersurfaces have long been recognized as being crucial tools in the study of the structure of spacetimes. The so-called " 3 + 1 formalism" in relativity is a decomposition of the fields on a spacetime by means of a spacelike hypersurface (see, e.g., [IN]). The 3 + 1 formalism is used to formulate the Cauchy problem (assume that the metric and second fundamental form on a spacelike hypersurface are prescribed, and that the Ricci curvature of the spacetime has the form of a classical matter/energy field—e.g., zero for vacuum; is the spacetime metric determined?); this can be found in [HE, ChoqY]. The notion of assigning a total mass to an isolated system is defined in terms of a spacelike hypersurface (one that is "asymptotically flat") (see [BJ, Ho]).

In what sense can a spacelike hypersurface be said to represent "the shape of space" in the sense of homeomorphism or diffeomorphism class? There are two aspects to this questions: What properties should a spacelike hypersurface have in order to qualify as reasonable exemplum of "spatial topology," and is it possible that in a given spacetime there could be more than one homeomorphism class of such exempla? This latter aspect is addressed in a general way in the survey article on global structure [GH]: A reasonable exemplum is assumed to be a spacelike hypersurface which is closed as a subset, and examples are given of classical spacetimes in each of which there are several different homeomorphism classes of closed spacelike hypersurfaces. That article also recalls Geroch's result in [G] that in a spacetime obeying the property called stable causality, if two compact spacelike hypersurfaces bound a compact subset of the spacetime, then they are diffeomorphic. A somewhat stronger result appeared in [BILY]: In a globally hyperbolic spacetime, any two compact spacelike hypersurfaces are homeomorphic (and are each Cauchy surfaces).

What might be the physical implications of two different spatial topologies in one spacetime? Possibly, this reflects nothing more than the differing views of the universe perceived by different classes of observers. For example, consider the Minkowski cylinder, $\mathbb{R}^1 \times \mathbb{S}^1$ with metric $-dt^2 + d\theta^2$: One spacelike hypersurface is the circle embedded as $\{(0, \theta)|0 \leq \theta \leq 2\pi\}$, another is the line embedded as the helix $\{(a\theta, \theta \pmod{2\pi})|\theta \in \mathbb{R}\}$, $|a| < 1$; the first is a restspace of observers standing still—such as $\{(t, \theta_0)|t \in \mathbb{R}\}$, θ_0 constant—while the second is a restspace of observers running around the circle—such as $\{(t, at + \theta_0 \pmod{2\pi})|t \in \mathbb{R}\}$, θ_0 constant.

However, other spacetimes may experience more drastic forms of competing spatial topologies. The quantum effects of a sudden change in spatial topology are explored in [AD, MCD], where the trousers space is considered: the Minkowski cylinder for $t < 0$ glued to two copies of the cylinder for $t > 0$ and with half the diameter of the first one (the gluing is done along a circle—$t = 0$—with two points removed, resulting in a smooth, flat spacetime). Although detailed calculations are somewhat murky, it appears that the topology change is responsible for an instability in the quantum evolution of a scalar field in this spacetime. Examination of the stability of the causal structure of the trousers space is made in [HD], where it is shown that the trousers space admits the smooth addition of one point at $t = 0$ (representing where "new information," for quantum purposes, enters the spacetime), that there can be no extension of the Lorentz causal structure to this point (there are, in effect, two light cones at the point), and that these properties are invariant under small C^0 changes in the metric.

Let us return to the first aspect of the question of the topological shape of space: What spacelike hypersurfaces ought to be admitted as carrying this information? One reasonable criterion is to insist that the hypersurface be *achronal*, i.e., that no two of its points be chronologically related; this would rule out, for instance, the helix in the Minkowski cylinder above. One problem, then, is to determine when this condition is satisfied. If a putative representative of the shape of space is locally well-behaved—i.e., is a spacelike hypersurface—does it have the requisite global properties, such as achronality? This question appears to have been first addressed in [CheY], where closed spacelike hypersurfaces in Minkowski space are considered and a rough argument given for why they should be achronal; the thrust of the paper is to show that closed spacelike hypersurfaces with zero mean curvature must be hyperplanes, and achronality turns out to be needed to show this.

To further refine the notion of topological shape of space, we need to be more careful in talking about a hypersurface. If (V^n, g) is our spacetime, with g the Lorentz metric, let us consider our domain of discourse to be spacelike immersions $f: M^{n-1} \to V$, i.e., immersions with f^*g Riemannian; we shall also need to place some completeness condition on f. For f to give us a reasonable shape of space, we ought to demand, besides that $f(M)$ be closed and achronal, that f be injective and proper: Thus f should be a closed achronal embedding.

What completeness condition is appropriate for f, i.e., strong enough to yield a closed achronal embedding in "appropriate" spacetimes, but weak enough to be widely applicable? It is not sufficient to ask merely that $f(M)$ be closed: [Ha1] gives an example of a closed and injective spacelike immersion of $\mathbb{R}^2$ into $\mathbb{L}^3$ that is not achronal and not proper. We could use proper as the required completeness, but that is rather strong. A weaker notion that fully captures the physical idea of "having no edge" is to insist only that f be proper when restricted to curves in M, i.e., that for any curve $\sigma: [0, 1) \to M$, if $f \circ \sigma$ has a continuous endpoint at 1, then so does σ. If f has this property, let us call it *curve-proper*.

Since f^*g is Riemannian, asking that it be a complete metric would seem to be a viable approach. However, all the properties mentioned so far have depended only on the causal structure or topology of V, so it would be desirable to maintain conformal invariance of the discussion; besides, any immersion $f: \mathbb{R} \to \mathbb{L}^2$ given by $f(x) = (x, \phi(x))$, $|\phi'| < 1$, ought to be included, but this is incomplete if φ' approaches 1 fast enough. What does the trick here is to ask that f be *conformally completable*: that for some positive scalar function $\Omega: V \to \mathbb{R}^+$, $f^*(\Omega g)$ be complete.

Let us call a spacelike immersion *edgeless* if it satisfies any of the three conditions: proper, curve-proper, or conformally completable. For locally injective maps (such as immersions), proper implies curve-proper. Proper also implies conformally completable; for Riemannian ambient spaces, even curve-proper implies conformally completable, but the question is open for Lorentzian ambient spaces.

The two aspects of the question "what is the shape of space" can now be precisely defined: For what spacetimes is it true that any edgeless spacelike immersion of codimension one must be a closed achronal embedding? And for what spacetimes is it true that all closed achronal embedded spacelike hypersurfaces are homeomorphic or diffeomorphic to one another?

In the following discussion of these questions, the references, where not otherwise explicated, are [**Ha1, Ha2, Ha3, Ha4, Ha5**].

2. When spacelike hypersurfaces are globally well-behaved

The question which forms the title of this section is to be interpreted as a quest for "good" spacetimes or, at least, for simple ways to characterize a "good" spacetime: one for which all edgeless spacelike immersions (codimension one) are closed achronal embeddings. As one might expect, Minkowski space is such a spacetime. With $f: M^{n-1} \to \mathbb{L}^n$ the immersion, one can show that f is injective and achronal by examining the intersection of $f(M)$ with a continuous family of timelike 2-planes P_t, $0 \le t \le 1$, where P_t contains $f(c(0))$ and $f(c(t))$ for an arbitrary curve c in M with distinct endpoints. For t small enough, this intersection contains a connected curve from $f(c(0))$ to $f(c(t))$; the edgelessness of f implies that this is true for all t. This yields a spacelike curve from $f(c(0))$ to $f(c(1))$ in P_1, a copy of $\mathbb{L}^2$. In $\mathbb{L}^2$, it is well known that a spacelike curve must be injective and achronal; therefore, $f(c(0))$ and $f(c(1))$ must be distinct and not timelike-related. Once f is known to be injective, being proper follows easily from edgelessness.

How can this process be generalized to other spacetimes? One needs a way of emulating the selection of a continuous family of timelike two-planes. Just about any simply-connected timelike two-submanifolds will serve—so long as they are "long enough." The main problem is that the intersection curves might "run off the edge" of the timelike two-submanifolds, so that one does not end up with a connected curve from $f(c(0))$ to $f(c(1))$.

That this is a live possibility is exemplified by spacetimes which "have an edge" at finite time (or, somewhat more precisely, have a spacelike boundary at timelike infinity). A typical example is $\mathbb{L}^n_+$: Minkowski space for $t > 0$. For $n = 3$, consider a closed planar curve σ immersed in $\mathbb{L}^3_+$ in the $t = 1$ plane. Let M be $\mathbb{R}^1 \times \mathbb{S}^1$ and let $f: M \to \mathbb{L}^3_+$ be the cone on σ with the origin as (imaginary) vertex; f will be a spacelike immersion so long as the normal to σ (as a planar curve) is never parallel to the radial vector and σ stays sufficiently outside the unit disk about the t-axis. In any event, f is edgeless. Clearly, σ can be self-intersecting and obey these restrictions, resulting in f being neither injective nor achronal. Alternatively, σ can be a nonclosed curve and we take $M = \mathbb{R}^1 \times \mathbb{R}^1$ (with f the corresponding cone); if σ is not proper, such as a spiral with a limit-circle, then f, also, will not be proper.

The simple connectivity referred to above is also a real problem: Consider the Minkowski cylinder with its spacelike helix: This fails to be achronal. Other helices fail to be proper (having a limit-circle) or fail to be injective.

The way to avoid these possibilities is to insist on the existence, for any two points x and y in the spacetime V, of a timelike two-submanifold P

containing x and y and also containing timelike curves through x and y which meet—both in the future and in the past—while still in P; and to insist that these submanifolds exist in a continuous manner as y approaches x. The effect of this is to preclude the possibility of the intersection curve in $f(M)$ running off the edge of P, since, being a spacelike curve, it lies within the region in P enclosed by the timelike curves through x and y.

If we consider all the timelike two-submanifolds P_t containing $x(0)$ and $x(t)$ for a closed curve $x(t)$, $0 \leq t \leq 1$, we end up with a disk spanning the closed curve and extending significantly into the future and the past. The whole idea can be encapsulated more simply by the notion of a timelike-contractible disk:

By a *disk* let us mean an immersion $\delta: B^2 \to V$, where B^2 is the closed unit two-ball in $\mathbb{R}^2$. A *timelike contraction* of δ is a continuous map $C: B^2 \times [-1, 1] \to V$ such that $C(-, 0) = \delta$, C is an immersion on $B^2 \times (-1, 1)$, $C(p, -)$ is a timelike curve for each p in B^2, and $C(B^2, 1)$ and $C(B^2, -1)$ are each a single point. A disk is *timelike-contractible* if it has a timelike contraction.

One further requirement is needed: In the Minkowski space proof, an essential idea was that for any point p in M, if f is restricted to a small enough neighborhood U of p, then it is well behaved, i.e., $f(U)$ is achronal. This is true for a wide variety of ambient spacetimes, but not, for instance, in ones with closed causal curves. In order for this "local achronality" of f to hold we need to assume something about the causal structure of the spacetime. The requisite property is known as *strong causality*: the existence, for each point in the spacetime, of a fundamental system of neighborhoods such that any causal curve which starts in one of those neighborhoods cannot exit and re-enter it. Almost all classical spacetimes are strongly causal (see [**HE, BE**]).

We can now state a theorem:

1. THEOREM. *Let V^n be a strongly causal spacetime such that for any loop in V, there is a timelike-contractible disk spanning it. Let $f: M^{n-1} \to V$ be an edgeless spacelike immersion. Then f is a closed achronal embedding; furthermore, $\pi_1(M) = 0$.*

Can anything be said in the case of a nonsimply connected spacetime? Indeed, yes: Note that in the case of the Minkowski cylinder, the "bad" helices all have smaller π_1 than the ambient spacetime. One can generate "bad" spacelike immersions from the circle, such as a circle wrapping twice around and (necessarily) intersecting itself; in this case, $f_*(\pi_1(\mathbb{S}^1))$ is still not all of the ambient fundamental group. An examination of universal covering spaces shows that this is perfectly general:

2. THEOREM. *Let V^n be a strongly causal spacetime such that for any null-homotopic loop in V, there is a timelike-contractible disk spanning it. Let*

$f: M^{n-1} \to V$ be an edgeless spacelike immersion. Then
(1) $f_*: \pi_1(M) \to \pi_1(V)$ is injective,
(2) if f_* is onto, then f is a closed, achronal embedding, and
(3) if f_* is not onto, then f is not injective or not achronal.

Thus, in the right kind of spacetime, it is purely the π_1 behavior of an edgeless spacelike "hypersurface" that determines whether those quotation marks are really warranted.

So what spacetimes are the right kind, i.e., obey the hypotheses of Theorem 2? For any Riemannian manifold (N, h), $(V, g) = (\mathbb{R}^1 \times N, -dt^2 + h)$ is the right kind; furthermore, since all definitions involved have been conformally invariant, anything conformal to such a product is the right kind. This includes a number of classical spacetimes, such as external Schwarzschild (simple black hole, outside the event horizon), Reissner-Nordstrøm (charged black hole), universal anti-de Sitter (constant negative curvature), external slow Kerr (rotating black hole, outside the "inner horizon" at $r = r_-$; $a \le m$), and many Robertson-Walker models (common cosmological models with perfect fluid). On the other hand, interior Schwarzschild has a spacelike boundary at timelike infinity (the singularity), as do de Sitter space and the remaining Robertson-Walker models; these spacetimes do not have the timelike-contractible disk property. Internal slow Kerr and all of fast Kerr are not even causal. (See [HE] for definitions and discussions of these spacetimes.)

One feature of these classical spacetimes that makes them easy to analyze is their invariance under time translations: They all have timelike Killing fields (or, in the cases of Robertson-Walker, de Sitter, and anti-de Sitter, are conformal to such). A spacetime with a timelike Killing field is called *stationary*; if the Killing field is hypersurface-orthogonal, then it is called *static* (Kerr is the best known stationary but not static spacetime). Let us call a spacetime *conformally stationary* or *conformally static* if it is conformal to one such.

In the list of spacetimes above with the timelike-contractible disk property, all have complete Killing fields; those with spacelike boundaries all have incomplete Killing fields. This is no coincidence. Let us call a spacetime *stationary-complete* (and *mutatis mutandis*) if the Killing field is complete; this makes it easy to construct timelike contractions of disks. Since the notion of timelike-contractible disks is conformally invariant, we have

3. PROPOSITION. *In a conformally stationary-complete spacetime, all disks are timelike-contractible.*

Thus, strongly causal, conformally stationary-complete spacetimes are among the right kind. Straightforward techniques yield the structure of such spacetimes:

4. PROPOSITION. *Let (V, g) be a strongly causal, conformally stationary-complete spacetime. Let P be the space of integral curves of the Killing field.*

Then P is a manifold, and V naturally has the structure of a principle $\mathbb{R}$-bundle over P. A choice of cross-section yields for (V, g) the conformal structure of $(\mathbb{R}^1 \times P, -(dt + \omega)^2 + h)$, where h is the Riemannian metric induced on P and ω is a one-form on P (depending on the cross-section). (V, g) is conformally static-complete iff $d\omega = 0$.

With no causal assumptions on V, P could be a very nasty topological space. If V is strongly causal but the Killing field is not complete, one still gets a generalized manifold structure for P, but it may not be hausdorff; in that case, there may very well be no cross-section (example: Minkowski space minus one point).

A conformally stationary spacetime has a timelike vector field which is "conformally Killing": its Lie derivative of the metric is a scalar function times the metric. In general, the existence of a timelike conformal Killing field does not yield a conformal factor for which the field is truly Killing, but the two notions are equivalent in the current context:

5. PROPOSITION. *Let V be a strongly causal spacetime with a complete timelike conformal Killing field. Then V is conformally stationary-complete.*

The reason for spending time on the structure of stationary spacetimes is that they are excellent laboratories in which to explore the second aspect of the question of the shape of space.

3. When spacelike hypersurfaces are all the same

For a stationary spacetime, the space of integral curves of the Killing field provides a ready model for the shape of space: One can project any spacelike hypersurface onto this space, and one might suppose that this is a diffeomorphism. Using Proposition 4 and Theorem 1, we can show this is indeed the case if we have all the usual conditions and the spacetime is simply connected:

6. THEOREM. *Let V^n be a simply connected, strongly causal, conformally stationary-complete spacetime, and let $\pi: V \to P$ be projection to the space of integral curves of the conformal Killing field. Let $f: M^{n-1} \to V$ be an edgeless spacelike immersion. Then f is a closed achronal embedding and $\pi \circ f$ is a diffeomorphism from M to P. If σ is any cross-section of π, then f is isotopic to $\sigma \circ \pi \circ f$. In particular, any two edgeless spacelike hypersurfaces in V have isotopic inclusions.*

As before, examination of covering spaces yields results in the multiply-connected case:

7. THEOREM. *Let V^n be a strongly causal, conformally stationary-complete spacetime, with $\pi: V \to P$ as above. Let $f: M^{n-1} \to V$ be an edgeless spacelike immersion. Then*

    $f_*: \pi_1(M) \to \pi_1(V)$ *is injective, and*
    $\pi \circ f: M \to P$ *is a covering map.*
Furthermore, the following are all equivalent:

f_* is onto;
f is injective and achronal;
$\pi \circ f: M \to P$ is a diffeomorphism.
If these are true, then for any cross-section σ of π, f is isotopic to $\sigma \circ \pi \circ f$.
In particular, any two edgeless spacelike hypersurfaces which are achronal and
non-self-intersecting have isotopic inclusions.

In a static spacetime, one might expect that a restspace—hypersurface
orthogonal to the Killing field—might be a good model for the shape of
space. This depends on how closely the restspace mirrors the topology of
the spacetime on the π_1 level (recall the helical restspace in the Minkowski
cylinder):

If V is strongly causal and conformally static-complete, let $i: N \to V$
be the inclusion of a "conformal restspace." Let $G = \pi_1(V)$ and $G_0 =
i_*\pi_1(N)$; then G_0 is normal in G (due to the interaction of the G-action
on the universal covering space of V and of the $\mathbb{R}$-action coming from the
conformal Killing field). Let $\overline{V}$ be the covering space of V with fiber $H =
G/G_0$. Then $\overline{V}$ is diffeomorphic to $\mathbb{R}^1 \times N$; let $\overline{\pi}: \overline{V} \to N$ be projection. If
M is any achronal and non-self-intersecting edgeless spacelike hypersurface
in V, then let $\overline{M}$ be its covering space with fiber H; $\overline{M}$ has a natural
inclusion into $\overline{V}$. Then $\overline{\pi}$ yields a diffeomorphism from $\overline{M}$ to N; in fact,
the inclusion of $\overline{M}$ into $\overline{V}$ is isotopic to the inclusion of N into $\mathbb{R} \times N$.

More generally, for any codimension-one edgeless spacelike immersion
$f: M \to V$, let $\overline{M}$ be the (regular) covering space of M with fiber
$\pi_1(M)/f_*^{-1}(G_0)$. Then f has a natural lift $\overline{f}: \overline{M} \to \overline{V}$, and $\overline{\pi} \circ \overline{f}: \overline{M} \to N$
is a covering projection.

In particular, if a conformal restspace is achronal, then (by Theorem 7) it
has the same fundamental group as the spacetime, and it will be a true model
for the (unique) topological shape of space.

References

[AD] A. Anderson and B. DeWitt, *Does the topology of space fluctuate?*, Found. Phys. **16** (1986), 91–105.

[BE] J. K. Beem and P. E. Ehrlich, *Global Lorentzian geometry*, Marcel Dekker, New York, 1981.

[BILY] R. Budic, J. Isenberg, L. Lindblom, and P. B. Yasskin, *On the determination of Cauchy surfaces from intrinsic properties*, Comm. Math. Phys. **61** (1978), 87–95.

[BJ] D. R. Brill and P. S. Jang, *Positive mass conjecture*, in [He], pp. 173–193.

[CheY] S.-Y. Cheng and S. T. Yau, *Maximal spacelike hypersurfaces in Lorentz-Minkowski space*, Ann. of Math. **104** (1976), 407–419.

[ChoqY] Y. Choquet-Bruhat and J. W. York, Jr., *The Cauchy problem*, in [He], pp. 99–172.

[F] F. J. Flaherty (ed.), *Asymptotic behavior of mass and spacetime geometry*, Lecture Notes in Physics, vol. 202, Springer-Verlag, Berlin, 1984.

[G] R. Geroch, *Topology in general relativity*, J. Math. Phys. **8** (1967), 782–786.

[GH] R. Geroch and G. Horowitz, *Global structure of spacetime*, in [HI], pp. 212–293.

[Ha1] S. G. Harris, *Closed and complete spacelike hypersurfaces in Minkowski space*, Class. Quantum Grav. **5** (1988), 111–119.

[Ha2] ____, *Complete codimension-one spacelike immersions*, Class. Quantum Grav. **4** (1987), 1577–1585.

[Ha3] ____, *Complete spacelike immersions with topology*, Class. Quantum Grav. **5** (1988), 833–838.

[Ha4] ____, *Conformally stationary spacetimes*, Class. Quantum Grav. **9** (1992), 1823–1827.

[Ha5] ____, *Global structure of static-complete spacetimes* (in preparation).

[HD] S. G. Harris and T. Dray, *The causal boundary of the trousers space*, Class. Quantum Grav. **7** (1990), 149–161.

[He] A. Held (ed.), *General relativity and gravitation*, Plenum, New York, 1980.

[HE] S. W. Hawking and G. F. R. Ellis, *The large scale structure of space-time*, Cambridge Univ. Press, Cambridge, 1973.

[HI] S. W. Hawking and W. Israel (eds.), *General relativity*, Cambridge Univ. Press, Cambridge, 1979.

[Ho] G. Horowitz, *The positive mass theorem and its extensions*, in [F], pp. 1–22.

[IN] J. Isenberg and J. Nester, *Canonical gravity*, in [He], pp. 23–98.

[MCD] C. A. Manogue, E. Copeland, and T. Dray, *The trousers problem revisited*, Pramāṇa **30** (1988), 279–292.

[O] B. O'Neill, *Semi-riemannian geometry*, Academic Press, New York, 1983.

St. Louis University

Proceedings of Symposia in Pure Mathematics
Volume **54** (1993), Part 2

The Kinematics of the Gravitational Field

ADAM D. HELFER

A celebrated recent result in differential geometry is the Positive Energy Theorem, which was proved by Schoen and Yau [**16, 17**], Witten [**27**] and others. The conjecture arose naturally in general relativity, and is an important result there, but it has also had significant consequences in Riemannian geometry. The quantity whose positivity is established is the total energy of a space-time, called the Arnowitt-Deser-Misner energy. This energy is the simplest of a number of kinematic quantities of interest in general relativity. The most ambitious aim is to give a quasilocal treatment of the kinematics of the gravitational field: that is, to quantify the amount of energy, momentum and angular momentum enclosed in a given box (spacelike two-surface of spherical topology) in space-time. Although we are far from having such an understanding, a recent twistorial program due to Penrose [**12**] has had remarkable successes to the extent it has been carried out. The ideas involved are new and deep. In this article, I shall survey this work, and explain too how it resolves some difficulties with kinematics at other levels in general relativity.

Let me begin by explaining why making such definitions is a difficult and even a profound problem. Recall that in mechanics conserved quantities arise from continuous symmetries. For example a component of momentum is the conserved quantity associated to (one says "conjugate to") translation in the corresponding direction. Similarly the components of angular momentum are conjugate to rotations, and energy is conjugate to temporal translations. Now, according to Einstein's theory of general relativity, the effects of gravitation are encoded in a Lorentzian metric on a four-manifold. In general, this metric will admit no Killing vectors, and hence no continuous symmetries. *Therefore the* foundations *for the usual discussion of conserved quantities are* *absent.*

1991 *Mathematics Subject Classification.* Primary 83–02; Secondary 83C40, 83C60.
This paper is in final form and no version will be submitted for a publication elsewhere.

It should be evident that the problem of defining the kinematic quantities of the gravitational field is not well-posed in the mathematical sense. The criteria for success are partly mathematical and partly physical consistency. And since we are trying to do something new physically, although we must be guided by our physical intuition, we must also be prepared to modify that intuition. One must expect, too, that there will be various competing definitions and each will have its own virtues and faults. I shall emphasize the twistorial one, which seems to me the best on physical grounds. However, some of the other proposed definitions may have some mathematical interest, or even some physical interest for a purpose not quite the same as that originally intended.

I shall consider the following desiderata for quasilocal kinematic quantities:

(a) No mass, momentum, or angular momentum should be ascribed to any region in Minkowski space.

(b) The definition should be robust. (If the box is crinkled slightly, the kinematic quantities of the field inside should not change much.)

(c) The definition should be broadly applicable.

(d) Reasonable results should be produced for reasonable surfaces in physically reasonable space-times. In particular, the momentum should reduce to the Arnowitt-Deser-Misner momentum as the surface recedes to spatial infinity, and to the Bondi momentum as the surface recedes to null infinity. Also, one should recover the correct answers in the "weak-field limit," where the metric tensor is taken to differ only infinitesimally from that of Minkowkski space.

A few comments are in order. First, it is embarrassing that a number of proposed definitions fail (a) if the box is not a round sphere. One can try to remedy this by considerably restricting the allowable boxes, but this would be in violation of (c) and of the spirit of (b). Second, it is sometimes argued that the mass should increase monotonically with the box. However, this is not clear since the gravitational binding energy is expected to be negative. For example, if the space-time admits a compact Cauchy surface, there are at least two independent arguments that the total energy should be zero. The only general program which seems to have a hope of fulfilling (a)–(d) is the twistorial one initiated by Penrose. Although it is not yet fully developed, it has given satisfactory answers in every case in which it can be checked. (There are now a large number of these examples, due mainly to Tod [22, 24].)

NOTATION AND CONVENTIONS. Our conventions are those of Penrose and Rindler [14]. We use the abstract index notation for vectors and tensors, so that indices do not take numerical values nor refer to components with respect to a frame. Lorentzian metrics have signature $+---$. The Riemann tensor satisfies $[\nabla_a, \nabla_b]v^d = R_{abc}{}^d v^c$. The Ricci curvature is $R_{ab} = R_{acb}{}^c$

and the scalar curvature is $R = R_a{}^a$. If V is a vector space, then V^* is the dual space. If G is a Lie group, then G_e means the component of G connected to the identity.

Hamiltonian kinematics

We recall that in Hamiltonian mechanics on $\mathbf{R}^3$, a system is specified by giving its *Hamiltonian* $H(p_a, q^a)$, a function of the system's position q^a and its momentum p_a. These evolve according to

$$\text{(1)} \qquad \dot{p}_a = -\frac{\partial H}{\partial q^a}, \qquad \dot{q}^a = \frac{\partial H}{\partial p_a}.$$

The value of H is constant along the trajectory determined by these equations, and this value is the system's energy.

A more general, axiomatic and invariant approach is common these days. We start with a manifold Γ equipped with a closed nondegenerate symplectic form ω_{ab}. The pair (Γ, ω_{ab}) is *phase space*. (In the example above, $\Gamma = \{(q^a, p_a) \in \mathbf{R}^6\}$ and $\omega = 2dp_a \wedge dq^a$.) A vector field V on Γ is called *Hamiltonian* if

$$\text{(2)} \qquad \mathscr{L}_V \omega = 0.$$

In this case, the one-form $\omega_{ab}V^b$ is closed, and (assuming Γ is simply-connected) must be dH for some function H on Γ, determined up to an additive constant. Notice that we must have $\mathscr{L}_V H = 0$, so H is a conserved quantity along any trajectory of V.

Now suppose that a Lie group G acts on M and preserves ω. For each ξ^i in the Lie algebra of G, let $\xi^i V_i^a$ be the vector field on M generating the action of ξ^i. Then, as before, we must have

$$\text{(3)} \qquad \omega_{ab} V_i^b \xi^i = dH(\xi^i).$$

We choose $H(\xi^i) = \xi^i H_i$ for some function H_i on M (which takes values in the dual of the Lie algebra of G). Then $\xi^i H_i$ is conserved along $\xi^i V_i^a$. It is called the *canonical momentum conjugate to* $\xi^i V_i^a$. This can be somewhat misleading, as for example when G is the group of isometries of $\mathbf{R}^3$, if $\xi^i V_i^a$ generates a rotation, then $\xi^i H_i$ is the associated *angular* momentum. We shall refer to each $\xi^i H_i$ as a *kinematic quantity*.

Relativity

Minkowski space. Let us begin with special relativity, which is Einstein's theory in the absence of gravitational forces. Physics is described on *Minkowski space*, which is a four-manifold $\mathbf{M} = \{(t, x, y, z) \in \mathbf{R}^4\}$ equipped with a Lorentzian metric

$$\text{(4)} \qquad \eta_{ab} = c^2 dt^2 - dx^2 - dy^2 - dz^2.$$

We shall use units in which c, the speed of light, is unity.

The trajectory of a particle is represented by a curve γ in M, and one sees from the previous equation that the speed of the particle is less than c iff $\eta_{ab}\dot{\gamma}^a\dot{\gamma}^b > 0$. All known particles travel at speeds less than that of light (or equal to it, if they are massless), and it is this restriction which leads to the relativistic definitions of causality. In general, one calls a vector v^a *timelike, null* or *spacelike* according to whether $\eta_{ab}v^a v^b$ is positive, zero or negative. Physical particles are represented by curves with timelike or null tangents; such curves are called *world-lines*.

The timelike vectors from the interior of a cone and are divided into two components: those with $v^t > 0$, which are *future-pointing*; and those with $v^t < 0$, which are *past-pointing*. The nonzero null vectors are divided into future- and past-pointing sets in the same way. The set of spacelike vectors is connected; there is no invariant meaning to a spacelike vector being future- or past-pointing.

Kinematics on Minkowski space. The identity-connected component of the group of isometries of Minkowski space is the *Poincaré group* P. It is the semidirect product of the group $T = \mathbf{R}^4$ of translations and the Lorentz group $L = SO(1, 3)_e$; one has an exact sequence

$$(5) \qquad\qquad 0 \to T \to P \to L \to 0$$

or, at the Lie algebra level

$$(6) \qquad\qquad 0 \to \mathscr{T} \to \mathscr{P} \to \mathscr{L} \to 0.$$

Structures like this will recur and will be important for understanding the difficulties associated with defining angular momentum, so it is worth spelling out what this group theory means. First, the translations form a canonical subgroup of the isometries. Also any isometry determines an element of L (by the projection $P \to L$). This group L can be thought of as the "rotations measured from infinity": each element of P induces a motion on Minkowski space whose asymptotics determine a unique rotation. (Here "rotation" means "element of $SO(1, 3)_e$.") The rotations about a particular origin in Minkowski space, on the other hand, form a certain (nonnormal) subgroup of P, and there is one of these subgroups for each point in Minkowski space. The group L_O of rotations about O is obviously identified with L (by the projection $P \to L$). That P be a semidirect product is the statement that P contain many copies of L as subgroups which project to L under the map $P \to L$; and that these subgroups (called *lifts* of L) are related to each other by conjugation with elements of T. In ther words, the rotations about different origins all project to L, and those about one origin may be identified with those about another by conjugating with the translation which takes the first origin to the second.

The Lie algebra $\mathscr{T}$ is identified with the space of (constant) tangent vectors; the algebra $\mathscr{P}$ with the Killing vector fields, and the algebra $\mathscr{L}$ with the skew bivectors. Since kinematic quantities are dual to one-parameter families of isometries, the kinematic quantities of a particular system are represented by an element of $A \in \mathscr{P}^*$. In this connection, consider the dual sequence

$$0 \leftarrow \mathscr{T}^* \leftarrow \mathscr{P}^* \leftarrow \mathscr{L}^* \leftarrow 0. \tag{7}$$

We see that the element A gives rise to an element of $\mathscr{T}^*$. This is the *momentum* of the system; it is evidently represented by a covector p_a. The recovery of the angular momentum is less direct; one must choose an origin O of Minkowski space. This determines a lift $L \to P$, whose dual map induces an image of A in $\mathscr{L}^*$, the angular momentum $M_{ab}(O)$. Note that the angular momentum is origin-dependent. When we come to define angular momentum in general relativity, one of the main problems will be to identify a suitable space of origins.

Finally, there are two invariants (called Casimir operators) associated with A. There is the system's *mass*, given by

$$m^2 = p_a p^a \tag{8}$$

and its *spin*, given by

$$s^2 = -S^a S_a / m^2 \tag{9}$$

where

$$S_a = (1/2)\varepsilon_{abcd} P^b M^{cd} \tag{10}$$

is the *Pauli-Lubanski spin vector*. (The origin-dependence of M_{ab} cancels out in S_a.) For physical particles, the squared-mass is always nonnegative, and so the momentum is timelike or null. For physical systems, p^a must always be future-pointing as well. This is equivalent to saying that $p_a t^a \geq 0$ for any future-pointing timelike vector t_a. Here $p_a t^a$ is the time-component of the four-momentum as measured by an observer whose world-line has tangent t^a, so the requirement is that all observers measure a positive energy for the system. It is in this sense that one wishes to establish positivity-of-energy theorems.

General relativity. In general relativity, we replace Minkowski space by a smooth four-dimensional manifold M together with a Lorentzian metric g_{ab}. Such a pair (M, g_{ab}) is called a *space-time*. A vector v^a is said to be *timelike, null,* or *spacelike* according to whether $g_{ab}v^a v^b$ is positive, zero or negative. Physical particles are represented by *worldlines*, curves with everywhere timelike tangents (or everywhere null tangents if the particle is massless). A curve with a spacelike tangent represents a particle moving faster than light and these have never been observed.

It is necessary to assume the space-time has more structure in order to have a meaningful treatment of kinematics. We shall assume that the space-time is *oriented*, and also that it is *time-orientable* and *time-oriented*. This means that it is possible to make a continuously-varying choice of one half of the open cone of timelike vectors over the manifold, and that such a choice has been made. The vectors in this cone, and the nonzero null-vectors on its boundary, are called *future-pointing*; their negatives are *past-pointing*. The world-lines of physical particles will always be parametrized so that their tangent vectors are future-pointing.

Einstein's field equation is

$$(11) \qquad R_{ab} - (1/2)Rg_{ab} = -8\pi G T_{ab}.$$

Here G is Newton's gravitational constant; and T_{ab}, the *stress-energy* tensor, is determined by the matter in the space-time. (If the matter is described by a Lagrangian, then T_{ab} is the variation of the Lagrangian with respect to the metric.) An observer with a world-line γ perceives a four-momentum density $T_{ab}\dot{\gamma}^b$. For physically realistic matter, this density must be a future-pointing null or timelike vector (or zero). This is the *dominant energy condition*.

If (M, g_{ab}) possesses a Killing field ξ^a, then the current $J_a = T_{ab}\xi^b$ is conserved (that is, $\nabla^a J_a = 0$) as a consequence of Killing's equation and the identity $\nabla^a T_{ab} = 0$. In this case, the integral

$$(12) \qquad A(\xi) = \int_\Sigma {}^*J$$

over a three-surface with suitable asymptotics (if J_a falls off quickly enough on Σ) will be conserved under compact deformations of Σ and suitable limits. With mild topological restrictions on M, we have ${}^*J = dQ$ for some two-form Q. Then

$$(13) \qquad A(\xi) = \lim_{S \to \infty} \int_S Q$$

where the limit is taken as the two-surface S approaches infinity on Σ in a suitable way. Thus one can recover conserved quantities associated to genuine symmetries of the space-time from asymptotic information. One approach to defining the kinematic quantities for the entire space-time would be to identify vector fields which were "asymptotically Killing" and from those construct an analog of $Q(\xi)$, which would then be integrated over large spheres as above. This approach was begun by Komar [9] and modified and extended by Winicour and Tamburino [26] (see also Geroch and Winicour [4]). We shall return to this.

The "weak-field limit" of general relativity is a theory of fields on Minkowski space which however represent weak gravitational fields. Let $(\mathbf{M}, \eta_{ab})$ be Minkowski space, and consider a tensor field

$$(14) \qquad g_{ab} = \eta_{ab} + \varepsilon h_{ab},$$

where ε is regarded as infinitesimal. The idea is to treat g_{ab} as a metric tensor on $\mathbf{M}$, but in all calculations to keep only terms of orders zero and one in ε. One calculates the curvature $\varepsilon K_{abc}{}^d$ and the weak-field stress-energy tensor εE_{ab} which are related by the weak-field Einstein equations

(15) $$\varepsilon[K_{ab} - (1/2)g_{ab}K] = \varepsilon[-8\pi G E_{ab}].$$

(In forming the trace εK it is irrelevant whether one uses g^{ab} or η^{ab}, since the difference is second order in ε.) Then the kinematic quantity conjugate to the Minkowski Killing vector ξ^a is

(16) $$\varepsilon A(\xi) = \varepsilon \int_\Sigma {}^*(E_{ab}\xi^a),$$

where Σ is a Minkowski Cauchy surface.

Twistor theory

Twistors are the spinors of the conformal group of Minkowski space. We begin by looking at twistors from a group-theoretical point of view, so as to understand how kinematics will be expressed twistorially. (See Penrose and Rindler [14] for further details.)

A vector field ξ^a on a space-time is called a *conformal Killing vector* if

(17) $$\mathscr{L}_\xi g_{ab} \propto g_{ab},$$

or equivalent

(18) $$\nabla_{(a}\xi_{b)} = (1/4)\nabla_r\xi^r g_{ab}.$$

The conformal Killing vectors on Minkowski space form a 15-dimensional Lie algebra, isomorphic to $so(2,4)$ and $su(2,2)$. This algebra does not exponentiate to a Lie group action on Minkowski space, quite, because flowing a finite increment along some of the vector fields carries one to infinity. The situation can be remedied by "conformally compactifying" Minkowski space to a manifold $\mathbf{M}^\#$ equipped with the conformal class of a metric. Then the conformal Killing fields are the Lie algebra of a connected Lie group $C(1,3)_e$ acting on $\mathbf{M}^\#$ and preserving the conformal class. The group $C(1,3)_e$ is the *conformal group* of Minkowkski space. It is generated by: (a) the Poincaré motions; (b) the dilations $x^a \mapsto kx^a$; and (c) the inversions $x^a \mapsto -(x^a - p^a)/(x - p)^2$.

The group $C(1,3)_e$ is four-to-one covered by $SU(2,2)$. The space on which the defining representation of $SU(2,2)$ acts is *twistor space* $\mathbf{T}$. It is evidently a four-complex-dimensional vector space, and its elements are usually written as Z^α, V^β, etc. It is equipped with a pseudo-Hermitian norm $H_{\alpha\bar{\alpha}}$ of signature $++--$ and an alternating form $\varepsilon_{\alpha\beta\gamma\delta}$. The norm provides a canonical isomorphism between the complex conjugate and the dual of $\mathbf{T}$, and this will be used without comment.

Evidently the Poincaré group is embedded in the conformal group. It turns out that (a four-to-one cover of) the Poincaré group in $SU(2,2)$ can be recovered as those elements preserving a certain skew bitwistor $I^{\alpha\beta}$ called the *infinity twistor*. The infinity twistor is *real* in the sense that

$$(19) \qquad\qquad I_{\alpha\beta} = \overline{I}_{\alpha\beta}$$

where $I_{\alpha\beta} := (1/2)\varepsilon_{\alpha\beta\gamma\delta}I^{\gamma\delta}$, and also simple:

$$(20) \qquad\qquad I^{[\alpha\beta}I^{\gamma]\delta} = 0.$$

As we noted above, the kinematic quantities of a system on Minkowksi space are encoded in an element of the dual to the Poincaré Lie algebra. In twistor terms, this turns out to be a symmetric dual bitwistor $A_{\alpha\beta}$ satisfying

$$(21) \qquad\qquad A_{\alpha\beta}I^{\beta\gamma} \text{ is Hermitian.}$$

One calls $A_{\alpha\beta}$ the *kinematic twistor* (or sometimes, a little misleadingly, the *angular momentum twistor*). Sorting through the group theory in a straight-forward way, one finds

$$(22) \qquad\qquad m^2 = -(1/2)A_{\alpha\beta}\overline{A}^{\alpha\beta}.$$

It is curious that the infinity twistor does not appear explicitly in this formula: since the mass is a Poincaré but not a conformal invariant, one would expect some manifestation of the breaking of conformal symmetry. But it turns out otherwise; the condition (21) is where this enters, not in the explicit formula for the mass. There are also unambiguous ways to recover the momentum and angular momentum (about any origin), and these involve the infinity twistor. We shall not need the details.

The treatment so far has been abstract, but can be made concrete. The essential thing is to have a representation of twistor space in space-time terms. It turns out that there is a natural one. The elements of **T** turn out to be identified with certain spinor fields. The spinor field representing Z^α will be denoted $\omega^A(x, Z^\alpha)$. (This is a two-component spinor field. For details, seen Penrose and Rindler [14]). The condition that a spinor field represent a twistor is that it satisfy the *twistor equation*,

$$(23) \qquad\qquad \nabla^{A'(A}\omega^{B)} = 0.$$

An understanding of the particulars of this equation is not necessary for this paper. One should simply realize that the twistors are identified with the spinor fields satisfying a certain system of first-order partial differential equations.

The norm, alternating symbol and infinity twistor can all be expressed in terms of twistor fields. For the moment, we note that the norm of Z^α is

$$(24) \qquad Z^\alpha\overline{Z}_\alpha = (-i/2)\omega^A\nabla_{AA'}\bar{\omega}^{A'} + \text{complex conjugate},$$

where $\omega^A = \omega^A(X, Z^\alpha)$. Again, the details are not important. What *is* important is that $Z^\alpha\overline{Z}_\alpha$ is a certain function on **M**, and this function turns out to be constant by virtue of the twistor equation.

Finally, one can recover all the usual structures of Minkowski space from twistor data. One finds that the complexification of conformally completed Minkowski space is naturally identified with the Grassmannian of (complex) two-planes in $\mathbf{T}$:

$$(25) \qquad\qquad \mathbf{CM}^{\#} = Gr(2, \mathbf{T}).$$

Real conformally completed Minkowski space is the set of two-planes in $\mathbf{T}$ which are *null*, i.e., every point on which satisfies $Z^{\alpha}\overline{Z}_{\alpha} = 0$. Note that one needs the norm on twistor space to determine which points of $\mathbf{CM}^{\#}$ are real. The points in $\mathbf{M}^{\#}$ which are finite are those whose two-planes in $\mathbf{T}$ meet the two-plane determined by the infinity twistor only at zero.

Now suppose $A_{\alpha\beta}$ is a kinematic twistor, and for simplicity suppose the system it represents is massive $(m > 0)$. The system's world-line can be recovered from the kinematic twistor as follows. Take twistor space projectively. Then $A_{\alpha\beta}$ defines a quadric hypersurface $A_{\alpha\beta}Z^{\alpha}Z^{\beta} = 0$. All two-dimensional quadrics have a canonical form $CP_1 \times CP_1$, so there are two families of projective lines on the quadric. One of these families contains $I^{\alpha\beta}$, and this family is the *complex world-line* of the system. (In general, the only projective line on the quadric representing a real point in $\mathbf{CM}^{\#}$ will be $I^{\alpha\beta}$.) The system's world-line is the real part of this. (One can show that the real part is invariantly defined. The imaginary part of the complex world-line turns out to measure of the system's spin.)

(Twistors can be used to give treatments of de Sitter and anti-de Sitter space without much change. For these spaces, the infinity twistor is not simple, but satisfies

$$(26) \qquad\qquad \varepsilon_{\alpha\beta\gamma\delta}I^{\alpha\beta}I^{\gamma\delta} = 4\lambda/3,$$

where λ is the cosmological constant. In this case, the kinematic twistor is an element of the dual of the de Sitter or anti-de Sitter Lie algebra.)

Two-surface twistors

Now let us try to develop a twistorial kinematics of the gravitational field enclosed by a space-like two-surface S of spherical topology in a general space-time. We begin by constructing the *two-surface twistor space* $\mathbf{T}(S)$ of S, sometimes called *superficial twistor space*. For this, we consider the twistor equation. It turns out that it makes sense to restrict the equation to S, in the sense that it is possible to isolate the components of the equation which involve only tangential derivatives. The result is a system of equations usually written as

$$(27) \qquad\qquad \eth'\omega^{0} = \sigma'\omega^{1}, \qquad \eth\omega^{1} = \sigma\omega^{0}.$$

Again, the details are not relevant. This system is rather like a Dirac equation (in fact, the operators $\eth$ and $\eth'$ are Dirac operators on certain line bundles on S; the functions σ and σ' are sections of certain line bundles and are determined by the geometry of the embedding of S in M). The Atiyah-Singer Index Theorem can be used to show that the system always has at least a four-complex-dimensional family of solutions, and moreover the dimension is four generically. We shall assume $\mathbf{T}(S)$, the space of solutions to (27), is a four-complex-dimensional vector space.

One would like now to construct the following: a norm on $\mathbf{T}(S)$, an alternating form, an infinity twistor, and a kinematic twistor. At present, in the most general situation, there is no candidate for any of these that one can have complete confidence in. The situation is best for the kinematic twistor. It has the form

$$
(28) \qquad A_{\alpha\beta}Z^{\alpha}Z^{\beta} = \int_{S} \text{quadratic form in } \omega^{A}(x, Z^{\alpha})
$$

$$
\text{with curvature components as coefficients} \cdot \eta,
$$

where η is a factor to be discussed below. This is arrived at by looking at the weak-field limit, in which the metric is taken to be Minkowskian plus an infinitesimal perturbation. In this case, the isometries of Minkowski space act on the perturbation, and one can deduce the correct form of the kinematic twistor. The general form, above, is the obvious generalization of this, apart from the factor η.

(The factor η is not well understood, but is motivated by two related investigations of surfaces in Schwarzschild space-time (Penrose [13], Tod [23]). The idea is this. Given $\mathbf{T}(S)$, one can construct an associated complexified compactified Minkowski space $\mathbf{CM}^{\#}(S)$ as the space of two-planes in $\mathbf{T}(S)$. The surface S naturally immerses in $\mathbf{CM}^{\#}(S)$: the image of $p \in S$ is the two-plane of two-surface twistors vanishing at p. It seems that the integral defining the kinematic twistor ought to be given more naturally over this immersed surface than over S. Then η is essentially the Jacobian of the immersion. It turns out however there is a scale ambiguity, so it is not clear how η ought to be normalized. If η is constant, one takes $\eta = 1$.)

The norm is the next thing we should like, for with it we can compute the mass. There is a sensible restriction of the Minkowski-space expression (24) to S, but it not satisfactory because it is not constant on the two-surface, in general. One could simply average it, and this might turn out to be the right thing, but the procedure is not obviously correct. There are certain situations in which there is a natural candidate for the norm, however. One of these is when S is uncontorted:

DEFINITION. A two-surface S in a space-time is *uncontorted* if it can be embedded in a space-time conformally equivalent to Minkowski space, so that its first and second fundamental forms are preserved.

REMARKS. (a) Some elementary counting shows that locally the condition that S be uncontorted is given by the vanishing of three real functions of two variables. (b) There are many interesting examples of uncontorted surfaces in space-times, however. (c) Clearly any two-surface in a conformally flat space-time is uncontorted.

If S is uncontorted, then the expression (24) for the norm (and a similar one for the alternating twistor) are constant on S. (Also the factor η is unity in this case.) Thus the mass can be calculated.

OPEN QUESTION. If the surface S is the boundary of a space-like three-surface on which the dominant energy condition holds, must we have $m^2 \geq 0$?

REMARK. The answer is not even known if (M, g_{ab}) is conformally flat, and this case should certainly be tractable.

If S is contorted, much less is known. In some space-times, there are natural candidates for the norm; in particular, Kelly, Tod and Woodhouse [7] have analyzed certain small contorted spheres in the Schwarzschild solution with satisfactory results. However, there is one broad class of contorted spheres on which there is a natural candidate for the norm: spheres which are "cuts" of null infinity. This class is important for physics, and also provides a useful testing-ground for quasilocal kinematic constructions. It will be discussed at length below.

To calculate the momentum and angular momentum, one requires an infinity twistor as well. Here the result is at least superficially negative: even for a generic two-surface in a generic conformally flat space-time, there is no skew bitwistor satisfying (21), even if we relax the requirements of reality and simplicity (Helfer [5]). It is not clear that this should be regarded as damning the superficial twistor program, however. One can argue plausibly (if, at the moment, vaguely) that perhaps the presence of the gravitational field should destroy the reduction of the conformal group to the Poincaré group. This is made less unreasonable, perhaps, by the results on the angular momentum at null infinity to be discussed shortly.

Successes of the quasilocal construction

Although many important difficulties with the twistorial quasilocal construction remain to be resolved, much has been achieved. Throughout this section, S is a spacelike two-surface of spherical topology.

(1) The twistorial definition gives the correct answers in the weak-field limit. In particular, $A_{\alpha\beta} = 0$ for any S in Minkowski space.

(2) Tod [24] has very nearly given a positive answer to the following

CONJECTURE. For any uncontorted S in the Schwarzschild solution, the mass is defined and is equal to the Schwarzschild mass parameter if the surface links the Schwarzschild throat once and vanishes if the surface does not link the throat.

For contorted spheres in Schwarzschild, there is a natural although not compelling candidate for the norm. The factor η in the integral for the kinematic twistor is not unity, and it is difficult to evaluate. For this reason, little is known. An investigation of contorted spheres in Schwarzschild would be an important test of the definition in new circumstances. (See Penrose [13], Tod [23], Kelly, Tod and Woodhouse [7].)

OPEN QUESTION. Is the mass zero or the Schwarzschild mass parameter for contorted surfaces in Schwarzschild?

(3) Satisfactory results have also been obtained for spheres of symmetry in Reissner-Nordstrom space-time (representing a charged spherically symmetric mass). See Tod [22]. There is a contribution due to the energy of the electromagnetic field. (Some previous definitions had given such a contribution but an incorrect one, as could be deduced by looking at the weak-field limit.)

(4) In the $k = 0$ Friedmann-Robertson-Walker space-times (spatially homogeneous universes filled with fluids), for S in a surface of constant "cosmological time" (but not necessarily round), the mass is the proper volume enclosed by S times the density (Tod [22]).

(5) Consider the time-symmetric initial value problem for n stars. Then the mass enclosed by S is the sum of the mass parameters of the starts within S minus a contribution for the gravitational binding energy (Tod [22]).

(6) de Sitter space is a space-time of constant curvature, in which there are space-like three-surfaces which are S^3's of constant curvature. Let S_θ be a round S^2 of constant colatitude θ on such an S^3. Then the mass enclosed by S_θ increases from zero at $\theta = 0$ to a maximum at $\theta = \pi/2$, and decreases to zero again as θ approaches π. Similarly, quite generally the total mass of a spatially closed universe is zero.

(7) The Bondi momentum and the Arnowitt-Deser-Misner momentum are obtained as S recedes to infinity in the appropriate senses. More asymptotic results will be given below (Penrose [12]).

(8) The Ashtekar-Hansen-Sommers angular momentum is obtained as S recedes to spatial infinity if the Weyl tensor is "asymptotically electric" (Shaw [19]).

Null infinity

If we think of an isolated gravitational system, there are two distinct regimes in which we can discuss its asymptotics. First, we might imagine going far away from the sources in a space-like sense (e.g., on a Cauchy surface). Second, we might imagine leaving the source region at the speed of light; in the limit, we end up then at null infinity. The first regime is the appropriate one for posing Einstein's equations as a Hamiltonian system. The second is more interesting for the study of gravitational and other radiation. Since these waves travel at the speed of light, they never escape to infinity on a Cauchy surface. There is an extensive theory of null asymptotics which is beyond the scope of this article. (See Penrose and Rindler [14].) We only outline the main ideas.

DEFINITION. A space-time (M, g_{ab}) is *future asymptotically simple* if it embeds in a manifold-with-boundary $\widehat{M} = M \cup \mathscr{I}^+$ (where $\mathscr{I}^+$ is the boundary of $\widehat{M}$) and:

(i) There exists a smooth metric $\hat{g}_{ab}$ on $\widehat{M}$ conformally related to g_{ab} on M. (So $\hat{g}_{ab} = \Omega^2 g_{ab}$ with Ω positive on M.)

(ii) The function Ω vanishes on $\mathscr{I}^+$ and $\widehat{\nabla}_a \Omega \neq 0$ on $\mathscr{I}^+$. (Here $\widehat{\nabla}_a$ is the Levi-Cività derivative with respect to $\hat{g}_{ab}$.)

(iii) Every null geodesic acquires a future end-point on $\mathscr{I}^+$.

REMARKS. (a) The symbol $\mathscr{I}$ is called *scri* (for "script I".) (b) The particular function Ω used here is not physically significant. Thus one much check that calculations of physical quantities are independent of the particular choice of Ω. (c) The notion of a null geodesic is conformally invariant (but not its affine parameterization). (d) One could treat the past null asymptotics of the space-time by changing "future" to "past" in (iii); then one writes $\mathscr{I}^-$ for $\mathscr{I}^+$. This would be relevant for studying incoming radiation. There is no essential change, since general relativity is, in the absence of asymmetric boundary conditions, a time-symmetric theory.

We shall also assume that the stress-energy tensor vanishes in a neighborhood of $\mathscr{I}^+$ and that $\mathscr{I}^+$ has only one component. (These conditions could be weakened without difficulty; one really only needs the stress-energy tensor to approach zero at a suitable rate.) In these circumstances, we have

THEOREM (Penrose). $\mathscr{I}^+$ *is a null hypersurface diffeomorphic to* $S^2 \times \mathbf{R}$, *and the "*$\mathbf{R}$*" factors can be taken to be null geodesics.*

In particular, this means that $\mathscr{I}^+$ has the structure of a bundle over S^2. A section of this bundle is called a *cut*. Any cut can be (locally) extended back into M to a null hypersurface meeting $\mathscr{I}^+$ transversely. Such a hypersurface is an "instant of retarded time," and one would like to quantify the momentum and angular momentum of the space-time at such an instant. Then, for example, the amount of momentum radiated between two such instants ought to be the difference in the momenta at the two instants.

Before discussing kinematics on $\mathscr{I}^+$, it is useful to give an account of the structure of this boundary. First, note that $\mathscr{I}^+$ is equipped with the conformal class of a degenerate metric, and that this metric is negative-definite on any cut. In fact, the conformal class passes down to the quotient S^2. (This is a consequence of Einstein's equations, which imply that $\widehat{\nabla}^a \Omega$ is tangent to the null generators of $\mathscr{I}^+$ and shear-free there.) Therefore the base S^2 of the fibration has naturally a conformal structure. In particular, any symmetry of $\mathscr{I}^+$ must induce a Möbius transformation on S^2.

Next, it is always possible to choose the conformal factor Ω so that $\widehat{\nabla}_a \Omega$ is divergence-free at $\mathscr{I}^+$. Once this is done, the remaining freedom in Ω is rather small, and in particular the notion of an affine parametrization of a generator of $\mathscr{I}^+$ is invariant (although not the scale or the origin of

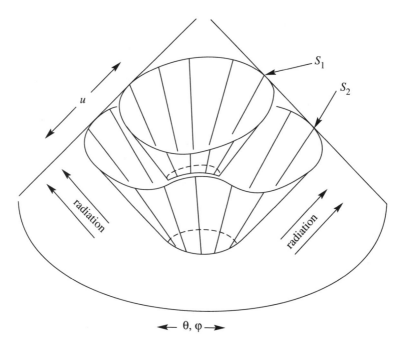

FIGURE 1. A space-time with boundary $\mathscr{I}^+$ at future null infinity. One spatial dimension has been suppressed. The boundary $\mathscr{I}^+$ is represented by a half-cone opening downwards; the interior is the region below this. Null vectors are at forty-five degrees; time increases generally upwards. Two cuts, S_1 and S_2 of $\mathscr{I}^+$ are indicated, as well as the null hypersurfaces extending inwards from them. The difference in the momenta and angular momenta between S_1 and S_2 ought to give the amount carried away by gravitational or other radiation.

the parametrization). Having made such a choice of Ω, it is customary to coordinatize $\mathscr{I}^+ = \{(u, \theta, \phi)\}$ where u is an affine parameter along the generators and (θ, ϕ) are polar coordinates on S^2. We shall assume for simplicity that the affine parameter attains all real values on each generator of $\mathscr{I}^+$. (That is, that the generators of $\mathscr{I}^+$ are infinitely long in a natural sense.) There are also other invariant structures at $\mathscr{I}^+$, which will not be needed explicitly here. Note that a cut of $\mathscr{I}^+$ is given by a function $u = u(\theta, \phi)$.

The group of motions of $\mathscr{I}^+$ preserving its invariant structures is called the *Bondi-Metzner-Sachs (BMS) group*. It is formally similar to the Poincaré group in that it is a semidirect product:

$$(29) \qquad 0 \to \text{Supertranslations} \to \text{BMS} \to \text{Lorentz Group} \to 0.$$

Here "Supertranslations" is an infinite-dimensional Abelian subgroup of motions of the form

$$(30) \qquad (u, \theta, \phi) \mapsto (u + \alpha(\theta, \phi), \theta, \phi)$$

and the quotient Lorentz group can be identified by looking at the Möbius motions induced on the base of the fibration $\mathscr{I}^+ \to S^2$.

REMARKS. (a) A general BMS motion will alter the scale as well as the origin of the affine parametrization of each generator in a generator-dependent way. The details are not needed here. (b) It is clear that Supertranslations acts simply transitively on the space of all cuts, so no cut is preferred above any other. (c) There is a lift of the Lorentz group to BMS for each cut of $\mathscr{I}^+$. Thus if one wants to exploit the analogy between the BMS group and the Poincaré group to define angular momentum, one has to consider the space of all possible cuts as the candidate for the space of origins about which angular momentum might be defined. (d) Supertranslations has a unique four-dimensional subgroup which is normal in BMS, and one can identify this with the subgroup of translations. Thus "asymptotic translations" are well-defined. (e) Combining the two previous observations, we see that every cut S of $\mathscr{I}^+$ determines a copy P_S of the Poincaré group embedded in BMS. This is the group generated by the Lorentz motions fixing S and the Lorentz motions fixing the translates of S.

It might be surprising that one gets an infinite-dimensional asymptotic symmetry group. What is the extra information in Minkowski space which allows one to reduce the BMS group to the Poincaré group? It turns out that in Minkowski space there is a preferred four-dimensional family of cuts of $\mathscr{I}^+$, each of which is the intersection of the future null-cone of a point in **M** with $\mathscr{I}^+$. These cuts are characterized as those which are shear-free. (That is, null hypersurface extending inwards from the cut is shear-free at $\mathscr{I}^+$.) This suggests looking, in a general space-time, for shear-free cuts of $\mathscr{I}^+$. In general, there will be none, but the shear structure of the cuts does play an important role. For our purposes, it is enough to remark that the shear is naturally described by a complex-valued function σ whose argument determines the principal axis of the shear. (This is the same σ as in (27). More properly, σ takes values in a certain line bundle.) This function is essentially the freely specifiable datum for the radiative part of the gravitational field. On the other hand, a cut is uncontorted iff $\sigma = \eth^2 \alpha$ for some *real-valued* function α. This represents a constraint, and so generally cuts of $\mathscr{I}^+$ are contorted.

Kinematics at null infinity

The identification of the kinematic quantities relative to a cut is not straightforward. There is no known Hamiltonian treatment of the system. This should not be surprising, since one expects the evolution of the field from one instant of retarded time to a future one to be irreversible, on domain-of-dependence arguments, and one expects the quantities not to be conserved, since some of the momentum and angular momentum between the two instants will have been carried off by gravitational waves.

OPEN PROBLEM. Can one modify the usual Hamiltonian formalism to treat such systems? (Perhaps semigroup techniques are relevant.)

Originally, Bondi [2] and Sachs [15] (see also Bondi, van der Burg and Metzner [3]) identified the energy of the gravitational field by comparing the asymptotics of the metric at $\mathscr{I}^+$ with those of the Schwarzschild solution. There are good reasons to believe that the Bondi momentum is the correct one. The positivity of the Bondi energy has been established by Ludvigsen and Vickers [10, 11] (using a Witten-type argument) and by Schoen and Yau [18]. (This actually implies positivity of the ADM energy for a wide class of space-times (Ashtekar and Magnon-Ashtekar [1]).) Also the Bondi energy carried by gravitational waves is positive (Bondi [2]).

At present, there are two calculations of the kinematical quantities in this situation which can claim to be based on first principles. The first is the *theory of linkages*, due to Winicour and Tamburino (see Geroch and Winicour [4]). This approach is based on an asymptotic version of Komar's integrals. The kinematic quantities at a particular cut S take values in the Poincaré group P_S. One recovers the Bondi momentum, but the angular momentum one recovers is not satisfactory unless the cut is shear-free. (In particular, nonzero angular momenta are ascribed to shearing cuts of $\mathscr{I}^+$ for Minkowksi space.) It is worthwhile emphasizing that the construction of the linkages is natural and reasonable. It is (at least at first) a surprise that they do not lead to an expression fulfilling our requirements.

The second approach based on first principles is the twistorial one. It turns out that there is no difficulty in taking the two-surface S to be a cut of $\mathscr{I}^+$, and that all the twistor quantities one would like (the kinematic twistor, the norm, the infinity twistor, and the alternating twistor) have unambiguous definitions there. (The factor η is unity there.) One can therefore define twistorially the momentum and angular momentum of the gravitational field at any retarded time. The momentum turns out to be the Bondi momentum. The angular momentum turns out to be different from expressions previously proposed, and more satisfactory. It vanishes identically for arbitrary cuts of $\mathscr{I}^+$ in Minkowski space.

REMARK. Interestingly, one needs positivity of the Bondi energy to define the norm.

So far so good. The next thing one would like is to compare the kinematic quantities at two different instants of retarded time. There is no difficulty in doing this for the momentum, since the Bondi momenta are elements of the same space (Translations)*. For angular momentum, however, the problem is more difficult.

The twistorial construction at $\mathscr{I}^+$ associates to each cut S a twistor space $\mathbf{T}(S)$. Since this twistor space comes equipped with a norm, an infinity twistor and an alternating symbol, one can construct from it the associated

Minkowski space $\mathbf{M}(S)$, which is the space of origins for the twistorially-defined angular momentum at S. In order to compare the angular momenta at S_1 and S_2, then, one must be able to identify $\mathbf{M}(S_1)$ with $\mathbf{M}(S_2)$. At the moment, there is no unambiguous prescription for this in the most general circumstances, but if S_1 and S_2 are both $u = $ constant cuts for the same coordinate u, there is a natural identification (Helfer [6]). In particular, one can compute the total emitted angular momentum by taking the limit that S_1 goes to $u = -\infty$ and S_2 to $u = +\infty$. In appropriate circumstances, one recovers the formula Ashtekar and Streubel [28] (1981) derived from completely different considerations. (These authors defined a symplectic structure on the space of purely radiative degrees of freedom of the gravitational field.) This is encouragement for the twistor program.

The success of the twistor definition of angular momentum at cuts of $\mathscr{I}^+$ is due to the fact that the cuts are no longer taken to be the origins, but the origins are constructed in a different and less direct way. The price paid for this is that it is no longer clear in what sense this angular momentum is conjugate to motions of the space-time. The structure one has is this. For any cut S, we have the Minkowski space $\mathbf{M}(S)$, and we can construct the future null infinity $\mathscr{I}^+(S)$ of $\mathbf{M}(S)$. It turns out that *there is a natural identification of the complexifications of $\mathscr{I}^+(S)$ and $\mathscr{I}^+$ but this identification does not preserve the real slice in general.* In particular, S is embedded in the complexification of $\mathscr{I}^+(S)$. The angular momentum at S is conjugate to genuine Poincaré motions in $\mathbf{M}(S)$, and these latter can be identified with certain elements of the BMS group of $\mathscr{I}^+(S)$, and these will be identified with certain elements of the *complexified* BMS group of $\mathscr{I}^+$. To summarize: the Poincaré group to which the twistorial kinematic quantities are conjugate is embedded in the complexification of the BMS group.

At the moment, the question of what these complex BMS motions mean (in other than a formal sense) is completely open. I know of no examples in classical Hamiltonian mechanics where one is naturally led to consider complexified Hamiltonian vector fields. Indeed, the concept is so new that one must be cautious about whether there will be any deep interpretation of it, or whether it will just be a useful formal device. But the evidence that there is something deep and as-yet not understood is this: that we anticipated that any discussion of the kinematics of gravity may lead to a radical reconsideration of the foundations of kinematics; that in the regime we are here considering, the system is not classically a Hamiltonian one, since the temporal evolution is irreversible and kinematic quantities are not conserved; that the twistor quasilocal construction has a remarkable number of successes, and we should consider seriously anything it suggests to us.

If we do take the twistorial angular momentum at $\mathscr{I}^+$ seriously, and we believe that it is the manifestation of a deeper structure not yet understood, then it seems that $\mathscr{I}^+$ is the edge of the regime in which the usual Hamiltonian interpretation of the kinematic quantities will be sufficient. The results on the nonexistence of an infinity twistor for S in the interior of a space-time then appear less to invalidate the program of twistorial kinematics than to indicate a limitation of the usual Hamiltonian understanding of momentum and angular momentum. At the moment, though, this is all speculative.

REMARKS. (a) The introduction of complex origins is closely related to Newman's *theory of heaven* (Ko et al. [8]). (b) The complexification of the motions is entirely in the u direction. (c) One might wonder to what extent the complexification is natural, since usually complex quantities on manifolds not endowed with complex or almost-complex structures are artificial conveniences. Although this question is not completely understood, complex structure enters more deeply into four-dimensional space-times than elsewhere. This is because the Lie algebra $so(1, 3) \approx sl(2, \mathbf{C})$ admits a natural complex structure; physically, this comes about because there is a natural identification between the axes of rotations and the directions in which one might boost.

OPEN QUESTION. Let the cut S in $\mathscr{I}^+(S)$ be represented by a complex function $u(\theta, \phi)$. Can one say anything about the sign of the imaginary part of this?

Further open questions

I summarize here some further questions which were not included in the exposition above.

Given the negative result on the existence of an infinity twistor for $\mathbf{T}(S)$ in general, one might look for more modest structures which still convey some kinematic interest. One can always construct from $\mathbf{T}(S)$ an associated complexified compactified Minkowski space $\mathbf{CM}^{\#}(S)$ (by taking the Grassmannian of two-planes in $\mathbf{T}(S)$). If the norm is known on $\mathbf{T}(S)$ (for example, if S is uncontorted), then one can invariantly distinguish a real compactified Minkowski space $\mathbf{M}^{\#}(S)$. As noted above, the kinematic twistor defines a quadric surface in $\mathbf{T}(S)$ (taken projectively), and there will be two families of lines on this quadric. Each of the lines represents a point in $\mathbf{CM}^{\#}(S)$, and in fact each of the families represents a conic curve in $\mathbf{CM}^{\#}(S)$. One would like to identify one of these as the complex world-line of the system. A little projective geometry shows that distinguishing one of the conics is equivalent to choosing a square root of $\det A$.

OPEN QUESTION. For an uncontorted surface in a space-time satisfying the dominant energy condition, is the real part of $\det A$ nonnegative?

REMARKS. (a) This is not even known for surfaces in conformally flat space-time. (b) If an infinity twistor with the usual properties exists, then $\det A = 4m^4$ is necessarily real and nonnegative.

This suggests the more general

OPEN QUESTION. What can one say about the invariants of $A_{\alpha\beta}$?

REMARK. A technical difficulty in attacking this problem is that the twistor group $SU(2, 2)$ is *pseudo*-unitary, and the extraction of invariants with respect to it is not quite so straightforward as for unitary groups.

Previous attempts to define the angular momentum at spatial infinity were successful only when the Weyl tensor was "asymptotically electric." In this case, the twistorial approach reproduces those definitions.

OPEN QUESTION. Can the twistor approach provide a satisfactory treatment of angular momentum at spatial infinity even if the Weyl tensor is not asymptotically electric?

REMARKS. (a) It is conceivable that such a treatment might help us to understand the structure of spatial infinity better, and this is an important problem. (b) It is also possible that this might help one attack the problem of what the complex BMS motions arising in the treatment of angular momentum at null infinity represent.

Also on the topic of angular momentum:

OPEN QUESTION. Is there a spin-down theorem for angular momentum at $\mathscr{I}^+$? That is, can one show that the total angular momentum as measured from $\mathscr{I}^+$ decreases with time?

REMARKS. (a) The analogous result for energy was established by Bondi in 1960. (b) It is not clear even what the appropriate measure of angular momentum should be here. One possibility is the spin.

One wants to know that the twistorial definitions of the kinematic quantities are robust in that they do not depend sensitively on the surface S or the metric of the space-time. (The point is that neither of these can be measured with absolute accuracy in practice.) It seems likely that something along these lines will be correct, given the ellipticity of the two-surface twistor equations. One would like to make this conjecture precise and to verify it.

REFERENCES

1. A. Ashtekar and A. Magnon-Ashtekar, *Energy-momentum in general relativity*, Phys. Rev. Lett. **43** (1979), 181–184.
2. H. Bondi, *Gravitational waves in general relativity*, Nature **186** (1960), 535.
3. H. Bondi, M. G. J. van der Burg, and A. W. K. Metzner, *Gravitational waves in general relativity*. VII, *Waves from axi-symmetric isolated systems*, Proc. Roy. Soc. Lond. Ser. A **269** (1962), 21–52.
4. R. Geroch and J. Winicour, *Linkages in general relativity*, J. Math. Phys. **22** (1981), 803–812.
5. A. D. Helfer, *Difficulties with quasilocal momentum and angular momentum* Classical Quantum Gravity **9** (1992) 1001–1008.
6. ____, *The angular momentum of gravitational radiation*, Phys. Lett. A. **150** (1990) 342–344.
7. R. M. Kelly, K. P. Tod, and N. M. J. Woodhouse, *Quasi-local mass for small surfaces*, Classical Quantum Gravity **3** (1986), 1151–1167.
8. M. Ko, M. Ludvigsen, E. T. Newman, and K. P. Tod, *The theory of $\mathscr{H}$-space*, Phys. Rep. **71** (1981), 51–139.
9. A. B. Komar, *Covariant conservation laws in general relativity*, Phys. Rev. **113** (1959), 934–936.
10. M. Ludvigsen and J. A. G. Vickers, *The positivity of the Bondi mass*, J. Phys. A **14** (1981), L389–391.

11. ___, *A simple proof of the positivity of the Bondi mass*, J. Phys. A **151** (1982), L67–70.

12. R. Penrose, *Quasi-local mass and angular momentum in general relativity*, Proc. Roy. Soc. Lond. Ser. A **381** (1982), 53–63.

13. ___, *New improved quasi-local mass and the Schwarzschild solution*, Twistor News. **18** (1984), 7–11.

14. R. Penrose and W. Rindler, *Spinors and space-time*, Cambridge Univ. Press, 1984/86.

15. R. K. Sachs, *Gravitational waves in general relativity. VIII, Waves in asymptotically flat space-time*, Proc. Roy. Soc. Lond. Ser. A **270** (1962), 103–126.

16. R. Schoen and S. T. Yau, *Positivity of the total mass of a general space-time*, Phys. Rev. Lett. **43** (1979), 1457–1459.

17. ___, *On the proof of the positive mass conjecture in general relativity*, Comm. Math. Phys. **65** (1979), 45–76.

18. ___, *Proof that the Bondi mass is positive*, Phys. Rev. Lett. **48** (1982), 369–371.

19. W. T. Shaw, *Twistor theory and the energy-momentum and angular momentum of the gravitational field at spatial infinity*, Proc. Roy. Soc. Lond. Ser. A **390** (1983), 191–215.

20. ___, *The asymptopia of quasi-local mass and momentum*: I. *General formalism and stationary space-times*, Classical Quantum Gravity **3** (1986), 1069–1104.

21. ___, *Quasi-local mass for "large" spheres*, Contemp. Math. (J. A. Isenberg, ed.), Amer. Math. Soc., Providence, RI, 1988, pp. 15–22.

22. K. P. Tod, *Some examples of Penrose's quasi-local mass construction*, Proc. Roy. Soc. Lond. Ser. A **388** (1983), 457– 477.

23. ___, *More on quasi-local mass*, Twistor News. **18** (1984), 3–6.

24. ___, *More on Penrose's quasi-local mass*, Classical Quantum Gravity **3** (1986), 1169–1189.

25. J. Winicour, *Angular momentum in general relativity*, General Relativity and Gravitation: One Hundred Years After the Birth of Albert Einstein (A. Held, ed.), Plenum, New York, 1980, pp. 71–96.

26. J. Winicour and L. Tamburino, *Lorentz-covariant gravitational energy-momentum linkages*, Phys. Rev. Lett. **15** (1965), 601–605.

27. E. Witten, *A new proof of the positive energy theorem*, Comm. Math. Phys. **80** (1981), 381–402.

28. A. Ashtekar and M. Streubel, *Symplectic geometry of radiative modes and conserved quantities at null infinity*, Proc. Roy. Soc. Lond. Ser A **376** (1981) 585–607.

UNIVERSITY OF MISSOURI

Proceedings of Symposia in Pure Mathematics
Volume **54** (1993), Part 2

Support Theorems in Integral Geometry
and Their Applications

SIGURDUR HELGASON

1. Introduction

The well-known paper [9] of Radon suggests, already by its title, the following general inversion problem in integral geometry. Let X be a manifold, Ξ a family of certain submanifolds $\xi \subset X$, each ξ carrying a specific measure dm.

A. Inversion problem. Determine a function f on a manifold X on the basis of its integrals

$$(1) \qquad \hat{f}(\xi) = \int_\xi f(x)\, dm(x), \qquad \xi \in \Xi.$$

The mapping $f \longrightarrow \hat{f}$ is often referred to as the *Radon transform*.

With this problem solved in a number of cases (e.g., k-planes in $\mathbf{R}^n$, totally geodesic submanifolds in a two-point homogeneous space, horocycles in a symmetric space) it is natural to pose additional problems:

B. Range problem. Determine $C_c^\infty(X)^\wedge$ as a function space on Ξ. Similar problems for other naturally defined function spaces on X.

A more restricted problem is the following:

C. Support problem. Show that the conditions:

   (a) f "small" at ∞,
   (b) $\hat{f}$ of compact support,

imply

   (c) f has compact support.

1991 *Mathematics Subject Classification.* Primary 44A12; Secondary 53C35, 53C65.
Supported by National Science Foundation Grant DMS–8805665.
This paper is in final form and no version of it will be submitted for publication elsewhere.

In case X is a Riemannian manifold one would naturally sharpen this problem as follows:

Let $B \subset X$ be a ball, f a function on X, "small" at ∞. Assuming

$$(2) \qquad\qquad \hat{f}(\xi) = 0 \qquad \text{for} \quad \xi \cap B = \varnothing$$

prove that

$$(3) \qquad\qquad f(x) = 0 \qquad \text{for} \quad x \notin B.$$

Roughly speaking, the support problem is a part of the range problem. Letting B shrink to a point in (2), (3) we see that a positive solution to Problem C implies that $f \longrightarrow \hat{f}$ is injective so Problem A is at least implicitly solved. However, a solution to Problem A, even in the form of an explicit formula

$$(4) \qquad\qquad f = I(\hat{f}),$$

does not imply a solution to Problem C. In fact, determining $f(x)$ for x outside a ball B by means of (4) will normally require knowledge of $\hat{f}(\xi)$ for all ξ through x including those ξ which intersect B. Thus the implication $(2) \Longrightarrow (3)$ would not follow from (4). In view of this it is not surprising that support theorems have had applications which could not have been obtained by means of the inversion formula. In this note we concentrate on solutions to Problem C and give an indication of some applications. Note that while the assumptions and the proofs of the theorems below are all different, the conclusion is always the same!

2. Totally geodesic submanifolds in constant curvature spaces

Let X be a simply connected Riemannian manifold of constant curvature and dimension n and Ξ the set of k-dimensional totally geodesic submanifolds of X. It is well known that a function $f \in C_c^\infty(X)$ (f assumed even in the case $X = S^n$) is explicitly determined by $\hat{f}(\xi)$ ($\xi \in \Xi$). However, this does not imply a support theorem as pointed out above.

(a) Let $X = \mathbf{R}^n$ so Ξ is the set of k-planes in $\mathbf{R}^n$ and dm the Euclidean measure. Here we have the following result [5c].

THEOREM 2.1. *Let f be continuous on $\mathbf{R}^n$ and*

$$(5) \qquad\qquad f(x) = O(|x|^{-m}) \quad \text{for each} \ \ m > 0.$$

Suppose $C \subset \mathbf{R}^n$ is a convex compact set. If

$$\hat{f}(\xi) = 0 \qquad \text{for } \xi \cap C = \varnothing$$

then

$$f(x) = 0 \qquad \text{for } x \notin C.$$

REMARKS. (i) The theorem fails to hold unless condition (5) holds for *each* $m > 0$.

(ii) For $n = 3$, $k = 1$ the theorem has the following consequence for tomography. Let $H \subset \mathbf{R}^3$ be a body (for example, a part of the human body) and let $f(x)$ denote its density at a point x. For a line ξ intersecting H consider a thin beam of X-rays along ξ. Let I_o and I, respectively, denote the intensity of the beam before entering H and after leaving H. From principles of physics one has (Cormack [3])

$$(6) \qquad \log\left(\frac{I_o}{I}\right) = \int_\xi f(x)\,dm(x) = \hat{f}(\xi).$$

Since the left-hand side is determined by the X-ray picture, the tomographic reconstruction of f amounts to the known inversion of the transform $f \longrightarrow \hat{f}$. If $H_o \subset H$ is a convex subset (for example, the heart) it may be of interest to determine the density f outside H_o using only X-rays ξ which do not intersect H_o. From the support theorem above we see that this is possible, at least theoretically.

A substantial generalization of Theorem 2.1, dropping the requirement that dm should be the Euclidean measure, was recently proved by Boman [1]. He used microlocal analysis methods developed by Boman and Quinto [2] for the case when f is assumed of compact support. For the so-called exponential X-ray transform in $\mathbf{R}^2$, more explicit results are proved in [6].

(b) Let X be the hyperbolic space $\mathbf{H}^n$ and Ξ the family of k-dimensional totally geodesic submanifolds ξ of $\mathbf{H}^n$ $(1 \le k \le n - 1)$. Let d denote the distance function on X and $o \in X$ some fixed origin, say 0 in the model of $\mathbf{H}^n$, which is the unit ball $|x| < 1$ in $\mathbf{R}^n$. We then have [5b, 5f].

THEOREM 2.2. *Let f be continuous on $\mathbf{H}^n$ and*

$$(7) \qquad f(x) = O(e^{-md(o,x)}) \quad \text{for each } m > 0.$$

Suppose $B \subset \mathbf{H}^n$ is a ball. If

$$\hat{f}(\xi) = 0 \qquad \text{for } \xi \cap B = \varnothing$$

then

$$f(x) = 0 \qquad \text{for } x \notin B.$$

REMARKS. (i) Condition (7) is a natural replacement for (5) since the volume element in $\mathbf{H}^n$ increases exponentially with the distance from 0.

(ii) The proof of Theorem 2.1 uses the vector space structure of $\mathbf{R}^n$ and thus does not extend to $\mathbf{H}^n$. The proof of Theorem 2.2 relies heavily on the fact that in the indicated model of $\mathbf{H}^n$, the spheres are Euclidean spheres. For this reason Theorem 2.2 has not been extended from $\mathbf{H}^n$ to the other noncompact two-point homogeneous spaces in spite of the fact that there an inversion formula for $f \longrightarrow \hat{f}$ is known [5e, III, §3].

The case $k = 1$ in which the ξ are the geodesics, is however an exception and then X can even by more general than two-point homogeneous [5e, III, §4].

THEOREM 2.3. *Let X be a symmetric space of the noncompact type, B a ball in X. Let f be continuous on X satisfying (7). Suppose*

$$\hat{f}(\xi) = 0 \qquad \text{for each geodesic } \xi \text{ outside } B.$$

Then

$$f(x) = 0 \qquad \text{for } x \notin B.$$

REMARK. If rank $X > 1$ condition (7) on f can be replaced by (5) with $|x|$ replaced by $d(o, x)$.

(c) The case $X = \mathbf{S}^n$ is a bit different from the cases $X = \mathbf{R}^n$ and $X = \mathbf{H}^n$ in that the inversion formula for f [5a, 5g] can only hold for f even $(f(-x) = f(x))$. With this assumed, the expected support theorem holds for the integration over great circles ξ.

THEOREM 2.4. *Let $B \subset \mathbf{S}^2$ be a closed spherical cap centered at the north pole o, and f an even C^∞ function on $\mathbf{S}^2$. Assume (ξ being a great circle)*

$$\hat{f}(\xi) = \int_\xi f(s)\, dm(s) = 0 \quad \text{for } \xi \cap B = \varnothing.$$

Then, if $f = 0$ on the equator along with all its derivatives, we have

$$f(s) = 0 \quad \text{for } s \notin B \cup B',$$

B' being the spherical cap antipodal to B.

REMARKS. In the case when f is *assumed* to vanish in some belt around the equator, Theorem 2.4 is contained in Proposition 4.1 in Quinto [8]. An example devised by P. Goodey and W. Weil shows that the assumption $f = 0$ cannot be dropped. This and the evenness of f show that the odd-order derivatives in $\theta = d(o, s)$ must vanish on the equator. As proved by A. Kurusa (to appear) this condition is needed for the even order derivatives too.

3. Horocycles and horocycle planes in a symmetric space

(a) Let $X = G/K$ be a symmetric space of the noncompact type. With an Iwasawa decomposition $G = KAN$ the horocycles ξ in X are the orbits of groups conjugate to N. Using the Fourier transform on X and the theory of generalized spherical functions (Eisenstein integrals) the following support theorem is proved in [5d].

THEOREM 3.1. *Let $B \subset X$ be a closed ball and $f \in C_c^\infty(X)$. If*

$$\hat{f}(\xi) = 0 \qquad \text{for } \xi \cap B = \varnothing$$

then

$$f(x) = 0 \qquad \text{for } x \notin B.$$

The assumption $f \in C_c^\infty(X)$ can be weakened; the proof works if one just assumes f in the L^2 Schwartz space in the sense of Eguchi [4].

The theorem can be used to prove that each G-invariant differential operator D on X is globally solvable: given $f \in C^\infty(X)$ there exists a $u \in C^\infty(X)$ such that $Du = f$. To indicate the method let $o = \{K\}$, the origin in X, and ξ_o the horocycle $N \cdot o$. Then

(8) $$\hat{f}(ka \cdot \xi_o) = \int_N f(kan \cdot o)\, dn, \qquad k \in K, a \in A,$$

where dn is a Haar measure on N. Also

(9) $$(Df)^\wedge = \Gamma(D)\hat{f},$$

where $\Gamma(D)$ is a constant coefficient differential operator on A. The surjectivity

$$DC^\infty(X) = C^\infty(X)$$

follows by combining general functional analysis methods with the following key lemma in whose proof Theorem 3.1 is the main ingredient.

LEMMA 3.2. *The following implication holds,* supp *denoting support*:

$$f \in C_c^\infty(X), \qquad \operatorname{supp}(Df) \subset B \implies \operatorname{supp}(f) \subset B.$$

In fact (9) implies $(\Gamma(D)\hat{f})(\xi) = 0$ for $\xi \cap B = \varnothing$. Taking B with center o and radius R this means

(10) $$\Gamma(D)_a(\hat{f}(ka \cdot \xi_o)) = 0 \quad \text{for } d(o, a \cdot o) > R.$$

By a special case of the Lions-Titchmarsh convexity theorem, (10) implies $\hat{f}(ka \cdot \xi_o) = 0$ for $d(o, a \cdot o) > R$ so by Theorem 3.1, $\operatorname{supp}(f) \subset B$ as desired.

(b) With the notation for (a) let $\mathfrak{g}$ and $\mathfrak{k}$ denote the Lie algebras of G and K and $\mathfrak{g} = \mathfrak{k} + \mathfrak{p}$ the corresponding Cartan decomposition. Then $\mathfrak{a}$, the Lie algebra of A, is contained in $\mathfrak{p}$ and we write $\mathfrak{p} = \mathfrak{a} + \mathfrak{q}$ (orthogonal decomposition). Then $\mathfrak{q}$ is naturally identified with the tangent space to the horocycle $N \cdot o$ at o. Let G_o denote the group of affine transformations of $\mathfrak{p}$ generated by the translations and the isotropy action of K. A *horocycle plane* in $\mathfrak{p}$ is by definition the image $\xi = g \cdot \mathfrak{q}$ of $\mathfrak{q}$ under some $g \in G_o$. Let Ξ_o denote the manifold of all horocycle planes. The following result from [5h] is a sharper version of Theorem 2.1 in that the vanishing condition (11) is only required for a lower-dimensional family of planes.

THEOREM 3.3. *Let f be a rapidly decreasing function on $\mathfrak{p}$ and $B \subset \mathfrak{p}$ a closed ball. Assume*

(11) $$\int_\xi f(x)\, dm(x) = 0$$

for each horocycle plane ξ, $\xi \cap B = \varnothing$. Then

$$f(x) = 0 \qquad \text{for } x \notin B.$$

The proof is based on the generalized Bessel transform on $\mathfrak{p}$. To indicate the tools needed consider the following case where we write the Fourier transform

$$\tilde{f}(\lambda\omega) = \int_{\mathbf{R}^n} f(x)e^{-i\lambda(x,\omega)}\,dx, \qquad |\omega| = 1, \lambda \in \mathbf{R},$$

on $\mathbf{R}^n$ in polar coordinates. Then the mapping $f \longrightarrow \tilde{f}$ maps $C_c^\infty(\mathbf{R}^n)$ onto the set of functions $\tilde{f}(\lambda\omega) = \phi(\lambda, \omega) \in C^\infty(\mathbf{R} \times \mathbf{S}^{n-1})$ satisfying:

(i) $\lambda \longrightarrow \phi(\lambda, \omega)$ is holomorphic and for some $A > 0$,

$$\sup_{\lambda,\omega} \big| \phi(\lambda, \omega)(1 + |\lambda|^N)e^{-A|\operatorname{Im}\lambda|} \big| < \infty$$

for all $N \in \mathbf{Z}^+$.

(ii) For each $k \in \mathbf{Z}^+$ and each isotropic vector $a \in \mathbf{C}^n$, the function

(12) $$\lambda \longrightarrow \lambda^{-k} \int_{\mathbf{S}^{n-1}} \phi(\lambda, \omega)(a, \omega)^k\,d\omega$$

is even and holomorphic on $\mathbf{C}$.

For Theorem 3.3 one requires an extension of this result to the polar coordinate $(k, H) \longrightarrow k \cdot H$ $(k \in K, H \in \mathfrak{a})$ decomposition of $\mathfrak{p}$. The generalization of condition (ii) and (12) then amounts to a description of the matrix-valued function

$$H \longrightarrow \int_K \tilde{f}(k \cdot H)\delta(k)\,dk, \qquad H \in \mathfrak{a},$$

δ being any irreducible representation of K : there is a matrix-valued function $J^\delta(H)$ whose entries are certain explicit polynomials such that

$$H \longrightarrow J^\delta(H)^{-1} \int_K \tilde{f}(k \cdot H)\delta(k)\,dk$$

extends to a holomorphic Weyl group invariant function on $\mathfrak{a} + i\mathfrak{a}$.

REMARK. For the case $\dim \mathfrak{a} = 1$ and dm in (11) replaced by any rotation-invariant measure, Theorem 3.3 has been proved by Orloff [7], in an entirely different fashion.

BIBLIOGRAPHY

1. J. Boman, *Helgason's support theorem for Radon transforms—a new proof and a generalization,* Mathematical Methods in Tomography. Proc. Oberwolfach Eds. G. P. Herman, A. K. Louis and F. Natterer, Springer Lecture Notes, 1947, New York, 1991.
2. J. Boman and E.T. Quinto, *Support theorems for real-analytic Radon transforms,* Duke Math. J. **55** (1987), 943–948.
3. A.M. Cormack, *Representation of a function by its line integrals, with some radiological applications,* I. J. Appl. Phys. **34** (1963), 2722–2727.
4. M. Eguchi, *Asymptotic expansions of Eisenstein integrals and Fourier transform on symmetric spaces,* J. Funct. Anal. **34** (1979), 167–216.
5a. S. Helgason, *Differential operators on homogeneous spaces,* Acta Math. **102** (1959), 239–299.

5b. ———, *A duality in integral geometry; some generalizations of the Radon transform*, Bull. Amer. Math. Soc. **70** (1964), 435–446.

5c. ———, *The Radon transform on Euclidean spaces, compact two-point homogeneous spaces and Grassmann manifolds*, Acta Math. **113** (1965), 153–180.

5d. ———, *The surjectivity of invariant differential operators on symmetric spaces*, Ann. of Math. (2) **98** (1973), 451–480.

5e. ———, *The Radon transform*, Birkhäuser, Boston, 1980.

5f. ———, *Support of Radon transforms*, Adv. in Math. **38** (1980), 91–100.

5g. ———, *The totally geodesic Radon transform on a constant curvature space*, Contemp. Math. **113** (1990), 141–149.

5h. ———, *The flat horocycle transform on a symmetric space*, Adv. in Math. **91** (1992), 232–251.

6. P. A. Kuchment and S. Y. L'vin, *Paley-Wiener theorem for exponential Radon transform*, Acta Appl. Math. **18** (1990), 251–260.

7. J. Orloff, *Invariant Radon transform for a symmetric space*, Trans. Amer. Math. Soc. **318** (1990), 581–600.

8. E. T. Quinto, *The invertibility of rotation-invariant Radon transforms*, J. Math. Anal. Appl. **91** (1983), 510–522.

9. J. Radon, *Über die Bestimmung von Funktionen durch ihre Integralwerte längs gewisser Mannigfaltigkeiten*, Ber. Verh. Sächs. Akad. Wiss. Leipzig **69** (1917), 262–277.

MASSACHUSETTS INSTITUTE OF TECHOLOGY

Proceedings of Symposia in Pure Mathematics
Volume **54** (1993), Part 2

Killing Spinors and Eigenvalues of the Dirac Operator

OUSSAMA HIJAZI

0. Introduction

In this paper, we give an overview on recent developments of the eigenvalues estimate problem for the Dirac operator on compact manifolds. The main ingredients are the Lichnerowicz formula, the modification of the Levi-Civita connection, and the conformal invariance of the Dirac operator. We refer to the original papers for the proofs.

1. Killing spinors

Let (M^n, g) be a complete Riemannian spin manifold of dimension $n \geq 2$. Denote by ∇ the Levi-Civita connection acting on sections of the spinor bundle $\Gamma(\Sigma M)$. In general relativity and supergravity theories, physicists constructed manifolds on which there are spinor fields ψ which are solutions of the differential equation

$$\forall X \in \Gamma(\Sigma M), \qquad \nabla_X \psi = \lambda X \cdot \psi,$$

where λ is a complex number and "$\cdot$" denotes Clifford multiplication. These solutions are called Killing spinors since the associated tangent vector fields are Killing fields or Killing conformal fields, depending on whether λ is real or purely imaginary.

In a different context, real Killing spinors appeared in the work of T. Friedrich [**Fr1**] as eigenspinor fields associated with the smallest eigenvalues of the Dirac operator. For our purpose, we allow the number λ to be any function, more precisely, take f to be any complex-valued function on M, define the modified connection on the spin bundle by

$$\nabla_X^f \psi := \nabla_X \psi + \frac{f}{n} X \cdot \psi.$$

DEFINITION 1. A Killing spinor is a spinor field $\psi \not\equiv 0$ such that for a complex valued function $f \neq 0$, $\nabla^f \psi \equiv 0$.

1991 *Mathematics Subject Classification.* Primary 58G25, 53C25, 53A50.
This paper is in final form and no version of it will be submitted for publication elsewhere.

PROPOSITION 2 [**Lic3**]. *A Killing spinor is either real (i.e., f real) or imaginary (i.e., f purely imaginary).*

2. Real Killing spinors

In this paragraph (M^n, g) is supposed to be a complete Riemannian spin manifold of dimension $n \geq 2$.

THEOREM 3 [**Hi2, Hi3**]. *If (M^n, g) admits a real Killing spinor, then*

i) (M^n, g) *is Einstein, compact with*

$$f^2 = \frac{n}{4(n-1)} S(:= \lambda_1^2),$$

where S is the scalar curvature of (M^n, g);

ii) *in dimensions* 4 *and* 8, *the manifold* (M^n, g) *is isometric to the standard sphere.*

LEMMA 4 [**Hi1**]. *Let (M^n, g) be complete with nontrivial real Killing spinor ψ; then any harmonic k-form $(k \neq 0, n)$ kills ψ.*

COROLLARY 5 [**Hi1**]. *In even dimensions, the existence of a real Killing spinor implies that the manifold is non-Kähler.*

COROLLARY 6 [**Li2**]. *The existence of a real Killing spinor on a complete manifold implies that there are no nontrivial parallel forms.*

Concerning real Killing spinors in low dimensions, there are classification results where real Killing spinors are in one-to-one correspondence with contact structures in dimension 5 and 7 (see [**FK1**] and [**FK2**]), with nonintegrable complex structure in dimension 6 (see [**Gr**]).

3. Eigenvalues estimate

Assume that (M^n, g) is a compact Riemannian spin manifold of dimension $n \geq 3$. We have the following fundamental result:

THEOREM 7 [**Lic1**]. *If the scalar curvature S is positive, then*

i) $\ker D = 0$, *where D is the Dirac operator, and*
ii) *any eigenvalue λ of D satisfies*

$$\lambda^2 > \frac{1}{4} \inf_M S.$$

PROOF. Consider the Lichnerowicz formula applied to a spinor field ψ and take its scalar product with ψ, to get

$$\int_M |D\psi|^2 v_g = \int_M |\nabla\psi|^2 v_g + \frac{1}{4} \int_M S|\psi|^2 v_g.$$

This equation implies Theorem 7.

Part i) of Theorem 7 gives topological obstructions for the existence on the manifold M of metrics with positive scalar curvature. More precisely, for $n = 4k$, the positivity of the scalar curvature implies that the $\hat{A}$-genus, which is equal to the index of the Dirac operator, should be zero. This result has been extended by Hitchin later on for any dimension (see [**Hit, GL**]). For the second part of Theorem 7, we have

THEOREM 8 [Fr1]. *Any eigenvalue λ of the Dirac operator satisfies*

$$(1) \qquad \lambda^2 \geq \frac{n}{4(n-1)} \inf_M S.$$

Equality in (1) *implies the existence of a real Killing spinor.*

The key point is to use the modified connection ∇^f in the Lichnerowicz formula instead of the Levi-Civita connection ∇.

THEOREM 9 [Hi2]. *Any eigenvalue λ of the Dirac operator on a compact Riemannian spin manifold of dimension $n \geq 3$, satisfies*

$$(2) \qquad \lambda^2 \geq \frac{n}{4(n-1)} \sup_{h>0} \inf_M \left[4\frac{n-1}{n-2} h^{-1}\Delta h + S \right] = \frac{n}{4(n-1)}\mu_1(g),$$

where $\mu_1(g)$ is the first eigenvalue of the conformal scalar Laplacian, i.e.,

$$4\frac{n-1}{n-2}\Delta + S.$$

Moreover, equality in (2) *implies the existence of a real Killing spinor.*

IDEA OF THE PROOF. Consider a conformal change of the metric $\bar{g} = e^{2u}g$, for some real function u on M. Use the isometry G_u between the spin bundles ΣM and $\Sigma \overline{M}$ in order to relate the associated Dirac operators (see [**Hit**])

$$\overline{D}(e^{-(n-1)u/2}\bar{\psi}) = e^{-(n+1)u/2}\overline{D\psi}.$$

Take ψ so that $D\psi = \lambda\psi$ and observe that $\overline{D}\bar{\varphi} = f\bar{\varphi}$ where

$$f = \lambda e^{-u} \quad \text{and} \quad \bar{\varphi} = e^{-(n-1)u/2}\bar{\psi}.$$

The Lichnerowicz formula applied to $\overline{\nabla}^f$ and $\bar{\varphi}$ gives

$$\int_M |\overline{\nabla}^f\bar{\varphi}|^2 v_{\bar{g}} + \int_M \left(\frac{\overline{S}}{4} - \frac{n-1}{n}f^2 \right) |\bar{\varphi}|^2 v_{\bar{g}} = 0.$$

Choose the metric $g_1 = e^{2u_1}g$ where u_1 is an eigenfunction associated with $\mu_1(g)$ to get the result.

THEOREM 10 [Ki1, Ki2, Ki3]. *Let (M^n, g) be a compact Kähler spin manifold of dimension $n \geq 2$. Then*

i) *Any eigenvalue λ of the Dirac operator satisfies*

$$(3) \qquad \lambda^2 \geq \frac{n+2}{4n}\inf_M S, \quad \text{if } n = 4k+2.$$

Equality in (3) *implies that the manifold is Einstein. In dimension* 6, M^6 *should be isometric to* $P^3(C)$ *or to the flag manifold* $F(1, 2)$.

ii) *Any eigenvalue* λ *of the Dirac operator satisfies*

(4) $$\lambda^2 \geq \frac{n}{4(n-2)} \inf_M S, \quad \text{if } n = 4k.$$

Equality in (4) *implies that the manifold is non-Einstein (except for* $n = 4$) *with constant scalar curvature.*

4. Imaginary Killing spinors

There is a complete classification by H. Baum [**Ba2**] of complete Riemannian spin manifolds of dimension $n \geq 2$, admitting imaginary Killing spinors (see also [**Li5**] and [**Ra**]). These manifolds are the hyperbolic space or the warped product of the real line with an $(n-1)$-dimensional manifold having a parallel spinor.

5. Twistor-spinors

The twistor operator plays an important role in the study of Killing spinors and in the eigenvalues estimate for the Dirac operator [**Li4**]. It should be pointed out that R. Penrose used this operator in his twistor program in a different context. So we will define this operator and will give some results obtained in this direction. We refer to A. Lichnerowicz [**Li4, Li5**] and to [**Fr3**] for recent results.

DEFINITION 11 [**Li4**]. The twistor operator P is defined, for all vector fields X and spinor fields ψ by

$$P_X \psi := \nabla_X \psi + \frac{1}{n} X \cdot D\psi,$$

where D is the Dirac operator.

DEFINITION 12. A spinor field ψ is called a twistor-spinor if and only if $P\psi \equiv 0$.

THEOREM 13 [**Li4**]. *Let* (M^n, g) *be a Riemannian spin manifold of dimension* $n \geq 3$. *We have* $\dim \ker P$ *is a conformal invariant, and if* M *is compact, then*

i) $\ker P = 0$, *if* $\mu_1(g) < 0$, *and*

ii) *up to a conformal deformation of the metric, a twistor-spinor is a real Killing spinor or the sum of two real Killing spinors, if* $\mu_1(g) > 0$.

DEFINITION 14. For any real function f, any tangent vector field X, and any spinor field ψ, define the operator L by

$$L_X \psi := \nabla_X D\psi + \frac{f}{n} X \cdot \psi.$$

THEOREM 15 [**HL**]. *If* (M^n, g) *is a compact spin manifold of dimension* $n \geq 3$, *with* $\mu_1(g) \geq 0$ *which is not conformally isometric to the sphere, then* *the existence of a spinor field ψ lying in $\ker(L)$, implies that the function f is constant $(= \lambda_1)$ and ψ is a twistor-spinor, the sum of two real Killing spinors.*

REFERENCES

[Ba1] H. Baum, *Spin-strukturen and Dirac-operator über pseudo-riemannachen Mannigfaltig-keiten*, Teubner-Texte zur Mathematik Band 41, Teubner-Verlag, Leipzig, 1981.

[Ba2] ____, *Complete Riemannian manifolds with imaginary Killing spinors*, Ann. Global Anal. Geom. **7**, (1989) 205–226.

[Fr1] Th. Friedrich, *Der erste Eigenwert des Dirac-operators einer Kompacten Riemannschen Mannigfaltigkeit nichtnegativer skalar krümmung*, Math. Nach. **97**, (1980) 117–146.

[Fr2] ____, *A remark on the first eigenvalue of the Dirac operator on 4-dimensional manifolds*, Math. Nach. **102**, (1981) 53–56.

[Fr3] ____, *On the conformal relation between twistors and Killing spinors*, Humboldt-Universität zu Berlin, Preprint No. 209.

[FK1] Th. Friedrich and I. Kath, *Einstein manifolds of dimension five with small eigenvalues of the Dirac operator*, J. Differential Geom. **29**, (1989) 263–279.

[FK2] ____, *7-dimensional compact Riemannian manifolds with Killing spinors*, Humboldt-Universität zu Berlin, Preprint No. 200.

[GL] M. Gromov and H. B. Lawson, *Positive scalar curvature and the Dirac operator on complete Riemannian manifolds*, Inst. Hautes Études Sci. Publ. Math. **58**, (1983) 295–408.

[Gr] R. Grunewald, *Six-dimensional compact Riemannian manifolds with Killing spinors*, Humboldt-Universität-zu Berlin, Preprint No. 208.

[Hi1] O. Hijazi, *Thèse de 3ème cycle*, École Polytechnique, France, 1984.

[Hi2] ____, *A conformal lower bound for the smallest eigenvalue of the Dirac operator and Killing spinors*, Comm. Math. Phys. **104**, (1986) 151–162.

[Hi3] ____, *Caractérisation de la sphère par les premières valeurs propes de l'opérateur de Dirac en dimension 3, 4, 7 et 8*, C. R. Acad. Sci. Paris, Sér. I, **303**, (1986) 417–419.

[HL] O. Hijazi and A. Lichnerowicz, *Spineurs harmoniques, spineurs-twisteurs et géométrie conforme*, C. R. Acad. Sci. Paris, Série I, **307**, (1988) 833–838.

[Hit] N. Hitchin, *Harmonic spinors*, Adv. in Math. **14**, (1974) 1–55.

[Ki1] K.-D. Kirchberg, *An estimation for the first eigenvalue of the Dirac operator on closed Kähler manifolds*, Ann. Global. Anal. Geom. **4**, (1986) 291–326.

[Ki2] ____, *Compact six-dimensional Kähler spin manifolds of positive scalar curvature with the smallest possible eigenvalue of the Dirac operator*, Math. Ann., (1986) 157–176.

[Ki3] ____, *Twistor-spinors on Kähler manifolds and the first eigenvalue of the Dirac operator*, Humboldt-Universität zu Berlin, Preprint No. 213.

[Li1] A. Lichnerowicz, *Spineurs harmoniques*, C. R. Acad. Sci. Paris Sér. AB, (1963) 7–9.

[Li2] ____, *Killing spinors according to O. Hijazi and applications*, Trieste, SISSA-ISAS, Italy, 1986.

[Li3] ____, *Spin manifolds, Killing spinors and universality of the Hijazi inequality*, Lett. Math. Phys. **13**, (1987) 331–344.

[Li4] ____, *Les spineurs-twisteurs sur une variété spinorielle compacte*, C. R. Acad. Sci. Paris, Sér. I **306**, (1988) 381–385.

[Li5] ____, *Sur les résultats de H. Baum et Th. Friedrich concernant les spineurs de Killing valeur propre imaginaire*, C. R. Acad. Sci. Paris Sér. I **309**, (1989) 41–45.

[Ra] H. B. Rademacher, *Generalized Killing spinors with imaginary Killing function and conformal Killing fields*, Preprint, Bonn.

UNIVERSITÉ DE NANTES, FRANCE
E-mail address: hijazi@narech.dnet.circe.fr

Proceedings of Symposia in Pure Mathematics
Volume **54** (1993), Part 2

Einstein Metrics on Circle Bundles

GARY R. JENSEN AND MARCO RIGOLI

ABSTRACT. We give a new method for constructing Riemannian metrics on fiber bundles. On the tangent bundle we obtain the Sasaki metric. If the structure group does not act transitively on the fiber, then the constructed metric depends on arbitrary functions on the orbit space. The Einstein condition on the constructed metric imposes differential equations with appropriate boundary values on these arbitrary functions. In some cases these equations can be solved. We show how it works for circle bundles over the two dimensional sphere.

Introduction

In 1978 T. Eguchi and A. J. Hanson [**4**] constructed a complete Einstein metric on T^*CP^1. In the same year D. Page [**7**] constructed an Einstein metric on the nontrivial S^2 bundle over S^2. In 1979 E. Calabi [**3**] found Einstein metrics on T^*CP^n for $n \geq 1$. In 1981 L. Berard-Bergery [**1**] and [**2**] constructed Einstein metrics on complex line bundles and S^2-bundles over Einstein Kähler manifolds in the context of a general construction on cohomogeneity one spaces. In 1987 D. Page and C. N. Pope [**8**] produced the same constructions from quite a different point of view.

In this paper we present a metric construction on fiber bundles. In the case where the structure group of the bundle is the circle this construction gives a new method for obtaining the above Einstein metrics. As the method works in much more generality, it may lead to the construction of compact Einstein spaces of higher cohomogeneity.

1. The metric construction

The constuction begins with the following data:

(1) M, ds^2 is a Riemannian manifold
(2) G is a Lie group with Lie algebra $\mathscr{G}$

1991 *Mathematics Subject Classification.* Primary 53C25.
Key words and phrases. Einstein metrics, fiber bundles.
This paper is in final form and no version of it will be submitted for publication later.

(3) $\pi : P \to M$ is a principal G-bundle
(4) ω is a $\mathscr{G}$-valued connection form on P
(5) F is a manifold on which G acts
(6) g is a G-invariant Riemannian metric on F
(7) h is a positive G-invariant C^∞ function on F.

The action of G on F, g induces a homomorphism $\hat{}$ of $\mathscr{G}$ into the Lie algebra of Killing vector fields $\mathscr{I}(F, g)$, given by: if $X \in \mathscr{G}$ and $x \in F$, then $\hat{X}(x) = \frac{d}{dt}|_0 e^{tX} x$. If we let $\hat{\omega} = \hat{} \circ \omega : TP \to \mathscr{I}(F, g)$, then for any 1-form φ on F it follows that $\varphi(\hat{\omega})$ is a 1-form on P.

If $\{\varphi^a\}$ is a local orthonormal coframe field in F, g, then we define a local symmetric quadratic form q on $P \times F$ by

$$q = h^2 \pi^* ds^2 + \sum_a (\varphi^a + \varphi^a(\hat{\omega}))^2.$$

As this form does not depend on the choice of orthonormal coframe field $\{\varphi^a\}$, it is globally defined on $P \times F$.

The group G acts on $P \times F$ by $a(p, x) = (pa^{-1}, ax)$. The quotient, denoted $PF = P \times_B F$, is the fiber bundle associated to P with standard fiber F. In [5] we proved the following result.

THEOREM. *The form q is G-invariant, horizontal and $q(v, v) = 0$ if and only if v is a vertical tangent vector of $P \times F$. Hence it descends to a Riemannian metric ds_{PF}^2 on PF such that $\sigma^* ds_{PF}^2 = q$, where $\sigma : P \times F \to PF$ is the projection map.*

REMARKS. 1. In our approach here the curvature computations take place on $P \times F$ which is the natural home of all the data. The major problem is to describe the G-invariant metrics and functions on F. This, of course, involves functions on the orbit space $G\backslash F$.

2. This construction, with $h = 1$ and g the standard metric on $\mathbf{R}^n$, produces the Sasaki metric on the tangent bundle TM.

3. In 1979 J. Nash [6] studied metrics defined on $P \times_G F$ so that σ is a Riemannian submersion. The metric he used on $P \times F$ is

$$\pi^* ds^2 + < \omega, \omega > + g,$$

where $< , >$ is a biinvariant metric on G. If G acts transitively on F and F, $g = G/H$ is a normal homogeneous space, then his metric and ours are the same. In general, say when G does not act transitively on F, the two metrics are different.

2. Bundles over the two-sphere

In order to simplify the exposition for this short note we will carry out the above construction for the case of the Hopf fibration $\pi : S^3 \to S^2$. As long as the structure group of the bundle is S^1 the details of the computations are not changed significantly.

On S^2 we take the standard metric ds^2 of constant curvature equal to four.

The Hopf fibration is a principal S^1-bundle, where $S^1 \subset \mathbf{C}$ as the complex numbers of modulus one, and where $S^3 \subset \mathbf{C}^2$ as the set of unit vectors with respect to the standard hermitian inner product $<, >$ on $\mathbf{C}^2$. Here S^1 acts on the right on S^3 by scalar multiplication from the right on $\mathbf{C}^2$. For the connection form on this bundle we take

$$\omega_{(p)} X = \langle p, X \rangle = \bar{p}^t X,$$

which takes values in $i\mathbf{R}$, the Lie algebra of S^1. It is convenient to set $\omega = iA$, so that A is a real valued 1-form on S^3.

On S^3 there is a complex valued 1-form defined at $p = \binom{a}{b} \in S^3$ on $X = \binom{u}{v} \in T_p S^3$ by

$$(\theta^1 + i\theta^2)_p X = av - bu.$$

It has the property that

$$\pi^* ds^2 = (\theta^1)^2 + (\theta^2)^2.$$

It is easily verified that

$$dA = 2\theta^1 \wedge \theta^2.$$

For each integer k we let $\rho_k : S^1 \to S^1$ be the representation $\rho_k(a) = a^k$.

Let F be either $\mathbf{C}$ or $S^2 \subset \mathbf{C} \times \mathbf{R}$. The action of S^1 on F is by multiplication of $S^1 \subset \mathbf{C}$ on $\mathbf{C}$. Then

$$E_k = S^3 \times_{\rho_k} \mathbf{C}$$

is a complex line bundle over S^2 with Chern number equal to $-k$; and

$$S^3 \times_{\rho_1} S^2$$

is the nontrivial S^2 bundle over S^2.

Using polar coordinates, we can write any complete G-invariant metric on F as

$$g = dr^2 + f(r)^2 d\theta^2.$$

An arbitrary G-invariant function on F is given by $h(r)$. Of course both $f(r)$ and $h(r)$ must be positive in the interior of their domains, I_F. The boundary conditions which they must satisfy depend on the two cases as follows.

In case $F = \mathbf{C}$, then $I_F = 0 \leq r < \infty$ and

(1) h is an even function at $r = 0$ in the sense that its odd order derivatives are zero there;

(2) f is an odd function at $r = 0$ in the sense that its even order derivatives are zero there;

(3) $\dot{f}(0) = 1$.

In case $F = S^2$, then $I_F = 0 \leq r \leq \pi$ and

(1) h is an even function at $r = 0$ and at $r = \pi$;
(2) f is an odd function at $r = 0$ and at $r = \pi$;
(3) $\dot{f}(0) = 1$ and $\dot{f}(\pi) = -1$.

Our symmetric bilinear form on $S^3 \times F$ becomes

$$q = h(r)^2 \pi^* ds^2 + dr^2 + f(r)^2 (d\theta + kA)^2.$$

By our theorem q descends to a complete metric ds^2_{PF} on E_k or $S^3 \times_{\rho_1} S^2$, as the case may be.

REMARK. It turns out to be sufficient to consider the form

$$\tilde{q} = h(r)^2 \pi^* ds^2 + dr^2 + k^2 f(r)^2 A^2$$

on $S^3 \times I_F$. This reduction is a crucial step in making the calculations on more complicated examples, say when the structure group is nonabelian.

3. The Einstein equations

The 1-forms on $S^3 \times F$

$$\psi^1 = h\theta^1, \qquad \psi^2 = h\theta^2, \qquad \psi^3 = dr, \qquad \psi^4 = f(d\theta + kA)$$

diagonalize q. If they are pulled back to PF by a local section of σ, they form a local orthonormal coframe field for ds^2_{PF}.

Using this coframe, we find the components of the Ricci tensor of ds^2_{PF} to be:

$$R_{ab} = 0, \quad \text{if} \quad a \neq b,$$

$$R_{11} = R_{22} = \frac{4}{h^2} - \left(\frac{\dot{h}}{h}\right)^2 - 2\left(\frac{kf}{h^2}\right)^2 - \frac{\ddot{h}}{h} - \frac{\dot{f}\dot{h}}{fh}$$

$$R_{33} = -2\frac{\ddot{h}}{h} - \frac{\ddot{f}}{f}$$

$$R_{44} = 2\left(\frac{kf}{h^2}\right)^2 - 2\frac{\dot{f}\dot{h}}{fh}.$$

Then ds^2_{PF} is Einstein if and only if

$$\frac{4}{h^2} - \left(\frac{\dot{h}}{h}\right)^2 - 2\left(\frac{kf}{h^2}\right)^2 + \frac{\ddot{h}}{h} + \frac{\ddot{f}}{f} - \frac{\dot{f}\dot{h}}{fh} = 0$$

$$\frac{\ddot{h}}{h} + \left(\frac{kf}{h^2}\right)^2 - \frac{\dot{f}\dot{h}}{fh} = 0.$$

We refer to these as the Einstein equations. The Einstein constant will be

$$\lambda = -2\frac{\ddot{h}}{h} - \frac{\ddot{f}}{f}.$$

Multiplying the second of the Einstein equations by $\frac{h}{f}$, rearranging and then multiplying by $2\frac{h}{f}$, we find that

$$\left[\left(\frac{\dot{h}}{f}\right)^2 - \frac{k^2}{h^2}\right]^{\cdot} = 0.$$

Hence we have a first integral which we write as

$$\left(\frac{\dot{h}}{f}\right)^2 - \frac{k^2}{h^2} = k^2 a,$$

where a is the constant of integration. Thus any solution of the Einstein equations must satisfy

$$f = \frac{|\dot{h}|h}{|k|\sqrt{1 + ah^2}}.$$

The solutions of these equations with each set of boundary conditions are derived in detail in §11 of [1]. Complete solutions restrict the allowable values of k.

4. Ricci flat, complete, Kähler metrics on $E_{\pm 2}$

Throughout this section we set $\varepsilon = \pm 1$. On $S^3 \times \mathbf{C}$ we set

$$\varphi^1 = \psi^1 + i\psi^2$$
$$\varphi^2 = \psi^3 + i\varepsilon\psi^4.$$

It is easy to see that these descend to a pair of integrable almost complex structures on PF which we shall refer to as J_ε. The Kähler form of J_ε is (the pull back by any section of σ of)

$$\kappa = h^2\theta^1 \wedge \theta^2 + \varepsilon f dr \wedge (d\theta + kA).$$

Thus

$$d\kappa = 2(h\dot{h} - \varepsilon k f)dr \wedge \theta^1 \wedge \theta^2,$$

from which we see that ds_{PF}^2, J_ε is Kähler if and only if

(1) $$h\dot{h} = \varepsilon k f.$$

THEOREM. *Let k be a nonzero integer and let $\varepsilon = \text{sign}(k)$ so that $\varepsilon k > 0$. Then the metric $ds_{E_k}^2$ on the complex line bundle E_k is Ricci flat, complete, and Kähler if and only if*

(2) $$h\ddot{h} + 2\dot{h}^2 = 2$$
(3) $$\varepsilon k f = h\dot{h},$$

where f and h are positive on $0 < r < \infty$ and satisfy the appropriate initial conditions described above. In particular, it follows that $\varepsilon k = 2$. This problem, with the normalization $h(0) = 1$, has a unique solution on $0 \le r < \infty$.

PROOF. Combining the Kähler condition (1) with the Einstein equations, we have

$$(4) \qquad\qquad h\dot{h} = \varepsilon k f$$

$$(5) \qquad\qquad 2\frac{\ddot{h}}{h} + \frac{\ddot{f}}{f} = 0$$

$$(6) \qquad\qquad 2\left(\frac{kf}{h^2}\right)^2 - 2\frac{\dot{f}\dot{h}}{fh} - \frac{\ddot{f}}{f} = 0$$

$$(7) \qquad \frac{4}{h^2} - \left(\frac{\dot{h}}{h}\right)^2 - 2\left(\frac{kf}{h^2}\right) - \frac{\ddot{h}}{h} - \frac{\dot{f}\dot{h}}{fh} = 0.$$

One can show, under our positivity and initial conditions, that functions f and h are solutions of equations (4)–(7) if and only if they are solutions of (2) and (3).

From (2) and the fact that $\dot{h}(0) = 0$, it follows that $h(0)\ddot{h}(0) = 2$. Substituting this into the derivative of (3) and using the condition that $\dot{f}(0) = 1$, it follows that $\varepsilon k = 2$.

Finally, it is an elementary exercise to show that the initial value problem $h\ddot{h} + 2\dot{h}^2 = 2$, $h(0) = 1$ and $\dot{h}(0) = 0$ has a unique solution. It is defined and strictly increasing on $0 \le r < \infty$, and it is an even function at 0.

It follows then that $ds^2_{E_k}$ is complete since h and f are both positive on $0 < r < \infty$ and thus g on $\mathbf{C}$ is complete. $\square$

REMARK. The bundle E_2 is the cotangent bundle, T^*S^2. The above Ricci flat, Kähler metric on E_2 was found in [4] and [1]. Our metric on $E_{-2} = TS^2$ seems to be new.

REFERENCES

1. L. Berard-Bergery, *Sur de nouvelles variétés riemanniennes d'Einstein*, Publications de l'Institut E. Cartan **4** (1982), 1–60, Nancy.
2. A. L. Besse, *Einstein Manifolds*, Springer-Verlag, Berlin, 1987.
3. E. Calabi, *Métriques kähleriennes et fibrés holomorphes*, Ann. Ecol. Norm. Sup. **12** (1979), 269–294.
4. T. Eguchi and A. J. Hanson, *Asymptotically flat self-dual solutions to Euclidean Gravity*, Phys. Lett. **237** (1978), 249–251.
5. G. R. Jensen and M. Rigoli, *Harmonic Gauss maps*, Pac. J. of Math. **136** (1989), 261–282.
6. J. C. Nash, *Positive Ricci curvature on fibre bundles*, J. Diff. Geom. **14** (1979), 241–254.
7. D. Page, *A compact rotating gravitational instanton*, Phys. Lett. **79 B** (1979), 235–238.
8. D. Page and C. Pope, *Inhomogeneous Einstein metrics on complex line bundles*, Classical Quantum Gravity **4** (1987), 213–225.

WASHINGTON UNIVERSITY

UNIVERSITÀ DI MILANO, ITALY

Proceedings of Symposia in Pure Mathematics
Volume **54** (1993), Part 2

Topological and Differentiable Structures of the Complement of an Arrangement of Hyperplanes

TAN JIANG AND STEPHEN S.-T. YAU

0. Introduction

An *arrangement of hyperplanes* is a finite collection of $\mathbf{C}$-linear subspace of dimension $(l-1)$ in $\mathbf{C}^l$. To such an arrangement $\mathscr{A}$, it is associated an open real $2\,l$-manifold, the *complement*

$$M(\mathscr{A}) = \mathbf{C}^l - U\{H : H \in \mathscr{A}\}.$$

Perhaps the central problem in this area is to find connections between the topology of $M(\mathscr{A})$ and the combinatorial geometry of $\mathscr{A}$. Specifically the most difficult unsolved problem is to decide to what extent the topology or differentiable structure of $M(\mathscr{A})$ is determined by the combinatorial geometry of $\mathscr{A}$.

Let M_l denote the braid space with l strands i.e. M_l is the complement of complexified braid arrangement $\mathscr{A}_l$ defined by $Q = \prod_{1 \le i < j \le l}(z_i - z_j)$. In 1969, Arnold [**Ar**] was able to calculate the Poincaré polynomial of the pure braid space M_l and the cohomology ring structure of $H^*(M_l)$. Arnold showed that

$$P(M_l, t) = (1 + t)(1 + 2t) \cdots (1 + (l - 1)t)$$

where $P(M_l, t)$ is the Poincaré polynomial of M_l. Arnold also showed that $H^*(M_l)$ is generated by the one-dimensional elements

$$\omega_{p,q} = \frac{1}{2\pi_i} \frac{dz_p - dz_q}{z_p - z_q}$$

and that all relations among these generators are consequences of the relations $\omega_{p,q}\omega_{q,r} + \omega_{q,r}\omega_{r,p} + \omega_{r,p}\omega_{p,q} = 0$ In general for an arbitrary arrangement $\mathscr{A}$, define holomorphic differential forms $\omega_H = (\frac{1}{2\pi_i})(d\alpha_H|\alpha_H)$

1991 *Mathematics Subject Classification.* Primary 05B35, 14B05, 57R55.
This research is partially supported by NSF.
This paper is in final form and no version of it will be submitted for publication elsewhere.

where α_H is the linear form defining the hyperplane H for $H \in \mathscr{A}$ and let $[\omega_H]$ denote the corresponding cohomology class. Let $R(\mathscr{A}) = \bigoplus_{p=0}^{l} R_p$ be the graded $\mathbf{C}$-algebra of holomorphic differential forms on $M(\mathscr{A})$ generated by the ω_H and 1. Arnold conjectured that the natural map $\eta \longrightarrow [\eta]$ of $R(\mathscr{A}) \longrightarrow H^*(M(\mathscr{A}), \mathbf{C})$ is an isomorphism of graded algebras. This was proved by Brieskorn [Br] in 1971, who showed in fact that the $\mathbf{Z}$-subalgebra of $R(\mathscr{A})$ generated by the forms ω_H and 1 is isomorphic to the singular cohomology $H^*(M(\mathscr{A}), \mathbf{Z})$. Although Brieskorn proved the Arnold conjecture that $R(\mathscr{A})$ is isomorphic to $H^*(M(\mathscr{A}), \mathbf{C})$ as graded algebra for arbitrary arrangement $\mathscr{A}$, it was not known whether the algebra R is determined by the combinatorial data of $\mathscr{A}$, since the linear forms enter the definition of $R(\mathscr{A})$. In 1980, Orlik and Solomon [Or–So 1] showed that for an arbitrary arrangement $\mathscr{A}$ the Poincaré polynomial of $M(\mathscr{A})$ equals the Poincaré polynomial of $\mathscr{A}$. Hence the betti number of $M(\mathscr{A})$ is combinatorially determined. They also introduced a graded algebra $A(\mathscr{A})$ in [Or–So 1]. It is a combinatorial invariant of $\mathscr{A}$. The main result of [Or–So 1] asserts that there is an isomorphism of algebras $A(\mathscr{A}) \approx R(\mathscr{A})$. This together with the Brieskorn's solution to Arnold's conjecture imply that the cohomological ring $H^*(M(\mathscr{A}), \mathbf{C})$ is a combinatorial invariant of $\mathscr{A}$. The investigation of the cohomological ring $H^*(M(\mathscr{A}), \mathbf{C})$ is therefore completed.

The next difficult unsolved problems involve the homotopy groups of $M(\mathscr{A})$. In a Bourbaki Seminar talk, Brieskorn [Br] generalized Arnold's results. He replaced the symmetric group and the braid arrangement by a Coxeter group W acting in $\mathbf{R}^l$. Then W acts as a reflection group in $\mathbf{C}^l$. Let $\mathscr{A} = \mathscr{A}(W)$ be its reflection arrangement. Brieskorn conjectured that $\mathscr{A}(W)$ is a $K(\pi, 1)$ arrangement for all Coxeter groups W. He proved this for some of the groups by representing M as the total space of a sequence of fibrations. Deligne [De] settled the question by proving that complement of complexification of real simplicial arrangement is $K(\pi, 1)$. This result proves Brieskorn's conjecture because the arrangement of a Coxeter group is simplicial. Shephard and Todd [Sh–To] classified finite irredicible complex reflection groups. Recall that real reflection groups are also called Coxeter groups because finite irreducible real reflection groups were classified by Coxeter [Co]. Every real reflection group may be viewed as a complex reflection group. There are examples of complex reflection groups which are not Coxeter groups. Orlik and Solomon [Or–So 2] conjectured that all the complex reflection arrangements are $K(\pi, 1)$. For a subclass of irreducible complex reflection groups, called Shephard groups this was proved by Orlik and Solomon [Or–So 3]. The conjecture is still open for the remaining irreducible complex reflection groups. Recently, Jambu and Terao [Ja–Te] have introduced the property of supersolvability of an arrangement. This property is combinatorial in nature. That is, they depend only on the pattern of intersection of the hyperplanes, i.e. on the lattice associated to the arrangement.

It turns out that complement M of a supersolvable arrangement is the total space of a fiber bundle in which the base and fiber are $K(\pi, 1)$ spaces. The long exact homotopy sequence of the bundle shows that M is $K(\pi, 1)$ also. On the other hand, one also knows certain conditions on an arrangement $\mathscr{A}$ for which $M(\mathscr{A})$ is not $K(\pi, 1)$. For example, Hattori [Ha] observed that the arrangement of 4 planes in $\mathbf{C}^3$ in general position does not yield a $K(\pi, 1)$ since in this case $\pi_2(M(\mathscr{A}))$ is the free $\mathbf{Z}[\pi_1(M^*)]$-module of rank 1 $(M^* = M(\mathscr{A})/\mathbf{C}^*$, $\pi_1(M^*) \cong \mathbf{Z}^3)$. More generally, if $\pi_1(M(\mathscr{A}))$ contains a subgroup isomorphic to $\mathbf{Z}^4$, then $M(\mathscr{A})$ cannot be a $K(\pi, 1)$ (for $M(\mathscr{A}) \subseteq \mathbf{C}^3$). In [Fa–Ra 1], a numerical relationship was established between $\pi_1(M(\mathscr{A}))$ and $H^*(M(\mathscr{A}))$ for the class of fiber-type arrangements. This LCS *formula* reads as follows:

$$\prod_{n \geq 1}(1 - t^n)^{\phi_n\left(M(\mathscr{A})\right)} = P\left(M(\mathscr{A}), -t\right)$$

where the $\phi_n(M(\mathscr{A}))$ are the ranks of successive quotients in the lower central series of $\pi_1(M(\mathscr{A}))$ and $P(M(\mathscr{A}), t)$ is the Poincaré polynomial of $H^*(M(\mathscr{A}))$. Because the sequence $\phi_n(M(\mathscr{A}))$ is related to the one-minimal $\mathscr{S}$ of $M(\mathscr{A})$, Falk and Randell conjectured in [Fa–Ra 2] that the LCS formula holds when $\mathscr{S}$ determines $H^*(M(\mathscr{A}))$. Recently Kohno [Ko 5] and Falk [Fa 1] has resolved independently this conjecture. Specifically, they showed that the LCS formula holds when $H^*(\mathscr{S})$ is isomorphic to $H^*(M(\mathscr{A}))$. Arrangements satisfying the latter condition have been called rational $K(\pi, 1)$, which is a deceptive terminology because a $K(\pi, 1)$ space is not necessarily rational $K(\pi, 1)$. In 1987, Salvetti [Sa] made a fundamental contribution to the understanding of the higher homotopy in the complement of an arrangement. Consider a union of real affine hyperplanes in $\mathbf{C}^l$ with complement M. In [Sa 1], Salvetti constructed explicitly a CW-complex $X \subset M$ of dimension l, and proved that X and M are homotopically equivalent. As an application to higher homotopy groups he considered the general position case, already consider in [Ha]. In this case the complex X is a union of cellularly decomposed l-tori $((S^1)^l)$ attached by $(l - 1)$-cells, and he showed that it can be obtained from its real part by suitable identifications. He obtained a more compact proof of the fact, that, if the number of hyperplanes is greater than the dimension l, then M is not a $K(\pi, 1)$. Because of the very explicit description of the regular complex X, it should be possible to compute the homotopy of Y in other not yet solved cases.

The most difficult unsolved problem is whether the topological or diffeomorphic type of complement $M(\mathscr{A})$ of an arrangement is combinatorial in nature. It seems to us that these major problems in the theory of arrangements remain untouched. The purpose of this paper has two folds. On the one hand, the study of the topology of $M(\mathscr{A})$ is important both in the theory of hypergeometric functions (see the work of Gelfand [Ge] and his subsequent papers, the work of Deligne and Mostow [De–Mo] and subsequent papers by Mostow) and in the singularity theory [Ar, Br, De and also Ca]. Moreover, it plays a role in some interesting problems in algebraic geometry (see especially the works of Hirzebruch [Hi] and Moishezon [Mo]). The first purpose of this paper is to give a rough survey of the theory of arrangements. Hopefully this will generate some interest among differential geometors to this subject. In fact the well-known Hadamard-Cartan criterion, which states that $M(\mathscr{A})$ is $K(\pi, 1)$ if its sectional curvature is nonpositive, may help in many unsolved problems in the theory of arrangements. The second purpose of this paper is to announce a new result (cf. Theorem 4.3) which gives a partial solution to the above problem. Let $\mathscr{A}$ be an arrangement of n lines in $\mathbf{CP}^2$. Then our theorem asserts that if the graph of $\mathscr{A}$ is not very complicated, then the diffeomorphic type of $M(\mathscr{A})$ is combinatorial in nature. The detail of the proof will appear elsewhere.

In §1 we recall some definitions in lattice theory as well as the definition of lattice $L(\mathscr{A})$ associated to arrangement $\mathscr{A}$. We show an example of arrangement $\mathscr{A}$ in $\mathbf{C}^3$ and show how to compute its Poincaré polynomial. In §2 we describe the beautiful work of Orlik and Solomon on cohomology ring of $M(\mathscr{A})$. The graded algebra $A(\mathscr{A})$ is constructed purely from the arrangement $\mathscr{A}$. It turns out that $A(\mathscr{A})$ is isomorphic to cohomology algebra $H^*(M(\mathscr{A}))$ as a graded algebra. In §3, we study homotopy properties of complement of arrangement. We recall the definition for a lattice to be supersolvable due to Stanley. We then discuss the beautiful results of Terao on $K(\pi, 1)$ property and of Falk and Randell on LCS formula for supersolvable arrangements. We also discuss the beautiful work of Falk and Kohno on LCS formula for rational $K(\pi, 1)$ arrangement. Due to lack of space, we regret that we cannot describe the Salvetti complex although it is by far one of the most important results in this direction. In §4, we discuss the diffeomorphic types of the complements of arrangements. We introduce the concepts complexity and graph of an arrangement. We show an example of two arrangements in $\mathbf{CP}^2$ such that they have the same cardinality and the same number of k-tuple points for all $k \geq 2$, but their associated lattices and graphs are different. We show further an example of two arrangements $\mathscr{A}_1$, $\mathscr{A}_2$ in $\mathbf{CP}^2$ such that there exists a bijection $\phi: \mathscr{A}_1 \longrightarrow \mathscr{A}_2$ in such a way that for every $H_1 \in \mathscr{A}_1$, the number of k-tuple points on H_1 and on $\phi(H_1)$ are the same for all k. However the lattices associated to $\mathscr{A}_1$ and $\mathscr{A}_2$ are not isomorphic. These two examples were communicated to us by Salvetti.

Those readers who want to have further detail of the subject should consult the work of [Or, Fa, Fa–Ra 2, Sa 1, Te]. In fact, this paper owe the existence to their works.

1. Arrangement $\mathscr{A}$ and its lattice $L(\mathscr{A})$

We begin by recalling some terminology in lattice theory.

DEFINITION 1.1. A poset is a set in which a binary relation $x \leq y$ is defined which satisfies for all x, y, z the following conditions:

P1 (Reflexive). For all x, $x \leq x$.

P2 (Antisymmetry). If $x \leq y$ and $y \leq x$, then $x = y$.

P3 (Transitivity). If $x \leq y$ and $y \leq z$, then $x \leq z$. If $x \leq y$, we shall say that x is less than or equal to y. If $x \leq y$ and $x \neq y$, one writes $x < y$.

An *upper bound* of a subset X of a point P is an element $a \in P$ such that $x \leq a$ for every $x \in X$. The least upper bound is an upper bound less than or equal to every other upper bound, it is denoted by sup X. By P2, sup X is unique if it exists. The notion of lower bound of X and greatest lower bound (inf X) of X are defined dually. Again by P_2, inf X is unique if it exists.

DEFINITION 1.2. A *lattice* is a poset P any two of whose elements have a greatest lower bound or "meet" denoted by $x \wedge y$, and a lowest upper bound or "join" denoted by $x \vee y$.

DEFINITION 1.3. An element y covers an element x in a lattice L if and only if $x < y$, but $x < z < y$ for no element z in L.

DEFINITION 1.4. A chain in a lattice L is any linearly ordered subset of L.

DEFINITION 1.5. A lattice having no infinite chains is said to be semimodular whenever it has the covering property : for all lattice elements x, y, if x and y cover $x \wedge y$, then $x \vee y$ covers x and y.

DEFINITION 1.6. Let L be a lattice with finite length. The rank of $a \in L$, denoted by $r(a)$, is the length of the longest chain in L below a. Let $\hat{0} = \inf L$ and $\hat{1} = \sup L$. Then $r(\hat{0}) = 0$. The rank of L (rank L) is defined to be $r(\hat{1})$. If a in L has rank 1, then a is called a point or an atom of the lattice.

DEFINITION 1.7. A *point lattice* (or atomic lattice) is a lattice in which every element is a join of points. A *geometric lattice* is a semimodular point lattice with no infinite chains. In this paper an arrangement $\mathscr{A}$ is a finite collection of hyperplanes $\{H_1, \ldots, H_n\}$ through the origin in $\mathbf{C}^l$. Arrangements in our sense are sometimes called central. If we view $\mathbf{C}^l$ as affine space and allow $\mathscr{A}$ to contain affine hyperplanes, we call $\mathscr{A}$ an affine arrangement. A projective arrangement is a finite set of projective hyperplanes in projective space $\mathbf{CP}^{l-1}$. Recall the canonical bundle $p : \mathbf{C}^l - \{0\} \longrightarrow \mathbf{CP}^{l-1}$ with fiber $\mathbf{C}^*$, which identifies z with λz for $\lambda \in \mathbf{C}^*$.

PROPOSITION 1.8. *Let $\mathscr{A}$ be a nonempty arrangement with complement $M = M(\mathscr{A})$ and let $M^* = p(M)$. The restriction $p : M \longrightarrow M^*$ is a trivial fibration so $M = M^* \times \mathbf{C}^*$. There is a projective $(l - 1)$-arrangement $\mathscr{A}^*$*

such that $M^* = M(\mathscr{A}^*)$ where $M(\mathscr{A}^*)$ is the complement of the projective arrangement in $\mathbf{CP}^{l-1}$.

PROOF. Let $H \in \mathscr{A}$. The restriction of p to $\mathbf{C}^l - H$ has base space $\mathbf{CP}^{l-1} - \mathbf{CP}^{l-2} \cong \mathbf{C}^{l-1}$. Thus $p : \mathbf{C}^l - H \longrightarrow \mathbf{C}^{l-1}$ is a trivial bundle and $p : M \longrightarrow M^*$ is a subbundle. The rest of the proposition follows easily. Q.E.D.

Since the complement of a hyperplane in projective space is affine space, a nonempty projective arrangement may be viewed as an affine arrangement and vice versa. Proposition 1.8 above gives a close connection between central l-arrangements and affine $(l - 1)$-arrangements.

Following Orlik-Solomon [**Or–So**], we define the lattice $L(\mathscr{A})$ of an arrangement. The set $L(\mathscr{A})$ is the set of all intersections of subsets of $\mathscr{A}$, partially ordered by reverse inclusion i.e. $X \leq Y \Longleftrightarrow Y \subseteq X$. Thus $\mathbf{C}^l$ is the minimal element.

Define a rank function r on $L(\mathscr{A})$ by $r(X) = \operatorname{codim} X = l - \dim_{\mathbf{C}} X$ for $X \in L(\mathscr{A})$. Call H_i an *atom* of $L(\mathscr{A})$. Define the *join* by $X \vee Y = X \cap Y$ and the *meet* by $X \wedge Y = \bigcap \{Z : Z \in L(\mathscr{A}), X \cup Y \subset Z\}$.

LEMMA 1.9. *Let $\mathscr{A}$ be an arrangement. Then*
(i) *for every $X \in L(\mathscr{A})$ all maximal linear ordered subsets*

$$X_0 = \mathbf{C}^l < X_1 < \cdots < X_p = X$$

have the same cardinality,
(ii) *every element of $L(\mathscr{A}) - \{\mathbf{C}^l\}$ is a join of atoms,*
(iii) *for all X, Y in $L(\mathscr{A})$ the rank function satisfies,*

$$r(X \wedge Y) + r(X \vee Y) \leq r(X) + r(Y).$$

Thus $L(\mathscr{A})$ is a geometric lattice.

DEFINITION 1.10. Let $L_p = L_p(\mathscr{A}) := \{X \in L(\mathscr{A}) : r(X) = p\}$. The Hasse diagram of $L(\mathscr{A})$ has vertices labeled by the elements of $L(\mathscr{A})$ and arranged on levels L_p, $p \geq 0$. Suppose $X \in L_p$ and $Y \in L_{p+1}$. An edge connects X with Y if $X < Y$.

EXAMPLE 1.1. Let $\mathscr{A}$ be an arrangement of hyperplanes in $\mathbf{C}^3$ consisting of the elements $\{(x, y, z) \in \mathbf{C}^3 : x = y\}$, $\{(x, y, z) \in \mathbf{C}^3 : x = -y\}$, $\{(x, y, z) \in \mathbf{C}^4 : x = z\}$, $\{(x, y, z) \in \mathbf{C}^3 : x = -z\}$ $\{(x, yz) \in \mathbf{C}^3 : y = -z\}$, $\{(x, y, z) \in \mathbf{C}^3 : y = z\}$. (See Figure 1.)

DEFINITION 1.11. The Möbius function μ of $L(\mathscr{A})$ is defined recursively by $\mu(\mathbf{C}^l) = 1$, and $\mu(X) = -\sum_{Y \in L(\mathscr{A}), Y < X} \mu(Y)$.

EXAMPLE 1.2. The values of $\mu(X)$ for the arrangement in Example 1.1 above are given by the diagram in Figure 2.

DEFINITION 1.4. Let $\mathscr{A}$ be an arrangement with lattice $L(\mathscr{A})$ and Möbius function μ. Let t be an indeterminate. Define the Poincaré polynomial of $\mathscr{A}$ by $\pi(\mathscr{A}, t) = \sum_{X \in L(\mathscr{A})} \mu(X)(-t)^{r(X)}$.

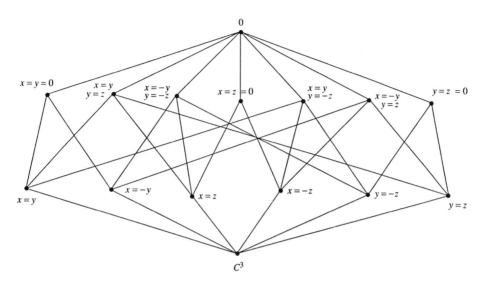

Figure 1.

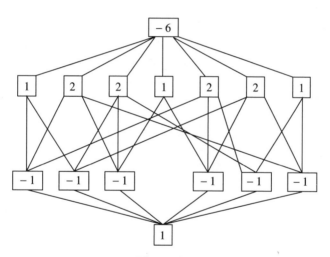

Figure 2.

EXAMPLE 1.3. The Poincaré polynomial of the arrangement in Example 1.1 above is $\pi(\mathscr{A}, t) = 1 + 6t + 8t^2 + 3t^3$.

2. Cohomology ring of complement of arrangement

In this section, we shall study the cohomology of $M(\mathscr{A})$.

DEFINITION 2.1. Let $\mathscr{A}$ be a complex arrangement in $\mathbf{C}^l$ and let $M = M(\mathscr{A})$ be the complement of its hyperplanes. The Poincaré polynomial of the complement is $P(M(\mathscr{A}), t) = \sum_{p \geq 0} \operatorname{rk} H^p(M(\mathscr{A})) t^p$.

The following beautiful theorem is due to Orlik and Solomon [**Or–So 1**]. It says that one can compute the betti numbers of the complement $M(\mathcal{A})$ easily from the lattice $L(\mathcal{A})$.

THEOREM 2.2. *Let $\mathcal{A}$ be a complex arrangement with complement $M(\mathcal{A})$. Then $P\big(M(\mathcal{A}), t\big) = \pi(\mathcal{A}, t)$.*

DEFINITION 2.3. Let Ω be the exterior algebra of the $\mathbf{C}$ vector space $(\mathbf{C}^l)^*$ graded by $\Omega = \bigoplus_{p=0}^{l} \Omega^p$ where

$$\Omega^p = \bigoplus_{1 \le i_1 < \cdots < i_p \le l} \mathbf{C} dz_{i_1} \wedge \cdots \wedge dz_{i_p}.$$

Let $\mathcal{A}$ be an arrangement in $\mathbf{C}^l$. For $H \in \mathcal{A}$, let α_H be the linear form defining H and let $\omega_H = d\alpha_H/\alpha_H \in \Omega^1$. Let $R = R(\mathcal{A})$ be the $\mathbf{C}$-subalgebra of Ω generated by 1 and ω_H for $H \in \mathcal{A}$.

Let $R_p = R \cap \Omega^p$. Since R is generated by the homogeneous elements ω_H, it is naturally graded $R = \bigoplus_{p=0}^{l} R_p$. This algebra was first computed by Arnold [**Ar**] for the braid arrangement. Brieskorn [**Br**] defined it for all arrangements and shows that it is isomorphic to the cohomology algebra of the complement, which solved affirmatively the Arnold conjecture. In order to show that $R(\mathcal{A})$ depends only on the geometric lattice $L(\mathcal{A})$, Orlik and Solomon [**Or–So 1**] introduced the following combinatorial algebra $A(\mathcal{A})$.

Let $E_1 = \bigoplus_{H \in \mathcal{A}} \mathbf{C} e_H$ and let $E = E(\mathcal{A}) = \bigwedge(E_1)$ be the exterior algebra of E_1. Note that E_1 has a $\mathbf{C}$-basis consisting of elements e_H in one-to-one correspondence with the hyperplanes of $\mathcal{A}$. Write $uv = u \wedge v$ and note that $e_H^2 = 0$, $e_H e_K = -e_K e_H$ for $H, K \in \mathcal{A}$. The algebra E is graded. If the cardinality of $\mathcal{A}$ is n, then $E = \bigoplus_{p=0}^{n} E_p$ where $E_0 = \mathbf{C}$, E_p is spanned over $\mathbf{C}$ by all $e_{H_1} \cdots e_{H_p}$ with $H_k \in \mathcal{A}$. Define a $\mathbf{C}$-linear map $\partial : E \longrightarrow E$ by $\partial 1 = 0$, $\partial e_H = 1$ and for $p \ge 2$,

$$\partial(e_{H_1} \cdots e_{H_p}) = \sum_{k=1}^{p} (-1)^{k-1} e_{H_1} \cdots \hat{e}_{H_k} \cdots e_{H_p}$$

for all $H_1, \ldots, H_p \in \mathcal{A}$.

Given a p-tuple of hyperplanes $S = (H_1, \ldots, H_p)$ write $|S| = P$, $e_S = e_{H_1} \cdots e_{H_p} \in E$ and $\bigcap S = H_1 \cap \cdots \cap H_p \in L(\mathcal{A})$. If $p = 0$, we agree that $S = (\ )$ is the empty tuple, $e_S = 1$ and $\bigcap S = \mathbf{C}^l$.

LEMMA 2.4. *The map $\partial : E \longrightarrow E$ satisfies* (i) $\partial^2 = 0$.

(ii) *If $u \in E_p$ and $v \in E$, then $\partial(uv) = (\partial u)v + (-1)^p u(\partial v)$. Since the map ∂ is homogeneous of degree -1, we see from (i) that (E, ∂) is a chain complex. Part (ii) says that ∂ is a derivation of the exterior algebra. It may be characterized as the unique derivation of E with $\partial e_H = 1$. Since the rank function on $L(\mathcal{A})$ is codimension, it is clear that if $|S| = p$ then $r(\bigcap S) \le p$.*

DEFINITION 2.5. Let $S = (H_1, \ldots, H_p)$. Call S independent if $r(\bigcap S) = p$ and dependent if $r(\bigcap S) < p$.

Clearly S is independent if the corresponding linear forms $\alpha_1, \ldots, \alpha_p$ are linearly independent. Equivalently, the hyperplanes of S are in general position.

DEFINITION 2.6. Let $\mathscr{A}$ be an arrangement. Let $I = I(\mathscr{A})$ be the ideal of E generated by ∂e_S for all dependent p-tuple S, $p \geq 0$. Define $A = A(\mathscr{A}) := E/I$.

Let $\phi: E \longrightarrow A$ be the natural homomorphism and let $A_p = \phi(E_p)$. If $H \in \mathscr{A}$ let $a_H = \phi(e_H)$ and if S is a p-tuple of hyperplanes let $a_S = \phi(e_S)$.

LEMMA 2.7. *If S is a p-tuple of hyperplanes and $H \in S$, then $e_S = e_H(\partial e_S)$. If S is dependent then $e_S \in I$.*

PROOF. If H is in S, then $e_H e_S = 0$ so that $0 = \partial(e_H e_S) = e_S - e_H(\partial e_S)$. The second assertion follows from the first.

REMARK 2.8. Since both E and I are graded, A is a graded anticommutative algebra. Since the elements of 1-tuple are independent we have $I_0 = 0$ and hence $A_0 = \mathbf{C}$. The only dependents of 2-tuple are of the form $S = (H, H)$. Since $e_S = e_H^2 = 0$ we have $I_1 = 0$. Thus the elements a_H are linearly independent over $\mathbf{C}$ and $A_1 = \bigoplus_{H \in \mathscr{A}} a_H$. If $p > l$, then every p-tuple of hyperplanes is dependent and it follows from Lemma 2.7 that $A_p = 0$. Thus $A = \bigoplus_{p=0}^{l} A_p$.

In what follows we shall recall some properties of the algebra A due to Orlik and Solomon.

LEMMA 2.9. $\partial_E I \subset I$.

PROOF. Recall that I is a $\mathbf{C}$-linear combination of elements of the form $e_T \partial e_S$ where S is dependent. We have

$$\partial(e_T \partial e_S) = (\partial e_T)(\partial e_S) \pm e_T(\partial^2 e_S) = (\partial e_T)(\partial e_S) \in I. \quad \text{Q.E.D.}$$

DEFINITION 2.10. Since $\partial_E I \subset I$ we may define $\partial_A: A \longrightarrow A$ by $\partial_A(\phi(u)) = \phi(\partial_E u)$ for $u \in E$ where $\phi: E \longrightarrow A$ is the natural homomorphism.

LEMMA 2.11. *The map $\partial_A: A \longrightarrow A$ satisfies* (i) $\partial_A^2 = 0$.
(ii) *If $a \in A_p$ and $b \in A$, then $\partial_A(ab) = (\partial_A a)b + (-1)^p a(\partial_A b)$.*

PROOF. (i) and (ii) follow from the corresponding facts for ∂_E. For (iii), it suffices to show that $\operatorname{im} \partial_A \supset \ker \partial_A$. Since $\mathscr{A}$ is not empty, we may choose $H \in \mathscr{A}$. Let $v = e_H$. Then $\partial_E v = 1$. Let $b = \phi(v)$. Then $\partial_A(b) = 1$ since $\phi \partial_E = \partial_A \phi$. It follows that if $a \in \ker \partial_A$, then

$$\partial_A(ba) = (\partial_A b)a - b(\partial_A a) = a$$

Thus $\operatorname{im} \partial_A \supset \ker \partial_A$ as claimed. Q.E.D.

COROLLARY 2.12. *Suppose $\mathscr{A}$ is not empty. Then $\sum_{p \geq 0}(-1)^p \dim A_p = 0$.*

PROOF. Since the chain complex (A, ∂_A) is acyclic, its Euler characteristic is zero. Q.E.D.

THEOREM 2.13. *Let $\mathscr{A}$ be an arrangement and let $A(\mathscr{A})$ be the associated algebra. Let $P(A(\mathscr{A}), t)$ be the Poincaré polynomial of the graded $\mathbf{C}$-algebra $A(\mathscr{A})$. Then $P(A(\mathscr{A}), t) = \pi(\mathscr{A}, t)$.*

The works of Arnold [Ar], Brieskorn [Br], Orlik and Solomon [Or–So 1] established that the cohomology algebra $H^*(M(\mathscr{A}))$ of an arrangement $\mathscr{A}$ is a combinatorial invariant of $\mathscr{A}$. Its combinatorial nature is a consequence of the following result of [Or–So 1] which establishes an algebra isomorphism between $A(\mathscr{A})$ and $R(\mathscr{A})$.

THEOREM 2.14. *Let $\mathscr{A}$ be an arrangement and let $R(\mathscr{A})$ be the algebra of differential forms generated by 1 and $\omega_H = d\alpha_H / \alpha_H$ where α_H is the linear form defining the hyperplane $H \in \mathscr{A}$. The map $\gamma: A(\mathscr{A}) \longrightarrow R(\mathscr{A})$ induces an isomorphism of graded $\mathbf{C}$-algebras such that $\gamma(e_H) = \omega_H$.*

Theorem 2.14 together with the Brieskorn solution to Arnold conjecture establish the following most successful investigations concerning the cohomology of $M(\mathscr{A})$.

THEOREM 2.15. *Let $\mathscr{A}$ be a complex arrangement. Then $A(\mathscr{A})$ is isomorphic as a graded algebra to the cohomology $H^*(M(\mathscr{A}))$ of the complement $M(\mathscr{A})$.*

3. Homotopy properties of complement of arrangement

In this section we consider homotopy properties of the space $M(\mathscr{A})$, complement of arrangement $\mathscr{A}$. Since $M(\mathscr{A})$ is the complement of a real codimension two embedding, it has a rich fundamental group structure. One may see that

(a) $H_1(M(\mathscr{A}), \mathbf{Z}) \cong \mathbf{Z}^n$, with generators loops meridional to the individual H_i,

(b) $\pi_1(M(\mathscr{A})) \cong \pi_1(M(\mathscr{A}) \cap P)$, where P is a generic three-space through the origin in $\mathbf{C}^l$ [Ha–Lê],

(c) $\pi_1(M(\mathscr{A}) \cap P)$ may be computed for a given arrangement with the standard "pencil" technique [Chéniot],

(d) Randell [Ra 2] gave a presentation for the fundamental group of the complement of a complexified real arrangement (i.e. equations of hyperplanes are defined by real equation). Salvetti [Sa 1, Part II] gave a different presentation of $\pi_1(M(\mathscr{A}))$ with one generator for each hyperplane, and one set of relations for each codimension 2 subspace. For the higher homotopy groups much less is known. One does know certain conditions for which

$M(\mathscr{A})$ is not $K(\pi, 1)$. In fact at it follows from a theorem of Hattori [**Ha**] which says that most arrangements are not $K(\pi, 1)$.

DEFINITION 3.1. The arrangement $\mathscr{A}$ is called a *general position* arrangement if for every subset $\{H_1, \ldots, H_p\} \subset \mathscr{A}$ with $p \leq l$, $r(H_1 \cap \ldots \cap H_p) = p$ and when $p > l$, $H_1 \cap \cdots \cap H_p = \varnothing$.

Note that if $\mathscr{A}$ is a central general position l-arrangement then $|\mathscr{A}| \leq l$. Thus the interesting general position arrangements are affine.

DEFINITION 3.2. Let $I \subseteq \{1, \ldots, n\}$ and $|I|$ be its cardinality. Define the subtorus T_I of T^n by $T_I = \{(z_1, \ldots, z_n) \in T^n : |z_j| = 1 \text{ for } j \notin I\}$.

Then Hattori [**Ha**] proved the following theorem

THEOREM 3.3. *Let $\mathscr{A}^*$ be an affine l-arrangement in general position and assume that $n = |\mathscr{A}^*| \geq l + 1$. Then $M^* = M(\mathscr{A}^*)$ has the homotopy type of $\bigcup_{|I|=l} T_I$.*

Hattori [**Ha**] also proved that $\pi_1(M^*)$ is free abelian of rank n, and that the universal covering space $\widetilde{M}^*$ of M^* has trivial homology in dimensions $\neq 0, l$. He also gave a free $\mathbf{Z}(\pi_1(M^*))$ resolution of $H_l(\widetilde{M}^*, \mathbf{Z})$. In particular if $n = l + 1$ then $H_l(\widetilde{M}^*, \mathbf{Z})$ is a free $\mathbf{Z}(\pi_1(M^*))$-module of rank 1.

DEFINITION 3.4. Let $\mathscr{A}$ be a central $(l+1)$-arrangement with $l \geq 2$. Let $\mathscr{A}^*$ be the corresponding affine l-arrangement. Call $\mathscr{A}$ a *generic* arrangement if $\mathscr{A}^*$ is a general position arrangement.

COROLLARY 3.5. *Generic arrangements are not $K(\pi, 1)$.*

PROOF. Hurewicz isomorphism theorem implies $\pi_i(\widetilde{M}^*) = H_i(\widetilde{M}^*, \mathbf{z}) = 0$ for $1 \leq i < l$ and $\pi_l(\widetilde{M}^*) = H_l(\widetilde{M}^*, \mathbf{Z})$. Thus $\pi_l(M^*) \neq 0$ and M^* is not a $K(\pi, 1)$-space. In view of Proposition 1.8, M is not a $K(\pi, 1)$-space. Q.E.D.

DEFINITION 3.6. An arrangement $\mathscr{A}$ is *fibered* if $M(\mathscr{A})$ is the total space of a fiber bundle $F \longrightarrow M(\mathscr{A}) \longrightarrow M'$, where F is a punctured surface, and M' is a $K(\pi, 1)$ space.

DEFINITION 3.7. (i) The arrangement $\{0\}$ in $\mathbf{C}^1$ is a *fiber-type* arrangement.

(ii) Suppose that, after suitable linear coordinate change, projection to the first $(l-1)$ coordinate is a fiber bundle projection $M(\mathscr{A}) \longrightarrow M'$, where M' is the complement of a fiber type arrangement in $\mathbf{C}^{l-1}$. Then $\mathscr{A}$ is a *fiber-type arrangement*.

REMARK 3.8. It follows by application of the homotopy exact sequence that fibered arrangements and fiber-type arrangements are $K(\pi, 1)$.

Let L be a finite geometric lattice with minimal element $\hat{0}$, the maximal element $\hat{1}$, and rank function r. We shall briefly review the notions of modular elements of L and supersolvable lattices, both of which are defined by Stanley [**St 1, St 2**].

DEFINITION 3.9. A pair $(X, Y) \in L \times L$ is said to be a *modular pair* when

$$Z = (Z \vee X) \wedge Y \quad \text{for all } Z \text{ with } X \wedge Y \leq Z \leq Y.$$

REMARK 3.10. In a geometric lattice $(X, Y) \in L \times L$ is a modular pair if and only if $r(X) + r(Y) = r(X \vee Y) + r(X \wedge Y)$ [**Bi**, p. 83, Theorem 2].

DEFINITION 3.11. An element $x \in L$ is called a *modular element* it forms a modular pair with every $y \in L$ i.e. if $y < z$ then $y \vee (x \wedge z) = (y \vee x) \wedge z$.

The following theorem is due to Stanley [**St 1**, Theorem 1].

THEOREM 3.12. *An element $x \in L$ is modular if and only if no two complements of x are comparable, i.e. if $x \wedge y = x \wedge z = \hat{0}$, $x \vee y = x \vee z = \hat{1}$ and $y \geq z$, then $y = z$.*

The following lemma due to Terao [**Te**] provides a characterization of modular elements.

LEMMA 3.13. *An element x of L is a modular element if and only if x forms a modular pair with any y satisfying $x \wedge y = \hat{0}$.*

In [**St 2**], Stanley defined a lattice to be supersolvable.

DEFINITION 3.14. A geometric lattice L is supersolvable if there exists a maximal modular chain $\hat{0} = x_0 < x_1 < \ldots < x_l = \hat{1}$ i.e. $l = \text{rank } L$ and each x_i is a modular element $(i = 0, \ldots, l)$.

In [**Ja–Te**], Jambu and Terao first studied those arrangements $\mathscr{A}$ whose lattices $L(\mathscr{A})$ are supersolvable. In [**Te**], Terao has shown the following theorem.

THEOREM 3.15. *An arrangement $\mathscr{A}$ is a fiber-type if and only if $L(\mathscr{A})$ is supersolvable.*

COROLLARY 3.16. *For any arrangement $\mathscr{A}$, if $L(\mathscr{A})$ is supersolvable, then $M(\mathscr{A})$ is a $K(\pi, 1)$ space.*

Supersolvability is the only known property on $L(\mathscr{A})$ that will imply that $M(\mathscr{A})$ is a $K(\pi, 1)$ space. In particular if $L(\mathscr{A})$ is supersolvable, then $\pi_i(M(\mathscr{A})) = 0$ for $i \geq 2$. Falk and Randell [**Fa–Ra 1**] studied $\pi_1(M(\mathscr{A}))$ when $\mathscr{A}$ is a fiber-type arrangement. Set $G = \pi_1(M(\mathscr{A}))$, $G_1 = G$, and $G_{n+1} = [G_n, G]$, where $[A, B]$ denotes the commutator subgroup of A and B. That is, $[A, B]$ is the subgroup generated by $a^{-1}b^{-1}ab$, $a \in A$, $b \in B$. Then the G_n are just the terms of the lower central series of G. We further set $G(n) = G_n/G_{n+1}$. By [**Ma–Ka–So**, Theorem 5.4], $G(n)$ is a finitely generated abelian group. We set $\phi_n = \text{rank } G(n)$. The following theorem is due to Falk and Randell [**Fa–Ra**].

THEOREM 3.17. *Let $\mathscr{A}$ be an arrangement such that $L(\mathscr{A})$ is supersolvable. Then*

$$\prod_{j=1}^{\infty}(1 - t^j)^{-\phi_j} = \frac{1}{P(M(\mathscr{A}), -t)} \quad \text{in } \mathbf{Z}[[t]].$$

Kohno [**Ko 1, Ko 2, Ko 5**] and Falk [**Fa**] were able to generalize the above result using the theory of minimal models of Sullivan. For the sake of convenience to the readers, we recall Sullivan's theory of minimal model briefly. In what follows, all vector spaces and algebras are over $\mathbf{Q}$. A *differential graded algebra* (DG algebra) A is a graded vector space $A = \bigoplus_{i \geq 0} A^i$ with a degree 1 coboundary operator $d: A^i \longrightarrow A^{i+1}$, and a product $\bigwedge: A^i \otimes A^j \longrightarrow A^{i+j}$. These satisfy

(i) $d^2 = 0$,

(ii) $d(x \wedge y) = dx \wedge y + (-1)^i x \wedge dy$ for $x \in A^i$,

(iii) $x \wedge y = (-1)^{ij} y \wedge x$ for $x \in A^i$ and $y \in A^j$,

(iv) $\bigwedge$ makes A an associative algebra with unit $1 \in A^\circ$.

If $A^\circ = \mathbf{Q}$, then A is called connected. Let (A^*, d) be a DG algebra. For each $n \geq 0$, let $A(n)$ be the subalgebra generated by A^k, $0 \leq k \leq n$, and dA^n. Construct subalgebras $A(n, q)$ of $A(n)$ for each $q \geq 0$ inductively as follows: $A(n, 0) = A(n - 1)$, and, for $q > 0$, $A(n, q)$ is the subalgebra generated by $A(n, q - 1)$ and $\{x \in A^n : dx \in A(n, q - 1)\}$.

DEFINITION 3.18. (cf. [**Bo–Gu**]). The DG algebra M is *minimal* provided

(i) $M^\circ \cong \mathbf{Q}$.

(ii) M is a free graded-commutative algebra.

(iii) $M(n) = \bigcup_{q \geq 0} M(n, q)$ for each $n \geq 1$.

Given a $\mathbf{Q}$-vector space V of finite dimension, we denote by $\bigwedge_r(V)$ the free graded-commutative algebra on V, with V homogeneous of degree r. Thus $\bigwedge_r(V)$ is an exterior or polynomial algebra depending on the parity of r.

Minimal algebras are usually built using the following construction.

DEFINITION 3.19. (cf. [**Gr–Mo**]). An inclusion $A \subseteq B$ of DG algebras is a *Hirsch extension* of degree r if, for some V, $B = A \otimes \bigwedge_r(V)$ as graded-commutative algebras, and $dV \subseteq A^{r+1}$.

In practice, minimal algebras are constructed inductively by setting $M(0, 0) = M(0) = \mathbf{Q}$ and $M(n) = \bigcup_{q \geq 0} M(n, q)$ for $n \geq 1$, where each $M(n, q)$, $q \geq 1$, is a degree n Hirsch extension of $M(n, q - 1)$.

Let A be a DG algebra. Let $A_+ = \bigoplus_{k \geq 1} A^k$. The complex $Q = A/A_+ \wedge A_+$ is the *complex of indecomposable* of A. The cohomology group $H^n(Q)$ is denoted by $\pi^n(A)$. There is a (pointed) homotopy theory of DG algebra [**Bo–Gu**, §6] in which $\pi^n(A)$ is the group of homotopy classes of maps from A to a fixed DG algebra ("the n-sphere") [**Bo–Gu**, 6.16]. The thrust of Sullivan's theory is that, with some restrictions, the homotopy category of DG algebras is equivalent to the homotopy category of a certain class of topological space [**Bo–Gu**, §9].

A minimal algebra M has the property that the differential is decomposable, that is, $dM \subseteq M_+ \wedge M_+$ [**Bo–Gu**, 7.3]. In this case $\pi^n(M) = Q^n$.

The next proposition says that each homotopy class of (connected) DG algebras contains a unique minimal algebra.

PROPOSITION 3.20 [Bo–Gu, 7.7 and 7.8]. *Let A be a DG algebra with $H^{\circ}(A) \cong \mathbf{Q}$. Then there is a minimal algebra M and a DG algebra map $f : M \longrightarrow A$ such that $f^* : H^*(M) \longrightarrow H^*(A)$ is an isomorphism. The algebra M is unique up to isomorphism, and the map $f : M \to A$ is unique up to homotopy of DG algebra maps.*

The algebra M of 3.20 is called the *minimal model* of A. The subalgebra $M(n)$ of M ($n \geq 1$) is called the *n-minimal model* of A. The n-minimal model is characterized by the following properties:

(i) $M(n)$ is a minimal algebra.

(ii) $M(n)$ is generated by elements of degree at most n.

(iii) $f/M(n) \colon M(n) \longrightarrow A$ induces isomorphisms in cohomology through degree n, and an injection in degree $n + 1$. (This is implied by [Bo–Gu, 7.9].)

Suppose X is a connected simplicial complex. Sullivan defined the DG algebra $A = A(X)$ of $\mathbf{Q}$ polynomial forms on X. The minimal (n-minimal) model of X is by definition the minimal (n-minimal) model of the DG algebra of $\mathbf{Q}$-polynomial forms on X. The main tool of [Fa 1] is the connection between the one-minimal model of X and the fundamental group of X. We shall describe it below (cf. [Mor] for detail).

Let A be DG algebra with $H^{\circ}(A) = \mathbf{Q}$ and $H^*(A)$ finitely generated. In this case the one-minimal model of A is an increasing union of degree one Hirsch extensions. Let $S = M(1)$ be the one-minimal model, and set $S(n) = M(1, n)$. Then $S(n-1) \subseteq S(n)$ is a degree one Hirsch extension for each $n \geq 1$; write $S(n) \cong S(n-1) \otimes \bigwedge_1(V_n)$. Set $W_n = S(n)^1 = V_1 \oplus \cdots \oplus V_n$. The dual of the map $d \colon W_n \longrightarrow W_n \wedge W_n$ give rise to a (graded) Lie algebra structure on the dual vector space $\operatorname{Hom}(W_n, \mathbf{Q}) = W_n^*$, the Jacobi identity being the dual of the equation $d^2 = 0$. The exact sequence $0 \longrightarrow W_{n-1} \longrightarrow W_n \longrightarrow V_n \longrightarrow 0$ dualizes to $0 \longrightarrow V_n^* \longrightarrow W_n^* \longrightarrow W_{n-1}^* \longrightarrow 0$. The fact that $dV_n \subseteq W_{n-1} \wedge W_{n-1}$ implies that the latter sequence is a central extension of the Lie algebras. Since $d/_{W_1} = d/_{V_1} = 0$, W_1^* is an abelian Lie algebra. It follows that W_n^* is a nilpotent Lie algebra for each $n \geq 1$. The tower of nilpotent rational Lie algebras

$$\cdots \longrightarrow W_n^* \longrightarrow W_{n-1}^* \longrightarrow \cdots \longrightarrow W_1^* \longrightarrow 0$$

is called the *dual Lie algebra* of S.

Let G be a finitely presented group. Construct the lower central series G_n of G by setting $G_0 = G$ and $G_n = [G_{n-1}, G]$ for $n \geq 1$. Setting $G(n) = G_{n-1}/G_n$ and $N_n = G/G_n$, we have for each $n \geq 1$ the central extension $1 \longrightarrow G(n) \longrightarrow N_n \longrightarrow N_{n-1} \longrightarrow 1$ of nilpotent groups. By a construction of Malcev, see [Ma, pp. 142–145], there is a method for tensoring those nilpotent groups by $\mathbf{Q}$ which preserves central extensions and agrees with ordinary tensor product for abelian groups. Thus we have a central extension $1 \longrightarrow G(n) \otimes \mathbf{Q} \longrightarrow N_n \otimes \mathbf{Q} \longrightarrow N_{n-1} \otimes \mathbf{Q} \longrightarrow 1$. The sequence $\cdots \longrightarrow N_n \otimes \mathbf{Q} \longrightarrow N_{n-1} \otimes \mathbf{Q} \longrightarrow \cdots N_1 \otimes \mathbf{Q} \longrightarrow 1$ is called the *rational nilpotent completion* of G; the inverse limit is the *Malcev completion* of G, denoted $G \otimes \mathbf{Q}$. The groups $N_n \otimes \mathbf{Q}$ are nilpotent rational Lie groups. The Campbell-Hausdorff formula yields a rational Lie algebra structure on each $N_n \otimes \mathbf{Q}$.

The topological significance of the 1-minimal model is the following result of Sullivan.

PROPOSITION 3.21 (cf. [Su] or [Gr–Mo]). *Let X be a connected simplicial complex with finitely generated rational cohomology. Let S be the one-minimal model of X, and let G be the fundamental group of X. Then the dual Lie algebra of S is isomorphic to the rational nilpotent completion of G.*

Under this isomorphism, the central extensions $0 \longrightarrow V_n^* \longrightarrow W_n^* \longrightarrow W_{n-1}^* \longrightarrow 0$ and $1 \longrightarrow G(n) \otimes \mathbf{Q} \longrightarrow N_n \otimes \mathbf{Q} \longrightarrow N_{n-1} \otimes \mathbf{Q} \longrightarrow 1$ correspond. Since $G(n)$ is finitely generated and abelian [Ma–Ka–So], $G(n) \otimes \mathbf{Q}$ is an ordinary tensor product with dimension equal to the rank of $G(n)$. This gives a useful corollary of Proposition 3.21.

COROLLARY 3.22. $\operatorname{rank}(G(n)) = \dim V_n$.

We denote rank $(G(n))$ by $\phi_n(X)$; these are the exponents appearing in the LCS formula.

The minimal model often contains information on the higher homotopy groups of a space. In particular, we point out another major result of Sullivan.

PROPOSITION 3.23 (cf. [Su] or [Gr–Mo]). *Let X be a connected simplicial complex with $H^*(X, \mathbf{Q})$ finitely generated and $\pi_1(X, *) = 1$. Let M be the minimal model of X. Then, for each $n \geq 2$, $\pi_n(X) \otimes \mathbf{Q}$ is isomorphic to $\operatorname{Hom}(\pi^n(M), \mathbf{Q})$.*

Now suppose $\mathscr{A}$ is an arrangement. Since $M(\mathscr{A})$ the complement of the arrangement is a formal space in the sense of Sullivan [Su, p. 315], we may replace the algebra $A(M)$ of $\mathbf{Q}$-polynomial forms on $M(\mathscr{A})$ with the algebra $A = A(\mathscr{A})$. Let $\rho: \mathscr{M} \longrightarrow A$ be a one-minimal model. We may view A as a DG algebra with zero differential so $H^*(A) = A$.

DEFINITION 3.24. Call $\mathscr{A}$ is a rational $K(\pi, 1)$-arrangement if $\rho^*: H^*(\mathscr{M}) \longrightarrow A$ is an isomorphism i.e. minimal model of $M(\mathscr{A})$ is generated by elements of degree ≤ 1.

Falk [Fa 1] and Kohno [Ko 5] used different methods to prove the following.

THEOREM 3.25. *Let $\mathscr{A}$ be a rational $K(\pi, 1)$-arrangement and write $\phi_n = \phi_n(\pi_1((\mathscr{A})))$. Then $\prod_{n \geq 1}(1 - t^n)^{\phi_n} = P(M(\mathscr{A}), -t)$.*

This theorem is stronger than Theorem 3.17 because Falk [Fa 1] proved that fiber-type arrangements are rational $K(\pi, 1)$ and there is an example

of arrangement which satisfies the lower central series formula in Theorem
3.25 but is not rational $K(\pi, 1)$.

Let $\mathscr{A}$ be the complexification of a real arrangement which may be affine.
In [**Sa 1**], Salvetti constructed explicitly a finite CW-complex $X \subseteq M(\mathscr{A})$
such that the inclusion is a homotopy equivalence. We refer the reader to
the original paper [**Sa 1**] of Salvetti for the construction.

4. Differentiable type of complement of arrangement

In this section, we denote $\mathscr{A}$ the central arrangement of hyperplanes in
$\mathbf{C}^3$ and $\mathscr{A}^*$ its associated projective arrangement of hyperplanes in $\mathbf{CP}^2$.
Associated with the arrangement is the lattice L of intersections. As we saw
in §2 , one can compute the cohomology ring of the complement from this
intersection lattice by a result of [**Or–So 1**], but it is unknown whether L
determines the homotopy type of the complement. In what follows, we shall
define a numerical invariant for any projective arrangement of hyperplanes
in $\mathbf{CP}^2$. We shall call this numerical invariant the complexity of the arrangement. Our main result is that if the complexity of the arrangement is not so
high, then L determines the differentiable structure of the complement.

DEFINITION 4.1. Let $\mathscr{A}$ be a central arrangement in $\mathbf{C}^3$ and $\mathscr{A}^*$ be
the corresponding projective arrangement in $\mathbf{CP}^2$. A point p in $\mathbf{CP}^2$ is
a k-tuple point of $\mathscr{A}^*$ if p is the intersection of exactly k lines in $\mathscr{A}^*$.
Let t_k be the number of k-tuple points in the arrangement $\mathscr{A}^*$. Then the
complexity $c(\mathscr{A}^*)$ of $\mathscr{A}^*$ (or $\mathscr{A}$) is defined to be $\sum_{k\geq3}(k-2)t_k$.

DEFINITION 4.2. Let $\mathscr{A}$ be a central arrangement in $\mathbf{C}^3$ and $\mathscr{A}^*$ be the
corresponding projective arrrangement in $\mathbf{CP}^2$. We define a graph $G(\mathscr{A}^*)$
associated to $\mathscr{A}^*$ as follows. The set of vertices of G is $V_G = \{p: p$ is a k-
tuple of $\mathscr{A}^*$ where $k \geq 3\}$ and the set of edges of G is $E_G = \{$lines in $\mathscr{A}^*$
which connect at least two points in $V_G\}$.

EXAMPLE 4.1. The graph $G(\mathscr{A}^*)$ of Example 1.1 in §1 is shown in Figure
3 where • denotes triple point.

FIGURE 3

The detail of the proof of the following theorem can be found in [**Ji–Ya**].

THEOREM 4.3. *Let $\mathscr{A}_1$ and $\mathscr{A}_2$ be two central arrangements in $\mathbf{C}^3$ and
$\mathscr{A}_1^*$, $\mathscr{A}_2^*$ be the corresponding projective arrangements in $\mathbf{CP}^2$. If the lattices
of $\mathscr{A}_1$ and $\mathscr{A}_2$ are isomorphic and there is a vertex v_0 in $G(\mathscr{A}_1^*)$ such that
there is no loop in $G(\mathscr{A}_1^*)\backslash\{v_0$ and all lines passing through $v_0\}$, then the
complements of the projective arrangements $\mathscr{A}_1^*$, $\mathscr{A}_2^*$ are diffeomorphic to
each other.*

REMARK 4.4. The condition in Theorem 4.3 implies that $c(\mathscr{A}_1^*) \le n - 2$ where n is the number of lines in $\mathscr{A}_1^*$. We next give two examples which are communicated to us by Salvetti. The first example shows two different lattices with same t_k for all $k \ge 2$.

EXAMPLE 4.2. We only show the affine part of the projective arrangements $\mathscr{A}_1^*$ and $\mathscr{A}_2^*$. It is understood that the lines at infinity are not in $\mathscr{A}_1^*$ and $\mathscr{A}_2^*$ (see Figure 4).

- denotes triple point.
○ denotes double point.

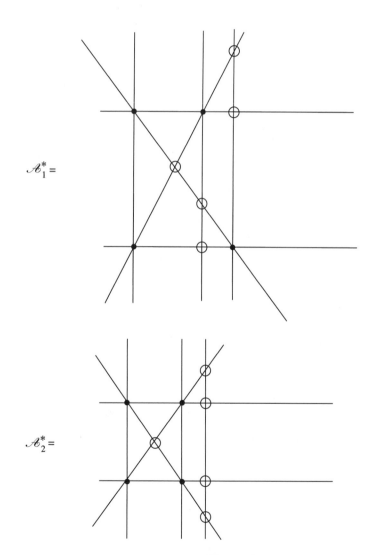

Figure 4

Here $|\mathscr{A}_1^*| = |\mathscr{A}_2^*| = 7$ and $t_2 = 9$, $t_3 = 4$, $t_4 = 0$.

The next example shows two arrangements $\mathscr{A}_1^*$, $\mathscr{A}_2^*$ in $\mathbf{CP}^2$ such that there exists a bijection $\phi: \mathscr{A}_1^* \longrightarrow \mathscr{A}_2^*$ in such a way that for every $H_1 \in \mathscr{A}_1$, the number of k-tuple points on H_1 and on $\phi(H_1)$ are the same for all k. However the lattices associated to $\mathscr{A}_1$ and $\mathscr{A}_2$ are not isomorphic.

EXAMPLE 4.3. We only show the affine part of the projective arrangements $\mathscr{A}_1^*$ and $\mathscr{A}_2^*$. It is understood that the lines at infinity are not in $\mathscr{A}_1^*$ and $\mathscr{A}_2^*$ (see Figure 5).

In both diagrams, we label each line by an ordered pair (i_1, i_2) where i_1 represents the number of triple points on the line while i_2 represents the number of double points on the line. The existence of the bijection ϕ from $\mathscr{A}_1^*$ to $\mathscr{A}_2^*$ with the desired properties is clear. We next claim that

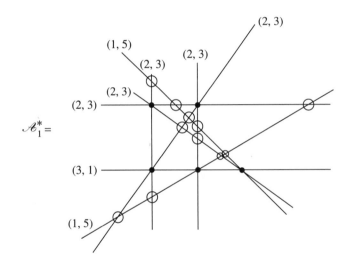

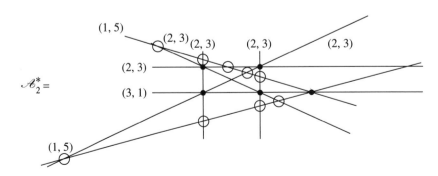

Figure 5

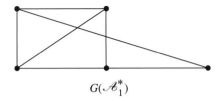

$$G(\mathscr{A}_1^*)$$

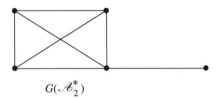

$$G(\mathscr{A}_2^*)$$

Figure 6

the two lattice $L(\mathscr{A}_1)$ and $L(\mathscr{A}_2)$ are not isomorphic. If there was a lattice isomorphism from $L(\mathscr{A}_1)$ to $L(\mathscr{A}_2)$, then it would send (1, 5) line to (1, 5) lines. However observe that with (1, 5) lines removed from both $L(\mathscr{A}_1)$ and $L(\mathscr{A}_2)$, we get the nonisomorphic lattices in Example 4.3 without (1, 5) lines.

REMARK 4.5. In Example 4.3, the graphs of $\mathscr{A}_1^*$ and $\mathscr{A}_2^*$ are given in Figure 6. Here • denotes triple point. We remark that $G(\mathscr{A}_1^*)$ (but not $G(\mathscr{A}_2^*)$) satisfies the condition of Theorem 4.3 and $G(\mathscr{A}_1^*)$ is different from $G(\mathscr{A}_2^*)$.

REFERENCES

[Ar] V. I. Arnold, *The cohomology ring of the colored braid group*, Mat. Zametki **5** (1969), 227–231: Math. Notes **5** (1969), 138–140.

[Bi] G. Birkhoff, Lattice theory, 3rd ed., Amer. Math. Soc. Colloq. Publ., vol. 25, Providence, R.I., 1967.

[Bo-Gu] A. Bousfield and V. Guggenheim, *On PL de Rham theory and rational homotopy type*, Mem. Amer. Math. Soc., vol. 8, no. 179, Amer. Math. Soc., Providence, R.I., 1976.

[Br] E. Brieskorn, *Sur les groups de tresses*, Séminaire Bourbaki 1971/1972, Lecture Notes in Math., vol. 317, Springer-Verlag, Berlin/Heidelberg/New York, 1973, pp. 21–44.

[Ca] P. Cartier, *Les arrangements d'hyperplans: un chapitre de géométrie combinatoire*, Séminaire Bourbaki 1980/1981, Lecture Notes in Math., vol. 901, Springer-Verlag, Berlin/Heidelberg/New York, 1981, pp. 1–22.

[Ch] D. Chéniot, *Le théorèm de Van Kampen sur le groupe fondamental du complémentaire d'une courbe algébrique projective plane*, Séminaire F. Norguet, 1970–1973: Fonctions de Plusiers Variables Complexes, Lecture Notes in Math., vol. 409, Springer-Verlag, 1974, pp. 394–417.

[Co] H. S. M. Coxeter, *Discrete groups generated by reflections*, Ann. of Math. (2) **35** (1934), 588–621.

[De] P. Deligne, *Les immeubles des groups de tresses génèralisés*, Invent. Math. **17** (1972), 273–302.

[De-Mo] P. Deligne and G. D. Mostow, *Monodromy of hypergeometric functions and non-lattice integral monodromy*, Publ. Math. Inst. Hautes Études Sci. **63** (1986), 5–89.

[Fa1] M. Falk, *The minimal model of the complement of an arrangement of hyperplanes*, Trans. Amer. Math. Soc. **309** (1988), 543–556.

[Fa2] ___, *The cohomology and fundamental group of a hyperplane complement*, Contemporary Math. **90** (1989), 55–72.

[Fa-Ra1] M. Falk and R. Randell, *The lower central series of a fiber-type arrangement*, Invent. Math. **82** (1985), 77–88.

[Fa-Ra2] ___, *On the homotopy theory of arrangements*, Complex Analytic Singularities, Advanced Studies in Pure Math., no. 8, North–Holland, 1987, pp. 101–124.

[Ge] I. M. Gelfand, *General theory of hypergeometric functions*, Soviet Math. Dokl. **33** (1986), 573–577.

[Gr-Mo] P. Griffiths and J. Morgan, *Rational homotopy theory and differential forms*, Birkhäuser, Boston, 1981.

[Ha] A. Hattori, *Topology of $\mathbf{C}^n$ minus a finite number of affine hyperplanes in general position*, J. Fac. Sci. Univ. Tokyo **22** (1975), 205–219.

[Ha-Le] H. Hamm and Lê Dũng Tráng, *Un Théorème de Zariski du type de Lefschetz*, Ann. Sci. École Norm. Sup. **6** (1973), 317–366.

[Hi] F. Hirzebruch, *Arrangements of lines and algebraic surfaces*, Arithmetic and Geometry, vol. II, Progress in Math., no. 36, Birkhäuser, Boston, 1983, pp. 113–140.

[Ja-Te] M. Jambu and H. Terao, *Arrangements libres d'hyperplans et treillies hyper-résolubles*, C. R. Acad. Sci. Paris **296** (1983), 623–624.

[Ko] T. Kohno, *On the minimal algebra and $K(\pi, 1)$-property of affine algebraic varieties* (preprint).

[Ko2] ___, *On the holonomy Lie algebra and the nilpotent completion of the fundamental group of the complement of hypersurfaces*, Nagoya Math. J. **92** (1983), 21–37.

[Ko3] ___, *Série de Poincaré-Koszul a associée aux groupes de tresses pures*, Invent. Math. **82** (1985), 57–75.

[Ko4] ___, *Homology of a local system on the complement of hyperplanes*, Proc. Japan Acad. Ser. A **62** (1986), 144–147.

[Ko5] ___, *Rational $K(\pi, 1)$ arrangements satisfy the LCS formula*, preprint.

[Ma-Ka-So] W. Magnus, A. Karrass and D. Solitar, *Combinatorial group theory*, Wiley, 1966.

[Moi] B. Moishezon, *Simply connected algebraic surfaces of general type*, Invent. Math. **89** (1987), 601–643.

[Mor] J. Morgan, *The algebraic topology of smooth algebraic varieties*, Publ. Math. Inst. Hautes Études Sci. **48** (1978), 137–204.

[Or] P. Orlik, *Introduction to arrangements*, CBMS Regional Conf. Ser. in Math. no. 72, Amer. Math. Soc., Providence, R.I. 1989.

[Or-So1] P. Orlik and L. Solomon, *Combinatories and topology of complements of hyperplanes*, Invent. Math. **56** (1980), 167–189.

[Or-So2] ___, *Unitary reflection groups and cohomology*, Invent. Math. **59** (1980), 77–94.

[Or-So3] ___, *Discriminants in the invariant theory of reflection groups*, Nagoya Math. J. **109** (1988), 23–45.

[Ra1] R. Randell, *On the fundamental group of the complement of a singular plane curve*, Quart. J. Math. Oxford Ser. (2) **31** (1980), 71–79.

[Ra2] ___, *The fundamental group of the complement of a union of complex hyperplanes*, Invent. Math. **69** (1982), 103–108. Correction Invent. Math. **80** (1985), 467–468.

[Ro] G.-C. Rota, *On the foundations of combinatorial theory. I, Theory of Möbius functions*, Z. Wahrscheinlichkeit-screchnung **2** (1969), 340–368.

[Sa1] M. Salvetti, *Topology of the complement of real hyperplanes in $\mathbf{C}^N$*, Invent. Math. **88** (1987), 603–618.

[Sa2] ____, *Arrangements of lines and monodromy of plane curves*. Compositio Math. **68** (1988), 103–122.

[Sh-To] G. C. Shephard and J. A. Todd, *Finite unitary reflection groups*, Canad. J. Math. **6** (1954), 274–304.

[St1] R. P. Stanley, *Modular elements of geometric lattices*, Algebra Universalis **1** (1971), 214–217.

[St2] ____, *Supersolvable lattices*, Algebra Universalis **2** (1972), 214–217.

[Su] D. Sullivan, *Infinitesimal computations in topology*, Publ. Math. Inst. Hautes Études. Sci. **47** (1977), 269–331.

[Te] H. Terao, *Modular elements of lattices and topological fibration*, Adv. in Math. **62** (1986), 135–154.

UNIVERSITY OF ILLINOIS AT CHICAGO

Proceedings of Symposia in Pure Mathematics
Volume **54** (1993), Part 2

The Relationship between the Moduli Spaces of Vector Bundles on K3 Surfaces and Enriques Surfaces

HOIL KIM

I. Introduction

In the construction of the moduli spaces of vector bundles, we need the concepts of simple bundles and stable bundles.

DEFINITION. (1) A vector bundle E on a variety X is simple if $H^0(X, \mathscr{E}nd\, E) = \mathbb{C}$.

(2) Let the H-slope of E be $c_1(E) \cdot H^{n-1}/\operatorname{rank}(E)$, where H is some ample divisor of X and n is the dimension of X. Then a vector bundle E on a variety X is H-stable (semistable) if the slope $\mathscr{F} < (\leq)$ slope of E for any proper subsheaf $\mathscr{F}$ of E.

S. Mukai [**Muk1**] showed that the moduli space of simple vector bundles on a K3 surface or an abelian surface is smooth and has a symplectic structure, i.e., there exists a nondegenerate global holomorphic 2-form which is closed.

Here we show that the image of the pull-back map from the space of stable vector bundles on an Enriques surface to the moduli space of simple vector bundles on its K3 covering is a Lagrangian subvariety. Recall that a Lagrangian subvariety in a symplectic variety of dimension $2n$ is a subvariety of dimension n where the symplectic form vanishes. Our main result is the following.

THEOREM. *Let Y be an Enriques surface and let X be the associated* K3 *cover. Then the pull-back map from the open subvariety which consists of stable vector bundles E satisfying $E \not\cong E(K)$ in the moduli space of stable vector bundles on Y to the moduli space of simple vector bundles on X is two to one and the image is a Lagrangian subvariety.*

1991 *Mathematics Subject Classification*. Primary 14J10, 14F05, 14J28.
The final version of this paper will be submitted for publication elsewhere.

In Chapter 2 we recall some preliminary facts and in Chapter 3 we prove the theorem. (In this paper we assume that the rank is two.)

II. Preliminaries

(1) K3 surfaces and Enriques surfaces. A K3 surface is a surface X with a trivial canonical class and $b_1(X) = 0$, where b_1 is the first betti number. Furthermore, X is simply connected. An Enriques surface is a surface Y with $2K_Y = 0$ but $K_Y \neq 0$, $q = h^1(Y, 0_Y) = 0$ and $p_g = h^2(Y, 0_Y) = 0$. The fundamental group of Y is $\mathbb{Z}/2\mathbb{Z}$ and its universal covering space is a K3 surface.

(2) simple bundles and stable bundles.

THEOREM 2.1. (a) *E is stable if and only if E^v, dual of E, is stable.*

(b) *E is stable if and only if $E(D)$ is stable, where D is any divisor.*

(c) *Let $f : E_1 \rightarrow E_2$ be a nontrivial sheaf homomorphism between two stable vector bundles E_1 and E_2 with the same ranks and the same slopes. Then f is an isomorphism.*

PROOF. See [OSS].

THEOREM 2.2. *Every stable bundle is simple.*

PROOF. See [OSS]. We denote by $\sum_{c_1,\dots,c_r}^{X}(S)$ the set of equivalent classes of families of stable r-bundles over X. This defines a contravariant functor

$$\sum_{c_1,\dots,c_r}^{X} : \quad An \rightarrow Ens$$

from the category of complex spaces to the category of sets.

DEFINITION 2.3. A complex space $M = M_X(c_1, \dots, c_r)$ is a coarse moduli space for $\sum_{c_1,\dots,c_r}^{X}$ if the following conditions are satisfied:

(i) there is a natural transformation of contravariant functors

$$\sum_{c_1,\dots,c_r}^{X} \rightarrow \mathrm{Hom}(-, M_X(c_1, \dots, c_r)),$$

which is bijective for any (reduced) point x_0;

(ii) for every complex space N and every natural transformation $\sum_{c_1,\dots,c_r}^{X} \rightarrow \mathrm{Hom}(-, N)$ there is a unique holomorphic mapping

$$f : M_X(c_1, \dots, c_r) \rightarrow N$$

for which the diagram

$$\sum_{c_1,\dots,c_r}^{X} \longrightarrow \mathrm{Hom}(-, M_X(c_1, \dots, c_r))$$

$$\mathrm{Hom}(-, N)$$

with maps labeled f^*, commutes.

REMARK. There exist coarse moduli spaces of stable vector bundles on curves [**Mu, Se**], surfaces and higher dimensional varieties [**Gi, Ma**]. By deformation theory, the dimension of the tangent space of M_X, T_{M_X} at E is

$$2rc_2 - (r-1)c_1^2 - r^2(1-q+p_g) + 1 + h^2(X, \mathscr{E}nd\, E)$$

when X is a surface.

(3) Symplectic structure on the moduli spaces of vector bundles on K3 surfaces.

THEOREM 2.4 [**Muk1**]. *Assume that X is an abelian or K3 surface. Let E_0 be a simple sheaf on X and T denote the tangent sheaf of the moduli space of simple sheaves on X, Spl_X. Then we have*

(1) Spl_X *is smooth and its dimension at E_0 is equal to $c_1(E_0)^2 - 2r(E_0)\chi(E_0) + r(E_0)^2\chi(O_X) + 2$, where χ denotes the Euler-Poincaré characteristic.*

(2) *There is a line bundle L isomorphic to O_{Spl_X} and a skew-symmetric bilinear form $B : T \times T \to L$ such that B at E_0 is nondegenerate and canonically isomorphic to the natural pairing $\mathrm{Ext}^1(E_0, E_0) \times \mathrm{Ext}^1(E_0, E_0) \to \mathrm{Ext}^2(E_0, E_0)$ for any simple sheaf E_0.*

III. Proof of the main theorem

Let us fix the notation.

$$
\begin{array}{ccc}
X & \xrightarrow{\ i\ } & X \\
& \searrow{\scriptstyle \pi} \quad {\scriptstyle \pi}\swarrow & \\
& Y &
\end{array}
\qquad
\begin{array}{l}
X : \text{K3 surface,} \\
Y : \text{Enriques surface,} \\
i : \text{fixed point free involution,} \\
\pi : \text{projection,}
\end{array}
$$

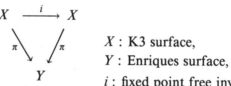

$$
\begin{array}{ccc}
M_X(2, \pi^*c_1, \pi^*c_2) & \xleftarrow{\ i^*\ } & M_X(2, \pi^*c_1, \pi^*c_2) \\
{\scriptstyle \pi^*}\nwarrow & & \nearrow{\scriptstyle \pi^*} \\
& M \subset M_Y(2, c_1, c_2) &
\end{array}
$$

$M_Y(2, c_1, c_2)$: the moduli space of stable rank 2 vector bundles E on an Enriques surface Y with $c_1(E) = c_1$ and $c_2(E) = c_2$, M: the open subvariety of M_Y which consists of bundles E satisfying $E \not\cong E(K)$,

$M_X(2, \pi^*c_1, \pi^*c_2)$: the moduli space of simple rank 2 vector bundles E on a K3 surface X with $c_1(E) = \pi^*c_1$ and $c_2(E) = \pi^*c_2$.

We divide the proof into five steps.

STEP (1). We show that the dimension of M is half of the dimension of M_X.

Dim M_X is equal to dim TM_X at $\pi^* E = 2r\pi^* c_2 - (r-1)(\pi^* c_1)^2 - 2r^2 + 2$ for any E since $h^2(\mathscr{E}nd\, E) = h^0(\mathscr{E}nd\, E) = 1$.

Dim TM_Y at $E = 2rc_2 - (r-1)c_1^2 - r^2 + 1 + h^0((\mathscr{E}nd\, E)(K))$.

However $H^0((\mathscr{E}nd\, E)(K)) = 0$ if and only if $E \not\cong E(K)$. (See Theorem 2.1.) So

$$\dim M = \dim TM_Y \text{ at } E = 2rc_2 - (r-1)c_1^2 - r^2 + 1 = \tfrac{1}{2}\dim M_X.$$

Furthermore M is smooth.

STEP (2). Next we show that if E is stable, then $\pi^* E$ is simple for E with $E \not\cong E(K)$, and that $\pi^* E$ is not simple for E with $E \cong E(K)$. Hence π^* is well defined from M to M_X.

By using $\pi_* O_X = O_Y \oplus O_Y(K)$, hence $\pi_* \pi^* E = E \oplus E(K)$, we have

$$H^0(\mathscr{E}nd\, \pi^* E) = H^0(\mathscr{E}nd\, E) \oplus H^0((\mathscr{E}nd\, E)(K)).$$

So $\pi^* E$ is simple if $E \not\cong E(K)$.

However $H^0(\mathscr{E}nd\, \pi^* E) \cong \mathbb{C} \oplus \mathbb{C}$, if $E \cong E(K)$. This implies that $\pi^* E$ is not simple.

REMARK. Note that π^* is a regular map. Indeed we can construct a family $\mathscr{E}_i$ of stable vector bundles on $U_i \times Y$, such that $\mathscr{E}_{i|E \times Y} \cong E$ for each E in U_i, where $M = \bigcup U_i$. (See [No].) Then $(\mathrm{id} \times \pi)^* \mathscr{E}_i$ is a family of simple bundles on $U_i \times X$ such that the restriction on $E \times X$ is $\pi^* E$. By using the universality of the coarse moduli space of M_X, we can define a morphism f_i from U_i to M_X. After gluing, we can define a morphism f from M to M_X, which is just π^*.

STEP (3). Here we show that the map π^* is 2 to 1.

If $\pi^* E \cong' \pi^* E'$, $H^0(\pi^*(E^v \otimes E')) \neq 0$, where E, E' are in M. However

$$H^0(\pi^*(E^v \otimes E')) = H^0(E^v \otimes E') \oplus H^0(E^v \otimes E' \otimes O_Y(K)).$$

So either $H^0(E^v \otimes E') \neq 0$, or $H^0(E^v \otimes E' \otimes O_Y(K)) \neq 0$. The property of stability implies $E \cong E'$ in the first case and $E \cong E'(K)$ in the second case. So we can conclude that the map π^* is 2 to 1 since $E \cong E(K)$.

STEP (4). We claim that $\pi^* M \subset M_X^{i^*}$, where $M_X^{i^*}$ is the fixed locus of the involution i^* on M_X.

This is obvious, since $i^*(\pi^* E) = (\pi \circ i)^* E = \pi^* E$, but $\pi^* M \neq M_X^{i^*}$ in many examples. (See (K2).)

STEP (5). Finally we show that $\omega_{M_X}(\pi^* E)|_{T\pi^* M} = 0$, where ω_{M_X} is the nonvanishing syplectic 2 form on M_X.

$$\omega_{M_X} : TM_X \times TM_X \to O_{M_X}.$$

Here $TM_{X(\pi^*E)} \cong \operatorname{Ext}^1(\pi^*E, \pi^*E)$, so,

$$\omega_{M_X}(\pi^*E): \operatorname{Ext}^1(\pi^*E, \pi^*E) \times \operatorname{Ext}^1(\pi^*E, \pi^*E)$$
$$\to \operatorname{Ext}^2(\pi^*E, \pi^*E) \cong \mathbb{C}.$$

We illustrate this with the following diagrams.

$$
\begin{array}{ccc}
\operatorname{Ext}^1(\pi^*E, \pi^*E) \times \operatorname{Ext}^1(\pi^*E, \pi^*E) & \xrightarrow{\ f\ } & \operatorname{Ext}^2(\pi^*E, \pi^*E) \cong \mathbb{C} \\
\big\uparrow {\scriptstyle (\pi^*,\pi^*)} & & \big\uparrow {\scriptstyle \pi^*} \\
\operatorname{Ext}^1(E, E) \times \operatorname{Ext}^1(E, E) & \xrightarrow{\ g\ } & \operatorname{Ext}^2(E, E) \\
& & \big\| \\
& & H^0((\operatorname{End} E)(K))
\end{array}
$$

where $TM_Y(E) \cong \operatorname{Ext}^1(E, E)$.

$$
\begin{array}{ccc}
(\pi^*G, \pi^*H) & \xrightarrow{\ f\ } & f(\pi^*G, \pi^*H) \\
\big\uparrow {\scriptstyle (\pi^*,\pi^*)} & & \big\uparrow {\scriptstyle \pi^*} \\
(G, H) & \xrightarrow{\ g\ } & g(G, H)
\end{array}
$$

However $g(G, H) = 0$ if $E \not\cong E(K)$, for G, $H \in \operatorname{Ext}^1(E, E) \cong TM_Y(E)$, since $\operatorname{Ext}^2(E, E) = H^0((\operatorname{End} E)(K)) = 0$ if $E \cong E(K)$.

So, $f(\pi^*G, \pi^*H) = 0$ by the commutativity of the diagram, for π^*G, $\pi^*H \in \operatorname{Ext}^1(\pi^*E, \pi^*E) \cong TM_X(\pi^*E)$. This implies that

$$\omega_{M_X}(\pi^*E)|_{T\pi^*M} = 0.$$

From these 5 steps we can conclude that π^*M is a Lagrangian subvariety in M_X fixed by involution. $\square$

REFERENCES

[BPS] W. Barth, C. Peters, and A. Van de ven, *Compact complex surfaces*, Springer-Verlag, 1984.

[CD] F. Cossec and I. Dolchev, *Enriques surfaces*. I, Birkhäuser, 1989.

[Gi] D. Gieseker, *On the moduli of vector bundles on an algebraic surface*, Ann. of Math. (2) **106** (1977), 45–60.

[K1] H. Kim, *Exceptional bundles on Enriques surfaces* (to appear in Manuscripta MetaMetica).

[K2] ____, *Moduli spaces of stable bundles on Enriques surfaces* (in preparation).

[Ma1] *Stable vector bundles on an algebraic surface*, Naggoya Math. J. **58** (1975), 25–68.

[Ma2] ____, *Moduli of stable sheaves*. I, J. Math. Kyoto Univ. **17** (1977), 91–126.

[Muk1] S. Mukai, *Symplectic structure of the moduli spaces of sheaves on an abelian or K3 surface*, Invent. Math. **77** (1984), 101–116.

[Muk2] ____, *On the moduli space of bundles on K3 surface*. I, Tata Inst. Fund. Res. Stud. in Math. **11** (1987), 341–413.

[Mu] D. Mumford, *Geometric invariant theory*, Springer-Verlag, 1965.

[No] V. A. Norton, *Analytic moduli of complex vector bundles*, Indiana Univ. Math. J. **28** (1979), 365–387.

[OSS] C. Okonek, M. Schneider, and H. Spindler, *Vector bundles on complex projective spaces*, Birkhäuser, 1980.

[Se] C. S. Seshadri, *Space of unitary vector bundles on a compact Riemann surface*, Ann. of Math. (2) **85** (1967), 303–336.

[Ty] A. N. Tyurin, *Cycles, curves and vector bundles*, Duke Math. J. **54** (1987), 1–26.

UNIVERSITÄT BAYREUTH, GERMANY

Proceedings of Symposia in Pure Mathematics
Volume **54** (1993), Part 2

Self-Dual Manifolds with Symmetry

C. LeBRUN AND Y. S. POON

ABSTRACT. An oriented Riemannian four-manifold is said to be *half-conformally-flat* if its conformal curvature W is either self-dual or anti-self-dual as a bundle-valued two-form. We review a construction [18] of compact half-conformally-flat manifolds with semifree isometric S^1-action, starting from the Green's function of a collection of points in a hyperbolic three-manifold. If the three-manifold in question is just hyperbolic space, the resulting four-manifolds are one-point conformal compactifications of scalar-flat Kähler surfaces. We then show that any asymptotically Euclidean scalar-flat Kähler surface with a nonzero conformal Killing field arises from this construction.

1. Overview

Imagine that you are given a smooth oriented compact four-manifold M, and, perhaps motivated by pleasant memories of the uniformization theory of surfaces, you try to find a conformally-flat metric g on it. You would thus be looking for a Riemannian metric g which, relative to some atlas of charts $\{x_\alpha : \mathscr{U}_\alpha \longrightarrow \mathscr{V}_\alpha \subset \mathbb{R}^4\}$ for M, is of the form

$$g = f_\alpha \sum_{j=1}^{4} (dx_\alpha^j)^{\otimes 2}$$

on each open set $\mathscr{U}_\alpha$, where f_α is some positive smooth function. Generally speaking, you would be out of luck, for the following reason: the conformal curvature[1] W of such a metric must vanish, whereas, for an arbitrary Riemannian metric g on M, the signature $\sigma = b_+ - b_-$ of M is given by

$$\sigma = \frac{1}{12\pi^2} \int_M (\|W_+\|^2 - \|W_-\|^2) \, d \, \text{vol},$$

1991 *Mathematics Subject Classification*. Primary 53C25; Secondary 53C55, 32C17, 32L25.
Partially supported by NSF grants # DMS 9003263 and DMS 8906809.
This paper is in final form and no version will be submitted for publication later.

[1] Weyl tensor: the totally-trace-free piece of the Riemann curvature.

so the existence of a conformally-flat metric on M implies that $\sigma = 0$. Here $W_{\pm}$ is the self-dual (respectively, anti-self-dual) piece of the Weyl tensor, as defined by $W_{\pm} := \frac{1}{2}(W \pm *W)$, where the Hodge-star operator $*$ treats W as a bundle-valued two-form.

Since the above problem is, generally speaking, hopeless, one should perhaps try to settle for a metric with as little conformal curvature W as allowed by the topology. A good measure of the the total amount of conformal curvature present is given by the L^2-norm

$$A(g) := \int_M \|W\|^2 \, d \operatorname{vol} = \int_M (\|W_+\|^2 + \|W_-\|^2) \, d \operatorname{vol}$$

of the Weyl tensor, because, unlike other norms you might be tempted to apply, this one has the virtue that it is *conformally invariant* — $A(g) = A(f \cdot g)$ for any smooth positive function f. But $A(g) \geq 12\pi^2 |\sigma|$, with equality iff either $W_+ \equiv 0$ or $W_- \equiv 0$, in which case the conformal Riemannian manifold $(M, [g])$ is said to be *half-conformally-flat*. The cases $W_+ \equiv 0$ and $W_- \equiv 0$ are more specifically called anti-self-dual and self-dual, respectively; they differ, of course, only by a choice of the orientation.

The two simplest examples of compact *self-dual* manifolds are provided by the Riemannian symmetric spaces S^4 and $\mathbb{CP}_2$, oriented in the usual manner. The half-conformal-flatness of the latter example has, fundamentally, nothing at all to do with its Kähler structure, to such an extent that the following general observation now seem rather surprising: a Kähler manifold of complex dimension 2 is *anti-self-dual* with respect to the standard orientation iff its scalar curvature is identically zero [8].

A particularly compelling reason for the study of half-conformally-flat four-manifolds comes from the Penrose twistor construction [2, 24]. Let (M, g) be an orientable Riemannian four-manifold, and let $F \to M$ be the principal $SO(4)$-bundle of orthonormal frames determining the same orientation on M. Let $Z = F/U(2)$, which is a bundle over M with typical fiber $S^2 = SO(4)/U(2)$. Then the smooth six-manifold Z carries a natural almost-complex structure $J : TZ \to TZ$, $J^2 = -1$, which leaves invariant both the tangent spaces of each fiber and the horizontal spaces of the metric connection of g. Indeed, let us notice that, by construction, Z is exactly the space of almost-complex structures $\jmath : TM \to TM$ compatible with the given metric and orientation, and so, thinking of the g-horizontal subspace of TZ as the pull-back of TM to Z, there is thus a tautological way to let J act on the horizontal sub-bundle of TZ. In the vertical directions, on the other hand, J will simply act as the standard complex structure on S^2, namely rotation by $+90°$. Provided that we give the fibers the correct orientation in defining this almost-complex structure J, the entire construction turns out, rather surprisingly, to be *conformally invariant*, meaning that J remains completely unchanged if the given Riemannian metric g is replaced by αg, where $\alpha : M \to \mathbb{R}^+$ is any smooth positive function. This construction of an almost-complex manifold for each conformal Riemannian manifold may thus be thought of as a higher-dimensional analogue of the correspondence between conformal Riemannian two-manifolds and complex one-manifolds. However, the almost-complex manifold (Z, J) will not, in general be a complex manifold— there need not be an atlas of charts for Z relative to which J identically becomes multiplication by i in $\mathbb{C}^3 = \mathbb{R}^6$. Instead, the relevant integrability condition turns out to $W_+ = 0$. When Z is the space of almost-complex structures on M compatible with the given metric and the *conjugate* orientation on M, the tautological almost complex structure on the twistor space is integrable if and only if $W_- = 0$. In short, every half-conformally-flat four-manifold determines a complex three-fold Z, called its *twistor space*, and this complex three-manifold in turn completely encodes the conformal geometry of the original manifold. For example, the twistor space of S^4 is $\mathbb{CP}_3$, whereas the twistor space of $\mathbb{CP}_2$ is the flag-manifold

$$\mathbb{F} = \{([z_1, z_2, z_3], [w_1, w_2, w_3]) \in \mathbb{CP}_2 \times \mathbb{CP}_2 | z \cdot w = 0\} .$$

Which smooth compact four-manifolds M admit half-conformally-flat metrics? Certainly not all, for example, neither $S^2 \times S^2$ nor $\mathbb{CP}_2 \# \overline{\mathbb{CP}}_2$ admit such metrics, since these manifolds have signature zero, which would force any putative half-conformally-flat metric to be conformally-flat — whereas [17] the only simply-connected conformally-flat manifold is S^4! There is a reasonable hope, however, that, for any M, its connected sum $M \# m\mathbb{CP}_2$, $m \gg 0$, with a sufficiently large number of complex projective planes might always admit such metrics. (Note added in proof: This has now been proved by C. H. Taubes, J. Differential Geom. **36** (1992), 163–253.) While this remains beyond the scope of current technology, a panoply of interlocking methods [**26, 6, 7, 18, 19, 20**] has evolved in the past several years for the construction of self-dual metrics on connected sums of self-dual manifolds. The present article will focus on a particularly elementary and explicit construction of this type, but which is limited to the case of metrics with an isometric S^1-action.

2. Constructing self-dual manifolds

Suppose that (M_{free}, g) is a self-dual four-manifold with a free isometric circle action generated by a Killing field ξ, and let $\xi^b = g(\xi, \cdot)$ be the corresponding one-form. We may then equip the three-manifold $X := M/S^1$ with the unique metric h for which the canonical projection π becomes a Riemannian submersion; i.e. such that

$$\pi^* h = \left(g - \frac{\xi^b \otimes \xi^b}{\|\xi\|^2} \right) .$$

Let β be the unique one-form on X such that

$$\pi^* \beta = \frac{-d\|\xi\|^2 + 2 * \xi^b \wedge d\xi^b}{2\|\xi\|^2} ,$$

and define a connection $\mathbf{D}$ on X by

$$\mathbf{D}_v w := \nabla_v w + \beta(v)w + \beta(w)v - g(v, w)\beta^{\#},$$

where $\beta = h(\beta^{\#}, \cdot)$ and ∇ is the Riemannian connection of h. If we replace g by αg, where α is any S^1-invariant function on M, then the connection $\mathbf{D}$ and the conformal class $[h]$ remain unchanged. By construction, the torsion-free connection $\mathbf{D}$ preserves the conformal structure $[h]$ in the sense that parallel transport preserves angles, and is thus a so-called *Weyl connection*

$$(2.1) \qquad\qquad \mathbf{D}h = \omega \otimes h,$$

where $\omega = -2\beta$. The hypothesis that (M, g) is self-dual then has the consequence that the symmetrization of the Ricci tensor $\mathbf{r_D}$ of $\mathbf{D}$ is a multiple of h; i.e. there is a function $\lambda : X \to \mathbb{R}$ such that

$$(2.2) \qquad\qquad \mathbf{r_D}(v, w) + \mathbf{r_D}(w, v) = \lambda h(v, w) \quad \forall v, w.$$

We will call a three-manifold X equipped with a conformal metric h and a connection $\mathbf{D}$ satisfying (2.1) and (2.2) an *Einstein-Weyl* manifold. Such geometries were first studied by Elie Cartan [4], but their relation to self-dual four-manifolds was first recognized by Hitchin [12]. Several critical further observations described below were then made by Jones and Tod [14].

In order to reconstruct a self-dual four-manifold from an Einstein-Weyl geometry, we need an extra piece of information, namely the function $V = \|\xi\|^{-1}$. If we think of $M \to X$ as a circle bundle, we may equip it with a connection θ whose horizontal spaces are the g-orthogonal complements of the fibers. The self-duality of g then implies that the curvature of θ is given by $d\theta = *(d - \beta)V$.

We may invert this construction as follows: let $(X, [h], \mathbf{D})$ be an Einstein-Weyl three-manifold, and let $V : X \to \mathbb{R}$ be a positive solution of the elliptic equation $d * (d - \beta)V = 0$. Assume, in addition, that the closed 2-form $\frac{1}{2\pi} * (d - \beta)V$ represents an integral class in the de Rham cohomology $H^2(X)$. Then, by the Chern-Weil theorem, there is a circle-bundle $\pi : M \to X$ which admits a connection θ whose curvature is $d\theta = *(d - \beta)V$. Then, for any positive function μ on M, the metric $g = \mu(\pi^* h + V^{-2}\theta \otimes \theta)$ is self-dual. Often one takes $\mu = V$, so that the above expression becomes $g = \pi^* V h + V^{-1}\theta \otimes \theta$.

EXAMPLE 2.3. Take X to be $\mathbb{R}^3$ punctured at n points $p_1, \ldots, p_n$, with h the Euclidean metric and $\mathbf{D}$ the usual flat connection, and let V be the sum of their Green's functions:

$$V = \sum_{j=1}^{n} \frac{1}{2r_j},$$

where r_j is the Euclidean distance from p_j. Then $\frac{1}{2\pi} * dV$ has integral -1 on a small sphere around any one of the puncture points p_j; since such spheres generate $H_2(\mathbb{R}^3 - \{p_1, \dots, p_n\})$, we conclude that $[\frac{1}{2\pi} * dV]$ is an integral de Rham class. We can therefore consider the circle bundle $M_{free} \rightarrow (\mathbb{R}^3 - \{p_1, \dots, p_n\})$ with connection one-form θ whose curvature is $*dV$. There is then a self-dual metric on M given by $g = Vh + V^{-1}\theta \otimes \theta$. This is the metric of Gibbons and Hawking [9]. If we add n points $\hat{p}_1, \dots, \hat{p}_n$ to M_{free} to obtain a new space M which comes equipped with a circle action having $\hat{p}_1, \dots, \hat{p}_n$ as its fixed points and a projection $M \rightarrow \mathbb{R}^3$, then M admits a unique smooth structure such that g extends to M as a smooth Riemannian metric:

$$
\begin{array}{ccccc}
M & = & M_{free} & \cup & \{\hat{p}_1, \dots, \hat{p}_n\} \\
\downarrow & & \downarrow & & \downarrow \\
\mathbb{R}^3 & = & (\mathbb{R}^3 - \{p_1, \dots, p_n\}) & \cup & \{p_1, \dots, p_n\}
\end{array}
$$

Moreover, the resulting Riemannian manifold is complete and, by virtue of special properties of the Einstein-Weyl space $\mathbb{R}^3$, actually Ricci-flat Kähler. With a little care, this construction can easily be generalized to the case of infinitely many (sparsely located) centers [1].

EXAMPLE 2.4. [18]. Let $(X, h, \mathbf{D})$ be hyperbolic 3-space $\mathscr{H}^3$ punctured at n points $p_1, \dots, p_n$, where $\mathbf{D}$ is the Riemannian connection. We again build V from the Green's functions of the given points

$$
(2.5) \qquad V = 1 + \sum_{j=1}^{n} \frac{1}{e^{2\rho_j} - 1},
$$

where ρ_j denotes the hyperbolic distance from p_j. Then $[\frac{1}{2\pi} * dV]$ is again an integral class, and we can define a circle bundle $M_{free} \rightarrow (\mathscr{H}^3 - \{p_1, \dots, p_n\})$ with connection one-form θ whose curvature is $*dV$. Let ρ denote the hyperbolic distance from any reference point. The metric

$$
(2.6) \qquad g = (\mathrm{sech}^2 \rho)\,(\pi * Vh + V^{-1}\theta \otimes \theta)
$$

is then self-dual, and, because of our choice of conformal gauge, may be smoothly compactified by adding a two-sphere and n points $\hat{p}_1, \dots, \hat{p}_n$. Indeed, let B denote the closed unit ball in $\mathbb{R}^3$, and identify the interior of B with $\mathscr{H}^3$ via the Poincaré conformal model. Then $M = M_{free} \cup S^2 \cup \{p_1, \dots, p_n\}$ can be made into a smooth four-manifold with circle-action in such a manner that $S^2 \cup \{p_1, \dots, p_n\}$ is the fixed point set and B is the orbit space, so that the projection to B is as follows:

$$
\begin{array}{ccccccc}
M & = & M_{free} & \cup & S^2 & \cup & \{\hat{p}_1, \dots, \hat{p}_n\} \\
\downarrow & & \downarrow & & \downarrow & & \downarrow \\
B & = & (\mathscr{H}^3 - \{p_1, \dots, p_n\}) & \cup & \partial B & \cup & \{p_1, \dots, p_n\}
\end{array}
$$

Calculations similar to those involved in the analysis of Example (2.3) then show

THEOREM 2.7. *The metric g of equation (2.6) has nonnegative scalar curvature, and extends to M to yield a compact self-dual four-manifold diffeomorphic to the n-fold connected sum* $\mathbb{CP}_2 \# \cdot s \# \mathbb{CP}_2$.

When $n = 0, 1$, this construction produces the standard metrics on S^4 and $\mathbb{CP}_2$, respectively. When $n = 2$, we instead get the self-dual metrics on $\mathbb{CP}_2 \# \mathbb{CP}_2$ first found in [26].

EXAMPLE 2.8 [19, 20, 15]. Let Y be a hyperbolic three-manifold with smooth conformal compactification $\overline{Y}$, meaning that we assume that $\overline{Y}$ is a smooth compact three-manifold-with-boundary such that $Y = \overline{Y} - \partial\overline{Y}$, and the hyperbolic metric h of Y is of the form $h = f^{-2}\hat{h}$ for $\hat{h}$ a smooth Riemannian metric on $\overline{Y}$ and f a nondegenerate defining function of the boundary $\partial\overline{Y}$; by Thurston's main theorem [29], the class of $\overline{Y}$ admitting structures of this kind includes "most" atoroidal three-manifolds-with-boundary. Let $p_1, \ldots, p_n \in Y$ be given, let G_j be the corresponding Green's functions, and set $X = Y - \{p_1, \ldots, p_n\}$. Then we can mimic the previous construction of compact self-dual four-manifolds by taking

$$V = 1 + \sum_{j=1}^{n} G_j ,$$

trying to find a circle-bundle with connection one-form θ whose curvature is $*dV$, setting

$$g = f^2 \left(\pi^* V h + V^{-1} \theta \otimes \theta \right) ,$$

and compactifying by adding a copy of $\partial\overline{Y}$ and n isolated fixed points $\{\hat{p}_1, \ldots, \hat{p}_n\}$. The only catch lies in showing that $[\frac{1}{2\pi} * dV]$ is an integral cohomology class and, indeed, this will usually only be true for some special configurations of points! Nonetheless, one can verify the integrality condition in many cases. For example, if Y is a handle-body, the integrality condition is automatically verified, and one may use this to construct explicit self-dual metrics on arbitrary connected sums of $S^1 \times S^3$'s and $\mathbb{CP}_2$'s. On the other hand, if $Y = S_{\mathbf{g}} \times \mathbb{R}$, where $S_{\mathbf{g}}$ is a compact surface of genus $\mathbf{g} \geq 2$, one finds that the integrability condition is nontrivial, but, by restricting ones choice of point-configurations, the construction can be made to yield self-dual metrics on $(S^2 \times S_{\mathbf{g}}) \# n\mathbb{CP}_2$ provided that $n \neq 1$.

3. Twistor spaces

All the self-dual manifolds described in the previous section of course are associated with complex three-manifolds, namely their twistor spaces, and these complex manifolds completely encode the conformal geometry of each self-dual four-manifold, providing a higher-dimensional analog of the familiar dictionary between Riemann surfaces and complex curves. The fact that the metrics in question have conformal Killing fields is then reflected by a $\mathbb{C}^*$-action on their twistor spaces. At least locally, one can then construct the quotient of the twistor space by this action, thereby producing a complex surface, called the *minitwistor space* [14], which corresponds to the Einstein-Weyl quotient geometry [12]. Let us now examine our key examples in this light. We begin with the Gibbons-Hawking metrics of Example (2.3), the twistor spaces of which were discovered by Hitchin [10]. The relevant Einstein-Weyl geometry is in this case that of Euclidean three-space, and the corresponding mini-twistor space [13] is $T\mathbb{CP}_1$. Let $\mathcal{O}(k) \to T\mathbb{CP}_1$ denote the pull-back of the degree k line-bundle over $\mathbb{CP}_1$ via the canonical projection. The data points $p_1, \ldots, p_n \in \mathbb{R}^3$ specify n sections of $T\mathbb{CP}_1 \to \mathbb{CP}_1$, and these are the zero loci of n sections $P_1, \ldots, P_n$ of $\mathcal{O}(2)$. In the total space of the rank 2 vector bundle $\mathcal{O}(n) \oplus \mathcal{O}(n)$, let $\tilde{Z}$ denote the hypersurface $xy = \prod_{j=1}^n P_j$, where x and y refer to the two factors of $\mathcal{O}(n) \oplus \mathcal{O}(n)$. The twistor space Z of the Gibbons-Hawking metric is then given by a "small resolution" of this three-dimensional complex algebraic variety, meaning that each singular point is replaced by a rational curve. For an important generalization of this class of twistor spaces, see [16].

We now turn to the manifolds given by Example (2.4). In this case, the relevant Einstein-Weyl geometry is that of hyperbolic three-space, and the corresponding minitwistor space is $\mathbb{CP}_1 \times \mathbb{CP}_1$. Let $\mathcal{O}(k, l)$ denote the unique holomorphic line-bundle over $\mathbb{CP}_1 \times \mathbb{CP}_1$ with degree k on the first factor and degree l on the second, and let the data points $p_1, \ldots, p_n \in \mathcal{H}^3$ correspond to the zero loci of n sections $P_1, \ldots, P_n \in \Gamma(\mathbb{CP}_1 \times \mathbb{CP}_1, \mathcal{O}(1, 1))$. Let $\mathcal{B}$ denote the total space of the $\mathbb{CP}_2$-bundle

$$(3.1) \qquad \mathcal{B} := \mathbb{CP}(\mathcal{O}(n - 1, 1) \oplus \mathcal{O}(1, n - 1) \oplus \mathcal{O}) \xrightarrow{\pi} \mathbb{CP}_1 \times \mathbb{CP}_1,$$

and define an algebraic variety $\tilde{Z} \subset \mathcal{B}$ by the equation

$$(3.2) \qquad xy = t^2 \prod_{j=1}^n P_j,$$

where $x \in \mathcal{O}(n-1, 1)$, $y \in \mathcal{O}(1, n-1)$, and $t \in \mathcal{O} := \mathcal{O}(0, 0)$. The twistor space Z of the metric constructed in Example (2.4) is then obtained from $\tilde{Z}$ by making small resolutions of the singular points and blowing down the surfaces $x = t = 0$ and $y = t = 0$ to $\mathbb{CP}_1$'s. Notice that Hitchin's twistor spaces are degenerations of these.

These twistor spaces thus turn out to be *Moishezon spaces*, meaning that they are bimeromorphic to smooth projective varieties; for example, when $n = 2$, the above twistor space is bimeromorphic to a resolution of the intersection of two hyper-quadrics in $\mathbb{CP}_5$ with four ordinary double points [26]. For $n > 3$ one can also show [5, 22] that their generic small deformations are not even bimeromorphic to *Kähler* manifolds, so that one observes from these twistor examples a rather unexpected phenomenon of broader interest: the class of compact complex manifolds bimeromorphic to Kähler is not stable under deformation of complex structure.

4. Scalar-flat Kähler surfaces with symmetry

As mentioned in §1, an interesting class of half-conformally-flat four-manifolds is given by the scalar-flat Kähler surfaces, i.e. complex two-manifolds with Kähler metrics with scalar curvature $\equiv 0$. A beautiful characterization of such metrics in terms of their twistor spaces was found by Pontecorvo [25]. Namely, the complex structure J, as well as the conjugate complex structure $-J$, are, by definition, sections of the twistor fibration. These are "conjugate," in the sense that they are interchanged by the antipodal map on each fiber of the twistor fibration. (This fiber-wise antipodal map is an anti-holomorphic involution of the twistor space, and will henceforth be called the "real structure.") The integrability condition on J then implies that the image of each of these two sections are complex hypersurfaces, say Σ and $\overline{\Sigma}$, in the twistor space Z. The fact that the metric is Kähler then implies that the line bundle $(\Sigma\overline{\Sigma})^2$ is isomorphic to the anti-canonical bundle K^{-1} of the twistor space Z; the crux of the argument is that the Kähler form, being parallel, corresponds by the Penrose transform [11] to a holomorphic section of $K^{-1/2}$, and the zero-locus of this section is, by inspection, $\Sigma \coprod \overline{\Sigma}$. Conversely, if a holomorphic section of $K^{-1/2}$ is invariant under the real structure of a twistor space, it defines a scalar-flat Kähler metric in the conformal class on the open subset of the four-manifold over which the given holomorphic section has two distinct zeroes on each twistor fiber.

A remarkable consequence of this is the following: a self-dual manifold arising as in Example (2.3) or (2.4) is automatically conformally isometric to a number of different scalar-flat Kähler surfaces! Indeed, as we saw in §3, the corresponding twistor spaces Z admit $\mathbb{C}^*$-actions for which the stable quotient is either $T\mathbb{CP}_1$ or $\mathbb{CP}_1 \times \mathbb{CP}_1$. But this means that a section of $K^{-1/2}$ on $T\mathbb{CP}_1$ or $\mathbb{CP}_1 \times \mathbb{CP}_1$, respectively, will pull back as section of $K^{-1/2}$ on Z. Since these complex surfaces admit many half-anti-canonical divisors, the claim follows. For applications of these ideas to the construction of *compact* scalar-flat Kähler surfaces via Example (2.5), see [20].

Let us now focus on the twistor spaces (3.2) arising from Example (2.4), in which there are two distinguished families of effective divisors, corresponding to the two factors of the projection $\pi : \tilde{Z} \to \mathbb{CP}_1 \times \mathbb{CP}_1$ in (3.1). These two families, henceforth denoted by $|D|$ and $|\overline{D}|$, are interchanged by the real structure. The intersection of any such divisor D with its conjugate $\overline{D}$ is just a fiber of the projection π; moreover, it is a *real twistor line*, corresponding to the twistor fiber of a point of the fixed two-sphere of the circle action on $n\mathbb{CP}_2$. When this twistor line is removed, the resulting space N is the blow-up of $\mathbb{C}^2$ at n-points along a complex line in $\mathbb{C}^2$. When the conformal factor $(\mathrm{sech}^2 \rho)$ in (2.6) is replaced by e^{2B}, where $B : \mathcal{H}^3 \to \mathbb{R}$ is the

Busemann function of any geodesic, the resulting metric

$$(4.1) \qquad e^{2B}(\pi^* V h + V^{-1} \theta \otimes \theta),$$

where V is given by (2.5), is a scalar-flat Kähler metric on N. (Of course, the orientation on N is opposite that of M.) Moreover, the S^1-action is carried over to N to a conformal action. It turns out that these properties completely characterize our metrics:

THEOREM 4.2. *Suppose that N is a complete scalar-flat Kähler surface. If the metric is asymptotically Euclidean and admits a nontrivial conformal Killing field, it is isometric to (4.1).*

PROOF. When N is asymptotically Euclidean, it can be conformally compactified by adding one point at infinity, henceforth called o. After a change of orientation, the conformal curvature of the resulting compact conformal Riemannian manifold M is self-dual. Let $\wp : Z \to M$, be the twistor fibration. Thus Z is a compact complex manifold.

Over $M \backslash \{o\}$, the twistor fibration has two sections, namely the complex structure J on N and its conjugate $-J$. The integrability of J implies that the images Σ and $\overline{\Sigma}$ of these sections are complex submanifolds of Z. Let D and $\overline{D}$ respectively denote the closures of Σ and $\overline{\Sigma}$. Since $D \backslash \Sigma$ is contained in the twistor line $L_o = \wp^{-1}(o)$ of the point at infinity, $D \cup L_o$ is *-analytic in the sense of Mumford [23], and hence a subvariety of Z. Thus $D \subset Z$ is a complex hypersurface. The same applies to $\overline{D}$.

We already know that the divisor D is nonsingular away from the twistor line L_o. On the other hand, a generic twistor fiber intersects D transversely in a single point, so the homological intersection number $D \cdot L_o$ is 1. Let $z \in Z$ be any point. Through z passes a two-complex-parameter family of $\mathbb{CP}_1$'s with normal bundle $\mathcal{O}(1) \oplus \mathcal{O}(1)$ and representing the fiber homology class. The tangent spaces of these curves fill out an open cone in $T_z Z$. In particular, through every point of $z \in D$ passes such a $\mathbb{CP}_1$ meeting D in precisely one point. Fix another point z' on such a curve, and consider the two-parameter family of nearby $\mathbb{CP}_1$'s through z'. Each such curve meets D in one point. This provides a holomorphic chart for a neighborhood of $z \in D$. Hence the divisor D is a nonsingular surface.

Let us consider the divisor $\Sigma + \overline{\Sigma}$ in the twistor space $Z \backslash L_o$ of our original noncompact complex surface. By Pontecorvo's theorem [25], the fact that the metric is scalar-flat Kähler implies that there is a line bundle isomorphism $\Sigma \overline{\Sigma} \cong \mathbf{K}^{-\frac{1}{2}}$. Let $\phi \in \Gamma(Z \backslash L_o, (\mathbf{D} \overline{\mathbf{D}})^* \otimes \mathbf{K}^{-\frac{1}{2}})$ be the section which realizes this isomorphism. Since $L_o \subset Z$ has complex codimension 2, this section extends across L_o by Hartog's theorem. But the extended section is nonzero away from the codimension 2 set L_o, and hence is everywhere nonzero. This shows that

$$(4.3) \qquad \mathbf{D} \overline{\mathbf{D}} \cong \mathbf{K}^{-\frac{1}{2}}.$$

After this observation, we can show that the surface D contains L_o. If not, both D and $\overline{D}$ intersect L_o transversely in one point each, and these points are interchanged by Σ. Thus $D \cap \overline{D} = \varnothing$, and we can apply Pontecorvo's theorem [25] to conclude that there is a scalar-flat Kähler metric on M in the fixed conformal class. This metric and our original Kähler metric g on $N = M\backslash\{o\}$ are thus conformally related and Kähler with respect to the same complex structure J. Since the complex dimension of N is > 1, the conformal factor is therefore constant. This would then imply that (M, g) is incomplete, which is a contradiction.

To finish our proof, we will need the following

PROPOSITION 2.5. *The surface D is isomorphic to $\mathbb{CP}_2$ blown up at a collinear collection of points.*

PROOF. With the isomorphism (4.3), we apply the adjunction formula on D to find the canonical class on D in terms of the restriction of the divisor classes of D and $\overline{D}$. Then applying the same formula to L_o as a divisor on D, we find that the self-intersection of this real twistor line is equal to one. It follows that D is a rational surface (see e.g. [3, Proposition V.4.3]). In particular, $h^1(D, \mathscr{O}) = 0$. With the last equality and the following exact sequence on D: $0 \to \mathscr{O} \to \mathscr{O}(\mathbf{L}_o) \to \mathscr{O}_{L_o}(1) \to 0$, we find that the twistor line L_o is in a net of rational curves. The associated map of this complete linear system exhibits the divisor D as the blow up of $\mathbb{CP}_2$.

Now the naturality of the twistor construction insures that any conformal automorphism of a self-dual manifold M induces a holomorphic automorphism of its twistor space. A conformal Killing field therefore lifts to a holomorphic vector field on Z. If this holomophic vector field is not tangent to D, we conclude that $\dim|D| \geq 1$; similarly for $|\overline{D}|$. In this case, we may then apply (3.1) in [27] to conclude our result. Otherwise, our one-parametergroup of conformal transformation of M induces holomorphic transformations of Z preserving on D.

Since a holomorphic transformation homotopic to the identity must leave any exceptional divisor fixed, the transformation of D descends to be a transformation of $\mathbb{CP}_2$ leaving the blown-up points fixed. As the twistor line at infinity is also invariant, there is an additional pair of fixed points, say, p and q. This is possible only if all these fixed points lie on two lines. As the line joining p and q does not contain any points of blowing up, all the blown-up points are collinear. $\square$

When all the blown-up points are collinear, the system $|D|$ on the surface D contains at least one member, namely the proper transform of the line containing all the blown up points. This is so because the isomorphism (4.3) implies that the divisor class of D is linearly equivalent to the proper transform of the line passing through all points of blowing up.

The restriction of the twistor fibration to D induces an isomorphism of first homotopy groups between D and M, since this map is a diffeomorphism away from the twistor line at infinity, which it contracts to a point. Thus M is simply-connected. But, by the Penrose transform, we therefore have $h^1(Z, \mathscr{O}) = b_1(M) = 0$ [11]. The exact sequence $0 \to \mathscr{O} \to \mathscr{D} \to \mathscr{O}_D(\mathbf{D}) \to 0$, on the twistor space now tells us that the linear system $|D|$ on Z has dimension one. According to theorem (3.1) in [27], the metric on M is therefore conformal to the metric (4.1). Moreover, the two complex structures agree. Since two Kähler metrics in the same conformal class on a connected complex surface can only differ by a homothety, the proof of our main theorem is finished.   □

REMARK 4.4. The first part of the above proof can easily be adapted to prove the following more general result: *Let M be an asymptotically Euclidean, scalar-flat Kähler surface, perhaps with many ends. Then M is biholomorphic to $\mathbb{C}^2$ blown up at a finite number of points.* In particular, such a manifold can only have one end, and is automatically simply connected. The reason for the "one end" conclusion is that any real twistor line "at infinity" will have self-intersection 1 in the surface D; yet two distinct real twistor lines must be disjoint! Surface classification thus excludes the possibility of the surface D containing two real twistor lines.

REMARK 4.5. If M is just anti-self-dual Hermitian instead of scalar-flat Kähler, theorem (4.2) will still hold provided one imposes the additional hypothesis that M be simply connected. Indeed, without any assumptions whatsoever, the isomorphism (4.3) generalizes to become $\mathbf{D}\overline{\mathbf{D}} \cong \mathbf{K}^{-\frac{1}{2}}\mathbf{F}$ where $\mathbf{F}$ is a holomorphic line bundle with vanishing Chern class. The assumption of simple connectivity then forces the torsion bundle $\mathbf{F}$ to be trivial. One can then once again apply Pontecorvo's theorem. However, scalar-flat anti-self-dual Hermitian counter-examples with fundamental group $\mathbb{Z}$ may be easily constructed by removing a point from a Hopf surface $S^3 \times S^1$.

REMARK 4.6. The Kähler metric (4.1) was originally found [18] by the method of Kähler reduction. The present theorem of course shows that this (or any other) method will not lead to other asymptotically Euclidean solutions. Nevertheless, there are many other complete Kähler surfaces of constant scalar curvature which may be found by this approach [20] and its generalizations [28].

REMARK 4.7. The metrics (4.1) can of course be conformally compactified so as to yield self-dual metrics on connected-sums of complex projective planes, and the corresponding compact four-manifolds automatically admit an isometric circle-action. An interesting generalization of the problem solved in this section would thus be that of classifying compact self-dual four-manifolds with nontrivial isometric S^1-action. The first author has recently succeeded in proving that, provided the action is assumed to be **semifree**, any such manifold with nonnegative scalar curvature must either arise as in Example (2.4), or else be conformally flat; however, it also seems certain that there will exist classes of examples for which the semifree hypothesis will fail.

References

1. M. Anderson, P. Kronheimer, and C. LeBrun, *Ricci-flat Kähler manifolds of infinite topological type*, Comm. Math. Phys. **125** (1989), 637–642.
2. M.F. Atiyah, N.J. Hitchin, and I.M. Singer, *Self-duality in four dimensional Riemannian geometry*, Proc. Roy. Soc. London Ser A. **362** (1978), 425–461.
3. W. Barth, C. Peters, and A. Van de Ven, *Compact complex surfaces*, Springer-Verlag, Berlin Heidelberg, 1984.
4. E. Cartan, *Sur une classe d'espaces de Weyl*, Ann. Sci. Ecole Norm. Sup. **60** (1943), 1–16.
5. F. Campana, *The class $\mathscr{C}$ is not stable by small deformations*, Math. Ann. **290** (1991), 19-30.
6. S.K. Donaldson and R. Friedman, *Connected sums of self-dual manifolds and deformations of singular spaces*, Nonlinearity **2** (1989), 197–239.
7. A. Floer, *Self-dual conformal structure on $l\mathbb{CP}_2$*, J. Differential Geom. **33** (1991), 551–573.
8. P. Gauduchon, *Surfaces kählériennes dont la courbure vérifie certaines conditions de positivité*, Géometrie Riemannienne en Dimension 4 (Séminaire A. Besse, 1978/1979), (Bérard-Bergery, Berger, and Houzel, eds.), CEDIC/Fernand Nathan, 1981.
9. G. W. Gibbons and S. W. Hawking, *Gravitational multi-instantons*, Phys. Lett **78B** (1978), 430–476.
10. N. J. Hitchin, *Polygons and gravitons*, Math. Proc. Cambridge Philos. Soc. **83** (1979), 465–476.
11. ____, *Linear field equations on self-dual spaces*, Proc. London. Math. Soc. **43** (1980), 133–150.
12. ____, *Complex manifolds and Einstein's equations*, Lecture Notes in Math., vol. 970 Springer-Verlag, Berlin and New York, 1982, pp. 73–99.
13. ____, *Monopoles and geodesics*, Commun. Math. Phys **83** (1982), 579–602.
14. P. E. Jones and K. P. Tod, *Minitwistor spaces and Einstein-Weyl spaces*, Classical Quantum Gravity **2** (1985), 565-577.
15. J.-S. Kim, *On a class of 4-dimensional minimum-energy metrics and hyperbolic geometry*, Ph.D. Thesis, SUNY Stony Brook, 1991.
16. P. Kronheimer, *The construction of ALE spaces as hyper-Kähler quotients*, J. Differential Geom. **29** (1989), 665-683.
17. H. N. Kuiper, *On conformally flat spaces in the large*, Ann. of Math. (2) **50** (1949), 916–924.
18. C. LeBrun, *Explicit self-dual metrics on $\mathbb{CP}_2\#...\#\mathbb{CP}_2$*, J. Differential Geom. **34** (1991), 223–253.
19. ____, *Anti-self-dual Hermitian metrics on blown-up Hopf surfaces*, Math. Ann. **289** (1991), 383–392.
20. ____, *Scalar-flat Kähler metrics on blown-up ruled surfaces*, J. Reine Angew. Math. **420** (1991), 161–177.
21. ____, *Twistors, Kähler manifolds, and bimeromorphic geometry. I*, J. Amer. Math. Soc. **5** (1992), 289–316.
22. C. LeBrun and Y.-S. Poon, *Twistors, Kähler manifolds, and bimeromorphic geometry. II*, J. Amer. Math. Soc. **5** (1992), 317–325.
23. D. Mumford, *Algebraic geometry I: Complex projective varieties*, Springer-Verlag, 1970.
24. R. Penrose, *Non-linear gravitons and curved twistor theory*, Gen. Relativity gravitation **7** (1976), 31-52.
25. M. Pontecorvo, *On twistor spaces of anti-self-dual Hermitian surfaces*, Trans. Amer. Math. Soc **331** (1992).
26. Y. S. Poon, *Compact self-dual manifolds with positive scalar curvature*, J. Differential Geom. **24** (1989), 97–132.
27. ____, *On the algebraic structure of twistor spaces*, J. Differential Geom. (to appear).

28. H. Pedersen and Y. S. Poon, *Hamiltonian constructions of Kähler-Einstein metrics and Kähler metrics of constant scalar curvature*, Comm. Math. Phys. **136** (1991), 309–326.
29. W. Thurston, *Three-dimensional manifolds, Kleinian groups, and hyperbolic geometry*, Bull. Amer. Math. Soc. (N.S.) **6** (1983), 357–381.

SUNY AT STONY BROOK

RICE UNIVERSITY

Proceedings of Symposia in Pure Mathematics
Volume **54** (1993), Part 2

Remarks on the Skyrme Model

ELLIOTT H. LIEB

One of the interesting constructs in theoretical physics was invented by
Skyrme [**S1,2**] in 1961. It carries with it some intriguing differential geometric
problems that have not been solved to this day, and the purpose of this note
is to make their existence a little better known to geometers.

The model is not fundamental in the sense of, say, Schrödinger's quantum
mechanics, or Maxwell's electromagnetism. Rather, it is phenomenological
in the sense that it is supposed to give a good account of the structure of
baryons and their interaction with mesons at low energies. The model posits
a simple energy functional of a simple "field" (i.e., a map from $\mathbf{R}^3$ to $SU(2)$)
and the problem is to understand the lowest energy fields and the fluctuations
around them. Only the energy minimization problem will be discussed here.
Physicists believe that the "correct theory" (quantum chromodynamics)—
if it is eventually understood—will have properties closely matched by the
Skyrme model. Therefore it is important to understand it in detail.

For mathematicians, a way to get started is to consult [**E1, LM**], and [**M1**].
Surveys of the physics literature are contained in [**SWHH**] and [**ZB**]. There,
one will find that the original model has since been elaborated in many ways,
especially by replacing $SU(2)$ by $SU(3)$. An insightful discussion of the relation of harmonic maps to physics is in [**MC**].

The objects of interest are maps

$$U : \mathbf{R}^3 \to SU(2) \approx \mathbf{S}^3.$$

This can be written as

$$U(x) = \exp[i f(x) \cdot \sigma]$$

1991 *Mathematics Subject Classification*. Primary 53C80, 81V05.
Work partially supported by U.S. National Science Foundation grant PHY-85-15288-A04.
This paper is in final form and no version of it will be submitted for publication elsewhere.

where $f : \mathbf{R}^3 \to \mathbf{R}^3$ and where σ^1, σ^2 and σ^3 are the Pauli matrices

$$\sigma^1 = \begin{pmatrix} 0 & 1 \\ 1 & 0 \end{pmatrix}, \quad \sigma^2 = \begin{pmatrix} 0 & -i \\ i & 0 \end{pmatrix}, \quad \sigma^3 = \begin{pmatrix} 1 & 0 \\ 0 & -1 \end{pmatrix}$$

and $f \cdot \sigma \equiv \sum_{i=1}^{3} f_i \sigma^i$. The function $x \mapsto f(x) \cdot \sigma$ is called the *meson field*. Associated with U is an energy

$$\mathscr{E}(U) = \mathscr{E}_1(U) + \mathscr{E}_2(U),$$

$$\mathscr{E}_1(U) = \int_{\mathbf{R}^3} \sum_{i=1}^{3} \operatorname{Tr} K_i^* K_i,$$

(1) $$\mathscr{E}_2(U) = -\frac{1}{4} \int_{\mathbf{R}^3} \sum_{i,j=1}^{3} \operatorname{Tr}(K_i K_j - K_j K_i)^2,$$

where $*$ denotes adjoint, K_i is the Lie algebra valued function $U^* \partial_i U$, and Tr denotes trace. Clearly $\mathscr{E}_1(U) \geq 0$ and $\mathscr{E}_2(U) \geq 0$ since $K_i^* = -K_i$. Assuming that $U(x) \to$ constant as $x \to \infty$, U defines a map from $\mathbf{S}^3$ to $\mathbf{S}^3$ with a well-defined integral degree k. We define

$$E_k = \inf\{\mathscr{E}(U) : \operatorname{degree}(U) = k\}.$$

Primarily $k = 1$ is interesting because the minimum energy $k = 1$ map defines a "Skyrmion" and this, essentially, is the simplest baryon.

The energy $\mathscr{E}_1$ is the usual harmonic map energy (as will be clearer in a moment). There is, unfortunately, no minimizer for $\mathscr{E}_1$ alone because, under the dilation $U(x) \to U(\lambda x)$, $\mathscr{E}_1 \to \frac{1}{\lambda}\mathscr{E}_1$. The energy $\mathscr{E}_2$ is introduced solely to stabilize the system because $\mathscr{E}_2 \to \lambda\mathscr{E}_2$ under dilation. Together $\mathscr{E}_1 + \mathscr{E}_2$ is stable, and stability under dilation yields $\mathscr{E}_1 = \mathscr{E}_2$ for any critical point of $\mathscr{E}$. Although $\mathscr{E}_1$ is "natural," we could consider altering $\mathscr{E}_2$ to the trace of some other invariant quartic form in first derivatives of U.

Esteban [E1] has proved (modulo technicalities) that a minimizing U exists when $k = 1$ (and for $k > 1$ if $E_k < E_{k_1} + E_{k_2}$ with $k_1 + k_2 = k$). (See note added in proof.) Physicists believe that this $k = 1$ Skyrmion is a *hedgehog*:

(2) $$U(x) = \exp\left[iF(|x|)\frac{x}{|x|} \cdot \sigma\right],$$

i.e., the Skyrmion has the maximum possible O(3) symmetry. The "boundary conditions" are $F(0) = \pi$, $F(\infty) = 0$. In general a hedgehog of degree k has $F(0) = k\pi$, $F(\infty) = 0$. Kapetansky and Ladyzenskaya [KL] proved the existence of a hedgehog minimizer for $k = 1$ and Esteban [E2] showed this for $k > 1$. This hedgehog ansatz leads to a very nonlinear ODE for $F : \mathbf{R}^+ \to [0, \pi]$. See [ANW, Y1, EL]. We then have two open problems—of which the first is the most interesting and important.

PROBLEM 1. Is the hedgehog ansatz correct for $k = 1$; i.e., does a minimizer have the form (2)?

PROBLEM 2. The ODE for the $k = 1$ hedgehog has been solved numerically, but can it be solved in closed form? Since the ODE arises from a "simple" geometric problem, it is not inconceivable that it has a "simple" solution.

Further insight into the problem is gained by rewriting U as a map $\rho : \mathbf{R}^3 \to \mathbf{S}^3$. More generally [LM, M1, MR, JMW, Y2] we can consider maps

$$\rho : (M, g) \to (N, h)$$

between two Riemannian manifolds, and an energy $\mathscr{E}(\rho) = \mathscr{E}_1(\rho) + \mathscr{E}_2(\rho)$ with

$$(3) \qquad \mathscr{E}_i(\rho) = \int_M \Gamma_i(g^{-1}\rho^*h)\, dV_g,$$

where V_g is the volume form on M, ρ^*h is the pullback of the metric h and, for an arbitrary $n \times n$ matrix A, $\Gamma_i(A)$ is the ith elementary symmetric function of the eigenvalues, λ_j, of A. In particular,

$$\Gamma_1(A) = \sum_{j=1}^{h} \lambda_j(A) = \operatorname{Tr} A,$$

$$\Gamma_2(A) = \sum_{1 \le i < j \le n} \lambda_i(A)\lambda_j(A) = \frac{1}{2}(\operatorname{Tr} A)^2 - \frac{1}{2}\operatorname{Tr}(A^2).$$

The energies $\mathscr{E}_1$ and $\mathscr{E}_2$ correspond exactly to (1). In coordinates

$$(g^{-1}\rho^*h)^i_j(x) = g^{ik}(x)(h_{\alpha\beta} \circ \rho)(x)\frac{\partial \rho^\alpha}{\partial x^k}\frac{\partial \rho^\beta}{\partial x^j}$$

and we see that $\mathscr{E}_1(\rho)$ is just the usual harmonic map energy. If dimension$(M) > 3$ we have to introduce some $\mathscr{E}_i$ with $i \ge n/2$ to have stability under dilation.

In order to understand the $k = 1$ hedgehog problem better, Manton and Ruback [MR] suggested studying the problem from $L\mathbf{S}^3$ to $\mathbf{S}^3$ (with the standard metrics). Here, $L\mathbf{S}^3 = \{x \in \mathbf{R}^4 : |x|^2 = L^2\}$ is the 3-sphere with radius L. As $L \to \infty$ one should recover the original $\mathbf{R}^3 \to \mathbf{S}^3$ problem. They proposed that when L is small, the minimizing $k = 1$ map should be the "identity'" map, i.e., $\rho(x) = L^{-1}x$. When L exceeds some critical value L_c the O(4) symmetry of the "identity" map should be broken down to the O(3) symmetry of the hedgehog map (the obvious analogue of (2)), and should remain with this O(3) symmetry all the way to $L = \infty$.

Manton and Ruback conjectured that $L_c = \sqrt{2}$ because they could easily prove that the "identity" map is locally unstable for $L > \sqrt{2}$. Subsequently, Loss and Lieb [LM] and Manton [MN] proved that the "identity" map is the unique minimizer for $L \le 1$. Loss [LM], by an ingenious argument, also showed that the "identity" map is a local minimum for $L < \sqrt{2}$. Several years later, Wirzba and Bang [WB] actually found all the eigenvalues of the linear second variation operator at the "identity" for all L.

The proof of the above $L \leq 1$ stability result is so simple that I cannot resist giving it here. In this lemma M and N are compact, oriented, 3-dimensional manifolds without boundary with volumes $|M|$ and $|N|$. Maps ρ of degree k are considered.

LEMMA. *If* $\mu^3 \equiv |k||N|/|M| \geq 1$ *then* $\mathscr{E}(\rho) \geq 3|M|(\mu^2 + \mu^4)$, *with equality iff* $\rho^* h = \mu^2 g$.

COROLLARY. *If* M *and* N *are homothetic (i.e.,* $\exists \psi : M \to N$ *with* $L^2 \psi^* h = g$), *then* $\mu^3 L^3 = |k|$ *and, when* $L \leq 1$,

$$\min\{\mathscr{E}(\rho) : \deg(\rho) \geq 1\} = 3|M|(L^{-2} + L^{-4})$$

with the "identity" map (i.e., the ψ *above) being the unique minimizer up to isometry.*

PROOF OF LEMMA. By Newton's theorem, $\Gamma_1(A)/3 \geq [\Gamma_2(A)/3]^{1/2} \geq [\Gamma_3(A)]^{1/3}$. If we define $I(p) \equiv |M|^{-1} \int_M \Gamma_3(g^{-1}\rho^* h)^p dV_g$, we have that

$$(4) \qquad \mathscr{E}(\rho) \geq 3|M|\{I(1/3) + I(2/3)\}.$$

Furthermore, $|M|I\left(\frac{1}{2}\right) = \int_M dV_{\rho^* h} \geq |k||N|$ and also, by Schwarz's inequality, $I(1/3)^2 \leq I(2/3)$ and $I(1/2)^2 \leq I(1/3)I(2/3)$. Thus, $\mu^6 \leq I(1/3)I(2/3)$ and

$$|M|^{-1}\mathscr{E}(\rho) \geq 2\min\{x + y : xy \geq \mu^6, y \geq x^2\}$$
$$= 3\min\{x + \max(x^2, \mu^6/x)\} = 3(\mu^2 + \mu^4)$$

if $\mu \geq 1$. $\square$

PROBLEM 3. Prove that for the $LS^3 \to S^3$, $k = 1$ problem the "identity" is the unique minimizer for $1 < L < \sqrt{2}$.

PROBLEM 4. Prove that every minimizer has O(3), but not O(4), symmetry when $L > \sqrt{2}$.

By a combination of analytic plus numerical methods, Wirzba and Bang [WB] assert that the hedgehog is a local minimum when $L > \sqrt{2}$. But the global minimum for $L > 1$ is still a mystery. It has not even been proved that there is a minimizer satisfying both $(L, 0, 0, 0) \to (1, 0, 0, 0)$ and $(-L, 0, 0, 0) \to (-1, 0, 0, 0)$.

Using the ideas of the above lemma, we can get another useful bound [FL, S1, S2] when degree $(\rho) = k$. From (4),

$$(5) \qquad \mathscr{E}(\rho) \geq 6|M|\sqrt{I(1/3)I(2/3)} \geq 6|k||N|.$$

This is known as the *topological bound*; the important point is that it is independent of $|M|$, and hence of L. (5) is an identity for the identity map when $L = 1$, and for the $\mathbf{R}^3 \to S^3$ hedgehog (corresponding to $L = \infty$) the right side of (5) is not far from the hedgehog energy.

Now let us return to the $\mathbf{R}^3 \to SU(2)$ problem and consider the $k > 1$ minimizers. If we accept that the $k = 1$ minimizer is a hedgehog then

$E_2 < 2E_1$ and hence, according to [E1] (see also [EL]) there is a $k = 2$ minimizer. The reason $E_2 < 2E_1$ is that there is a simple ansatz U_2 [S1, S2, KI] with $\mathscr{E}(U_2) < 2\mathscr{E}(k = 1,$ hedgehog) and $\deg(U_2) = 2$.

$$(6) \qquad U_2(x) = \exp\left\{ iF(|x|)\frac{x}{|x|}\cdot\sigma \right\} W \exp\left\{ iF(|x-y|)\frac{x-y}{|x-y|}\cdot\sigma \right\} W^{-1}$$

with $|y| \gg 1$. Here F is the $k = 1$ hedgehog in (2) and $W = iv\cdot\sigma \in \mathrm{SU}(2)$ with $|v|^2 = 1$ and $v\cdot(x - y) = 0$. One finds $\mathscr{E}(U_2) - 2E_1 \approx -C|y|^{-3}$. A minimizer for $k = 2$ cannot be a $k = 2$ hedgehog because, as Esteban has shown by a brief, but elegant variational argument in [E2, Proposition 6],

$$\mathscr{E}(k = d, \text{ hedgehog}) > \mathscr{E}(k = d - l, \text{ hedgehog}) + \mathscr{E}(k = l, \text{ hedgehog})$$

for any $d > l \geq 1$.

Thus, Skyrmions bind—just as baryons do. Given two $k = 1$ skyrmions, the energy can be lowered by bringing them from infinity to some finite separation. But they have to be "twisted" relative to each other with the W above; this is poetic, of course, because the true $k = 2$ minimizer could be a complicated beast that cannot be easily identified as two skyrmions, as in (6). In fact, more sophisticated analyses of the $k = 2$ minimizer [BC,VJ] have led to the conjecture that the object has the shape of a torus. See also [M2].

PROBLEM 5. Assuming $E_2 < 2E_1$, what are the qualitative properties of the $k = 2$ minimizer? What about many Skyrmions?

Klebanov [KI] has analyzed the very-many-Skyrmion problem (i.e., nuclear matter) by generalizing (6) to a periodic array. Later, more detailed analyses of this kind of periodic structure are in [WBJ] and [GM]. If many Skyrmions really do bind, what kind of pattern is formed in $\mathbf{R}^3$? There is no evident reason that it has to be periodic.

The many-Skyrmion problem appears, for reasons that are unclear, to be very different from

(1) harmonic maps from $\mathbf{R}^3$ to $\mathbf{S}^2$,

(2) Yang-Mills $\mathrm{SU}(2)$ monopoles (with large λ),

(3) Ginzburg-Landau vortices (with $\lambda > 1$).

See [AL] for (1) and [JT] for (2), (3). In these three cases it is believed (and proved for (1) by Almgren-Lieb [AL]) that *only* $k = 1$ globally stable solutions exist; i.e., the elementary $k = 1$ "excitations" (or "particles") repel each other. Somehow, the quartic term in (1) plays a role whose significance has not been sufficiently understood.

NOTE ADDED IN PROOF. The technical restrictions in [E1] have been resolved in [EM].

REFERENCES

[AL] F. J. Almgren, Jr. and E. H. Lieb, *Singularities of energy minimizing maps from the ball to the sphere: Examples, counterexamples, and bounds*, Ann. of Math. (2) **128** (1988), 483–530.

[ANW] G. S. Adkins, C. R. Nappi, and E. Witten, *Static properties of nucleons in the Skyrme model*, Nuclear Phys. B **228** (1983), 552–566.

[BC] E. Braaten and L. Carson, *Deutron as a toroidal Skyrmion*, Phys. Rev. D **38** (1988), 3525–3539.

[E1] M. J. Esteban, *A direct variational approach to Skyrme's model for meson fields*, Comm. Math. Phys. **105** (1986), 571–591.

[E2] ____, *Existence of symmetric solutions for the Skyrme's problem*, Ann. Math. Pura Appl. (4) **147** (1987), 187–195.

[EL] M. J. Esteban and P. L. Lions, *Skyrmions and symmetry*, Asymp. Anal. **1** (1988), 187–192.

[EM] M. J. Esteban and S. Müller, *Sobolev maps with integer degree and applications to Skyrme's problem*, Proc. Roy. Soc. London A **436** (1992), 197–201.

[FL] L. D. Fadeev, *Some comments on the many-dimensional solitons*, Lett. Math. Phys. **1** (1976), 289–293.

[GM] A. S. Goldhaber and N. S. Manton, *Maximal symmetry of the Skyrme crystal*, Phys. Lett. B **198** (1987), 231–234.

[JJP] A. Jackson, A. D. Jackson, and V. Pasquier, *The Skyrmion-Skyrmion interaction*, Nuclear Phys. A **432** (1985), 567–609.

[JMW] A. D. Jackson, N. S. Manton, and A. Wirzba, *New Skyrmion solutions on a 3-sphere*, Nuclear Phys. A **495** (1989), 499–522.

[JT] A. Jaffe and C. Taubes, *Vortices and monopoles*, Birkhäuser, Boston, 1980.

[KI] I. Klebanov, *Nuclear matter in the Skyrme model*, Nuclear Phys. B **262** (1985), 133–143.

[KL] L. B. Kapitansky and O. A. Ladyzenskaya, *On Coleman's principle concerning the stationary points of invariant functionals*, Zap. Nauchn. Sem. Leningrad. Otdel. Mat. Inst. Steklov. (LOMI) **127** (1983), 84–102.

[LM] M. Loss, *The Skyrme model on Riemannian manifolds*, Lett. Math. Phys. **14** (1987), 149–156.

[MC] C. W. Misner, *Harmonic maps as models for physical theories*, Phys. Rev. D **18** (1978), 4510–4528.

[M1] N. S. Manton, *Geometry of Skyrmions*, Comm. Math. Phys. **111** (1987), 469–478.

[M2] ____, *Monopole and Skyrmion bound states*, Phys. Lett. B **198** (1987), 226–230.

[MR] N. S. Manton and P. J. Ruback, *Skyrmions in flat space and curved space*, Phys. Lett. B **181** (1986), 137–140.

[S1] T. H. R. Skyrme, *A non-linear field theory*, Proc. Roy. Soc. London Ser. A **260** (1961), 127–138.

[S2] ____, *A unified field theory of mesons and baryons*, Nuclear Phys. **31** (1962), 556–569.

[SWHH] B. Schwesinger, H. Weigel, G. Holzworth, and A. Hayashi, *The Skyrme soliton in pion, vector and scalar-meson fields: πN-scattering and photo production*, Phys. Rep. **173** (1989), 173–255.

[VJ] J. J. M. Verbaarschot, *Axial symmetry of bound baryon number-two solution of the Skyrme model*, Phys. Lett. B **195** (1987), 235–239.

[WB] A. Wirzba and H. Bang, *The mode spectrum and the stability analysis of Skyrmions on a 3-sphere*, Nuclear Phys. A **515** (1990), 571–598.

[WBJ] E. Wüst, G. E. Brown, and A. D. Jackson, *Topological chiral bags in a baryonic environment*, Nuclear Phys. A **468** (1987), 450–472 .

[Y1] Y. Yang, *On the global behavior of symmetric Skyrmions*, Lett. Math. Phys. **19** (1990), 25–33.

[Y2] ____, *Generalized Skyrme model on higher dimensional Riemannian manifolds*, J. Math. Phys. **30** (1989), 824–828.

[ZB] I. Zahed and G. E. Brown, *The Skyrme model*, Phys. Rep. **142** (1986), 1–102.

PRINCETON UNIVERSITY

Proceedings of Symposia in Pure Mathematics
Volume **54** (1993), Part 2

Geometric Settings for Quantum Systems with Isospin

E. B. LIN

1. Introduction

Classical dynamics of particles with internal degrees of freedom can be described in terms of a principal bundle over the cotangent bundle of the space-time manifold [5]. In contrast to this internal symmetry, we intend to develop quantum theory of a system of particles with external symmetry. Namely, we are interested in a molecule which is a system of particles or atomic nuclei in the Born-Oppenheimer approximation.

The physical properties of a quantum mechanical system depend upon the types of molecules out of which the system is built, but as critically upon the overall geometrical way in which the various constituents are put together. On the other hand, symmetry group often greatly simplifies complex problems. In fact, it offers one an understanding of the way in which mathematical properties of wave functions of a quantum system depend upon the physical symmetry of the system.

In a series of papers [2, 3], Iwai first developed a gauge theory for the quantum planar three-body system. He provided a mathematical meaning of nonrigidity of molecules. As an application of the connection theory due to Guichardet [1], he also established a gauge theory for nonrigid molecules on the basis of the observation that the vector bundle associated with the principal fibre bundle gives rise to a setting for quantum mechanics of the internal molecular motion. From the view point of Berry's phase, Wu also pointed out that the rotational and vibrational motions are not seperable in the planar three-body problem [7]. In fact, he showed that the corrected symplectic structure gives the correct quantization in the planar three-body system [8]. On the basis of Guichardet's work, Iwai demonstrated that the internal motion of the nonrigid molecule can be well described in terms of the gauge theory or the connection theory in differential geometry [2, 3].

1991 *Mathematics Subject Classification.* Primary 53C80,53C05,53B50.

This work was partially supported by a summer fellowship, University of Toledo.

This paper is in final form and no version of it will be submitted for publication elsewhere.

In this paper, we shall establish quantum theory of nonrigid molecule with isospin as external symmetry. A quantum system with external symmetry is made into a principal fiber bundle with SU(2) as the structure group, and is equipped with a connection. The base manifold of this bundle is called the internal space. A gauge theory for nonrigid molecules on the basis of the observation that the vector bundle associated with the principal fiber bundle provides a setting for quantum mechanics of the internal motion. This also illustrates that the connection has nonvanishing curvature gives rise to the nonseparability of motions.

2. The principal fiber bundle

2.1 SU(2)-**Action.** For simplicity, we consider a principal fiber bundle with the structure group SU(2). The total space is $\mathscr{C}^4 \cong \mathscr{R}^8$. The system is in $\mathscr{R}^8 = Q$ with an orthonormal system $\{f_i\}_{i=0}^7$.

The left action of SU(2) on Q is expressed with respect to the basis $\{f_j\}$ in the form

(2.1)
$$\begin{pmatrix} g & 0 \\ 0 & g \end{pmatrix}$$

where

$$g = \begin{pmatrix} u_1 + iu_2 & u_3 + iu_4 \\ -u_3 + iu_4 & u_1 - iu_2 \end{pmatrix}$$

$$u_1 = \cos\alpha, \qquad u_2 = \sin\alpha\cos\beta$$
$$u_3 = \sin\alpha\sin\beta\cos\gamma, \qquad u_4 = \sin\alpha\sin\beta\sin\gamma.$$

PROPOSITION. *The system* $\dot{Q} = \mathscr{R}^8 - \{0\}$ *is made into a principal fiber bundle with structure group* SU(2),

$$\pi: \dot{Q} \to M \simeq \dot{\mathscr{R}}^5, \qquad M := \dot{Q}/SU(2).$$

PROOF. We introduce in $\dot{Q}$ the structure of the algebra by setting the following operations on basis.

$$f_0^2 = f_0, \qquad f_j^2 = -f_0, \qquad j \neq 0,$$
$$f_j f_0 = f_0, \qquad f_j = f_j,$$
$$f_{ij} = -f_{ji}, \qquad i \neq j, \qquad i, j \neq 0,$$
$$f_1 f_2 = -f_3, \qquad f_4 f_5 = f_1, \qquad f_4 f_6 = f_2,$$
$$f_1 f_3 = f_2, \qquad f_2 f_3 = -f_1, \qquad f_4 f_7 = f_3,$$
$$f_1 f_4 = f_5, \qquad f_2 f_4 = f_6, \qquad f_3 f_4 = f_7,$$
$$f_1 f_5 = -f_4, \qquad f_2 f_5 = f_7, \qquad f_3 f_5 = -f_6,$$
$$f_1 f_6 = -f_7, \qquad f_2 f_6 = -f_4, \qquad f_3 f_6 = -f_5,$$
$$f_1 f_7 = f_6, \qquad f_2 f_7 = -f_5, \qquad f_3 f_7 = -f_4.$$
$$f_5 f_6 = -f_3, \qquad f_5 f_7 = f_2, \qquad f_6 f_7 = -f_1.$$

For each x in Q, $x = \xi_0 f_0 + \sum_{j=1}^{7} \xi_j f_j$, $\xi_j \in \mathcal{R}$, we set $\epsilon = (\cos \alpha) f_0 + (\sin \alpha \cos \beta) f_1 - (\sin \alpha \sin \beta \cos \gamma) f_2 + (\sin \alpha \sin \beta \sin \gamma) f_3$ Then the SU(2) action on Q, given by (2.1), is written as the left action:

$$(2.2) \qquad\qquad\qquad x \to \epsilon x.$$

2.2 **The connection.** From (2.2), we have the corresponding fundamental vector field $\{F_i\}_{i=1}^{3}$, the infinitesimal generator of the SU(2) action on $\dot{Q}$ as

$$F_1 = \frac{1}{3}\{(-\xi_1 + \xi_2 - \xi_3) f_0 + (\xi_0 + \xi_3 + \xi_2) f_1 + (\xi_3 - \xi_0 - \xi_1) f_2$$

$$+ (-\xi_2 - \xi_1 + \xi_0) f_3 + (-\xi_5 + \xi_6 - \xi_7) f_4 + (\xi_4 + \xi_7 + \xi_6) f_5$$

$$+ (\xi_7 - \xi_4 - \xi_5) f_6 + (-\xi_6 - \xi_5 + \xi_4) f_7\}$$

$$F_2 = \frac{1}{2}\{(\xi_2 - \xi_3) f_0 + (\xi_3 + \xi_2) f_1 + (-\xi_0 - \xi_1) f_2 + (-\xi_1 + \xi_0) f_3$$

$$+ (\xi_6 - \xi_7) f_4 + (\xi_7 + \xi_6) f_5 + (-\xi_4 - \xi_5) f_6 + (-\xi_5 + \xi_4) f_7\}$$

$$F_3 = -\xi_3 f_0 + \xi_2 f_1 - \xi_1 f_2 + \xi_0 f_3 - \xi_7 f_4 + \xi_6 f_5 - \xi_5 f_6 + \xi_4 f_7$$

where f_k are naturally identified with $\partial / \partial \xi_k$. In fact, the vector fields F_i are vertical, and the vector fields Y orthogonal to F_i are horizontal [4]; i.e.,

$$K_x(Y_x, F_x) = 0, \qquad x \in \dot{Q},$$

when K_x is the inner product naturally induced in the tangent space $T_x(\dot{Q})$. We choose the horizontal vector fields V_k, $k = 1, 2, 3, 4, 5$, as follows:

$$V_1 = \xi_0 f_0 + \xi_1 f_1 + \xi_2 f_2 + \xi_3 f_3 + \xi_4 f_4 + \xi_5 f_5 + \xi_6 f_6 + \xi_7 f_7$$

$$V_2 = \xi_4 f_0 + \xi_5 f_1 + \xi_6 f_2 + \xi_7 f_3 + \xi_0 f_4 + \xi_1 f_5 + \xi_2 f_6 + \xi_3 f_7$$

$$V_3 = -\xi_5 f_0 + \xi_4 f_1 - \xi_7 f_2 + \xi_6 f_3 + \xi_1 f_4 - \xi_0 f_5 + \xi_3 f_6 - \xi_2 f_7$$

$$V_4 = -\xi_6 f_0 + \xi_7 f_1 + \xi_4 f_2 - \xi_5 f_3 + \xi_2 f_4 - \xi_3 f_5 - \xi_0 f_6 + \xi_1 f_7$$

$$V_5 = \xi_7 f_0 + \xi_6 f_1 - \xi_5 f_2 - \xi_4 f_3 - \xi_3 f_4 - \xi_2 f_5 + \xi_1 f_6 + \xi_0 f_7$$

The linear subspace $W_{x, hor}$ of $T_x(\dot{Q})$ spanned by all the horizontal vectors at x is called the horizontal subspace.

PROPOSITION. *The connection forms are Lie algebra valued one-forms:* (*in terms of the basis of Lie algebra of* SU(2))

$$\omega_1 = \| \xi \|^{-2} [-\xi_1 d\xi^0 + \xi_0 d\xi^1$$
$$+ \xi_3 d\xi^2 - \xi_2 d\xi^3 - \xi_5 d\xi^4 + \xi_4 d\xi^5 + \xi_7 d\xi^6 - \xi_6 d\xi^7]$$

$$\omega_2 = \| \xi \|^{-2} [\xi_2 d\xi^0 + \xi_3 d\xi^1$$
$$- \xi_0 d\xi^2 - \xi_1 d\xi^3 + \xi_6 d\xi^4 + \xi_7 d\xi^5 - \xi_4 d\xi^6 - \xi_5 d\xi^7]$$

$$\omega_3 = \| \xi \|^{-2} [-\xi_3 d\xi^0 + \xi_2 d\xi^1$$
$$- \xi_1 d\xi^2 + \xi_0 d\xi^3 - \xi_7 d\xi^4 + \xi_6 d\xi^5 - \xi_5 d\xi^6 + \xi_4 d\xi^7]$$

$$where \; \| \xi \|^2 = \sum_{j=0}^{7} (\xi_j)^2 \; .$$

2.3 The curvature. Let e_1, e_2, e_3 be a basis for the Lie algebra $\mathscr{G}$ of SU(2) and c_{jk}^m, $m, j, k = 1, 2, 3$, the structure constants of $\mathscr{G}$ with respect to e_1, e_2, e_3, that is $[e_j, e_k] = \sum_m c_{jk}^m e_m$, $c_{jk}^m = 2i\epsilon_{jkm}$, $j, k, = 1, 2, 3$ where

$$\epsilon_{jkm} = \begin{cases} \mathrm{sgn} \begin{pmatrix} 1 & 2 & 3 \\ j & k & m \end{pmatrix}, & \\ 0, & \text{in other cases.} \end{cases}$$

Let $\omega = \sum_i \omega_i e_i$ and $\Omega = \sum_m \Omega^m e_m$. One can obtain the curvature Ω by applying the structure equation

$$\Omega^m = d\omega_m + \frac{1}{2} \sum_{j,k} c_{jk}^m \omega_j \wedge \omega_k, \qquad m = 1, 2, 3$$

where ω_m is obtained in §2.2.

To express Ω^m in terms of coordinates on internal space M, one can express $\pi : \dot{Q} \to M$ explicitly in terms of basis elements in Q. To avoid lengthy pages, we omit the expressions here.

3. Internal molecular motions

To set up quantum mechanics for internal motion, we associate complex vector bundles to the principal SU(2) bundle $\pi : \dot{Q} \to M$ as follows: [6]

For each j, $j = 0, \frac{1}{2}, 1, \ldots$.

Let D^j denote unitary representation of SU(2) and $\mathscr{C}^{2j+1}$ its representation space. We have $(D^j(g)z)_k = \sum_{m=j}^{-j} D_{km}^j(g)z_j$, $z \in \mathscr{C}^{2j+1}$, $g \in$ SU(2), $D_{km}^j(g)$: matrix elements. For a basis $|jm >$ with $J_3|jm >= m|jm >$, Casimir operator J^2, we have

$$J^2 = \sum_i (J_i)^2 = \frac{1}{2}(J_+ J_- + J_- J_+) + (J_3)^2$$

$$[J^2, J_i] = [J^2, J_\pm] = 0, \qquad J^2|jm >= j(j+1)|jm > .$$

Define a left action of SU(2) on $\dot{Q} \times \mathscr{C}^{2j+1}$ by $(x, z) \to (gx, D^j(g)z)$ which gives an equivalence relation in $\dot{Q} \times \mathscr{C}^{2j+1}$. We denote the quotient manifold by $\dot{Q} \times_{SU(2)} \mathscr{C}^{2j+1}$ which is made into a complex vector bundle $V_j = (\dot{Q} \times_{SU(2)} \mathscr{C}^{2j+1}, \pi_j, M)$ via the following commutative diagram:

$$\begin{array}{ccc} \dot{Q} \times \mathscr{C}^{2j+1} & \xrightarrow{\ q_j\ } & \dot{Q} \times_{SU(2)} \mathscr{C}^{2j+1} \\ \text{pr} \downarrow & & \pi_j \downarrow \\ \dot{Q} & \xrightarrow{\ \pi\ } & M \end{array}$$

Projection maps are related by $\pi_j \circ q_j = \pi \circ pr$. The internal states of the system are described by the cross-sections in the complex vector bundle V_j. The V_j is a trivial bundle, and hence the cross-sections become $\mathscr{C}^{2j+1}$-valued functions on the internal space M. To illustrate this idea, we proceed as follows:

A $\mathscr{C}^{2j+1}$-valued function f on $\dot{Q}$ is said to be D^j-equivariant if it satisfies $f(gx) = D^j(g)f(x)$. For each D^j-equivariant function, there corresponds a cross-section in the complex vector bundle V_j, and vice versa. Let $q_j^\#$ be the one-to-one correspondence from the cross-section to the D^j-equivariant function.

On the other hand, let σ denote the cross-section of $\pi : \dot{Q} \to M$. Then any point x of $\dot{Q}$ is of the form $g\sigma(u)$, $u \in M$. Therefore we have $f(x) = D^j(g)f(\sigma(u))$. One can then identify $\Phi = f \circ \sigma$ with a cross-section. The components of f are $f_k(x) = \sum_{m=-j}^{-j} D_{km}^j(g)\Phi_m(u)$. More precisely, the correspondence can be traced by the following commutative diagram:

$$\begin{array}{ccc} V_j & \xrightarrow{\ q^\#\ } & V_j \\ f \uparrow & & \uparrow \\ \dot{Q} & \xrightarrow{\ \epsilon\ } & \dot{Q} \\ \sigma \uparrow & & \uparrow \\ M & \xrightarrow{\ \ \ \ } & M \end{array}$$

On the other hand, we apply Casimir operator,

$$\hat{J}^2 D^j(g) = j(j+1)D^j(g), \qquad \hat{J}^2 = \sum \hat{J}_k^2,$$

to cross-sections in the complex vector bundle V_j which can also be regarded as $\mathscr{C}^{2j+1}$-valued functions on the internal space M.

Furthermore, one can construct Hilbert space of square integrable sections, and construct connection, curvature on the complex vector bundle V_j. To set up quantum mechanics for internal states of the quantum system, it is to obtain the internal Hamiltonian operator acting on cross-sections in V_j by direct calculations or by applying the technique of geometric quantization.

4. Remarks

One of the crucial parts of the above geometric setting of quantum system is the SU(2) action can be described by introducing an amazing algebraic structure on the operations of basis elements of the total space of the principal fiber bundle. It would be interesting to know how the parameter spaces setting in Berry's phase situation [8]. It is also interesting to set up the SU(3) case or higher degree of freedom for the internal motion of the quantum system by applying Bott's periodicity formula. We will work out quantization and reduction of the quantum system in a future paper.

REFERENCES

1. A. Guichardet, Ann. Inst. H. Poincare **40** (1984), 329.
2. T. Iwai, J. Math. Phys. **28** (1987), 1315.
3. _____, J. Math. Phys. **29** (1988), 1325.
4. S. Kobayashi and Nomizu, *Foundations of differential geometry*, Interscience, New York, (1969).
5. E. B. Lin, *Geometric quantization of particles in Quark model*, Lecture Notes in Physics **278** (1987), 383.
6. J. M. Normand, *A Lie group: Rotations in quantum mechanics*, North-Holland (1980).
7. Y. Wu, J. Phys. A:Math. Gen **22** (1989), L117.
8. _____, J. Math. Phys. **31** (1990), 294.

UNIVERSITY OF TOLEDO

Proceedings of Symposia in Pure Mathematics
Volume 54 (1993), Part 2

Heat Kernels on Covering Spaces
and Topological Invariants

JOHN LOTT

1. Introduction

It is well known that there are relationships between the heat flow, acting on differential forms on a closed oriented Riemannian manifold M, and the topology of M. For example, let Δ_p denote the Laplacian on p-forms. Then it follows from Hodge theory that the pth Betti number of M is given by

$$(1) \qquad b_p(M) = \lim_{T \to \infty} \operatorname{Tr} \exp(-T\Delta_p).$$

Another spectral invariant of M is the analytic torsion $\mathcal{T}(M)$ of Ray and Singer [RS]. It turns out to be equal to the classical Reidemeister torsion of M [Ch, Mü].

We will be interested in the heat flow on the universal cover $\widetilde{M}$ of M. As the Laplacian on $\widetilde{M}$ generally does not have discrete spectrum, its heat flow is somewhat different from the closed case. Nevertheless, one can define an analog of the Betti numbers, the L^2-Betti numbers $b_p^{(2)}(M)$, using the $T \to \infty$ limit of the heat flow [At, Si]. We review this theory in §2.

It was pointed out by Novikov and Shubin that there is more topological information in the large-time heat flow on $\widetilde{M}$ than just the L^2-Betti numbers [NS]. Roughly speaking, one finds that if one measures the heat at a point $x \in \widetilde{M}$ at time T which was induced by a point source at x at time 0, then for large T this decays with a power-law behaviour. The exponent of this decay, $\alpha_p(M)/2$, is independent of the Riemannian metric on M [NS]. This invariant has no counterpart in heat flow on closed manifolds. In §3 we discuss the topological properties of $\alpha_p(M)$ and give some results on its values [Lo1].

1991 *Mathematics Subject Classification.* Primary 58GXX.
The final version of this paper has been submitted for publication elsewhere.

In §4 we consider the covering space analog of the analytic torsion [**Lo1**]. This L^2-analytic torsion, $\mathcal{T}_\Gamma(M)$, has the same relationship to the L^2-cohomology as the Ray-Singer analytic torsion bears to de Rham cohomology. It turns out to be a smooth invariant of manifolds with vanishing L^2-Betti numbers. For odd-dimensional manifolds which admit a hyperbolic structure, $\mathcal{T}_\Gamma(M)$ is proportionate to the hyperbolic volume. Thus, one can view $\mathcal{T}_\Gamma(M)$ as an odd-dimensional analog of the L^2-Euler characteristic $\chi^{(2)}(M)$. (By the L^2-index theorem (6), $\chi^{(2)}(M) = \chi(M)$. If M^n is even-dimensional and admits a hyperbolic metric g then $\chi(M) = c(n) \, \mathrm{Vol}(M, g)$ for some nonzero constant $c(n)$.) Most of the results of this section have been obtained independently by V. Mathai [**Ma**].

I wish to thank the organizers of the UCLA conference for the stimulating atmosphere that they provided.

2. Type II traces of heat kernels

Let $\Lambda^p(\widetilde{M})$ denote the Hilbert space of L^2 p-forms on $\widetilde{M}$. The Laplacian is defined by $\widetilde{\Delta}_p = dd^* + d^*d$, acting on smooth p-forms of compact support on $\widetilde{M}$. It has a unique selfadjoint extension to $\Lambda^p(\widetilde{M})$, which we will also denote by $\widetilde{\Delta}_p$. Using the spectral theorem, one defines $\exp(-T\widetilde{\Delta}_p)$ for $T \geq 0$ by

$$(2) \qquad \exp(-T\widetilde{\Delta}_p) = \int_{\mathbb{R}} \exp(-T\lambda) \, dE_p(\lambda),$$

where $E_p(\lambda)$ is the spectral projection operator of $\widetilde{\Delta}_p$. It follows that $\exp(-T\widetilde{\Delta}_p)$ is a bounded operator on $\Lambda^p(\widetilde{M})$. For $T > 0$, the Schwartz kernel $\exp(-T\widetilde{\Delta}_p)(x, y)$ is a smooth form on $\widetilde{M} \times \widetilde{M}$.

Recall that for the Laplacian on M, we have

$$(3) \qquad \mathrm{Tr} \exp(-T\Delta_p) = \int_M \mathrm{tr} \, \exp(-T\Delta_p)(x, x) \, dx,$$

where tr denotes the finite-dimensional trace on $\mathrm{End}(\Lambda^p(T_xM))$. If $\pi_1(M)$ is infinite, the analogous integral of tr $\exp(-T\widetilde{\Delta}_p)(x, x)$ over $\widetilde{M}$ would be infinite, as the integral over each translation of a fundamental domain $\mathfrak{F}$ would be the same as over $\mathfrak{F}$. However, it makes sense to define the following "type II" trace:

$$(4) \qquad \mathrm{Tr}_\Gamma \exp(-T\widetilde{\Delta}_p) = \int_{\mathfrak{F}} \mathrm{tr} \, \exp(-T\widetilde{\Delta}_p)(x, x) \, dx.$$

The reason for calling Tr_Γ a type II trace is that $\exp(-T\widetilde{\Delta}_p)$ is an element of a certain type II_∞ von Neumann algebra, namely the algebra $\mathfrak{A}$ of all bounded operators on $\Lambda^p(\widetilde{M})$ which commute with the deck transformations. Denoting the trace on $\mathfrak{A}$ by Tr_Γ, (4) becomes an identity [**At**].

DEFINITION. $b_p^{(2)}(M) = \lim_{T \to \infty} \mathrm{Tr}_\Gamma \exp(-T\widetilde{\Delta}_p)$.

The L^2-Betti numbers are real numbers which in a sense measure the size of the space of harmonic L^2 p-forms on $\widetilde{M}$, relative to $\pi_1(M)$. The L^2-index theorem [At] states that for all $T > 0$,

(5) $$\sum_p (-1)^p \operatorname{Tr}_\Gamma \exp(-T\widetilde{\Delta}_p) = \chi(M),$$

implying that

(6) $$\sum_p (-1)^p b_p^{(2)}(M) = \chi(M).$$

EXAMPLE.

(7) $$b_p^{(2)}(\mathbb{R}^n/\Gamma) = 0 \quad \text{for all } p.$$

(8) $$b_p^{(2)}(H^{2n+1}/\Gamma) = 0 \quad \text{for all } p.$$

(9) $$b_p^{(2)}(H^{2n}/\Gamma) = \begin{cases} 0 & \text{for } p \neq n, \\ (-1)^n \chi(H^{2n}/\Gamma) & \text{for } p = n. \end{cases}$$

It follows from the definition that $b_p^{(2)}(M)$ is multiplicative under finite covers. One can show that $b_p^{(2)}(M)$ is a homotopy invariant of M [Do1]. For review articles on L^2 cohomology, see [Do2, An] and articles in these proceedings.

3. Novikov-Shubin invariants

Let us look at how $\operatorname{Tr}_\Gamma \exp(-T\widetilde{\Delta}_p)$ behaves for large T. Let $\widetilde{\Delta}_p'$ denote the Laplacian acting on $(\operatorname{Ker}\widetilde{\Delta}_p)^\perp$, so

(10) $$\operatorname{Tr}_\Gamma \exp(-T\widetilde{\Delta}_p') = \operatorname{Tr}_\Gamma \exp(-T\widetilde{\Delta}_p) - b_p^{(2)}(M).$$

EXAMPLE. If $M = \mathbb{R}^n/\Gamma$, we can use the explicit form of the heat kernel on $\mathbb{R}^n$ to find that

(11) $$\operatorname{Tr}_\Gamma \exp(-T\widetilde{\Delta}_p) = \binom{n}{p} \operatorname{Vol}(\mathbb{R}^n/\Gamma)(4\pi T)^{-n/2}.$$

Note that $\lim_{T\to\infty} \operatorname{Tr}_\Gamma \exp(-T\widetilde{\Delta}_p) = 0$, proving (7), and that $\operatorname{Tr}_\Gamma \exp(-T\widetilde{\Delta}_p')$ is $O(T^{-n/2})$ as $T \to \infty$. This is completely unlike what happens in the case of heat flow on a closed manifold, where

(12) $$\operatorname{Tr} \exp(-T\Delta_p) - b_p(M) = O(e^{-\lambda T}) \quad \text{as } T \to \infty,$$

λ being the smallest nonzero eigenvalue of Δ_p.

Let us generalize this example by the following definition [NS].

DEFINITION. $\alpha_p(M) = \sup\{\beta_p : \operatorname{Tr}_\Gamma \exp(-T\widetilde{\Delta}_p') = O(T^{-\beta_p/2}) \text{ as } T \to \infty\}$.

One can think of $\alpha_p(M)$ as an effective dimension of M as felt by diffusing p-forms. A priori, $\alpha_p(M)$ depends on the Riemannian metric on M. We give three results, in order of increasing strength, on the invariance properties of $\alpha_p(M)$.

PROPOSITION 1. a. $\alpha_p(M)$ *is independent of the Riemannian metric, and so is a smooth invariant of* M **[NS]**.

b. $\alpha_p(M)$ *is defined for all closed oriented topological manifolds* M, *and is a homeomorphism invariant* **[Lo1]**.

c. $\alpha_p(M)$ *is a homotopy invariant* **[GS]**.

The methods of proof of Proposition 1 involve spectral analysis. Taking the type II trace of both sides of (2), we have

$$(13) \qquad \mathrm{Tr}_\Gamma \exp(-T\widetilde{\Delta}_p) = \int_\mathbb{R} \exp(-T\lambda)\, dN_p(\lambda),$$

where $N_p(\lambda) = \mathrm{Tr}_\Gamma E_p(\lambda)$ is a monotonically nondecreasing function of λ. Integrating by parts, we have

$$(14) \qquad \mathrm{Tr}_\Gamma \exp(-T\widetilde{\Delta}'_p) = \int_\mathbb{R} T \exp(-T\lambda)(N_p(\lambda) - N_p(0))\, d\lambda.$$

The large-T behaviour of $\mathrm{Tr}_\Gamma \exp(-T\widetilde{\Delta}'_p)$ is clearly linked to the small-λ behaviour of $N_p(\lambda)$. Let us keep track of the dependence on the metric g by writing $N_p(\lambda, g)$. Part a of Proposition 1 follows from showing that, under change of metric, the change of $N_p(\lambda)$ is relatively bounded in the sense that for two metrics g and g', there is a constant $C > 0$ such that

$$(15) \qquad N_p(C\lambda, g) \geq N_p(\lambda, g') \geq N_p(C^{-1}\lambda, g).$$

Part b is proven by using a theorem of Sullivan that any topological manifold of dimension $\neq 4$ has a Lipschitz structure which is unique up to a Lipschitz homeomorphism **[Su1]**. If $\dim(M) \neq 4$, one defines $\alpha_p(M)$ using analysis on Lipschitz manifolds and proves the invariance as in part a. (If $\dim(M) = 4$, define $\alpha_p(M)$ to be $\alpha_p(M \times S^6)$.) The proof of part c involves changing homotopies to become submersions.

EXAMPLE.

$$(16) \qquad \alpha_p(\mathbb{R}^n/\Gamma) = n \quad \text{for all } p.$$

$$(17) \qquad \alpha_p(H^{2n}/\Gamma) = \infty \quad \text{for all } p.$$

$$(18) \qquad \alpha_p(H^{2n+1}/\Gamma) = \begin{cases} \infty & \text{for } p \neq n,\, n+1, \\ 1 & \text{for } p = n,\, n+1 \text{ [Lo1]}. \end{cases}$$

$$(19)$$
$$\alpha_0(M) = \begin{cases} \infty & \text{if } \pi_1(M) \text{ is not of polynomial growth}, \\ \text{the degree of growth if } \pi_1(M) \text{ is of polynomial growth [Va]}. \end{cases}$$

The reason for the ∞'s that appear in (17) and (18) is that, in these cases, the spectrum of $\widetilde{\Delta}'_p$ does not extend down to zero. More generally, it follows from the method of proof of Proposition 1 that the basic question of whether the spectrum of $\widetilde{\Delta}'_p$ is bounded away from zero only depends on the topology of M. It is known that the spectrum of $\widetilde{\Delta}_0$ is bounded away from zero if and only if $\pi_1(M)$ is not amenable [**Br**]. It seems reasonable to conjecture that if M is a $K(\pi, 1)$ manifold then 0 is in the spectrum of $\tilde{\Delta}_p$ for some p, and that if π is an amenable group then 0 is in the spectrum of $\tilde{\Delta}'_p$ for all p. (If π is amenable then a $K(\pi, 1)$ manifold has vanishing L^2-Betti numbers [**CG1**].)

EXAMPLE. If $M = \Gamma \backslash G/K$ for an irreducible symmetric space G/K of noncompact type, then if $\text{rank}(G) = \text{rank}(K)$, $\alpha_p(M) = \infty$ for all p. If $\text{rank}(G) \neq \text{rank}(K)$, put $d = \text{rank}(G) - \text{rank}(K)$. Then $\alpha_p(M) = \infty$ if $p \notin [(\dim(M)-d)/2, (\dim(M)+d)/2]$, and is finite if $p \in [(\dim(M)-d)/2, (\dim(M)+d)/2]$ and independent of p in this region.

EXAMPLE. If G^{2n+1} denotes the Heisenberg group of dimension $2n+1$ then

$$(20) \qquad \alpha_p(G^{2n+1}/\Gamma) \leq \begin{cases} 2(n+1) & \text{for } p \neq n, n+1, \\ n+1 & \text{for } p = n, n+1. \end{cases}$$

If $n = 1$ then these inequalities are equalities.

One would like to be able to compute $\alpha_p(M)$ in terms of standard topological invariants. One class of manifolds for which this can be done is that of manifolds with abelian fundamental group. As $\alpha_p(M)$ is invariant under finite covers, we can assume that $\pi_1(M) = \mathbb{Z}^l$ for some l. The reason for the simplification in the abelian case is that the regular representation ρ_{reg} of $\mathbb{Z}^l$ is highly reducible. Using Fourier transforms, we can decompose ρ_{reg} as an integral of one-dimensional representations, the integral being over the dual torus T^l. This decomposition can be used to write the heat kernel on $\widetilde{M}$ in terms of twisted heat kernels on M. More precisely, given $\theta \in T^l$, let E_θ be the corresponding flat unitary line bundle on M and let $\Delta_{p,\theta}$ denote the Laplacian on $\Lambda^p(M, E_\theta)$. Then

$$(21) \qquad \text{Tr}_\Gamma \exp(-T\tilde{\Delta}_p) = \int_{T^l} \text{Tr} \exp(-T\Delta_{p,\theta}) \, d^l\theta/(2\pi)^l.$$

Denoting the eigenvalues of $\Delta_{p,\theta}$ by $\{\lambda_{p,i}(\theta)\}_{i=1}^\infty$, it follows that

$$(22) \qquad \text{Tr}_\Gamma \exp(-T\tilde{\Delta}_p) = \int_{T^l} \sum_i \exp(-T\lambda_{p,i}(\theta)) \, d^l\theta/(2\pi)^l.$$

Now each $\lambda_{p,i}(\theta)$ is a locally analytic function of θ, and so $\{\lambda_{p,i}(\theta)\}_{i=1}^\infty$ gives a sequence of multivalued analytic functions on T^l. (It is necessary to allow for multivaluedness, as it is possible that there is a rearrangement of eigenvalues when going around a closed curve in T^l.)

The large-T behaviour in (22) is clearly determined by the asymptotics of the functions among $\{\lambda_{p,i}(\theta)\}_{i=1}^\infty$ which touch zero for some value of θ. By Hodge theory, for a given value of θ, $\lambda_{p,i}(\theta)$ is zero for some i if and only if $H^p(M, E_\theta) \neq 0$. If for some i, $\lambda_{p,i}(\theta)$ is zero for all θ then its contribution to (22) is independent of T, and so is part of $b_p^{(2)}(M)$. Thus, if $\{\lambda'_{p,i}(\theta)\}_{i=1}^{\infty}$ denotes the eigenvalues which are not identically zero in θ then

$$(23) \qquad \mathrm{Tr}_{\Gamma} \exp(-T\widetilde{\Delta}'_p) = \int_{T^l} \sum_i \exp(-T\lambda'_{p,i}(\theta)) \, d^l\theta/(2\pi)^l.$$

For any $\varepsilon > 0$, the large-T behaviour of (23) is determined by the functions $\{\lambda'_{p,i} : \lambda'_{p,i}(\theta) < \varepsilon \text{ for some } \theta\}$, up to an exponentially small error. Using a partition of unity, it follows that the asymptotics of (23) is that of a finite sum of integrals of the form

$$(24) \qquad \int_{\mathbb{R}^l} \rho(\theta) \exp(-T\lambda(\theta)) \, d^l\theta/(2\pi)^l,$$

where $\rho \in C_0^{\infty}(\mathbb{R}^l)$, and λ is analytic on $\mathbb{R}^l$ and nonnegative on $\mathrm{supp}(\rho)$. The large-T asymptotics of such integrals is known [**Ma1**], giving

PROPOSITION 2. *If $\pi_1(M)$ is abelian then $\alpha_p(M)$ is positive and rational.*

If $l = 1$, it is possible to be more explicit about the asymptotics of (24). If θ_0 is an isolated zero of λ and $\rho(\theta_0) > 0$ then (24) is $O(T^{-1/(2n)})$ as $T \to \infty$, where n is the smallest integer such that $\lambda^{(2n)}(\theta_0) \neq 0$. Thus we have

PROPOSITION 3. *If $\pi_1(M) = \mathbb{Z}$ then $\alpha_p(M) = 1/n_p$ for some nonnegative integer n_p.*

In order to compute n_p, it is necessary to know the degree of vanishing of the eigenvalue function $\lambda(\theta)$ around a zero θ_0. This can be computed using the perturbation theory of operators. One finds that n_p is determined by products in the cohomology ring of M, such as Massey triple products and their higher order generalizations. In particular, $\alpha_p(M)$ is determined explicitly by the minimal model of M [**Su2**]. We refer to [**Lo1**] for the precise statement, and only give the following examples.

EXAMPLE. Let P be the total space of a nontrivial principal SU(2) bundle over $S^1 \times S^3$.

$$(25) \qquad \alpha_3(S^1 \times S^6) = \infty.$$

$$(26) \qquad \alpha_3(S^1 \times S^3 \times S^3) = 1.$$

$$(27) \qquad \alpha_3(P) = 1/2.$$

These examples show that in general $\alpha_p(M)$ depends on more than just the dimension and fundamental group of M. By taking repeated principal SU(2) bundles, one can construct examples in which $\alpha_3(M)$ is arbitrarily close to zero.

One can also use the perturbation theory of operators to get some results in the case of general abelian fundamental group.

PROPOSITION 4. *If $\pi_1(M)$ is abelian then $\alpha_0(M) = \alpha_1(M) = \mathrm{rank}(\pi_1(M))$.*

This result seems to indicate that, for general M, $\alpha_1(M)$ may only depend on $\pi_1(M)$. By (19), this is true for $\alpha_0(M)$.

A class of manifolds for which the minimal model has nice properties is that of manifolds with a nilpotent fundamental group which acts nilpotently on the higher homotopy groups [Su2]. It should be possible to read off $\alpha_p(M)$ from the minimal model for such a manifold.

It is not known whether $\alpha_p(M) > 0$ for all p and all M.

4. L^2-analytic torsion

In order to motivate the Ray-Singer analytic torsion, suppose that one tries to find a topological invariant of an oriented closed Riemannian manifold by the ansatz

$$(28) \qquad I = \int_0^\infty \mathrm{Tr}\exp(-T\Delta_p)f(T)\,dT$$

for some smooth function f. As a uniform rescaling of the metric multiplies Δ_p by a constant, thereby effectively rescaling T, if I is to be metric independent then it is necessary that $f(T)\,dT$ be proportionate to dT/T. It turns out that to get an expression which is completely metric independent, one must take the combination

$$(29) \qquad \int_0^\infty \sum_p (-1)^p p\, \mathrm{Tr}\exp(-T\Delta_p)T^{-1}\,dT.$$

Two qualifications must be made about this integral.

1. The integrand is not integrable near $T = 0$. This is due to the small-T divergence of the heat kernel traces. One can get around this problem by the method of zeta-function regularization [RS].

2. The integrand is not integrable near $T = \infty$. This is due to the presence of nonzero real cohomology, which causes the heat kernel traces to not decay for large T. One can get around this problem by effectively killing the real cohomology, by twisting the differential forms by an acyclic flat unitary bundle E over M.

The precise definition of the analytic torsion can be taken to be

$$(30) \qquad \mathscr{T}_E(M) = \frac{d}{ds}\Big|_{s=0} \frac{1}{\Gamma(s)} \int_0^\varepsilon \sum_p (-1)^p p\, \mathrm{Tr}\exp(-T\Delta_p)T^{s-1}\,dT$$
$$+ \int_\varepsilon^\infty \sum_p (-1)^p p\, \mathrm{Tr}\exp(-T\Delta_p)T^{-1}\,dT.$$

Here ε is an arbitrary positive number and the Laplacian Δ_p acts on $\Lambda_p(M, E)$.

The covering space analog of (29) would be

$$(31) \qquad \int_0^\infty \sum_p (-1)^p p\, \mathrm{Tr}_\Gamma \exp(-T\widetilde{\Delta}_p)T^{-1}\,dT.$$

Again, two qualifications are necessary.

1. The integrand is not integrable near $T = 0$. This is because the type II heat kernel traces have the same small-T asymptotics as the usual heat kernel traces. One can again get around this problem by zero-function regularization.

2. The integrand may not be integrable near $T = \infty$. First, we should assume that $b_p^{(2)}(M) = 0$ for all p. Second, even with this assumption, the large-T falloff of $\mathrm{Tr}_\Gamma \exp(-T\widetilde{\Delta}_p)$ could conceivably be slow enough to make (31) still not integrable near $T = \infty$. However, if we assume that $\alpha_p(M) > 0$ for all p then this problem does not arise.

With these assumptions, we define the L^2-analytic torsion as

$$
\begin{aligned}
\mathscr{T}_\Gamma(M) = \ & \frac{d}{ds}\bigg|_{s=0} \frac{1}{\Gamma(s)} \int_0^\varepsilon \sum_p (-1)^p p \, \mathrm{Tr}_\Gamma \exp(-T\widetilde{\Delta}_p) T^{s-1} \, dT \\
& + \int_\varepsilon^\infty \sum_p (-1)^p p \, \mathrm{Tr}_\Gamma \exp(-T\widetilde{\Delta}_p) T^{-1} \, dT.
\end{aligned}
$$
(32)

PROPOSITION 5. $\mathscr{T}_\Gamma(M)$ is independent of the Riemannian metric used in the definition.

Thus $\mathscr{T}_\Gamma(M)$ is a smooth invariant of such M. The proof of Proposition 5 is similar to the analogous statement for the Ray-Singer torsion [RS], and involves differentiating $\mathscr{T}_\Gamma(M)$ along a curve in the space of Riemannian metrics. As the integrands of $\mathscr{T}_\Gamma(M)$ and $\mathscr{T}_E(M)$ have the same small-T asymptotics, there is no need for zeta-function regularization in the expression for $\mathscr{T}_\Gamma(M) - \mathscr{T}_E(M)$, and one can simplify the proof of Proposition 5 by using the known metric invariance of $\mathscr{T}_E(M)$. One must still be careful when dealing with the second integral of (32), due to the possible slow decay of $\mathrm{Tr}_\Gamma \exp(-T\widetilde{\Delta}_p)$.

$\mathscr{T}_\Gamma(M)$ has some properties analogous to those of the ordinary analytic torsion. For example, if $\dim(M)$ is even then $\mathscr{T}_\Gamma(M) = 0$. Also one has

$$
\mathscr{T}_\Gamma(M_1 \times M_2) = \mathscr{T}_\Gamma(M_1)\chi(M_2) + \mathscr{T}_\Gamma(M_2)\chi(M_1).
$$
(33)

Unlike the ordinary analytic torsion, $\mathscr{T}_\Gamma(M)$ is multiplicative under finite covers. It follows that $\mathscr{T}_\Gamma(M) = 0$ if M is finitely covered by $S^1 \times N$ for some closed N.

If M admits a locally homogeneous structure then it follows from the definition of $\mathrm{Tr}_\Gamma \exp(-T\widetilde{\Delta}_p)$ that $\mathscr{T}_\Gamma(M) = c \, \mathrm{Vol}(M, g)$, where g is a locally homogeneous metric on M which is normalized, for example, by its curvature tensor. The constant of proportionality c only depends on the homogeneous geometry of $\widetilde{M}$. Of course, c could be zero, as happens when M is flat. If $\widetilde{M}$ is an odd-dimensional irreducible symmetric space of noncompact type then $b_p^{(2)}(M) = 0$ and $\alpha_p(M) > 0$ for all p. It follows from [MS] that in this case, $\mathscr{T}_\Gamma(M) = 0$ if $\widetilde{M}$ is not $\mathrm{SO}(p+q)/\mathrm{S}(\mathrm{O}(p) \times \mathrm{O}(q))$ with p

and q odd or $\mathrm{SL}(3, \mathbb{R})/\mathrm{SO}(3, \mathbb{R})$. (These are the noncompact irreducible symmetric spaces G/K with $\mathrm{rank}(G) = \mathrm{rank}(K) + 1$.)

PROPOSITION 6. *If M is a three-dimensional manifold which admits a hyperbolic structure then $\mathcal{T}_\Gamma(M) = -\mathrm{Vol}(M, g)/(3\pi)$, where g is a hyperbolic metric on M.*

By Mostow rigidity, the hyperbolic metric g is unique up to isometry, and so Proposition 6 is consistent with Proposition 5. The proof of Proposition 6 is done by explicit computation of the heat kernels on the diagonal of $H^3 \times H^3$.

Just as the Reidemeister torsion is a combinatorial equivalent of the analytic torsion, there is an L^2-Reidemeister torsion $\mathcal{T}_\Gamma^c(M)$ defined combinatorially [CM, LR]. One expects that $\mathcal{T}_\Gamma^c(M) = \mathcal{T}_\Gamma(M)$. This would imply that $\mathcal{T}_\Gamma(M)$ vanishes if M is a Seifert 3-manifold with infinite fundamental group. The reason is such that a Seifert 3-manifold has a locally homogeneous structure based on one of five 3-manifold geometries [Sc]. Each of these 3-manifold geometries has a compact quotient M' which is a circle bundle. One can show that $\mathcal{T}_\Gamma^c(M')$ vanishes [LR], from which the vanishing of $\mathcal{T}_\Gamma(M)$ would follow.

One can also define L^2 analogs of other spectral invariants, such as the L^2-eta invariant [CG2] and the L^2-holomorphic torsion. In fact, the left-hand side of (5) is itself a L^2-spectral invariant, and there is a similar result for the signature operator which equates the L^2-index of the signature operator on $\widetilde{M}$ with $\sigma(M)$. It is known from work on the Novikov conjecture that one can extend the notion of the L^2-index of the signature operator to become an element of $K_0(C_r^* \pi_1(M))$, where $C_r^* \pi_1(M)$ is the reduced group C^*-algebra of $\pi_1(M)$ [MF]. There should be similar extensions for the other L^2 spectral invariants. In [Lo2] such an extension is proposed for the L^2-eta invariant, taking value in the noncommutative de Rham homology [Ka] of $C_r^* \pi_1(M)$.

BIBLIOGRAPHY

[An] M. Anderson, *L^2 harmonic forms on complete Riemannian manifolds*, Geometry and Analysis on Manifolds (Katata/Kyoto 1987), Lecture Notes in Math., vol. 1339, Springer-Verlag, 1988, 1–19.

[At] M. Atiyah, *Elliptic operators, discrete groups and von Neumann algebras*, Astérisque, no. 32, Soc. Math. France, Paris, 1976, 43–72.

[Br] R. Brooks, *The fundamental group and the spectrum of the Laplacian*, Comment. Math. Helv. **56** (1981), 581–598.

[CG1] J. Cheeger and M. Gromov, *L^2 cohomology and group cohomology*, Topology **25** (1986), 189–215.

[CG2] ——, *Bounds on the von Neumann dimension of L^2 cohomology and the Gauss-Bonnet theorem for open manifolds*, J. Differential Geom. **21** (1985), 1–34.

[Ch] J. Cheeger, *Analytic torsion and the heat equation*, Ann. of Math. (2) **109** (1979), 259–322.

[CM] A. Carey and V. Mathai, *L^2 acyclicity and L^2-torsion invariants*, in "Geometric and Topological Invariants of Elliptic Operators", Amer. Math. Soc. (1990), p. 91–118.

[Do1] J. Dodziuk, *De Rham-Hodge theory for* L^2 *cohomology of infinite coverings*, Topology **16** (1977), 157–165.

[Do2] _____, L^2 *Harmonic forms on complete manifolds*, Seminar on Differential Geometry (S. T. Yau, ed.), Ann. of Math. Stud., no. 102, Princeton Univ. Press, Princeton, NJ, 1982.

[GS] M. Gromov and M. Shubin, *von Neumann spectra near zero*, Geom. and Funct. Anal. **1** (1991), pp. 375–404.

[Ka] M. Karoubi, *Homologie cyclique et K-théorie*, Astérisque, no. 149, Soc. Math. France, Paris, 1987.

[Lo1] J. Lott, *Heat kernels on covering spaces and topological invariants*, J. of Diff. Geom. **35** (1992).

[Lo2] _____, *Superconnections and noncommutative de Rham homology*, Preprint (1990).

[LR] W. Lück and M. Rothenberg, PL *torsion and the* K *theory of von Neumann algebras*, K-Theory **5** (1991), pp. 213–264.

[Ma] V. Mathai, L^2 *analytic torsion*, Preprint (1990).

[Ma1] B. Malgrange, *Intégrales asymptotiques et monodromie*, Ann. Sci. École Norm. Sup. (4) **7** (1974), 405–430.

[MF] A. Miscenko and A. Fomenko, *The index of elliptic operators over* C^*-*algebras*, Izv. Akad. Nauk SSSR Ser. Mat. **43** (1979), 831–859.

[MS] H. Moscovici and R. Stanton, *R-torsion and zeta functions for locally symmetric manifolds*, Inv. Math. **105** (1991), pp. 185–216.

[Mü] W. Müller, *Analytic torsion and the R-torsion of Riemannian manifolds*, Adv. in Math. **28** (1978), 233–305.

[NS] S. Novikov and M. Shubin, *Morse inequalities and von Neumann invariants of non-simply-connected manifolds*, Uspekhi Mat. Nauk **41** (1986), 289–292.

[RS] D. Ray and I. Singer, *R-torsion and the Laplacian on Riemannian manifolds*, Adv. in Math. **7** (1971), 145–210.

[Sc] P. Scott, *The geometries of 3-manifolds*, Bull. London Math. Soc. **15** (1983), 401.

[Si] I. Singer, *Some remarks on operator theory and index theory*, K-Theory and Operator Algebras, Proc. Conf. Univ. Georgia, Athens, Ga. 1975, Lecture Notes in Math., vol. 575, Springer-Verlag, 1977.

[Su1] D. Sullivan, *Hyperbolic geometry and homeomorphisms*, Geometric Topology, Proc. Georgia Topology Conf., Athens, Ga. 1977, Academic Press, 1979, pp. 543–555.

[Su2] _____, *Infinitesimal computations in topology*, Inst. Hautes Études Sci. Publ. Math. **47** (1977), 269–331.

[Va] N. Varapoulos, *Random walks and Brownian motion on manifolds*, Sympos. Math. **29** (1988), 97–109.

UNIVERSITY OF MICHIGAN

Proceedings of Symposia in Pure Mathematics
Volume 54 (1993), Part 2

The Heat Kernels of Symmetric Spaces

QI-KENG LU

The symmetric spaces that we discuss here are the nonexceptional and irreducible Riemannian globally symmetric spaces (NIRGSS). Each of the NIRGSS can be identified as a space composed from certain kind of matrices [**LU2** or **LU5**]). Using such formations, one can in particular express the invariant Riemann metrics, the curvatures, the geodesics etc. of NIRGSS explicitly by matrix notation as we have done on classical domains (nonexceptional bounded symmetric domains) and classical manifolds (the complex manifolds which are transitive under certain classical groups respectively). Moreover, we have constructed the Bergman kernels, the Cauchy kernels, and the Poisson kernels of classical domains [**HUA, LU1**]. The Morse functions and Jacobian fields for compact NIRGSS can be constructed in general [**YAO, YE** and **LIU**]).

Recently we have become interested in constructing the heat kernels of NIRGSS [**LU4, LUH, HO, HOW1, HOW2** and **LUL**]. The problem arises from the intention to solve the Poisson equation

$$\Delta\phi = f,$$

where Δ is the Laplace-Beltrami operator of a NIRGSS. If the Green function (or Green form) $G(p, q)$ of such a space M is known, the solution, roughly speaking is

$$\phi(p) = \int_{q \in M} G(p, q) * f(q).$$

If M is of rank 1, $G(p, q)$ is known ([**HEL**] in case that f is a function and [**LU3**] in case that f is one-form). However, when M is of rank > 1, it seems that the Green function $G(p, q)$ is hard to construct even for the simple case that M is a bi-disc (private conversation with Professor E. M. Stein).

1991 *Mathematics Subject Classification.* Primary 53C35.
Partially supported by the Chinese NSF.
This paper is in final form and no version of it will be submitted for publication elsewhere.

But if we know the heat kernel $H(p, q, t)$ of the heat equation $\partial\phi/\partial t = \Delta\phi$ then the Green function is obviously

$$G(p, q) = \int_0^\infty H(p, q, t)\,dt.$$

Moreover if $H_1(p_1, q_1, t)$ and $H_2(p_2, q_2, t)$ are heat kernels of two Riemannian manifolds M_1 and M_2 respectively, then

(1) $$H((p_1, q_1), (p_2, q_2), t) = H_1(p_1, q_1, t)H_2(p_2, q_2, t)$$

is the heat kernel of Riemannian manifold $M_1 \times M_2$ and the Green function of $M_1 \times M_2$ is

$$\int_0^\infty H_1(p_1, q_1, t)H_2(p_2, q_2, t)\,dt.$$

For the study of the heat kernels of symmetric spaces, it is sufficient to study those of irreducible symmetric spaces.

At first the author [LU4] constructed the heat kernel of a ball $B^n = \{Z \in C^n \| Z| < 1\}$ with respect to the Bergman metric. It is

(2)
$$H_{B^n}(Z, W, t) = C_1 \frac{e^{-n^2 t}}{\sqrt{t}} \int_{-\infty}^{+\infty} \left[\frac{1}{2ch\sigma} \left(\frac{-1}{sh\sigma} \frac{\partial}{\partial\sigma} \right)^n e^{-\frac{\sigma^2}{4t}} \right]_{ch^2\sigma=ch^2 r(z,w)+\tau^2} d\tau$$

where

$$C_1 = \frac{1}{2^{3n-2}(n-1)!n!\sqrt{\pi}\pi^n}$$

and

$$r(Z, W) = \frac{1}{2}\log \frac{\sqrt{1 - W\overline{Z}^T} + \sqrt{(Z-W)(I-\overline{W}^T Z)^{-1}(\overline{Z}-\overline{W})^T}}{\sqrt{1 - W\overline{Z}^T} - \sqrt{(Z-W)(I-\overline{W}^T Z)^{-1}(\overline{Z}-\overline{W})^T}}$$

is the geodesic distance between the points Z and W of B^n. Notice that the heat kernel depends only upon $r(Z, W)$ and t. So we can write $h_n(r(Z, W), t) = H_{B^n}(Z, W, t)$. Afterwards in a joint work [LUH] we constructed the heat kernel of the NIRGSS

$$SU(m, n)/S(U(m) \times U(n)) \simeq R_I(m, n) = \{Z \in C^{m\times n}|I - Z\overline{Z}^T > 0\},$$

where $Z \in C^{m\times n}$ means that Z is a complex $m \times n$ matrix. The rank of $R_I(m, n)$ is $\min\{m, n\}$. In the meantime we discovered that our method can be applied to other NIRGSS if we use the mentioned matrix representation of such spaces (see Appendix). Of course we have not yet constructed all the heat kernel of all NIRGSS, but for each type of NIRGSS (according to E. Cartan's classification [HEL]) we can give an example to illustrate what expression the heat kernel is. The general routine is that, from the matrix representation of M (cf. Appendix), we can introduce in a neighborhood of a fixed point $o \in M$ such that any point p in this neighborhood possesses a coordinate $(\lambda_1, \ldots, \lambda_n, \theta_1, \ldots, \theta_N)$, where n is the rank of M and $\theta_1, \ldots, \theta_N$

is the local coordinate of the isotropic subgroup $I_0(M)$ at O and $\lambda_1, \ldots, \lambda_n$ is called the radial part of the matrix polar-coordinate. In fact $(\lambda_1, \ldots, \lambda_n)$ is the coordinate of a flat and totally geodesic submanifold passing through O and usually we can choose a monotonously increasing function ρ defined in an interval of R such that $(\lambda_1, \ldots, \lambda_n) = (\rho(r_1), \ldots, \rho(r_n))$ and the number

$$r = r(o, p) = [r_1^2 + \cdots + r_n^2]^{1/2}$$

is exactly the geodesic distance of M between the points O and p. We call $(r_1, \ldots, r_n, \theta_1, \ldots, \theta_N)$ the pseudo-geodesic coordinate of p. Since M is transitive under the isometric group $I(M)$, for any two points p_1 and p_2 of M there is a transformation $\sigma_{p_1} \in I(M)$ such that $\sigma_{p_1}(p_1) = O$. Let $p = \sigma_{p_1}(p_2)$. Since the geodesic distance $r(p_1, p_2)$ of p_1 and p_2 is invariant under $I(M)$, we have especially

$$r(p_1, p_2) = r(\sigma_{p_1}(p_1), \sigma_{p_1}(p_2)) = r(O, p).$$

Suppose that the pseudo-geodesic coordinate of p is $(r_1(p_1, p_2), \ldots, r_k(p_1, p_2), \theta_1(p_1, p_2), \ldots, \theta_N(p_1, p_2))$. By definition

$$r(p_1, p_2) = [(r_1^2(p_1, p_2) + \cdots + r_n^2(p_1, p_2))]^{1/2}$$

does not depend upon $\theta_1(p_1, p_2), \ldots, \theta_n(p_1, p_2)$. This implies that there would be a suitably chosen σ_{p_1} such that the pseudo-geodesic coordinate of $p = \sigma_{p_1}(p_2)$ is $(r_1(p_1, p_2), \ldots, r_n(p_1, p_2), 0, \ldots, 0)$. Since the heat kernel $H(p_1, p_2, t)$ of M must be invariant under $I(M)$, i.e.,

$$H(p_1, p_2, t) = H(\sigma(p_1), \sigma(p_2), t), \qquad \forall \sigma \in I(M),$$

we can take $\sigma = \sigma_{p_1}$ and see that $H(p_1, p_2, t)$ depends only upon $r_1(p_1, p_2)$, $\ldots, r_n(p_1, p_2)$. Writing the Laplace-Beltrami operator Δ into pseudo-geodesic coordinate we have in general $\Delta = \Delta_0 + \Delta_1$ where Δ_0 depends only upon $r_1, \ldots, r_n$ and Δ_1 is a differential operator of the form

$$\Delta_1 = \frac{1}{f_0} \sum_{\partial \theta_\alpha}^N \frac{\partial}{\partial \theta_\alpha} \left[f_0 g^{\alpha\beta} \frac{\partial}{\partial \theta_\beta} \right].$$

Then to find the heat kernel is reduced to find the fundamental solution of the following equation $\partial H/\partial t = \Delta_0 H$.

We use the notation and classification of Riemannian globally symmetric spaces as those used in [HEL] and the matrix representation of such spaces can be seen in the Appendix.

1. NIRGSS of Type III

EXAMPLE.

$$SU(m, n)/S(U(m) \times U(n)) \cong R_I(m, n) = \{Z \in C^{m \times n} | I - Z\overline{Z}^T > 0\}.$$

Without loss of generality we assume $m \leq n$. The heat kernel of $R_I(m, n)$ is [LUH]

$$H(Z, W, t) = \frac{c_1}{\omega(Z, W)} e^{-c_0 t} \det[h_{n-m+1}^{(2j-2)}(r_k(Z, W), t)]_{1 \leq j, k \leq m}$$

where c_1 is a well-defined constant, $c_0 = 2m(m-1)(3n-m+2)/3$,

$$\omega(Z, W) = \prod_{j>k}[sh^2 r_j(Z, W) - sh^2 r_k(Z, W)]$$

and

$$h_l^{(k)}(r, t) = \frac{e^{-l^2 t}}{\pi^l \sqrt{t}} \int_{-\infty}^{+\infty} \left[\frac{1}{2ch\sigma} \left(\frac{-1}{sh\sigma} \frac{2}{\partial \sigma} \right)^l \left(-\frac{\partial}{\partial \sigma} \right)^k e^{-\frac{\sigma^2}{4t}} \right]_{ch^2 \sigma = ch^2 r + \tau^2} d\tau.$$

2. NIRGSS of Type I

These are compact spaces dual to those of Type III.

EXAMPLE. $SU(m+n)/S[U(m) \times U(n)] \cong$ complex Grassmann manifold.
Its heat kernel [HOW] is formally analogous to that of $R_I(m, n)$ i.e.,

$$H(P, Q, t) = \frac{c_1}{\omega(P, Q)} e^{c_0 t} \det[h_{n-m+1}^{(2j-2)}(r_k(P, Q), t)]_{1 \leq j, k \leq m}$$

where c_0 is the same constant as in the last example but

$$c_1 = \frac{2^m \pi^{m(n+1)}}{(n-1)! \cdots (n-m)!(m-1)! \cdots 1!}$$

$$\omega(P, Q) = \prod_{j>k}[\sin^2 r_j(P, Q) - \sin^2 r_k(P, Q)]$$

and

$$h_l^{(k)}(r, t) = \frac{\Gamma\left(\frac{k+1}{2}\right) e^{l^2 t}}{2^{n-1} \Gamma\left(l + \frac{k}{2}\right) \Gamma\left(\frac{1}{2}\right)} \int_r^{\pi-r} \left[\left(\frac{-1}{\sin\phi} \frac{\partial}{\partial\phi} \right)^l \left(-\frac{\partial}{\partial\phi} \right)^k h_1(\phi, t) \right]$$

$$\cdot \sin\phi[\cos^2\phi - \cos^2 r]^{-\frac{1}{2}} d\phi$$

where

$$h_1(\phi, t) = \frac{1}{\pi} \sum_{l=0}^{\infty} e^{-l^2 t} \cos l\phi = \frac{1}{2\sqrt{\pi t}} \sum_{l=-\infty}^{+\infty} e^{-\frac{(\phi+2l\pi)^2}{4t}}$$

is the heat kernel of $U(1)$. Substituting this series into $H(P, Q, t)$ Hong obtained the development (private communication),

$$H(P, Q, t) = \sum_{j=0}^{\infty} e^{-\lambda_j t} H_{\lambda_j}(P, Q)$$

where $\lambda_1, \lambda_2, \ldots$, are eigenvalues which are of the form

$$4 \sum_{k=1}^{m} N_k(N_k + 2k - n - m + 1)$$

with integers $N_1, \ldots, N_m$ satisfying $N_1 \leq N_2 \leq \cdots \leq N_m$.

3. NIRGSS of Type II

EXAMPLES. The unitary group $U(n)$. The heat kernel is [LUL],

$$H_{U(n)}(U_1, U_2, t) = \frac{C_1}{\omega(U_1, U_2)} e^{C_0 t} \det[h_1^{(j-1)}(r_k(U_1, U_2), t)]_{1 \leq j, \, k \leq n}$$

where

$$C_0 = \tfrac{1}{3} n(n^2 - 1),$$

$$C_1^{-1} = \lim_{t \to 0} \int_{U_1 \in U(n)} \frac{e^{C_0 t}}{\omega(U_1, U_2)} \det[h_i^{(j-1)}(r_k(U_1, U_2), t)]_{1 \leq j, k \leq n} \dot{U}_1$$

which is independent of U_2,

$$\omega(U_1, U_2) = \prod_{j > k} \sin^2[r_j(U_1, U_2) - r_k(U_1, U_2)]$$

and $h_1^{(j)}(r, t) = (-\partial/\partial r)^j h_1(r, t)$ where $h_1(r, t)$ is the heat kernel of $U(1)$ as in the last example.

However $U(n)$ is not irreducible but $U(n) \cong U(1) \times SU(n)$ where $SU(n)$ is irreducible. By formula (1) the heat kernel of $SU(n)$ is

$$H_{SU(n)}(U_1, U_2, t) = \frac{H_{U(n)}(e^{i\theta_1} U_1, e^{i\theta_2} U_2, t)}{H_{U(1)}(e^{i\theta_1}, e^{i\theta_2}, t)}.$$

4. NIRGSS of Type IV

EXAMPLE. $GL(n, C)/U(n) \cong \mathscr{H} = \{$Space of $n \times n$ Hermitian matrix $H > 0\}$.

As an illustration, we give a more detailed sketch of the process for constructing the heat kernel of $\mathscr{H}$.

The map $T_A \colon \mathscr{H} \to \mathscr{H}$ defined by

$$H \mapsto A H \overline{A}^T, \qquad A \in GL(n, C),$$

is an isometry of the Riemann metric $ds^2 = \operatorname{tr}(H^{-1} dH H^{-1} dH)$ defined in $\mathscr{H} = \mathscr{H}(n)$.

For any $H \in \mathscr{H}$ there is a $U \in U(n)/[U(1)]^n$ such that

$$H = U \wedge \overline{U}^T, \qquad \wedge = \begin{pmatrix} \lambda_1 & & \\ & \ddots & \\ & & \lambda_n \end{pmatrix}.$$

Then $\lambda_1, \dots, \lambda_n$ and the local coordinate of $U(n)/[U(1)]^n$ together form the matrix polar-coordinate of H and

$$ds^2 = \operatorname{tr}[\Lambda^{-1}(d\Lambda + \delta U \Lambda + \Lambda \delta \overline{U}^T) \Lambda^{-1}(d\Lambda + \delta \overline{U} \Lambda + \Lambda \delta U^T)^T]$$
$$= \operatorname{tr}(\Lambda^{-1} d\Lambda \Lambda^{-1} d\Lambda) + \operatorname{tr}[(\delta U \Lambda - \Lambda \delta U)(\delta \overline{U} \Lambda - \Lambda \delta \overline{U})^T],$$

where $\delta U = U^{-1}dU = (\delta u_{jk})_{1 \le j, k \le n}$ and $\delta u_{jk} = -\delta \bar{u}_{kj}$ are left invariant one-forms of $U(n)$. Let $\theta_1, \ldots, \theta_{n(n-1)}$ be the local coordinate of $U(n)/[U(1)]^n$ and set $r_j = \log \lambda_j$, $\delta u_{jk} = \phi_{jk} + i\psi_{jk}$ with real ϕ_{jk}, ψ_{jk}.

Then

$$ds^2 = \sum_{j=1}^{n} dr_j^2 + 2 \sum_{j>k} sh^2 \tfrac{1}{2}(r_j - r_k)(\phi_{jk}^2 + \psi_{jk}^2),$$

by which we know the invariant volume element is

$$\dot{H} = f_0(\theta)\omega^2 dr_1 \cdots dr_n d\theta_1 \cdots d\theta_{n(n-1)}$$

where $f_0(\theta)$ depends only upon $\theta = (\theta_1, \ldots, \theta_{n(n-1)})$ and

$$\omega = \prod_{j>k} sh^2 \tfrac{1}{2}(r_j - r_k) \qquad (\omega = 1 \text{ when } n = 1)$$

and the heat equation is

$$\frac{\partial h}{\partial t} = \Delta h = \frac{1}{\omega^2} \sum_{j=1}^{n} \frac{\partial}{\partial r_j}\left(\omega^2 \frac{\partial h}{\partial r_j}\right) + \frac{1}{f_0} \sum_{\alpha, \beta=1}^{n(n-1)} \frac{\partial}{\partial \theta_\alpha}\left[f_0 g^{\alpha\beta} \frac{\partial h}{\partial \theta_\beta}\right]$$

or

$$\frac{\partial(\omega h)}{\partial t} = \sum_{j=1}^{n} \frac{\partial^2(\omega h)}{\partial r_j^2} - \left[\frac{1}{\omega} \sum_{j=1}^{n} \frac{\partial^2 \omega}{\partial r_j^2}\right](\omega h) + \sum_{\alpha, \beta=1}^{n(n-1)} \frac{1}{f_0} \frac{\partial}{\partial \theta_\alpha}\left[f_0 g^{\alpha\beta} \frac{\partial(\omega h)}{\partial \theta_\beta}\right],$$

where $(r_1, \ldots, r_n, \theta_1, \ldots, \theta_{n(n-1)})$ is the pseudo-geodesic coordinate. Since the heat kernel h depends only upon $r_1, \ldots, r_n$ and t, it must satisfy

(3) $$\frac{\partial(\omega h)}{\partial t} = \sum_{j=1}^{n} \frac{\partial^2(\omega h)}{\partial r_j^2} - c_0(\omega h)$$

where

$$c_0 = \frac{1}{\omega} \sum_{j=1}^{n} \frac{\partial^2 \omega}{\partial r_j^2} = \frac{1}{12}n(n^2 - 1)$$

by direct calculation.

When $n = 1$ the heat kernel of $\mathcal{H}(1)$ (the half axis) is

$$h_1(r, t) = \frac{1}{2\sqrt{\pi t}} e^{-r^2/(4t)}$$

and the functions

$$h_1^{(k)} = \left(-\frac{\partial}{\partial r}\right)^k h_1(r, t), \qquad k = 0, 1, 2, \ldots,$$

also satisfy the heat equation

$$\frac{\partial}{\partial t} h_1^{(k)}(r, t) = \frac{\partial^2}{\partial r^2} h_1^{(k)}(r, t).$$

Hence the function

$$H(r_1, \ldots, r_n, t) = \frac{C_1}{\omega} e^{-C_0 t} \sum_{j_1, \ldots, j_n = 1}^{n} \delta_{j_1 \cdots j_n}^{1 \cdots n} h_1^{(j_1 - 1)}(r_1, t) \cdots h_1^{(j_n - 1)}(r_n, t)$$

must satisfy equation (3) for arbitrary constant C_1. We choose

$$C_1^{-1} = \lim_{t \to 0} \int_{\mathcal{H}} \frac{1}{\omega} e^{-C_0 t} \det[h_1^{(j-1)}(r_h, t)]_{1 \le j, k \le n} \dot{H}.$$

It can be proved this limit C_1^{-1} exists. Then the heat kernel is

$$H_{\mathcal{H}(n)}(H_1, H_2, t) = h(r_1(H_1, H_2), \ldots, r_n(H_1, H_2), t).$$

Again $\mathcal{H}(n)$ is not an irreducible symmetric space but the space

$$\mathcal{H}_1(n) = \{H \in \mathcal{H}(n) | \det H = 1\}$$

is. So the heat kernel of $\mathcal{H}_1(n) \cong SL(n, C)/SU(n)$ (see Appendix) is

$$H_{\mathcal{H}_1(n)}(H_1, H_2, t) = H_{\mathcal{H}(n)}(\lambda_1 H_1, \lambda_2 H_2, t)/H_{\mathcal{H}(1)}(\lambda_1, \lambda_2, t).$$

REFERENCES

[HEL] S. Helgason, *Differential geometry and symmetric spaces*, Academic Press, New York, 1962.

[HO] Y. Hong, *The heat kernels of some symmetric spaces*, Intern. Symp., in Memory of Hua Loo Keng, Vol. II, Springer-Verlag and Science Press, Hong Kong. (1991), 143–155.

[HOW1] Y. Hong and G. Wang, *The heat kernels of some classical domains and classical manifolds*, Preprint.

[HOW2] ____, *The heat kernel of the symplectic unitary group $Sp(n)$*, Preprint. (Chinese)

[HUA] L. K. Hua, *The harmonic analysis of functions of several complex variables in classical domains*, Science Press, Beijing, 1958. (Chinese)

[LIU] W. Liu, *The conjugate points of the compact classical symmetric manifolds*, Dissertation, 1990. (Chinese)

[LU1] Qi-Keng Lu, *The classical manifolds and classical domains, scientific and technical publisher*, Shanghai, 1963. (Chinese)

[LU2] ____, *The matrix representation of the non-exceptional and irreducible Riemannian globally symmetric spaces*, Unpublished manuscript, 1964. (Chinese)

[LU3] ____, *The Green form to invariant metric of a ball*, Sci. China Ser. A (= Scientia Sinica) **32** (1989), 129–141.

[LU4] ____, *The heat kernel of a ball in C^n*, Chinese Ann. Math. Ser. B **11** (1990), 1–14.

[LU5] ____, *The theory of functions of several complex variables in China from 1949 to 1989*, Contemporary Geometry J.-Q. Zhong Memorial Volume, Plenum Press, New York, (1991) 53–93.

[LUL] Qi-keng Lu and K. Lu, *The heat kernel of the unitary group*, Preprint.

[LUH] Qi-keng Lu and Y. Hong, *The heat kernel of the classical domain $R_I(m, n)$*, Proc. Several Complex Variables, at Mittiag-Leffler Inst., '87–'88, Princeton Univ. Press Math. Notes, 1992.

[YAO] J. Yao, *the Morse functions of some symmetric spaces of Type* I, Dissertation, 1990. (Chinese)

[YE] F. Ye, *The conjugate points of the complex Grassmann manifolds*, Acta Math. Sinica **21** (1978), 367–374. (Chinese)

Appendix (a): NIRGSS of Type III

Quotient Space	Matrix Realization
$SL(n, R)/SO(n)$	$S > 0$, $\det S = 1$, $S^{(n)} = S' = \overline{S}$ or $I - TT' > 0$, $T^{(n)} = T' = \overline{T}$, $\det(I + T) = \det(I - T)$.
$SU^*(2n)/Sp(n)$	$H > 0$, $\det H = 1$, $H^{(2n)} = \overline{H}'$, $\overline{H}J = JH$, $J = \begin{pmatrix} 0 & I^{(n)} \\ -I^{(n)} & 0 \end{pmatrix}$ or $I - T\overline{T}' > 0$, $T^{(2n)} = -T'$, $\overline{T}J = JT$, $\det(J + T) = \det(J - T)$.
$SU(p, q)/S(U(p) \times U(q))$	$I - Z\overline{Z}' > 0$, $Z^{(p, q)}$ being complex matrix.
$SO(p, q)/S(O(p) \times O(q))$	$I - XX' > 0$, $X^{(p, q)}$ being real matrix.
$SO^*(2n)/U(n)$	$I - Z\overline{Z}' > 0$, $Z^{(n)} = -Z'$ being complex.
$Sp(n, R)/U(n)$	$I - Z\overline{Z}' > 0$, $Z^{(n)} = Z'$ being complex.
$Sp(p, q)/Sp(p) \times Sp(q)$	$I - Z\overline{Z}' > 0$, $Z^{(2p, 2q)}$ being complex, $K_0^{(2p)} Z = \overline{Z}K_0^{(2q)}$, $K_0^{(2p)} = I^{(p)} \times \begin{pmatrix} 0 & 1 \\ -1 & 0 \end{pmatrix}$ or $I - QQ^* > 0$, Q being an $p \times q$ matrix of quaternions, Q^* its conjugate and transposed matrix.

Appendix (b): NIRGSS of Type I

Quotient Space	Matrix Realization
$SU(n)/SO(n)$	$S^{(n)} = S'$, $S\overline{S}' = I$, $\det S = 1$.
$SU(2n)/Sp(n)$	$K^{(2n)} = -K'$, $K\overline{K}' = I$, $\det K = 1$.
$SU(p + q)/S(U(p) \times U(q))$	The space $G_I(p, q)$ (the complex Grassmann manifold) formed from all $p \times (p + q)$ complex matrices of rank p as its homogeneous coordinates.
$SO(p + q)/S(O(p) \times O(q))$	The space $G_I^R(p, q)$ (the real Grassmann manifold) formed from all $p \times (p + q)$ real matrices of rank p as its homogeneous coordinates.
$SO(2n)/U(n)$	The space $G_{III}(n)$ formed from all $n \times 2n$ complex matrices $\mathscr{Z}$ of rank n as its homogeneous coordinates and satisfying $\mathscr{Z}\mathscr{Z}' = 0$.
$Sp(n)/U(n)$	The space $G_{II}(n)$ formed from all $n \times 2n$ complex matrices $\mathscr{Z}$ of rank n as its homogeneous coordinates and satisfying $\mathscr{Z}J\mathscr{Z}' = 0$, $J = \begin{pmatrix} 0 & I^{(n)} \\ -I^{(n)} & 0 \end{pmatrix}$.
$Sp(p + q)/Sp(p) \times Sp(q)$	The space $G_I^Q(p, q)$ formed from all $2p \times (2p + 2q)$ complex matrices $\mathscr{Z}$ of rank n as its homogeneous coordinates and satisfying $\overline{Z}K_0^{(2p+2q)} = K_0^{(2p)}\mathscr{Z}$, $K_0^{(2p)} = I^{(p)} \times \begin{pmatrix} 0 & 1 \\ -1 & 0 \end{pmatrix}$ (the quaternion Grassmann manifold).

Appendix (c): NIRGSS of Type IV

Quotient Space	Matrix Realization
$SL(n+1, C)/SU(n+1)$	$H > 0$, $H^{(n+1)} = \overline{H}'$, $\det H = 1$.
$SO(2n+1, C)/SO(2n+1)$	All $(2n+1) \times (2n+1)$ matrices $$H = RP_0 \left[\begin{pmatrix} e^{-\lambda_1} & 0 \\ 0 & e^{\lambda_1} \end{pmatrix} \dotplus \cdots \dotplus \begin{pmatrix} e^{-\lambda_n} & 0 \\ 0 & e^{\lambda_n} \end{pmatrix} \dotplus 1 \right] \overline{P}_0' R'$$ where R is any real orthogonal matrix, $$P_0 = I^{(n)} \times \begin{pmatrix} 2^{-\frac{1}{2}} & 2^{-\frac{1}{2}} \\ i2^{-\frac{1}{2}} & -i2^{-\frac{1}{2}} \end{pmatrix} \dotplus 1$$ and $\lambda_1 \geq \cdots \geq \lambda_n \geq 0$.
$Sp(n, C)/Sp(n)$	All $2n \times 2n$ matrices $$H = U \begin{pmatrix} \Lambda & 0 \\ 0 & \Lambda^{-1} \end{pmatrix} \overline{U}, \quad U \text{ being unitary}, \qquad \overline{U}J = JU,$$ $$J = \begin{pmatrix} 0 & I^{(n)} \\ -I^{(n)} & 0 \end{pmatrix}, \quad \Lambda = \begin{pmatrix} \lambda_1 & & \\ & \ddots & \\ & & \lambda_n \end{pmatrix}$$ and $\lambda_1 \geq \cdots \geq \lambda_n > 0$.
$SO(2n, C)/SO(2n)$	All $2n \times 2n$ matrices $$H = RP_0 \left[\begin{pmatrix} e^{-\lambda_1} & 0 \\ 0 & e^{\lambda_1} \end{pmatrix} \dotplus \cdots \dotplus \begin{pmatrix} e^{-\lambda_n} & 0 \\ 0 & e^{\lambda_n} \end{pmatrix} \right] \overline{P}_0' R'$$ where R is real orthogonal matrix, $$P_0 = I^{(n)} \times \begin{pmatrix} 2^{-\frac{1}{2}} & 2^{-\frac{1}{2}} \\ i2^{-\frac{1}{2}} & -i2^{-\frac{1}{2}} \end{pmatrix}$$ and $\lambda_1 \geq \cdots \geq \lambda_n \geq 0$.

Notice that $S' = S^T$. Here $A \dotplus B$ of two matrices A and B means the matrix $\begin{pmatrix} A & 0 \\ 0 & B \end{pmatrix}$ and

$$A \times B = \begin{pmatrix} a_{11}B & \cdots & a_{1n}B \\ \cdots & \cdots & \cdots \\ a_{1m}B & \cdots & a_{mn}B \end{pmatrix} \quad \text{where } A = (a_{jk}).$$

The NIRGSS of the groups spaces $SU(n+1)$, $SO(2n)$, $SO(2n+1)$ and $Sp(n)$ are themselves matrices.

ACADEMIA SINICA, PEOPLE'S REPUBLIC OF CHINA

Proceedings of Symposia in Pure Mathematics
Volume **54** (1993), Part 2

CR-Geometry and Deformations
of Isolated Singularities

JOHN J. MILLSON

This paper is a greatly expanded version of my lecture at the A.M.S. Summer Institute in Differential Geometry in July 1990. It is an exposition of the paper [**BM**] which shows how to compute the parameter space $(\mathcal{X}, x)$ of the versal deformation of an isolated singularity $(V, 0)$ from the CR-structure on a link M of the singularity. The paper [**BM**] had two sources. The first was provided by the fundamental work [**K2**] of Kuranishi in 1977. In this paper, Kuranishi wrote down a first-order nonlinear system of partial differential equations on a link of the singularity with a finite-dimensional formal solution space assuming $\dim V \geq 3$ (see Theorem 5.2 for a precise statement). This space had the key property that it did not depend on the choice of sphere used to define the link. In 1983, in [**Mi1**], Miyajima proved that Kuranishi's space had a complex analytic structure to be denoted $(\mathcal{X}_M, 0)$ henceforth. His work was based on earlier work of Akahori, [**Ak1–Ak4**]. Our theorem below relating $(\mathcal{X}_M, 0)$ to the parameter space of the versal deformation of $(V, 0)$ was obtained independently by Miyajima in [**Mi2**]. His result is based in part on [**Fu**].

The second source for our paper was the principle that deformation problems are controlled by differential graded Lie algebras which we learned from Deligne [**D**]—similar ideas are to be found in [**SS**]. According to this principle, in order to prove that two deformation spaces are isomorphic one finds controlling differential graded Lie algebras (see §2) for the deformation theories and then proves that these algebras are one-quasi-isomorphic. We recall that two differential graded Lie algebras L and $\overline{L}$ are *one-quasi-isomorphic* if there is a sequence

$$L = L_1 \to L_2 \leftarrow L_3 \to \cdots \to L_{n-1} \leftarrow L_n = \overline{L}$$

1991 *Mathematics Subject Classification.* Primary 32G07, 32G11.
The author was supported by the National Science Foundation grant DMS-90-02116.
The detailed version of this paper, [**BM**], will be submitted for publication elsewhere.

of homomorphisms of differential graded Lie algebras such that each arrow above induces an isomorphism on first cohomology and an injection on second cohomology; see [**GM1**] and [**M1**] for other applications of this principle (one says that L and $\overline{L}$ are *quasi-isomorphic* if the above arrows induce isomorphisms of all cohomology groups). This is what we do in this paper. Our main theorem goes as follows:

THEOREM. *Suppose $(V, 0)$ is a normal isolated singularity and satisfies*
(1) $\dim V \geq 4$;
(2) $\mathrm{depth}_{\{0\}} V \geq 3$.
Then the base space of the versal deformation of $(V, 0)$ is isomorphic to $(\mathscr{K}_M, 0)$.

REMARKS. The assumption (2) is equivalent to the assumption that $H^1(U, \mathscr{O}) = \{0\}$ where $U = V - \{0\}$. If we do not assume (2) we show that the base space for the versal deformation of $(V, 0)$ is isomorphic to a closed subgerm of $(\mathscr{K}_M, 0)$. In Chapter 10 of [**BM**] we give examples such that $(V, 0)$ is normal but the deformation space of $(V, 0)$ is a proper subgerm of $(\mathscr{K}_M, 0)$.

Our proof involves many technical details but the basic idea is simple. It is to prove that the arrows in the following diagram of *formal* deformation theories

$$\mathrm{Def}(V, 0) \leftarrow \mathrm{Def}(V) \rightarrow \mathrm{Def}(U) \rightarrow \mathrm{Def}(M)$$

are isomorphisms. The middle arrow is the map of Schlessinger ([**Sc2**] or [**Ar2**, Part I, §9]). A formal deformation of V is a sheaf S on V satisfying certain axioms. Then the middle arrow restricts S to U. In fact we study the corresponding diagram of maps of controlling differential Lie algebras and prove they are all one-quasi-isomorphisms. A critical intermediate step is Theorem D below which allows us to replace the tangent complex L_U of U with the Kodaira-Spencer algebra $\mathscr{L}(U) = \mathscr{A}^{0,\cdot}(U, T^{1,0}(U))$ which is needed to compare with Kuranishi's theorem. Our proof divides into six theorems, some (especially Theorems C and D) of interest in their own right. In what follows $L_{V,0}$ denotes the tangent complex of the germ $(V, 0)$, L_V the tangent complex of V and $\overline{\mathscr{K}}$ the differential graded Lie algebra on M constructed in §6. Our six theorems are then the following.

THEOREM A. *The local tangent complex $L_{V,0}$ controls the deformation theory of an analytic germ $(V, 0)$.*

THEOREM B. *L_V is one-quasi-isomorphic to $L_{V,0}$.*

THEOREM C. *The global tangent complex L_V controls the deformation theory of a complex analytic space.*

COROLLARY. *The complete local $\mathbb{C}$-algebras $R_{L_{v,0}}$ and R_{L_U} are isomorphic (assuming $\mathrm{depth}_{\{0\}} V \geq 3$).*

PROOF. By [Sc2] the map of *formal* deformation theories

$$\mathrm{Def}(V) \to \mathrm{Def}(U)$$

is an isomorphism. Hence R_{L_V} and R_{L_U} are isomorphic. But by Theorem B and the Comparison Theorem of §2 the complete local $\mathbb{C}$-algebras R_{L_V} and $R_{L_{V,0}}$ are isomorphic. □

THEOREM D. *Let U be a complex manifold. Then the tangent complex L_U and the Kodaira-Spencer algebras $\mathscr{L}(U) = \mathscr{A}^{0,\cdot}(U, T^{1,0}(U))$ are quasi-isomorphic.*

COROLLARY 1. *The Kodaira-Spencer algebra controls the (formal) deformation theory of U (even if U is not compact).*

COROLLARY 2. *Assuming now that U is compact, then Kuranishi's space $(\mathscr{K}_U, 0)$ is analytically-versal for all deformations of U including those with nonreduced base.*

PROOF. By [Gr] there exists an analytically-versal family with base $(\mathscr{X}, x)$. But by Theorem C we have $\widehat{\mathscr{O}}_{\mathscr{X},x} \cong R_{L_U}$. By Theorem D we have $R_{L_U} \cong R_{\mathscr{L}(U)}$ and by [GM2, §3] or [BM, Theorem 1.4] we have $R_{\mathscr{L}(U)} \cong \widehat{\mathscr{O}}_{\mathscr{K}_U,0}$. The corollary follows from Artin's Theorem (see Theorem 1.1). □

THEOREM E. *The Kodaira-Spencer algebra $\mathscr{L}(U)$ and the differential graded Lie algebra $\overline{\mathscr{K}}^{\cdot}(M)$ on the link are one-quasi-isomorphic (assuming $\dim V \geq 4$).*

THEOREM F. *$\overline{\mathscr{K}}^{\cdot}(M)$ controls Kuranishi's CR-deformation theory.*

COROLLARY. *$(\mathscr{K}_M, 0)$ is isomorphic to the parameter space $(\mathscr{X}, x)$ of the versal deformation of $(V, 0)$.*

PROOF. The Comparison Theorem and Theorem E give $R_{\mathscr{L}(U)} \cong R_{\overline{\mathscr{K}}(M)}$. The first corollary above then gives $R_{L_{V,0}} \cong R_{\overline{\mathscr{K}}(M)}$. By Theorems A and F we obtain $\widehat{\mathscr{O}}_{(\mathscr{X},x)} \cong \widehat{\mathscr{O}}_{\mathscr{K}_M,0}$. By Artin's Theorem (see Theorem 1.1) $(\mathscr{X}, x) \cong (\mathscr{K}_M, 0)$. □

This paper could not have been written without many conversations with other mathematicians, among them, Bill Goldman, Steve Halperin and Mike Schlessinger. I would especially like to thank Madhav Nori for providing us with some important ideas concerning affine varieties in infinite-dimensional vector spaces, Jack Lee for many conversations about CR-manifolds and H. Flenner for help with the counterexamples mentioned above. Finally, I would like to thank Pierre Deligne for outlining the connection between differential graded Lie algebras and deformation theory in [D] five years ago and my collaborator Ragnar Buchweitz for patiently explaining many basic facts about complexes of sheaves, the derived category and the tangent complex.

1. Preliminaries on analytic germs

Let $(\mathscr{X}, x)$ be an analytic germ in $\mathbb{C}^n$ defined by analytic equations

$$f_1(z_1, z_2, \ldots, z_n) = 0,$$

$$\vdots$$

$$f_k(z_1, z_1, \ldots, z_n) = 0.$$

The geometry of $(\mathscr{X}, x)$ is then encoded in each of the three following algebraic objects:

(i) the analytic local ring

$$\mathscr{O}_{\mathscr{X}, x} = \frac{\mathbb{C}\{z_1, z_2, \ldots, z_n\}}{(f_1, f_2, \ldots, f_k)};$$

(ii) the complete local ring

$$\widehat{\mathscr{O}}_{\mathscr{X}, x} = \frac{\mathbb{C}[[z_1, z_2, \ldots, z_n]]}{(f_1, f_2, \ldots, f_k)};$$

(iii) the functor of points $F_{\mathscr{X}, x}$.

In the above $(f_1, f_2, \ldots, f_k)$ denotes the ideal generated by $f_1, f_2, \ldots, f_k$. We now define the functor of points, $F_{\mathscr{X}, x}$.

We recall that an Artin local $\mathbb{C}$-algebra A is a commutative $\mathbb{C}$-algebra which is finite dimensional as a complex vector space. Thus if $\mathfrak{m}$ denotes the maximal ideal of A, then every element of $\mathfrak{m}$ is nilpotent. Let $\mathscr{A}$ be the category of Artin local $\mathbb{C}$-algebras. We define

$$F_{\mathscr{X}, x} : \mathscr{A} \to \text{Sets}$$

by

$$F_{\mathscr{X}, x}(A) = \{(m_1, m_2, \ldots, m_n) \in \mathfrak{m}^n : f_j(m_1, m_2, \ldots, m_n) = 0, \ 1 \le j \le k\}.$$

REMARK. An invariant definition of $F_{\mathscr{X}, x}$ is given by

$$F_{\mathscr{X}, x}(A) = \text{Hom}_{\mathbb{C}\text{-alg}}(\widehat{\mathscr{O}}_{\mathscr{X}, x}, A).$$

Here the Hom denotes local homomorphisms of local $\mathbb{C}$-algebras. We will often write $F_{\mathscr{X}, x}$ as $F_{\widehat{\mathscr{O}}_{\mathscr{X}, x}}$.

We then have the following theorem.

THEOREM 1.1. *Let (X, x) and (Y, y) be analytic germs. Then the following are equivalent.*

(i) $(X, x) \approx (Y, y)$,

(ii) $\mathscr{O}_{X, x} \approx \mathscr{O}_{Y, y}$,

(iii) $\widehat{\mathscr{O}}_{X, x} \approx \widehat{\mathscr{O}}_{Y, y}$,

(iv) $F_{X,x} \approx F_{Y,y}$.

REMARK. For a discussion of this theorem see [GM1, §3.1]. The critical step (iii) implies (ii) is a deep theorem of M. Artin [Ar1].

2. The deformation theory associated to a differential graded Lie algebra

We begin with the following definitions which are basic for all that follows.

DEFINITIONS. Let **k** be a field. A graded Lie algebra over **k** is a pair $(L, [\ , \ ])$ where L is a graded vector space over **k**

$$L = \bigoplus_{i \geq 0} L^i$$

and $[\ , \ ]: L \oplus L \to L$ is a graded bilinear map of degree zero satisfying (graded) skew-commutativity

$$[\alpha, \beta] + (-1)^{ij}[\beta, \alpha] = 0$$

and the (graded) Jacobi identity

$$(-1)^{ki}[\alpha, [\beta, \gamma]] + (-1)^{ij}[\beta, [\gamma, \alpha]] + (-1)^{jk}[\gamma, [\alpha, \beta]] = 0$$

where $\alpha \in L^i$, $\beta \in L^j$, $\gamma \in L^k$.

A differential graded Lie algebra over **k** is a triple, $(L, [\ , \ ], d)$ such $(L, [\ , \ ])$ is a graded Lie algebra over **k**, (L, d) is a complex (and d has degree $+1$) and

$$d[\alpha, \beta] = [d\alpha, \beta] + (-1)^i[\alpha, d\beta]$$

where $\alpha \in L^i$.

We now let $(L, [\ , \ ], d)$ be a differential graded Lie algebra over $\mathbb{C}$ such that $\dim H^1(L) < \infty$. We wish to associate a complete local Noetherian $\mathbb{C}$-algebra R_L to L. The definition of R_L is indirect.

The definition of R_L. Instead of defining R_L we define the associated functor

$$F_{R_L} : A \to \text{Sets} .$$

Choose a complement C^1 to the one-coboundaries $B^1 \subset L^1$ whence

$$L^1 = B^1 + C^1 .$$

Then if $A \in \text{Obj}\,\mathscr{A}$ and A has maximal ideal $\mathfrak{m}$ we define

$$Y_L(A) = \{\eta \in C^1 \otimes \mathfrak{m} : d\eta + \tfrac{1}{2}[\eta, \eta] = 0\}.$$

THEOREM. *There exists a complete local Noetherian $\mathbb{C}$-algebra R_L such that*

$$Y_L = F_{R_L} .$$

PROOF. Apply Schlessinger's Theorem [Sc1] characterizing those functors $F: \mathscr{A} \to$ Sets that are pro-representable (see below) by complete local Noetherian $\mathbb{C}$-algebras. □

EXAMPLE. Suppose $(L, [\ ,\ ], d)$ satisfies $d = 0$. Then $R_L = \mathscr{O}_{Q_L}$ where

$$Q_L = \{\eta \in L^1 : [\eta, \eta] = 0\}.$$

DEFINITION. Suppose (X, x) is the versal deformation space [GM2, p. 364] for some deformation theory. Suppose L is a differential graded Lie algebra such that

$$R_L \approx \hat{\mathscr{O}}_{X,x}.$$

Then we call L a *controlling differential graded Lie algebra* for the deformation theory.

EXAMPLES.

(i) **The twisted de Rham algebra.** Let M be a compact manifold with fundamental group $\pi_1(M)$ and G be a Lie group with Lie algebra $\mathfrak{g}$. Let $\rho: \pi_1(M) \to G$ be a homomorphism. Define

$$P = \widetilde{M} \times_{\pi_1(M)} G,$$
$$\mathrm{ad}\, P = \widetilde{M} \times_{\pi_1(M)} \mathfrak{g},$$
$$L = (\mathscr{A}^*(M, \mathrm{ad}\, P), [\ ,\ ], d_\nabla).$$

Here $[\ ,\ ]$ is the pointwise bracket ($\mathrm{ad}\, P$ is a bundle of Lie algebras) and d_∇ is the exterior covariant derivative associated to the canonical flat connection ∇ on $\mathrm{ad}\, P$. Then L is a controlling differential graded Lie algebra for deforming the flat connection ∇ (this is proved in [GM2, §3]).

(ii) **The Kodaira-Spencer algebra.** Let M be a compact complex manifold. We define L by

$$L = \{\mathscr{A}^{0,\,\cdot}(M, T^{1,0}(M)), [\ ,\ ], \overline{\partial}\}.$$

Here, $[\ ,\ ]$ is the Nijenhuis bracket [FN1]. Then L is a controlling differential graded Lie algebra for deforming the complex structure on M (this is proved in [GM2, §3]).

REMARK. We will give more examples of controlling differential graded Lie algebras in §§3, 4, and 6.

We now state the critical invariance property of R_L.

THE COMPARISON THEOREM. *Suppose* $f: L \to \overline{L}$ *is a homomorphism of differential graded Lie algebras with*

(i) $H^1(f)$ *an isomorphism,*

(ii) $H^2(f)$ *an injection.*

Then

$$R_L \approx R_{\overline{L}}.$$

REMARK. We do not require $f(C^1) \subset \overline{C}^1$.

COROLLARY 1. *The isomorphism class of R_L is independent of the choice of the complement C^1.*

COROLLARY 2. *Suppose there exists a chain*

$$L = L_1 \to L_2 \leftarrow L_3 \to \cdots \to L_n = \overline{L}$$

with each arrow as above. Then

$$R_L \approx R_{\overline{L}}.$$

COROLLARY 3. *Suppose in the above that $\overline{L}$ has a zero differential. Then $R_L \approx \widehat{\mathscr{O}}_{Q_H}$ where*

$$Q_H \approx \{\eta \in H^1(L) : [\eta, \eta] = 0\}.$$

The last corollary explains the ubiquity of quadratic singularities in deformation theory. The map $\eta \to [\eta, \eta]$ is called the *first obstruction* to integrating η.

Definitions. In case L and $\overline{L}$ are related by a chain of arrows as above we will say L and $\overline{L}$ are *one-quasi-isomorphic*.

If L is one-quasi-isomorphic to $\overline{L}$ and $\overline{L}$ has zero differential we will say that L is *formal*.

In fact there is a "deformation theory" canonically associated to L, (see [**GM1**, §2], for details). Let A be an Artin local $\mathbb{C}$-algebra with maximal ideal $\mathfrak{m}$. We define a groupoid $\mathscr{C}(L; A)$ (i.e., a small category such that every morphism is an isomorphism) as follows. We define $\operatorname{Obj}\mathscr{C}(L; A)$, the set of objects, by

$$\operatorname{Obj}\mathscr{C}(L; A) = \{\eta \in L^1 \otimes \mathfrak{m} : d\eta + \tfrac{1}{2}[\eta, \eta] = 0\}.$$

We define $\operatorname{Mor}\mathscr{C}(L; A)$, the set of morphisms, to be the nilpotent group $\exp(L^0 \otimes \mathfrak{m})$. This group has underlying space $L^0 \otimes \mathfrak{m}$ and is equipped with the Campbell-Baker-Hausdorff multiplication

$$(X, Y) \to \log(\exp(X)\exp(Y)).$$

The morphisms act on the objects by the affine action ρ. This action is determined by the infinitesimal action

$$d\rho(\lambda) \cdot \eta = [\lambda, \eta] - d\lambda, \qquad \lambda \in L^0 \otimes \mathfrak{m}, \ \eta \in L^1 \otimes \mathfrak{m}.$$

We let $\operatorname{Iso}\mathscr{C}(L; A)$ be the set of isomorphism classes. The ring R_L is then a hull for the functor $\operatorname{Iso}\mathscr{C}(L; \cdot)$ in the sense of [**Sc1**, Definition 2.7], [**BM**, Theorem 1.3].

We now recall the definition of *hull* since it will be important in what follows. Let $\mathscr{A}$ be the category of Artin local $\mathbb{C}$-algebras and F and G be functors on $\mathscr{A}$ with values in the category **Sets** and let $\eta: F \to G$ be a natural transformation. Then η is smooth if for any surjection $A \to B$ in $\mathscr{A}$ the induced map

$$F(A) \to F(B) \times_{G(B)} G(A)$$

is surjective. The natural transformation η is *minimally smooth* if, for any Artin local $\mathbb{C}$-algebra A with maximal ideal $\mathfrak{m}$ satisfying $\mathfrak{m}^2 = 0$, $\eta_A: F(A) \to G(A)$ is an isomorphism.

Now let $\mathscr{C}$ be the category of complete local $\mathbb{C}$-algebras. Let R be an object of $\mathscr{C}$ and define $h_R: \mathscr{C} \to \textbf{Sets}$ by

$$h_R(S) = \text{Hom}_{\mathbb{C}\text{-alg}}(R, S).$$

Now suppose F is as above and extend F to $\mathscr{C}$ by the formula

$$F(R) = \varprojlim F(R/\mathfrak{m}^n)$$

where $\mathfrak{m}$ is the maximal ideal of R. We observe that if $u \in \text{Hom}_{\mathbb{C}\text{-alg}}(R, A)$ there is an induced map $F(u): F(R) \to F(A)$. Hence given $\xi \in F(R)$ we obtain a natural transformation of functors $e_\xi: h_R \to F$ given by

$$e_\xi(u) = F(u)(\xi).$$

We now give two important definitions from [**Sc1**]. A complete local Noetherian $\mathbb{C}$-algebra R together with an element $\xi \in F(R)$ is said to *pro-represent* F if the natural transformation e_ξ is an isomorphism of functors. A complete local Noetherian $\mathbb{C}$-algebra R together with an element $\xi \in F(R)$ is said to be a *hull* of F if the natural transformation $e_\xi: h_R \to F$ is minimally smooth.

It is easy to see that if (R, ξ) pro-represents Y_L then it is a hull for $\text{Iso}\,\mathscr{C}(L; \cdot)$—see [**BM**, Chapter 1]. Since Y_L is described by polynomial equations on a vector space it is intuitively clear that it is pro-representable. A proof along these lines of the pro-representability of Y_L is given in [**BM**, Theorem 1.2] based on ideas of Madhav Nori. The condition $\dim H^1(L) < \infty$ corresponds to the condition that R_L be Noetherian; see [**BM**, Theorem 1.2].

3. The local tangent complex

Let B be an analytic local $\mathbb{C}$-algebra and A be an Artin local $\mathbb{C}$-algebra with maximal ideal $\mathfrak{m}$. We define the groupoid $\text{Def}(B; A)$ of deformations of B with base A. We define an object of $\text{Def}(B; A)$ to be a flat A-algebra B' which reduces to B modulo $\mathfrak{m}$. If B' and B'' are objects then a morphism from B' to B'' is a homomorphism of A-algebras such that $f \equiv \text{id} \mod \mathfrak{m}$. We let $\text{Iso}\,\text{Def}(B; A)$ denote the set of isomorphism classes of $\text{Def}(B; A)$.

We would like to find a complete local Noetherian $\mathbb{C}$-algebra S and an element $\xi \in \text{Def}(B; S)$ such that (S, ξ) pro-represents $\text{Iso}\,\text{Def}(B; \cdot)$. Usually this is not possible but $\text{Iso}\,\text{Def}(B; \cdot)$ has a hull S [**Ar**, Part I, §8]. We now show we can construct S by the construction of §2.

Since B is an analytic local $\mathbb{C}$-algebra, we may write $B = \mathbb{C}\{z_1, \dots, z_n\}/ (f_1, f_2, \dots, f_k)$. Let $R(1)$ be the differential graded algebra obtained by adding free (exterior) algebra generators $y_1, y_2, \ldots, y_k$ of degree -1 to $\mathbb{C}\{z_1, \ldots, z_n\}$ so

$$R(1) = \mathbb{C}\{z_1, \ldots, z_n\}[y_1, \ldots, y_k].$$

Define ∂ on $R(1)$ by $\partial p = 0$, $p \in \mathbb{C}\{z_1, \ldots, z_n\}$ and $\partial y_j = f_j$. Now kill the first cohomology of $R(1)$ by adding free (polynomial) generators $x_1, x_2, \ldots, x_l$ of degree -2 to get $(R(2), \partial)$. Continuing in this way we obtain $R = \varinjlim R(k)$ which is a free graded-commutative algebra over $\mathbb{C}\{z_1, \ldots, z_n\}$ and a surjective homomorphism $\varepsilon: R \to B$ which makes R a resolution of B. The differential graded algebra R is called a *resolvent* for B.

REMARK. The resolvent R is not functorial in B because it involves a choice of presentation of B. Thus the construction of R does not "sheafify."

We now define the local tangent complex L_B by $L_B = \mathrm{Der}^+(R, R)$ when R is a resolvent for B (thus L_B depends upon the choice of resolvent). The symbol $\mathrm{Der}^+(R, R)$ means the Lie algebra of graded derivations of nonnegative degree of the graded-commutative algebra R. We recall that a derivation D of degree i satisfies $D: R^k \to R^{k+i}$, all k, and

$$D(r_1 r_2) = (Dr_1)r_2 + (-1)^{ij} r_1 (Dr_2)$$

where $\deg r_1 = j$. Then L_B is a graded Lie algebra under the graded bracket

$$[D_1, D_2] = D_1 \circ D_2 - (-1)^{ij} D_2 \circ D_1$$

where $\deg D_1 = i$, $\deg D_2 = j$. We note that ∂ is a graded derivation of R of degree 1. Thus we can define $d: L_B \to L_B$ by

$$dD = [\partial, D].$$

It is immediate that with these definitions $(L, [\,,\,], d)$ is a differential graded Lie algebra which will be seen below to control the deformation theory of B. It is not hard to see, [**F**] or [**P**], that *up to quasi-isomorphism* of differential graded Lie algebras L_B does not depend on the choice of R.

Now let $(V, 0)$ be an analytic germ and let B be the analytic local ring $\mathscr{O}_{V,0}$ of $(V, 0)$. Since the category of analytic local $\mathbb{C}$-algebras is the opposite of that of analytic germs (and deformations of the latter are defined dually to the above) we see that Theorem A is equivalent to the following theorem.

THEOREM 2.1. *The complete local Noetherian $\mathbb{C}$-algebra R_{L_B} is a hull for* $\mathrm{Iso\,Def}(B, \cdot)$.

PROOF. We describe a natural transformation $p: \mathscr{C}(L_B; A) \to \mathrm{Def}(B; A)$ which induces the required natural isomorphism of isomorphism classes. Choose a resolvent R for B and let $R' = (R \otimes A, \partial \otimes 1)$. Then $B' = H^0(R') = B \otimes A$ is the trivial deformation of B. Now let $\eta \in \mathrm{Obj}\,\mathscr{C}(L; A)$. Hence $\eta \in \mathrm{Der}^1(R, R) \otimes \mathfrak{m}$ and $d\eta + \frac{1}{2}[\eta, \eta] = 0$. It is immediate that this latter equation is equivalent to the identity $(\partial \otimes 1 + \eta) \circ (\partial \otimes 1 + \eta) = 0$ in $\text{Der}(R, R) \otimes A$. Now let $\tilde{\eta}$ be the image of η under the natural map $\text{Hom}(R, R) \otimes A \to \text{Hom}_A(R \otimes A, R \otimes A)$. Put $\partial'' = \partial \otimes 1 + \tilde{\eta}$. Then $\partial'' \circ \partial'' = 0$ whence $R'' = (R \otimes A, \partial'')$ is a differential graded algebra. We define p by

$$p(\eta) = H^0(R'').$$

The definitions of p on morphisms is clear and we obtain the required functor.   □

REMARKS. The functor p "perturbs the differential" of the trivial deformation of a resolvent for B. Theorem 2.1 is equivalent to the statement that L_B controls the deformation theory of B.

We can now prove Theorem A. Indeed if $(\mathscr{X}, x)$ is the base of the versal deformation of $(V, 0)$ then $\widehat{\mathscr{O}}_{\mathscr{X}, x}$ is also a hull for $\text{Iso Def}(B;)$. But hulls are unique [Sc1, Proposition 2.9]. Hence $R_{L_B} \approx \widehat{\mathscr{O}}_{\mathscr{X}, x}$.

4. The global tangent complex

Let X be a complex analytic space and A be an Artin local $\mathbb{C}$-algebra with maximal ideal $\mathfrak{m}$. We define the groupoid $\text{Def}(X; A)$ of deformations of X with base $\text{Spec} A$. An object of $\text{Def}(X; A)$ is defined to be a complex analytic space X' flat over $\text{Spec} A$, whose restriction to $\text{Spec} \mathbb{C} \subset \text{Spec} A$ is X. Thus as a topological space $X' = X$ but the extra information defining X' is a sheaf $B' = \mathscr{O}_{X'}$ of flat A-algebras whose reduction modulo $\mathfrak{m}$ is $\mathscr{O}_X$. Intuitively X' is an "infinitesimal thickening" of X. If X' and X'' are deformations of X over $\text{Spec} A$ we define a morphism $f: X' \to X''$ to be a morphism of complex analytic spaces whose restriction to X is the identity. All morphisms are isomorphisms and we may form the set of isomorphism classes $\text{Iso Def}(X; A)$. It is a standard application of Theorem 2.11 of [Sc1] (see [Be, Theorem 2.13]) that if the vector space $T^1(X)$ of infinitesimal deformations of X (i.e., $\text{Def}(X; \mathbb{C}[\varepsilon])$ where $\mathbb{C}[\varepsilon]$ is the dual numbers) is finite dimensional then the functor $\text{Iso Def}(X; \cdot)$ has a hull S. We now show that we can construct S by the construction of §2.

Suppose first that X is a "model space"; that is, X is the intersection of an analytic subset of a Stein domain V with another Stein domain W such that $\overline{W} \subset V$ and $\overline{W}$ has a fundamental system of Stein neighborhoods in V. We let $\varphi: X \to W$ be the inclusion. Using Cartan's Theorems A and B, we may repeat the construction of the previous section to obtain a sheaf R of differential (negatively) graded algebras over W such that R is a free graded-commutative $\mathscr{O}_W$-algebra in the sense that R is freely generated as an algebra by certain global sections (a finite number in each degree). Furthermore there is a surjective homomorphism $\varepsilon: R \to i_* \mathscr{O}_X$ which makes R a resolution of $i_* \mathscr{O}_X$. We say that R is a resolvent for X.

Now let X be an arbitrary complex analytic space. An analytic polyhedron in X is any relatively compact subset P of X together with a proper holomorphic imbedding $\varphi: U \to V \subset \mathbb{C}^N$ which is defined in some neighborhood $U \supset \overline{P}$ so that V is a Stein neighborhood of the closure of the unit coordinate poly-disk $W = D^N$ and $P = \varphi^{-1}(D^N)$. With each polyhedron there is associated the complex space $(\varphi(P), \varphi_*(\mathscr{O}_X|P))$ which is a model space. We call φ the barrier map for P.

Now let $\mathscr{P} = \{P_i\}_{i \in I}$ be a locally finite cover of X by polyhedra as above such that if $\mathscr{U} = \{U_i\}_{i \in I}$ is the corresponding cover (here $U_i \supset P_i$ as above) we have $\mathrm{Nerve}(\mathscr{P}) = \mathrm{Nerve}(\mathscr{U})$. This is possible by [**Bi**, Chapter IV, Lemma 3.7]. We let $\varphi_i: U_i \to V_i \subset \mathbb{C}^{N_i}$ by the corresponding barrier maps and put $\mathscr{N} = \mathrm{Nerve}(\mathscr{P})$.

We now construct a pair of simplicial complex spaces X_* and W_* with underlying simplicial complex $\mathscr{N}$. By this we mean that X_* and W_* are contravariant functors from $\mathscr{N}$ considered as a category (the objects are the simplices, the morphisms the face inclusions) to the category of complex spaces. If $\alpha = (i_0, i_1, \ldots, i_p) \in \mathscr{N}$ we define $X_\alpha = P_{i_0} \cap \cdots \cap P_{i_p}$. If $\alpha, \beta \in \mathscr{N}$ with $\alpha \subset \beta$ we define the corresponding morphism $j_{\alpha\beta}: X_\beta \to X_\alpha$ to be the inclusion. We define W_* as follows. If $\alpha = (i_0, i_1, \ldots, i_p) \in \mathscr{N}$ then $W_\alpha = W_{i_0} \times \cdots \times W_{i_p}$ and if $\alpha, \beta \in \mathscr{N}$ with $\alpha \subset \beta$ then we define $p_{\alpha\beta}: W_\beta \to W_\alpha$ to be the projection on the factors corresponding to α. We obtain an embedding (natural transformation) $\varphi: X_* \to W_*$ by defining φ_α to be the composition $X_\alpha \to X_\alpha \times \cdots \times X_\alpha \to W_{i_0} \times \cdots \times W_{i_p}$ where the first map is the diagonal imbedding and the second the product of the restrictions of the φ_{i_j}'s, $0 \leq j \leq p$. Thus we obtain φ_α as the restriction of the proper map $\prod_j \varphi_{i_j}$ to the closed subset of $X_{i_0} \times \cdots \times X_{i_p}$ obtained by intersecting with the diagonal $\Delta \subset X \times \cdots \times X$. Thus φ_α is proper.

We now explain the key "smoothness" property possessed by the simplicial complex space W_*. We observe that $\mathscr{O}_{W_*}$ is a simplicial sheaf of algebras on $\mathscr{N}$ (for each simplex $\alpha \in \mathscr{N}$ we have $\mathscr{O}_{W_\alpha}$ and for each $\alpha, \beta \in \mathscr{N}$ with $\alpha \subset \beta$ we have a ring homomorphism $\mu_{\beta\alpha}: p_{\alpha\beta}^{-1}\mathscr{O}_{W_\alpha} \to \mathscr{O}_{W_\beta}$). We define $\mathrm{Mod}(\mathscr{O}_{W_*})$, the category of simplicial $\mathscr{O}_{W_*}$-modules on $\mathscr{N}$ in the obvious way (a module is now a *covariant* functor on $\mathscr{N}$) and $\mathrm{Coh}(\mathscr{O}_{W_*})$ to be those $\mathscr{O}_{W_*}$-modules M_* so that M_α is a coherent $\mathscr{O}_{W_\alpha}$-module for all $\alpha \in \mathscr{N}$. There is a functor [**F**, p. 29] $p_\alpha^*: \mathrm{Mod}(\mathscr{O}_{W_\alpha}) \to \mathrm{Mod}(\mathscr{O}_{W_*})$ defined by

$$(p_\alpha^* M_\alpha)_\beta = \begin{cases} p_{\alpha\beta}^* M_\alpha, & \beta \supseteq \alpha, \\ 0, & \text{otherwise}. \end{cases}$$

The functor p_α^* has a right adjoint res_α where $\mathrm{res}_\alpha(M_*) = M_\alpha$ and consequently p_α^* carries projective objects to projective objects. The next lemma is the key point in the previous construction of W_*.

LEMMA 4.1. *The module $\Omega^1_{W_*}$ of differential 1-forms on W_* is a projective object in $\mathrm{Coh}(\mathscr{O}_{W_*})$.*

PROOF. Clearly $\Omega^1_W = \bigoplus_{i \in I} p_i^* \Omega^1_{W_i}$.   □

REMARK. The previous lemma plays a key role in the deformation theory of X. Indeed let $B' \in \mathrm{Obj}\,\mathrm{Def}(X; A)$ as above. We let B'_* be the corresponding simplicial sheaf on X_* defined by $B'_\alpha = B'|X_\alpha$. Then the previous lemma has as a consequence that the sheaf $\varphi_* B'_*$ on W_* is a quotient of $\mathcal{O}_{W_*} \otimes A$ and any morphism $f: \varphi_* B' \to \varphi_* B''$ can be covered by an algebra homomorphism $F: \mathcal{O}_{W_*} \otimes A \to \mathcal{O}_{W_*} \otimes A$.

We now explain the notion of a resolvent of a complex analytic space X. We need some preliminary definitions.

DEFINITIONS. Let $G_* \in \mathrm{Mod}(\mathcal{O}_{W_*})$. Then we define G_* to be free if there exist free $\mathcal{O}_{W_\alpha}$-modules F_α, $\alpha \in \mathcal{N}$, such that $G_* = \bigoplus_\alpha p_\alpha^* F_\alpha$. We note that free coherent modules are projective. A graded commutative $\mathcal{O}_{W_*}$-algebra $R_*^{\cdot}$ is said to be free if there exists a free negatively-graded $\mathcal{O}_{W_*}$-module $M_*^{\cdot}$ such that $R_*^{\cdot} = S_{\mathcal{O}_{W_*}}(M_*^{\cdot})$. This last symbol denotes the graded symmetric product. Finally if B_* is an $\mathcal{O}_{W_*}$-algebra then a resolvent for B_* over $\mathcal{O}_{W_*}$ is a differential graded $\mathcal{O}_{W_*}$-algebra R_* which is free as a commutative graded algebra and a surjective homomorphism $\varepsilon: R_* \to B_*$ which makes R_* a resolution of B_* (i.e., over each simplex $\alpha \in N$ we have $H^i(R_\alpha) = 0$, $i > 0$, and $\varepsilon_\alpha: H^0(R_\alpha) \to B_\alpha$ is an isomorphism). A resolvent of the complex space X is a triple $\mathfrak{g} = (X_*, W_*, R_*)$. Here $\mathcal{P}$ is a cover of X by analytic polyhedra with nerve $\mathcal{N}$, X_* and W_* are as above and R_* is a resolvent of $\varphi_* \mathcal{O}_{X_*}$ over $\mathcal{O}_{W_*}$. If $\alpha, \beta \in \mathcal{N}$ with $\alpha \subset \beta$ we let $r_{\beta\alpha}: p_{\alpha\beta}^* R_\alpha^{\cdot} \to R_\beta^{\cdot}$ be the structure map. It is proved in [F, Lemma 2.11] that every complex space X admits a resolvent (X_*, W_*, R_*).

We now define the tangent complex $L_X^{\cdot}$ of X, depending upon the choice of resolvent (X_*, W_*, R_*), to be the differential graded Lie algebra over $\mathbb{C}$ of global derivations of $R_*^{\cdot}$ of nonnegative degree. Thus an element of $L_X^{\cdot}$, $i \geq 0$, consists of a collection $\{D_\alpha\}_{\alpha \in \mathcal{N}}$ such that D_α is a global derivation of degree i of the graded algebra $R_\alpha^{\cdot}$ and such that for $\alpha, \beta \in \mathcal{N}$ with $\alpha \subset \beta$ we have a commutative diagram

$$
\begin{array}{ccc}
p_{\alpha\beta}^* R_\alpha & \xrightarrow{\ r_{\beta\alpha}\ } & R_\beta \\
{\scriptstyle p_{\alpha\beta}^* D_\alpha}\downarrow & & \downarrow{\scriptstyle D_\beta} \\
p_{\alpha\beta}^* R_\alpha & \xrightarrow{\ r_{\beta\alpha}\ } & R_\beta
\end{array}
$$

THEOREM 4.1. $L_X^{\cdot}$ controls the deformation theory of X.

PROOF. Again we just describe a natural transformation $p: \mathcal{C}(L; A) \to \mathrm{Def}(X; A)$. Choose a resolvent (R_*, ∂) for X as above and let $R'_* = (R_* \otimes A, \partial \otimes 1)$. Now let $\eta \in \mathrm{Obj}\,\mathcal{C}(L; A)$. As in the local case we can twist $\partial \otimes 1$ by $\tilde{\eta} \in \mathrm{Der}^1(R \otimes A, R \otimes A)$ and obtain $R''_* = (R_* \otimes A, \partial \otimes 1 + \tilde{\eta})$.

We put $C'_* = H^0(R''_*)$. Hence C'_* is an $\mathcal{O}_{W_*} \otimes A$-module which is easily seen to be supported on $\varphi(X_*)$. We put $B'_* = \varphi^{-1}C'_*$. Then B'_* is a simplicial sheaf on X_* with the property that *all its structure maps* $\mu_{\beta\alpha}: B'_\alpha|X_\beta \to X_\alpha$ *are isomorphisms*. It is then immediate [**BM**, Lemma 6.1] that we can glue together the B'_α's to get a global sheaf B' on X. We define $p(\eta) = B'$. □

REMARK. It is a basic result that $T^1(X) = H^1(L_X^{\cdot})$ (see [**F**, Lemma 3.13]). Thus the functor $\mathrm{Iso}\,\mathscr{C}(L_X^{\cdot}; \cdot)$ has a hull if and only if the functor $\mathrm{Iso}\,\mathrm{Def}(X; \cdot)$ does.

We conclude this section with some indications of the proof of Theorem D. Let X be a complex manifold. We have seen that the Kodaira-Spencer algebra $\mathscr{L}^{\cdot}(X)$ controls the deformation theory of X if X is compact [**GM2**,§3]. However, the proof of the latter theorem does not seem to generalize to the noncompact case. Thus it is somewhat surprising that even in the noncompact case $\mathscr{L}^{\cdot}(X)$ and $L_X^{\cdot}$ are quasi-isomorphic. We prove Theorem D by comparing each of the latter two with the differential graded Lie algebra $\mathrm{Der}(\mathscr{A}^{0,\cdot}(X))$ of graded derivations of the Dolbeault algebra. We have an injective homomorphism $\Phi: \mathscr{L}^{\cdot}(X) \to \mathrm{Der}(\mathscr{A}^{0,\cdot}(X))$ that assigns to $\omega \in \mathscr{A}^{0,\cdot}(X, T^{1,0}(X))$ the Lie derivative operator $\mathscr{L}_\omega$ (denoted d_ω in [**FN1**]) on $\mathscr{A}^{0,\cdot}(X)$. There are other graded derivations, for example of interior multiplication by an element of $\mathscr{A}^{0,\cdot}(X, T^{0,1}(X))$. However, it is proved in [**BM**, Theorem 2.2] that Φ is a quasi-isomorphism.

In order to construct a quasi-isomorphism from $L_X^{\cdot}$ to $\mathrm{Der}(\mathscr{A}^{0,\cdot}(X))$ we proceed as follows. We let $\mathscr{A}^{0,\cdot}$ be the sheaf associated to $\mathscr{A}^{0,\cdot}(X)$ and $\mathscr{A}_*^{0,\cdot}$ be the $\mathcal{O}_{X_*}$-module given by

$$\mathscr{A}_\alpha^{0,\cdot} = \mathscr{A}^{0,\cdot}|X_\alpha, \qquad \alpha \in \mathcal{N}.$$

We define $\gamma: R_* \to \varphi_*\mathscr{A}_*^{0,\cdot}$ as the composition of $\varepsilon: R_* \to \varphi_*\mathcal{O}_{X_*}$ and $\varphi_*(\iota)$ where $\iota: \mathcal{O}_{X_*} \to \mathscr{A}_*^{0,\cdot}$ is the inclusion. Clearly γ is a quasi-isomorphism.

We obtain induced maps $\gamma_*: \mathrm{Der}^+(R_*, R_*) \to \mathrm{Der}^+(R_*, \mathscr{A}_*^{0,\cdot})$ and $\gamma^*: \mathrm{Der}(\mathscr{A}_*^{0,\cdot}, \mathscr{A}_*^{0,\cdot}) \to \mathrm{Der}^+(R_*, \mathscr{A}_*^{0,\cdot})$. It is proved in [**BM**, Propositions 7.1 and 7.2] that γ_* and γ^* are quasi-isomorphisms of complexes. Hence $\mathscr{L}^{\cdot}(X)$ and $L_X^{\cdot}$ are quasi-isomorphic as *complexes*. It is a considerably harder result, proved in [**BM**], that they are quasi-isomorphic as *differential graded Lie algebras*.

5. Kuranishi's CR-deformation theory

Let V be an analytic subset of $\mathbb{C}^N$ of dimension n which has a unique singular point which we assume is located at the origin. We let $U = V - \{0\}$ be the regular part of V. Let S_ε be a small sphere around the origin and put $M = S_\varepsilon \cap U$. Then M inherits a strongly pseudo-convex CR-structure from U. The horizontal distribution H on M is defined as follows. Let

$x \in M$. Then

$$H_x = T_x(M) \cap J_x T_x(M).$$

Here J denotes the complex structure on U. Then H_x is J_x-invariant by definition and the pair (H, J) is a strongly pseudo-convex CR-structure on M. Let $T^{1,0}(M)$ be the bundle of $+i$-eigenspaces for J acting on the complexification of $H \otimes \mathbb{C}$. We let $T^{0,1}(M) = \overline{T}^{1,0}(M)$ be the complex conjugate of $T^{0,1}(M)$. Then

$$T^{0,1}(M) = T^{0,1}(U)|M \cap T(M) \otimes \mathbb{C}.$$

Thus $T^{0,1}(M)$ is an integrable subbundle of $T(M) \otimes \mathbb{C}$. A pair (H, J) as above on an odd-dimensional manifold M such that the $\pm i$-eigenspaces of J on $H \otimes \mathbb{C}$ are integrable constitutes an (abstract) CR-structure on M (see [**Ta**, Chapter I]).

Let ρ be a defining function for M so $\rho: V \to \mathbb{R}$ with $\rho^{-1}(0) = M$, $d\rho|T(M)$ nowhere zero and ρ negative on the inside of M in V. We let $M_t = \rho^{-1}(t)$, for $t \in \mathbb{R}$. Then M is strongly pseudo-convex means that the Hermitian form $L(\rho)$ on $T^{1,0}(M)$ given by

$$L(\rho)(Z, W) = \partial \overline{\partial} \rho(Z, \overline{W})$$

is positive definite at each point of M. Here Z and W are smooth sections of $T^{1,0}(M)$. In fact we will take $\rho = r^2 - \varepsilon$ whence $L(\rho)$ is positive definite on $T^{1,0}(U)$. We note that $\ker \partial \rho|(T^{1,0}(U)|M_t) = T^{1,0}(M_t)$ and $\ker \overline{\partial} \rho|(T^{1,0}(U)|M_t) = T^{0,1}(M_t)$.

We will need some auxiliary vector fields and 1-forms associated to ρ. We define $\theta = J d\rho|T(M)$ where $J d\rho(X) = -d\rho(JX)$ for $X \in T(M)$. Then $\ker \theta = H$. We now construct a smooth vector field T on M everywhere transverse to H.

LEMMA 5.1. *There exists a unique vector field ξ of type $(1, 0)$ on U such that*
(i) $\partial \rho(\xi) = 1$.
(ii) $\iota_\xi \partial \overline{\partial} \rho = c \overline{\partial} \rho$, *for some smooth complex-valued function c on U.*

PROOF. The lemma follows from the observation that the annihilator $T^{0,1}(M_t)^\perp$ of $T^{0,1}(M_t)$ in $T^{1,0}(U)|M_t$ for the bilinear form $\partial \overline{\partial} \rho$ gives rise to a complex line field Λ on U which is everywhere transverse to $T^{1,0}(M_t)$. Equation (ii) states that $\xi \in \Lambda$ and (i) determines a unique vector in Λ.   □

REMARK. In fact c is real-valued because $c = \partial \overline{\partial} \rho(\xi, \overline{\xi})$.

We define $\widetilde{N} = \frac{1}{2}(\xi + \overline{\xi})$ and $\widetilde{T} = JN = \frac{i}{2}(\xi - \overline{\xi})$. Then $d\rho(\widetilde{T}) = 0$ whence $\widetilde{T}$ is tangent to the level sets of ρ. We let T be the restriction of $\widetilde{T}$ to M.

LEMMA 5.2. *The vector field T is the unique solution to the equations*
(i) $\theta(T) = 1$.

(ii) $\iota_T \, d\theta = 0$.

PROOF. The lemma follows immediately from the observation that $2i\partial\overline{\partial}\rho|T(M)\otimes\mathbb{C} = d\theta$ by taking real parts in (ii) of Lemma 5.1. $\quad\square$

Let F be the complementary vector bundle to H in $T(M)$ generated by T. We define $E \subset T(M)\otimes\mathbb{C}$ by

$$E = T^{1,0}(M)\oplus(F\otimes\mathbb{C}).$$

We next observe (following [K2]) that there is a natural isomorphism of complex vector bundles

$$\tau\colon T^{1,0}(U)|M \to E.$$

Indeed the map inverse to τ is induced by the composition

$$T(M)\otimes\mathbb{C} \to T(U)\otimes\mathbb{C}|M \to T^{1,0}(U)|M$$

which is a surjection with kernel $T^{0,1}(M)$. We observe that $\tau(\xi) = -2iT$ and $\tau|T^{1,0}(M) = \mathrm{id}$.

We recall that we can form a complex $(\mathscr{A}^{0,\cdot}(M), \overline{\partial}_b)$ where

$$\mathscr{A}^{0,i}(M) = \Gamma(M, \Lambda^i T^{0,1}(M)^*)$$

and $\overline{\partial}_b$ is defined as follows. Given $\varphi \in \mathscr{A}^{0,1}(M)$ extend it to $\tilde{\varphi} \in \mathscr{A}^{0,1}(U)$. Then $\overline{\partial}_b\varphi$ is defined to be the restriction of $\overline{\partial}\tilde{\varphi}$ to M. The integrability of $T^{0,1}(M)$ implies that $\overline{\partial}_b$ is well defined. We say $f \in \mathscr{A}^{0,0}(M)$ is a CR-function if $\overline{\partial}_b f = 0$. We say that a complex vector bundle M is CR if it has an atlas with CR-transition functions. If E' is such a bundle then we may twist the above complex by E' to form a complex $(\mathscr{A}^{0,\cdot}(M, E'), \overline{\partial}_b)$. If E' is the restriction to M of a holomorphic bundle E'' on U then E' is CR and the above complex may be obtained as the boundary complex associated to the twisted Dolbeault complex $(\mathscr{A}^{0,\cdot}(U, E''), \overline{\partial})$ as in the case of trivial coefficients above. Thus we obtain a complex $(\mathscr{A}^{0,\cdot}(M, T^{1,0}(U)|M), \overline{\partial}_b)$ and by transport of structure via τ a complex $(\mathscr{A}^{0,\cdot}(M, E), \overline{\partial}_b)$. We will abbreviate this complex to $(\mathscr{K}^{\cdot}(M), \overline{\partial}_b)$. Thus

$$\mathscr{K}^i(M) = \Gamma(M, \Lambda^i T^{0,1}(M)^*)\otimes E).$$

We now give a formula for $\overline{\partial}_b$ following [Ak1]. Let $P\colon T(M)\otimes\mathbb{C} \to E$ be the projection with kernel $T^{0,1}(M)$, $\omega \in \mathscr{A}^{0,q}(M, E)$ and $Z_1, \ldots, Z_{q+1}$ be locally defined C^∞-sections of $T^{1,0}(M)$. Then

$$\overline{\partial}_b\omega(\overline{Z}_1, \ldots, \overline{Z}_{q+1}) = \sum_{i=1}^{q+1}(-1)^{i+1}P([\overline{Z}_i, \omega(\overline{Z}_1, \ldots, \widehat{\overline{Z}}_i, \ldots, \overline{Z}_{q+1})])$$

$$+ \sum_{1\le i<j\le q+1}(-1)^{i+j}$$

$$\times \omega([\overline{Z}_i, \overline{Z}_j], \overline{Z}_1, \ldots, \widehat{\overline{Z}}_i, \ldots, \widehat{\overline{Z}}_j, \ldots, \overline{Z}_{q+1}).$$

We extend τ to a map $\tau \colon \mathscr{A}^{0,q}(U, T^{1,0}(U)) \to \mathscr{A}^{0,q}(M, E)$ by

$$\tau \varphi(\overline{Z}_1, \ldots, \overline{Z}_q) = \tau(\varphi(\overline{Z}_1, \ldots, \overline{Z}_q))$$

and $\overline{Z}_1, \ldots, Z_q$ as above. One can now check [**BM**, Lemma 3.3] that Akahori's formula for $\overline{\partial}_b$ agrees with that induced by the $\overline{\partial}$-operator on U as described above. Thus we have $\tau \circ \overline{\partial} = \overline{\partial}_b \circ \tau$.

In what follows we will need the analogue of this fact for scalar-valued $(0, q)$-forms. Let $j \colon M \to U$ be the inclusion. We define the restriction $j^* \colon \mathscr{A}^{0,q}(U) \to \mathscr{A}^{0,q}(M)$, $q \geq 0$, by

$$j^* \mu(\overline{Z}_1, \ldots, \overline{Z}_q) = \mu(dj(\overline{Z}_1), \ldots, dj(\overline{Z}_{q+1})).$$

We have defined $\overline{\partial}_b$ above such that if $\mu \in \mathscr{A}^{0,q}(U)$ then

$$j^* \overline{\partial} \mu = \overline{\partial}_b j^* \mu.$$

The proof of the next lemma is left to the reader.

LEMMA 5.3. *Let* $\mu \in \mathscr{A}^{0,q}(M)$. *Then*

$$\overline{\partial}_b \mu(\overline{Z}_1, \ldots, \overline{Z}_{q+1}) = \sum_{i=1}^{q+1} (-1)^{i-1} \overline{Z}_i \mu(\overline{Z}_1, \ldots, \widehat{\overline{Z}}_i, \ldots, \overline{Z}_q)$$

$$+ \sum_{1 \leq i < j \leq (-1)} (-1)^{i+j}$$

$$\times \mu([\overline{Z}_i, \overline{Z}_j], \ldots, \widehat{\overline{Z}}_i, \ldots, \widehat{\overline{Z}}_j, \ldots, \overline{Z}_{q+1}).$$

We have obtained maps of complexes $j^* \colon \mathscr{A}^{0,\cdot}(U) \to \mathscr{A}^{0,\cdot}(M)$ and $\tau \colon \mathscr{L}^{\cdot}(U) \to \mathscr{K}^{\cdot}(M)$. We will often abbreviate $\mathscr{L}^{\cdot}(U)$ to $\mathscr{L}$ and $\mathscr{K}^{\cdot}(U)$ to $\mathscr{K}$. The next theorem is of central importance to us. It is proved in pages 81 and 82 of [**Y**]. We leave the reader to check that the argument there applies when the coefficients Ω^p in [**Y**] are replaced by $T^{1,0}(U)$.

THEOREM 5.1. *The map* τ *induces isomorphisms on cohomology groups of degree* i, $1 \leq i \leq n - 2$.

COROLLARY. (i) *Suppose* $\dim V = 3$. *Then* τ *induces an isomorphism on first cohomology.*

(ii) *Suppose* $\dim V \geq 4$. *Then* τ *induces an isomorphism on first and second cohomology.*

We take this opportunity to observe that it makes no difference "how big U is." In fact we have the following lemma.

LEMMA 5.4. *Let* U_r *be the intersection of* U *with the open ball centered at* 0 *of radius* r. *Then the restriction map* $\rho \colon \mathscr{L}^{\cdot}(U) \to \mathscr{L}^{\cdot}(U_r)$ *induces isomorphisms of cohomology groups in positive degrees.*

PROOF. Let V_r be the intersection of the above ball with V. We compare the long exact cohomology sequences with coefficients in Θ of the pairs

(V, U) and (V_r, U_r). We obtain a diagram

$$\begin{array}{ccccccc}
\longrightarrow & H^i_{\{0\}}(V, \Theta) & \longrightarrow & H^i(V, \Theta) & \longrightarrow & H^i(U, \Theta) & \longrightarrow \\
& \downarrow & & \downarrow & & \downarrow & \\
\longrightarrow & H^i_{\{0\}}(V_r, \Theta) & \longrightarrow & H^i(V_r, \Theta) & \longrightarrow & H^i(U_r, \Theta) & \longrightarrow
\end{array}$$

By excision for local cohomology the left arrow is an isomorphism and since V and V_r are Stein the middle groups vanish if $i \geq 1$. $\square$

COROLLARY. *Let $\mathscr{L}^{\cdot}(\overline{U}_\varepsilon)$ be the differential graded Lie algebra of forms which extend smoothly over M. Then the restriction maps $\mathscr{L}^{\cdot}(U) \to \mathscr{L}^{\cdot}(\overline{U}_\varepsilon)$ and $\mathscr{L}^{\cdot}(\overline{U}_\varepsilon) \to \mathscr{L}^{\cdot}(U_\varepsilon)$ induce isomorphisms of cohomology groups of positive degree.*

PROOF. We have $\mathscr{L}(\overline{U}_\varepsilon) = \lim_{r > \varepsilon} \mathscr{L}(U_r)$. The corollary follows because cohomology commutes with direct limits of complexes and on the cohomology level the above limit system is constant by the previous lemma. $\square$

REMARK. We see then that we arrive at the same result (namely Kuranishi's theory on the link) whether we take U to be large, i.e., a neighborhood of M, and take τ to be the restriction map as we have done or take $U = \overline{U}_\varepsilon$ and take τ to be the boundary value map as is more traditional in complex variable theory, e.g., [Ak4].

We next explain the connection between Kuranishi's complex $(\mathscr{K}^{\cdot}(M), \overline{\partial}_b)$ and deformation theory. If (H', J') is a CR-structure near (H, J) then its $(0, 1)$-distribution may be represented as the graph of a bundle map $\varphi: T^{0,1}(M) \to E$. The form φ will satisfy the integrability condition [Ak1]

$$P(\varphi) = \overline{\partial}_b \varphi + \frac{1}{2} R_2(\varphi) + R_3(\varphi) = 0$$

where $R_2(\varphi)$ and $R_3(\varphi)$ are elements of $\mathscr{K}^2(M)$ given by the formulas

$$\begin{aligned}
R_2(\varphi)(\overline{Z}, \overline{W}) &= P([\varphi(\overline{Z}), \varphi(\overline{W})]) - P([\varphi(\overline{W}), \varphi(\overline{Z})]) \\
&\quad + 2\{\varphi(Q([\varphi(\overline{Z}), \overline{W}])) - \varphi(Q([\varphi(\overline{W}), \overline{Z}]))\}
\end{aligned}$$

and

$$R_3(\varphi)(\overline{Z}, \overline{W}) = -\varphi(Q([\varphi(\overline{Z}), \varphi(\overline{W})])).$$

Here $Q: T(M) \otimes \mathbb{C} \to T^{0,1}(M)$ denotes the projection with kernel E.

REMARK. It is important to observe that by defining

$$\mathscr{K}^0(M) = \Gamma(M, E) \cong \Gamma(M, T^{1,0}(U)|M)$$

we are dividing out by isotopies of M in U; hence, we are taking care of the dependence of M on the choice of sphere S_ε used to form the link. Kuranishi refers to such isotopies as "wiggles" in [K2].

We now describe the formal deformation functor $\mathscr{K}_M$ associated to Kuranishi's theory. Choose a Hermitian metric on $T(M) \otimes \mathbb{C}$. We can then obtain a formal adjoint $\overline{\partial}_b^*$ to $\overline{\partial}_b$ on $\mathscr{K}^{\cdot}(M)$. Let A be an Artin local $\mathbb{C}$-algebra with maximal ideal $\mathfrak{m}$. We define $\mathscr{K}_M(A)$ by

$$\mathscr{K}_M(A) = \{\eta \in \mathscr{K}^1 \otimes \mathfrak{m} : P(\eta) = \overline{\partial}_b^* \eta = 0\}.$$

THEOREM 5.2. *The functor $\mathscr{K}_M$ is pro-representable by a complete local Noetherian $\mathbb{C}$-algebra provided* $\dim M \geq 5$.

PROOF. The proof is identical to that of the Representability Theorem for R_L of §2. We need to know that $\dim H^1(\mathscr{K}^{\cdot}(M)) < \infty$. This is proved in [**Ta**, Theorem 6.5]. $\square$

6. A controlling differential graded Lie algebra for Kuranishi's CR-deformation theory

Unfortunately it turns out to be extremely difficult to prove directly that the functor $\mathscr{K}_M$ of the previous section can be "represented" by an analytic local $\mathbb{C}$-algebra. Although considerable progress in this direction was made in [**K2**] a complete proof of this fact was first given in [**Mi1**]. Miyajima's proof was based on the previous work of Akahori, especially [**Ak3**]. In [**Ak3**], Akahori modified Kuranishi's theory by keeping the underlying contact structure fixed. He was led to a complex $(\mathscr{B}^{\cdot}(M), \overline{\partial}_b)$ on M where

$$\mathscr{B}^i(M) = \{\mu \in \mathscr{K}^q : \mu \rfloor \theta = 0, \ \mu \rfloor d\theta = 0\}.$$

We define $\mathscr{B}^0(M)$ by

$$\mathscr{B}^0(M) = \{\lambda \in \mathscr{K}^0 : \overline{\partial}_b \lambda \rfloor \theta = 0\}.$$

THEOREM 6.1. *The inclusion $\mathscr{B}^{\cdot}(M) \to \mathscr{K}^{\cdot}(M)$ is a quasi-isomorphism.*

PROOF. This theorem is essentially Theorem 2.4 of [**Ak3**]. The necessary modifications are explained in Proposition 8.2 of [**BM**]. $\square$

We next review the definition and properties of Akahori's interior complex $(\mathscr{E}^{\cdot}, \overline{\partial})$ [**Ak4**]. Let ρ be the defining function of §5. We define $\mathscr{E}^q \subset \mathscr{A}^{0,q}(U, T^{1,0}(U))$, $q > 0$, by

$$\mathscr{E}^q = \{\mu \in \mathscr{A}^{0,q}(U, T^{1,0}(U)) : j^*(\mu \rfloor \partial \rho) = j^*(\mu \rfloor \partial\overline{\partial}\rho) = 0\}.$$

We define $\mathscr{E}^0$ by

$$\mathscr{E}^0 = \{\lambda \in \mathscr{A}^0(U, T^{1,0}(U)) : j^*(\overline{\partial}\lambda \rfloor \partial \rho) = 0\}.$$

The following lemma is now clear (it is Lemma 8.7 of [**BM**]).

LEMMA 6.1. *Let $\mu \in \mathscr{L}^q(U)$. Then $\mu \in \mathscr{E}^q(U)$ if and only if $\tau\mu \in \mathscr{B}^q(M)$.*

The analogue of Theorem 6.1 holds for the inclusion $\mathscr{E}^{\cdot}(U) \subset \mathscr{L}^{\cdot}(U)$. This is essentially Theorem 3.4 of [**Ak4**]; see Proposition 8.4 of [**BM**].

THEOREM 6.2. *The inclusion* $(\mathscr{E}^{\cdot}(U), \overline{\partial}) \to (\mathscr{L}^{\cdot}(U), \overline{\partial})$ *is a quasi-isomorphism of complexes.*

We can combine Theorems 6.1, 6.2 and 5.1 with the commutative diagram

$$
\begin{array}{ccc}
\mathscr{L}^{\cdot}(U) & \xrightarrow{\ \tau\ } & \mathscr{K}^{\cdot}(M) \\
\uparrow & & \uparrow \\
\mathscr{E}^{\cdot}(U) & \xrightarrow{\ \tau\ } & \mathscr{B}^{\cdot}(M)
\end{array}
$$

to deduce the following theorem.

THEOREM 6.3. *The restriction map* $\tau\colon \mathscr{E}^{\cdot}(U) \to \mathscr{B}^{\cdot}(M)$ *induces an isomorphism on first and second cohomology provided* $\dim U \geq 4$.

We next prove that the spaces of elements of positive degree in $\mathscr{E}^{\cdot}(U)$ and $\mathscr{B}^{\cdot}(M)$ have the structure of a differential graded Lie algebra such that τ is a homomorphism of differential graded Lie algebras. The general case is a little complicated (see [**BM**, Chapter 9]), so we will prove only a special case which contains the main point.

LEMMA 6.2. *Suppose* $\mu \in \mathscr{E}^{p}(U)$, $\eta \in \mathscr{E}^{q}(U)$ *with* $p > 0$, $q > 0$. *Then the Nijenhuis bracket* $[\mu, \eta] \in \mathscr{E}^{p+q}(U)$.

PROOF. We will prove that if $\mu \in \mathscr{E}^{1}(U)$ and $\eta \in \mathscr{E}^{1}(U)$ then $j^{*}([\mu, \eta] \rfloor \partial\rho) = 0$. Let Z_{1}, Z_{2} be smooth sections of $T^{1,0}(U)$. Then

$$
\begin{aligned}
\partial\rho([\mu, \eta])(\overline{Z}_{1}, \overline{Z}_{2}) = {} & \partial\rho([\mu(\overline{Z}_{1}), \eta(\overline{Z}_{2})] - [\mu(\overline{Z}_{2}), \eta(\overline{Z}_{1})]) \\
& - [\partial\rho(\mu([\overline{Z}_{1}, \eta(\overline{Z}_{2})]) - \mu([\overline{Z}_{2}, \eta(\overline{Z}_{1})]))] \\
& - [\partial\rho(\eta([\overline{Z}_{1}, \mu(\overline{Z}_{2})]) - \eta([\overline{Z}_{2}, \mu(\overline{Z}_{1})]))].
\end{aligned}
$$

The first term is zero because $T^{1,0}(M)$ is integrable and μ and η both take values in $T^{1,0}(M)$ by definition of $\mathscr{E}^{\cdot}(U)$. The second and third terms are symmetrical. We show that the second is zero.

We may write

$$
[\overline{Z}_{1}, \eta(\overline{Z}_{2})] = \overline{\partial}\rho([\overline{Z}_{1}, \eta(\overline{Z}_{2})])\overline{\xi} + W + \overline{Z}
$$

where W is a smooth section of $T^{1,0}(U)|M$ and $\overline{Z}$ is a smooth section of $T^{0,1}(M)$. Since μ is of type $(0, 1)$ we have $\mu(W) = 0$ whence

$$
\mu([\overline{Z}_{1}, \eta(\overline{Z}_{2})]) = \overline{\partial}\rho([\overline{Z}_{1}, \eta(\overline{Z}_{2})])\mu(\overline{\xi}) + \mu(\overline{Z}).
$$

Since $\mu \in \mathscr{E}^{1}(U)$ we have $\partial\rho(\mu(\overline{Z})) = 0$ whence

$$
\partial\rho(\mu([\overline{Z}_{1}, \eta(\overline{Z}_{2})])) = \overline{\partial}\rho([\overline{Z}_{1}, \eta(\overline{Z}_{2})])\partial\rho(\mu(\overline{\xi})).
$$

We next observe that

$$
\partial\overline{\partial}\rho(\overline{Z}_{1}, \eta(\overline{Z}_{2})) = d\overline{\partial}\rho(\overline{Z}_{1}, \eta(\overline{Z}_{2})) = -\overline{\partial}\rho([\overline{Z}_{1}, \eta(\overline{Z}_{2})]).
$$

Hence

$$\partial\rho(\mu[\overline{Z}_1, \eta(\overline{Z}_2)]) = -\partial\overline{\partial}\rho(\overline{Z}_1, \eta(\overline{Z}_2))\partial\rho(\mu(\overline{\xi})).$$

Repeating the above procedure with $\overline{Z}_1$ and $\overline{Z}_2$ interchanged we obtain

$$\begin{aligned}
\partial\rho(\mu([\overline{Z}_1, \eta(\overline{Z}_2)]) &- \mu([\overline{Z}_2, \eta(\overline{Z}_1)])) \\
&= -[\partial\overline{\partial}\rho(\overline{Z}_1, \eta(\overline{Z}_2)) - \partial\overline{\partial}\rho(\overline{Z}_2, \eta(\overline{Z}_1))]\partial\rho(\mu(\overline{\xi})) \\
&= -(\eta\,\lrcorner\,\partial\overline{\partial}\rho)(\overline{Z}_1, \overline{Z}_2)\partial\rho(\mu(\overline{\xi})).
\end{aligned}$$

But $(\eta\,\lrcorner\,\partial\overline{\partial}\rho)(\overline{Z}_1, \overline{Z}_2) = 0$ because $\eta \in \mathscr{E}^1$. $\quad\square$

The reader should be able to generalize the above to prove that $\mathscr{E}^+(U) = \bigoplus_{q>0}\mathscr{E}^q(U)$ is closed under the Nijenhuis bracket on $\mathscr{L}^{\boldsymbol{\cdot}}(U)$.

We next define $\mathscr{I}^{\boldsymbol{\cdot}}(U) \subset \mathscr{E}^{\boldsymbol{\cdot}}(U)$ by $\mathscr{I}^q(U) = \ker(\tau\colon \mathscr{E}^q(U) \to \mathscr{B}^q(M))$. We define $\mathscr{I}^+(U) = \bigoplus_{q>0}\mathscr{I}^q(U)$.

LEMMA 6.3. $\mathscr{I}^+(U)$ *is an ideal in* $\mathscr{E}^+(U)$.

PROOF. Once again we prove only the special case $q = 1$ referring the reader to [**BM**, Lemma 9.3] for the general case. Let $\mu \in \mathscr{E}^1(U)$, $\eta \in \mathscr{E}^1(U)$ and Z_1 and Z_2 be as above. We have

$$\begin{aligned}
[\mu, \eta](\overline{Z}_1, \overline{Z}_2) = {}&[\mu(\overline{Z}_1), \eta(\overline{Z}_2)] - [\mu(\overline{Z}_2), \eta(\overline{Z}_1)] \\
&- (\mu([\overline{Z}_1, \eta(\overline{Z}_2)]) - \mu([\overline{Z}_2, \eta(\overline{Z}_1)])) \\
&- (\eta([\overline{Z}_1, \mu(\overline{Z}_2)]) - \eta([\overline{Z}_2, \mu(\overline{Z}_1)])).
\end{aligned}$$

The first and third terms are clearly zero since they involve tangential derivatives of the restriction of μ to M which is identically zero. It remains to check that the second term is zero. We have as before

$$[\overline{Z}_1, \eta(\overline{Z}_2)] = \overline{\partial}\rho([\overline{Z}_1, \eta(\overline{Z}_2)])\overline{\xi} + W + \overline{Z}$$

with W a smooth section of $T^{1,0}(U)|M$ and $\overline{Z} \in T^{0,1}(M)$.

Since μ annihilates $T^{0,1}(M)$ by assumption we have

$$\mu([\overline{Z}_1, \eta(\overline{Z}_2)]) = \overline{\partial}\rho([\overline{Z}_1, \eta(\overline{Z}_2)])\mu(\overline{\xi})$$

and as before

$$\mu([\overline{Z}_1, \eta(\overline{Z}_2)] - [\overline{Z}_2, \eta(\overline{Z}_1)]) = -(\eta\,\lrcorner\,\partial\overline{\partial}\rho)(\overline{Z}_1, \overline{Z}_2)\mu(\overline{\xi}).$$

Once again the lemma follows by definition of $\mathscr{E}^1(U)$. $\quad\square$

COROLLARY. *Define* $\mathscr{B}^+(M)$ *by* $\mathscr{B}^+(M) = \bigoplus_{q>0}\mathscr{B}^q(M)$. *Then* $\mathscr{B}^+(M)$ *has the structure of a differential graded Lie algebra.*

We need to relate the bracket $[\ ,\ ]\colon \mathscr{B}^1(M) \otimes \mathscr{B}^1(M) \to \mathscr{B}^2(M)$ to the operator $R_2(\varphi)$ appearing in the integrability condition for φ in §5.

LEMMA 6.4. *Let* $\mu, \eta \in \mathscr{B}^1(M)$ *and* Z_1 *and* Z_2 *be smooth sections of* $T^{1,0}(M)$. *Then*

$$[\mu, \eta](\overline{Z}_1, \overline{Z}_2) = [\mu(\overline{Z}_1), \eta(\overline{Z}_2)] - [\mu(\overline{Z}_2), \eta(\overline{Z}_1)]$$
$$- [\mu(Q([\overline{Z}_1, \eta(\overline{Z}_2)] - [\overline{Z}_2, \eta(\overline{Z}_1)]))]$$
$$- [\mu(Q([\overline{Z}_1, \mu(\overline{Z}_2)] - [\overline{Z}_2, \mu(\overline{Z}_1)]))].$$

COROLLARY. *If* $\varphi \in \mathscr{B}^1(M)$ *then*
(i) $R_2(\varphi) = \frac{1}{2}[\varphi, \varphi]$,
(ii) $R_3(\varphi) = 0$.

We can now construct a controlling differential graded Lie algebra for Kuranishi's CR-deformation theory. Indeed let $\overline{C}^1$ be a vector space complement to $\overline{\partial}_b \mathscr{B}^0(M)$ in $\mathscr{B}^1(M)$ and define $\overline{\mathscr{K}}^{\cdot}(M)$ by

$$\overline{\mathscr{K}}^{\cdot}(M) = \overline{C}^1 \oplus \bigoplus_{i \geq 2} \mathscr{B}^i(M).$$

Then $(\overline{\mathscr{K}}^{\cdot}(M), [,], \overline{\partial}_b)$ is a differential graded Lie algebra.

LEMMA 6.5. *There exists a complement* C^1 *to* $\overline{\partial}\mathscr{E}^0(U)$ *in* $\mathscr{E}^1(U)$ *such that* $\tau C^1 \subset \overline{C}^1$.

PROOF. Choose subspaces B'' and C'' of $\mathscr{E}^1(U)$ such that τ maps B'' isomorphically onto $\overline{\partial}_b \mathscr{B}^0(M)$ and C'' isomorphically onto $\overline{C}^1$. Since τ maps $\mathscr{E}^0(U)$ surjectively onto $\mathscr{B}^0(M)$ we may choose B'' such that $B'' \subset \overline{\partial}\mathscr{E}^0(U)$. We have $\mathscr{E}^1(U) = \ker \tau \oplus B'' \oplus C''$. Let $B' = \ker \tau \cap \overline{\partial}\mathscr{E}^0(U)$ and let C' be a complement to B' in $\ker \tau$ whence

$$\mathscr{E}^1(U) = B' \oplus C' \oplus B'' \oplus C''.$$

But clearly $B' \oplus B'' = \overline{\partial}\mathscr{E}^0(U)$ since $\tau(B'') = \tau(\overline{\partial}\mathscr{E}^0(U))$. Put $C^1 = C' \oplus C''$. $\square$

We now define $\mathscr{L}^{\cdot}_{\tan}(U) \subset \mathscr{L}^{\cdot}(U)$ by

$$\mathscr{L}^{\cdot}_{\tan}(U) = C^1 \oplus \bigoplus_{i \geq 2} \mathscr{E}^i(U).$$

From Theorem 6.3 we obtain the following theorem.

THEOREM 6.1. *Suppose* $\dim U \geq 4$. *Then the restriction map* $\tau : \mathscr{L}^{\cdot}_{\tan}(U) \to \overline{\mathscr{K}}^{\cdot}(M)$ *is a one-quasi-isomorphism of differential graded Lie algebras.*

We then obtain Theorem E from the one-quasi-isomorphism

$$\mathscr{L}^{\cdot}(U) \leftarrow \mathscr{L}^{\cdot}_{\tan}(U) \to \overline{\mathscr{K}}^{\cdot}(M).$$

Finally Theorem F follows from [Mi1] as explained in Theorem 9.3 of [BM]. This is rather clear from the corollary to Lemma 6.4.

References

[Ak1] T. Akahori, *Intrinsic formula for Kuranishi's* $\overline{\partial}_b$, Publ. Res. Inst. Math. Sci. **14** (1978), 615–641.

[Ak2] ___, *Complex analytic construction of the Kuranishi family on a normal strongly pseudo convex manifold*, Publ. Res. Inst. Math. Sci. **14** (1978), 789–847.

[Ak3] ___, *The new estimate for the sub-bundles* E_j *and its application to the deformation of the boundaries of strongly pseudo-convex domains*, Invent. Math. **63** (1981), 311–344.

[Ak4] ___, *The new Neumann operator associated with deformation of strongly pseudo-convex domains and its application to deformation theory*, Invent. Math. **68** (1982), 317–352.

[AM] T. Akahori and K. Miyajima, *Complex analytic construction of the Kuranishi family on a normal strongly pseudo-convex manifold. II*, Publ. Res. Inst. Math. Sci. **18** (1980), 811–834.

[Ar1] M. Artin, *On solutions to analytic equations*, Invent. Math. **5** (1968), 277–291.

[Ar2] ___, *Lectures on deformations of singularities*, Lectures on Math. and Phys., vol. 54, Tata Institute, Bombay, 1976.

[Be] A. Beauville, *Foncteurs sur les anneaux Artiniens et applications aux deformations verselles*, Astérisque, no. 16, Soc. Math. France, Paris, 1974, pp. 82–104.

[Bi] J. Bingener, *Lokale Modulräume in der analytischen Geometrie*, Aspects of Math., vol. D2, Vieweg, Braunschweig, 1987.

[Bo] N. Bourbaki, *Commutative algebra*, Hermann, Paris, 1972.

[BM] R. Buchweitz and J. J. Millson, *CR-geometry and deformations of isolated singularities*, preprint.

[D] P. Deligne, Letter to J. J. Millson, April 24, 1986.

[Fu] A. Fujiki, *Flat Stein completion of a flat* (1, 1)-*convex concave map*, Preprint.

[F] H. Flenner, *Über Deformationen holomorpher Abildungen*, Math. Institut der Universität Göttingen Nachdruck, 1988.

[Fi] G. Fischer, *Complex analytic geometry*, Lecture Notes in Math., vol. 538, Springer-Verlag, 1976.

[FN1] A. Frölicher and A. Nijenhuis, *Theory of vector-valued forms. I. Derivations in the graded ring of differential forms*, Nederl. Akad. Wetensch. Proc. Ser. A **59** (1956), 338–359.

[FN2] ___, *Some new cohomology invariants for complex manifolds*, Nederl. Akad. Wetensch. Proc. Ser. A **59** (1956), 540–564.

[Go] R. Godement, *Topologie algébrique et Théorie des faisceaux*, Hermann, Paris, 1964.

[Gr] H. Grauert, *Der Satz von Kuranishi für kompakte complexe Räume*, Math. **25** (1974), 107–142.

[GR Mo] P. Griffiths and J. Morgan, *Rational homotopy theory and differential forms*, Prog. in Math., vol. 16, Birkhäuser, Boston, 1981.

[GM1] W. M. Goldmann and J. J. Millson, *The deformation theory of representations of fundamental groups of compact Kähler manifolds*, Inst. Hautes Études Sci. Publ. **67** (1988), 43–96.

[GM2] ___, *The homotopy invariance of the Kuranishi space*, Illinois J. Math. **34** (1990), 337–367.

[I] L. Illusie, *Complexe cotangent et déformations. I*, Lecture Notes in Math., vol. 239, Springer-Verlag, 1970.

[K1] M. Kuranishi, *Deformations of compact complex manifolds*, Presses Univ. Montréal, Montréal, 1971.

[K2] ___, *Application of* $\overline{\partial}_b$ *to deformation of isolated singularities*, Proc. Sympos. Pure Math., vol. 30, Amer. Math. Soc., Providence, RI, 1977, pp. 97–106.

[M1] J. Millson, *Rational homotopy theory and deformation problems from algebraic geometry*, Proc. Internat. Congr. Math. (Kyoto, 1990).

[M2] ___, *CR-geometry and deformations of cones*, in preparation.

[Mi1] K. Miyajima, *Completion of Akahori's construction of the versal family of strongly pseudo-convex CR structures*, Trans. Amer. Math. Soc. **277** (1983), 163–172.

[Mi2] ___, *Deformation of a complex manifold near a strongly pseudo-convex real hypersurface and a realization of the Kuranishi family of strongly pseudo-convex CR-structures*, Math. Z. **205** (1990), 593–602.

[P] V. Palamodov, *Deformations of complex spaces*, Russian Math. Surveys **31** (1976), no. 3, 129–197.

[Sc1] M. Schlessinger, *Functors of Artin rings*, Trans. Amer. Math. Soc. **130** (1966), 208–222.

[Sc2] ____, *On rigid singularities*, Conf. on Complex Analysis, Rice Univ. Stud., vol. 59 (1972), 147–162.

[Sc3] ____, *Infinitesimal deformation of singularities*, Ph.D. thesis, Harvard University.

[Schu] H. Schuster, *Infinitesimale Erweiterungen Komplexer Räume*, Comment. Math. Helv. **45** (1970), 265–286.

[SS] M. Schlessinger and J. Stasheff, *Deformation theory and rational homotopy type*, Preprint.

[Ta] N. Tanaka, *A differential-geometric study on strongly pseudo-convex manifolds*, Lectures in Math., vol. 9, Kyoto Univ., 1975.

[Y] Stephen S.-T Yau, *Kohn-Rossi cohomology and its application to the complex Plateau problem*. I, Ann. of Math. (2) **113** (1981), 67–110.

UNIVERSITY OF MARYLAND

Proceedings of Symposia in Pure Mathematics
Volume **54** (1993), Part 2

Generalized Symplectic Geometry
on the Frame Bundle of a Manifold

L. K. NORRIS

ABSTRACT. In this paper we develop the fundamentals of the generalized symplectic geometry on the bundle of linear frames LM of an n-dimensional manifold M that follows upon taking the $\mathbb{R}^n$-valued soldering one-form θ on LM as a generalized symplectic potential. The development is centered around generalizations of the basic structure equation $df = -X_f \, \lrcorner \, \omega$ of standard symplectic geometry to LM when the symplectic two-form ω is replaced by the closed and nondegenerate $\mathbb{R}^n$-valued two-form $\beta = d\theta = d\theta^i r_i$. The fact that $d\theta$ is $\mathbb{R}^n$-valued necessitates generalizing from $\mathbb{R}$-valued observables to vector-valued observables on LM, and there is a corresponding increase in the number of Hamiltonian vector fields assigned to each observable. We show that the algebras of symmetric and antisymmetric contravariant tensor fields on the base manifold have natural interpretations in terms of symplectic geometry on LM. For the analysis we consider in place of each rank p contravariant tensor field on the base manifold the uniquely related $\bigotimes^p \mathbb{R}^n$-valued tensorial function on LM. For symmetric contravariant tensor fields on M we show that the associated algebra $(ST, \otimes_s)$, where $ST = \sum_{p=1}^{\infty} ST^p$ is the vector space of all $\bigotimes_s^p \mathbb{R}^n$-valued tensorial functions on LM, becomes a Poisson algebra under a generalized Poisson bracket. In addition the associated set of locally defined $\bigotimes_s^{p-1} \mathbb{R}^n$-valued Hamiltonian vector fields $\widehat{X}_{\hat{f}}$ forms a Lie algebra under a generalized Lie bracket. In the case of antisymmetric contravariant tensor fields on M we show that the corresponding vector space $AT = \sum_{p=1}^{\infty} AT^p$ of $\bigotimes_a^p \mathbb{R}^n$-valued functions on LM becomes a Poisson super algebra under a naturally defined bracket. The associated set of locally defined $\bigotimes_a^{p-1} \mathbb{R}^n$-valued Hamiltonian vector fields $\widehat{X}_{\hat{f}}$ forms a super algebra under a generalized super bracket. The naturally defined brackets of the tensorial functions on LM give the Schouten differential concomitants when reinterpreted on the base manifold. Two applications of the geometry to physics are presented. First the dynamics of free inertial observers in spacetime is shown to follow upon taking the metric tensor as the Hamiltonian for free observers. We then show that the Dirac equation arises in a natural way as an eigenvalue equation for a naive prequantization operator assigned to the spacetime metric tensor Hamiltonian.

1991 *Mathematics Subject Classification.* Primary 53C15; Secondary 53C80.

This paper is in final form and no version of it will be submitted for publication elsewhere.

1. Introduction

The methods introduced by W. R. Hamilton more than a century and a half ago have since been used fundamentally in physics in the development of classical and quantum mechanics and the classical and quantum theory of fields. Although many physicists continue to use the classical coordinate dependent form of the Hamilton equations, mathematicians have given the entire subject a beautiful, invariant and geometrical formulation in the theory of symplectic geometry. The fundamentals of the subject can be found in the works of Sternberg [1], Hermann [2], Arnol'd [3], Abraham and Marsden [4], and Guillemin and Sternberg [5].

Since its inception symplectic geometry has served to motivate a number of new developments. In particular the formulation of Hamiltonian dynamics in terms of symplectic geometry is the starting point for the theory of geometric quantization due to Kostant [6] and Souriau [7]. Generalizations of symplectic geometry have led to the study of canonical manifolds [8], Poisson manifolds [9, 10, 11], and to the idea of Poisson algebras [11, 12]. A recent account of the applications of symplectic techniques in a wide variety of physical problems, including Yang-Mills theory, can be found in the book by Guillemin and Sternberg [13].

This paper introduces, and develops aspects of, a generalized symplectic geometry on the bundle of linear frames of a manifold. Mathematical motivation for this generalization may be provided in the following way. The canonical model of a symplectic manifold is the pair $(T^*M, d\tilde{\theta})$, where T^*M is the cotangent bundle of an n-dimensional differentiable manifold M, and $\tilde{\theta}$ is the *canonical one-form* on T^*M that plays the role of a globally defined symplectic potential. The point to be emphasized here is that one obtains $\tilde{\theta}$ "for free" from the differential structure of T^*M.

Now consider T^*M as the fiber bundle $LM \times_{GL(n)} (\mathbb{R}^n)^*$ associated to the bundle of linear frames LM of M and the standard action of $GL(n) \equiv GL(n, \mathbb{R})$ on $(\mathbb{R}^n)^*$. From this point of view it is clear that T^*M inherits its differential structure from LM. This being the case one is led to ask if the canonical symplectic structure on T^*M has its roots in a more general structure on LM. The natural candidate for a symplectic potential on LM is the $\mathbb{R}^n$-valued *soldering one-form* θ on LM which also comes "for free" from the differential structure of LM. Here we note that the exact two-form $d\theta$ is non-degenerate in the sense that

$$X \rfloor d\theta = 0 \Leftrightarrow X = 0$$

for X a vector field on LM. Moreover, if $\dim(M) = n$ then the dimension of the frame bundle LM is the even number $n(n+1)$. Thus we would have the necessary ingredients for a symplectic manifold if it were not for the fact that $d\theta$ is $\mathbb{R}^n$-valued. Nonetheless we make the following

DEFINITION. The $\mathbb{R}^n$-valued two form $d\theta$ on LM is a *generalized symplectic structure.* The pair $(LM, d\theta)$ will be referred to as a *generalized symplectic manifold*.

The geometry that one can build up from this definition is more general and at the same time more special than standard symplectic geometry. It is more general in the sense that the $\mathbb{R}$-valued observables on T^*M are replaced by *vector-valued* functions on LM, together with an increase in the number of associated Hamiltonian vector fields. On the other hand the geometry is more special in the sense that while in principle an arbitrary $\mathbb{R}$-valued function on T^*M is an allowable observable, not all vector-valued functions on LM are compatible with the geometry. Both of these features will be seen to be due to the fact that the general linear group $GL(n)$ acts on the manifold LM, and that the soldering one-form θ transforms *tensorially* under this action. Consequently the set of all allowable observables contains the vector-valued tensorial functions on LM that correspond uniquely to tensor fields on M. Although more general observables may be compatible with the geometry, in this paper we will restrict attention to the geometry associated with vector-valued observables related to the standard actions of $GL(n)$ on $\mathbb{R}^n$ and $\mathbb{R}^{n*}$.

Physical motivation for this generalized geometry comes from the observation that the bundle of linear frames LM is the bundle that is fundamentally related to spacetime observations. This fact coupled with the inescapable interaction of observer and object that is a basic feature of quantum mechanics supports the idea that symplectic geometry on LM may be useful in quantum theory. In fact we show in §8 that the Dirac equation arises in a natural way as the eigenvalue equation for a generalized pre-quantization operator assigned to the spacetime metric tensor Hamiltonian on LM. This new result lends strong support to this study of generalized symplectic geometry.

In the first few sections of this paper we develop the fundamentals of this symplectic geometry on LM. After providing a motivational algorithm for the fundamental structure equation in §2, we show in §3 that symmetric contravariant tensor fields on the base manifold M give rise to a *Poisson algebra* on LM. We show that the Poisson bracket of two tensorial functions $\hat{f}$ and $\hat{g}$ on LM, corresponding to tensor fields $\vec{f}$ and $\vec{g}$ on M, is the tensorial function corresponding to the differential concomitant of $\vec{f}$ and $\vec{g}$ discovered by Schouten and Nijenhuis [14, 15]. The corresponding sets of locally defined vector-valued Hamiltonian vector fields define Lie algebras on LM.

Extending the analysis to anti-symmetric contravariant tensor fields on M we show in §4 that such fields on M give rise to a *Poisson super-algebra* on LM. The super algebra bracket again reproduces the differential concomitants of Schouten and Nijenhuis [14, 15] for two antisymmetric contravariant tensor fields. The corresponding sets of locally defined vector-valued Hamiltonian vector fields define super algebras on LM.

In §5 we use a result from §2, that the *natural lift* of a vector field on M to

LM is a rank $p = 1$ Hamiltonian vector field on LM, to give an extension of the definition to contravariant rank > 1 tensor fields. We also show in §5 how the Poisson bracket introduced in §§2 and 3 leads to a natural definition of Killing tensors associated with a given Riemannian metric tensor field. In §6 we define *locally Hamiltonian vector fields* in the context of generalized symplectic geometry and study the associated integrability conditions. We show that the most general $\bigotimes_s^p \mathbb{R}^n$-valued functions on LM that are compatible with the geometry are generalized *polynomial observables*, that is $\bigotimes_s^p \mathbb{R}^n$-valued polynomials in the generalized momentum coordinates π_j^i with coefficients in the set of $\mathbb{R}$-valued functions on the base space M. The analogous result in the antisymmetric case is obtained by replacing "polynomial in π_j^i" with "exterior products of the π_j^i".

In §§7 and 8 we provide two applications of the geometry to physics. The equations of motion of free inertial observers are derived in §7, while in §8 we derive the Dirac equation from a generalized geometric prequantization argument. Finally in §9 we present a summary and conclusions.

It is convenient to introduce here a portion of the notation that will be needed in the remainder of the paper. Let U be an open subset of M with $n = \dim(M)$. Set $\hat{U} = \pi^{-1}(U)$ where $\pi \colon LM \to M$ is the projection map. All actions of $GL(n) \equiv GL(n, \mathbb{R})$ on the spaces $\bigotimes^p \mathbb{R}^n$, $p = 1, 2, \ldots,$ indicted below by a central dot "$\cdot$", are the standard tensorial actions. We use the abbreviated notation $\bigotimes_s^p \mathbb{R}^n$ for the p-fold symmetric tensor product $\mathbb{R}^n \otimes_s \mathbb{R}^n \otimes_s \cdots \otimes_s \mathbb{R}^n$, and $\bigotimes_a^p \mathbb{R}^n$ for the p-fold anti-symmetric tensor product $\mathbb{R}^n \otimes_a \mathbb{R}^n \otimes_a \cdots \otimes_a \mathbb{R}^n$.

- $ST^p = \{\hat{f} \colon LM \to \bigotimes_s^p \mathbb{R}^n | \hat{f}(u \cdot g) = g^{-1} \cdot \hat{f}(u) \forall g \in GL(n)\}$ is the vector space of symmetric $\bigotimes_s^p \mathbb{R}^n$-valued tensorial functions on LM. An element of ST^p corresponds to a unique rank p symmetric contravariant tensor field on M.
- $AT^p = \{\hat{f} \colon LM \to \bigotimes_a^p \mathbb{R}^n | \hat{f}(u \cdot g) = g^{-1} \cdot \hat{f}(u) \forall g \in GL(n)\}$ is the vector space of antisymmetric $\bigotimes_a^p \mathbb{R}^n$-valued tensorial functions on LM. An element of AT^p corresponds to a unique rank p antisymmetric contravariant tensor field on M.
- $ST = \sum_{p=1}^{\infty} ST^p$.
- $AT = \sum_{p=1}^{\infty} AT^p$.
- $\mathscr{X}(N)$ denotes the vector space of smooth vector fields on a differentiable manifold N.
- $S\mathscr{X}^p \equiv S\mathscr{X}^p(N)$ denotes the vector space of smooth symmetric contravariant tensor fields on N of rank p.
- $A\mathscr{X}^p \equiv A\mathscr{X}^p(N)$ denotes the vector space of smooth anti-symmetric contravariant tensor fields on N of rank p.
- $S\mathscr{X} = \sum_{p=1}^{\infty} S\mathscr{X}^p$.
- $A\mathscr{X} = \sum_{p=1}^{\infty} A\mathscr{X}^p$.

2. Generalized symplectic geometry

Let M be an n-dimensional manifold and LM the principal fiber bundle of linear frames of M. The dimension of LM is the even number $n(n+1)$. A point $u \in LM$ will be denoted by the pair (p, e_i) where $p \in M$ and $(e_i) \equiv (e_1, e_2, \ldots, e_n)$ denotes a linear frame at p. The projection map $\pi: LM \to M$ is defined by $\pi(p, e_i) = p$. The structure group of LM is the general linear group $GL(n)$, which acts freely on the right of LM by $R_g(p, e_i) \equiv (p, e_i) \cdot g = (p, e_j g_i^j)$ for $g = (g_j^i) \in GL(n)$. In this definition and throughout this paper the summation convention on repeated indices is employed.

Local coordinates on LM may be defined as follows. If (U, x^i) is a chart on M, then define local coordinates $(x^i, \pi_k^j): \pi^{-1}(U) \to \mathbb{R}^n \times \mathbb{R}^{n^2}$ by

$$(2.1) \qquad x^i(p, e_j) = x^i(p), \qquad \pi_k^j(p, e_i) = e^j\left(\frac{\partial}{\partial x^k}\right).$$

In this definition (e^j), $j = 1, 2, \ldots, n$ denotes the coframe dual to the linear frame (e_j). Moreover we follow the standard practice of using x^i to denote coordinates on both $U \subset M$ and $\pi^{-1}(U) \subset LM$.

The structure of LM is special in the sense that it supports a globally defined $\mathbb{R}^n$-valued one-form, the *soldering one-form* $\theta = \theta^i r_i$. Here $r_1, r_2, \ldots, r_n$ denotes the standard basis of $\mathbb{R}^n$. For each point $u \in LM$ let u also denote the linear map $u: \mathbb{R}^n \to T_{\pi(u)}M$ defined by [16],

$$(2.2) \qquad u(\xi^i r_i) \equiv (p, e_j)(\xi^i r_i) \overset{\text{def}}{=} \xi^i e_i,$$

with inverse

$$(2.3) \qquad u^{-1}(X) \equiv (p, e_i)^{-1}(X) = e^i(X)r_i, \qquad X \in T_p M.$$

Then the soldering one-form θ may be defined by

$$(2.4) \qquad \theta(Y) \overset{\text{def}}{=} u^{-1}(d\pi Y), \qquad \forall Y \in T_u LM.$$

In local coordinates (x^i, π_k^j) the soldering one-form has the local expression

$$(2.5) \qquad \theta^i r_i = (\pi_j^i dx^j)r_i.$$

One may compare this form to the expression $\tilde{\theta} = \pi_j dx^j$ for the canonical one-form on T^*M in canonical coordinates.

The basic properties of θ that follow from its definition are

$$(2.6) \qquad \begin{array}{ll} \text{(a)} & \theta(Y) = 0 \Leftrightarrow d\pi(Y) = 0, \\ \text{(b)} & R_g^*\theta = g^{-1} \cdot \theta \equiv (g^{-1})_j^i \theta^j r_i. \end{array}$$

The one-form θ is the basic element needed to define the torsion Θ of a linear connection. If ω denotes the $gl(n)$-valued one-form of a linear connection on LM, then the torsion of ω may be defined by [16],

$$(2.7) \qquad \Theta = d\theta + \omega \wedge \theta = (d\theta^i + \omega_j^i \wedge \theta^j)r_i.$$

In particular, if a linear connection ω is torsion-free then

$$(2.8) \qquad d\theta^i = -\omega^i_j \wedge \theta^j .$$

By the Frobenius theorem the n-dimensional codistribution spanned globally by the one-forms θ^i is integrable, and the integral submanifolds of the codistribution are clearly the fibers of LM.

Consider now the exact $\mathbb{R}^n$-valued two-form $\beta \overset{\text{def}}{=} d\theta$. By (2.5) it has the local coordinate expression

$$(2.9) \qquad \beta = \beta^i r_i = (d\pi^i_j \wedge dx^j) r_i .$$

Using this last equation (or equation (2.8)) it is easy to show that β is *nondegenerate* in the sense that

$$(2.10) \qquad X \lrcorner \beta = 0 \Leftrightarrow X = 0 .$$

In standard symplectic geometry on T^*M one uses the canonical one-form $\tilde{\theta}$ to assign a unique Hamiltonian vector field X_f to each observable $f : T^*M \to \mathbb{R}$ via the equation

$$(2.11) \qquad df = -X_f \lrcorner d\tilde{\theta} .$$

If we attempt to transcribe this equation to LM using the soldering one-form we have

$$df = -X_f \lrcorner d\theta^i r_i ,$$

and it is clear that this expression makes no sense for f an $\mathbb{R}$-valued function and X_f a vector field on LM. However, if we replace f with an $\mathbb{R}^n$-valued function $\hat{f} = \hat{f}^i r_i : LM \to \mathbb{R}^n$, then the equation

$$(2.12) \qquad d\hat{f} = -X_{\hat{f}} \lrcorner d\theta^i r_i$$

defines a unique vector field $X_{\hat{f}}$ given $\hat{f}$. In order to facilitate the derivation of other generalizations of equation (2.11) to LM it is convenient to introduce the following geometrical derivation of equation (2.12).

Consider the problem of finding a torsion-free linear connection ω on LM with respect to which a given vector field $\vec{f}$ on the base manifold M is covariant constant. Let $\hat{f} = \hat{f}^i r_i$ denote the unique $\mathbb{R}^n$-valued tensorial zero-form on LM determined by $\vec{f}$, defined invariantly by $\hat{f}(u) = u^{-1}(\vec{f}(\pi(u)))$. The covariant derivative of $\vec{f}$ on M is uniquely determined [16] by the exterior covariant derivative $D\hat{f} = (d\hat{f}^i + \omega^i_j \cdot \hat{f}^j) r_i$ of $\hat{f}$ on LM. Let (B_i), $i = 1, 2, \ldots, n$ denote the standard horizontal vector fields on LM determined by ω. Then these vector fields satisfy

$$(2.13) \qquad \omega(B_i) = 0, \qquad \theta^i(B_j) = \delta^i_j ,$$

for $i, j = 1, 2, \ldots, n$.

From equations (2.8) and (2.13) we find for a torsion-free linear connection the relationship

$$(2.14) \qquad \omega^i_j = B_j \lrcorner \beta^i.$$

If this expression for ω^i_j is substituted into the formula for the exterior covariant derivative of $\hat{f}$ then the $\mathbb{R}^n$ components of $D\hat{f}$ may be expressed as

$$(2.15) \qquad D\hat{f}^i = d\hat{f}^i + (B_j \hat{f}^j) \lrcorner \beta^i.$$

Defining $X_{\hat{f}} \overset{\text{def}}{=} B_j \hat{f}^j$ and assuming $D\hat{f} = 0$ we obtain the equation

$$(2.16) \qquad d\hat{f}^i = -X_{\hat{f}} \lrcorner \beta^i.$$

Given the functions $\hat{f}^i$ the vector field $X_{\hat{f}}$ is uniquely determined by equation (2.16) since $\beta = \beta^i r_i$ is nondegenerate. Thus solutions to the original problem can be sought by first solving equation (2.16) for $X_{\hat{f}}$ and by then solving $X_{\hat{f}} = B_i \hat{f}^i$ for the vector fields (B_i), which would certainly not be unique. If the vector fields (B_i) can be made to satisfy certain additional conditions then they would define a linear connection with the required property.

The purpose of this example is not to discuss the existence or uniqueness of such a linear connection, but rather to bring to light equation (2.16). If we now disregard the method of derivation of equation (2.16) and the definition of $X_{\hat{f}}$, then it is clear that $\beta = \beta^i r_i$ plays the role of a generalized symplectic structure on LM. We may think of $X_{\hat{f}}$ as the *generalized Hamiltonian vector field* determined by the $\mathbb{R}^n$-valued tensorial function $\hat{f}$, and we may consider the flow of $X_{\hat{f}}$ as generating local one-parameter families of *generalized canonical transformations*. These transformations are canonical in the sense that $\mathscr{L}_{X_{\hat{f}}}(\beta) = 0$, which follows in the standard way from equation (2.16) and the general formula $\mathscr{L}_X \Psi = X \lrcorner d\Psi + d(X \lrcorner \Psi)$.

When the functions $\hat{f}^i$ are determined as above from a vector field $\vec{f}$ on M, then the Hamiltonian vector field $X_{\hat{f}}$ determined from equation (2.16) will be shown below to be the *natural lift* of $\vec{f}$ to LM. This point is important because, since the natural lift is independent of any connection on LM, it shows the basic independence of equation (2.16) from ideas of covariant differentiation based on linear connections. Explicitly, let $\vec{f}$ be given in local coordinates on M by

$$(2.17) \qquad \vec{f} = f^i \frac{\partial}{\partial x^i}$$

so that the corresponding function $\hat{f}$ on LM is given by

$$(2.18) \qquad \hat{f} = \hat{f}^i r_i = ((f^j \circ \pi) \pi^i_j) r_i.$$

Note that under right translation on LM the functions $\hat{f}^i$ transform according to the rule $\hat{f}^i(u \cdot g) = (g^{-1})^i_j \hat{f}^j(u)$ for $u \in LM$ and $g = (g^i_j) \in GL(n)$. This is the *tensorial* transformation law [16] for $\mathbb{R}^n$-valued functions on LM.

Solving equation (2.16) locally for $X_{\hat{f}}$ with $\hat{f}$ as in (2.18) yields

$$(2.19) \qquad X_{\hat{f}} = (f^i \circ \pi)\frac{\partial}{\partial x^i} - \left(\frac{\partial(f^i \circ \pi)}{\partial x^j}\pi^k_i\right)\frac{\partial}{\partial \pi^k_j}.$$

Using this result it is easy to show that $X_{\hat{f}}$ has the following three properties:

$$(2.20) \qquad \begin{array}{ll} (1) & dR_a(X_{\hat{f}}) = X_{\hat{f}} \quad \text{for every } a \in GL(n), \\ (2) & \mathscr{L}_{X_{\hat{f}}}(\theta) = 0, \\ (3) & d\pi(X_{\hat{f}}) = \vec{f}. \end{array}$$

Properties (1) and (3) follow from (2.19), while property (2) follows from equation (2.16). These three properties uniquely characterize [16] the *natural lift* of a vector field on M to LM. It follows that the canonical transformations generated by the flow of X_f on LM represent the natural lift to LM of the local diffeomorphisms of M generated by the flow of $\vec{f}$. We have the result that the natural dynamics of vector fields on a manifold M is *Hamiltonian dynamics with respect to the symplectic structure* $d\theta$ *on* LM. We formalize these results in the following

THEOREM 2.1. *Let* $\hat{f}: LM \to \mathbb{R}^n$ *be the tensorial zero-form on* LM *determined by a vector field* $\vec{f}$ *on* M. *Then the Hamiltonian vector field* $X_{\hat{f}}$ *determined by* $d\hat{f} = -X_{\hat{f}} \rfloor \beta$ *is the natural lift of* $\vec{f}$ *to* LM.

Modifications of standard symplectic geometry begin to appear when equation (2.16) is examined more closely. The first thing to notice is that while there are no restrictions placed on the $\mathbb{R}$-valued functions f on T^*M by equation (2.11), not every $\mathbb{R}^n$-valued function on LM is compatible with equation (2.16). Let

- $HF^1 \equiv HF^1(LM, \mathbb{R}^n)$ denote the set of $\mathbb{R}^n$-valued functions on LM that satisfy equation (2.16) for some vector field $X_{\hat{f}}$,
- HV^1 denote the set of Hamiltonian vector fields determined by HF^1 and equation (2.16),
- $T^1 \equiv T^1(LM, \mathbb{R}^n)$ denote the set of $\mathbb{R}^n$-valued tensorial zero-forms on LM relative to the standard action of $GL(n)$ on $\mathbb{R}^n$.
- $LHF^1 \equiv HF^1(\pi^{-1}(U), \mathbb{R}^n)$, U an open subset of M, denotes the set of locally defined $\mathbb{R}^n$-valued functions on LM that satisfy equation (2.16) for some vector field $X_{\hat{f}}$ on $\pi^{-1}(U)$.

An analysis of equation (2.16) (see §6) shows that the locally defined set of $\mathbb{R}^n$-valued functions LHF^1 consists of functions of the form

$$(2.21) \qquad \hat{f} = \hat{f}^i r_i = \{(f^i \circ \pi)\pi^j_i + \xi^j \circ \pi\}r_j,$$

where f^i and ξ^i are functions defined on $U \subset M$. Thus, upon comparing (2.21) with (2.18) we have

$$(2.22) \qquad LHF^1 = T^1(\pi^{-1}(U), \mathbb{R}^n) + C^\infty(U, \mathbb{R}^n).$$

The Hamiltonian vector field $X_{\hat{f}}$ determined locally by such an element of LHF^1 has the local expression

$$(2.23) \qquad X_{\hat{f}} = (f^i \circ \pi)\frac{\partial}{\partial x^i} - \left(\frac{\partial(f^i \circ \pi)}{\partial x^k}\pi_i^j + \frac{\partial(\xi^j \circ \pi)}{\partial x^k}\right)\frac{\partial}{\partial \pi_k^j}.$$

It is straightforward to show that HF^1 is a *Lie algebra* under the bracket defined by

$$(2.24) \qquad \{\hat{f}, \hat{g}\} \stackrel{\text{def}}{=} X_{\hat{f}}(\hat{g}).$$

Moreover, by direct calculation one may show that HV^1 is a *Lie algebra* under Lie bracket, and that $[X_{\hat{f}}, X_{\hat{g}}] = X_{\{\hat{f}, \hat{g}\}}$.

The explicit local expression for the Poisson bracket of

$$\hat{f} = \{(f^i \circ \pi)\pi_i^j + \xi^j \circ \pi\}r_j$$

and $\hat{g} = \{(g^i \circ \pi)\pi_i^j + \eta^j \circ \pi\}r_j$ is

$$(2.25) \qquad \{\hat{f}, \hat{g}\} = \left\{f^i\frac{\partial g^k}{\partial x^i} - g^i\frac{\partial f^k}{\partial x^i}\right\}\pi_k^j r_j + \left\{f^i\frac{\partial \eta^j}{\partial x^i} - g^i\frac{\partial \xi^j}{\partial x^i}\right\}r_j.$$

Therefore the sum in (2.22) is a semidirect sum.

The center of LHF^1 consists of the constant functions $\pi^{-1}(U) \to \mathbb{R}^n$, so that as Lie algebras $LHV^1 \simeq LHF^1/\mathbb{R}^n$. LHF^1 can thus be regarded as a *central extension* of LHV^1, in general agreement with the standard theory on T^*M.

3. The Poisson algebra ST on LM

As shown above the Lie bracket of vector fields on M is equivalent to the Lie bracket of tensorial $\mathbb{R}^n$-valued functions on LM defined in equation (2.24). Now the Lie bracket of vector fields is a derivation on the space of vector fields, and it is well known to have an extension to derivations of arbitrary tensor fields. However, the Lie derivative $\mathscr{L}_{\hat{f}}(T)$ of a rank $p > 1$ contravariant tensor field T with respect to a vector field $\hat{f}$ is not a Lie bracket, and hence $\mathscr{L}_{\hat{f}}(T)$ would not seem to have any relationship to a Poisson bracket. However we will show that there is a relationship for certain classes of irreducible tensors.

A symmetric rank p contravariant tensor field $\vec{g} \in S\mathscr{X}^p$ has the local coordinate expression

$$(3.1) \qquad \vec{g} = g^{i_1 \cdots i_p}\partial_{i_1} \otimes \cdots \otimes \partial_{i_p}, \qquad g^{i_1 \cdots i_p} = g^{(i_1 \cdots i_p)}.$$

In this equation and in the following "round" brackets on indices denotes symmetrization. The corresponding function $\hat{g} \in ST^p$ is given in local coordinates (x^i, π^j_k) by

$$(3.2) \qquad \hat{g} = \hat{g}^{i_1 \cdots i_p} r_{i_1} \cdots r_{i_p} = (g^{j_1 \cdots j_p} \circ \pi) \pi^{i_1}_{j_1} \cdots \pi^{i_p}_{j_p} r_{i_1} \otimes \cdots \otimes r_{i_p}.$$

Now consider $\hat{f} \in ST^1$ and $\hat{g} \in ST^2$. Let $X_{\hat{f}}$ denote the Hamiltonian vector field associated with $\hat{f}$. By analogy with equation (2.24) one is led to try the definition

$$(3.3) \qquad \{\hat{f}, \hat{g}\} \stackrel{\text{def}}{=} X_{\hat{f}}(\hat{g}) = X_{\hat{f}}(\hat{g}^{ij}) r_i \otimes r_j$$

for the Poisson bracket of $\hat{f}$ with $\hat{g}$. It is straightforward to check using (2.19) that the term $X_{\hat{f}}(\hat{g})$ on the right-hand side of this definition is in fact the element of ST^2 corresponding to the Lie derivative of the associated tensor field $\vec{g}$ on M with respect to the vector field $\vec{f}$. This is the LM form of the extension of the Lie derivative mentioned above. The problem of course with the definition (3.3) is how to make sense of

$$(3.4) \qquad \{\hat{g}, \hat{f}\} = -\{\hat{f}, \hat{g}\}$$

so that we actually have a Poisson bracket.

So far we have only associated Hamiltonian vector fields with elements of ST^1. To give meaning to the left-hand side of equation (3.4) we need to associate Hamiltonian vector fields with all elements of ST. A method for doing this can be found by considering again the derivation of equation (2.16), which we illustrate for an element $\hat{g} \in ST^2$. The derivation starts with the problem of finding a torsion-free linear connection that leaves $\hat{g}$ covariant constant. The generalization of equation (2.16) for this problem is

$$d\hat{g} = (d\hat{g}^{ij}) r_i \otimes r_j = (-X^j_{\hat{g}} \rfloor \beta^i - X^i_{\hat{g}} \rfloor \beta^j) r_i \otimes r_j,$$

or simply

$$(3.5) \qquad d\hat{g}^{ij} = -2X^{(i}_{\hat{g}} \rfloor \beta^{j)}.$$

Thus a symmetric $\mathbb{R}^n \otimes \mathbb{R}^n$-valued tensorial function on LM has associated with it a *set of Hamiltonian vector fields* $X^i_{\hat{g}}$, $i = 1, 2, \ldots, n$ rather than a single Hamiltonian vector field. Taken together these vector fields $X^i_{\hat{g}}$ define the $\mathbb{R}^n$-valued vector field $\widehat{X}_{\hat{g}} = X^i_{\hat{g}} \otimes r_i$.

Now although β is nondegenerate in the sense of equation (2.10), because of the symmetrization in equation (3.5) $\widehat{X}_g$ is not uniquely determined by $\hat{g}$ and equation (3.5). However, the nonuniqueness is easily characterized, at least locally. In particular, an element $\hat{g} \in ST^2$ determines n vector fields X^i_g via equation (3.5) up to addition of vector fields Y^i on LM satisfying the kernel equation $Y^{(i} \rfloor \beta^{j)} = 0$. More generally, an element $\hat{g} \in ST^p$ as in (3.2) above determines

$$N_S(p) = \binom{n+p-2}{p-1}$$

vector fields $X_g^{i_1 \cdots i_{p-1}}$ via the generalized symplectic structure equation

$$(3.6) \qquad d\hat{g}^{i_1 \cdots i_p} = -p! X_{\hat{g}}^{(i_1 \cdots i_{p-1}} \rfloor \beta^{i_p)}$$

up to addition of vector fields $Y^{i_1 \cdots i_{p-1}}$ satisfying the kernel equation

$$(3.7) \qquad Y^{(i_1 \cdots i_{p-1}} \rfloor \beta^{i_p)} = 0.$$

The nonuniqueness can be characterized locally as follows. For $\hat{g}$ as in (3.2) the associated Hamiltonian vector fields $X_{\hat{g}}^{i_1 \cdots i_{p-1}}$ determined by equation (3.6) have the local coordinate expressions

$$(3.8) \qquad \begin{aligned} X_{\hat{g}}^{i_1 \cdots i_{p-1}} = {} & \frac{1}{(p-1)!}(g^{j_1 \cdots j_{p-1}k} \circ \pi)\pi_{j_1}^{i_1} \cdots \pi_{j_{p-1}}^{i_{p-1}} \frac{\partial}{\partial x^k} \\ & - \frac{1}{p!}\left\{ \frac{\partial}{\partial x^l}(g^{j_1 \cdots j_p} \circ \pi)\pi_{j_1}^{i_1} \cdots \pi_{j_{p-1}}^{i_{p-1}}\pi_{j_p}^k + T_l^{i_1 \cdots i_{p-1}k} \right\} \frac{\partial}{\partial \pi_l^k}. \end{aligned}$$

The nonuniqueness is contained completely in the vertical component

$$(3.9) \qquad Y^{i_1 \cdots i_{p-1}} = T_j^{i_1 \cdots i_{p-1}k} \frac{\partial}{\partial \pi_j^k},$$

where the coefficients $T_j^{i_1 \cdots i_{p-1}k}$ must satisfy

$$(3.10) \qquad T_j^{(i_1 \cdots i_{p-1}k)} = 0$$

but are otherwise arbitrary.

There are special cases (see, for example, §7) in which the nonuniqueness can be removed globally by placing additional invariantly defined conditions on the vector fields $X_{\hat{g}}^{i_1 \cdots i_{p-1}}$ in addition to equation (3.6). Here, however, we will resolve the nonuniqueness by working locally as follows.

We assign to $\hat{g} \in ST^p$ the $N_S(p)$ Hamiltonian vector fields $X_{\hat{g}}^{i_1 \cdots i_{p-1}}$ determined by equation (3.6) and the conditions

$$(3.11) \qquad d\pi_j^k(X_{\hat{g}}^{i_1 \cdots i_{p-1}}) = d\pi_j^{(k}(X_{\hat{g}}^{i_1 \cdots i_{p-1})}).$$

This condition is clearly a local condition and accordingly for the moment we restrict attention to objects defined locally on $\pi^{-1}(U) \subset LM$, where $U \subset M$ is the domain of the chart (x^i). Since the undetermined elements of $X_{\hat{g}}^{i_1 \cdots i_{p-1}}$ given in (3.8) satisfy (3.10), the auxiliary conditions (3.11) now provides uniqueness locally, with

$$(3.12) \qquad \begin{aligned} X_{\hat{g}}^{i_1 \cdots i_{p-1}} = {} & \frac{1}{(p-1)!}(g^{j_1 \cdots j_{p-1}k} \circ \pi)\pi_{j_1}^{i_1} \cdots \pi_{j_{p-1}}^{i_{p-1}} \frac{\partial}{\partial x^k} \\ & - \frac{1}{p!}\frac{\partial}{\partial x^l}(g^{j_1 \cdots j_p} \circ \pi)\pi_{j_1}^{i_1} \cdots \pi_{j_{p-1}}^{i_{p-1}}\pi_{j_p}^k \frac{\partial}{\partial \pi_l^k}. \end{aligned}$$

Returning now to equation (3.3) we see that

$$(3.13) \quad \begin{aligned} \{\hat{f}, \hat{g}\}^{ij} &= X_{\hat{f}}(\hat{g}^{ij}) = d\hat{g}^{ij}(X_{\hat{f}}) = -4\beta^{(i}(X_{\hat{g}}^{j)}, X_{\hat{f}}) \\ &= +4\beta^{(i}(X_{\hat{f}}, X_{g}^{j)}) = -2d\hat{f}^{(i}(X_{g}^{j)}) = -2X_{\hat{g}}^{(i}(\hat{f}^{j)}). \end{aligned}$$

Thus the definition

$$(3.14) \quad \{\hat{g}, \hat{f}\} \stackrel{\text{def}}{=} 2! X_{\hat{g}}^{(i}(\hat{f}^{j)}) r_i \otimes r_j$$

has the right properties to make sense of equation (3.4).

More generally, let $\hat{f} \in ST^p$ and $\hat{g} \in ST^q$, and denote their corresponding Hamiltonian vector fields, determined uniquely but locally by equations (3.6) and (3.11), by $X_{\hat{f}}^{i_1 \cdots i_{p-1}}$ and $X_{\hat{g}}^{i_1 \cdots i_{q-1}}$, respectively. Then one may show that the two definitions

$$(3.15) \quad \{\hat{f}, \hat{g}\} \stackrel{\text{def}}{=} p! X_{\hat{f}}^{(i_1 \cdots i_{p-1}}(\hat{g}^{i_p \cdots i_{p+q-1})}) r_{i_1} \otimes \cdots r_{i_{p+q1}}$$

and

$$(3.16) \quad \{\hat{g}, \hat{f}\} \stackrel{\text{def}}{=} q! X_{\hat{g}}^{(i_1 \cdots i_{q-1}}(\hat{f}^{i_q \cdots i_{q+p-1})}) r_{i_1} \otimes \cdots r_{i_{p+q-1}}$$

are related by

$$\{\hat{g}, \hat{f}\} = -\{\hat{f}, \hat{g}\}.$$

The fact that if $\hat{f} \in ST^p$ and $\hat{g} \in ST^q$, then $\{\hat{f}, \hat{g}\} \in ST^{p+q-1}$ follows from the following argument. Writing out the explicit formula for $\{\hat{f}, \hat{g}\}$ using (3.2), (3.12) and (3.15) we obtain

$$(3.17)$$

$$\{\hat{f}, \hat{g}\}^{i_i \cdots i_{p+q-1}} = \begin{pmatrix} p f^{k(j_2 \cdots j_p} \partial_k g^{|a_2 \cdots a_q)} \\ -q g^{k(a_2 \cdots a_q} \partial_k f^{|j_2 \cdots j_p)} \end{pmatrix} \pi_l^{(i_1} \pi_{j_2}^{i_2} \cdots \pi_{j_p}^{i_p} \pi_{a_2}^{i_{p+1}} \cdots \pi_{a_q}^{i_{p+q-1})}.$$

The right-hand side of this equation is the $\bigotimes_s^{p+q-1} \mathbb{R}^n$-valued function on LM corresponding to the *differential concomitant*, due to Schouten and Nijenhuis [14, 15], of the tensor fields $\vec{f}$ and $\vec{g}$ on M. It is easy to show that ∂_k on the right-hand side of (3.17) can be replaced by ∇_k for an arbitrary symmetric linear connection, and hence the right-hand side of (3.17) is independent of coordinates. Thus $\{\hat{f}, \hat{g}\}$ is an element of ST^{p+q-1}.

This is a rather remarkable result. It shows that although the vector-valued vector fields $X_{\hat{f}}$ given in (3.12) depend explicitly on the choice of coordinate chart, the bracket $\{\hat{f}, \hat{g}\}$ calculated explicitly in terms of $X_{\hat{f}}$ itself does not depend on the choice of coordinates. We can see why this is true by noting that

$$\begin{aligned} \{\hat{f}, \hat{g}\}^{i_1 \cdots i_{p+q-1}} &= p! X_{\hat{f}}^{(i_1 \cdots i_{p-1}}(g^{i_p \cdots i_{p+q-1})}) \\ &= -p! q! d\theta^{(i_1}(X_{\hat{g}}^{i_2 \cdots i_q}, X_{\hat{f}}^{i_{q+1} \cdots i_{p+q-1})}). \end{aligned}$$

It is clear from this last equation and (3.10) that the undetermined components $T_j^{i_1 \cdots i_{p-1} k}$ do not contribute to the bracket $\{\hat{f}, \hat{g}\}$.

For $\hat{f} \in ST^p$ we get from (3.6) and (3.11) a unique set of locally defined Hamiltonian vector fields $X_{\hat{f}}^{i_1 \cdots i_{p-1}}$. We introduce the multi-index notation $I = (i_1 \cdots i_{p-1})$ and view the set of vector fields $X_{\hat{f}}^{i_1 \cdots i_{p-1}}$ as the $\bigotimes_s^{p-1} \mathbb{R}^n$-valued vector field $\hat{X}_{\hat{f}}$ defined by

$$(3.18) \qquad \hat{X}_{\hat{f}} = X_{\hat{f}}^I \otimes r_I, \qquad r_I = r_{i_1} \otimes_s r_{i_2} \otimes_s \cdots \otimes_s r_{i_{p-1}} .$$

For two *arbitrary* vector fields $\hat{X} = X^I \otimes r_I$ and $\hat{Y} = Y^J \otimes r_J$ with values in $\bigotimes_s^{p-1} \mathbb{R}^n$ and $\bigotimes_s^{q-1} \mathbb{R}^n$, respectively, we define a bracket by

$$(3.19) \qquad [\hat{X}, \hat{Y}] = [X^I, Y^J] \otimes r_I \otimes_s r_J .$$

The bracket on the right-hand side of this last equation is the ordinary Lie bracket of vector fields. One shows directly that the bracket defined in (3.19) has all the properties of a Lie bracket. Thus the infinite dimensional vector space $S\widehat{\mathscr{X}}(LM) = \sum_{p=1}^{\infty} S\widehat{\mathscr{X}^p}(LM)$, where $S\widehat{\mathscr{X}^p}(LM)$ denotes the vector space of $\bigotimes_s^{p-1} \mathbb{R}^n$-valued vector fields on LM, is a *Lie algebra* under the bracket defined in (3.19).

We introduce the notation
- $LSHV^p =$ the set of all $\bigotimes_s^{p-1} \mathbb{R}^n$-valued Hamiltonian vector fields $\hat{X}_{\hat{f}}$ determined locally by elements $\hat{f} \in ST^p$ by equations (3.6) and (3.11),
- $LSHV = \sum_{p=1}^{\infty} LSHV^p$

LEMMA. *Let $X_{\hat{g}}^J$ be the Hamiltonian vector fields determined locally by $\hat{g} \in ST^q$ where J denotes the multi-index $J = i_1 i_2 \cdots i_{q-1}$. Then $L_{X_{\hat{g}}^{(J}} \beta^{i)} = 0$ where the notation $(J$ and $i)$ indicates symmetrization over the indices.*

PROOF. Using the identity $L_X \omega = X \lrcorner d\omega + d(X \lrcorner \omega)$ we obtain

$$L_{X_{\hat{g}}^{(J}} \beta^{i)} = X_{\hat{g}}^{(J} \lrcorner d\beta^{i)} + d(X_{\hat{g}}^{(J} \lrcorner \beta^{i)}) .$$

The first term on the right-hand side vanishes since $\beta^i = d\theta^i$, and the second term vanishes because by (3.8) $X_{\hat{g}}^{(J} \lrcorner \beta^{i)} = (-1/q!)d(\hat{g}^{Ji})$. $\square$

The next two theorems will establish that the space ST of symmetric tensorial zero-forms, and the local spaces $LSHV$ of the corresponding locally defined Hamiltonian vector fields, are Lie algebras under multiplications defined by the Poisson bracket (3.15) and Lie bracket (3.19), respectively.

THEOREM 3.1. *Let $\hat{f} \in ST^p$ and $\hat{g} \in ST^q$, and denote the corresponding locally Hamiltonian vector fields by $\hat{X}_{\hat{f}}$ and $\hat{X}_{\hat{g}}$. Then*

$$[\hat{X}_{\hat{f}}, \hat{X}_{\hat{g}}] = \frac{(p+q-1)!}{p!q!} \hat{X}_{\{\hat{f}, \hat{g}\}} .$$

PROOF. Using the identity $L_X(Y \lrcorner \beta) = X \lrcorner (L_Y \beta) + [X, Y] \lrcorner \beta$ we have

$$[X_{\hat{f}}^{(I}, X_{\hat{g}}^{J]} \lrcorner \beta^{i)} = L_{X_{\hat{f}}^{(I}}(X_{\hat{g}}^{J} \lrcorner \beta^{i)}) - X_{\hat{f}}^{(I} \lrcorner (L_{X_{\hat{g}}^{J}} \beta^{i)})$$

with symmetrization over the indices I, J, i. The second term on the right-hand side of this equation vanishes by the lemma. Using the identity $L_X \omega = X \rfloor d\omega + d(X \rfloor \omega)$ the equation now reduces to

$$[X_{\hat{f}}^{(I}, X_{\hat{g}}^{J)}] \rfloor \beta^{i)} = X_{\hat{f}}^{(I} \rfloor d(X_{\hat{g}}^{J} \rfloor \beta^{i)}) + d(X_{\hat{f}}^{(I} \rfloor X_{\hat{g}}^{J} \rfloor \beta^{i)})$$
$$= d(X_{\hat{f}}^{(I} \rfloor X_{\hat{g}}^{J} \rfloor \beta^{i)})$$

where the second line follows from the lemma. If we now use equation (3.6) and the definition (3.15) we obtain

$$[X_{\hat{f}}^{(I}, X_{\hat{g}}^{J)}] \rfloor \beta^{i)} = d(X_{\hat{f}}^{(I} \rfloor X_{\hat{g}}^{J} \rfloor \beta^{i)})$$
$$= d\left(X_{\hat{f}}^{(I} \rfloor \left(\frac{-1}{q!} d\hat{g}^{Ji)} \right) \right)$$
$$= \frac{-1}{q!} d(X_{\hat{f}}^{(I}(\hat{g}^{Ji)}))$$
$$= -d\left(\frac{1}{p!q!} \{\hat{f}, \hat{g}\}^{(IJi)} \right).$$

The result now follows from equation (3.6). □

Since the bracket used in this theorem is a (generalized) Lie bracket of vector-valued vector fields we have the

COROLLARY 3.1. *The vector spaces of locally defined Hamiltonian vector fields LSHV determined by elements of ST are Lie algebras under the bracket defined in* (3.19).

The formula

(3.20) $\{\hat{f}, \hat{g}\}^{iIJ} = 2p!q! \beta^{(i}(X_g^I, X_g^{J)}),$

which follows easily from (3.6) and (3.15), will be useful in the following.

THEOREM 3.2. *The bracket defined in* (3.15) *for elements of ST satisfies the Jacobi identity. That is, for* $\hat{f} \in ST^p$, $\hat{g} \in ST^q$ *and* $\hat{h} \in ST^r$

$$\{\hat{f}, \{\hat{g}, \hat{h}\}\} + \{\hat{h}, \{\hat{f}, \hat{g}\}\} + \{\hat{g}, \{\hat{h}, \hat{f}\}\} = 0.$$

PROOF. Let X_f^I, X_g^J and X_h^K denote the Hamiltonian vector fields determined by $\hat{f}$, $\hat{g}$ and $\hat{h}$, where I, J, K denote the multi-indices $I = i_2 i_3 \cdots i_p$, $J = i_{p+1} i_{p+2} \cdots i_{p+q-1}$ and $K = i_{p+q} i_{p+q+1} \cdots i_{p+q+r-2}$. Then by using the standard identity for evaluating $d\omega(X, Y, Z)$ for ω a two-form we obtain

$$0 = 3d\beta^{(i_1}(X_{\hat{f}}^I, X_{\hat{g}}^J, X_{\hat{h}}^K)$$
$$= X_{\hat{f}}^{(I} \beta^{i_1}(X_{\hat{g}}^J, X_{\hat{h}}^K) + X_{\hat{g}}^{(J} \beta^{i_1}(X_{\hat{h}}^K, X_{\hat{f}}^I) + X_{\hat{h}}^{(K} \beta^{i_1}(X_{\hat{f}}^I, X_{\hat{g}}^J)$$
$$- \beta^{(i_1}([X_{\hat{f}}^I, X_{\hat{g}}^J], X_{\hat{h}}^K) - \beta^{(i_1}([X_{\hat{h}}^K, X_{\hat{f}}^I], X_{\hat{g}}^J) - \beta^{(i_1}([X_{\hat{g}}^J, X_{\hat{h}}^K], X_{\hat{f}}^I).$$

Using the formula (3.20) and Theorem 3.1 in this equation we obtain

$$
0 = X_{\hat{f}}^{(I}\left(\frac{1}{2q!r!}\{\hat{g},\,\hat{h}\}^{JKi_1)}\right) + X_{\hat{h}}^{(J}\left(\frac{1}{2p!q!}\{\hat{f},\,\hat{g}\}^{IJi_1)}\right)
$$
$$
+ X_{\hat{g}}^{(K}\left(\frac{1}{2p!r!}\{\hat{h},\,\hat{f}\}^{KIi_1)}\right) - \beta^{(i_1}\left(\frac{(p+q-1)!}{p!q!}X_{\{\hat{f},\,\hat{g}\}}^{IJ},\,X_{\hat{h}}^{K)}\right)
$$
$$
- \beta^{(i_1}\left(\frac{(p+r-1)!}{p!r!}X_{\{\hat{h},\,\hat{f}\}}^{KI},\,X_{\hat{g}}^{J)}\right) - \beta^{(i_1}\left(\frac{(q+r-1)!}{q!r!}X_{\{\hat{g},\,\hat{h}\}}^{JK},\,X_{\hat{f}}^{I)}\right).
$$

Next we use the definition (3.15) in the first three terms and formula (3.18) and Theorem 3.1 in the last three terms to obtain

$$
0 = \frac{1}{2p!q!r!}\{\hat{f},\,\{\hat{g},\,\hat{h}\}\}^{L} + \frac{1}{2p!q!r!}
$$
$$
\times\{\hat{h},\,\{\hat{f},\,\hat{g}\}\}^{L} + \frac{1}{2p!q!r!}\{\hat{g},\,\{\hat{h},\,\hat{f}\}\}^{L}
$$
$$
- \frac{(p+q-1)!}{p!q!}\left(\frac{1}{2(p+q-1)!r!}\{\{\hat{f},\,\hat{g}\},\,\hat{h}\}^{L}\right)
$$
$$
- \frac{(p+r-1)!}{p!r!}\left(\frac{1}{2(p+r-1)!q!}\{\{\hat{h},\,\hat{f}\},\,\hat{g}\}^{L}\right)
$$
$$
- \frac{(q+r-1)!}{q!r!}\left(\frac{1}{2(q+r-1)!p!}\{\{\hat{g},\,\hat{h}\},\,\hat{f}\}^{L}\right),
$$

where the multi-index L denotes (i_1IJK). Cancelling the common factor $\frac{1}{p!q!r!}$ we obtain

$$
0 = \tfrac{1}{2}\{\hat{f},\,\{\hat{g},\,\hat{h}\}\}^{L} + \tfrac{1}{2}\{\hat{h},\,\{\hat{f},\,\hat{g}\}\}^{L}
$$
$$
+ \tfrac{1}{2}\{\hat{g},\,\{\hat{h},\,\hat{f}\}\}^{L} - \tfrac{1}{2}\{\{\hat{f},\,\hat{g}\},\,\hat{h}\}^{L}
$$
$$
- \tfrac{1}{2}\{\{\hat{h},\,\hat{f}\},\,\hat{g}\}^{L} - \tfrac{1}{2}\{\{\hat{g},\,\hat{h}\},\,\hat{f}\}^{L}
$$
$$
= \{\hat{f},\,\{\hat{g},\,\hat{h}\}\}^{L} + \{\hat{h},\,\{\hat{f},\,\hat{g}\}\}^{L} + \{\hat{g},\,\{\hat{h},\,\hat{f}\}\}^{L}. \quad \square
$$

The symmetrized tensor product $\otimes_s$ makes ST into a commutative algebra. If we now consider again elements $\hat{f} \in ST^p$, $\hat{g} \in ST^q$ and $\hat{h} \in ST^r$, then by using definition (3.15) and techniques as in the proofs above one may show that

$$(3.21) \qquad \{\hat{f},\,\hat{g}\otimes_s\hat{h}\} = \{\hat{f},\,\hat{g}\}\otimes_s\hat{h} + \hat{g}\otimes_s\{\hat{f},\,\hat{h}\}.$$

Thus the bracket defined in (3.15) acts as a derivation on the commutative algebra, and as a result we have

THEOREM 3.3. *The space ST of symmetric tensorial 0-forms on LM is a Poisson algebra with respect to the Poisson bracket defined in (3.15).*

4. The Poisson super algebra AT on LM

The analysis presented in §3 can be modified to define a symplectic geometry for antisymmetric contravariant tensor fields. We will find that the natural generalization of the Poisson bracket to the space $AT = \sum_{p=1}^{\infty} AT^p$ of antisymmetric $\otimes_a^p \mathbb{R}^n$-valued tensorial functions on LM will make AT into a Poisson super algebra, while the corresponding sets of locally defined vector-valued Hamiltonian vector fields will form super algebras. Because of a uniqueness problem of the type encountered in §3 the Hamiltonian vector fields will be defined locally on $\pi^{-1}(U)$ where (x^i, U) is a chart domain on M.

To $\hat{f} \in AT^p$ one assigns an anti-symmetric $\otimes_a^p \mathbb{R}^n$-valued Hamiltonian vector field

$$\widehat{X}_{\hat{f}} = X_{\hat{f}}^{i_1 i_2 \cdots i_{p-1}} r_{i_1} \otimes_a \cdots \otimes_a r_{i_{p-1}} := X_{\hat{f}}^I \otimes r_{[I]},$$

where the multi-index I is defined by $I = i_1 i_2 \cdots i_{p-1}$, and where $r_{[I]} := r_1 \otimes_a r_2 \otimes_a \cdots \otimes_a r_{p1}$. Here square brackets on indices denotes anti-symmetrization. We will refer to $\widehat{X}_{\hat{f}}$ with values in $\otimes_a^{p-1} \mathbb{R}^n$ as a rank p Hamiltonian vector field. The $N_A(p) = \binom{n}{p-1}$ component vectors fields $X_{\hat{f}}^{[i_1 i_2 \cdots i_{p-1}]}$ are determined by the generalized symplectic structure equation

$$(4.1) \qquad d\hat{f}^{i_1 \cdots i_p} = -p! X_{\hat{f}}^{[i_1 \cdots i_{p-1}} \lrcorner \beta^{i_p]}.$$

The map $\hat{f} \to \widehat{X}_f$ is also not unique because of the antisymmetrization in (4.1). The nonuniqueness is also contained completely in a vertical component $T_j^{i_1 i_2 \cdots i_{p-1} k} \frac{\partial}{\partial \pi_j^k}$ that is arbitrary except for the condition

$$(4.2) \qquad T_j^{[i_1 i_2 \cdots i_{p-1} k]} = 0.$$

As in §3 we can therefore determine $\widehat{X}_{\hat{f}}$ uniquely, given $\hat{f}$, by equation (4.1) and the locally defined auxillary condition

$$(4.3) \qquad d\pi_j^k (X_{\hat{f}}^{i_1 i_2 \cdots i_{p-1}}) = d\pi_j^{[k} (X_{\hat{f}}^{i_1 i_2 \cdots i_{p-1}]}).$$

The explicit local formula for $X_{\hat{f}}$ determined by $\hat{f} \in AT^p$ from (4.1) and (4.3) is

$$(4.4) \qquad \begin{aligned} X_{\hat{g}}^{i_1 \cdots i_{p-1}} &= \frac{1}{(p-1)!} (g^{j_1 \cdots j_{p-1} k} \circ \pi) \pi_{j_1}^{i_1} \cdots \pi_{j_{p-1}}^{i_{p-1}} \frac{\partial}{\partial x^k} \\ &\quad - \frac{1}{p!} \frac{\partial}{\partial x^l} (g^{j_1 \cdots j_p} \circ \pi) \pi_{j_1}^{i_1} \cdots \pi_{j_{p-1}}^{i_{p-1}} \pi_{j_p}^k \frac{\partial}{\partial \pi_l^k}. \end{aligned}$$

Now let $LAHV = \sum_{p=1}^{\infty} LAHV^p$, where $LAHV^p$ denotes the vector space of locally defined $\otimes_a^{p-1} \mathbb{R}^n$-valued vector fields determined uniquely by elements of AT^p from equations (4.1) and (4.3). Using these Hamiltonian vector fields we define a map $\{\ ,\ \}: AT^p \times AT^q \to AT^{p+q-1}$ by

$$(4.5) \qquad \{\hat{f}, \hat{g}\}^{i_1 \cdots i_{p+q-1}} = p! X_{\hat{f}}^{[i_1 \cdots i_{p-1}} (\hat{g}^{i_p \cdots i_{p+q-1}]}).$$

Working out the right-hand side of this equation using (4.4) we obtain
(4.6)

$$\{\hat{f}, \hat{g}\}^{i_i \cdots i_{p+q-1}} = \begin{pmatrix} p f^{k[j_2 \cdots j_p} \partial_k g^{la_2 \cdots a_q]} \\ -q g^{k[a_2 \cdots a_q} \partial_k f^{lj_2 \cdots j_p]} \end{pmatrix} \pi_l^{[i_1} \pi_{j_2}^{i_2} \cdots \pi_{j_p}^{i_p} \pi_{a_2}^{i_{p+1}} \cdots \pi_{a_q}^{i_{p+q-1}]}.$$

The right-hand side of this equation is the $\bigotimes_a^{p+q-1} \mathbb{R}^n$-valued function on
LM corresponding to the differential concomitant, due to Schouten and Nijenhuis [14, 15], of the antisymmetric tensor fields $\vec{f}$ and $\vec{g}$ on M. It is
easy to show that ∂_k on the right-hand side of (4.6) can be replaced by ∇_k
for an arbitrary symmetric linear connection, and hence the right-hand side
of (4.6) is independent of coordinates. Thus for $\hat{f} \in AT^p$ and $\hat{g} \in AT^q$ the
bracket $\{\hat{f}, \hat{g}\}$ is an element of AT^{p+q-1}.

We also define a bracket $[\widehat{X}_{\hat{f}}, \widehat{X}_{\hat{g}}]$ for Hamiltonian vector fields $\widehat{X}_{\hat{f}}$ by

(4.7) $$[\widehat{X}_{\hat{f}}, \widehat{X}_{\hat{g}}] := [X_{\hat{f}}^I, X_{\hat{g}}^J] \otimes r_{[IJ]}.$$

As we will see the brackets in (4.5) and (4.7) are not Lie brackets, but rather
are brackets appropriate for super-algebras.

The following theorems establish basic facts about the spaces AT and
$LAHV$. The proofs are omitted since most are simply modifications of the
corresponding proofs given in §3.

THEOREM 4.1. *For all $\hat{f} \in AT^p$, $\hat{g} \in AT^q$, and $\hat{h} \in AT^r$, the bracket
defined in (4.5) has the following properties*:

(4.8)
$$\begin{aligned}
&\text{(a)} \quad \{\hat{f}, \hat{g}\} = -(-1)^{(p-1)(q-1)}\{\hat{g}, \hat{f}\}, \\
&\text{(b)} \quad 0 = (-1)^{(p-1)(r-1)}\{\hat{f}, \{\hat{g}, \hat{h}\}\} + (-1)^{(p-1)(q-1)} \\
&\qquad\qquad \times \{\hat{g}, \{\hat{h}, \hat{f}\}\} + (-1)^{(q-1)(r-1)}\{\hat{h}, \{\hat{f}, \hat{g}\}\}, \\
&\text{(c)} \quad \{\hat{f}, \hat{g} \wedge \hat{h}\} = \{\hat{f}, \hat{g}\} \wedge \hat{h} + (-1)^{(p-1)q} \hat{f} \wedge \{\hat{g}, \hat{h}\}.
\end{aligned}$$

An algebra with properties (a) and (b) above has been given the name
super-algebra by physicists [17, 18]. When the derivation property (c) is
included the resulting algebra has been given the name Schouten algebra [11].
In keeping with the general philosophy of generalizing standard symplectic
geometry we will refer to the algebra AT as a Poisson super algebra.

THEOREM 4.2. *Let $\widehat{X}_{\hat{f}}$, $\widehat{X}_{\hat{g}}$ and $\widehat{X}_{\hat{h}}$ be elements of LHV of rank p, q
and r, respectively. Then under the bracket defined in (4.5) above*
(a) $[\widehat{X}_{\hat{f}}, \widehat{X}_{\hat{g}}] = \frac{(p+q-1)!}{p!q!} \widehat{X}_{\{\hat{f}, \hat{g}\}}$,
(b) $[\widehat{X}_{\hat{f}}, \widehat{X}_{\hat{g}}] = -(1)^{(p-1)(q-1)}[\widehat{X}_{\hat{g}}, \widehat{X}_{\hat{f}}]$,
(c)

$$\begin{aligned}
0 = &(-1)^{(p-1)(r-1)}[\widehat{X}_{\hat{f}}, [\widehat{X}_{\hat{g}}, \widehat{X}_{\hat{h}}]] + (-1)^{(p-1)(q-1)}[\widehat{X}_{\hat{g}}, [\widehat{X}_{\hat{h}}, \widehat{X}_{\hat{f}}]] \\
&+ (-1)^{(q-1)(r-1)}[\widehat{X}_{\hat{h}}, [\widehat{X}_{\hat{f}}, \widehat{X}_{\hat{g}}]].
\end{aligned}$$

Thus each local algebra $(LHV, [\ ,\ ])$, *with bracket for* $\bigotimes_a^{p-1} \mathbb{R}^n$-*valued vector fields defined as in* (4.5) *above, is a local super algebra.*

5. Geometrical applications

In this section we consider two applications of the algebras introduced in §§3 and 4. As a first example we show how to define the natural lift of a symmetric or antisymmetric contravariant tensor field on M to LM. As discussed in §2 the Hamiltonian vector field $X_{\hat{f}}$ determined by $\hat{f} \in ST^1$ is the natural lift to LM, as defined for example in [16], of the corresponding vector field $\vec{f}$ on M. What has been shown in §2 is that the definition of natural lift of a vector field to LM, in terms of the three properties given in equation (2.20), can be replaced by the

DEFINITION. The natural lift of a vector field $\vec{f}$ on M to LM is the Hamiltonian vector field $X_{\hat{f}}$ determined by the equation

$$d\hat{f} = -X_{\hat{f}} \rfloor \beta$$

where $\hat{f}$ is the element of ST^1 determined by $\vec{f}$.

With this fact in hand it is natural to use the symplectic structure to extend the definition of natural lift to arbitrary symmetric contravariant tensor fields. We illustrate the extension for a rank 2 tensor field. Thus let $\hat{g} \in ST^2$ corresponding to the rank 2 tensor field $\vec{g}$ on M. Then the Hamiltonian vector fields $X_{\hat{g}}^i$ determined locally by $\hat{g}$ are (from equation (3.8)),

$$(5.1) \qquad X_{\hat{g}}^i = (g^{ab} \circ \pi) \pi_a^i \frac{\partial}{\partial x^b} - \frac{1}{2} \left\{ \frac{\partial}{\partial x^j} (g^{ab} \circ \pi) \pi_a^i \pi_b^k \frac{\partial}{\partial \pi_j^k} \right\}.$$

It is a simple matter to check that the vector fields $X_{\hat{g}}^i$ have, instead of property (1) listed in equation (2.20), the transformation property

$$dR_a(X_{\hat{g}}^i) = (a^{-1})_j^i X_{\hat{g}}^j \quad \forall a \in GL(n).$$

Recalling definition (2.2) we define a map

$$u \to d_u\pi \otimes u: T_u(LM) \otimes \mathbb{R}^n \to T_{\pi(u)}M \otimes T_{\pi(u)}M$$

by

$$(5.2) \qquad (d_u\pi \otimes u)(X \otimes \eta) \stackrel{\text{def}}{=} d_u\pi(X) \otimes u(\eta) \quad \text{for } X \in T_uLM, \ \eta \in \mathbb{R}^n.$$

It is then easy to check that if $\widehat{X}_{\hat{g}} = X_{\hat{g}}^i \otimes r_i$, with the $X_{\hat{g}}^i$ as given above, then

$$(5.3) \qquad d_u\pi \otimes u(\widehat{X}_{\hat{g}}) = \vec{g}(\pi(u)).$$

Two remarks are in order. First it is clear that the domain $T_uLM \otimes \mathbb{R}^n$ in the above may be replaced by $T_uLM \otimes (\bigotimes^p \mathbb{R}^n)$ by using (2.2) on each factor of $\bigotimes^p \mathbb{R}^n$. Second, we note that although the vector-valued vector fields $\widehat{X}_{\hat{g}}$

are not uniquely determined for $p > 1$, the nonunique components are, by (3.9), vertical and hence do not contribute to the right-hand side of (5.2). It is therefore natural to make the following

DEFINITION. The natural lift of a symmetric contravariant rank p tensor field $\vec{f}$ on M to LM is the symmetric $\bigotimes_s^{p-1} \mathbb{R}^n$-valued vector field

$$\widehat{X}_{\hat{f}} = X_{\hat{f}}^{i_1 \cdots i_{p-1}} r_{i_1} \otimes_s \cdots \otimes_s r_{i_{p-1}}$$

where the component vector fields $X_{\hat{f}}^{i_1 \cdots i_{p-1}}$ are determined locally by equations (3.6) and (3.11).

It is clear that this definition may be modified to give a definition of the natural lift of an antisymmetric contravariant tensor field $\vec{f}$ on M to LM as the corresponding $\bigotimes_a^{p-1} \mathbb{R}^n$-valued vector field $\widehat{X}_{\hat{f}}$.

As a second geometrical application we consider the definition of Killing tensors. Suppose that $\vec{g}$ is a contravariant metric tensor field on a manifold M, and denote by $\hat{g}$ the corresponding element of ST^2. Then we may define pth *order Killing tensors* as elements $\widehat{K} \in ST^p$ that *commute with $\hat{g}$ under the Poisson bracket*. The analogy from mechanics is that if we consider $\hat{g}$ as a generalized Hamiltonian (see §7), then the equation $\{\widehat{K}, \hat{g}\} = 0$ is *the constants of the motion* equation.

For $p = 1$ we have the well-known definition of Killing vector fields since, as argued in §2, for $\widehat{K} \in ST^1$ and $\hat{g} \in ST^2$ the Poisson bracket $\{\widehat{K}, \hat{g}\}$ corresponds to the Lie derivative of $\vec{g}$ with respect to $\vec{K}$.

For $p = 2$ the vanishing of the bracket $\{\widehat{K}, \hat{g}\}$ can easily be shown to reduce, using (3.17), to the equation

$$(5.4) \qquad\qquad \nabla^{(i}(K^{jk)}) = 0,$$

where covariant differentiation is with respect to the Levi-Civita connection defined by $\vec{g}$. This definition (5.4) of rank 2 Killing tensors has been used, for example, by Sommers [19].

6. Local spaces of allowable observables on LM

As remarked in §2, not all $\bigotimes_s^p \mathbb{R}^n$-valued functions on LM are compatible with the equation

$$d\hat{g}^{i_1 \cdots i_p} = -p! X_{\hat{g}}^{(i_1 \cdots i_{p-1}} \lrcorner \beta^{i_p)}.$$

In order to find the general form of allowable observables on LM we proceed as follows. Generalizing a definition from standard symplectic geometry we have

DEFINITION. A $\bigotimes_s^{p-1} \mathbb{R}^n$-valued vector field $X^I \otimes r_I$ on LM is *locally Hamiltonian* if

$$(6.1) \qquad\qquad d(X^{(i_1 \cdots i_{p-1}} \lrcorner \beta^{i_p)}) = 0.$$

The corresponding condition $d(X \lrcorner \omega) = 0$ on T^*M implies that locally on $U \subset T^*M$ there is a function $f: U \to \mathbb{R}$ such that $X = X_f$, but places no further restrictions on f. On the other hand equation (6.1) asserts that (A) there exists on $\widehat{U} \subset LM$ a function $\hat{g} : \widehat{U} \to \bigotimes_s^p \mathbb{R}^n$ such that $X = X_{\hat{g}}$, and (B) that $\hat{g}$ must be a polynomial of degree p in the momentum coordinates π_i^i with coefficients in the set of real-valued functions on $\pi(\widehat{U}) \subset M$.

To see this consider (6.1) for the case $p = 2$. With the vector fields X^i expressed in local coordinates as

$$(6.2) \qquad X^i = X^{ij} \frac{\partial}{\partial x^j} + X_k^{ij} \frac{\partial}{\partial \pi_k^j}$$

equation (6.1) splits up into the three sets of equations

$$(6.3) \qquad \frac{\partial X_b^{(ij)}}{\partial x^a} - \frac{\partial X_a^{(ij)}}{\partial x^b} = 0,$$

$$(6.4) \qquad \delta_k^{(i} \frac{\partial X^{j)b}}{\partial \pi_s^r} - \delta_r^{(i} \frac{\partial X^{j)s}}{\partial \pi_b^k} = 0,$$

and

$$(6.5) \qquad \frac{\partial X_a^{(ij)}}{\partial \pi_s^r} + \delta_r^{(i} \frac{\partial X^{j)s}}{\partial x^a} = 0.$$

Working first with the set of equations (6.4) one shows by a series of contractions and resubstitutions that

$$(6.6) \qquad \frac{\partial X^{ib}}{\partial \pi_s^r} = \frac{1}{n} \frac{\partial X^{jb}}{\partial \pi_s^j} \delta_r^i.$$

Computing a second derivative of this last equation with respect to π_b^a and resubstituting (6.6) on the right-hand side after permuting the order of differentiation, one can show by contraction of the resulting equation that

$$(6.7) \qquad \frac{\partial^2 X^{ij}}{\partial \pi_b^a \pi_s^r} = 0.$$

Hence the components X^{ij} are linear in the momentum coordinates π_j^i. We conclude that

$$(6.8) \qquad X^{ij} = A^{jk}(x)\pi_k^i + B^{ij}(x), \qquad A^{ij} = A^{ji}.$$

The symmetry of the coefficients A^{ij} follows from (6.4).

Now using (6.8) in the right-hand side of (6.5) one can show that

$$(6.9) \qquad X_a^{(ij)} = \frac{\partial}{\partial x^a}\left(\frac{-1}{2}(A^{kl}(x)\pi_k^{(i}\pi_l^{j)} + 2\pi_k^{(i}B_k^{j)}(x)) \right) + C_a^{ij}(x).$$

Finally, using (6.3) and (6.9) one concludes that

$$(6.10) \qquad C_a^{ij}(x) = \frac{\partial C^{ij}}{\partial x^a}, \qquad C^{ij} = C^{ji}.$$

Hence the vector fields X^i must be of the form

(6.11)
$$X^i = (A^{jk}(x)\pi_k^i + B^{ij}(x))\frac{\partial}{\partial x^j}$$
$$+ \left(\frac{\partial}{\partial x^a}\left(\frac{-1}{2}A^{kl}(x)\pi_k^{(i}\pi_l^{j)} + 2\pi_k^{(i}B^{j)k}(x) + 2C^{ij)}(x)\right)\right)\frac{\partial}{\partial \pi_a^j}.$$

If we now write $d\hat{g}^{ij} = -2X^{(i} \rfloor \beta^{j)}$ with $\hat{g}^{ij} = \hat{g}^{ji}$, then we find

(6.12)
$$\hat{g}^{ij} = A^{kl}(x)\pi_k^i\pi_l^j + \pi_k^{(i}B^{j)k}(x) + C^{ij}(x).$$

The following theorem is the general result that one can prove using methods patterned after those in the above discussion.

THEOREM 6.1. *If* $\hat{X} = X^I \otimes r_I$ *is a* $\bigotimes_s^{p-1}\mathbb{R}^n$-*valued vector field on* LM *satisfying*

(6.13)
$$d(X^{(i_1 i_2 \cdots i_{p-1}} \rfloor \beta^{i_p)}) = 0,$$

then locally there exist $\bigotimes_s^p \mathbb{R}^n$-*valued functions* $\hat{g}$ *such that* $\hat{X} = \hat{X}_{\hat{g}}$ *and*

(6.14)
$$\hat{g}^{i_1 i_2 \cdots i_p} = \frac{1}{p}\pi_{l_1}^{(i_1}\cdots\pi_{l_p}^{i_p)}A^{l_1\cdots l_p}(x)$$
$$+ \frac{1}{p-1}\pi_{L_{p-2}}^{J_{p-2}}\pi_{l_p}^{(i_p}B_{1,J_{p-2}}^{i_1\cdots i_{p-1})l_p L_{p-2}}(x)$$
$$+ \frac{1}{p-2}\pi_{L_{p-3}}^{J_{p-3}}\pi_{l_p}^{(i_p}B_{2,J_{p-3}}^{i_1\cdots i_{p-1})l_p L_{p-3}}(x) + \cdots$$
$$+ \cdots + \pi_{l_p}^{(i_p}B_{p-1}^{i_1\cdots i_{p-1})l_p}(x) + B_p^{(i_1\cdots i_p)}(x)$$

where $L_{p-k} = l_1\cdots l_{p-k}$, $J_{p-k} = j_1\cdots j_{p-k}$ *and where* $\pi_{L_{p-k}}^{J_{p-k}} = \pi_{l_1}^{j_1}\pi_{l_2}^{j_2}\cdots\pi_{l_{p-k}}^{j_{p-k}}$.

The analogous result that one can prove for antisymmetric $\bigotimes_a^{p-1}\mathbb{R}^n$-valued locally Hamiltonian vector fields is

THEOREM 6.2. *If* $\hat{X} = X^I \otimes r_I$ *is a* $\bigotimes_a^{p-1}\mathbb{R}^n$-*valued vector field on* LM *satisfying*

(6.15)
$$d(X^{[i_1 i_2 \cdots i_{p-1}} \rfloor \beta^{i_p]}) = 0,$$

then locally there exist $\bigotimes_a^p \mathbb{R}^n$-*valued functions* $\hat{g}$ *such that* $\hat{X} = \hat{X}_{\hat{g}}$ *and*

(6.16)
$$\hat{g}^{i_1 i_2 \cdots i_p} = \frac{1}{p}\pi_{l_1}^{[i_1}\cdots\pi_{l_p}^{i_p]}A^{l_1\cdots l_p}(x) + \frac{1}{p-1}\pi_{L_{p-2}}^{J_{p-2}}\pi_{l_p}^{[i_p}B_{1,J_{p-2}}^{i_1\cdots i_{p-1}]l_p L_{p-2}}(x)$$
$$+ \frac{1}{p-2}\pi_{L_{p-3}}^{J_{p-3}}\pi_{l_p}^{[i_p}B_{2,J_{p-3}}^{i_1\cdots i_{p-1}]l_p L_{p-3}}(x) + \cdots$$
$$+ \cdots + \pi_{l_p}^{[i_p}B_{p-1}^{i_1\cdots i_{p-1}]l_p}(x) + B_p^{[i_1\cdots i_p]}(x)$$

where $L_{p-k} = l_1\cdots l_{p-k}$, $J_{p-k} = j_1\cdots j_{p-k}$ *and where* $\pi_{L_{p-k}}^{J_{p-k}} = \pi_{l_1}^{j_1}\pi_{l_2}^{j_2}\cdots\pi_{l_{p-k}}^{j_{p-k}}$.

7. The metric tensor as a generalized Hamiltonian tensor
for free inertial observers

A metric tensor field $\vec{g}$ on a spacetime M defines a real-valued function $\tilde{g}$ in canonical coordinates (x^i, π_j) on T^*M by $\tilde{g}(x, \pi) = g^{ij}(x)\pi_i\pi_j$. The free-particle Hamiltonian on T^*M for this spacetime is then $\mathscr{H} = \frac{1}{2}\tilde{g}$, and the solutions of the associated Hamilton equations on T^*M are the linear geodesics of the unique Levi-Civita connection Γ_g defined by $\vec{g}$. One may then build up parallel transport of linear frames in terms of Γ_g, but geometrical ideas of this type are only indirect consequences of the Hamiltonian dynamics of $\tilde{g}$ on T^*M. We now show that the generalized Hamiltonian dynamics of $\hat{g}$ on LM gives the full Levi-Civita connection geometry directly and explicitly.

Let $\vec{g} = g^{ij}\partial_i \otimes \partial_j$ be the local coordinate form of the metric tensor on spacetime, and let $\hat{g} = (g^{ij} \circ \pi)\pi_i^a\pi_j^b r_a \otimes r_b$ denote the corresponding tensorial function in ST^2 on LM. Then from equation (3.8) the associated Hamiltonian vector fields $X_{\hat{g}}^i$ determined by equation (3.5) have the local expressions

$$(7.1) \qquad X_{\hat{g}}^i = (g^{ab} \circ \pi)\pi_a^i \frac{\partial}{\partial x^b} - \frac{1}{2}\left\{ \frac{\partial(g^{ab} \circ \pi)}{\partial x^j}\pi_a^i\pi_b^k + T_j^{ik} \right\} \frac{\partial}{\partial \pi_j^k}.$$

It is not difficult to show that if for the arbitrary functions T_j^{ik} we take smooth functions that transform under right translations on LM according to the law

$$(7.2) \qquad T_j^{ik}(u \cdot h) = (h^{-1})_m^i T_j^{mk}(u), \qquad \forall h \in GL(n),$$

then the distribution Δ on LM spanned by the vector fields

$$(7.3) \qquad\qquad\qquad B_i \stackrel{\text{def}}{=} \hat{g}_{ij}X_{\hat{g}}^j$$

defines a linear connection on LM. Here $\hat{g}_{ij} = (g_{ab} \circ \pi)(\pi_i^a)^{-1}(\pi_j^b)^{-1}$ where the functions g_{ij} are the components of the matrix inverse of (g^{ij}). Roughly, from equations (7.1), (7.3) and the nonsingularity of $\vec{g}$, one checks that the vector fields B_i never vanish on LM and form a complement to the vertical subspace of T_uLM at each $u \in LM$. The condition given in equation (7.2) is then sufficient to guarantee that the smooth distribution Δ spanned by the vector fields B_i is invariant by right translation. These properties taken together show that Δ satisfies the distributional definition [16] of a linear connection on LM.

From the method used to derive equation (3.5) ("Find a linear connection that leaves $\vec{g}$ covariant constant and...") it is clear that the set of all connections defined in this way contains the set of "metric linear connections" defined by $\vec{g}$. Contained in this set is the Levi-Civita connection, i.e. the unique torsion-free metric linear connection defined by $\vec{g}$. This unique connection can be defined in terms of the generalized symplectic structure as follows. Require the generalized Hamiltonian vector fields to satisfy, in addition to

$$(7.4) \qquad\qquad (a) \quad d\hat{g}^{ij} = -2X_{\hat{g}}^{(i} \, \lrcorner \, \beta^{j)},$$

the invariantly defined constraint equations

$$(7.5) \qquad\qquad (b) \quad \beta^i \, \lrcorner \, X_{\hat{g}}^j \, \lrcorner \, X_{\hat{g}}^k = 0 \quad \forall i, j, k = 1, \dots, 4.$$

These last equations are in fact just the "torsion free" condition in a different form. Equations (7.4) and (7.5) uniquely determine the arbitrary functions T_k^{ij} so that the resulting Hamiltonian vector fields are

$$(7.6) \qquad X_{\hat{g}}^i = (g^{ab} \circ \pi)\pi_a^i \frac{\partial}{\partial x^b} + (\Gamma_{jc}^b \circ \pi)g^{ac}\pi_a^i\pi_b^k \frac{\partial}{\partial \pi_j^k}.$$

The functions Γ_{jc}^b are the Christoffel symbols of the Levi-Civita connection defined by $\vec{g}$. In the following it will be convenient to drop the composition "$\circ \pi$" whenever there is no possibility of confusion.

It is straightforward to check that the distribution spanned by the vector fields

$$(7.7) \qquad B_k = \hat{g}_{ki}X_g^i = (\pi_k^j)^{-1}\left(\frac{\partial}{\partial x^j} + \Gamma_{ij}^a\pi_a^b \frac{\partial}{\partial \pi_i^b}\right)$$

is the horizontal distribution of the Levi-Civita connection. The vector fields B_i are easily seen to be the "standard horizontal vector fields" [16] determined by the connection.

For simplicity we work with the Hamiltonian vector fields $X_{\hat{g}}^i$ defined in equation (7.6). The first question we ask is: *What dynamics is determined by the four Hamiltonian vector fields $X_{\hat{g}}^i$?* When there is only a single Hamiltonian vector field $X_{\hat{f}}$, as in the case for $\hat{f} \in T^1$ as well as in standard symplectic geometry on T^*M, then the dynamics is given by the integral curves of $X_{\hat{f}}$. One can ask if the distribution spanned by the $X_{\hat{g}}^i$ is integrable, but it is well known that only flat connections have integrable distributions. On the other hand the vector fields B_k, and hence also the vector fields $X_{\hat{g}}^i$, are tangent to the subbundle of orthonormal linear frames $O_{\hat{g}}(M)$ determined by $\hat{g}$. Thus we may define an "integral" of the set of Hamiltonian vector fields $X_{\hat{g}}^i$ to be $O_{\hat{g}}(M)$. The subbundle $O_{\hat{g}}(M)$ is thus the analogue of the "constant energy surfaces" in standard symplectic geometry.

Since a section of $O_{\hat{g}}(M)$ represents a local orthonormal linear frame field on M we conclude that *the dynamics defined by the four Hamiltonian vector fields is the dynamics of orthonormal frames, and hence the dynamics of local observers on spacetimes.* More explicitly, consider the integral curves of the Hamiltonian vector field $X_{\hat{g}}^1$ with "time-like" initial conditions, i.e.,

$u = (p, e_i) \in LM$ with e_1 a time-like vector in $T_p M$. The differential equations for the integral curve of $X_{\hat{g}}^1$ through u are, from equation (7.6),

(7.8)
$$\text{(a)} \quad \frac{dx^i}{dt} = g^{ij} \pi_j^1 ,$$

$$\text{(b)} \quad \frac{d\pi_j^k}{dt} = \Gamma_{jc}^b g^{ac} \pi_a^1 \pi_b^k .$$

These two equations decouple into two sets of equations. For $k = 1$ we obtain

(7.9)
$$\text{(a}') \quad \frac{dx^i}{dt} = g^{ij} \pi_j^1 ,$$

$$\text{(b}') \quad \frac{d\pi_j^1}{dt} = \Gamma_{jc}^b g^{ac} \pi_a^1 \pi_b^1 ,$$

and for $k = \alpha = 2, 3, 4$,

(7.10)
$$\text{(b}'') \quad \frac{d\pi_j^\alpha}{dt} = \Gamma_{jc}^b g^{ac} \pi_a^1 \pi_b^\alpha .$$

The pair of equations (7.9-(a$'$)) and (7.9-(b$'$)) combine into the second order *geodesic equation*

(7.11)
$$\frac{d^2 x^i}{dt^2} + \Gamma_{jk}^i \frac{dx^j}{dt} \frac{dx^k}{dt} = 0 ,$$

while the equation (7.10-(b$''$)) can be rewritten as

$$\frac{D\pi_j^\alpha}{Dt} = \frac{d\pi_j^\alpha}{dt} - \Gamma_{ij}^k \frac{dx^i}{dt} \pi_k^\alpha = 0 , \qquad \alpha = 2, 3, 4 .$$

These last equations are just the equations for parallel transport of the two, three, and four legs of a coframe along the geodesic determined by equation (7.11). The result is that $X_{\hat{g}}^1$ *generates parallel transport of linear frames and coframes along time-like geodesic of* Γ_g. If we repeat this discussion for, say $X_{\hat{g}}^2$, then again we obtain parallel transport of linear frames along geodesics, but these geodesics will generally be spacelike.

The four Hamiltonian vector fields $X_{\hat{g}}^i$ associated with the spacetime metric tensor can therefore be used to construct the local Lorentzian coordinate systems carried by a freely-falling observer. Let $p_0 \in M$ and let (e_i) be an orthonormal frame at p_0. By integrating $X_{\hat{g}}^1$ with initial condition $(p_0, e_i) \in LM$ we obtain the time-like geodesic $\gamma(s)$ through p_0 determined by e_1, and a parallelly propagated orthonormal spatial triad $(e_\alpha(s))$, $\alpha = 2, 3, 4$, determined by e_2, e_3 and e_4. At each point along $\gamma(s)$ we can then fill in the spatial coordinate axes locally by integrating the vector fields $X_{\hat{g}}^\alpha$ with initial conditions $(\gamma(s), e_i(s))$. Each of these integrations will produce a spatial geodesic through $\gamma(s)$ and the parallel propogation of a triad of vectors along that geodesic. A local coordinate system determined in this way is referred to as *the local Lorentzian coordinate system carried by a freely-falling observer* [**20, 21**]. Because it takes all four Hamiltonian vector fields $X_{\hat{g}}^i$ to determine a coordinate system in this way, it seems appropriate to refer to $\hat{g}$ as the *generalized Hamiltonian tensor for free inertial observers*.

The existence of such local coordinate systems on spacetime is one aspect of Einstein's original correspondence principle. What we have shown above is that the dynamics of such local inertial observers does not have to be postulated, but rather is derivable from generalized Hamilton equations on LM.

8. The Dirac equation

We recast the fundamentals of the Kostant-Souriau theory of geometric quantization [**6, 7**], taking now for the symplectic manifold the bundle of linear frames LM of spacetime M with the generalized symplectic form $\beta = d\theta$. We restrict attention to the essentials of the initial *prequantization* procedure without concern here for the details of a more complete development of a full theory on LM.

In the naive prequantization program one assigns to each observable $f: T^*M \to \mathbb{R}$ a Hermitian operator

(8.1) $$f \to \mathscr{P}_f = i\hbar X_f.$$

The operator $\mathscr{P}_f$ operates as a linear operator on the set of square integrable functions $\psi: T^*M \to \mathbb{C}$, square integrability being defined with respect to the natural Liouville volume element on T^*M. Although the assignment (8.1) is not the completely correct assignment in the full geometric quantization theory (see [**6, 7**]) it will suffice for our purposes here.

We consider a spacetime $(M, \vec{g})$ which admits a spin structure [**22**], and ask for the prequantization operator assignments that one can make for the metric tensor Hamiltonian observable $\hat{g}$ on LM. The natural analogue of (8.1) is

(8.2) $$\hat{g} \to \mathscr{P}_{\hat{g}} = i\hbar \widehat{X}_{\hat{g}} = i\hbar X_{\hat{g}}^i \hat{r}_i,$$

with the $X_{\hat{g}}^i$ given in (7.6). We consider three *representations* of this operator $\mathscr{P}_{\hat{g}}$. The first and simplest representation is as the *scalar operator*

(8.3) $$\mathscr{P}_{\hat{g}} \to \mathscr{P}_{\hat{g}}^2 \equiv -\hbar^2 \hat{g}_{ij} X_{\hat{g}}^i \circ X_{\hat{g}}^j$$

acting on functions that are invariant on fibers of LM. In this case the operator $\mathscr{P}_{\hat{g}}^2$ is proportional to the d'Alembertian operator

(8.4) $$\mathscr{P}_{\hat{g}}^2(\Psi) = (-\hbar^2)\nabla^2(\Psi).$$

The eigenvalue equation for this operator is then the Klein-Gordon equation

(8.5) $$(-\hbar^2)\nabla^2\Psi = \mu\Psi.$$

A second representation of the operator $\mathscr{P}_{\hat{g}}$, which might be called the *vector representation*, can be defined as follows. Let $\vec{t}$ denote a timelike vector field on spacetime M, and let $\hat{t}$ be the corresponding element in T^1 on LM. We consider the operator

$$(8.6) \qquad \mathscr{P}_{\hat{g}} \to \hat{g}_{ij}\hat{t}^i\mathscr{P}_{\hat{g}}^j = i\hbar\hat{g}_{ij}\hat{t}^i X_{\hat{g}}^j = i\hbar\hat{t}^i B_i,$$

where the last equality follows from (7.7). It is easy to show that this operator is $i\hbar$ times the *horizontal lift* of $\vec{t}$ to LM relative to the Levi-Civita connection defined by the metric tensor $\vec{g}$. Since $\vec{t}$ is time-like we can think of the operator defined in (8.6) as a relativitic analogue of the Schrödinger energy operator $i\hbar\frac{d}{dt}$.

Finally we consider the *spinor representation* of the operator $\mathscr{P}_{\hat{g}}$, which we define by

$$(8.7) \qquad \mathscr{P}_{\hat{g}} \to \gamma_i\mathscr{P}_{\hat{g}}^i = i\hbar\gamma_i X_{\hat{g}}^i = i\hbar\gamma^i B_i,$$

where the four γ_i is a set of appropriate Dirac matrices. In writing (8.7) we are assuming that the vector fields $X_{\hat{g}}^i$ have been lifted to the spin bundle $SP(M)$ over the orthonormal frame subbundle $O_{\hat{g}}(M)$ of LM; that is, the B_i in (8.7) are the standard horizontal vector fields defined by the Levi-Civita connection on $SP(M)$. It follows that (8.7) is the Dirac operator on $SP(M)$ [22].

Let $\Psi: SP(M) \to \mathbb{C}^4$ be a Dirac 4-spinor transforming under $SL(2, \mathbb{C})$ transformations on the spin bundle as

$$(8.8) \qquad \Psi(u \cdot a) = \rho(a^{-1}) \cdot \Psi(u), \quad \forall u \in SP(M), \ \forall a \in SL(2, \mathbb{C}),$$

where ρ denotes the 4-spinor representation of $SL(2, \mathbb{C})$. It is straightforward to show that

$$(8.9) \qquad \gamma_i\mathscr{P}_{\hat{g}}^i(\Psi)(u \cdot a) = \rho(a^{-1}) \cdot \gamma_i\mathscr{P}_{\hat{g}}^i(\Psi)(u).$$

Thus the eigenvalue equation

$$(8.10) \qquad \gamma_i\mathscr{P}_{\hat{g}}^i(\Psi) = \mu\Psi$$

for the prequantization operator $\gamma_i\mathscr{P}_{\hat{g}}^i$ is equivariant on $SP(M)$ and is in fact just the Dirac equation

$$(8.11) \qquad i\hbar\gamma^i B_i(\Psi) = \mu\Psi.$$

9. Conclusions

In this paper we have developed the fundamentals of the generalized symplectic geometry on the bundle of linear frames $L\mathrm{M}$ of an n-dimensional manifold M that follows upon taking the $\mathbb{R}^n$-valued soldering one-form θ on LM as a generalized symplectic potential. This study was motivated by the following two points:

I. An essential feature of quantum mechanics is the unavoidable interaction of observer and object, and in relativistic physics observers are modeled as points in the bundle of linear frames LM of spacetime M.

II. The Kostant-Souriau theory of geometric quantization takes symplectic geometry on phase space as a starting point for the theory.

It is reasonable to suppose that the fundamental quantum mechanical phenomenon of observer-object interaction ought to be able to be studied well on the *manifold of observers*, namely the frame bundle LM of spacetime M. To then carry through a generalization of the Kostant-Souriau theory one would need a symplectic geometry on LM, and we have developed the fundamentals of the geometry with this long range goal in mind. Indeed, as a preliminary application of the generalized symplectic geometry to quantum theory we showed in §8 that the Dirac equation arises in a natural way as an eigenvalue equation for a naive prequantization operator assigned to the spacetime metric tensor Hamiltonian.

The heart of standard symplectic geometry is the assignment $f \to X_f$ of a Hamiltonian vector field X_f to each real-valued observable f. This is done via the equation

$$(9.1) \qquad\qquad df = -X_f \, \lrcorner \, \omega,$$

where ω is the symplectic two-form. Once the assignments are made for each observable one may then proceed to compute Poisson brackets, integrate the equations of motion (find integral curves!), and in general do symplectic geometry. Everything flows from the basic equation (9.1) which may be considered as a *structure equation* of symplectic geometry. The development of generalized symplectic geometry on LM presented in this paper has been centered around generalizations of (9.1) to LM when ω is replaced by $\beta = d\theta = d\theta^i r_i$ with $\theta = \theta^i r_i$ the $\mathbb{R}^n$-valued *soldering one-form*. The fact that $d\theta$ is $\mathbb{R}^n$-valued necessitated generalizing from $\mathbb{R}$-valued observables to vector-valued observables on LM, and we found in §§2, 3 and 4 that the algebras of symmetric and antisymmetric contravariant tensor fields on the base manifold have natural interpretations in terms of symplectic geometry on LM.

For a rank p symmetric contravariant tensor field $\vec{f}$ on M we used the uniquely related $\bigotimes_s^p \mathbb{R}^n$-valued function $\hat{f} = \hat{f}^{i_1 \cdots i_p} r_{i_1} \otimes_s \cdots \otimes_s r_{i_p}$ on LM to assign a set of Hamiltonian vector fields $X_{\hat{f}}^{i_1 \cdots i_{p-1}}$ via the generalized structure equation

$$(9.2) \qquad\qquad d\hat{f}^{i_1 \cdots i_p} r_{i_1} = -p! X_{\hat{f}}^{(i_1 \cdots i_{p-1}} \, \lrcorner \, d\theta^{i_p)}.$$

For $p > 1$ the $\bigotimes_s^{p-1} \mathbb{R}^n$-valued Hamiltonian vector fields $X_{\hat{f}} = X_{\hat{f}}^I r_I$ where $I = i_1 \cdots i_{p-1}$, were determined uniquely only locally on LM using the condition (3.11) that depends explicitly on a choice of coordinates. However, the generalized bracket defined by

$$(9.3) \qquad \{\hat{f}, \hat{g}\}^{i_1 \cdots i_{p+q-1}} = p! X_{\hat{f}}^{(i_1 \cdots i_{p-1}}(\hat{g}^{i_p \cdots i_{p+q-1})})$$

for $\hat{f} \in ST^p$, $\hat{g} \in ST^q$, produces an element of ST^{p+q-1} which is independent of the choice of coordinates. The result proved in §3 is that the algebra $(ST, \otimes_s)$ where $ST = \sum_{p=1}^{\infty} ST^p$ is the vector space of all $\otimes_s^p \mathbb{R}^n$-valued tensorial functions on LM, becomes a *Poisson algebra* under the Poisson bracket defined in (9.3). In addition the set of all locally defined $\otimes_s^{p-1} \mathbb{R}^n$-valued vector fields $\widehat{X}_{\hat{f}}$ forms a *Lie algebra* under the Lie bracket defined by

$$(9.4) \qquad [\widehat{X}_{\hat{f}}, \widehat{X}_{\hat{g}}] = [X_{\hat{f}}^{i_1 \cdots i_{p-1}}, X_{\hat{g}}^{i_p \cdots i_{p+q-1}}] \otimes r_{i_1} \otimes_s r_{i_2} \otimes_s \cdots \otimes_s r_{i_{p+q-1}}.$$

In the case of antisymmetric contravariant tensor fields on M we found that the corresponding vector space $AT = \sum_{p=1}^{\infty} AT^p$ of $\otimes_a^p \mathbb{R}^n$-valued functions on LM becomes a *Poisson super algebra* under the bracket defined by

$$(9.5) \qquad \{\hat{f}, \hat{g}\}^{i_1 \cdots i_{p+q-1}} = p! X_{\hat{f}}^{[i_1 \cdots i_{p-1}}(\hat{g}^{i_p \cdots i_{p+q-1}]})$$

for $\hat{f} \in AT^p$ and $\hat{g} \in AT^q$. The component vector fields $X_{\hat{f}}^{i_1 \cdots i_{p-1}}$ of the $\otimes_a^{p-1} \mathbb{R}^n$-valued vector field $\widehat{X}_{\hat{f}} = X_{\hat{f}}^I r_I$, $I = i_1 \cdots i_{p-1}$, are determined locally by the equation

$$(9.6) \qquad d\hat{f}^{i_1 \cdots i_p} r_{i_1} = -p! X_{\hat{f}}^{[i_1 \cdots i_{p-1}} \rfloor d\theta^{i_p]},$$

and the set of all such vector fields forms a *super algebra* under the super bracket defined by

$$(9.7) \qquad [\widehat{X}_{\hat{f}}, \widehat{X}_{\hat{g}}] = [X_{\hat{f}}^{i_1 \cdots i_{p-1}}, X_{\hat{g}}^{i_p \cdots i_{p+q-1}}] \otimes r_{i_1} \otimes_a r_{i_2} \otimes_a \cdots \otimes_a r_{i_{p+q-1}}.$$

In both of these special cases the brackets, defined in (9.3) and (9.5), are related to differential invariants discovered by Schouten [14] and studied by Nijenhuis [15]. More specifically, expressing the right-hand side of (9.3) in coordinates on the base manifold one finds that the Poisson bracket $\{\hat{f}, \hat{g}\}$ corresponds to the *differential concomitant* of the corresponding symmetric contravariant tensor fields $\vec{f}$ and $\vec{g}$ on M due to Schouten and Nijenhuis. Similarly, the rank $r = p + q - 1$ anti-symmetric tensor field on M represented by (9.5) is the Schouten-Nijenhuis differential concomitant of the corresponding tensor fields $\vec{f}$ and $\vec{g}$ on M. The generalized symplectic geometry on LM thus provides a natural and unified treatment of the Schouten-Nijenhuis concomitants.

Bhaskara and Viswanath [11] have studied the *symmetric* and *alternating products* of contravariant tensor fields, introduced by Schouten, from an algebraic point of view. Bloore and Assimakopoulos [23], in a study of the Schouten product for symmetric contravariant tensor fields $S\mathscr{X}(M)$ on a manifold M, discovered a natural *one-cochain* that they then used to define the Schouten concomitant "...in exactly the same way as the Poisson bracket is defined using" the canonical one-form on T^*M. Bloore and Assimakopoulos remarked that they were forced into studying derivations because $S\mathscr{X}(M)$ "...is not an algebra of functions on a manifold and so we cannot set up homology chains involving vector fields." We point out that by working on LM we have replaced $S\mathscr{X}(M)$ with the algebra (under $\otimes_s$) of *functions ST* on LM, which we feel is the essential ingredient in setting up the Poisson algebra $(ST, \{ , \})$ on LM. The conclusion to be drawn is that the differential geometry of symmetric (antisymmetric) contravariant tensor fields on a manifold has a natural formulation on LM as a Poisson algebra (Poisson super algebra). The naturally defined brackets then give the Schouten differential concomitants when reinterpreted on the base manifold.

It should be pointed out that the Schouten concomitant for symmetric contravariant tensor fields has long been known [23] to be related to the standard Poisson bracket on T^*M. If $\vec{f} \in S\mathscr{X}^r(M)$, then $\vec{f}$ defines a homogeneous *polynomial observable* [23],

$$(9.8) \qquad \tilde{f} = f^{i_1 \cdots i_r}(x) p_{i_1} \cdots p_{i_r}$$

on T^*M in standard canonical coordinates (x^i, p_j). Let $\tilde{f}$ and $\tilde{g}$ be polynomial observables induced on T^*M as in (9.8) by rank r and s symmetric contravariant tensor fields $\vec{f}$ and $\vec{g}$, respectively, on M. Then the Poisson bracket $\{\tilde{f}, \tilde{g}\}$ defined with respect to the canonical symplectic two-form on T^*M gives the Schouten concomitant of $\vec{f}$ and $\vec{g}$ when re-expressed on M. Note, however, that there is no possibility of obtaining the Schouten concomitant for antisymmetric tensor fields on T^*M in this way since the right-hand side of (9.8) vanishes identically if $f^{i_1 \cdots i_r}(x)$ is antisymmetric. On the other hand the antisymmetric case was handled quite satisfactorily in terms of the generalized symplectic geometry on LM. Generalized symplectic geometry on the frame bundle of a manifold thus unifies and clarifies the many different approaches to the differential concomitants of Schouten.

The homogeneous polynomial observables mentioned above are special cases of the *polynomial observables* that one may define [23, 24] on T^*M. We found in §6 that the *locally defined allowable observables* on LM for $\otimes_s^p \mathbb{R}^n$-valued functions are polynomials in the generalized momentum coordinates π_j^i with coefficients in the set of functions on LM that are invariant on fibers. *The locally defined allowable observables* on LM for $\otimes_a^p \mathbb{R}^n$-valued functions are exterior products of the π_j^i with coefficients also in the set of functions that are invariant on fibers in LM. Knowledge of these locally defined allowable observables would be important, for example, in setting up canonical commutation relations for a generalized canonical quantization scheme on LM. Although we will not go into the details here, we point out that the natural *canonically conjugate* variables on $(LM, d\theta)$ that generalize the q^i and p_j on T^*M are the $\mathbb{R}^n$-valued coordinate functions

$$(9.9) \qquad \hat{x}^i = x^i r_i = x^i \delta_i^j r_j, \quad \text{(no sum on } i\text{)},$$

and the $\mathbb{R}^n$-valued conjugate momentum coordinates

$$(9.10) \qquad \hat{\pi}_i = \pi_i^j r_j.$$

Observe that the momentum coordinate $\hat{\pi}_i$, by (9.10) and (2.21), corresponds to the locally defined vector field $\frac{\partial}{\partial x^i}$ on M. On the other hand the canonical coordinate $\hat{x}^i$, by (9.9) and (2.21), *does not* correspond to a vector field on M, but rather corresponds to a locally defined allowable observable in LHF^1.

The generalized Poisson brackets, calculated using $X_{\hat{\pi}_i} = \frac{\partial}{\partial x^i}$ and $X_{\hat{x}^i} = -\frac{\partial}{\partial \pi_i^i}$, are

$$(9.11) \qquad \begin{aligned} &\{\hat{x}^i, \hat{x}^j\} = 0, \qquad \{\hat{\pi}_i, \hat{\pi}_j\} = 0, \\ &\{\hat{\pi}_j, \hat{x}^i\} = \delta_j^i r_i, \qquad \text{(no sum on } i\text{)}. \end{aligned}$$

These are the commutation relations that could serve as a starting point for a generalized canonical quantization scheme on LM.

Finally, we mention the obvious fact that the discussions in this paper have just scratched the surface on the complete theory of generalized symplectic geometry on the frame bundle of a manifold. It is hoped that future developments of the subject will validate the conjecture that a deeper understanding of the quantum theory can be obtained by studying the natural Hamiltonian geometry of the manifold of observers.

Acknowledgments

The author wishes to express his appreciation to Professor R. O. Fulp for many helpful discussions during the preparation of this manuscript.

References

1. S. Sternberg, *Lectures on differential geometry*, Prentice-Hall, Englewood Cliffs, NJ, 1964.
2. R. Hermann, *Differential geometry and the calculus of variations*, Academic Press, New York, 1968.
3. V. I. Arnol'd, *Mathematical methods of classical mechanics*, Springer-Verlag, New York, 1978.
4. R. Abraham and J. E. Marsden, *Foundations of mechanics*, Benjamin Cummings, Reading, MA, 1978.
5. V. Guillemin and S. Sternberg, *Geometric asymptotics*, A.M.S., Providence, RI, 1977.
6. B. Kostant, *Quantization and unitary representations*, Lecture Notes in Math., vol. 170, Springer-Verlag, New York, 1970.
7. J.-M. Souriau, *Structures des systèmes dynamiques*, Dunod, Paris, 1970.
8. W. M. Tulczyjew, *Poisson brackets and canonical manifolds*, Bull. Acad. Polon. Sci. Ser. Sci. Math. Astronom. Phys. XXII, no. 9, (1974), 931–935.
9. A. Lichnerowicz, *Les variétiés de Poisson et leurs algèbres de Lie associées*, J. Differential Geom. **12** (1977), 253–300.

10. R. Hermann, *Toda lattices; cosymplectic manifolds, Backlund transformations and kinks*, Interdisciplinary Math. vol. XV, Math. Sci. Press, Brookline, MA, 1977.

11. K. H. Bhaskara and K. Viswanath, *Poisson algebras and Poisson manifolds*, Pitman Research Notes in Mathematics, vol. 174, Longman Scientific and Technical, Essex, 1988.

12. V. Guillemin and S. Sternberg, *The moment map and collective motion*, Ann. Physics **127** (1980), 220–253.

13. ____, *Symplectic techniques in physics*, Cambridge Univ. Press, Cambridge, 1984.

14. J. A. Schouten, *Über differentialkomitanten zweier kontravarianter grossen*, Proc. Kon. Ned. Akad. Wet. Amsterdam **43** (1940), 449–452.

15. A. Nijenhuis, *Jacobi-type identities for bilinear differential concomitants of certain tensor fields*, Indag. Math. **17** (1955), 390–403.

16. S. Kobayashi and K. Nomizu, *Foundations of differential geometry*. vol. 1, Interscience, New York, 1963.

17. B. Zumino, *Supersymmetry*, Gauge Theories and Modern Field Theory, (R. Arnowitt and P. Nath, eds.) MIT Press, Cambridge, MA, 1975.

18. L. Corwin, Y. Ne'eman, and S. Sternberg, *Graded Lie algebras in mathematics and physics (Bose-Fermi symmetry)*, Rev. Modern Phys. **47** (1975), 573–603.

19. P. Sommers, *On Killing tensors and constants of motion*, J. Math. Phys. **14** (1973), 787–790.

20. C. W. Misner, K. S. Thorne, and J. A. Wheeler, *Gravitation*, Freeman, San Francisco, 1970.

21. J. L. Synge, *Relativity: the general theory*, North-Holland, Amsterdam, 1971.

22. D. Bleecker, *Gauge theory and variational principles*, Addison-Wesley, Reading, MA, 1981.

23. N. Woodhouse, *Geometric quantization*, Clarendon Press, Oxford, 1980.

24. B. Kostant, *Géométrie symplectique et physique mathématique* (J.-M. Souriau, ed.) CNRS, Paris, 1974.

NORTH CAROLINA STATE UNIVERSITY

Proceedings of Symposia in Pure Mathematics
Volume **54** (1993), Part 2

Complex Geometry and String Theory

D. H. PHONG

I. Introduction

General relativity and quantum mechanics are fundamental laws of physics, and yet they have so far resisted unification. String theories are very compelling in this regard, since they incorporate a quantum theory of gravity, and are expected to be finite [**42, 43, 100**]. There is still no proof of this basic property. In fact explicit rules for evaluating string scattering amplitudes— say, comparable to Feynman rules in quantum field theory—are not even available [**30**].

These issues raise extremely interesting mathematical questions, since string amplitudes are integrals of forms of maximal rank over the moduli space of Riemann surfaces. These forms are built from differential geometric notions such as determinants of Laplacians and Green's functions. It is however crucial for the construction of string theories that they be realized also as pairings of holomorphic and antiholomorphic sections of vector bundles over moduli space. Finiteness of string amplitudes is dictated by the boundary behavior of these forms, that is, when the underlying Riemann surfaces degenerate. Thus string theory unifies seemingly unrelated areas of differential geometry, algebraic geometry, and the theory of modular forms. Its least understood aspects presently have to do with supersymmetry, and we expect further investigation to bring even greater and more surprising rewards.

In this paper we provide a survey of recent progress in

- formulating rules for evaluating string amplitudes;

1991 *Mathematics Subject Classification.* Primary 81Txx, 81T30, 81T40, 81T60, 30Fxx, 30F60, 32G15, 32G20, 58D17, 58D27, 58A20, 32C11, 11Fxx, 11F11, 11F20, 11F72, 14H42.

Contribution to Proceedings of the Geometry Festival, Los Angeles, AMS Summer Session, July 1990 (by R. Greene and S. T. Yau, eds.).

This paper is in final form and no version of it will be submitted for publication elsewhere. Work supported in part by NSF Grant DMS-90-04062.

- developing the geometry of super Riemann surfaces, which provides the framework;
- establishing the equivalence between the Polyakov and Mandelstam formulations of string theory.

These topics are discussed in §§IV, V, VI below. In II and III we give an account of the well-understood case of the bosonic string, as well as of some underlying physical principles of string and conformal field theory. Conventions, and calculus with Grassmann variables are gathered in the Appendices.

This work is a long term collaboration with Eric D'Hoker, and also with Ken Aoki on the unitarity problem.

II. The bosonic string

A. Physical principles. In string theory, elementary particles arise as excitations of a more fundamental object, a *string*, which is a one-dimensional curve without thickness. The mass and spin of a particle correspond to the level of excitation. The length of the string is expected to be of the order of the Planck scale, which is $10^{-33} cm$. In this paper we shall restrict our attention to theories of closed strings.

Strings interact by joining and splitting (Figure 1). Such a process is described by the *world—sheet*, which is the surface spanned by the strings. Its boundaries correspond to the initial and final strings, and the handles to the creation and annihilation of virtual pairs. In a fixed Lorentz frame, the interaction is local, and thus causality is preserved. However, there is no interaction point independent of the choice of frame (Figure 2). From this point of view, the presence of interactions is indicated only by the world-sheet, while locally the motion is free. This is in sharp contrast with quantum field theories of point particles, where interactions are introduced through well-defined vertices. Vertices approaching one another can lead to uncontrollable divergences. This difference underlies the soft ultraviolet behavior of strings.

In path integral quantization, scattering amplitudes are given by sums over paths, weighted by the classical action. Paths of strings are surfaces. A key

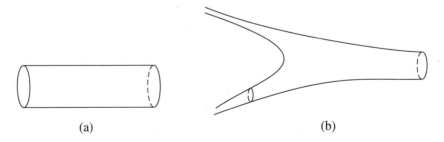

(a)   (b)

FIGURE 1. (a) Free propagation of a closed string;
(b) String interaction.

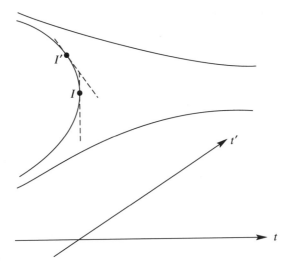

FIGURE 2. The interaction point I depends on the choice of Lorentz frame.

requirement of string theory is *conformal invariance*, which means that the contribution of each surface depends only on its conformal class, and not on its metric. Thus string scattering amplitudes are sums over all genera h of integrals of forms of maximal rank on the moduli space of Riemann surfaces of genus h.

B. The action and vertex operators. Although the theory of greatest physical interest is the superstring to be described in V–VIII, we begin by implementing the above ideas in the simpler case of the bosonic string propagating in Minkowski space-time. At fixed order h of perturbation theory, the world-sheet is a surface of genus h imbedded in Minkowski space-time. Fixing a surface M of genus h, the imbedding is given by d scalar functions $M \ni \xi \to x^\mu(\xi)$ on M. In the Polyakov formulation of string theory [77, 78], the geometry of the world-sheet requires an auxiliary metric g_{mn} on M, and the action is the harmonic map action

$$(2.1) \qquad I(g, x) = \frac{1}{8\pi} \int_M d^2\xi \sqrt{g} g^{mn} \partial_m x^\mu \partial_n x^\nu \eta_{\mu\nu}.$$

Here $\eta_{\mu\nu}$ is the metric of Minkowski space-time. The sum over histories of the quantum theory becomes the functional integral over g_{mn} and x^μ. For example the partition function, or vacuum-to-vacuum amplitude, is given by

$$(2.2) \qquad Z_{\text{BOS}} = \int Dg\, Dx\, e^{-I(g, x)}.$$

More general scattering amplitudes require the insertion in (2.2) of *vertex* *operators,* [*] i.e., expressions of the form

(2.3) $$V(k^\mu) = \int d^2\xi \sqrt{g} P(Dx) e^{ik^\mu x_\mu}.$$

Here $P(Dx)$ is a polynomial in the covariant derivatives of x^μ, and k^μ is a d-dimensional vector in Minkowski space with norm $-k^2 = -2, 0, 2, 4, \ldots$. The expression (2.3) should be viewed as a random variable on the probability space of the (g_{mn}, x^μ). Its expectation values are functions of the external parameters k^μ. Each particle in the string spectrum corresponds to such a vertex operator, k^μ to its momentum, and $m^2 = -k^2$ to the square of its mass. Of particular importance are the lowest mass levels

$$tachyons \qquad V_{-2}(k^\mu) = \int d^2\xi \sqrt{g} e^{ik^\mu x_\mu(\xi)} \qquad m^2 = -2,$$

(2.4)
$$graviton \quad V_0(k^\mu) = \varepsilon_{\nu\kappa} \int d^2\xi \sqrt{g} g^{mn} \partial_m x^\nu \partial_n x^\kappa e^{ik^\mu x_\mu(\xi)} \quad m^2 = 0.$$
$$multiplet$$

Often the term "vertex operator" will just refer to the integrand in the above expressions. For example the scattering amplitude of N tachyons with respective momenta $k_1^\mu, \ldots, k_N^\mu$ is given by the functional integral (Figure 3)

(2.5) $$A(k_1^\mu, \ldots, k_N^\mu) = \int Dg\, Dx\, e^{-I(g,x)} \prod_{i=1}^{N} \int d^2\xi_i \sqrt{g}(\xi_i) e^{ik_i^\mu x_\mu(\xi_i)}.$$

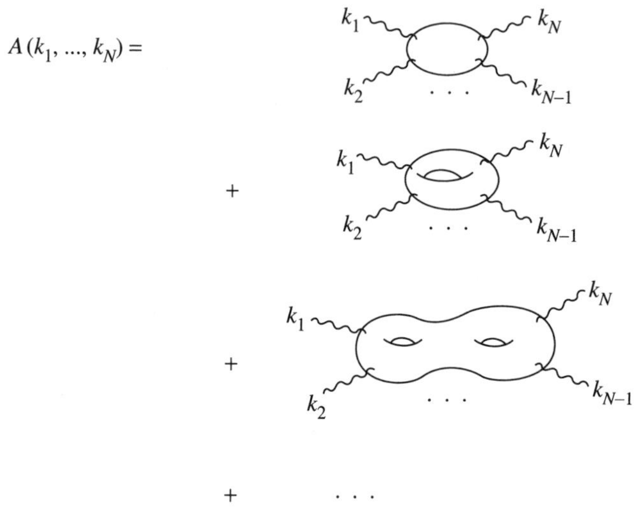

FIGURE 3. The loop expansion for string scattering amplitudes.

[*] The terminology "operator" comes from the Hamiltonian formalism, where the functional integrals of the Lagrangian formalism are interpreted as matrix elements of operators on the Hilbert space of states.

Symmetries and anomalies. The action (2.1) is invariant under diffeomorphisms and conformal transformations (or "Weyl scalings") of the worldsheet metric ($g_{mn} \to e^{2\sigma(\xi)} g_{mn}$). Naively the Dx^μ integral should first produce a functional of g which depends only the complex structure of g, and the integral over Dg should reduce next to an integral over the moduli space of complex structures, after factoring out the volume of the symmetry group. There is however a hidden dependence on the trace of g in the measures Dg and Dx^μ because of the metrics on the spaces of x^μ and g,

$$
\begin{aligned}
\| \delta x^\mu \|^2 &= \int d^2\xi \sqrt{g} (\delta x^\mu)^2 \\
\| \delta g_{mn} \|^2 &= \int d^2\xi \sqrt{g} g^{pq} g^{mn} (\delta g_{pm} \delta g_{qn} + c \delta g_{pq} \delta g_{mn}).
\end{aligned}
$$

(2.6)

This loss of a symmetry of the classical action is a quantum effect known as the *conformal anomaly*. The critical dimension $d = 26$ is the one where the anomaly from the Dg integral cancels that of the Dx^μ integral. This will be explained in some detail below. Similarly, although general vertex operators seemingly depend on the trace of g_{mn}, the ones corresponding to physical particles are those for which this dependence ultimately cancels. Thus the Polyakov formulation of string theory gives a cogent explanation of string scattering amplitudes as integrals over moduli space.

We shall consider separately the integrals in x^μ and g in (2.2) and (2.5). Physically the integrals in x describe a theory of massless scalar fields fluctuating in a fixed background g. This theory is a key example of a "conformal field theory" [14, 36, 38]. By integrating in turn in g, we are incorporating the fluctuations of the background as well. In this sense string theory is a theory of "conformal fields coupled to two-dimensional gravity."

C. String amplitudes as integrals over moduli space. We give now explicit formulas for string amplitudes as integrals over moduli space. Such formulas are referred to as "gauge-fixed" in the physics literature, in the sense that we have chosen a "gauge" for each orbit of the symmetry group, and factored out its volume.

We need some facts about Riemann surfaces and their moduli space, which will be explained more fully in IV.A below. For each metric g it is convenient to introduce complex coordinates z so that $ds^2 = 2g_{z\bar{z}} dz\, d\bar{z}$. Let $\bar{\partial}_n$ the Cauchy-Riemann operator on $(n, 0)$ forms

$$
(2.7) \qquad \bar{\partial}_n : \phi(dz)^n \to \partial_{\bar{z}} \phi\, d\bar{z}(dz)^n.
$$

Its zero modes are the holomorphic tensors of type $(n, 0)$, which we denote by $\phi_a^{(n)}$, $a = 1, \ldots, N_n$. In genus h, $N_1 = h$ and $N_n = 1$ for $h = 1$, $N_n = (2n - 1)(h - 1)$ for $n > 1$, $h > 1$. Of particular importance are the holomorphic tensors of type $(1, 0)$ and $(2, 0)$, called "abelian differentials" and "quadratic differentials." It is convenient to denote them respectively by ω_I, $I = 1, \ldots, h$, and ϕ_a, $a = 1$ for $h = 1$, $a = 1, \ldots, 3(h - 1)$ for

$h > 1$. Finally we denote by $\mathcal{M}_h$ the moduli space of Riemann surfaces of genus h,

(2.8) $\mathcal{M}_h = \{\text{Metrics}\}/\text{Diff} \times \text{Weyl}$.

A fundamental propery of $\mathcal{M}_h$ is that its cotangent space at each point is exactly the space of quadratic differentials of the corresponding Riemann surface. In particular a (local) section of its canonical bundle κ is just a choice of basis ϕ_a for each Riemann surface. The gauge-fixed formulas for the string partition function and tachyon scattering (2.2), (2.5) in $d = 26$ dimensional space-time are then given by

(2.9)

$$Z_{\text{BOS}} = \int_{\mathcal{M}_h} \bigwedge_{a=1}^{3h-3} \phi_a \otimes \bigwedge_{a=1}^{3h-3} \overline{\phi_a} \, \frac{\det'_g 4\overline{\partial}_2^\dagger \overline{\partial}_2}{\det\langle \phi_a | \phi_b \rangle_g} \left(\frac{8\pi^2 \det'_g 2\overline{\partial}_0^\dagger \overline{\partial}_0}{\int d^2\xi \sqrt{g}} \right)^{-13}$$

$$A(k_1, \ldots, k_N) = \delta\left(\sum_{i=1}^N k_i \right) \int_{\mathcal{M}_h} \bigwedge_{a=1}^{3h-3} \phi_a$$

$$\otimes \bigwedge_{a=1}^{3h-3} \overline{\phi_a} \, \frac{\det'_g 4\overline{\partial}_2^\dagger \overline{\partial}_2}{\det\langle \phi_a | \phi_b \rangle_g} \left(\frac{8\pi^2 \det'_g 2\overline{\partial}_0^\dagger \overline{\partial}_0}{\int d^2\xi \sqrt{g}} \right)^{-13} \int_M \cdots \int_M \prod_{i<j} F(z_i, z_j)^{k_i k_j}$$

$$= \delta\left(\sum_{i=1}^N k_i \right) \int_{\mathcal{M}_h} \bigwedge_{a=1}^{3h-3} \phi_a$$

$$\otimes \bigwedge_{a=1}^{3h-3} \overline{\phi_a} \, \frac{\det'_g 4\overline{\partial}_2^\dagger \overline{\partial}_2}{\det\langle \phi_a | \phi_b \rangle_g} \left(\frac{8\pi^2 \det'_g 2\overline{\partial}_0^\dagger \overline{\partial}_0}{\int d^2\xi \sqrt{g} \det \text{Im} \, \Omega} \right)^{-13}$$

$$\times \int_M \cdots \int_M \int dp_I^\mu \left| \exp\left(i\pi p_I^\mu \Omega_{IJ} p_J^\mu + 2\pi i p_I^\mu \sum_{i=1}^N k_i^\mu \int_P^{z_i} \omega_I \right) \right|^2$$

$$\times \prod_{i<j} |E(z_i, z_j)|^{2k_i k_j}$$

where

• The adjoints of $\overline{\partial}_n$ are taken with respect to a metric g in the conformal class determined by each Riemann surface. The determinants of Laplacians are regularized by the zeta function method (cf. Appendix B). The inner products $\langle \phi_a | \phi_b \rangle$ are taken with respect to (2.6). The final expressions appearing in (2.9) are however independent of the choice of g within its conformal class;

• $F(z, w)$ is a positive $(-1/2, -1/2)$ form in each variable z and w, which also depends only on the complex structure of the surface, and not on the metric. In particular the integrand in the second line of (2.9) is a $(1, 1)$

form in each z_i (recall that $k_i^2 = 2$), and can be integrated without a metric on M.

• The above formulas are manifestly independent of the choice of bases ϕ_a, and thus patch together into a smooth section of $\kappa \otimes \overline{\kappa}$ over moduli space.

• The second formula for tachyon scattering amplitudes $A(k_1, \ldots, k_N)$ exhibits the integrand in terms of sections of holomorphic vector bundles over $\mathcal{M}_h$. The key ingredients are the period matrix Ω and the prime form $E(z, w)$ which will be described in IV and V below.

• Although we have now expressed string amplitudes as well-defined finite-dimensional integrals, we shall see in IV.D that these integrals diverge because the integrand blows up near the boundary of $\mathcal{M}_h$. This is to be expected on physical grounds, because the bosonic string spectrum contains a tachyon. Nevertheless the integrands in (2.9) are of great intrinsic mathematical interest. Defined in terms of determinants and Green's functions, they can however be expressed in terms of modular forms, zeta functions, and sections of holomorphic vector bundles, leading to rather unexpected identities between very different notions.

The derivation of (2.9) will be given in IV.B. The key ingredients are provided by the conformal field theories of the "matter fields" x^μ and "ghost system" b, c to be discussed next.

III. Conformal field theory

A. Scalar fields. Throughout this section we shall fix a metric g and consider the Dx^μ integrals in (2.2), (2.5). It suffices to treat each x^μ separately, and we drop the index μ. Thus string theoretic interpretations can be set aside temporarily, and the main issue is the quantum field theory of a scalar field $x(z)$. The mathematical theory of functional integrals is still not fully developed at the present time. Nevertheless we can assign a completely unambiguous meaning to the Dx integrals because they are gaussian. Gaussian integrals are completely determined by the inverse of the covariance matrix and its determinant

$$Integrals\ on\ \mathbf{R}^n$$

$$\int \frac{d^d x}{(2\pi)^{d/2}} e^{-\langle Ax|x\rangle/2} = (\det A)^{-1/2}$$

(3.1)
$$\int \frac{d^d x}{(2\pi)^{d/2}} e^{-\langle Ax|x\rangle/2} \exp(ikx) = (\det A)^{-1/2} e^{-\langle A^{-1}x|x\rangle/2}$$

$$\int \frac{d^d x}{(2\pi)^{d/2}} e^{-\langle Ax|x\rangle/2} x^i x^j = (\det A)^{-1/2} (A^{-1})^{ij}.$$

Their infinite-dimensional analogues with the matrix A replaced by the Laplacian are given similarly in terms of a regularized determinant and Green's function

Functional Integrals

$$\int Dx e^{-\langle \Delta_g x | x \rangle / 2} = \left(\frac{8\pi^2 \det' \Delta}{\int d^2 z \sqrt{g}} \right)^{-1/2}$$

(3.2)

$$\int Dx e^{-\langle \Delta_g x | x \rangle / 2} \exp \left(\sum_{i=1}^{N} ik_i x(z_i) \right)$$

$$= \delta \left(\sum_{i=1}^{N} k_i \right) \left(\frac{8\pi^2 \det' \Delta}{\int d^2 z \sqrt{g}} \right)^{-1/2} \exp \left(-\frac{1}{2} \sum_{i,j=1}^{N} k_i k_j G(z_i, z_j) \right)$$

$$\int Dx e^{-\langle \Delta_g x | x \rangle / 2} x(z_i) x(z_j) = \left(\frac{8\pi^2 \det' \Delta}{\int d^2 z \sqrt{g}} \right)^{-1/2} G(z_i, z_j).$$

There are two subtleties to be clarified in (3.2). The first is that the Laplacian on scalars admits a zero mode, namely the constant functions. If 1 is the constant function, we may split $x = x_0 1 + x'$ with x' orthogonal to constants. The measure Dx splits accordingly as

$$Dx = Dx' dx_0 \parallel 1 \parallel_g = Dx' dx_0 \left(\int d^2 z \sqrt{g} \right)^{1/2}.$$

On the x' space the Laplacian is nondegenerate and the Gaussian integral produces $(8\pi^2 \det' \Delta)^{-1/2}$, while the dx_0 produces either an infinite factor (independent of metrics!) to be ignored in the case of the partition function, or a δ function in presence of insertions. The second subtlety is a feature of quantum field theory. Since the Green's function is singular along the diagonal, the second line of (3.2) does not make sense as it stands. Thus we consider only the "normal ordered" vertex operator : $e^{ikx(z)}$:, where by convention the divergent Green's function $G(z, z)$ is replaced systematically by

(3.3) $$G_R(z, z) = \lim_{w \to z} (G(z, w) + \ln d(z, w)^2).$$

We now discuss the dependence of Dx integrals on the background metric. Evidently they depend on g and not just its conformal class. However for a particular class of insertions, called "primary fields" in the language of conformal field theory, this dependence resides entirely in a single factor called the Liouville action. It is convenient to adopt also the language of conformal field theory, so the Dx integral without any insertion

(3.4) $$Z_x(g) \equiv \int Dx\, e^{-I(g, x)}$$

is called the partition function Z_x of the x theory, and expressions of the form

$$(3.5) \qquad \left\langle \prod_{i=1}^{N} \phi_i(z_i) \right\rangle_x \equiv \int Dx \, e^{-I(g,x)} \prod_{i=1}^{N} \phi_i(z_i)$$

are called the "correlation functions of the fields ϕ_i." In view of (3.2),

$$(3.6) \qquad Z_x(g) = \left(\frac{8\pi^2 \det' \Delta}{\int d^2 z \sqrt{g}} \right)^{-1/2}.$$

Its transformation under a conformal change of background metric $\hat{g} \to g = e^{2\sigma(z)} \hat{g}$ is given by the famous calculation of Polyakov [77] (see Appendix B)

$$(3.7) \qquad Z_x(g) = e^{S_L(\sigma)} Z_x(\hat{g}).$$

Here $S_L(\sigma)$ is the Liouville action

$$(3.8) \qquad S_L(\sigma) = \frac{1}{12\pi} \int d^2 z (\partial_z \sigma \partial_{\bar{z}} \sigma + \hat{g}_{z\bar{z}} \hat{R}\sigma + \mu^2(e^{2\sigma(z)} - 1)).$$

Now the primary fields are $\partial_z x$ and $g_{z\bar{z}}^{k^2/2} : e^{ikx(z)} :$. Indeed

$$(3.9) \qquad \langle \partial_z x(z) \partial_w x(w) \rangle_x = Z_x(g) \partial_z \partial_w G(z,w)$$

is conformally invariant up to the factor $Z_x(g)$. The same statement holds for

$$(3.10) \qquad \left\langle \prod_{i=1}^{N} g_{z_i \bar{z}_i}^{k_i^2/2} : e^{ik_i x(z_i)} : \right\rangle_x = Z_x(g) \delta \left(\sum_{i=1}^{N} k_i \right) \prod_{i<j} F(z_i, z_j)^{k_i k_j}.$$

Here the symmetric $(-1/2, -1/2)$ form $F(z,w)$ is given by

$$(3.11) \qquad \begin{aligned} F(z,w) = 2^{-1} &(g_{z\bar{z}} g_{w\bar{w}})^{-1/2} \\ &\times \exp(-G(z,w) + \tfrac{1}{2} G_R(z,z) + \tfrac{1}{2} G_R(w,w)) \end{aligned}$$

and can be checked to be invariant under a conformal change. We note that primary fields are tensors. Their tensor type is called their "conformal dimension." Thus $\partial_z x$ has dimension $(1,0)$, while $g_{z\bar{z}}^{k^2/2} : e^{ikx(z)} :$ has dimension $(k^2/2, k^2/2)$.

It is now easy to explain the mass spectrum in (2.4). We revert to d scalar fields x^μ. In string theory, the locations z_i of vertex operators are to be integrated over. The conformal dimensions should be $(1,1)$. For example

$$(3.12) \qquad \begin{aligned} &\left\langle \prod_{i=1}^{N} \int_M d^2 z_i \sqrt{g}(z_i) : e^{ik_{i\mu} x^\mu(z_i)} : \right\rangle_{x^\mu} \\ &= (Z_x)^{26} \prod_{i=1}^{N} \int_M d^2 z_i (g_{z_i \bar{z}_i})^{1-k_i^2/2} \prod_{i<j} F(z_i, z_j)^{k_i k_j}. \end{aligned}$$

We must impose $k_i^2 = 2$ to concentrate all dependence on $g_{z\bar{z}}$ in Z_x. This is the mass shell condition for tachyons. More generally the mass shell condition for higher level physical particles is that $k^2/2 + \#\partial_z$ derivatives in $P(Dx)$ be 1. In particular for the second multiplet in (2.4), we obtain $k_i^2 = 0$, so these describe the emission and absorption of massless particles.

B. Ghost fields. Ghost fields are another key example of a conformal field theory. We shall see in IV.B how to express the bosonic string measure on moduli space in terms of their correlation functions. Furthermore they are essential to fermion emission processes in superstring theory.

Ghost fields are tensor fields $b\,dz^n$ and $c(dz)^{1-n}$ of type $(n,0)$ and $(1-n,0)$ respectively. We discuss the case of anticommuting fields, which is the one required for the bosonic string. For our purposes commuting fields are similar. In Appendix C, we provide the elements of Grassmann calculus needed. The ghost action is

$$(3.13) \qquad I_{gh}(b,c) = \frac{1}{2\pi} \int_M d^2 z (b_{zz}\partial_{\bar{z}}c^z + \bar{b}_{\bar{z}\bar{z}}\partial_z \bar{c}^{\bar{z}}).$$

The main correlation functions of interest are those of the fields b, c themselves

$$\left\langle \left| \prod_{i=1}^N b(z_i) \prod_{j=1}^M c(w_j) \right|^2 \right\rangle = \int D(b\bar{b}c\bar{c}) e^{-I_{gh}(b,c)-\overline{I_{gh}(b,c)}} \left| \prod_{i=1}^N b(z_i) \prod_{j=1}^M c(w_j) \right|^2$$

which are valued in $(\bigotimes_{i=1}^N T_{z_i}^{(n,n)}(M)) \otimes (\bigotimes_{j=1}^M T_{w_j}^{(1-n,1-n)}(M))$. Naively the correlation function with no insertions should give the determinant of the Laplacian $\bar{\partial}_n^\dagger \bar{\partial}_n$,

$$\langle 1 \rangle = \int D(b\bar{b}c\bar{c}) e^{-I_{gh}} = \det \bar{\partial}_n^\dagger \bar{\partial}_n.$$

This assumes however that the modes of the b, c fields (i.e. eigenvectors of the $\bar{\partial}_n^\dagger \bar{\partial}_n$ and $\bar{\partial}_n \bar{\partial}_n^\dagger$) occur in pairs. Whether they do is measured by the index of the $\bar{\partial}_n$ operator

$$\Upsilon_n = \dim \operatorname{Ker} \bar{\partial}_n - \dim \operatorname{Ker} \bar{\partial}_n^\dagger = N_n - N_{1-n}.$$

Thus the only nonvanishing correlation functions are of the form

$$\int D(b\bar{b}c\bar{c}) e^{-I_{gh}-\overline{I_{gh}}} \left| \prod_{a=1}^{\Upsilon_n+M} b(z_a) \prod_{b=1}^M c(w_b) \right|^2.$$

Assume for simplicity that $N_{1-n} = 0$, and let $\phi_a^{(n)}$, $a = 1, \ldots, N_n$ be a orthonormal basis of zero modes for $\bar{\partial}_n$. We can break b as $b = b_a\phi_a + b'$ where b' is orthogonal to zero modes. The $\bar{\partial}_n$ operator restricted to the b'

space is nondegenerate, and we obtain

(3.14)

$$\int D(b\overline{b}c\overline{c})\, e^{-I_{gh}-\overline{I}_{gh}} \left| \prod_{a=1}^{N_n} b(z_a) \right|^2$$

$$= \int D(b'\overline{b}'c\overline{c})\, e^{-I_{gh}-\overline{I}_{gh}} \int \left| \prod_{a=1}^{N_n} db_a \prod_{a=1}^{N_n} \left(\sum_{p=1}^{N_n} b_p \phi_p(z_a) + b'(z_a) \right) \right|^2$$

$$= \int D(b'\overline{b}'c\overline{c})\, e^{-I_{gh}-\overline{I}_{gh}} \int \left| \prod_{a=1}^{N_n} db_a \prod_{a=1}^{N_n} \sum_{p=1}^{N_n} b_p \phi_p(z_a) \right|^2$$

$$= (\det{}' \overline{\partial}_n^\dagger \overline{\partial}_n) |\det \phi_p(z_a)|^2.$$

In particular for fields b, c of type $(2,0)$ and $(-1,0)$ on a surface of genus $h > 1$, $N_1 = 0$, $N_2 = 3(h-1)$, and we have

(3.15)

$$\int D(b\overline{b}c\overline{c}) e^{-I_{gh}-\overline{I}_{gh}} \left| \prod_{a=1}^{3(h-1)} \langle \mu_a | b \rangle \right|^2$$

$$= \frac{\det{}' \overline{\partial}_2^\dagger \overline{\partial}_2}{\det \langle \phi_a | \phi_b \rangle} |\det \langle \mu_a | \phi_b \rangle|^2.$$

It is readily seen that the same arguments produce in the general case

(3.16)

$$\int D(b\overline{b}c\overline{c})\, e^{-I_{gh}-\overline{I}_{gh}} \left| \prod_{a=1}^{\Upsilon_n+M} b(z_a) \prod_{b=1}^{M} c(w_b) \right|^2$$

$$= \left| \sum \det \phi_a^{(n)}(z_i) \det \phi_b^{(1-n)}(w_j) \det_{l \neq i,\, m \neq j} G(z_l, w_m) \right|^2$$

$$\times \frac{\det \overline{\partial}_n^\dagger \overline{\partial}_n}{\det \langle \phi_a^{(n)} | \phi_b^{(n)} \rangle \det \langle \phi_a^{(1-n)} | \phi_b^{(1-n)} \rangle}$$

where $\phi_a^{(n)}$, $\phi_b^{(1-n)}$ are now arbitrary bases of holomorphic $(n,0)$ and $(1-n,0)$ tensors, and the sum runs over all selections of N_n and N_{1-n} points z_i and w_j from the points z_a and w_b respectively. The expression $G(z,w)$ denotes the Green's function of the operator $\overline{\partial}_n$ orthogonal to zero modes.

We discuss now the dependence of the above correlation functions on the metric g. Again the action (3.13) is conformally invariant, but a metric dependence arises due to the measure $D(b\overline{b}c\overline{c})$ and manifests itself in the determinant of the Laplacian and the Green's function. The scaling law for $\det{}' \overline{\partial}_n^\dagger \overline{\partial}_n$ under $\hat{g} \to g = e^{2\sigma(z)} \hat{g}$ is given by Polyakov [77] and Alvarez [1]

(cf. Appendix B)

$$\frac{\det'_g \overline{\partial}_n^\dagger \overline{\partial}_n}{\det\langle\phi_a^{(n)}|\phi_b^{(n)}\rangle_g \det\langle\phi_a^{(1-n)}|\phi_b^{(1-n)}\rangle_g}$$

(3.17)

$$= \frac{\det'_{\hat{g}} \overline{\partial}_n^\dagger \overline{\partial}_n}{\det\langle\phi_a^{(n)}|\phi_b^{(n)}\rangle_{\hat{g}} \det\langle\phi_a^{(1-n)}|\phi_b^{(1-n)}\rangle_{\hat{g}}} e^{-2(6n^2-6n+1)S_L(\sigma)}$$

where $S_L(\sigma)$ is the Liouville action (3.8). As for the expression involving Green's functions, it can be checked that it is conformally invariant although $G(z, w)$ is not. Thus b, c are primary fields for the ghost theory, in the same sense as $\partial_z x$ and $g_{z\bar{z}}^{k^2/2}: e^{ikx(z)}$ were primary fields for the scalar field theory: the only dependence on the metric itself is through the Liouville action. The coefficient $c = -2(6n^2 - 6n + 1)$ is called the central charge. The $-$ sign is due to quantization of anticommuting b, c fields. The corresponding charge for commuting fields is $2(6n^2 - 6n + 1)$.

A final property of correlation functions in a conformal field theory can be verified from (3.16). Ignoring the determinant factor, the remaining expression is the modulus squared of a *meromorphic* tensor in each of the z_i, w_j variables, with poles as $z_i \rightarrow w_j$.

C. The stress tensor and the Virasoro algebra. Quantum field theory is so mathematically rich partly because of its dual formalisms. The Lagrangian formalism with functional integrals is geometric in character, while the Hamiltonian formalism in terms of operators and Hilbert space is algebraic. We shall show here that conformal invariance leads to a Virasoro algebra acting on the Hilbert space of states. The coefficient of the conformal anomaly corresponds to the central charge of the algebra. We shall also take the opportunity to discuss some principles of conformal field theory.

Our setting is a field theory on a surface equipped with a metric g. The quantities of interest are the fields φ_i, and their correlation functions $\langle\prod_{i=1}^N \varphi_i(z_i)\rangle$, which are tensors of type $(h_i, \bar{h}_i)$ in each z_i. The dependence on g of the correlation functions is described by a field called the "stress tensor" T_{mn},

(3.18)
$$-\frac{4\pi}{\sqrt{g}}\frac{\delta}{\delta g^{mn}}\left\langle\prod_{i=1}^N \varphi_i(z_i)\right\rangle = \left\langle T_{mn}\prod_{i=1}^N \varphi_i(z_i)\right\rangle.$$

Under an infinitesimal diffeomorphism δv^m, the variation of the metric is $\delta g_{mn} = \nabla_m \delta v_n + \nabla_n \delta v_m$, and thus the variation of the correlation functions is

$$-\frac{1}{2\pi}\int d^2z\sqrt{g}\delta v^m \nabla^n \left\langle T_{mn}\prod_{i=1}^N \varphi_i(z_i)\right\rangle.$$

On the other hand the general transformation law under infinitesimal diffeomorphisms for tensors $\varphi_{\alpha_1 \cdots \alpha_p}$ of rank p is

$$\delta\varphi_{\alpha_1 \cdots \alpha_p} = \sum_{k=1}^{p} (\nabla_{\alpha_k} \delta v^m) \varphi_{\alpha_1 \cdots m \cdots \alpha_p} + \delta v^m \nabla_m \varphi_{\alpha_1 \cdots \alpha_p}.$$

Comparing the two transformation laws yields

$$\nabla^n \left\langle T_{mn}(z) \prod_{i=1}^{N} \varphi_{\alpha_1 \cdots \alpha_p}(z_i) \right\rangle = \sum_{j=1}^{N} \left(\delta(z, z_j) \nabla_m \left\langle \prod_{i=1}^{N} \varphi_{\alpha_1 \cdots \alpha_p}(z_i) \right\rangle \right.$$

$$\left. + \sum_{k=1}^{p} \nabla_{\alpha_k} \delta(z, z_j) \left\langle \prod_{i=1}^{N} \varphi_{\alpha_1 \cdots m \cdots \alpha_p}(z_i) \right\rangle \right)$$

where the covariant derivatives on the right-hand side act on the variables z_j. In particular for tensors of type $(h_i, \bar{h}_i)$ on a Riemann surface we obtain the conservation law

(3.19)
$$\frac{1}{2\pi} \nabla^n \left\langle T_{zn} \prod_{i=1}^{N} \varphi_i(z_i) \right\rangle$$

$$= \sum_{j=1}^{N} (h_i \nabla_{z_j} \delta(z, z_j) + \delta(z, z_j) \nabla_{z_j}) \left\langle \prod_{i=1}^{N} \varphi_i(z_i) \right\rangle.$$

Conformal field theories are characterized by the fact that the Hilbert space of states is generated by primary fields and their descendants. The dependence of the correlation functions of primary fields on the background metric is concentrated entirely on the Liouville factor (cf. (3.9), (3.10), (3.16)). In terms of the stress tensor, this translates into

(3.20)
$$\left\langle T_{z\bar{z}} \prod_{i=1}^{N} \varphi_i(z_i) \right\rangle = -\frac{c}{6} R g_{z\bar{z}} \left\langle \prod_{i=1}^{N} \varphi_i(z_i) \right\rangle$$

where c is a constant characteristic of each theory, called the "conformal anomaly". The equation (3.19) becomes

(3.21)
$$\frac{1}{2\pi} \nabla^z \left\langle T_{zz} \prod_{i=1}^{N} \varphi_i(z_i) \right\rangle - \frac{c}{12\pi} \nabla_z R \left\langle \prod_{i=1}^{N} \varphi_i(z_i) \right\rangle$$

$$= \sum_{j=1}^{N} (h_i \nabla_{z_j} \delta(z, z_j) + \delta(z, z_j) \nabla_{z_j}) \left\langle \prod_{i=1}^{N} \varphi_i(z_i) \right\rangle.$$

This equation can be integrated using the Green's function G_{zz}^w for $\bar{\partial}$ on $(2, 0)$ tensors

$$\nabla^z G_{zz}^w \stackrel{*}{=} 2\pi \delta(z, w)$$

$$\nabla^w G_{zz}^w = -2\pi \delta(z, w) + 2\pi \sum_{a=1}^{3h-3} g^{w\bar{w}} \mu_{a,\bar{w}}^{} \phi_{a, zz}$$

where ϕ_a is a basis of holomorphic quadratic differentials, and μ_b is the dual basis of Beltrami differentials. We have assumed $h > 1$, the other cases being similar. The result is the first "Ward identity for reparametrization invariance"

$$
\left\langle T_{zz} \prod_{i=1}^{N} \varphi_i(z_i) \right\rangle - \sum_{a=1}^{3h-3} \phi_{a,zz} \int d^2 y \mu_{a,\bar{y}}^{y} \left\langle T_{yy} \prod_{i=1}^{N} \varphi_i(z_i) \right\rangle
$$

(3.22)
$$
= \sum_{j=1}^{N} (h_j \nabla_{z_j} G_{zz}^{z_j} + G_{zz}^{z_j} \nabla_{z_j}) \left\langle \prod_{i=1}^{N} \varphi_i(z_i) \right\rangle
$$

$$
- \frac{c}{12\pi} \int d^2 y \sqrt{g} G_{zz}^{y} \nabla_y R \left\langle \prod_{i=1}^{N} \varphi_i(z_i) \right\rangle.
$$

Differentiating (3.21) and again integrating gives the second Ward identity

$$
\left\langle T_{zz} T_{ww} \prod_{i=1}^{N} \varphi_i(z_i) \right\rangle - \sum_{a=1}^{3h-3} \phi_{a,zz} \int d^2 y \mu_{a,\bar{y}}^{y} \left\langle T_{yy} T_{ww} \prod_{i=1}^{N} \varphi_i(z_i) \right\rangle
$$

$$
= \frac{c}{12} \nabla_w^3 G_{zz}^{w} \left\langle \prod_{i=1}^{N} \varphi_i(z_i) \right\rangle
$$

(3.23)
$$
+ \left[\frac{c}{12\pi} \int d^2 y \sqrt{g} G_{zz}^{y} \nabla_y R + 2\nabla_w G_{zz}^{w} + G_{zz}^{w} \nabla_w \right.
$$

$$
\left. + \sum_{j=1}^{N} (h_j \nabla_{z_j} G_{zz}^{z_j} + G_{zz}^{z_j} \nabla_{z_j}) \right] \left\langle T_{ww} \prod_{i=1}^{N} \varphi_i(z_i) \right\rangle.
$$

We give now the interpretation of the Ward identities in the operator language. Here we need to introduce a coordinate system and restrict our discussion to a disk D with center at the origin. The radius $|z|$ serves as time. The fields in the theory become operator valued distributions, and their correlation functions are interpreted as the matrix elements on the vacuum vector $|0\rangle$ and a suitable vector $|v\rangle$ of their time-ordered product, e.g.,

$$
(3.24) \qquad \left\langle \prod_{i=1}^{N} \varphi_i(z_i) \right\rangle = \left\langle v \left| \prod_{|z_{i_N}| > |z_{i_{N-1}}| > \cdots > |z_{i_1}|} \varphi_{i_j}(z_{i_j}) \right| 0 \right\rangle
$$

with $|v\rangle$ determined by $M \backslash D$. Since the Green's function G_{zz}^{w} is equal to $(z - w)^{-1}$ up to regular terms, we can read off the singularities as insertion points come close

$$
T(T_{zz}\phi_i(z_i)) \sim \left(\frac{h_i}{(z - z_i)^2} + \frac{1}{z - z_i} \partial_{z_i} \right) \phi_i(z_i)
$$

$$(3.25) \quad \begin{aligned} T(T_{zz}T_{ww}) &\sim \left(\frac{2}{(z-w)^2} + \frac{1}{z-w}\partial_w \right) T_{ww} \\ &+ \frac{c}{12}(\partial_w - \Gamma^w_{ww})\partial_w(\partial_w + \Gamma^w_{ww})\frac{1}{z-w} \end{aligned}$$

where T denotes time-ordering. Such developments are called "operator product expansions." To obtain a *metric independent* operator product expansion, we are led to modify the stress tensor to the "chiral (or *holomorphic*) stress tensor"

$$(3.26) \quad T^{\text{chi}}_{ww} = T_{ww} + \frac{c}{6}\left(\partial_w \Gamma^w_{ww} - \frac{1}{2}(\Gamma^w_{ww})^2 \right).$$

We note that unlike T_{zz}, T^{chi}_{zz} is holomorphic, as follows from (3.19) and (3.20),

$$(3.27) \quad \partial_{\bar{z}}T^{\text{chi}}_{zz} = \frac{c}{6}g_{z\bar{z}}\partial_z R + \frac{c}{6}\partial_{\bar{z}}\left(\partial_z \Gamma^z_{zz} - \frac{1}{2}(\Gamma^z_{zz})^2 \right) = 0.$$

Furthermore the operator product expansion (3.25) can now be expressed solely in terms of the complex structure

$$(3.28) \quad \begin{aligned} T(T^{\text{chi}}_{zz}\phi_i(z_i)) &\sim \left(\frac{h_i}{(z-z_i)^2} + \frac{1}{z-z_i}\partial_{z_i} \right)\phi_i(z_i) \\ T(T^{\text{chi}}_{zz}T^{\text{chi}}_{ww}) &\sim \frac{c/2}{(z-w)^4} + \left(\frac{2}{(z-w)^2} + \frac{1}{z-w}\partial_w \right)T^{\text{chi}}_{ww}. \end{aligned}$$

Due to the noncovariant counterterm in (3.26) the stress tensor is no longer a tensor, but transforms with a Schwarzian derivative

$$(3.29) \quad \begin{aligned} T^{\text{chi}}_{zz} &= T^{\text{chi}}_{ww}\left(\frac{dw}{dz} \right)^2 + \frac{c}{12}S(w,z) \\ S(w,z) &= \frac{d^3w/dz^3}{dw/dz} - \frac{3}{2}\left(\frac{d^2w/dz^2}{dw/dz} \right)^2. \end{aligned}$$

Thus we see a key feature of conformal field theory: it is not possible to maintain covariance and holomorphicity simultaneously. To obtain the Virasoro algebra we adopt the *holomorphic* stress tensor, and henceforth drop its superscript. We can now introduce the operators L_n as

$$(3.30) \quad L_n = \oint \frac{dz}{2\pi i}z^{n+1}T_{zz}.$$

The matrix elements of the commutator $[L_n, T_{zz}]$ can be obtained as follows

(Figure 4)

$$[L_n, T_{zz}] = \oint_{C_{0,z}} \frac{dw}{2\pi i} w^{n+1} T_{ww} T_{zz} - \oint_{C_0} \frac{dw}{2\pi i} w^{n+1} T_{zz} T_{ww}$$

(3.31)
$$= \left(\oint_{C_{0,z}} - \oint_{C_0} \right) \frac{dw}{2\pi i} w^{n+1} T(T_{ww} T_{zz})$$

$$= \oint_{C_z} \frac{dw}{2\pi i} w^{n+1} T(T_{ww} T_{zz}).$$

The singularities for the time-ordered product $T(T_{zz} T_{ww})$ are given by (3.28). Substituting in (3.31) yields

(3.32) $\qquad [L_m, L_n] = (m - n)L_{m+n} + \dfrac{c}{12} m(m^2 - 1)\delta_{m, -n}.$

This is a Virasoro algebra with the central charge given by the conformal anomaly.

In practice the holomorphic stress tensor is often obtained rather by regularizing the classical stress tensor $-4\pi g^{-1/2} \delta I / \delta g^{mn}$ in a holomorphic but not covariant manner. For example the classical stress tensor for scalar fields is $-(\partial_z x \partial_z x)/2$. The quantum stress tensor is defined by the following

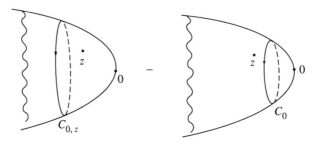

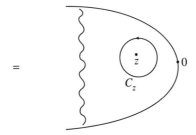

FIGURE 4. Deformation of contours arguments for commutators of the stress tensor.

normal ordering prescription

(3.33)
$$T_{zz} = : -\frac{1}{2}\partial_z x(z)\partial_z x(z) :$$
$$\equiv \lim_{w \to z}\left(-\frac{1}{2}\partial_z x(z)\partial_w x(w) - \frac{1}{2}\frac{1}{(z-w)^2}\right).$$

For ghost fields (3.13) the classical stress tensor is $-nb\partial_z c + (1-n)(\partial_z b)c$ normal ordered as

(3.34) $$T_{zz} = \lim_{w \to z}\left(-nb(z)\partial_w c(w) + (1-n)(\partial_z b(z))c(w) + \frac{1}{(w-z)^2}\right).$$

We can check directly that this regularization procedure causes T_{zz} to transform with a Schwarzian exactly as in (3.29). Through operator product expansions the stress tensor gives a very simple way of recapturing the conformal anomaly. Thus in the case of scalar fields

(3.35)
$$T_{zz}T_{ww} = : -\tfrac{1}{2}\partial_z x(z)\partial_z x(z) : : -\tfrac{1}{2}\partial_w x(w)\partial_w x(w) :$$
$$\sim \tfrac{1}{2}\langle\partial_z x(z)\partial_w x(w)\rangle^2 \sim \frac{1/2}{(z-w)^4}$$

by Wick contraction rules, giving the value 1 for the conformal anomaly. Similarly for ghost fields

(3.36)
$$T_{zz}T_{ww} =: -nb(z)\partial_z c(z) + (1-n)(\partial_z b(z))c(z) :$$
$$\times : -nb(w)\partial_w c(w) + (1-n)(\partial_w b(w))c(w) :$$
$$\sim 2n(1-n)\langle b(z)c(w)\rangle\langle\partial_w b(w)\partial_z c(z)\rangle$$
$$- n^2\langle b(z)\partial_w c(w)\rangle\langle b(w)\partial_z c(z)\rangle$$
$$- (1-n)^2\langle\partial_w b(w)c(z)\rangle\langle\partial_z b(z)c(w)\rangle \sim \frac{-(6n^2-6n+1)}{(z-w)^4}$$

where we have used the expansion $\langle b(z)c(w)\rangle \sim (z-w)^{-1}$ for the Green's function. The $-$ sign is due to the anticommuting nature of the ghost fields.

The Hilbert space of states must decompose into irreducible representations of the two Virasoro algebras generated by the components T_{zz} and $T_{\overline{z}\overline{z}}$ of the stress tensor. Highest weight vectors $|h, \overline{h}\rangle$ are defined by

(3.37) $$L_n|h, \overline{h}\rangle = \overline{L}_n|h, \overline{h}\rangle = 0 \quad \text{for } n > 0.$$

They correspond to primary fields under $|h, \overline{h}\rangle = \phi_{h,\overline{h}}(0)|0\rangle$ where $|0\rangle$ is the vacuum. The highest weight condition is equivalent to the condition that $T_{zz}\phi_{h,\overline{h}}(w)$ have no pole higher than second order, which is exactly the definition (3.28) of primary fields in the manifestly holomorphic (rather than covariant) formulation. The partition function of the theory is defined as a trace in the Hamiltonian formalism

(3.38) $$Z = \text{Tr}(q^{L_0-c/24}\overline{q}^{\overline{L}_0-c/24}) = \sum_{h,\overline{h}} N_{h\overline{h}}\chi_h(q)\overline{\chi_{\overline{h}}(q)}$$

where $\chi_h(q)$ is the character of the representation h, and $N_{h\overline{h}}$ is the multiplicity of $h\overline{h}$. For example the partition function for scalar fields is

(3.39)

$$Z = \mathrm{Tr}\left(q^{\sum\limits_{n=1}^{\infty} \alpha_{-n}\alpha_n - 1/24} \; \overline{q}^{\sum\limits_{n=1}^{\infty} \overline{\alpha}_{-n}\overline{\alpha}_n - 1/24} \right) \int_{-\infty}^{\infty} dp |q|^{p^2}$$

$$= \int_{-\infty}^{\infty} dp \, \frac{q^{p^2/2}}{\eta(q)} \cdot \frac{\overline{q}^{p^2/2}}{\eta(\overline{q})}$$

where α_{-n}, α_n obey the relation $[\alpha_n, \alpha_{-n}] = n$, and

$$\eta(q) \equiv q^{1/24} \prod_{n=1}^{\infty} (1 - q)^n$$

is the Dedekind eta function (see also (4.15), (4.16) below). In the Lagrangian formulation Z is given by a path integral on the torus with moduli parameter $q = e^{2\pi i \tau}$. Since $SL(2, \mathbf{Z})$ is the mapping class group for the torus, this forces Z to transform as a modular form. This requirement is a powerful tool in the classification of conformal field theories. At the present time only theories with $c \leq 1$ have been completely classified. For $c > 1$ the problem remains extremely complex, although the classification of all modular covariant combinations of Kac-Moody SU(2) characters (which are a particularly important class of Virasoro characters) is known.

The Hamiltonian-Lagrangian correspondence is most complete for the torus because of the existence of a global time. It is a challenging problem to develop the Hamiltonian formalism for higher genus surfaces. Work still relying on the Virasoro algebra includes [3, 72, 85, 90]. In a different direction, Krichever and Novikov [56] have constructed more general algebras adapted to higher genus. Progress in understanding this algebra can be found in [53, 76].

D. Holomorphic structure of conformal field theory. We have seen that the dependence of a conformal field theory on the background metric g on the Riemann surface is governed by the stress tensor (3.18). In particular if we deform the background metric g by a Beltrami differential μ,

(3.40) $$2g_{z\overline{z}}|dz|^2 \to 2g_{z\overline{z}}|dz + \mu_{\overline{z}}{}^z d\overline{z}|^2$$

the second order variation in the partition function is given by

(3.41) $$\delta_\mu \delta_{\overline{\mu}} \ln Z = \frac{1}{16\pi^2} \int d^2z \, d^2w \, \mu_{\overline{z}}{}^z \overline{\mu}_{\overline{w}}{}^w \langle T_{zz} T_{\overline{w}\overline{w}} \rangle_{\mathrm{conn}}.$$

Thus the independence of T_{zz} and $T_{\overline{z}\overline{z}}$ implies that Z splits as a product of a holomorphic by an antiholomorphic function of moduli parameters τ. More generally, we expect the correlation functions to split as

(3.42) $$\left\langle \prod_{i=1}^{N} \phi_i(z_i) \right\rangle = \sum_{K\overline{L}} h_{K\overline{L}} C_{i_1 \cdots i_N}^K (z_1, \ldots, z_N) \overline{C_{i_1 \cdots i_N}^L (z_1, \ldots, z_N)}$$

where the "conformal blocks" $C^K_{i_1 \cdots i_N}(z_1, \ldots, z_N)$ [14] are holomorphic functions of τ and of the insertion points z_i. From the operator theoretic standpoint, this holomorphic splitting is dictated by the presence of two independent Virasoro algebras.

This naive discussion is however spoiled by the conformal anomaly. Indeed the *covariant* stress tensor develops at the quantum level a trace which links the components T_{zz} and $T_{\bar{z}\bar{z}}$ (3.19). This difficulty is eliminated by using instead the *chiral* stress tensor (3.26). Now quadratic differentials can be viewed as one-forms on the moduli space $\mathscr{M}_h$ of Riemann surfaces, and hence as connections on line bundles on $\mathscr{M}_h$. The chiral stress tensor transforms with a Schwarzian and is rather a connection on a *projective* line bundle with Chern class given by the central charge. If we let λ be the Hodge bundle over $\mathscr{M}_h$, i.e., the maximum wedge power of the bundle of abelian differentials, the relevant projective line bundle may be identified with $\lambda^{c/2}$. This suggests that a conformal field theory is described by a geometric notion, which is a section $\Gamma(\tau)$ of $\lambda^{c/2} \otimes \overline{\lambda^{c/2}}$, covariantly constant with respect to the connection defined by the chiral stress tensor. Holomorphic splitting now means that $\Gamma(\tau)$ can be written as [38],

$$\Gamma(\tau) = \sum_{K\bar{L}} h_{K\bar{L}} s^K(\tau) \overline{s^{\bar{L}}(\tau)}$$

where $s^K(\tau)$, $s^{\bar{L}}(\tau)$ are covariantly constant sections of $\lambda^{c/2} \otimes W$, with W a flat holomorphic vector bundle over $\mathscr{M}_h$ and $h_{K\bar{L}}$ a hermitian metric on W. The partition function of the theory can be recaptured by taking the norm of $\Gamma(\tau)$ with respect to the Quillen metric (3.44), $Z = \|\Gamma(\tau)\|_Q^2$. String theories arise from coupling conformal field theories to two-dimensional gravity. Conformal invariance requires that the total conformal anomalies, including ghost fields, cancel. Thus we may neglect the contributions from the projective line bundles in this case, and string amplitudes are given by hermitian pairings of holomorphic sections of genuine vector bundles over the moduli space of Riemann surfaces. This powerful constraint is indispensable for the Gliozzi-Scherk-Olive projection [41] in superstring theory (cf. VII.A).

Curvature of determinant line bundles. The above physical discussion can be given a careful mathematical treatment in the case of the matter and ghost fields of III.A and III.B. The key observation is that the Cauchy-Riemann operator depends holomorphically on moduli parameters. In fact, under the deformation (3.40), the Cauchy-Riemann operator $\bar{\partial}$ is deformed to $\partial_{\bar{z}} - \mu \partial_z$, which involves only μ and not $\bar{\mu}$; furthermore, the zero modes of the $\bar{\partial}$ operator on $(n, 0)$ tensors also depend holomorphically on moduli, in the sense that they form a holomorphic vector bundle over moduli space for each n. Naively it can be expected that the determinants of the Laplacians $\det' \bar{\partial}_n^\dagger \bar{\partial}_n$ be the absolute value squared of a holomorphic function of moduli, except that nonholomorphicity has been introduced through the removal of the zero modes (otherwise the determinants would just vanish), and the choice of a metric (in order to take adjoints). The net outcome is the Belavin-Knizhnik formula [12, 17, 21]

$$
(3.43) \qquad \delta_\mu \delta_{\bar\mu} \ln \frac{\det' \bar\partial_n^\dagger \bar\partial_n}{\det\langle \phi_a^{(1-n)} | \phi_b^{(1-n)} \rangle \, \det\langle \phi_a^{(n)} | \phi_b^{(n)} \rangle}
$$
$$
= -\frac{6n^2 - 6n + 1}{12\pi} \int d^2 z \, \nabla_{\bar z} \mu \nabla_z \bar\mu
$$

where $\phi_a^{(n)}$ is a basis of holomorphic tensors of type $(n, 0)$, depending holomorphically on moduli.

Note that by Serre duality the holomorphic tensors of type $(1-n, 0)$ are in one-to-one correspondence with the zero modes of the adjoint $\bar\partial_n^\dagger$ $(\phi_a^{(1-n)} \to g_{z\bar z}^n \overline{\phi_a^{(1-n)}})$.

Formally the equation (3.43) can be integrated to

$$
\frac{\det' \bar\partial_n^\dagger \bar\partial_n}{\det\langle \phi_a^{(1-n)} | \phi_b^{(1-n)} \rangle \, \det\langle \phi_a^{(n)} | \phi_b^{(n)} \rangle} = e^{-2(6n^2 - 6n + 1) S_L(g_{z\bar z})} |Z_n|^4
$$

where $S_L(g_{z\bar z})$ is the noncovariant form of the Liouville action

$$
S_L(g_{z\bar z}) = \frac{1}{48\pi} \int d^2 z \, \partial_z \ln g_{z\bar z} \partial_{\bar z} \ln g_{z\bar z}
$$

and Z_n depends holomorphically on moduli. In string theories determinants occur in combinations for which the central charges $\pm 2(6n^2 - 6n + 1)$ cancel. Thus we may ignore the Liouville terms, and Z_n can be viewed as the partition function of an effective *chiral* field. For example for $n = 0$, Z_0 is just the Dedekind eta function $\eta(\tau)$ (3.39), (4.16).

The Belavin-Knizhnik formula admits a simple geometric interpretation. As in index theory for families of Dirac operators depending on parameters [8, 2, 16], we introduce the determinant line bundle

$$
\mathrm{DET}(\bar\partial_n) = \left(\bigwedge^{\max} \mathrm{Ker}\, \bar\partial_n \right)^{-1} \otimes \bigwedge^{\max} \mathrm{Ker}\, \bar\partial_n^\dagger
$$

over $\mathcal{M}_h$. In view of Serre duality we can identify $\mathrm{DET}(\bar\partial_n)$ with the holomorphic line bundle $(\bigwedge^{\max} \mathrm{Ker}\, \bar\partial_n)^{-1} \otimes (\bigwedge^{\max} \mathrm{Ker}\, \bar\partial_{1-n})^{-1}$ over $\mathcal{M}_h$. In particular $(\mathrm{DET}\, \bar\partial_0)^{-1}$ and $(\mathrm{DET}\, \bar\partial_2)^{-1}$ are readily recognized as the Hodge bundle λ and the canonical bundle κ over $\mathcal{M}_h$. A local section s of $\mathrm{DET}\, \bar\partial_n$ is of the form $(\bigwedge_a^{\max} \phi_a^{(n)})^{-1} \otimes (\bigwedge_b^{\max} \phi_b^{(1-n)})^{-1}$. If we choose a metric g_{mn} on the surface to represent each complex structure (e.g., the constant curvature metric), we may define the Quillen metric [80] on $\mathrm{DET}\, \bar\partial_n$ by

$$
(3.44) \qquad \| s \|_Q^2 = \frac{\det' \bar\partial_n^\dagger \bar\partial_n}{\det\langle \phi_a^{(1-n)} | \phi_b^{(1-n)} \rangle \, \det\langle \phi_a^{(n)} | \phi_b^{(n)} \rangle}.
$$

Now vectors on moduli space are just Beltrami differentials μ. Thus the Belavin-Knizhnik formula gives the curvature of the determinant line bundle with respect to the Quillen metric. In this sense the Belavin-Knizhnik formula is a local refinement of the family's index theorem, since it measures the nontriviality of determinant bundles in terms of curvature and not just characteristic classes. The holonomy of the determinant line bundle can be evaluated in terms of eta invariants [96, 16, 23].

We turn to holomorphic properties of correlation functions, say of the vertex operators $g_{z\bar{z}}^{k^2/2} : e^{ikx(z)} :$. They are given by the conformally invariant expression (3.10). On the sphere, the Green's function with respect to the constant curvature metric is

$$G(z, w) = -\ln\frac{|z - w|^2}{(1 + |z|^2)(1 + |w|^2)}$$

and we obtain

$$\left\langle \prod_{i=1}^{N} g_{z_i\bar{z}_i}^{k_i^2/2} : e^{ik_ix(z_i)} : \right\rangle_x = \delta\left(\sum_{i=1}^{N}k_i\right)\prod_{i<j}^{N}|z_i - z_j|^{2k_ik_j}$$

up to a constant independent of the insertion points z_i. This is manifestly the absolute value squared of a holomorphic function of the z_i's, as expected. On surfaces of nontrivial homology, the situation is more subtle due to the presence of Abelian differentials ω_I. However holomorphic splitting can be restored by introducing "loop momenta" p_I [92],

$$\left\langle \prod_{i=1}^{N} g_{z_i\bar{z}_i}^{k_i^2/2} : e^{ik_ix(z_i)} : \right\rangle_x$$

$$(3.45) \qquad = \delta\left(\sum_{i=1}^{N}k_i\right)\int dp_I \left|\exp\left(i\pi p_I\Omega_{IJ}p_J + 2\pi ip_I\sum_{i=1}^{N}k_i\int_P^{z_i}\omega_I\right)\right|^2$$

$$\times |E(z_i, z_j)|^{2k_ik_j}\left(\frac{8\pi^2 \det_g' 2\bar{\partial}_0^\dagger\bar{\partial}_0}{\int d^2\xi\sqrt{g}\det\mathrm{Im}\,\Omega}\right)^{-1/2} .$$

IV. Modular forms and zeta functions in string theory

We give now the derivation of the expressions (2.9) for string scattering amplitudes (2.2), (2.5) as integrals over moduli space. The integrals in Dx have already been treated in II.A as correlation functions of quantum scalar fields. It remains to analyze the integrals Dg over all metrics. For this we begin by describing the moduli space $\mathcal{M}_h$ of Riemann surfaces of genus h in greater detail in IV.A. The arguments for (2.9) are in IV.B together with the ghost formulation which will be crucial for superstrings. Explicit formulas in terms of modular forms and zeta functions are in IV.C. The bosonic string in Minkowski space-time is ill-behaved because of the presence in its spectrum of tachyons, i.e., particles moving faster than light. Indeed the vertex for emission and absorption of such particles is given in (2.4). With explicit formulas for string scattering amplitudes we can see another manifestation of their presence: the integrands in the amplitudes develop a pole near the boundary of moduli space. This is explained in IV.D in both hyperbolic geometry and algebraic geometry formalisms.

A. The moduli space of Riemann surfaces. We recall some basic facts about moduli space, viewed as the space $\{\text{Metrics}\}/\text{Diff}(M) \times \text{Weyl}$. An infinitesimal deformation of a metric g_{mn} is a symmetric tensor δg_{mn}. In local complex coordinates $(z, \overline{z})$ on M where the metric g_{mn} becomes $ds^2 = 2g_{z\overline{z}}dz\,d\overline{z}$, we can decompose the deformation δg_{mn} as a trace $\delta g_{z\overline{z}}$, and a traceless part $\delta g_{zz}, \delta g_{\overline{z}\overline{z}}$. The only part which may change the complex structure is the traceless part, so deformations of complex structures should be parametrized by tensors of type $(2, 0)$, or rather by the corresponding "Beltrami differential"

$$(4.1) \qquad\qquad \mu_{\overline{z}}{}^z = \tfrac{1}{2}g^{z\overline{z}}\delta g_{\overline{z}\overline{z}}$$

which is independent of the choice of g within its conformal class. Now not all Beltrami differentials change the complex structure. Those of the form $\partial_{\overline{z}}v^z$ do not, since they just measure the change in the metric induced by an infinitesimal diffeomorphism parametrized by the vector field v^z. Thus the tangent space to moduli space at g is given by

$$(4.2) \qquad\qquad T_g(\mathcal{M}_h) = \left\{ \frac{\text{Beltrami differentials } \mu_{\overline{z}}{}^z}{\text{Range } \partial_{\overline{z}} \text{ on } (-1, 0) \text{ tensors}} \right\}.$$

The pairing

$$(4.3) \qquad\qquad \langle \mu | \phi \rangle = \int d^2 z\, \mu_{\overline{z}}{}^z \phi_{zz}$$

between $(2, 0)$ and $(-1, 1)$ tensors allows us to identify the cotangent space of moduli space at g as the space of (holomorphic) quadratic differentials

$$(4.4) \qquad\qquad T_g^*(\mathcal{M}_h) = \{\phi_{zz}dz^2, \ \partial_{\overline{z}}\phi_{zz} = 0\}.$$

Since the dimension of the space of holomorphic vector fields is $3, 1$, and 0 for genus $h = 0, h = 1$, and $h \geq 2$, it follows that $\mathcal{M}_h$ is a complex manifold of dimension $0, 1$, and $3h - 3$ respectively. More precisely, it is an orbifold, with singular points at the surfaces with discrete automorphisms.

We shall need the parametrization of moduli space by period matrices. For genus $h = 1$, the surface is a torus $\mathbf{C}/\omega_1\mathbf{Z}+\omega_2\mathbf{Z}$. The ratio of the generators $\tau = \omega_2/\omega_1$ characterizes the complex structure. We may evidently choose the generators so that $\operatorname{Im}\tau > 0$. Since $(\omega_1, \omega_2 + \omega_1)$ and $(\omega_2, -\omega_1)$ are two equally good sets of generators, the transformations $T : \tau \to \tau + 1$ and $S : \tau \to -1/\tau$ leave the complex structure unchanged. Thus

$$(4.5) \qquad\qquad \mathcal{M}_1 = \{\tau\,; \operatorname{Im}\tau > 0\}/PSL(2, \mathbf{Z}).$$

A different way of viewing τ is more suitable for generalization to higher genus. If we represent the torus as the parallelogram with vertices at $z = 0, 1, \tau, \tau + 1$ and opposite sides identified, the sides A going from 0 to 1 and B going from 0 to τ constitute a homology basis. The parameter τ can be viewed as the period around B of the unique holomorphic one-form $\omega = dz$ which is normalized to have period 1 around A. For higher genus $h \geq 2$ we begin then by choosing a homology basis A_I, B_I with the following intersection numbers

$$(4.6) \qquad \#(A_I, A_J) = 0, \quad \#(A_I, B_J) = \delta_{IJ}, \quad \#(B_I, B_J) = 0.$$

A basis ω_I of abelian differentials is singled out by requiring that it be dual to the A_I's. Their periods around the B_I's define the "period matrix" Ω,

$$(4.7) \qquad \oint_{A_I} \omega_J = \delta_{IJ}, \qquad \oint_{B_I} \omega_J = \Omega_{IJ}.$$

It is not difficult to show that (Riemann's bilinear relations)

$$(4.8) \qquad \Omega_{IJ} = \Omega_{JI}, \qquad \operatorname{Im} \Omega_{IJ} = \langle \omega_I | \omega_J \rangle$$

so that Ω is in the Siegel upper half-space. Evidently transformations of Ω resulting from changes of homology bases preserving the intersection relations (4.7) do not affect the complex structure. These generate the symplectic group $\operatorname{Sp}(2h, \mathbf{Z})$. Since the dimension of the space of symmetric matrices is $h(h + 1)/2$, it is evident that the space of period matrices of Riemann surfaces is in general only a subvariety of the quotient of the Siegel upper half-space by $\operatorname{Sp}(2h, \mathbf{Z})$.

B. Integrals over metrics and integrals over moduli space. The space of metrics can be parametrized locally by diffeomorphisms f, conformal scalings $e^{2\sigma}$ and a $3h - 3$ dimensional slice $\mathscr{S} = \{\hat{g}(\tau_i); i = 1, \ldots, 3h - 3\}$ transversal to the orbits of the diffeomorphism and Weyl groups

$$(4.9) \qquad (\hat{g}(\tau_i), f, e^{2\sigma}) \to g = f_*(e^{2\sigma} \hat{g}(\tau_i)).$$

We wish to compute the Jacobian of the change of variables (4.9). Since we are dealing with manifestly reparametrization invariant expressions, it suffices to calculate the Jacobian of the above transformation at $g = e^{2\sigma} \hat{g}(\tau_i)$. Now infinitesimal variations of the left-hand side of (4.8) are $(\mu_i \delta \tau_i, \delta v^m, 2\delta \sigma)$ with $\mu_i = \frac{1}{2} \hat{g}^{z\bar{z}} \partial \hat{g}_{\bar{z}\bar{z}}(\tau)/\partial \tau_i$. The Jacobian is the volume of their images under the mapping (4.9). To compute this volume, we decompose the tangent space to the space of metrics at g as a direct sum

$$T_g(\text{Metrics}) = \{2\sigma g_{z\bar{z}}\} \oplus \operatorname{Range} \nabla_z \oplus \operatorname{Ker} \nabla_z^\dagger \oplus \overline{\operatorname{Range} \nabla_z \oplus \operatorname{Ker} \nabla_z^\dagger}$$

where ∇_z is the covariant derivative on $(1, 0)$ tensors. Its adjoint can be identified with the operator $\bar{\partial}_2$ on $(2, 0)$ tensors, and its kernel with the space ϕ_a of quadratic differentials. The Jacobian can now be easily seen to be (see Figure 5, next page)

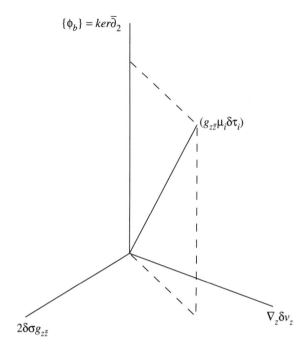

$$\{\phi_b\} = \ker \bar\partial_2$$

$$(g_{z\bar z}\mu_i\delta\tau_i)$$

$$2\delta\sigma g_{z\bar z}$$

$$\nabla_z\delta v_z$$

FIGURE 5.

$$\mathrm{Vol}_g(2\delta\sigma g_{z\bar z}; \, 2\nabla_z\delta v_z, \, 2\nabla_{\bar z}\delta v_{\bar z}; \, g_{z\bar z}\mu_i\delta\tau_i, \, g_{z\bar z}\overline{\mu_i\delta\tau_i})$$

(4.10)

$$= \delta\sigma\delta v\delta\bar v \prod_{i=1}^{3h-3} \delta\tau_i\delta\bar\tau_i \frac{|\det\langle\mu_a|\phi_b\rangle|^2}{\det\langle\phi_a|\phi_b\rangle_g}\det'_g 4\bar\partial_2^\dagger\bar\partial_2.$$

We can use the scaling law (3.17) for determinants of Laplacians in order to rewite (4.10) in terms of metrics $\hat g$. The result is

(4.11) $$Dg = Dv\, D\sigma \prod_{i=1}^{3h-3} d\tau_i\, d\bar\tau_i \frac{|\det\langle\mu_a|\phi_b\rangle|^2}{\det\langle\phi_a|\phi_b\rangle_{\hat g}}(\det'_{\hat g} 4\bar\partial_2^\dagger\bar\partial_2)\exp(-26S_L(\sigma)).$$

Returning to the bosonic string, we can put (3.6), (3.7) and (4.11) together to obtain the partition function. In dimension $d = 26$ the Liouville terms cancel, and we obtain

(4.12) $$Z_{\mathrm{BOS}} = \int_{\mathscr{S}} \prod_{i=1}^{3h-3} d\tau_i d\bar\tau_i \frac{|\det\langle\mu_a|\phi_b\rangle|^2}{\det\langle\phi_a|\phi_b\rangle}(\det'_g 4\bar\partial_2^\dagger\bar\partial_2)\left(\frac{\det' 2\bar\partial_0^\dagger\bar\partial_0}{\int d^2z\sqrt g}\right)^{-13}$$

where we have dropped the $\hat{}$ notation for metrics in the slice $\mathscr{S}$. This is clearly equivalent to (2.9). The formulas for vertex operators follow from (3.10) and (3.45). We can also rewrite them in terms of ghost fields, using (3.15). Thus

(4.13) $$Z_{\mathrm{BOS}} = \int_{\mathscr{S}} D(b\bar b c\bar c x^\mu)e^{-I(g,\,x^\mu)-I_{gh}(b,\,c)-\overline{I_{gh}(b,\,c)}} \prod_{i=1}^{3h-3} |\langle\mu_a|b\rangle|^2.$$

In this paper we deal only with string theories in their critical dimensions, which are candidates for grand unification including gravity. Noncritical strings are expected to be equally important, for example as an effective theory of color flux tubes in quantum chromodynamics [78]. Many models seem amenable to a nonperturbative analysis [19, 28, 46], which has led recently to intense research. From the Polyakov viewpoint noncritical strings are governed by the quantum Liouville theory, which is a fascinating issue in its own right. On the mathematical side the Liouville theory has sparked some beautiful work on the variational problems associated with the determinants of Laplacians in two and higher dimensions [73, 22, 74, 10, 18].

C. Explicit formulas for low genus. For the torus there is a complex one-dimensional group $\mathrm{Aut}(M)$ of automorphisms and the formula (4.12) has to be modified to

(4.14)

$$Z_{\mathrm{BOS}} = \int_{\mathcal{M}_h} d\tau d\bar{\tau} \frac{|\det\langle \mu_a | \phi_b \rangle|^2}{\det\langle \phi_a | \phi_b \rangle} \det' 4\bar{\partial}_2^\dagger \bar{\partial}_2 \left(\frac{\det' 2\bar{\partial}_0^\dagger \bar{\partial}_0}{\int d^2 z \sqrt{g}} \right)^{-13} \frac{1}{\mathrm{Vol}_g(\mathrm{Aut}(M))}.$$

The reason is that the parametrization (4.9) for metrics requires only vector fields v^m orthogonal to the infinitesimal generators of $\mathrm{Aut}(M)$, while we still factor out the full group of diffeomorphisms. The integrand in (4.14) is readily calculated. Represent each complex structure by a torus with parameter τ and metric $ds^2 = dz\, d\bar{z}$. The eigenvalues of $2\bar{\partial}_0^\dagger \bar{\partial}_0$ are $4\pi^2 |n + m\tau|^2/\tau_2^2$, n, m integers, and

$$\det'(2\bar{\partial}_0^\dagger \bar{\partial}_0) = \prod_{(m,n)\neq(0,0)} \frac{4\pi^2}{\tau_2^2} |n + m\tau|^2.$$

From the product expansion formula for $\sin$

$$\prod_{n=-\infty}^{\infty} (n + z) = 2i \sin \pi z$$

and the special values $\zeta(0) = -1/2$, $\zeta(-1) = -1/12$, $\zeta'(0) = -(\ln 2\pi)/2$ for the Riemann zeta function $\zeta(s) = \sum_{n=1}^{\infty} n^{-s}$ we deduce that

(4.15)

$$\det'(2\bar{\partial}_0^\dagger \bar{\partial}_0) = \left(\prod_{(m,n)\neq(0,0)} \frac{4\pi^2}{\tau_2^2} \right) \left(\prod_{n=1}^{\infty} n \right)^4 \prod_{m=1}^{\infty} \prod_{n=-\infty}^{\infty} |(n - m\tau)(n + m\tau)|^2$$

$$= \left(\frac{4\pi^2}{\tau_2^2} \right)^{4\zeta(0)(1+\zeta(0))} e^{-4\zeta'(0)} \prod_{m=1}^{\infty} \prod_{n=-\infty}^{\infty} |(n - m\tau)(n + m\tau)|^2$$

$$= \tau_2^2 \prod_{m=1}^{\infty} |e^{i\pi m\tau} - e^{-i\pi m\tau}|^4$$

$$= \tau_2^2 |\eta(\tau)|^4$$

where $\eta(\tau)$ is the Dedekind eta function

$$(4.16) \qquad \eta(\tau) = e^{i\pi\tau/12} \prod_{n=1}^{\infty} (1 - e^{2i\pi n\tau}).$$

This agrees with (3.39) of the Hamiltonian formalism. Evidently $\det' 4\bar{\partial}_2^\dagger \bar{\partial}_2$ is given by the same expression, up to a factor of $1/2$ due to the removal of the zero mode (cf. Appendix B). Next $\text{Aut}(M)$ consists of translations, and can be identified with M itself. Since the tangent space to $\text{Aut}(M)$ can be identified with holomorphic vector fields, $\text{Aut}(M)$ inherits a metric $\tau_2 dz d\bar{z}$ from the torus. Thus the volume of $\text{Aut}(M)$ is τ_2^2. The Beltrami differentials of this slice are $\mu = (2i\tau_2)^{-1} d\bar{z}(dz)^{-1}$. Choosing $\phi = (dz)^2$, we arrive at

$$(4.17) \qquad Z_{\text{BOS}} = \int_{H/PSL(2,\mathbb{Z})} \frac{d\tau \, d\bar{\tau}}{64\pi^2 \tau_2^2} (4\pi^2 \tau_2)^{-12} |\eta(\tau)|^{-48}.$$

In higher genus we can obtain an equally explicit formula for (2.9) by choosing constant curvature metrics to represent complex structures. The corresponding metric on moduli space is the Weil-Petersson metric. The measure $d\tau_i \, d\bar{\tau}_i |\det\langle \mu_a | \phi_b \rangle|^2 / \det\langle \phi_a | \phi_b \rangle$ can be identified up to a constant with ω_{WP}^{3h-3} where ω_{WP} is the Weil-Petersson Kaehler form. Furthermore $\det' \bar{\partial}_0^\dagger \bar{\partial}_0$ and $\det' \bar{\partial}_2^\dagger \bar{\partial}_2$ can be evaluated to be $Z'(1)$ and $Z(2)$ where $Z(s)$ is the Selberg zeta function

$$Z(s) = \prod_l \prod_{k=1}^{\infty} (1 - e^{-(k+s)l}).$$

Here l denotes the length of closed geodesics. Thus (2.9) reduces to [29],

$$(4.18) \qquad Z_{\text{BOS}} = c_h \int_{\mathcal{M}_h} \omega_{WP}^{3h-3} Z(2) Z'(1)^{-13}.$$

Another way which exploits to the fullest the complex geometry of moduli space is the following. By an earlier result of Mumford [70] the line bundle $\kappa \otimes \lambda^{-13}$ admits a global holomorphic section s over $\mathcal{M}$. Now the bundle of abelian differentials over moduli space admits a natural inner product

$$\langle \omega_I | \omega_J \rangle = \int_M \omega_I \bar{\omega}_J$$

which is independent of the choice of metric within each conformal class. The norm $\| s \|_{\lambda^{-13}}^2$ of s with respect to the induced inner product on λ^{-13} is a section of the canonical bundle $\kappa \otimes \bar{\kappa}$, and the bosonic string partition function can be expressed as [12],

$$(4.19) \qquad Z_{\text{BOS}} = c_h \int_{\mathcal{M}_h} \| s \|_{\lambda^{-13}}^2.$$

Indeed the Belavin-Knizhnik formula (3.43) shows that the expression

$$(4.20) \qquad \left(\frac{\det' \bar{\partial}_0^\dagger \bar{\partial}_0}{\int d^2 z \sqrt{g} \, \det\langle \omega_I | \omega_J \rangle} \right)^{-13} \frac{\det' \bar{\partial}_2^\dagger \bar{\partial}_2}{\det\langle \phi_a | \phi_b \rangle}$$

is a holomorphic function of moduli parameters, for

$$s = \bigwedge_{a=1}^{3h-3} \phi_a \otimes \left(\bigwedge_{I=1}^{h} \omega_I \right)^{-13}.$$

a holomorphic section of $\kappa \otimes \lambda^{-13}$. Since s is a global section, (4.20) is a global holomorphic function over $\mathscr{M}_h$. Now $\mathscr{M}_h$ admits a codimension 2 compactification (e.g., the Satake compactification by period matrices), so (4.20) must reduce to a constant, and (2.9) reduces to (4.19).

This holomorphic structure leads easily to an explicit expression for Z_{BOS} in terms of modular forms for genus 2 and 3 [13, 65, 60, 69],

$$Z_{\mathrm{BOS}} = c_2 \int \left| \prod_{I \leq J} d\Omega_{IJ} \right|^2 \left| \prod_{\delta \text{ even}} \vartheta[\delta](0, \Omega) \right|^{-2} (\det \mathrm{Im}\,\Omega)^{-13}, \qquad h = 2,$$

$$= c_3 \int \left| \prod_{I \leq J} d\Omega_{IJ} \right|^2 \left| \prod_{\delta \text{ even}} \vartheta[\delta](0, \Omega) \right|^{-1} (\det \mathrm{Im}\,\Omega)^{-13}, \qquad h = 3.$$

In fact if we parametrize moduli space by period matrices Ω and apply (4.8) the partition function can be written as

$$Z_{\mathrm{BOS}} = \int \left| \prod_{I \leq J} d\Omega_{IJ} \right|^2 (\det \mathrm{Im}\,\Omega)^{-13} |\Phi(\Omega)|^{-2}$$

with $\Phi(\Omega)$ holomorphic. Under $\mathrm{Sp}(2h, \mathbf{Z})$ transformations, the determinant transforms as

$$\det(\mathrm{Im}\,\Omega) \to (\det \mathrm{Im}\,\Omega) |\det(C\Omega + D)|^{-2}$$

and the invariant measure is $|\prod_{I \leq J} d\Omega_{IJ}|^2 (\det \mathrm{Im}\,\Omega)^{h+1}$. Thus $\Phi(\Omega)$ must be a modular form of weight $12 - h$, that is,

$$\Phi(\Omega) \to \Phi(\Omega)(\det(C\Omega + D))^{12-h}.$$

In genus 2 the weight of $\Phi(\Omega)$ is 10. Now the ring of modular forms in genus 2 has been identified by Igusa [52] as a polynomial ring with generators $\Psi_4, \Psi_6, \Psi_{10}, \Psi_{12}$ of weights $4, 6, 10, 12$ respectively. Furthermore the behaviors of the two weight 10 forms $\Psi_4 \Psi_6$ and Ψ_{10} differ near the boundary of the space of period matrices. The boundary consists of the components $\Delta_0 = \{\Omega_{11} \to \infty\}$ and $\Delta_1 = \{\Omega_{12} \to 0\}$, with transversal coordinates $\Omega_{11}^{-1/2}$ and Ω_{12} respectively. Now $\Psi_4 \Psi_6$ is known not to vanish along Δ_1 while $\Phi(\Omega)$ does. In the next section we shall see that the string integrand must have a pole. This identifies $\Phi(\Omega)$ with Ψ_{10}, which is given by

$$\Psi_{10} = \prod_{\delta \text{ even}} \theta[\delta](0, \Omega).$$

(For the theta function $\theta[\delta](\zeta, \Omega)$, see V.D below.) This gives the above formula for the bosonic string partition function in genus 2. The analysis for genus 3 is similar.

D. Degeneration behavior of the string integrand. The bosonic string is ill-behaved because of the presence of a tachyon. This manifests itself in the divergence near the boundary of moduli space of the partition function Z_{BOS}. Physically, as we approach the boundary of $\mathcal{M}_h$ a collar stretches out. If $T \to \infty$ is the Euclidean length of the collar, canonical quantization shows that the leading contribution to the string propagation along the collar is $e^{-E_0 T}$ where E_0 is the ground state energy. But $E_0 = -2$ because of the tachyon.

A precise mathematical analysis can be provided using either (4.18) or (4.19). The asymptotic behavior of special values of the Selberg zeta function as we pinch a geodesic of length l is [**99, 50, 39**],

$$Z(s) \sim l^{-2s+1} e^{-\pi^2/(3l)}, \qquad s > 1,$$
$$Z'(1) \sim l^{-1} e^{-\pi^2/(3l)} \prod_{0 < \lambda_n < 1/4} \lambda_n, \qquad s = 1.$$

Here λ_n are the eigenvalues of the hyperbolic Laplacian. By [**82, 37**] there are at most $4h - 2$ eigenvalues less than $1/4$, and λ_n is of the order of the sum of the lengths of closed geodesics disconnecting the surface into $n + 1$ components. Furthermore the asymptotic behavior of the Weil-Petersson metric is also known [**64**],

$$\omega_{WP}^{3h-3} \sim |dt|^2 |t|^{-2} \left(\ln \frac{1}{|t|} \right)^{-3}$$

where t is the complex coordinate transversal to the boundary of moduli space defined by the "plumbing fixture." In terms of hyperbolic geometry $|t| \sim \exp(-2\pi^2/l)$. Altogether we obtain for the string integrand (4.18) the asymptotic behavior

(4.21) $|dt|^2 |t|^{-4}$

up to smaller factors. Alternatively, we can exploit the fact that the bundles κ, l can be extended to the Deligne-Mumford compactified moduli space $\overline{\mathcal{M}}_h$, the first as the canonical bundle of $\overline{\mathcal{M}}_h$, the second as the bundle of dualizing differentials,i.e., meromorphic differentials on surfaces with nodes with poles only at the nodes, and opposite residues. The divisor of $\kappa \otimes l^{-13}$ is known to be $[-2\Delta]$, so the presence of the tachyon is indicated here by a pole of order 2 for the section s of (4.19) [**12**]. Yet another way of deriving the singularity of the string integrand is by analyzing the behavior of the determinants and Green's functions of the Laplacian with respect to the Arakelov metric [95, 35, 54].

V. Scalar superfields in a supergeometry background

In this section we shall describe the quantum theory of scalar superfields (x, ψ_α) in a fixed supergeometry background (g, χ_m^α). This theory is the supersymmetric generalization of the theory of quantum scalar fields x on a surface with metric g (cf. III.A) which was one of the building blocks of the bosonic string. Scalar superfields will play an equally central role in the construction of superstrings. It is also important in its own right as the simplest superconformal field theory.

A. The action and vertex operators. In a supersymmetric extension of a field theory, each field acquires a "supersymmetric partner." The partner of a boson is a fermion, and the partner of a commuting field is an anticommuting one. Supersymmetry is a symmetry with Grassmann generator which rotates the partners into one another. Bosons are given by tensor fields and fermions by spinor fields.

On a Riemann surface, spinors can be identified with tensors of half-integer weights. To define these, we need a square root of the canonical bundle $\kappa(M)$. It is not difficult to see that there are 2^{2h} square roots which we denote by S_δ. A concrete way of describing them is by choosing a homology basis, which singles out a spin bundle $S_0(M)$. Spin bundles differ from $S_0(M)$ by a bundle with zero Chern class and can be viewed as elements of the Jacobian. As such they are given by half-integer characteristics $\delta = (\delta', \delta'') \in \frac{1}{2}\mathbf{Z}^{2h}$. Each δ is called a "spin structure." A spin bundle of $U(1)$ weight $\frac{1}{2} + n$ and spin structure δ is given by $S_\delta(M) \otimes \kappa^n(M)$. Locally we write sections of $S_\delta(M)$ as $\psi = \psi_+(dz)^{1/2}$, $\overline{\psi} = \psi_-(d\overline{z})^{1/2}$. The transition functions are square roots of those of $\kappa(M)$, with the correct phases dictated by the choice of spin structure. Spinor indices $\pm$ are denoted generally by Greek letters, and tensor indices $z, \overline{z}$ by Latin letters.

We can now describe the supersymmetric extension of the theory of scalar fields of III.A. We fix a spin structure δ. The partner of x is a "Majorana spinor" which corresponds in Euclidean formalism to a section ψ of $S_\delta(M)$ and its conjugate $\overline{\psi}$. The partner of g is a "gravitino field" χ_m^α. We may represent g by a frame (or zweibein) $e^a = d\xi^m e_m^a$. The components of the gravitino consist of $\chi_{\overline{z}}^+$ of $U(1)$ weight 3/2, $\chi_{\overline{z}}^-$ of $U(1)$ weight 1/2, and their conjugates χ_z^-, and χ_z^+. Equivalently $\chi_{\overline{z}}^+$ and χ_z^- are tensors of types $(-1/2, 1)$ and $(1, -1/2)$ respectively

$$\chi = \chi_{\overline{z}}^+ (dz)^{-1/2} d\overline{z}, \qquad \overline{\chi} = \chi_z^- dz (d\overline{z})^{-1/2}$$

which we can view as supersymmetric partners of Beltrami differentials. The χ_z^+, $\chi_{\overline{z}}^-$ are partners of the trace of g and are usually referred to as the "γ

trace". The action is

$$(5.1) \qquad \begin{aligned} I_\delta = \frac{1}{4\pi} \int d^2z (\partial_z x^\mu \partial_{\bar{z}} x^\mu - \psi_+^\mu \partial_{\bar{z}} \psi_+^\mu - \psi_-^\mu \partial_z \psi_-^\mu \\ + \chi_{\bar{z}}^+ \psi_+^\mu \partial_z x^\mu + \chi_z^- \psi_-^\mu \partial_{\bar{z}} x^\mu - \tfrac{1}{2} \chi_{\bar{z}}^+ \chi_z^- \psi_+^\mu \psi_-^\mu). \end{aligned}$$

It is super Weyl invariant in the sense that it is independent of the γ trace of χ. In addition it is invariant under local supersymmetry. Supersymmetry is defined by the following infinitesimal transformation

$$(5.2) \qquad \begin{aligned} \delta e_m^a = \zeta^\alpha (\gamma^a)_{\alpha\beta} \chi_m^\beta, \qquad \delta x^\mu = \zeta^\alpha \psi_\alpha^\mu \\ \delta \chi_m^\alpha = -2\nabla_m \zeta^\alpha, \qquad \delta \psi_\alpha^\mu = -(\gamma^m)_{\alpha\beta} \zeta^\beta (\partial_m x^\mu - \tfrac{1}{2} \chi_m^\delta \psi_\delta^\mu) \end{aligned}$$

where ζ is a spinor. We shall call (x, ψ) a "scalar superfield" and (g, χ) a (two-dimensional) "supergeometry."

The quantum theory of scalar superfields in a fixed supergeometry background can be described in terms of functional integrals $Dx \, D\psi_\pm$. Besides the partition function

$$(5.3) \qquad Z_\delta = \int D(x\psi_+\psi_-) e^{-I(g,\chi;x,\psi_\pm)},$$

we shall also be interested in the correlation functions of the vertex operators which describe the simplest "primary fields" in this theory as well as graviton scattering in superstring theory. It is convenient to introduce anticommuting variables θ, $\bar{\theta}$, and "superfields" $X(x, \theta, \bar{\theta})$,

$$(5.4) \qquad X(x, \theta, \bar{\theta}) = x(z) + \theta \psi_+ + \bar{\theta} \psi_-$$

which are functions of θ, $\bar{\theta}$ without $\theta\bar{\theta}$ coefficient. The "super covariant derivatives" $\mathscr{D}$ is defined by

$$(5.5) \qquad \begin{aligned} \mathscr{D}_- X = \psi_- + \bar{\theta}(\partial_{\bar{z}} x + \tfrac{1}{2} \chi_{\bar{z}}^+ \psi_+) \\ + \theta\bar{\theta}(-\tfrac{1}{4} \chi_{\bar{z}}^+ \chi_z^- \psi_- - \tfrac{1}{2} \chi_{\bar{z}}^+ \partial_z x - \partial_{\bar{z}} \psi_+). \end{aligned}$$

This expression is dictated by supersymmetry, in the same way as covariant derivatives are dictated by transformation laws for tensors. The vertex operators of interest are now given by

$$(5.6) \qquad V(z, \theta, \bar{\theta}, \zeta, \bar{\zeta}; k) = \exp(ikX + \zeta\sqrt{E}\mathscr{D}_+ X + \bar{\zeta}\sqrt{E}\mathscr{D}_- X)$$

where E is the super determinant $E = \sqrt{g}(1 + \tfrac{1}{4} \theta\bar{\theta} \chi_z^- \chi_{\bar{z}}^+)$. Their correlation functions are given by the functional integrals

$$(5.7) \qquad \begin{aligned} \left\langle \prod_{i=1}^N V(z_i, \theta_i, \bar{\theta}_i, \zeta_i, \bar{\zeta}_i, k_i) \right\rangle_X \\ = \int DX \prod_{i=1}^N e^{-I_\delta} V(z_i, \theta_i, \bar{\theta}_i, \zeta_i, \bar{\zeta}_i, k_i). \end{aligned}$$

As in the case of bosonic scalar fields on a Riemann surface (cf. III.A), the functional integrals (5.7) are gaussian and can be defined rigorously in terms of regularized determinants of Laplacians and Green's functions on scalars and spinors. The action (5.1) is independent of traces, but the determinants and Green's functions, and hence the integrals (5.7) are not. However all dependence of (5.7) on traces is concentrated in Liouville terms which ultimately cancel. The main issues are rather the following:

• Ignoring the Liouville terms, the correlation functions (5.7) depend only on the complex structure of g and the γ traceless components $\chi_{\bar{z}}^{+}$, χ_{z}^{-} of the gravitino. The main problem is to decompose these functions as pairings of holomorphic vs. antiholomorphic forms in (g, χ), with holomorphicity in χ meaning dependence on $\chi_{\bar{z}}^{+}$ only and not on χ_{z}^{-}. This task is referred to as "holomorphic (or chiral) splitting." We shall see later that it is these holomorphic forms rather than the correlation functions (5.7) which are needed to build the superstring.

• Let a "super Riemann surface" be an equivalence class of (g, χ) under conformal transformations, diffeomorphisms, super Weyl scalings, and local supersymmetry. We need to develop complex supersymmetry invariants and function theory on super Riemann surfaces, and relate them to the invariants dictated by (5.7).

B. Green's functions, zero modes, and supersymmetry. There are several obstructions to holomorphic splitting:

• The action (5.1) and the covariant derivatives (5.5) mix intricately $\psi_{+}, \chi_{\bar{z}}^{+}$ with ψ_{-}, χ_{z}^{-}. This is required by local supersymmetry and is to be expected, since covariant derivatives on tensors already do not respect holomorphic structures;

• The Green's function on scalars is neither conformally invariant nor holomorphically dependent on moduli. This is evident from its defining equations

$$-\partial_z\partial_{\bar{z}}G(z, w) = 2\pi\delta(z, w) - \frac{2\pi g_{z\bar{z}}}{\int d^2 z\sqrt{g}}$$

(5.8)

$$\partial_z\partial_{\overline{w}}G(z, w) = 2\pi\delta(z, w) - \pi\sum_{I,J=1}^{h}\omega_I(\mathrm{Im}\,\Omega)_{IJ}^{-1}\omega_J.$$

The tensor $F(z, w)$ of (3.11) is conformally invariant, but does not split holomorphically;

• The functional integrals in $\psi_{\pm}$ will introduce the Green's function $S_{\delta}(z, w)$ on spinors and its conjugate (cf. Appendix C). Whether $S_{\delta}(z, w)$ varies holomorphically in z, w and moduli depends on the existence of zero modes for $\bar{\partial}$. For generic even spin structures δ, there is no zero mode, so that

(5.9) $$\partial_{\bar{z}}S_{\delta}(z, w) = 2\pi\delta(z, w)$$

and S_{δ} is meromorphic in z, w, holomorphic in moduli. However for odd spin structures there are always zero modes. Generically there is one zero mode denoted by h_δ, and the Green's function satisfies now

$$(5.10) \qquad \partial_{\bar{z}} S_\delta(z, w) = 2\pi\delta(z, w) - \frac{h_\delta(z)\overline{h_\delta(w)}}{\langle h_\delta | h_\delta \rangle}.$$

As in (4.9) the presence of h_δ clearly destroys the meromorphicity of $S_\delta(z, w)$.

All this suggests that naive holomorphic splitting with conformal blocks as just $(1/2, 0)$ tensors in each z_i is not possible. However, it can be accomplished by introducing conformal blocks as $(1/2, 0)$ forms tensored with sections of auxiliary vector bundles, as described in the Chiral Splitting Theorem below.

C. The chiral splitting theorem [32].

THEOREM 1. *Up to Liouville terms, the correlation functions* (5.7) *can be expressed as*

$$(5.11) \qquad \left\langle \prod_{i=1}^{N} V(z_i, \bar{z}_i, \theta_i, \bar{\theta}_i, \zeta_i, \bar{\zeta}_i; k_i) \right\rangle_X$$
$$= (2\pi)^{10} \delta\left(\sum_{i=1}^{N} k_i\right) \int dp_I^\mu |\mathscr{F}_\delta(z_i, \theta_i, \zeta_i, \Omega, \chi_{\bar{z}}^+; k_i, p_I^\mu)|^2$$

where
- *The "conformal block" $\mathscr{F}_\delta$ is unique up to a constant phase;*
- *$\mathscr{F}_\delta$ is holomorphic in ζ_i, Ω_{IJ}, and $\chi_{\bar{z}}^+$;*
- *$\mathscr{F}_\delta$ is a meromorphic $(1/2, 0)$ form in each of the insertion points z_i;*
- *$\mathscr{F}_\delta$ has nontrivial monodromy.*

$$\mathscr{F}_\delta(z_i + \delta_{ij} A_K, \theta_i, \zeta_i, \Omega, \chi_{\bar{z}}^+; k_i, p_I^\mu) = \mathscr{F}_\delta(z_i, \theta_i, \zeta_i, \Omega, \chi_{\bar{z}}^+; k_i, p_I^\mu)$$

$$(5.12) \qquad \begin{aligned} &\mathscr{F}_\delta(z_i + \delta_{ij} B_K, \theta_i, \zeta_i, \Omega, \chi_{\bar{z}}^+; k_i, p_I^\mu) \\ &\quad = \mathscr{F}_\delta(z_i, \theta_i, \zeta_i, \Omega, \chi_{\bar{z}}^+; k_i, p_I^\mu + \delta_{IK} k_j^\mu). \end{aligned}$$

Here $z_j + B_K$ means that the point z_j has been transported around the B_K homology cycle.

D. Effective Green's functions and effective rules [32]. The conformal blocks $\mathscr{F}_\delta$ can actually be written down explicitly. The idea is to introduce "effective Green's functions" which are manifestly holomorphic in all parameters. They cannot be single-valued on the Riemann surface, but the final expression for the conformal blocks and correlation functions will be.

The main ingredient is the prime form $E(z, w)$ which plays the role of $z - w$ in the plane [34, 71]. Fix a homology basis A_I, B_I satisfying (4.6) and the corresponding basis of abelian differentials ω_I. The Jacobian $J(M)$ of the surface M is the space of its line bundles of zero Chern class. It can be identified with the torus $\mathbf{C}/\mathbf{Z}^h + \Omega\mathbf{Z}^h$. On $J(M)$ one can construct the "ϑ line bundle" by the transition functions

$$s(\zeta + M + \Omega N) = \exp(-\pi i N^t \Omega N - 2\pi i N^t \zeta) s(\zeta).$$

Its unique (up to constants) holomorphic section is given by the theta function

$$(5.13) \qquad \theta(\zeta, \Omega) = \sum_{n \in \mathbf{Z}^h} \exp(\pi i n^t \Omega n + 2\pi i n^t \zeta).$$

The Riemann surface M can be imbedded into its Jacobian variety by the Abel map

$$(5.14) \qquad M \ni z \to \int_P^z \omega_I$$

where P is a fixed reference point. In particular we have the map α,

$$(z, w) \to \int_z^w \omega_I$$

from $M \times M$ into the Jacobian. The ϑ divisor is the variety of zeroes of the theta function $\theta(\zeta, \Omega)$. To each of its 2^{2h} symmetric (i.e., invariant under $\zeta \to -\zeta$) translates corresponds a spin bundle [71]. Let S_0 be the spin bundle corresponding to the ϑ divisor itself. The prime form $E(z, w)$ is a section of the line bundle

$$(5.15) \qquad (\pi_z)_*(S_0) \times (\pi_w)_*(S_0) \times (\alpha)_*(\vartheta)$$

on $M \times M$ obtained in the following way. Let

$$\theta[\delta](\zeta, \Omega) = \sum_{n \in \mathbf{Z}^h} \exp(\pi i (n + \delta')^t \Omega(n + \delta') + 2\pi i (n + \delta')(\zeta + \delta''))$$

be the theta function with characteristics $[\delta] = (\delta', \delta'') \in \frac{1}{2}\mathbf{Z}^{2h}$. Its zero set is a symmetric translate of the theta divisor, and thus characterizes a spin bundle S_δ. Its parity is the one of $4\delta'\delta''$. For odd spin bundles, there is generically exactly one holomorphic spinor, given explicitly by

$$(5.16) \qquad h_\delta(z) = \left(\sum_{I=1}^h \partial_I \theta[\delta](0, \Omega) \omega_I \right)^{1/2}.$$

The prime form is defined as

$$(5.17) \qquad E(z, w) = \frac{\theta[\delta](\int_z^w \omega_I, \Omega)}{h_\delta(z) h_\delta(w)}$$

and is actually independent of the choice of odd spin structure $[\delta]$. It vanishes exactly for $z = w$, and depends holomorphically on all variables z, w, Ω. Its relation with the Green's function (3.11) for the scalar Laplacian is

$$(5.18) \quad F(z, w) = \exp\left(-2\pi \mathrm{Im} \int_z^w \omega_I (\mathrm{Im}\,\Omega)_{IJ}^{-1} \mathrm{Im} \int_z^w \omega_J \right) |E(z, w)|^2.$$

TABLE 2.

	Original	Effective
Bosons	$x(z)$	$x_+(z)$
Fermions, δ even	$\psi_+(z)$	$\psi_+(z)$
Fermions, δ odd	$\psi_+(z)$	$\psi_+(z) + \lambda h_\delta(z)$
Derivatives	$\mathscr{D}_+$	$\partial_+ = \partial_\theta + \theta\partial_z$
x-Propagator	$G(z, w)$	$-\ln E(z, w)$
ψ-Propagator	$S_\delta(z, w)$	$\langle \psi_+(z)\psi_+(w) \rangle$
Determinants	$\dfrac{\det' \overline{\partial}_n^\dagger \overline{\partial}_n}{\det\langle \phi_a^{(n)} \vert \phi_b^{(n)} \rangle \det\langle \phi_a^{(1-n)} \vert \phi_b^{(1-n)} \rangle}$	$\vert Z_n(\Omega) \vert^4$
Loop momenta	None	$\exp(p_I \oint_{B_I} dz\, \partial_z x_+)$

THEOREM 2. *The conformal blocks $\mathscr{F}_\delta$ can be obtained by replacing the fields x, ψ, their Green's functions, and determinants in (5.7) by their effective counterparts as given in the table below, and carrying out the same rules of gaussian integration. For odd spin structure, we also have to integrate in the odd parameter λ introduced here in Table 2. Here the propagator for the effective fermionic field ψ_+ is given by*

$$(5.19)$$

$$-\langle \psi_+(z)\psi_+(w) \rangle = \frac{1}{E(z, w)} \theta[\delta]\left(\int_z^w \omega_I, \Omega \right)$$

$$= \frac{1}{E(z, w)} \left(\partial_I \theta[\delta]\left(\int_w^z \omega_I \right) \frac{\omega_I(z_0)}{h_\delta(z_0)^2} + \frac{\theta[\delta](\int_w^z \omega_I)}{h_\delta(z_0)^2} \partial_{z_0} \ln \frac{E(z, z_0)}{E(w, z_0)} \right)$$

for even and odd spin structures respectively. The terms $Z_n(\Omega)^2$ are the holomorphic roots of the determinant factors (3.44), defined up to an anomaly which ultimately cancels.

A comment on the effective Green's function on spinors is in order. For even spin structures, the genuine Green's function $S_\delta(z, w)$ with a single pole at $z = w$ exists and is given by (5.19), which is also known as the Szegö kernel. For odd spin structure, there is no meromorphic Green's function with a single pole. The effective Green's function has an additional pole at z_0 and is also denoted by $S_\delta(z, w; z_0)$,

$$(5.20) \qquad \partial_{\bar{z}} S_\delta(z, w; z_0) = 2\pi \delta(z, w) - 2\pi \frac{h_\delta(w)}{h_\delta(z_0)} \delta(z, z_0)$$

and is antisymmetric in z and w. The additional pole cancels in the final answer for $\mathscr{F}_\delta$. As far as the conformal blocks $\mathscr{F}_\delta$ are concerned, we could have adopted other effective Green's functions, for example Green's functions without additional poles but with additional monodromy. The choice of (5.19) is suggested by the most attractive properties for the super period matrix to be discussed next.

VI. Supergeometry and super Riemann surfaces

A. The super period matrix. Geometric invariants and function theory on super Riemann surfaces will be based on supersymmetric generalizations of abelian differentials and period matrices. For even spin structure, this can be done in parallel with Riemann surfaces. For odd spin structures, however, the presence of a holomorphic spinor leads to features having no analogue. It is remarkable that in both cases the notions which impose themselves are also the ones dictated by the conformal blocks (5.11), (5.12) [**32, 4**].

We begin by defining super abelian differentials. A meromorphic super abelian differential with first and second order poles is a rank $1/2$ tensor $\hat{\omega} = \hat{\omega}_+ + \theta\hat{\omega}_z$ satisfying

(6.1)
$$\partial_{\bar{z}}\hat{\omega}_+ + \frac{1}{2}\chi_{\bar{z}}^+\hat{\omega}_z = 2\pi\sum_{r=1}^{p}\alpha_r\theta_{z_r}\delta(z, z_r) + 2\pi\sum_{\rho=1}^{q}\nu_\rho\delta(z, z_\rho)$$

$$\partial_{\bar{z}}\hat{\omega}_z + \frac{1}{2}\partial_z(\chi_{\bar{z}}^+\hat{\omega}_+) = 2\pi\sum_{r=1}^{p}\alpha_r\delta(z, z_r) - 2\pi\sum_{\rho=1}^{q}\nu_\rho\theta_\rho\partial_z\delta(z, z_\rho).$$

For such differentials we may define the super line integral between "super points" $\mathbf{z} = (z, \theta_z)$ and $\mathbf{w} = (w, \theta_w)$,

(6.2)
$$\int_{\mathbf{w}}^{\mathbf{z}} d\mathbf{z}\,\hat{\omega} = \int_{w}^{z}\left(dz\hat{\omega}_z - \frac{1}{2}d\bar{z}\chi_{\bar{z}}^+\hat{\omega}_+\right) + \theta_w\hat{\omega}_+(w) - \theta_z\hat{\omega}_+(z).$$

When $\chi_{\bar{z}}^+ = 0$ dz reduces to $dz\,d\theta$ and $\hat{\omega} = \hat{\omega}_+ + \theta\hat{\omega}_z$ is meromorphic in the usual sense with poles at z_r and z_ρ. In the supergeometry framework, $\mathbf{z}_r$ should be viewed as a pole of first order, and $\mathbf{z}_\rho$ as a pole of second order. Indeed the $d\theta$ factor in the super line integral (6.2) produces nonvanishing residues only for singularities of the form $\theta_z(z - w - \theta_z\theta_w)^{-1}$.

It is helpful (although not necessary) to know that the above set-up corresponds to a covariant equation $\mathcal{D}_-$ on a $U(1)$ bundle over a two-dimensional supermanifold

(6.3)
$$\mathcal{D}_-\hat{\omega} = 2\pi\sum_{r=1}^{p}\alpha_r\delta(\mathbf{z}, \mathbf{z}_0) - 2\pi\sum_{\rho=1}^{q}\mathcal{D}_+(\nu_\rho\delta(\mathbf{z}, \mathbf{z}_0))$$

which provides a closer analogy with the equations for the usual abelian differentials.

The second equation in (6.1) yields immediately the necessary condition,

(6.4)
$$\sum_{r=1}^{p}\alpha_r = 0.$$

To solve the first one we must consider separately the cases of even and odd spin structures.

Generic even spin structures.

THEOREM 3. *For generic even spin structures there exist super abelian differentials of the*

- *First kind,* $\mathscr{D}_-\hat{\omega}_I = 0$, $I = 1, \dots, h$ *dual to the cycles* A_I *in the sense of super line integrals,*

$$\oint_{A_I} \hat{\omega}_J = \delta_{IJ}.$$

- *Second kind,* $\mathscr{D}_-\hat{\omega}_\mathbf{w} = \mathscr{D}_+\delta(\mathbf{z}, \mathbf{w})$.
- *Third kind,* $\mathscr{D}_-\hat{\omega}_{\mathbf{uv}} = \delta(\mathbf{z}, \mathbf{u}) - \delta(\mathbf{z}, \mathbf{v})$.

The super period matrix can be defined next as the periods of the $\hat{\omega}_I$ around the B_J cycles. To relate it to the conformal blocks of (5.11) and the usual period matrix Ω_{IJ}, we introduce a modified Cauchy-Riemann operator and its Green's function $\hat{S}_\delta(z, w)$,

$$(6.5) \quad \partial_{\bar{z}}\hat{S}_\delta(z, w) + \frac{1}{8\pi}\chi_{\bar{z}}^+ \int d^2 u \chi_{\bar{u}}^+ \partial_z \partial_u \ln E(z, u)\hat{S}_\delta(u, w) = 2\pi\delta(z, w)$$

which can be solved perturbatively since the Grassmann variables $\chi_{\bar{z}}^+$ depend on a large but finite number of generators.

THEOREM 4. *The following are equivalent definitions of the super period matrix* $\hat{\Omega}_{IJ}$,

- *as the periods of super abelian differentials,*

$$(6.6) \qquad\qquad \hat{\Omega}_{IJ} = \oint_{B_I} \hat{\omega}_J,$$

- *as the covariance matrix of the conformal blocks of correlators of scalar superfields,*

$$(6.7) \qquad\qquad \hat{\Omega}_{IJ} = \frac{1}{2\pi i}\partial_{p_I p_J}^2 \ln \mathscr{F}_\delta,$$

- *as a supersymmetric modification of the period matrix*

$$(6.8) \qquad \hat{\Omega}_{IJ} = \Omega_{IJ} - \frac{i}{8\pi} \int d^2 z d^2 w \omega_I(z)\chi_{\bar{z}}^+ \hat{S}_\delta(z, w)\chi_{\bar{w}}^+ \omega_J(w).$$

Generic odd spin structures.

The first key difference with even spin structures is the existence of an *even* holomorphic super abelian differential $\hat{\omega}_0 = (\hat{\omega}_+)_0 + \theta(\hat{\omega}_z)_0$ which is a supersymmetric extension of the holomorphic spinor h_δ [32],

$$(6.9) \quad \begin{aligned} (\hat{\omega}_+)_0(z) &= h_\delta + \frac{1}{4\pi} \int d^2 y \hat{S}_\delta(z, y)\chi_{\bar{y}}^+ (\hat{\omega}_z)_0(y) \\ (\hat{\omega}_z)_0(z) &= \frac{1}{4\pi} \int d^2 w \partial_z \partial_w \log E(z, w)\chi_{\bar{w}}^+ (\hat{\omega}_+)_0(w). \end{aligned}$$

Remarkably this differential has zero A_I periods. We set

$$(6.10) \qquad \oint_{A_I} \hat{\omega}_0 = 0, \qquad \oint_{B_I} \hat{\omega}_0 = \hat{\Omega}_{I0}.$$

The second difference is that there are no holomorphic *odd* super abelian differentials which we may have naively expected to generate perturbatively from the ω_I. In general we have to allow for a double pole at an arbitrary point z_0. In fact (6.3) implies

$$(6.11) \qquad 2\pi i \sum_{r=1}^{p} \alpha_r \int_{z_0}^{z_r} \hat{\omega}_0 + \sum_{\rho=1}^{q} \nu_\rho \hat{\omega}_0(z_\rho) = -\sum_{I=1}^{h} \oint_{B_I} \hat{\omega}_0 \oint_{A_I} \hat{\omega}$$

which combines with (6.4) as the necessary and sufficient condition for solvability. This means that the positions of the poles can be chosen arbitrarily, but the coefficients ν_ρ have to be solved for. In particular

THEOREM 5. *For generic odd spin structures, there exist super abelian differentials of the*

- *First kind, even, $\mathscr{D}_-\hat{\omega}_0 = 0$, vanishing A_I periods;*
- *First kind, odd, $\mathscr{D}_-\hat{\omega}_I = -\nu_I \mathscr{D}_+ \delta(z, z_0)$, $I = 1, \ldots, h$, dual to the A_I cycles;*
- *Second kind, $\mathscr{D}_-\hat{\omega}_w = \mathscr{D}_+ \delta(z, w) + \nu_w \mathscr{D}_+ \delta(z, z_0)$;*
- *Third kind, $\mathscr{D}_-\hat{\omega}_{uv} = \delta(z, u) - \delta(z, v) - \nu_{uv} \mathscr{D}_+ \delta(z, z_0)$. The "residues" of the double poles are*

$$\nu_I = \frac{1}{\hat{\omega}_0(z_0)} \oint_{B_I} \hat{\omega}_0, \qquad \nu_w = \frac{\hat{\omega}_0(w)}{\hat{\omega}_0(z_0)}, \qquad \nu_{uv} = \frac{1}{\hat{\omega}_0(z_0)} \int_u^v \hat{\omega}_0.$$

THEOREM 6. *The following are equivalent definitions of the even components $\hat{\Omega}_{IJ}$ of the super period matrix (the odd component is given by (6.10)),*
- *as the periods of super abelian differentials*

$$(6.12) \qquad \hat{\Omega}_{IJ} = \oint_{B_I} \hat{\omega}_J,$$

- *as the covariance matrix of the conformal blocks of correlators of scalar superfields*

$$(6.13) \qquad \hat{\Omega}_{IJ} = \frac{1}{2\pi i} \partial^2_{p_I p_J} \ln \mathscr{F}_\delta,$$

- *as a supersymmetric modification of the period matrix*

$$\begin{aligned}
(6.14) \qquad \hat{\Omega}_{IJ} = {} & \Omega_{IJ} - \frac{1}{h_\delta(z_0)} \hat{\Omega}_{I0} \theta_{z_0} \omega_I(z_0) - \frac{1}{h_\delta(z_0)} \hat{\Omega}_{I0} \theta_{z_0} \omega_J(z_0) \\
& + \frac{1}{16\pi^2 h_\delta(z_0)} \langle [\hat{\Omega}_{J0} \theta_{z_0} \omega_I \chi \hat{S}_\delta \chi \\
& \qquad + \Omega_{I0} \theta_{z_0} \omega_J \chi \hat{S}_\delta \chi] \partial_z \partial_{z_0} \ln E(z, z_0) \rangle \\
& - \frac{i}{8\pi} \langle \omega_I \chi \hat{S}_\delta \chi \omega_J \rangle
\end{aligned}$$

where $\hat{S}_\delta$ is viewed as an operator, and $\langle \ , \ \rangle$ denotes the natural pairing between $(1/2, 1)$ and $(1/2, 0)$ differential. Both $\hat{\Omega}_{IJ}$ and $\hat{\Omega}_{I0}$ depend on the location of the additional pole z_0.

B. The super prime form. The conformal blocks can now be written in a remarkably simple form by introducing the super prime form $\mathscr{E}_\delta(\mathbf{z}, \mathbf{w})$,

(6.15)

$$
\ln \mathscr{E}_\delta(\mathbf{z}, \mathbf{w}) = \ln E(z, w) - \theta_z \theta_w \hat{S}_\delta(z, w)
$$
$$
- \frac{1}{4\pi} \theta_z \int d^2 y \chi_{\bar{y}}^+ \partial_y \ln \frac{E(y, w)}{E(y, z)} \hat{S}_\delta(z, y)
$$
$$
- \frac{1}{4\pi} \theta_w \int d^2 y \chi_{\bar{y}}^+ \partial_y \ln \frac{E(y, z)}{E(y, w)} \hat{S}_\delta(w, y)
$$
$$
- \frac{1}{32\pi^2} \int d^2 x d^2 y \chi_{\bar{x}}^+ \partial_x \ln \frac{E(x, z)}{E(x, w)} \hat{S}_\delta(x, y) \chi_{\bar{y}}^+ \partial_y \ln \frac{E(y, w)}{E(y, z)}
$$
$$
+ \left\{ \hat{c}_{\mathbf{w}} \int_{\mathbf{z}}^{\mathbf{z}} \hat{\omega}_0 + \hat{c}_{\mathbf{z}} \int_{\mathbf{w}}^{\mathbf{w}} \hat{\omega}_0 \right\}
$$

where terms between brackets are only necessary in the case of odd spin structures. Recall that for even spin structures $\hat{S}_\delta(z, w)$ can be obtained perturbatively from $S_\delta(z, w)$ using (6.5). For odd spin structures we define $\hat{S}_\delta(z, w)$ by the same perturbative expansion but starting instead from the effective Green's function $S_\delta(z, w; z_0)$ of (5.19). The coefficients $\hat{c}_{\mathbf{z}}$, $\hat{c}_{\mathbf{w}}$ are uniquely determined by requiring the usual relation between the prime form and abelian differentials. Then

THEOREM 7. *The super prime form is an antisymmetric* $(-1/2, 0)$ *form in* $\mathbf{z}$, $\mathbf{w}$. *For even spin structures it is superholomorphic in these variables as well as in* g *and* χ. *For odd spin structures it has a pole at* $\mathbf{z}_0$. *Its "zeroes"* $\mathscr{E}_\delta(\mathbf{z}, \mathbf{w}) = 0$, $\mathscr{D}_+ \mathscr{E}_\delta(\mathbf{z}, \mathbf{w}) = 0$ *are on the diagonal* $\mathbf{z} = \mathbf{w}$. *Its monodromy is given by*

(6.16)

$$
\mathscr{E}_\delta(\mathbf{z} + A_I, \mathbf{w}) = \mathscr{E}_\delta(\mathbf{z}, \mathbf{w})
$$
$$
\mathscr{E}_\delta(\mathbf{z} + B_I, \mathbf{w}) = \mathscr{E}_\delta(\mathbf{z}, \mathbf{w}) \exp\left(-i\pi \hat{\Omega}_{II} + 2\pi i \int_{\mathbf{z}}^{\mathbf{w}} \hat{\omega}_I - \left\{ 2\pi i c_I \int_{\mathbf{z}}^{\mathbf{w}} \hat{\omega}_0 \right\} \right)
$$

with $c_I = -\theta_{z_0}(\hat{\omega}_z)_I(z_0)/h_\delta(z_0)$.

THEOREM 8. *The conformal blocks* $\mathscr{F}_\delta$ *admit the manifestly supersymmetric form*

(6.17)

$$
\mathscr{F}_\delta = Z_{\mathscr{D}, \delta}^{-10} \left\{ \int d\lambda \right\} \exp\left(i\pi p_I^\mu \hat{\Omega}_{IJ} p_J^\mu + \left\{ 2\pi i p_I \oint_{B_I} \hat{\omega}_0 \lambda \right\} \right.
$$
$$
\left. + 2\pi \sum_{i=1}^n (i k_i^\mu + \zeta_i^\mu \partial_+^i) \int_P^{\mathbf{z}_i} (p_I \hat{\omega}_I + \{\lambda \hat{\omega}_0\}) \right)
$$
$$
\times \left\langle \exp\left(\sum_{i=1}^n (i k_i^\mu X_+(\mathbf{z}_i) + \zeta_i^\mu \partial_+ X_+^\mu(\mathbf{z}_i)) \right) \right\rangle_{X_+}.
$$

Here X_+ is the "effective scalar superfield" with propagator

$$(6.18) \qquad \langle X_+(\mathbf{z}) X_+(\mathbf{w}) \rangle_{X_+} = -\ln \mathscr{E}_\delta(\mathbf{z}, \mathbf{w})$$

and $Z_{\mathscr{D},\delta}$ is the chiral determinant defined up to Liouville terms by

$$(6.19) \qquad Z_{\mathscr{D},\delta} = Z_0^{-1} Z_{1/2} \left\langle \exp\left(-\frac{1}{4\pi} \int d^2 z \chi_{\bar{z}}^+ \psi_+^\mu \partial_z x_+^\mu \right) \right\rangle.$$

The precise relation between $Z_{\mathscr{D},\delta}$ and the determinants of the super Laplacian on scalars is

$$(6.20) \qquad \left(\frac{s \det' \mathscr{D}_+ \mathscr{D}_-}{\int d\mathbf{z} E \det \operatorname{Im} \hat{\Omega}} \right)^{-1/2} = |Z_{\mathscr{D},\delta}|^{-2} \left\{ \oint_{B_I} \hat{\omega}_0 (\operatorname{Im} \hat{\Omega})_{IJ}^{-1} \overline{\oint_{B_J} \hat{\omega}_0} \right\}$$

where again terms relevant only to odd spin structures have been put between brackets.

C. The superzweibein formalism for two-dimensional supergravity. We have seen how the pair $x, \psi_\pm$ can be unified into a single superfield $X(z, \theta, \bar{\theta})$. The superzweibein formalism of supergravity accomplishes the same thing for the pair e_m^a, χ_m^α. It maintains manifest supersymmetry and also explains the complicated covariant derivatives of (5.5). Let $z^M = (z, \bar{z}, \theta, \bar{\theta})$ be viewed as local coordinates for a supermanifold. A supergeometry in the superfield formalism is a superzweibein $E^A = dz^M E_M^A$ and a $U(1)$ super connection $\Omega = dz^M \Omega_M$ satisfying the Wess-Zumino torsion constraints

$$(6.21) \qquad T_{ab}^c = T_{\alpha\beta}^\gamma = 0, \qquad T_{\alpha\beta}^c = 2(\gamma^c)_{\alpha\beta}.$$

Here γ^c are the Dirac matrices. As usual we have denoted Lorentz indices by early letters (a, A, α), Einstein indices by middle letters (m, M, μ), commuting coordinates by Latin letters, anticommuting coordinates by Greek letters, by T_{AB}^C the torsion tensor and by R_{AB} the curvature tensor as defined from commutation relations of covariant derivatives on Lorentz tensors of $U(1)$ weight n,

$$(6.22) \qquad \begin{aligned} \mathscr{D}_A &= E_A^M (\partial_M + in\, \Omega_M) \\ [\mathscr{D}_A, \mathscr{D}_B] &= T_{AB}^C \mathscr{D}_C + in\, R_{AB}. \end{aligned}$$

In (6.22) the bracket $[\ ,\ ]$ is a commutator unless both A, B are spinor indices, in which case it becomes an anticommutator. On scalars there is no curvature term in (6.22), and the second torsion constraint (6.21) just says that $\mathscr{D}_+^2 = \partial_z$. The other constraints are conditions for the almost-complex structure

$$(6.23) \qquad J_M^N = E_M^a \varepsilon_a^b E_b^N + E_M^\alpha (\gamma_5)_\alpha^\beta E_\beta^N$$

to be integrable. The Wess-Zumino constraints are more severe than torsion constraints in differential geometry. The structure group is $U(1)$ instead of

$O(2|2)$. Indeed there are only 16 degrees of freedom for E_M^A, 4 for Ω_M, while there are 14 constraints from (6.21). This implies that one cannot solve for the connection given *any* superzweibein.

The simplest example is flat superspace, as given by

$$(6.24) \qquad E_m^a = \delta_m^a, \qquad E_m^\alpha = 0, \qquad E_\mu^\alpha = (\gamma^\alpha)_\mu^\beta \theta_\beta, \qquad E_\mu^\alpha = \delta_\mu^\alpha$$

and the covariant derivatives take the familiar form $\mathscr{D}_+ = \partial_\theta + \theta \partial_z$. More generally there is a correspondence between superzweibeins satisfying the Wess-Zumino constraints and the pair g, χ we have used so far. By analogy with standard geometry, the groups of super diffeomorphisms and super Lorentz transformations act on supergeometries. Their generators are respectively super vector fields V^M and super scalars Φ. By suitable transformations, any superzweibein satisfying the Wess-Zumino constraints is locally equivalent to

$$E_\mu^\alpha = \delta_\mu^\alpha + \theta^\nu e^{*\alpha}_{\nu\mu}, \qquad E_\mu^a = \theta^\nu e^{**a}_{\nu\mu}$$

up to higher order terms in θ. Such a form is called a "Wess-Zumino gauge." Expanding the other components of the superzweibein gives

$$(6.25) \qquad E_m^a = e_m^a + \theta^\alpha (\gamma^a)_\alpha^\beta \chi_{m\beta} - i\theta\bar{\theta} e_m^a A/2.$$

The scalar A can be set to 0 in practice, giving the desired correspondence with g, χ. The derivative $\mathscr{D}_-$ of (5.5) is just the covariant derivative with respect to the connection Ω in Wess-Zumino gauge. On $U(1)$ rank n tensors $V = V_0 + \theta V_+ + \bar{\theta} V_- + i\theta\bar{\theta} V_1$ it is given by

$$(6.26) \qquad \begin{aligned} \mathscr{D}_-^{(n)} V &= V_- + \theta(-iV_1) + \bar{\theta}(\partial_{\bar{z}} V_0 + \tfrac{1}{2}\chi_{\bar{z}}^+ V_+) \\ &+ \theta\bar{\theta}(-\tfrac{1}{4}\chi_{\bar{z}}^+\chi_z^- V_- - \tfrac{1}{2}\chi_{\bar{z}}^+\partial_z V_0 - \partial_{\bar{z}} V_+ - n(\partial_z\chi_{\bar{z}}^+)V_0). \end{aligned}$$

Next the action (5.1) in superfield language becomes

$$(6.27) \qquad I_\delta = \frac{1}{8\pi} \int d^2 z \, d\theta \, d\bar{\theta} \, E \mathscr{D}_+ X \mathscr{D}_- X$$

which is the exact analogue of the bosonic action (2.1).

Finally we have been parametrizing supergeometries by $\chi_{\bar{z}}^+$ and the moduli of g. In the superzweibein formalism it is natural to parametrize deformations of supergeometries by

$$(6.28) \qquad H_A^B = E_A^M \delta E_M^B.$$

The Wess-Zumino constraints imply that all components of H_A^B can actually be written in terms of $H_-^-, H_-^{\bar{z}}, H_-^z$ and their conjugates only. The components $H_-^-, H_-^{\bar{z}}$ can be set to 0 by super Weyl. Thus a deformation of super Riemann surface is parametrized by the sole H_-^z which can be expanded as

$$(6.29) \qquad H_-^z = \bar{\theta}(e_{\bar{z}}^m \delta e_m^z - \theta \delta \chi_{\bar{z}}^+).$$

Evidently $\mu_{\bar{z}}^{\ z} = e_{\bar{z}}^m \delta e_m^z$ is a Beltrami differential, so that a variation of superzweibein corresponds indeed to a change of complex structure and of gravitino $\chi_{\bar{z}}^+$.

VII. Superstrings

A. Physical principles. The bosonic string model previously described is unrealistic because it contains a tachyon and carries no fermionic degrees of freedom. A way out of these difficulties is to supersymmetrize the theory. In II we have introduced fermionic partners $\psi_{\pm}^{\mu}$, χ_m^{α} to the bosonic fields x^{μ}, g_{mn} respectively. Geometrically (g, χ) describe a super surface, and (x^{μ}, ψ^{μ}) can be identified with d scalar superfield X^{μ} which provide an imbedding of the super surface in space-time. Conformal invariance at the quantum level will require that $d = 10$. The action is (5.1), or equivalently (6.27). Quantization leads to a sum over histories

$$(7.1) \qquad \left\langle \prod_{i=1}^N V_i \right\rangle_{\delta} = \int DE_M^A DX^{\mu} e^{-I_{\delta}} \prod_{i=1}^N V_i.$$

This is the quantum theory of scalar superfields coupled to two-dimensional supergravity, and is not yet superstring theory. It is a fundamental requirement of superstrings that the so-called Gliozzi-Scherk-Olive projection be performed. The reason is that fermionic degrees of freedom have been introduced, but undesirable states such as the tachyon are still present. The GSO projection is a suitable truncation of the spectrum which eliminates the tachyon and insures **space-time supersymmetry**. Now the GSO projection is an operator which can be given a rigorous definition in the Hamiltonian formalism. In the functional formalism, it is believed to correspond to the following rough prescription: separate the holomorphic from the antiholomorphic blocks in correlation functions with respect to a fixed spin structure, sum over spin structures separately for each side, and group them back together. This is why conformal blocks of scalar superfields in V are needed. In the case of genus 1, the Hamiltonian formalism is also available, and the functional prescription of summing over spin structures" agrees with the operator theoretic prescription. It is our goal presently to give a precise and workable prescription for the GSO projection on surfaces of any genus.

B. Supermoduli and gauge-fixing. In the same way as bosonic string integrands reduce to moduli space, (7.1) can be expressed as integrals over the supermoduli space $s\mathcal{M}_h$, which is the space of equivalence classes of supergeometries under superdiffeomorphisms and super Weyl transformations. The arguments leading to the dimension of moduli space outlined in II.D can be duplicated here to give the dimensions of $s\mathcal{M}_h$ as

$$\dim s\mathcal{M}_h = \begin{pmatrix} 0 & h = 0 \\ (1|0) & h = 1, \delta \ even \\ (1|1) & h = 1, \delta \ odd \\ (3h - 3|2h - 2) & h \geq 2 \end{pmatrix}$$

where $(m|n)$ denotes m even and n odd dimensions. Since in the superfield and superzweibein formalism the theory is formally identical to the bosonic string, we can write down immediately the gauge-fixed version of (7.1) in analogy with (4.12),

$$(7.2) \quad Z_\delta = \int \prod_{J=1}^{5h-5} d\tau_J d\bar\tau_J |s \det\langle\mu_J|\Phi_K\rangle|^2 \frac{s \det \mathscr{D}_2^\dagger \mathscr{D}_2}{s \det\langle\Phi_J|\Phi_K\rangle} \left(\frac{s \det \mathscr{D}_0^\dagger \mathscr{D}_0}{\int dz d\theta d\bar\theta E}\right)^{-5}.$$

Here we have parametrized supermoduli by a $(3h - 3|2h - 2)$ dimensional slice $\mathscr{S}$ of superzweibeins with coordinates τ_J, denoted by μ_J the "super Beltrami differentials" tangent to $\mathscr{S}$,

$$(7.3) \quad (\mu_J)_-^z = (-)^{m+1} E_-^M \partial E_M^z / \partial \tau_J$$

and chosen a basis Φ_J of zero modes for $\mathscr{D}_2$. For superdeterminants, see Appendix D. The operator $\mathscr{D}_2$ is the covariant derivative of (6.26). Again it is convenient to reformulate (7.2) in terms of ghost fields. Thus we introduce superghost fields B and C of $U(1)$ weights $3/2$ and -1 respectively, with action

$$(7.4) \quad I_{sgh}(B, C) = \frac{1}{2\pi} \int d^2\mathbf{z} E(B\mathscr{D}_- C + \bar{B}\mathscr{D}_+ \bar{C}).$$

The same reasoning leading to (4.13) gives

$$(7.5) \quad \begin{aligned} &\int D(B\bar{B}C\bar{C})e^{-I_{sgh}(B,C)} \prod_{J=1}^{5h-5} |\delta(\langle\mu_J|B\rangle)|^2 \\ &= \frac{|s \det\langle\mu_J|\Phi_K\rangle|^2}{s \det\langle\Phi_J|\Phi_K\rangle} s \det \mathscr{D}_2^\dagger \mathscr{D}_2. \end{aligned}$$

The Dirac delta function is appropriate here, since $\langle\mu_J|B\rangle$ can be both even or odd. For $\langle\mu_J|B\rangle$ odd, we have $\delta(\langle\mu_J|B\rangle) = \langle\mu_J|B\rangle$, as can be readily verified from integration rules of Grassmann quantities (cf. Appendix C). Altogether we obtain

$$(7.6) \quad Z_\delta = \int \prod_{J=1}^{5h-5} d\tau_J d\bar\tau_J \int D(B\bar{B}C\bar{C}X^\mu)e^{-I_\delta - I_{sgh}(B,C)} \left|\prod_{J=1}^{5h-5} \delta(\langle\mu_J|B\rangle)\right|^2.$$

We may also insert vertex operators and carry out the X integrals
(7.7)

$$\left\langle\prod_{i=1}^N V_i\right\rangle_\delta = \int \prod_{J=1}^{5h-5} d\tau_J d\bar\tau_J \int D(B\bar{B}C\bar{C})e^{-I_{sgh}(B,C)} |\delta(\langle\mu_J|B\rangle)|^2 \left\langle\prod_{i=1}^N V_i\right\rangle_X.$$

Finally we turn to the GSO projection to arrive at superstring amplitudes. The X correlation functions can be holomorphically split in view of the Theorem 1. Furthermore up to ultimately irrelevant super Liouville terms, the contribution of the ghosts splits as

$$(7.8) \qquad \frac{|s \det\langle \mu_J | \Phi_K \rangle|^2}{s \det\langle \Phi_J | \Phi_K \rangle} s \det \mathscr{D}_2^\dagger \mathscr{D}_2 = |\mathscr{F}_\delta^{gh}(\Omega, \chi_{\bar{z}}^+)|^2.$$

This can established by superfield arguments as in [30, 9]. We arrive now at our most important formulas, giving scattering amplitudes for gravitons in superstring theory [32],

$$(7.9) \qquad \left\langle \prod_{i=1}^{N} V_i \right\rangle_{II} = \int \prod_{J=1}^{5h-5} d\tau_J \, d\bar{\tau}_J \int dp_I^\mu \left| \sum_\delta (-1)^\delta \mathscr{F}_\delta^{gh}(\Omega, \chi_{\bar{z}}^+) \right.$$
$$\left. \times \mathscr{F}_\delta(\mathbf{z}_i, \zeta_i^\mu, k_i^\mu; \Omega, \chi_{\bar{z}}^+; p_I^\mu) \right|^2.$$

Here $(-1)^\delta$ is a relative sign between odd and even spin structures. It is $+$ in the Type IIA superstring, and $-$ in the Type IIB superstring.

C. Component formalism for the superstring.

We have seen in V–VI that the quantum theory of scalar superfields as well as the geometry of super Riemann surfaces can be described equally well in the superfield formalism with X^μ and E_M^A, or in the "component formalism" with (x^μ, ψ_α^μ) and (g_{mn}, χ_m^α). To integrate over supergeometries however, we have relied so far on the superfield formalism, and written string scattering amplitudes as integrals over *supermoduli* space, with the correct measure expressed in terms of the superghosts (B, C). To make contact ultimately with *moduli* space, we need now a component formalism for the superghosts.

This can be obtained by expanding the superfields $B(z, \theta)$, $C(z, \theta)$

$$(7.10) \qquad B(z, T) = \beta + \theta b, \qquad C(z, \theta) = c + \theta \gamma.$$

The fields b and c are the familiar anticommuting ghost fields of III.B. They have $U(1)$ weight 2 and -1. The new fields β and γ are commuting, and have weight $3/2$ and $-1/2$ respectively. The ghost action becomes

$$(7.11) \qquad \begin{aligned} I_{gh} = &\frac{1}{2\pi} \int d^2 z (b\partial_{\bar{z}} c + \bar{b}\partial_z \bar{c} + \beta \partial_{\bar{z}} \gamma + \bar{\beta} \partial_z \bar{\gamma}) \\ &- \frac{1}{2\pi} \int d^2 z \chi_{\bar{z}}^+ \left(-(\partial_z \beta)c - \frac{3}{2}\beta \partial_z c + \frac{1}{2} b\gamma + \text{c.c.} \right). \end{aligned}$$

Combining the matter and ghost actions give

$$I_\delta + I_{gh}(B, C) = I_{\text{free}} - \frac{1}{2\pi} \int d^2 z (\chi_{\bar{z}}^+ S + \chi_z^- \bar{S})$$

where I_{free} is the *total* free action and $S(z)$ is the total "super current"

$$I_{\text{free}} = \frac{1}{2\pi} \int d^2 z \left(\frac{1}{2} \partial_z x \partial_{\bar{z}} x - \frac{1}{2} \psi_+^\mu \partial_{\bar{z}} \psi_+^\mu - \frac{1}{2} \psi_-^\mu \partial_z \psi_-^\mu + b \partial_{\bar{z}} c \right.$$

(7.12)
$$\left. + \bar{b} \partial_z \bar{c} + \beta \partial_{\bar{z}} \gamma + \bar{\beta} \partial_z \bar{\gamma} \right)$$

$$S(z) = -\frac{1}{2} \psi_+^\mu \partial_z x^\mu - (\partial_z \beta) c - \frac{3}{2} \beta \partial_z c + \frac{1}{2} b \gamma.$$

By the effective rules for chiral splitting, we have dropped mixed terms such as $\chi_{\bar{z}}^+ \chi_z^-$. This requires insertion of loop momenta p_I^μ which we do not write explicitly. Separating the super Beltrami differentials μ_J into the even μ_i and the odd μ_a, we can rewrite (7.5) as

$$Z_\delta = \int_{s\mathcal{M}_h} \int D(b\bar{b}c\bar{c}\beta\bar{\beta}x^\mu \psi_\pm^\mu) \prod_{i=1}^{3h-3} d\tau_i \, d\bar{\tau}_i$$

(7.13)
$$\times \prod_{a=1}^{2h-2} d\tau_a \, d\bar{\tau}_a |\langle \mu_i | b \rangle|^2 \prod_{a=1}^{2h-2} |\delta(\langle \mu_a | \beta \rangle)|^2$$

$$\times \exp(-I_{\text{free}}(x, b, c, \beta, \gamma) + \langle \chi_{\bar{z}}^+ | S \rangle + \langle \chi_z^- | \bar{S} \rangle).$$

This is the fundamental gauge-fixed formula for superstrings in component language.

D. The case of the torus, the Jacobi identity, and space-time supersymmetry. We work out the partition function in the case of the torus explicitly and discuss the underlying physics. For the odd spin structure the fields $\psi_\pm$ have a zero mode and the functional integral $D\psi$ vanishes. It suffices to consider then the three even spin structures. In this case there are no supermoduli and we can set $\chi_{\bar{z}}^+ = 0$. The contributions of the x, ψ are given by

(7.14)
$$\int D(x\psi_+\psi_-)e^{-I_\delta} = \left(\frac{\det' \bar{\partial}_0^\dagger \bar{\partial}_0}{\int d^2 z \sqrt{g}} \right)^{-5} (\det'_\delta \bar{\partial}_{1/2}^\dagger \bar{\partial}_{1/2})^5.$$

For the ghosts and superghosts the measure is given by

(7.15)
$$\int D(b\bar{b}c\bar{c}\beta\bar{\beta}\gamma\bar{\gamma})e^{-I_{gh}} bc\delta(\beta)\delta(\gamma) = \frac{\det' \bar{\partial}_2^\dagger \bar{\partial}_2}{4\tau_2^2} (\det' \bar{\partial}_{1/2}^\dagger \bar{\partial}_{1/2})^{-1}$$

with the insertions due to the zero modes (cf. (3.14)). Now the determinant of the Laplacian on scalars has already been calculated in (4.15). Similar considerations give

$$\det_\delta \bar{\partial}_{1/2}^\dagger \bar{\partial}_{1/2} = \left| \frac{\theta[\delta](0, \tau)}{\eta(\tau)} \right|^2.$$

Evidently the determinants for ghosts and superghosts are given by the same formulas, since we are on the torus. Altogether the string integrand is for each fixed spin structure

$$(7.16) \qquad \left| \frac{\theta[\delta](0, \tau)^4}{\eta(\tau)^{12}} \right|^2.$$

Now we have to enforce the GSO projection, beginning by taking the holomorphic root of (7.16). Evidently there is an arbitrariness in the phase, to be resolved by requiring that the sum over spin structures be modular invariant. From transformation properties of theta functions with characteristics, we see that up to a global phase the only possible choice is

$$(7.17) \qquad \frac{1}{\eta(\tau)^{12}}(\theta_3^4(0, \tau) - \theta_4^4(0, \tau) - \theta_2^4(0, \tau)) = 0.$$

This is 0 according to the Jacobi identity [71]! This has the following important physical interpretation. The partition function is the generating function of the number of states. The term $q^{-1/2}(\theta_3^4 - \theta_4^4)/(2\eta^{12})$ counts the bosonic states, while the term $q^{-1/2}\theta_2^4/(2\eta^{12})$ counts the fermionic ones. That (7.15) vanishes identically means that the number of bosonic and fermionic states are the same at each mass level, which is a manifestation of space-time supersymmetry. The integral of (7.17) with respect to τ can also be interpreted as a first order loop correction to the space-time cosmological constant. To tree level, the cosmological constant is 0, and (7.17) strongly hints that it will remain 0 to all orders of perturbation theory. This is a desirable feature for phenomenology.

VIII. Current issues in perturbation theory

This section is devoted to problems of current interest in superstring perturbation theory. This part is not yet on as rigorous a mathematical footing as the previous parts, and our discussion will be more physics oriented. Because of lack of space we shall also assume more quantum field theory background than heretofore, and be briefer. We hope however to provide enough motivation for geometers to look into the many challenging problems in this area.

A. The light-cone gauge and unitarity. We have based our treatment of superstring theory on the formulation of Polyakov. There are other formulations which have each their own advantages and disadvantages. The Green-Schwarz formulation [43] is one with manifest space-time supersymmetry. It arises from a covariant action with the elegant interpretation as a sigma model with Wess-Zumino term in superspace. However its quantization is notoriously difficult and has not been developed nearly as far as the Polyakov formulation. The light-cone gauge formulation pioneered by Mandelstam [58, 59, 15] has the advantage of introducing only physical degrees of freedom, so that the interpretation of strings as an evolving quantum mechanical system and unitarity are manifest. Its main drawback is that Lorentz covariance in space-time is obscured. In both light-cone gauge and Polyakov formulations, space-time supersymmetry is a difficult issue.

It is an important problem to establish the equivalence of all these formulations to all orders of perturbation theory. Here we shall sketch an argument for the equivalence of the Mandelstam and Polyakov formulations [4]. Although the argument is not yet at the level of rigor of a mathematical theorem, it should be possible (and interesting!) to construct the mathematical framework to make it into one.

The light-cone gauge is based on the existence of a meromorphic form ω with purely imaginary periods on a Riemann surface with at least two punctures. In fact it is given by

$$(8.1) \qquad\qquad \omega = \sum_{r=1}^{p} \beta_r \omega_{P_r P_{r-1}}$$

where P_r, $r = 1, \ldots, p$ are the punctures. The zeroes z_a, $a = 0, \ldots, 2h - 3 + p$ are called the "interaction points." Integrating ω gives a multiple-valued coordinate system z on the surface. The real part of z defines however an unambiguous global time τ (Figure 6)

$$(8.2) \qquad\qquad \tau = \mathrm{Re} \int_{z_0}^{z} \omega.$$

The complex structure of the Riemann surface can be read off from ω.

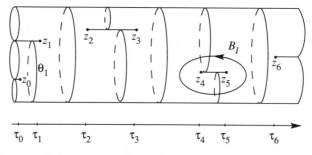

FIGURE 6. Representation of a punctured Riemann surface by its Mandelstam diagram.

Define
(8.3)

$$\tau_a = \tau(z_a), \qquad a = 1, \ldots, 2h-3+p, \qquad \theta_I = 2\pi \oint_{B_I} \omega \Big/ \oint_{A_I} \omega$$

$$I = 1, \ldots, h,$$

$$\alpha_I = \int_{A_I} \omega, \qquad I = 1, \ldots, h, \qquad \theta_a = 2\pi \int_{z_{a-1}}^{z_a} \mathrm{Im}\,\omega \Big/ \oint_{C_a} \omega,$$

$$a = 1, \ldots, 2h-3+p,$$

with C_a closed curves around the cylinder starting at z_{a-1} and ending at z_a. The parameters τ_a, α_I, (θ_I, θ_a) are called respectively "interaction times," "internal momenta," and "twist angles." As they vary over their natural range, we obtain a copy of moduli space up to a set of measure 0 [40, 58].

A similar parametrization of supermoduli space can be obtained by constructing for each super Riemann surface a superholomorphic $1/2$ form λ with purely imaginary periods. We discuss only the case of even spin structures for the sake of simplicity. The case of odd spin structures is treated in detail in [4]. As in (8.1) λ is constructed using super abelian differentials of the third kind which exist by Theorem 3. The interaction points z_a are more subtle, and are defined by the equations of Mandelstam

$$(8.4) \qquad \partial_+\lambda + \frac{1}{2}\frac{\lambda\partial_+^4\lambda}{\partial_+^3\lambda} = 0, \qquad \partial_+^2\lambda - \frac{1}{6}\frac{\lambda\partial_+^5\lambda}{\partial_+^3\lambda} = 0.$$

Although the equations are not covariant, their solutions are! We can now duplicate (8.3) with super line integrals of the $1/2$-form λ replacing the standard line integrals of the one-form ω and obtain invariants $(\hat\tau, \hat\alpha, \hat\theta)$ for the super Riemann surface. These will form the $(3h-3)$ even coordinates for supermoduli. The remaining $2h-2+p$ odd coordinates are taken to be η_a, $a = 1, \ldots, 2h-2+p$ where $(w_a|\eta_a)$ are the Mandelstam coordinates of the interaction points

$$(8.5) \qquad w_a = \int_{z_0}^{z_a} \lambda, \qquad \eta_a = (\mathscr{D}_+\lambda)^{-1/2}\lambda.$$

In the Mandelstam-Berkovits formulation [58, 59, 15], superstring propagation is described by transverse degrees of freedom only, so that the number of scalar superfields is now eight instead of ten. On the other hand, there is no gauge symmetry. The super Riemann surface is represented by its Mandelstam diagram. Scattering amplitudes are given by

$$(8.6) \qquad A(k_1, \ldots, k_N) = \int \prod d(\hat\tau\hat\alpha\hat\theta\eta\bar\eta) \int DX^i e^{-I} \prod_{r=1}^{p} V_r.$$

Here $i = 1, \ldots, 8$, the action I_δ is that of (6.27),

$$(8.7) \qquad I_\delta = \frac{1}{4\pi} \int d^2z E \mathscr{D}_- X^i \mathscr{D}_+ X^i$$

and V_r denotes vertex operators for emission of physical particles, written in the light-cone gauge

$$(8.8) \qquad V_r = \int DP_r^i \phi_r(P_r^i) e^{-P_r^- \tau_r + i \oint P_r^i X^i}.$$

In particular the partition function is given by

$$(8.9) \qquad Z_\delta = \int \prod d(\hat{\tau} \hat{\alpha} \hat{\theta} \eta \bar{\eta}) \left(\frac{\det' \mathscr{D}_+ \mathscr{D}_-}{\int d^2 \mathbf{z} E} \right)^{-4}.$$

The expressions (8.6),(8.9) are *before* the GSO projection, which still has to be performed as in the Polyakov formulation.

It suffices then to establish the equivalence of the Mandelstam and Polyakov formulations for the nonsplit amplitudes. Now the Mandelstam diagrams can be viewed as a slice of supergeometries representing supermoduli space, and we can apply the gauge-fixed formula (7.2). The problem reduces to evaluating the super Beltrami differentials μ_A for the Mandelstam slice, and show that

$$(8.10) \qquad \frac{|s \det \langle \mu_A | \phi_B \rangle|^2}{s \det \langle \phi_A | \phi_B \rangle} s \det \mathscr{D}_+^{(-3/2)} \mathscr{D}_-^{(-1)} = \frac{s \det \mathscr{D}_- \mathscr{D}_+^{(0)}}{\int d\mathbf{z} E}.$$

A major difficulty in (8.10) is a tractable choice of super 3/2 ("super quadratic") differentials Φ_A. In view of Theorem 3, it suffices to construct a one-differential from the canonical 1/2 differential λ. This task is peculiar to supergeometry, since in standard geometry, the canonical differential ω is already a one-form. It is natural to try $\hat{\omega} = \partial_+ \lambda$, but the coordinates $(w|\eta)$ obtained by integrating λ are singular near the zeroes $\mathbf{z}_a$. The following regularized form for $\varepsilon \to 0$ has been suggested by Mandelstam

$$(8.11) \qquad \hat{\omega}_\varepsilon = \partial_+ \lambda - \frac{\partial_+^2 \lambda}{\partial_+ \lambda} \lambda - \frac{\varepsilon^2}{\partial_+ \lambda}.$$

Its singularities are of the form

$$(8.12) \qquad \hat{\omega}_\varepsilon \sim \frac{(\mathbf{z} - \mathbf{z}_a^+)(\mathbf{z} - \mathbf{z}_a^-)}{(\mathbf{z} - \mathbf{z}_a^0)}$$

where the zeroes $\mathbf{z}_a^\pm$ and pole $\mathbf{z}_a^0$ are regularization dependent and have the same body. Now if $\hat{\omega}$ were a holomorphic one-differential with zeroes at $\mathbf{z}_a$, a basis of 3/2 differentials would be given by

$$(8.13) \qquad \begin{aligned} \Phi_I &= \hat{\omega} \hat{\omega}_I, & I &= 1, \ldots, h, \\ \Phi_a &= \hat{\omega} \hat{\omega}_{\mathbf{z}_0 \mathbf{z}_a}, & a &= 1, \ldots, 2h - 3 + p, \\ \Phi_\alpha &= \hat{\omega} \hat{\omega}_{\mathbf{z}_\alpha}, & \alpha &= 0, 1, \ldots, 2h - 3 + p. \end{aligned}$$

The presence of the complicated zero and pole structure in the regularized $\hat{\omega}_\varepsilon$ is troublesome, and it is remarkable that they can be cancelled by suitable choice of the poles of the abelian differentials, e.g.,

$$(8.14) \qquad \Phi_\alpha = \hat{\omega}_\varepsilon (\hat{\omega}_{\mathbf{z}_\alpha^+} + \hat{\omega}_{\mathbf{z}_\alpha^-} - \hat{\omega}_{\mathbf{z}_\alpha^0})$$

is holomorphic. Similar modifications yield a full basis of $3/2$-differentials Φ_A from the construction (8.13). The volume of the Φ_A's and the relative volume of the Beltrami differentials for the slice of Mandelstam diagrams are found to be

(8.15)
$$s \det \langle \phi_A | \phi_B \rangle = (2\pi)^{-1} 2^h s \det G \det \text{Im} \, \hat{\Omega}$$
$$|s \det \langle \mu_A | \Phi_B \rangle|^2 = (4\pi)^h (2\pi)^{4h-6+2p} \det \text{Im} \, \hat{\Omega}$$

where G is the matrix of super Green's functions

(8.16)
$$G = \begin{pmatrix} \langle X(\mathbf{z}_a) X(\mathbf{z}_b) \rangle & \text{Re} \langle X(\mathbf{z}_a) \mathscr{D}_+ X(\mathbf{z}_\beta) \rangle \\ \text{Re} \langle \mathscr{D}_+ X(\mathbf{z}_\alpha) X(\mathbf{z}_b) \rangle & -\langle \mathscr{D}_+ X(\mathbf{z}_\alpha) \mathscr{D}_+ X(\mathbf{z}_\beta) \rangle \end{pmatrix}.$$

Next vector fields V^z may be set in correspondence with scalars X by letting $V^z = (\hat{\omega})^{-1} X$. For $\hat{\omega}$ holomorphic form with zeroes at $\mathbf{z}_a$, the condition for regularity of V is $X(\mathbf{z}_\alpha) = \text{Re} \, \mathscr{D}_+ X(\mathbf{z}_\alpha) = 0$, and we may write

(8.17)
$$(s \det \mathscr{D}_+^{(-3/2)} \mathscr{D}_-^{(-1)})^{-1/2} = \int DX \, e^{-I_\delta} \prod_{a=0}^{2h-3+p} \delta(X(\mathbf{z}_a)) \delta(\text{Re} \, \mathscr{D}_+ X(\mathbf{z}_a))$$
$$= \int \prod_{a=0}^{2h-3+p} \frac{dk_a d\kappa_a}{2\pi} \int DX \exp\left(-I_\delta + \sum_{a=0}^{2h-3+p} ik_a X(\mathbf{z}_a) + \kappa_a \text{Re} \, \mathscr{D}_+ X(\mathbf{z}_a)\right).$$

With the regularized form $\hat{\omega}_\varepsilon$ instead of a holomorphic $\hat{\omega}$ we have to modify in the same way as (8.14) the conditions on X and the Dirac δ functions in (8.17). The resulting functional integrals are of the form (5.7). Applying Theorem 8 gives the remarkably simple answer

$$s \det \mathscr{D}_+^{(-3/2)} \mathscr{D}_-^{(-1)} = \frac{s \det \mathscr{D}_- \mathscr{D}_+^{(0)}}{\int d\mathbf{z} \, E} s \det G.$$

The desired identity (8.10) follows immediately.

B. BRST Invariance and reduction to moduli space. We return now to the problem of explicit evaluation of superstring scattering amplitudes. In VII we have provided a formula for these quantities as an integral over supermoduli space. At the present time supermoduli space appears intractable, and most efforts have rather been in the direction of reducing these formulas to more familiar integrals over moduli space. The problem is how to integrate out the odd supermoduli parameters.

In the gauge-fixed form the original gauge symmetries of a quantum field theory are no longer recognizable. A powerful discovery of Becchi-Rouet-Stora-Tyutin is that the gauge-fixed theory incorporating the ghosts acquires instead a new symmetry with Grassmann parameter, called the BRST symmetry. This symmetry constrains severely the possible expressions for the gauge-fixed amplitudes. For the superstring Friedan, Martinec, and Shenker

[37] have proposed the following formula

$$(8.18) \qquad Z_\delta = \int_{\mathscr{M}_h} \prod_{i=1}^{3h-3} d\tau_i \, d\overline{\tau}_i |\langle \mu_i | b \rangle|^2 \prod_{a=1}^{2h-2} |Y(z_a)|^2 e^{-I_{\text{free}}(x,\psi,b,c,\beta,\gamma)}.$$

Here $b = b_{zz} dz^2$, $c = c^z (dz)^{-1}$ are the familiar ghost fields for diffeomorphism invariance, and $\beta = \beta_{z+} (dz)^{3/2}$, $\gamma = \gamma^+ (dz)^{-1/2}$ are the analogous (commuting) ghost fields for local supersymmetry (cf. VII.C). The action in (8.18) is the free action of (7.12), and $Y(z)$ is the "picture changing operator"

$$(8.19) \qquad \qquad Y(z) = e^{-i\sigma(z)} S(z)$$

with $S(z)$ the super current (7.12), and σ comes from bosonization of the supersymmetry ghosts

$$\beta(z) = \partial_z \xi(z) e^{i\sigma(z)}, \qquad \gamma(z) = \eta(z) e^{-i\sigma(z)}.$$

Here we must introduce fermion fields ξ and η because β, γ are commuting fields. The operators $Y(z)$ and S are BRST transforms of $H(\beta(z))$ and $\beta(z)$ respectively, and hence are BRST invariant. As for b we have $\oint_{C_z} dw \, j_{\text{BRST}}(w) b(z) = T_{zz}^{\text{total}}$ where T^{total} denotes the stress tensor of matter and ghost fields combined. It is now easy to check the BRST invariance of the Ansatz (8.18). An insertion of the BRST charge corresponds to an insertion of a contour integral of the BRST current. We can deform this contour across the insertions $Y(z_a)$ and $\langle \mu | b \rangle$. There is no contribution from the $Y(z_a)$ since they are BRST invariant, but the $\langle \mu | b \rangle$ produce a factor of

$$(8.20) \qquad \qquad \langle \mu_i | T_{zz} \rangle = \int d^2 z (\mu_i)_{\overline{z}}^{\ z} T_{zz}^{\text{total}}.$$

In view of (3.18), we can identify an insertion of the form (8.20) as a derivative over moduli space. In particular a change of points z_a can be viewed as a change of gauge slice and hence a BRST change. Thus the expression (8.18) is independent of the choice of insertion points z_a up to derivatives on moduli space. How harmful are these derivatives? First moduli space has a boundary, and hence they may very well contribute. Worse still, they are defined only over the overlaps of coordinate patches where the insertion points z_a are chosen. This makes the integrals (8.18) ill-defined, a situation reminiscent of the Wu-Yang analysis of the action for an electric charge in the field of a magnetic monopole.

One way out of this difficulty is to add "by hand" to (8.18) Čech cohomology type correction terms [91]. Another way is to trace the origin of these difficulties by going back to the general gauge-fixed partition function over supermoduli space (7.13). Parametrize now supermoduli by choosing a slice for moduli, with Beltrami differentials μ_i, and gravitino fields

$\chi_{\bar{z}}^{+} = \sum_{a=1}^{2h-2} \tau_a \delta(z, z_a)$. The integral over the odd supermoduli τ_a gives

(8.21)

$$Z_\delta = \int_{\mathcal{M}_h} \prod_{i=1}^{3h-3} d\tau_i \, d\bar{\tau}_i |\langle \mu_i | b \rangle|^2$$

$$\times \prod_{a=1}^{2h-2} |\delta(\langle \mu_a | \beta \rangle)|^2 \prod_{a=1}^{2h-2} |\delta(\beta(z_a)) S(z_a)|^2 \exp(-I_{\text{free}}(x, b, c, \beta, \gamma)).$$

From operator product expansions, we recognize that $\delta(\beta(z))$ actually coincides with $\exp(-i\sigma(z))$, so we seem to recapture the expression suggested by BRST invariance. How did the slice-independent formula (7.13) lead to a slice-dependent one (8.18)? A close look shows that we have relied in the above derivation on the seemingly natural projection

$$\begin{array}{ccc}
(g_{mn}, \chi_m^\alpha) & \to & (g_{mn} + \delta g_{mn}, \chi_m^\alpha + \delta \chi_m^\alpha) \\
\downarrow & & \downarrow \\
g_{mn} & \not\mapsto & g_{mn} + \delta g_{mn}
\end{array}$$

which is however not invariant under local supersymmetry. This suggests that the correct way of integrating over odd supermoduli requires a projection to a supersymmetric, conformally invariant modification of the metric g_{mn}. Although there is no candidate for a modification of g_{mn} as a *tensor*, we can identify g with its period matrix, and the desired notion is provided by the super period matrix. The prescription is then to parametrize supermoduli space by the space of super period matrices and gravitino fields. The gravitino fields can be integrated out without ambiguities, leaving an integral over the space of period matrices. Since these have the same body as period matrices, we can effectively deform the integral over this space to the integral over the space of standard period matrices of Riemann surfaces. In effect, it means that after the gravitino integration, we can treat $\hat{\Omega}$ exactly as Ω. The integrands of scattering amplitudes are then built out of modular forms and sections of holomorphic vector bundles. This program is being carried out, and we shall report on it elsewhere.

Other solutions to this problem are proposed in [6, 57].

C. The $i\varepsilon$ prescription in string theory. So far we have discussed only the problem of formulating rules for writing the scattering amplitudes $A(k_1, \ldots, k_N)$ of string theory as integrals over moduli space depending on the external momenta $k_1, \ldots, k_N$. In general we have to expect convergence only for values of the k_i's in some wedge region in momenta space, and physical values require analytic continuation. Even for the torus, where completely explicit formulas for the scattering of gravitons are available, this problem has not been addressed. In fact except for the cosmological constant to one loop which vanishes because of supersymmetry, there is no loop amplitude which is explicitly known to be finite [45]. In analogy with the $i\varepsilon$ prescription for the propagators of field theory, we refer to the problem of analytic continuation in the external momenta as the " $i\varepsilon$ prescription for string theory."

That analytic continuation is needed already occurs at tree level. In this case however there is no moduli, the amplitudes are represented by hypergeometric integrals over the sphere, so the problem is solved by the Γ and beta functions (see e.g., [30, III.L]). For the torus the presence of moduli poses new problems as can be seen from the scattering amplitude for 4 gravitons ([43, 30, III.M]),

(8.22)
$$A(k_1, \ldots, k_4) = \delta\left(\sum_{i=1}^{4} k_i\right) \int_{H/PSL(2,\mathbf{Z})} \frac{d^2\tau}{2\tau_2^2} \frac{1}{\tau_2^4}$$
$$\times \int \prod_{i=1}^{4} d^2 z_i \prod_{i<j} \exp(-k_i k_j G(z_i, z_j))$$

up to kinematic factors. It is convenient to introduce the Mandelstam variables $s = -(k_1 + k_2)^2$, $t = -(k_2 + k_3)^2$, $u = -(k_1 + k_3)^2$, and the form $F(z, w)$ of (5.18) which reduces in this case to

$$F(z, w) = \left|\frac{\theta_1(z - w, \tau)}{\theta_1'(0, \tau)}\right|^2 \exp(-2\pi(\mathrm{Im}(z - w))^2/\tau_2).$$

The amplitude (8.22) can be rewritten as
(8.23)
$$A(k_1, \ldots, k_4) = \delta\left(\sum_{i=1}^{4} k_i\right) \int_{\mathcal{M}_1} \frac{d^2\tau}{2\tau_2^2} \frac{1}{\tau_2^4} \int \prod_{i=1}^{4} d^2 z_i |F(z_1, z_2)F(z_3, z_4)|^{-s/2}$$
$$\times |F(z_2, z_3)F(z_1, z_4)|^{-t/2} |F(z_1, z_3)F(z_2, z_4)|^{-u/2}.$$

As noted in [43] the divergences in the region where all the points z_1, z_2, z_3, z_4 come close together with $|z_i - z_j|$ of uniform size are mild. This ultraviolet behavior can be attributed to the "smearing" of string interactions, as discussed in II.A. However severe divergences arise from the region where the points come close in pairs, with a specific distance between the two pairs, say $|z_1 - z_2| \sim |z_3 - z_4| \ll |z_1 - z_3|$, $z_1 - z_3 \sim (1 + \tau)/2$. This is due to the asymptotic behavior of $F((1 + \tau)/2, 0)$ as τ tends to $i\infty$,

$$F((1 + \tau)/2, 0) \sim \frac{1}{4\pi^2} e^{\pi\tau_2/2}.$$

It is tempting to try to solve this problem by rewriting the integrals in τ as integrals in the j function. Since τ as a function of j is given by a ratio of hypergeometric functions, this brings us closer to the situation for the sphere. There are however difficulties coming from the fact that $F(z, w)$ is not holomorphic, and work in this direction (jointly with E. D'Hoker) is in progress. The $i\varepsilon$ prescription for higher genus seems to be an even more daunting task, requiring probably a deep understanding of the cell structure [**48, 49, 75**] of moduli space.

VII. Concluding remarks

We conclude by listing some directions warranting further investigation:

• The most urgent task is to complete the above program for writing string scattering amplitudes as integrals over moduli rather than over supermoduli. The key intermediate step is to write them in terms of the super period matrix $\hat{\Omega}$. The Chiral Splitting Theorem does this for correlation functions of matter fields, but the ghost system remains, especially the commuting β, γ which pose some subtle problems. The BRST formalism should also be re-examined in the context of $\hat{\Omega}$;

• An $i\varepsilon$ prescription needs to be found, beginning with the torus;

• We need to establish nonrenormalization theorems, which would imply for example that the cosmological constant, the graviton mass,etc. receive no perturbative corrections. These can be viewed as yet unknown generalizations of the Jacobi identity. Nonexistence theorems for modular forms or sections of vector bundles over moduli space will probably play a major role here, as suggested in [**68**];

• We need to develop two-dimensional supergeometry further, to the point where we can read off the conformal blocks of vertex operators for higher mass level bosonic particles as well as fermion vertices. This last problem is especially challenging, since the fermion emission vertex admits no superfield formulation and it couples to the ghosts [**37**]. In a different direction we have described supergeometry by the pair g, $\chi_{\bar{z}}^{+}$, which is analogous to describing a Riemann surface by a metric. It is important to develop it as well from flat coordinate systems (z, θ), $\mathscr{D}_{+} = \partial_{\theta} + \theta\partial_{z}$ with transition functions, in the same way as we would characterize a Riemann surface by its holomorphic coordinate charts. Finally it would be very helpful to make contact with the more algebraic approach to supermanifolds proposed by Berezin-Kostant-Leites-Manin [**24, 25, 61**];

• The arguments for unitarity in VIII should be made mathematically precise. For example they make use of determinants for noncompact surfaces. Such determinants have been proposed in [**31, 86**], and should be investigated further. Their interest may go beyond this problem, since they also play a role in the "picture changing" mechanism central to topological gravity [**98, 94, 26**];

• The path integrals on the torus admit a Hamiltonian interpretation. It would be useful to relate the functional integral approach to the operator methods [**3, 72, 85, 90**] for higher genus. In the same vein, it would be interesting to develop further the representation theory of the Krichever-Novikov algebras [**56**] which generalize to higher genus the Virasoro algebra on the twice-puncture sphere. Efforts in this direction are in [**53, 76**];

• We have discussed the equivalence between the Mandelstam and the

Polyakov formulations. It is of course important to establish their equivalence with the Green-Schwarz formulation. This can be expected to be challenging, since the Green-Schwarz formulation does not involve summation over spin structures and is notoriously difficult to quantize covariantly.

• Finally we mention for the sake of completeness other areas of strings besides perturbation theory. We have discussed strings propagating in ten-dimensional Minkowski space-time. Now this is only one example of manifolds where strings can propagate consistently, the consistency requirement being that the coordinates of the string (x^μ for flat Minkowski) describe a conformal field theory. This requirement together with phenomenological considerations such as unbroken supersymmetry have led to the condition that the ambiant manifold be the product of $4 - d$ Minkowski space-time by a Calabi-Yau three-fold [20]. Vigorous work has recently produced many Calabi-Yau manifolds [88, 44], and it is important to determine the one selected by physics. This issue can be determined only by nonperturbative considerations. Divergence in perturbation theory would indicate tunneling effects. Thus we need to be able to sum over all genera. There has been recently a breakthrough in this seemingly intractable problem for noncritical string theories [19, 28, 46]. For noncritical strings, the Liouville theory has to be quantized, a well-known obstacle in the last decade [77], although there has been some important progress [79, 55, 84]. The unexpected discovery has been the relation with the double scaling limits of matrix models and topological gravity [98, 94]. In particular recursive relations between different genera and an integrable structure emerge. We can speculate whether anything similar can take place for the critical string.

Appendices

A. Conventions. Real coordinates for the surface M are denoted $\xi = (\xi^1, \xi^2)$. In isothermal coordinates z, the metric becomes $ds^2 = 2g_{z\bar{z}}dz\,d\bar{z}$, and the corresponding real coordinates are defined by $z = (\xi^1 + i\xi^2)/\sqrt{2}$. The scalar curvature R is given by $R = -g^{z\bar{z}}\partial_z\partial_{\bar{z}}\ln g_{z\bar{z}}$. With these conventions, the metric $ds^2 = 4(1 + |z|^2)^{-2}|dz|^2$ on the sphere has curvature $R = 1$. The covariant derivatives ∇ on $(n, 0)$ tensors are $\nabla_{\bar{z}} = \partial_{\bar{z}}$ and $\nabla_z = g_{z\bar{z}}^n\partial_z(g^{z\bar{z}})^n$. Under a variation of the metric δg_{mn}, the covariant derivatives vary by [36]

$$(A.1)\quad \begin{aligned} \delta\nabla^z &= -\frac{1}{2}g^{z\bar{z}}\delta g_{z\bar{z}} + \frac{1}{2}\delta g^{zz}\nabla_z + \frac{n}{2}\nabla_z(\delta g^{zz}) \\ \delta\nabla_z &= -\frac{n}{2}\partial_z(g^{z\bar{z}}\delta g_{z\bar{z}}) - \frac{1}{2}\delta g_{zz}\nabla^z + \frac{n}{2}\nabla^z(\delta g_{zz}) \end{aligned}$$

with $\delta g^{zz} = -(g^{z\bar{z}})^2\delta g_{\bar{z}\bar{z}}$. The Dirac γ matrices in two dimensions satisfy

$$\{\gamma^a, \gamma^b\} = -\delta^{ab}, \qquad [\gamma^a, \gamma^b] = -\varepsilon^{ab}\gamma_5.$$

We choose the following representation

$$(A.2) \quad (\gamma^z)_{++} = (\gamma^{\bar{z}})_{--} = -(\gamma^z)_+^- = (\gamma^{\bar{z}})_-^+ = 1$$
$$(\gamma^a)_{\alpha\beta} = 0, \qquad \alpha \neq \beta$$

where we have used the same notation for Einstein and $U(1)$ indices, which is possible in conformally flat coordinates. The antisymmetric tensor is given by

$$\varepsilon^{\bar{z}z} = -\varepsilon^{z\bar{z}} = \varepsilon_{z\bar{z}} = -\varepsilon_{\bar{z}z} = i, \qquad \varepsilon_{zz} = \varepsilon_{\bar{z}\bar{z}} = 0.$$

B. Determinants of Laplacians. The determinant of a Laplacian Δ can be defined by zeta function regularization. If λ_n are the nonzero eigenvalues of Δ and N the dimension of its kernel, the zeta function

$$(A.3) \quad \begin{aligned} \zeta(s) &= \sum_n (\lambda_n)^{-s} \\ &= \frac{1}{\Gamma(s)} \int_0^\infty dt((\mathrm{Tr}\, e^{-t\Delta}) - N)t^{s-1} \end{aligned}$$

admits a meromorphic extension to the whole s plane which is regular at $s = 0$. Indeed the above integral converges for all s near $t \to \infty$, while at $t = 0$ we may check explicitly that its singularities are poles, located away from $s = 0$, by using a short time asymptotic expansion for the heat kernel. For surfaces the heat kernels $K_n^\pm(z, z; t)$ of the Laplacians $2\bar{\partial}_n^\dagger \bar{\partial}_n (= 2\nabla_{\bar{z}}^\dagger \nabla_{\bar{z}}^{(n)})$ and $2\nabla_z^\dagger \nabla_z^{(n)}$ are given by

$$(A.4) \quad \begin{aligned} K_n^\pm(z, z; t) &= \frac{1}{4\pi t} + \frac{1 \pm 3n}{12\pi} R + O(t) \\ \int d^2z \sqrt{g} K_n^\pm(z, z; t) &= \frac{1}{4\pi t} \int d^2z \sqrt{g} + \frac{1 \pm 3n}{6} \chi(M) + O(t) \end{aligned}$$

and the corresponding zeta functions have a unique pole as $s = 1$. Another closely related way of defining the logarithm of the determinant is by taking the finite part in the small ε expansion of

$$(A.5) \quad -\int_\varepsilon^\infty \frac{dt}{t}((\mathrm{Tr}\, e^{-t\Delta}) - N).$$

The zeta function definition of determinants differs from (A.5) by $-(c_0 - N)(s\Gamma(s))'(0)$, where c_0 is the constant coefficient in the short time expansion for the trace of the heat kernel. It is also easy to see that $\zeta(0) = c_0 - N$. Substituting in the value of c_0 from (A.4) we find

$$(A.6) \quad \det'(2\kappa\bar{\partial}_n^\dagger \bar{\partial}_n) = \kappa^{\zeta(0)} \det'(2\bar{\partial}_n^\dagger \bar{\partial}_n) = \kappa^{-N_n + (1-3n)\chi(M)/6} \det'(2\bar{\partial}_n^\dagger \bar{\partial}_n).$$

In string theory factors of the form $\kappa^{c\chi(M)}$ can be absorbed into a finite correction to the string coupling constant.

We describe now how to obtain the crucial scaling law (3.17) for determinants. Under a variation $2\delta\sigma g_{z\bar{z}}$ the Laplacian varies by (cf. (A.1)),

$$\delta(\bar{\partial}_n^\dagger \bar{\partial}_n) = 2(n-1)\delta\sigma\bar{\partial}_n^\dagger \bar{\partial}_n - 2n\bar{\partial}_n^\dagger \delta\sigma\bar{\partial}_n.$$

The cyclicity of the trace and the operator identity $Be^{AB}A = e^{BA}BA$ imply (A.7)

$$-\delta \operatorname{lndet}' \overline{\partial}_n^\dagger \overline{\partial}_n = -\int_\varepsilon^\infty dt \operatorname{Tr}(\delta(\overline{\partial}_n^\dagger \overline{\partial}_n)e^{-t\overline{\partial}_n^\dagger \overline{\partial}_n})$$

$$= -\int_\varepsilon^\infty dt \operatorname{Tr}(2(n-1)\delta\sigma \overline{\partial}_n^\dagger \overline{\partial}_n e^{-t\overline{\partial}_n^\dagger \overline{\partial}_n} - 2n\delta\sigma\, e^{-t\overline{\partial}_n \overline{\partial}_n^\dagger} \overline{\partial}_n \overline{\partial}_n^\dagger)$$

$$= \operatorname{Tr}(2(n-1)\delta\sigma\, e^{-t\overline{\partial}_n^\dagger \overline{\partial}_n})\big|_\varepsilon^\infty - \operatorname{Tr}(2n\delta\sigma\, e^{-t\overline{\partial}_n \overline{\partial}_n^\dagger})\big|_\varepsilon^\infty.$$

As $t \to \infty$ the only contribution to the heat kernel comes from the zero modes, and we can readily check that

$$\lim_{t\to\infty} \operatorname{Tr}(2(n-1)\delta\sigma\, e^{-t\overline{\partial}_n^\dagger \overline{\partial}_n}) = 2(n-1)\sum_a \langle \delta\sigma\phi_a^{(n)}|\phi_a^{(n)}\rangle = \delta \ln \det\langle \phi_a^{(n)}|\phi_b^{(n)}\rangle$$

and similarly for the zero modes of $\overline{\partial}_n^\dagger$. As $\varepsilon \to 0$ we may expand as in (A.4) and get

$$\delta \ln \frac{\det' \overline{\partial}_n^\dagger \overline{\partial}_n}{\det\langle \phi_a^{(1-n)}|\phi_b^{(1-n)}\rangle \det\langle \phi_a^{(n)}|\phi_b^{(n)}\rangle}$$

$$= -\frac{6n^2 - 6n + 1}{6\pi}\int d^2z\sqrt{g}\delta\sigma R - \frac{1}{2\pi\varepsilon}\int d^2z\sqrt{g}\delta\sigma.$$

Integrating in σ gives the scaling law in terms of the Liouville action (3.8), (3.17). We can also recapture (A.6) by setting $\delta\sigma$ to a constant and use the values for N_n given in II.C.

C. Grassmann numbers. The rules for Grassmann calculus are as follows. If θ is an odd Grassmann number $\theta^2 = 0$, a function f of θ can be expanded as $f(\theta) = x + \theta\psi$ and differentiation and integration are defined by

$$\partial_\theta f(\theta) = \psi = \int d\theta f(\theta).$$

In particular $\delta(\theta) = \theta$. This can be readily generalized to N anticommuting Grassmann generators θ_i, $i = 1, \ldots, N$, $\{\theta_i, \theta_j\} = 0$,

$$\{\partial_{\theta_i}, \theta_j\} = \delta_{ij}, \qquad \int d\theta_i \theta_i = 1, \qquad \int d\theta_i = 0.$$

In particular the only functions with nonvanishing integrals are those with a term $\prod_{i=1}^N \theta_i$ in their Taylor expansions. By expanding the exponential in power series we can verify that

(A.8)

$$\int \prod_{i=1}^N d\theta_i e^{\theta_i M_{ij}\theta_j/2} = (\det M)^{1/2}$$

$$\int \prod_{i=1}^N d\theta_i e^{\theta_i M_{ij}\theta_j/2} e^{i\zeta_i\theta_i} = (\det M)^{1/2}e^{-\zeta_i(M^{-1})_{ij}\zeta_j/2}$$

for any $N \times N$ antisymmetric matrix M_{ij}. Thus we can only have non-vanishing answers for even N. For functional integrals over anticommuting fields $\psi(z)$, $\overline{\psi}(z)$ this generalizes to

$$
(A.9) \quad
\begin{aligned}
&\int D\psi\, D\overline{\psi} \exp\left(\int d^2 z(\psi\partial_{\overline{z}}\psi + \overline{\psi}\partial_z\overline{\psi} + \eta\psi + \overline{\eta\psi})\right) \\
&= |\det(\partial_{\overline{z}}\partial_z)|^{1/2}\left|\exp\left(\frac{1}{2}\int d^2z\, d^2w\, \eta(z)S(z,w)\eta(w)\right)\right|^2.
\end{aligned}
$$

Ghost fields arise rather in pairs $b(z)$, $c(z)$, so it is convenient to have the corresponding version of (A.8)

$$
(A.10) \quad \int \prod_{i=1}^{N/2} db_i\, dc_i\, e^{c_i M_{ij} b_j} e^{\eta_i b_i + \zeta_i c_i} = (\det M) e^{\zeta_i (M^{-1})_{ij}\eta_j}
$$

where M is now a $(N/2) \times (N/2)$ matrix. The functional integral case together with the subtleties associated with zero modes and a nonvanishing index for the $\overline{\partial}$ operator are discussed in III.B.

D. Superdeterminants.

$$
(A.11) \quad s\det\begin{pmatrix} A & \beta \\ \gamma & D \end{pmatrix} = \frac{\det A}{\det D}\det(I - A^{-1}\beta D^{-1}\gamma).
$$

Here A, D denote even elements, while β, γ denote odd ones.

E. Superderivatives. In presence of a complex variable and a Grassmann variable θ, the key operator is the square root of the ∂_z operator defined by

$$
(A.12) \quad \partial_+ = \partial_\theta + \theta\partial_z.
$$

It can be viewed as the covariant derivative $\mathscr{D}_+$ for the flat supergeometry (6.24), characterized by $\chi_{m\alpha} = 0$. In general a supergeometry described by g_{mn}, $\chi_{m\alpha}$ can be brought locally to the flat one by successive diffeomorphisms, supersymmetry transformations, conformal and superconformal transformations. As in the case of Riemann surfaces, this cannot be done globally. We obtain in this way the super analogue of isothermal coordinates on a Riemann surface. We can either work in these "super-isothermal coordinates" (also referred to as the "superconformal gauge" in the physics literature) use the flat operator ∂_+ of (A.12), and the global geometry is encoded in the super transition functions, or work with a nonvanishing χ and complicated but globally well-defined covariant derivatives (6.26). We have adopted the second formalism, which is differential geometric in character and also at this point the most reliable one for the global geometry of supermanifolds.

REFERENCES

1. O. Alvarez, *Theory of strings with boundary*, Nucl. Phys. B **216** (1983), 125–184.
2. O. Alvarez, I. M. Singer, and B. Zumino, *Gravitational anomalies and the family's index theorem*, Comm. Math. Phys. **96** (1984), 409–417.

3. L. Alvarez-Gaume, C. Gomez, P. Nelson, G. Sierra, and C. Vafa, *Fermionic strings in the operator formalism*, Nucl. Phys. B **311** (1988), 333–400.

4. K. Aoki, E. D'Hoker, and D. H. Phong, *Unitarity of closed superstring perturbation theory*, Nucl. Phys. B **342** (1990), 149–230.

5. E. Arbarello, C. DeConcini, V. Kac, and C. Procesi, *Moduli spaces of curves and representation theory*, Comm. Math. Phys. **117** (1988), 1–36.

6. A. A. Atick, G. Moore, and A. Sen, *Catoptric tadpoles*, Nucl. Phys. B **307** (1988), 221–273.

7. J. J. Atick, J. Rabin, and A. Sen, *An ambiguity in fermionic string perturbation theory*, Nucl. Phys. B **299** (1988), 279–294.

8. M. Atiyah and I. M. Singer, *Dirac operators coupled to vector potentials*, Proc. Nat. Acad. Sci. U.S.A. **81** (1984), 2597–2600.

9. M. Baranov, Yu. I. Manin, I. Frolov, and A. S. Schwarz, *A superanalog of the Selberg trace formula and multiloop contributions to fermonic strings*, Comm. Math. Phys. **111** (1987), 373–392.

10. W. Beckner, *Moser-Trudinger inequalty in higher dimensions*, Int. Math. Res. Notices **7** (1991), 83–91.

11. A. Beilinson and Yu. I. Manin, *The Mumford form and the Polyakov measure in string theory*, Comm. Math. Phys. **107** (1986), 359–376.

12. A. A. Belavin and V. G. Knizhnik, *Algebraic geometry and the theory of quantized strings*, Phys. Lett. B **168** (1986), 201–206.

13. A. Belavin, V. Knizhnik, A. Morozov, and A. Perelomov, *Two and three-loop amplitude in the bosonic string theory*, Phys. Lett. B **177** (1986), 324–328.

14. A. Belavin, A. M. Polyakov, and A. Zamolodchikov, *Infinite conformal symmetry and two-dimensional quantum field theory*, Nucl. Phys. B **241** (1984), 333–380.

15. N. Berkovits, *Supersheet functional integration and the interacting Neveu-Schwarz string*, Nucl. Phys. B **304** (1988), 537–556.

16. J. M. Bismut and D. Freed, *The analysis of elliptic families I and II*, Comm. Math. Phys. **106** (1986), **159, 107** (1986), 103–163.

17. J. B. Bost and T. Jolicoeur, *A holomorphy property and the critical dimension of string theory from an index theorem*, Phys. Lett. B **174** (1986), 273–276.

18. T. Branson, A. Chang, and P. Yang, *Estimates and extremals for zeta functions on four-manifolds*, to appear in Comm. Math. Phys.

19. E. Brezin and V. Kazakov, *Exactly solvable field theories of closed strings*, Phys. Lett. B **236** (1990), 144–149.

20. P. Candelas, G. Horowitz, A. Strominger, and E. Witten, *Vacuum configurations for superstrings*, Nucl. Phys. B **258** (1985), 46–74.

21. R. Catenacci, M. Cornalba, M. Martinelli, and C. Reina, *Algebraic geometry and path integrals for closed strings*, Phys. Lett. B **172** (1986), 328–332.

22. A. Chang and P. Yang, *Prescribing Gaussian curvature on S^2*, Acta Math. **159** (1987), 215–259.

23. J. Cheeger, η *invariants, the adiabatic approximation, and conical singularities*, J. Diff. Geom. **26** (1987), 175–221.

24. P. Deligne, letter to Yu. I. Manin, unpublished, 1988.

25. B. DeWitt, *Supermanifolds*, Cambridge Univ. Press, 1984.

26. J. Distler, *Two-dimensional quantum gravity, topological field theory, and the multicritical matrix models*, Nucl. Phys. B **342** (1990), 523–538.

27. J. Dodziuk, T. Pignataro, B. Randol, and D. Sullivan, *Estimating small eigenvalues of Riemann surfaces*, Contemp. Math. **64** (1987), 93–121.

28. M. Douglas and S. Shenker, *Strings in less than one dimension*, Nucl. Phys. B **335** (1990), 635–654.

29. E. D'Hoker and D. H. Phong, *Multiloop amplitudes for the bosonic Polyakov string*, Nucl. Phys. B **269** (1986), 205–234.

30. ____, *The geometry of string perturbation theory*, Rev. Mod. Phys. **60** (1988), 917–1065.

31. ____, *Functional determinants on Mandelstam diagrams*, Comm. Math. Phys. **124** (1989), 629–645.

32. ____, *Conformal scalar fields and chiral splitting on super Riemann surfaces*, Comm. Math. Phys. **125** (1989), 469–513.

33. H. Farkas and I. Kra, *Riemann surfaces*, Springer-Verlag, Berlin, 1980.

34. J. Fay, *Theta functions on Riemann surfaces*, Lecture Notes in Math., vol. 352, Springer-Verlag, 1983.

35. ____, *Kernel functions, analytic torsion, and moduli spaces*, Memoirs of the Amer. Math. Soc. No. 464 (1992).

36. D. Friedan, *Introduction to Polyakov's string theory*, in "Recent Advances in Field Theory and Statistical Mechanics", Les Houches 1982, ed. J. B. Zuber and R. Stora (North-Holland, Amsterdam, 1984) 839–867.

37. D. Friedan, E. Martinec, and S. Shenker, *Conformal invariance, supersymmetry, and string theory*, Nucl. Phys. B **271** (1986), 93–165.

38. D. Friedan and S. Shenker, *The integrable analytic geometry of quantum strings*, Phys. Lett. B **175** (1986), 287–296; *The analytic geometry of two-dimensional conformal field theory*, Nucl. Phys. B **281** (1987), 509–545.

39. E. Gava, R. Iengo, T. Jayaraman, and R. Ramachandran, *Multiloop divergences in the closed bosonic string theory*, Phys. Lett. B **168** (1896), 207–211.

40. S. Giddings and S. Wolpert, *A triangulation of moduli space from light-cone strong theory*, Comm. Math. Phys. **109** (1987), 177–190.

41. F. Gliozzi, J. Scherk, and D. Olive, *Supersymmetry, supergravity theories, and the dual spinor model*, Nucl. Phys. B **122** (1976), 253–290.

42. M. Green and D. Gross, eds., *Unified String Theories*, World Scientific, 1986.

43. M. Green, J. Schwarz, and E. Witten, *Superstring Theory*, Cambridge Univ. Press, 1987.

44. B. R. Greene, K. Kiklin, and P. Miron, *Topology and geometry in superstring-inspired phenomenology*, in "Mathematical Aspects of string theory", ed. by S. T. Yau, 441–487, World Scientific, 1987.

45. D. Gross, *Superstrings and unification*, XXIV International Conference on High Energy Physics, Munich, 1988, Princeton Preprint PUPT-1108.

46. D. Gross and A. Migdal, *Non-perturbative two-dimensional quantum gravity*, Phys. Rev. Lett. **64** (1990), 127–130.

47. R. Gunning, *Lectures on Riemann surfaces*, Princeton Lecture Notes Series, 1966.

48. J. Harer, *Stability of the homology of the mapping class group of orientable surfaces*, Ann. of Math. (2) **121** (1985), 215–249.

49. J. Harer and D. Zagier, *The Euler characteristic of the moduli space of curves*, Invent. Math. **85** (1986), 457–485.

50. D. Hejhal, *Regular b-groups, degenerating Riemann surfaces, and spectral theory*, Memoirs of the Amer. Math. Soc. No. 437 (1990).

51. P. Howe, *Super Weyl transformations in two dimensions*, J. Phys. A **12** (1979), 393–402.

52. I. Igusa, *Theta functions*, Springer-Verlag, 1972.

53. A. Jaffe, S. Klimek, and A. Lesniewski, *Representations of the Heisenberg algebra on a Riemann surface*, Comm. Math. Phys. **126** (1989), 421–431.

54. J. Jorgenson, *Asymptotic behavior of Faltings' δ function*, Duke Math. J. **61** (1990), 221–254.

55. V. Knizhnik, A. M. Polyakov, and A. Zamolodchikov, *Fractal structure of two-dimensional gravity*, Mod. Phys. Lett. A **3** (1988), 819–826.

56. I. Krichever and S. P. Novikov, *Algebras of Virasoro type, Riemann surfaces, and structures in the theory of solitons*, Funct. Anal. Appl. **21** (1987), 126–146, *Virasoro-type algebras, Riemann surfaces, and strings in Minkowski space*, Func. Anal. Appl. **21** (1988), 294–307.

57. H. S. La and P. Nelson, *Unambiguous fermionic string amplitudes*, Phys. Rev. Lett. **63** (1989), 24–27.

58. S. Mandelstam, *The integrating-string picture and functional integration*, in *Unified String Theories*, ed. by M. Green and D. Gross, 46–102, World Scientific, 1986.

59. ____, *The n-loop string amplitude*, in *Second Nobel Symposium on Elementary Particle Physics*, Marstrand, 1986.

60. Yu. I. Manin, *The partition function of the Polyakov string can be expressed in terms of theta functions*, Phys. Lett. B **172** (1986), 185–186.

61. ____, *New directions in geometry*, Russian Math. Surveys **39** (1984), 51–83.

62. ____, *Quantum strings and algebraic curves*, in Proc. 1986 Int. Congress of Math., Berkeley, 1286–1295.

63. E. Martinec, *Superspace geometry of fermionic strings*, Phys. Rev. D **28** (1983), 2604–2613.

64. H. Masur, *The extension of the Weil-Petersson metric to the boundary of Teichmueller space*, Duke Math. J. **43** (1976), 623–635.

65. G. Moore, *Modular forms and two-loop string physics*, Phys. Lett. B **176** (1986), 369–379.

66. G. Moore and A. Morozov, *Some remarks on two-loop string calculations*, Nucl. Phys. B **306** (1988), 387–404.

67. G. Moore, P. Nelson, and J. Polchinski, *Strings and supermoduli* Phys. Lett. B **169** (1986), 47–53.

68. G. Moore, J. Harris, P. Nelson, and I. M. Singer, *Modular forms and the cosmological constant*, Phys. Lett. B **178** (1986), 167–173.

69. A. Morozov and A. Perelomov, *Partition function in superstring theory: the case of genus 2*, Phys. Lett. B **197** (1987), 115–118.

70. D. Mumford, *Stability of projective variables*, Ens. Math. **23** (1977), 39–110.

71. ____, *Tata Lectures on Theta* I *and* II, Birkhäuser, 1983.

72. A. Neveu and P. West, *Group theoretic approach to the perturbative string S-matrix*, Phys. Lett. B **193** (1987), 187–194.

73. E. Onofri, *On the positivity of the effective action in a theory of random surfaces*, Comm. Math. Phys. **86** (1982), 321–326.

74. B. Osgood, R. S. Phillips, and P. Sarnak, *Extremals of determinants of Laplacians*, J. Funct. Anal. **79** (1988), 148–211.

75. R. Penner, *The decorated Teichmueller space of punctured surfaces*, Comm. Math. Phys. **113** (1987), 299–339.

76. H. Pinson, *On the geometry of minimal conformal models*, Ph.D. Thesis, Dept. of Mathematics, Columbia University, 1992.

77. A. M. Polyakov, *Quantum geometry of bosonic strings*, Phys. Lett. B **103** (1981), 207–210, *Quantum geometry of fermionic strings*, Phys. Lett. B **103** (1981), 211–214.

78. ____, *Gauge fields and strings*, Harcourt, 1987.

79. ____, *Quantum gravity in two dimensions*, Mod. Phys. Lett. A **2** (1987), 893–898.

80. D. Quillen, *Determinants of Cauchy-Riemann operators over a Riemann surface*, Funct. Anal. Appl. **19** (1984), 31–34.

81. A. Restuccia and J. G. Taylor, *Light-cone gauge analysis of superstrings*, Phys. Reports **174** (1989), 285–407.

82. R. Schoen, S. Wolpert, and S. T. Yau, *Geometric bounds on the low eigenvalue of the compact surface*, Proc. AMS Symposium Pure Math. **36** (1980), 279–285.

83. J. Schwinger, *On gauge invariance and vacuum polarization*, Phys. Rev. **82** (1951), 664–679.

84. N. Seiberg, *Notes on quantum Liouville theory and quantum gravity*, Rutgers Preprint RU-90-29, 1990.

85. J. R. Sidenius, J. L. Petersen, and A. K. Tollsten, *Covariant super-reggeon calculus for superstrings*, Nucl. Phys. B **317** (1989), 109–146.

86. H. Sonoda, *Functional determinants on punctured Riemann surfaces and their application to string theory*, Nucl. Phys. B **294** (1987), 157–192.

87. L. Takhtajan and P. Zograf, *A local index theorem for $\bar{\partial}$ operators on punctured Riemann surfaces and a new Kaehler metric on their moduli space*, Comm. Math. Phys. **137** (1991), 399–426.

88. G. Tian and S. T. Yau, *Three-dimensional algebraic manifolds with $c_1 = 0$ and $\chi = -6$*, in *Mathematical aspects of string theory*, ed. by S. T. Yau, 543–628, World Scientific, 1987.

89. A. Tromba, *On a natural algebraic affine connection on the space of almost-complex structures and the curvature of Teichmueller space with respect to the Weil-Petersson metric*, Manuscript Math. **56** (1986), 475–497.

90. C. Vafa, *Operator formulation on Riemann surfaces*, Phys. Lett. **190** B (1987), 47–54.

91. H. Verlinde, *A note on the integral over fermionic supermoduli*, Utrecht Preprint No. THU-87/26 (1987) unpublished.

92. E. Verlinde and H. Verlinde, *Chiral bosonization, determinants, and the string partition function*, Nucl. Phys. B **288** (1987), 357–396.

93. ____, *Multiloop calculations in covariant superstring theory*, Phys. Lett. B **192** (1987), 95–102.

94. ____, *A solution of two-dimensional topological quantum gravity*, Nucl. Phys. B **348** (1991), 457–489.

95. R. Wentworth, *The asymptotics of the Arakelov-Green's function and Faltings's delta invariant*, Comm. Math. Phys. **137** (1991), 427–459.

96. E. Witten, *Global anomalies in string theory*, in Conference on *Anomalies, Topology, and Geometry*, ed. by A. White, 61–99, World Scientific, 1985.

97. ____, *Non-commutative geometry and string field theory*, Nucl. Phys. B **268** (1986), 253–294; *Interacting field theory of open superstrings*, Nucl. Phys. B. **276** (1986), 291–324.

98. ____, *Two-dimensional gravity and intersection theory on moduli space*, Surveys in Diff. Geometry **1** (1991), 243–310.

99. S. Wolpert, *Asymptotics of the spectrum and the Selberg zeta function on the space of Riemann surfaces*, Comm. Math. Phys. **112** (1987), 283–315.

100. S. T. Yau, ed., *Mathematical Aspects of String Theory*, World Scientific, 1987.

COLUMBIA UNIVERSITY

Proceedings of Symposia in Pure Mathematics
Volume **54** (1993), Part 2

Constructing Non-Self-Dual Yang-Mills Connections on S^4 with Arbitrary Chern Number

LORENZO SADUN AND JAN SEGERT

A connection A on a principal bundle over a 4-manifold M is called *Yang-Mills* if it is a critical point of the Yang-Mills (YM) action

$$(1) \qquad S(A) = \int_M |F_A|^2 d\mathrm{Vol} = \int_M -\mathrm{Tr}(*F_A \wedge F_A),$$

where $F_A = dA + [A, A]$ is the curvature of the connection A and $*$ is the Hodge dual. Equivalently, Yang-Mills connections are solutions of the *Yang-Mills equations*,

$$(2) \qquad\qquad d_A * F_A = 0,$$

where d_A denotes the covariant exterior derivative. These are the variational equations of the YM action, and constitute a system of second-order PDE's in A.

For a given second Chern number C_2, the YM action is bounded below by $8\pi^2 |C_2|$. To see this, let

$$F_\pm = \tfrac{1}{2}\left(F_A \pm *F_A\right)$$

be the self-dual and anti-self-dual parts of the curvature. We can then express the action and the Chern number as

$$S(A) = \int_M |F_A|^2 = \int_M |F_+|^2 + |F_-|^2$$

$$C_2 = \frac{-1}{8\pi^2} \int_M \mathrm{Tr}(F_A \wedge F_A) = \frac{1}{8\pi^2} \int_M |F_+|^2 - |F_-|^2.$$

The absolute minima of the action, connections with action attaining the topological bound $8\pi^2 |C_2|$, are thus characterized as having either self-dual

1991 *Mathematics Subject Classification.* Primary 81T13; Secondary 34B15, 53C05, 58E30.
The first author was partially supported by NSF Grant DMS-8806731. The second author was partially supported by a Bantrell Fellowship and NSF Grant DMS-8801918.
The detailed version of this paper has been submitted for publication elsewhere.

curvature $(*F_A = F_A, F_- = 0)$ or anti-self-dual curvature $(*F_A = -F_A, F_+ = 0)$. These (anti) self-dual connections have been well understood for some time. The first nontrivial example was the self-dual $SU(2)$ instanton on S^4, discovered in 1975 [BPST]. Three years later, all self-dual connections on S^4 were classified [ADHM], not only for $SU(2)$ but for all classical groups. The study of self-dual $SU(2)$ connections over arbitrary 4-manifolds led to spectacular progress in topology, including the discovery of exotic differentiable structures on $\mathbb{R}^4$ (see [FU] for an overview).

A natural question is whether any *non*-self-dual (NSD) Yang-Mills connections exist. Several classes of NSD YM connections on four-manifolds are known. One class are the homogeneous connections of Itoh [I] on bundles over S^4 with structure group $SU(4)$, and other large groups. Other solutions on S^4 have been constructed by 'twistor' methods, Buchdal [Bu] has produced solutions with the noncompact structure group $SL(2, \mathbb{C})$, and Manin [Ma] has produced solutions with various compact groups of very high dimension (as well as with supergroups). Parker [P] has constructed a solution on S^4 with a nonstandard Riemannian metric and structure group $SU(2)$. Some solutions are also known on other four-manifolds, namely $S^2 \times S^2$ [Ur], and $S^1 \times S^3$ [P, Ur].

Until recently it appeared that for $SU(2)$, NSD YM connections over the standard four-sphere might not exist. Indeed, analogies with harmonic maps from S^2 to S^2 appeared to indicate that none exist [AJ], and NSD YM connections with certain simple symmetries were ruled out [T2, J]. Moreover, it was shown that other than the (anti) self-dual connections, no *local* minima of the YM action exist [BLS, T1]; YM connections are either global minima (hence (anti) self-dual) or saddle points.

Sibner, Sibner and Uhlenback [SSU] recently showed that NSD YM connections on the trivial bundle $S^4 \times SU(2)$ do exist. As Lesley Sibner explained in this meeting, their construction involves using minmax theory to generate monopoles on hyperholic space H^3, which correspond to Yang-Mills connections on $\mathbb{R}^4$ with a certain $U(1)$ symmetry.

In this talk we would like to explain an alternate and somewhat simpler method for constructing NSD YM connections. This work is described in the papers [SS], and is based extensively on the work of Urakawa [Ur] and the work of Bor and Montgomery [BoMo]. Our construction produces examples not only on the trivial bundle, but on all $SU(2)$ bundles over S^4, except those with second Chern number equal to ± 1. We still do not know whether any NSD YM connections exist with Chern number ± 1.

The strategy is as follows:

(1) Pick a symmetry on S^4. This reduces all calculations from a four-dimensional space, S^4, to a much smaller space, the space of group orbits. In our case the symmetry group is the rotation group $SO(3)$ (more precisely its cover $SU(2)$), and the space of orbits is isomorphic to the interval $[0, \pi/3] \subset \mathbb{R}$.

(2) Consider $SU(2)$ connections on S^4 that are equivariant under this symmetry. Since connections live on bundles, we need to lift the symmetry group action from the base manifold S^4 to $SU(2)$ bundles over S^4. The equivariant connections fall into distinct classes, corresponding to different lifts of the group action. In our case these classes are indexed by two positive odd integers $n_\pm$. The bundle corresponding to (n_+, n_-) has second Chern number $C_2 = (n_+^2 - n_-^2)/8$.

(3) In each class of equivariant connections, look for *minima* of the action. By the principle of symmetric criticality [**Pal**], such minima must be stationary points of the action in the space of all connections, i.e. must be Yang-Mills. However, they need not be minima in the space of all connections, as the second variation in some nonequivariant directions may be negative. In our case, we show that minima exist for all classes with $n_+ \neq 1$ and $n_- \neq 1$.

(4) Finally, show that some classes do not contain (anti) self-dual connections. In our case we find that self-dual connections can only exist for $n_- = 1$ and anti-self-dual connections only exist for $n_+ = 1$.

These results give NSD YM connections in every class (n_+, n_-) with $n_\pm \geq 3$. Since every integer N except ± 1 can be written as $N = (n_+^2 - n_-^2)/8$ with $n_\pm \geq 3$ in at least one way, this gives examples with every Chern number except ± 1. For some Chern numbers we get several solutions (e.g. $5 = (7^2 - 3^2)/8 = (11^2 - 9^2)/8$), and for the trivial bundle we have a countably infinite number of solutions (just take $n_+ = n_-$).

1. The symmetry

Let $V \simeq \mathbb{R}^5$ be the space of symmetric, traceless, real 3×3 matrices Q, with inner product $\langle Q, Q' \rangle = \frac{1}{2}\mathrm{Tr}(QQ')$. It's convenient to work with an explicit orthonormal basis

$$Q_0 = \frac{1}{\sqrt{3}} \begin{pmatrix} -1 & 0 & 0 \\ 0 & -1 & 0 \\ 0 & 0 & 2 \end{pmatrix}; \qquad Q_1 = \begin{pmatrix} 0 & 0 & 1 \\ 0 & 0 & 0 \\ 1 & 0 & 0 \end{pmatrix}; \qquad Q_2 = \begin{pmatrix} 0 & 0 & 0 \\ 0 & 0 & 1 \\ 0 & 1 & 0 \end{pmatrix};$$

$$Q_3 = \begin{pmatrix} -1 & 0 & 0 \\ 0 & 1 & 0 \\ 0 & 0 & 0 \end{pmatrix}; \qquad Q_4 = \begin{pmatrix} 0 & 1 & 0 \\ 1 & 0 & 0 \\ 0 & 0 & 0 \end{pmatrix}.$$

We let $SO(3)$ act on V by conjugation, $g(Q) = gQg^{-1}$. Restricting ourselves to the unit sphere in V, this gives an action of $SO(3)$ on S^4. Since $SU(2)$ is the double cover of $SO(3)$, this also gives an action of $SU(2)$ on S^4.

Since all matrices in V are diagonalizable, it is not hard to check

PROPOSITION 1. *Every $Q \in S^4$ is related by the group action to a unique $Q_\theta = \cos(\theta)Q_0 + \sin(\theta)Q_3$ with $0 \leq \theta \leq \pi/3$. The orbits of Q_0 and $Q_{\pi/3}$ are two-dimensional, while all other orbits are three-dimensional.*

As a result, equivariant connections forms are determined by their values on the path Q_θ, $0 \leq \theta \leq \pi/3$. We put coordinates (θ, y^1, y^2, y^3) on a neighborhood of this path by $(\theta, \vec{y}) \mapsto \exp(\vec{y} \cdot \vec{\sigma})(Q_\theta)$, where σ_1, σ_2, and σ_3 are the usual (antihermitian) generators of $SU(2)$, and the action of $SU(2)$ on Q_θ is as above. On the path the tangent vectors $\partial_\theta \equiv \partial/\partial\theta$ and $\partial_i \equiv \partial/\partial y^i$ are orthogonal but not orthonormal. The vector ∂_θ is normalized, but the length of the vector ∂_i at Q_θ is $f_i(\theta)$, where

$$f_1(\theta) = 2\sin(\pi/3 + \theta); \qquad f_2(\theta) = 2\sin(\pi/3 - \theta); \qquad f_3(\theta) = 2\sin(\theta).$$

Note that f_3 vanishes at $\theta = 0$, as Q_0 is invariant under rotations about the z-axis. Similarly, f_2 vanishes at $\pi/3$.

From this it is easy to see how the Hodge dual operator $*$ acts on two-forms:

$$(3) \qquad\qquad *(d\theta \wedge dy^i) = \frac{f_j f_k}{f_i} dy^j \wedge dy^k,$$

where (i, j, k) are cyclic permutations of $(1, 2, 3)$. To simplify the notation we define the functions

$$G_1 = \frac{f_2 f_3}{f_1}; \qquad G_2 = \frac{f_1 f_3}{f_2}; \qquad G_3 = \frac{f_1 f_2}{f_3}.$$

G_1 and G_2 have zeroes at $\theta = 0$, while G_3 has a pole. Similarly, G_1 and G_3 have zeroes at $\theta = \pi/3$, while G_2 has a pole.

2. Equivariant connections

We next look at $SU(2)$ connections on S^4 that are equivariant under the above action of $SU(2)$. Such equivariant connections appeared in a study on non-Abelian Berry's phase [ASSS], and were classified by Bor and Montgomery [BoMo], who took Urakawa's general theory of equivariant connections with one-dimensional orbit spaces [Ur] and applied it to this particular symmetry. Much of this section is due to [BoMo].

An equivariant connection is of course determined by its values on the path $\{Q_\theta\}$. The most general Lie algebra valued one-form on the path is

$$A = -\sum_{i,j=1}^{3} \alpha_{ij}(\theta) dy^i \otimes \sigma^j - \sum_{i=1}^{3} \beta_i(\theta) d\theta \otimes \sigma^i,$$

where α_{ij} and β_i are real-valued functions. However, while rotations by $180°$ about the x, y, or z axes send Q_θ to itself, they do not preserve a general A of this form. For example, rotation by $180°$ about the z-axis flips the signs of dy^1, dy^2, σ^1, and σ^2, but not the signs of dy^3, $d\theta$, and σ^3. As a result, for an equivariant connection the coefficients α_{13}, α_{23}, α_{31}, α_{32}, β_1, and β_2 must be identically zero. Similarly, invariance under $180°$ rotations about the x-axis forces α_{12}, α_{21} and β_3 to be zero. Thus equivariant connections may be described by only three real-valued functions, $a_i = \alpha_{ii}$, and we write

$$A = -\sum_{i=1}^{3} a_i(\theta)dy^i \otimes \sigma^i.$$

We call such a triplet of functions $a = (a_1, a_2, a_3)$ a *reduced connection*.

Given an equivariant connection A, the curvature F_A is easily computed:

$$F_A = ((a_1 + a_2 a_3)dy^2 \wedge dy^3 - a_1' d\theta \wedge dy^1) \otimes \sigma^1 + (cyclic),$$

where $'$ denotes $d/d\theta$, and $(cyclic)$ denotes the other cyclic permutations of the indices $(1, 2, 3)$. By equation (3) the (anti) self-duality equations are then

(4)
$$-a_1' = \pm\frac{(a_1 + a_2 a_3)}{G_1}, \qquad -a_2' = \pm\frac{(a_2 + a_1 a_3)}{G_2},$$
$$-a_3' = \pm\frac{(a_3 + a_1 a_2)}{G_3},$$

where $+$ denotes self-duality and $-$ denotes anti-self-duality.

From F_A and $*$ we compute the action (1). The result is

(5) $$S(A) = \pi^2 \int_0^{\pi/3} d\theta \Big[(a_1')^2 G_1 + (a_2')^2 G_2 + (a_3')^2 G_3$$
$$+ \frac{(a_1 + a_2 a_3)^2}{G_1} + \frac{(a_2 + a_1 a_3)^2}{G_2} + \frac{(a_3 + a_1 a_2)^2}{G_3} \Big].$$

Finite action connections must have well-defined boundary values $r = a_3(0)$ and $t = a_2(\pi/3)$. Also, since G_1 and G_2 have zeroes at $\theta = 0$, they must have $a_1(0) + a_2(0)a_3(0) = a_2(0) + a_1(0)a_3(0) = 0$. If $r \neq \pm 1$, then these conditions imply that both $a_1(0)$ and $a_2(0)$ equal zero. Similarly, if $t \neq \pm 1$ then $a_1(\pi/3) = a_2(\pi/3) = 0$.

Not all finite-action reduced connections correspond to connections on all of S^4. A holonomy condition for infinitesimal paths around Q_0 forces $r \equiv -1 \pmod 4$, and a similar condition at $Q_{\pi/3}$ forces $t \equiv -1 \pmod 4$. If these conditions are met we define the positive odd integers $n_+ = |r|$, $n_- = |t|$. (If the holonomy conditions are not met, then our reduced connection corresponds to a singular connection with noninteger Chern number [FHP, SiSi], with the singularities occurring at the orbits of Q_0 and $Q_{\pi/3}$.)

Finally, we compute the Chern number of a connection. On our path the Chern form is

$$\frac{-1}{8\pi^2}\mathrm{Tr}(F_A \wedge F_A) = \frac{-1}{8\pi^2}(a_1'(a_2 + a_2 a_3) + cyclic)d\theta \wedge dy^1 \wedge dy^2 \wedge dy^3$$
$$= \frac{-1}{16\pi^2}d(a_1^2 + a_2^2 + a_3^2 + 2a_1 a_2 a_3) \wedge dy^1 \wedge dy^2 \wedge dy^3,$$

which we integrate, first over the group and then over $[0, \pi/3]$, to get a Chern number of $(r^2 - t^2)/8 = (n_+^2 - n_-^2)/8$.

3. Nonexistence of self-dual connections

Before showing that Yang-Mills connections do exist, we would like to prove that in certain classes (anti) self-dual connections do not exist. We prove this not only for the nonsingular classes (n_+, n_-) with $n_\pm \geq 3$, but also for a large number of singular classes (r, t). Specifically,

THEOREM 2. *There are no finite-action self-dual reduced connections with* $|t| > 1$. *There are no finite-action anti-self-dual reduced connections with* $|r| > 1$.

We prove the second statement, the first being similar. An anti-self-dual connection has nonpositive Chern number, so $|t| \geq |r| > 1$. Since $|r|$ and $|t|$ both differ from 1, finite action implies that $a_1(0) = a_2(0) = a_1(\pi/3) = a_3(\pi/3) = 0$. We will show that a solution to the anti-self-dual equations with $a_1(0) = a_2(0) = 0$ must have $a_3(\pi/3) \neq 0$, contradicting this.

Suppose $r > 1$ (the case $r < -1$ is similar). Then a_3 is positive and greater than 1 on a neighborhood $I_\epsilon = (0, \epsilon)$. If at some point in this neighborhood both a_1 and a_2 are nonnegative, then by the anti-self-duality equations (4) all three derivatives will be nonnegative, and the signs will persist. In particular, $a_3(\pi/3)$ will be positive, not zero. Similarly, if at some point in I_ϵ both a_1 and a_2 are nonpositive, then $a_1', a_2' \leq 0 \leq a_3'$ and again the signs persist. Thus it suffices to find a single point $\theta \in I_\epsilon$ at which a_1 and a_2 have the same sign (or where one is zero).

Suppose there is no such point, so a_1 and a_2 have opposite signs on all of I_ϵ. Then $|a_1 - a_2| < |a_1 + a_2|$. Equations (3) yield

$$\frac{d(a_1 - a_2)^2}{d\theta} = 2(a_1 - a_2)(a_1 - a_2)' = -\left(\frac{1}{G_1} + \frac{1}{G_2}\right)(a_3 - 1)(a_1 - a_2)^2$$
$$+ \left(\frac{1}{G_1} - \frac{1}{G_2}\right)(a_3 + 1)(a_1 - a_2)(a_1 + a_2).$$

The first term dominates in a neighborhood of $\theta = 0$, as $G_1^{-1} + G_2^{-1}$ has a pole at zero, while $G_1^{-1} - G_2^{-1}$ does not, and as $(a_3 - 1)$ is bounded away from zero on I_ϵ. Thus $[(a_1 - a_2)^2]'$ is strictly negative, and $(a_1 - a_2)^2$ is a strictly decreasing nonnegative function. However, at $\theta = 0$ we have $(a_1 - a_2)^2 = (0 - 0)^2 = 0$, and cannot decrease further. We have a contradiction, and so are done.

4. Existence of minima

What remains is to show that in each class (n_+, n_-) the action achieves its minimum. This is to be expected [BoMo], since the symmetry should prevent any bubbling-off phenomena, as in the equivariant Sobolev theorems of Parker [**P**]. By symmetry, such bubbling would have to occur on a complete orbit. But each orbit contains an infinite number of points, and Uhlenbeck's theorem only allows bubbling at a finite number of points.

This is in fact true, and as before, we prove our result for both singular and nonsingular connections.

THEOREM 3. *On each class* (r, t) *with* $|r| > 1$, $|t| > 1$ *the action achieves its minimum.*

The proof, which is only sketched here (the details being rather grungy), is by the direct method of the calculus of variations. We first define a Hilbert space $\mathscr{H}$ with norm

$$\|a\|^2 = \int_0^{\pi/3} d\theta \left[(a_1')^2 G_1 + (a_2')^2 G_2 + (a_3')^2 G_3 + \frac{(a_1)^2}{G_1} + \frac{(a_2)^2}{G_2} + \frac{(a_3)^2}{G_3} \right].$$

This norm resembles the action, only with the cubic and quartic terms removed. $\mathscr{H}$ is the direct sum of three weighted Sobolev spaces, one for each a_i.

We next show that, for fixed (r, t), sets of bounded action have bounded norm. This implies that any minimizing sequence lies in a finite radius ball in $\mathscr{H}$, which is weakly compact. Finally we show that the action is weakly lower-semicontinuous, hence that the weak limit of a minimizing sequence achieves the minimum action.

The difficulty is in showing that bounded action implies bounded norm. Away from the boundaries we have no problem, but near 0 and $\pi/3$ various functions G_i and G_i^{-1} diverge, complicating the analysis. The biggest difficulty is in bounding $a_1^2/G_1 + a_2^2/G_2$ near $\theta = 0$ (with a similar problem at $\theta = \pi/3$). This is done using the fact that both G_1 and G_2 go as $1/\theta$, and noting that

$$(a_1 + a_2 a_3)^2 + (a_2 + a_1 a_3)^2$$
$$= (a_3 + 1)^2 (a_1 + a_2)^2 + (a_3 - 1)^2 (a_1 - a_2)^2 \geq 2(|a_3| - 1)^2 (a_1^2 + a_2^2).$$

For $|r| > 1$ we can bound $|a_3| - 1$ away from zero in some neighborhood of $\theta = 0$, and so can bound $a_1^2/G_1 + a_2^2/G_2$ by a multiple of $(a_1 + a_2 a_3)^2/G_1 + (a_2 + a_1 a_3)^2/G_2$.

This whole approach breaks down for $r = \pm 1$ or $t = \pm 1$. In those cases bounded action does *not* imply bounded norm, and we have no proof that the minimum is achieved.

5. Regularity

For each pair (r, t) with $|r| > 1$ and $|t| > 1$ we have found solutions to the one-dimensional variational problem that do not satisfy the (anti) self-duality equations. These correspond to equivariant non-self-dual Yang-Mills connections on S^4 with the two exceptional orbits removed. The only remaining question is whether these connections may be extended smoothly across these two orbits.

For $r \not\equiv -1$ (mod 4) they cannot be extended across the orbit of Q_0 due to a holonomy obstruction, and for $t \not\equiv -1$ (mod 4) they cannot be extended across the orbit of $Q_{\pi/3}$. For r and t congruent to -1 (mod 4), however, the extension is possible. The proof is straightforward but lengthly, and we only sketch the main ideas here.

We first choose a particular connection (call it B), which is known to be smooth by the theorems of [BoMo]. We then show that the difference between our finite-action connection A and the reference connection approaches zero as $\theta \to 0$, and hence that A can be extended *continuously* across the orbit of Q_0.

We then compute the L_1^2 norm of $A - B$ in a particular gauge and show it to be finite. Since B is smooth, this means that A is in L_1^2. By Uhlenbeck's regularity theorem [Uh], there exists a gauge in which $d * \tilde{A} = 0$, where $\tilde{A}$ is gauge equivalent to A. The Yang-Mills equations together with this gauge condition form an elliptic system of equations, which implies that $\tilde{A}$ is smooth. Finally, since A was continuous to begin with, the gauge transformation must be C^1, and so cannot change the topology of the bundle.

Acknowledgment

We wish to thank Gil Bor, Percy Deift, Richard Montgomery, Lesley Sibner, Robert Sibner, Barry Simon, Cliff Taubes, and particularly Jalal Shatah for their help with this work.

NOTE ADDED IN PROOF: Since the time of this conference, we received the following papers: G. Bor, *SO(3) invariant Yang-Mills fields which are not self-dual*, Commun. Math. Phys. **145** (1992), 393–410; and T. H. Parker, *A Morse Theory for equivariant Yang-Mills*, preprint 1991.

REFERENCES

[ADHM] M.F. Atiyah, V.G. Drinfeld, N.J. Hitchin, and Y.I. Manin, *Construction of instantons*, Phys. Lett. **65A** (1978), 185–187.

[AJ] M.F. Atiyah and J.D.S. Jones, *Topological aspects of Yang-Mills theory*, Comm. Math. Phys. **61** (1978), 97–118.

[ASSS] J.E. Avron, L. Sadun, J. Segert, and B. Simon, *Chern numbers, quaternions, and Berry's phases in Fermi systems*, Commun. Math. Phys **124** (1989), 595–627.

[BLS] J.P Bourguignon, H.B. Lawson, and J. Simons, *Stability and gap phenomena for Yang-Mills fields*, Proc. Nat. Acad. Sci. U.S.A. **76** (1979), 1550–1553; *Stability and isolation phenomena for Yang-Mills equations*, Comm. Math. Phys. **79** (1982), 189–230.

[BoMo] G. Bor and R. Montgomery, *SO(3) invariant Yang-Mills fields which are not self-dual*, Hamiltonian Systems, Transformation Groups and Spectral Transform Methods, University of Montreal, Montreal, 1990, pp. 191–198.

[BPST] A.A. Belavin, A.M. Polyakov, A.S. Schwartz, and Yu. Tyupkin, *Pseudoparticle solutions of the Yang-Mills equations*, Phys. Lett. B **59** (1975), 85–87.

[Bu] N.P. Buchdal, *Analysis on analytic spaces and non-self-dual Yang-Mills fields*, Trans. Amer. Math. Soc. **288** (1985), 431–469.

[FHP] P. Forgacs, Z. Horvath, and L. Palla, *An exact fractionally charged self-dual solution*, Phys. Rev. Lett. **46** (1981), 392;*One can have noninteger topological charge*, Z. Phys. C. **12** (1982), 359–360.

[I] M. Itoh, *Invariant connections and Yang-Mills solutions*, Trans. Amer. Math. Soc. **267** (1981), 229–236.

[JT] A. Jaffe and C. Taubes, *Vortices and monopoles*, Birkhäuser, Boston, 1980.

[Ma] Yu. Manin, *New exact solutions and cohomology analysis of ordinary and supersymmetric Yang-Mills equations*, Proc. Steklov Inst. of Math. **165** (1984), 107–124.

[Ma] Yu. Manin, *Gauge field theory and complex geometry*, Springer-Verlag, Berlin, 1988.

[P] T. Parker, *Unstable Yang-Mills fields*, preprint;*Non-minimal Yang-Mills fields and dynamics*, Invent. Math. **107** (1992), 397–402.

[Pal] R.S. Palais, *The principle of symmetric criticality*, Comm. Math. Phys. **69** (1979), 19–30.

[SS] L. Sadun and J. Segert, *Non-self-dual Yang-Mills connections with nonzero Chern number*, Bull. Amer. Math. Soc. (N.S.) **24** (1991), 163–170; *Non-self-dual Yang-Mills connections with quadrupole symmetry*, Comm. Math. Phys. **145** (1992), 363–391.

[SiSi] L.M. Sibner and R.J. Sibner, *Singular Sobolev connections with holonomy*, Bull. Amer. Math. Soc. (N.S.) **19** (1988), 471–473;*Classification of singular Sobolev connections by their holonomy*, Comm. Math. Phys. **144** (1992), 337–350.

[SSU] L.M. Sibner, R.J. Sibner, and K. Uhlenbeck, *Solutions to Yang-Mills equations which are not self-dual*, Proc. Nat. Acad. Sci. U.S.A. **86** (1989), 8610–8613.

[T1] C.H. Taubes, *Stability in Yang-Mills theories*, Comm. Math. Phys. **91** (1983), 235–263.

[T2] C.H. Taubes, *On the equivalence of the first and second order equations for Gauge theories*, Comm. Math. Phys. **75** (1980), 207–227.

[Uh] K. Uhlenbeck, *Connections with L^p bounds on curvature*, Comm. Math. Phys. **83** (1982), 31–42.

[Ur] H. Urakawa, *Equivariant theory of Yang-Mills connections over Riemannian manifolds of cohomogeneity one*, Indiana Univ. Math. J. **37** (1988), 753–788.

NEW YORK UNIVERSITY
Current address: University of Texas, Austin
E-mail address: sadun@math.utexas.edu

UNIVERSITY OF MISSOURI
E-mail address: jan@segert.cs.missouri.edu

Proceedings of Symposia in Pure Mathematics
Volume **54** (1993), Part 2

The Yang-Mills Measure for the Two-Sphere

AMBAR SENGUPTA

1. Introduction

The study of Yang-Mills fields and the Yang-Mills functional has been an active area in mathematics for many years. Much of this interest, at least from a purely mathematical point of view, has been focused on classical (smooth) Yang-Mills fields. However, the fields of principal interest in physics are *quantum fields*. Quantization of gauge theories, in the so-called Euclidean formulation, involves the construction of a probability measure, the Yang-Mills measure, on an infinite-dimensional "manifold" whose points represent connections on some principal bundle. Thus a quantum theory of gauge fields would combine geometry with probability theory and infinite-dimensional analysis. An outline of problems in this area appears in [Si] and also in Singer's lecture at this conference.

In this report we outline the construction and study of the Yang-Mills measure for gauge fields over the sphere S^2. Details, including precise definitions and proofs, are in [Se]. The problem and the method of solution may be described in very general informal terms as follows. A probability measure (the Yang-Mills measure) μ is to be constructed on an infinite-dimensional manifold $\mathscr{C}$ (of gauge field configurations) starting from a heuristic expression describing μ. The measure μ is realized by embedding $\mathscr{C}$ in a Hilbert space $\mathscr{H}$ and setting μ equal to the restriction of Gauss measure on $\mathscr{H}$ to the submanifold $\mathscr{C}$. More precisely, $\mathscr{C}$ may be identified with $f^{-1}(e)$ for some function f on $\mathscr{H}$ taking values in a group G, and μ is realized as a Gauss measure on $\mathscr{H}$ *conditioned* to satisfy $f(\cdot) = e$. Each point of $\mathscr{C}$ describes a differential-geometric structure (or gauge field) over the two-sphere S^2. The embedding of $\mathscr{C}$ in $\mathscr{H}$ requires a splitting of S^2 into two hemispheres N and S. We prove that the measure μ is, in a sense, independent of the specific choice of hemispheres. Every area-preserving diffeomorphism

1991 *Mathematics Subject Classification.* Primary 81T13.
This paper is in final form and no revision will be submitted for publication elsewhere.

of S^2 induces a mapping of $\mathscr{C}$ into itself and we prove results demonstrating that (as expected from the specific form of the heuristic expression for μ) the measure μ is invariant under such transformations. This is done by computing explicitly the expectation values of important random variables (corresponding to configurations of Wilson loops in S^2) on $\mathscr{C}$.

The approach of the present work has been motivated by the works [**GKS, Dr**] which study the quantized Yang-Mills field over the plane. The Yang-Mills functional over the sphere and other compact Riemann surfaces has been studied in the classical sense (as distinct from the quantized formulation) by Atiyah and Bott [**AB-1, AB-2**] and more specifically over the sphere by several authors: [**F, FH, G**]. A study of the Yang-Mills measure for the two-sphere appears in Fine [**F**] by a method different from the one used in this work. A treatment of quantized gauge fields on the sphere has also been announced by Uhlenbeck and Nahm [**NU**]. The results in [**F, NU**] are consistent with our results but the principal difference between our results and theirs is that we are able to compute expectations for configurations of Wilson loops broad enough to specify connections completely up to gauge equivalence and thus it is clear in what sense our Yang-Mills measure lives on the configuration space $\mathscr{C}$. Moreover, we are able to prove essentially that the Yang-Mills measure is invariant under area-preserving diffeomorphisms of S^2. This seems to appear as a hypothesis in [**NU**]. To summarize,

(1) we construct a probability measure space for the Yang-Mills measure,

(2) we calculate Wilson loop expectation values for a class of configurations of loops broad enough to specify connections completely,

(3) we prove results demonstrating the invariance of the theory under area-preserving diffeomorphisms,

(4) our method seems to extend to the case of other surfaces (this extension is not carried out in this paper), and

(5) our use of the idea (see [**GKS, AHH**]) of defining parallel-transport by a stochastic differential equation provides the first example of a stochastic geometry over a compact surface and is of interest as a model in probability theory.

A Yang-Mills measure defines a stochastic differential geometry in the sense of [**Fr**]. That is, it can be viewed as a study of parallel-translation along (deterministic) paths using stochastic connections. Several works of S. Albeverio, R. Hoegh-Krohn, et al. (for example [**AHH**]) are devoted to the study of stochastic connections. The present work provides an example of stochastic differential geometry in this sense over a compact surface.

Objectives. (1) Construction of a probability measure μ_{YM}, called the Yang-Mills (YM) measure, on the infinite-dimensional manifold $\mathscr{C}$ of gauge-field configurations over the sphere S^2, which is given heuristically by the expression

$$d\mu_{\mathrm{YM}}(\omega) = \frac{1}{Z} e^{-S(\omega)} [\mathscr{D}\omega]$$

where $S : \mathscr{C} \to \mathbf{R}$ is the Yang-Mills functional, "$[\mathscr{D}\omega]$" is "infinite-dimensional Lebesgue measure" on $\mathscr{C}$ and Z is a normalizing constant.

(2) Description of invariance of μ_{YM} under area-preserving diffeomorphisms of S^2 (a diffeomorphism f of S^2 induces a map $\mathscr{C} \mapsto \mathscr{C}$ and the claim is that μ_{YM} is preserved if f preserves area).

(3) If μ_{YM}^T where $T > 0$ is defined as μ_{YM} with the YM functional scaled to $S(\cdot)/T$ then $\mu_{\mathrm{YM}}^T \to \mu_{\mathrm{YM}}^0$ where the latter is a probability measure on the set of minima of S (i.e. on Yang-Mills connections).

Notation and definitions

We summarize the concepts needed to make the statement (1) in the Objectives meaningful.

$$G = \text{compact connected Lie group,}$$

$$\mathfrak{g} = \text{Lie algebra of } G \text{ equipped with a bi-invariant inner-product,}$$

$$\pi : P \to S^2 : \text{principal } G \text{ bundle over } S^2,$$

$$\mathscr{A} = \{\text{connection forms on } P\}.$$

Fact. $\mathscr{A}$ is an infinite-dimensional affine space and there is an inner-product on the space of vectors in $\mathscr{A}$ induced by the metric on S^2 and the metric on $\mathfrak{g}$.

The **Yang-Mills functional** is given by

$$S : \mathscr{A} \to \mathbf{R} : \omega \mapsto \frac{1}{2} \int_{S^2} \|\Omega^\omega\|^2 d\sigma$$

where Ω^ω is the curvature form of ω, $\|\cdot\|$ is gotten from the metric on S^2 and the inner-product on $\mathfrak{g}$, and $d\sigma$ is the area-form on S^2.

The group of global gauge-transformations is given by

$$\mathscr{G} = \{\text{automorphisms of the bundle } P \text{ which cover the identity on } S^2\}.$$

The group of global gauge-transformations is given by

$$\mathscr{G} = \{\text{automorphisms of the bundle } P \text{ which cover the identity on } S^2\}.$$

The group $\mathscr{G}$ acts on $\mathscr{A}$: $\omega \mapsto \phi^*\omega$, where $\phi \in \mathscr{G}$ and $\omega \in \mathscr{A}$.

The configuration space is $\mathscr{C} = \mathscr{A}/\mathscr{G}$.

NOTE. The YM functional S is well defined on the quotient $\mathscr{C}$.

We recall the statement of the problem: Construct a probability measure μ_{YM} on $\mathscr{C}$ given heuristically by

$$d\mu_{\mathrm{YM}}(\omega) = Z^{-1} e^{-S(\omega)} [\mathscr{D}\omega].$$

We have explained the meanings of all the terms in this expression. The term $[\mathscr{D}\omega]$ describes the "measure" on $\mathscr{C}$ obtained by pushing down the "Lebesgue measure" from $\mathscr{A}$.

Outline of results

Method of construction. For simplicity of exposition we will assume G is simply connected.

Fix a point $n \in S^2$. Let $\mathscr{G}_n$ denote the subgroup of all elements of $\mathscr{G}$ which fix the fiber over n. Set $\mathscr{C}_n = \mathscr{A}/\mathscr{G}_n$.

It is more convenient *to construct* μ_{YM} on $\mathscr{C}_n$ and then one can get μ_{YM} on the quotient $\mathscr{C}$.

[Remark. If the base-space is the plane $\mathbf{R}^2$ then $\mathscr{C}_n$ is in bijective correspondence with a linear space of $\mathfrak{g}$-valued functions on $\mathbf{R}^2$ and under this correspondence $d\mu_{\text{YM}}$ looks like $Z^{-1}e^{-\frac{1}{2}\|x\|^2_{L^2(\mathbf{R}^2\,;\,\mathfrak{g})}}\mathscr{D}x$ and this is Gaussian (white-noise) measure on $L^2(\mathbf{R}^2\,;\,\mathfrak{g})$ which can be defined rigorously.]

We split the sphere S^2 into two closed hemispheres $S^2 = N \cup S$ where the hemispheres N and S intersect along the equator $\mathscr{E} = N \cap S$.

Then there is a one-one map

$$i : \mathscr{C}_n \hookrightarrow X = L^2(N\,;\,\mathfrak{g}) \oplus L^2(S\,;\,\mathfrak{g}) : [\omega] \mapsto (F^N(\omega), F^S(\omega))$$

where $F^N(\omega)\,d\sigma$ describes the curvature of ω over N expressed in "radial gauge" over N and F^S is a similar object over S. It is seen informally that under the correspondence between $\mathscr{C}_n$ and $i(\mathscr{C}_n) \subset X$:

μ_{YM} on $\mathscr{C}_n$ corresponds to Gauss measure on X *conditioned* to lie on $i(\mathscr{C}_n) \subset X$.

In general one does not expect probability measures on infinite-dimensional function spaces to be concentrated on smooth functions. So throwing smoothness aside we can take

$$i(\mathscr{C}_n) = \{(F^N, F^S) \in X : f(F^N, F^S) = e\}$$

where $f : X \to G$ is a function such that the condition that $i([\omega]) \in f^{-1}(e)$ insures that the holonomy around $\mathscr{E}$ calculated by $\omega|_{\pi^{-1}(N)}$ is the same as that calculated by ω over S.

To summarize: The YM measure μ_{YM} is taken as the product of Gaussian (white-noise) measures on $L^2(N\,;\,\mathfrak{g})$ and $L^2(S\,;\,\mathfrak{g})$ *conditioned* so that the equatorial holonomy is well defined. (To define the holonomy in this stochastic setting one defines stochastic parallel-translation with respect to a random connection by a stochastic differential equation.)

We write $(\overline{\mathscr{C}}_n, \mu_{\text{YM}})$ for this probability space.

Stochastic holonomy. Let C be a "nice" closed curve in S^2 based at n. For a connection $\omega \in \mathscr{A}$ we write

$$g_u(C\,;\,\omega) = \text{holonomy around } C \text{ starting at } u \in \pi^{-1}(n) \text{ using } \omega.$$

When u is a fixed point on the fiber $\pi^{-1}(n)$ then $g_u(C\,;\,\omega)$ depends only on $[\omega] \in \mathscr{C}_n$. Henceforth we drop reference to u.

Thus for each C as above, we have a function on $\mathscr{C}_n : [\omega] \mapsto g(C\,;\,\omega)$.

Under the probability measure μ_{YM} we have a corresponding random variable on $\overline{\mathscr{C}}_n$ which we denote again by $[\omega] \mapsto g(C; \omega)$. Thus this describes the *stochastic holonomy*. The precise definition of stochastic parallel-translation, in this sense, was used in [GKS] (see also [AHH]); briefly, one interprets the usual differential equation defining parallel-translation with respect to a connection as a stochastic differential equation where the randomness comes in through the randomness in the connection.

We then have

THEOREM. *Let $\mathscr{S} = \{C_1, \ldots, C_k\}$ be an "admissible" collection of closed curves in S^2 and let f be a bounded measurable function on G^k which satisfies*

$$f(gx_1 g^{-1}, \ldots, gx_k g^{-1}) = f(x_1, \ldots, x_k)$$

for all relevant values of the variables. Then

$$\int_{\overline{\mathscr{C}}_n} f(g(C_1; \omega), \ldots, g(C_k; \omega)) \, d\mu_{\text{YM}}(\omega)$$

$$= \frac{\int f(x(C_1), \ldots, x(C_k)) Q^{\Gamma \mathscr{S}}(x) \, dx}{Q_{|S^2|}(e)}$$

where on the right-hand side $Q_t(\cdot)$ is the fundamental solution of the heat equation on G, $\Gamma_{\mathscr{S}}$ is the (finite) graph formed by the system of curves $\mathscr{S}$, x runs over $G^{\text{edges of } \Gamma_{\mathscr{S}}}$, $x(C_i)$ is defined in the natural way by taking the ordered product of the values of x on the bonds appearing in C_i, and $Q^{\Gamma \mathscr{S}}(x)$ is a quantity defined using $Q_t(\cdot)$, x and the areas of the regions in the complement of the graph $\Gamma_{\mathscr{S}}$ in S^2.

COROLLARY. *If $R: S^2 \to S^2$ is an area-preserving diffeomorphism then*

$$\int_{\overline{\mathscr{C}}_n} f(g(C_1; \omega), \ldots, g(C_k; \omega)) \, d\mu_{\text{YM}}(\omega)$$

$$= \int_{\overline{\mathscr{C}}_n} f(g(R \circ C_1; \omega), \ldots, g(R \circ C_k; \omega)) \, d\mu_{\text{YM}}(\omega).$$

That is, informally speaking, μ_{YM} is invariant under area-preserving diffeomorphisms of S^2.

In particular μ_{YM} does not really depend on the choice of the hemispheres N, S used in the construction.

A limiting property. Consider the probability measure specified heuristically by the expression

$$d\mu_{\text{YM}}^T(\omega) = Z^{-1} e^{-S(\omega)/T} [\mathscr{D}\omega].$$

Then

THEOREM. *As $T \to 0$ the probability measure μ_{YM}^T converges weakly to a probability measure μ_{YM}^0 which is concentrated on the Yang-Mills connections (more precisely, on the set of minima of S on $\mathscr{C}$).*

The proof uses a similar property of the Brownian Bridge process proved in [M] (see also [H]).

Details of the results (along with more references) can be found in [Se]. The works [Dr, GKS] describe the Yang-Mills measure for the plane in a spirit similar to the present work. The works [F, NU] are also relevant.

Related questions. The methods used in proving the results described in this report seem to extend to the case of gauge fields over other compact surfaces where the interaction with the surface topology adds an extra element of interest. From a probabilistic point of view one could study questions of ergodicity of the Yang-Mills measure under area-preserving diffeomorphisms (or under isometries) and one may be able to carry out a more refined (for example, Large Deviations) study of the "Limiting Property" mentioned above.

The problem of constructing gauge fields in higher dimensions and studying holonomies in such set-ups is of great interest but the problems involved are more difficult.

Acknowledgement

I am grateful to Chris King and Leonard Gross for helpful conversations. A Sloan Dissertation Fellowship provided financial support.

Note added in proof: Since this talk was given several works ([F2, Fo, Se2, Se3, and Wi1, Wi2]) have appeared pertaining to the questions raised above.

REFERENCES

[AHH] S. Albeverio, R. Hoegh-Krohn, and H. Holden, J. Functional Anal. **78** (1988), 154–184.

[AB-1] M. Atiyah and R. Bott, Proc. Indian Acad. Sci. (Math. Sci.) **90** (1981), 11–20.

[AB-2] ____, Phil. Trans. Roy. Soc. London A **308** (1982), 523–615.

[Dr] B. Driver, Commun. Math. Phys. **123** (1989), 575–616.

[F] D. Fine, *Quantum Yang-Mills over the two-sphere*, Ph.D. Thesis, M.I.T., 1989.

[F2] D. Fine, Commun. Math. Phys. **140**, 321–338 (1991).

[FH] Th. Friedrich and L. Habermann, Commun. Math. Phys. **100** (1985), 231.]

[Fo] R. Forman, *Small volume limits of 2d Yang-Mills*, preprint (Rice University, 1992), to appear in Commun. Math. Phys.

[Fr] J. Frohlich, *On the construction of quantized gauge fields*, Field Theoretical Methods in Particle Physics (W. Rühl, ed.), Plenum Press, 1980.

[G] J. Graveson, Commun. Math. Phys. **127** (1990), 597.

[GKS] L. Gross, C. King, and A. Sengupta, Ann. Physics **194** (1989), 65–112.

[H] P. Hsu, *Probability theory and related fields* **84** (1990), 103–118.

[M] S. A. Molchanov, Russian Math. Surveys **30** (1975), 1–63.

[NU] W. Nahm, and K. Uhlenbeck, *The equivalence of quantized gauge fields on S^2 and the quantum mechanics of a particle moving on the group manifold*, preprint.

[Se] A. Sengupta, *The Yang-Mills measure for the two-sphere*, Ph.D. Thesis, Cornell Univ., 1990; to appear in J. Functional Anal.

[Se2] ____, Commun. Math. Phys. **141**, (1992) 153–209.

[Se3] ____, *Quantum Gauge Theory on Compact Surfaces*, to appear in Ann. Physics.

[Si] I. M. Singer, *Some problems in the quantization of gauge theories and string theories*, The Mathematical Heritage of Hermann Weyl, Amer. Math. Soc., Providence, R.I., 1987.

[Wi1] E. Witten, Commun. Math. Phys. **141**, (1992) 153–209.

[Wi2] E. Witten, *Two-Dimensional Gauge Theories Revisited*, preprint IASSNS-HEP92/15, March 1992.

LOUSIANA STATE UNIVERSITY

Proceedings of Symposia in Pure Mathematics
Volume **54** (1993), Part 2

Examples of Nonminimal Critical Points
in Gauge Theory

L. M. SIBNER

1. Introduction

Let P be a principal $SU(2)$ bundle over S^4. A connection $D_A = d + A$, where A is an $su(2)$-valued one form. The curvature of the connection is the two form $F_A = dA + A \wedge A$. We want to study the Yang-Mills functional

$$\text{YM}(A) = \int_{S^4} |F_A|^2 \, dV.$$

This functional is minimized by the self-dual (and antiself-dual) connections satisfying

$$(1) \qquad F_A = \pm * F_A.$$

The more general Euler-Lagrange equations satisfied by critical points of YM are the Yang-Mills equations

$$(2) \qquad D_A^* F_A = 0.$$

Principal bundles P are classified by their second Chern class

$$k = c_2 = -\frac{1}{8\pi^2} \int_{S^4} \text{tr}(F_A \wedge F_A).$$

By completing the square,

$$\text{YM}(A) = \frac{1}{2} \int_{S^4} |F_A \pm *F_A|^2 \, dV + 8\pi^2 |k|$$

which shows that $\text{YM}(A) \geq 8\pi^2 |k|$, with equality if and only if equations (1) are satisfied.

1991 *Mathematics Subject Classification*. Primary 35-XX, 53-XX.
Research partially supported by NSF Grant DMS-900270.
This paper is in final form and no version will be submitted for publication elsewhere.

An open question for some time has been the existence (or nonexistence) of critical points of YM which satisfy (2) but not (1) and hence are not minimal. Evidence against existence was provided by the corresponding situation for harmonic maps from S^2 to S^2 and questions of conformal invariance. Evidence for existence was contained in the work of Taubes [T] on the Higgs model in R^3.

Recently [SSU], using the general program of Taubes, we showed that nonminimal solutions do indeed exist. Previous results [BL] show that all of these critical points are unstable.

In this article, we outline Taubes's program and describe how this is applied in order to produce nonminimal solutions of the Yang-Mills equations. Taubes's main idea is to build a homotopically nontrivial loop of connections from more basic objects by reflecting, twisting and pasting. Crucial to the min-max argument is that this construction does not create too much energy. For this reason, the standard t'Hooft instanton is not a suitable basic object since it does not distinguish directions. Instead, it is necessary to exploit a fundamental relationship between *equivariant connections* over S^4 and *monopoles* over the hyperbolic ball $\mathbf{H}^3$. This is called *dimensional reduction* and is described more completely in Atiyah [A] and Braam [B].

Briefly, using coordinates on $R^4 = R^2 \times R^2$ given by (u, v, r, θ), there is a conformal equivalence between $R^4 - R^2$ and $\mathbf{H}^3 \times S^1$. Roughly speaking, a connection is called S^1-equivariant if the one form A has components which are independent of θ. In this case, D_A may be considered as a connection over $S^4 - S^2$ which extends over S^4 provided $\lim_{r \to 0} |A_\theta| = m$ is an integer [SS]. In this situation, the bundle has a limit as $r \to 0$ which splits as a direct sum of line bundles over S^2 having first Chern classes $\pm k$. The second Chern class of the original bundle can be seen to be $c_2 = -2km$.

2. Dimensional reduction

An equivariant D_A over $S^4 - S^2$ corresponds to a Higgs configuration $c = (B, \phi)$ over $\mathbf{H}^3$ by setting $B = A_u\, du + A_v\, dv + A_r\, dr$ and $\phi = A_\theta$. We denote by F_B the curvature of the connection $D_B = d + B$ over $\mathbf{H}^3$.

Note that

$$\int_{S^4} |F_A|^2\, d^4x = 2\pi \int_{\mathbf{H}^3} (|F_B|^2 + |D_B \phi|^2)\, d^3x$$

and, hence, that finite action connections over $S^4 - S^2$ become finite energy Higgs configurations over $\mathbf{H}^3$. The topological invariants associated with c are

(a) mass $= m = \lim_{|x| \to \infty} |\phi(x)|$

(b) monopole number $= (1/4\pi m) \int_{\mathbf{H}^3} \mathrm{tr}(F_B \wedge D_B \phi) = k$,

where the integers m and k are the same as those defined in §1 for equivariant connections on $S^4 - S^2$, with $c_2 = -2km$. Note that k, which may be regarded as the winding number of ϕ at ∞, is always an integer, but that m may be any positive real number.

We illustrate the above considerations with the simplest example, namely the equivariant *flat* $(F_A = 0)$ connection $A = m\hat{\imath}d\theta$ where $\hat{\imath} = \begin{pmatrix} i & 0 \\ 0 & -i \end{pmatrix}$. By dimensional reduction, on $\mathbf{H}^3$, $B = 0$ and $\phi \equiv m\hat{\imath}$. The action is zero and the monopole number $k = 0$. The codimension two singularity on S^2 is removable if and only if m is an integer [SS] in which case, D_A is gauge equivalent to d via the continuous gauge transformation $g = e^{-m\hat{\imath}\theta}$. Otherwise, D_A does *not* extend across S^2 and A itself is not very well behaved. In fact, it is an exercise to check that $\|A\|_{L^2(S^4)} = \infty$.

For more general connections, say with finite action, it can be shown that the limit m exists almost everywhere on S^2 and that if m is an integer, D_A extends as a connection over S^4 (see [SS]). For *equivariant connections*; it is not hard to show that F_A is a solution of the *Yang-Mills equations* on S^4 if and only if $c = (B, \phi)$ is a solution of the *Higgs equations* on $\mathbf{H}^3$. Moreover, F_A is a solution of the *self-dual* (or antiself-dual) *equations* if and only if c is a solution of the *Bogomolnyi equations*.

3. Construction of the loop

The basic object from which we construct the loop is the *Prasad-Sommerfield monopole* $c_0 = (B_0, \phi_0)$ on $\mathbf{H}^3$ (see [B] for an explicit formula) having monopole number $k = \pm 1$ and arbitrary fixed mass $m > 1$. Using half space coordinates (z, ρ, θ), one can show that for $z^2 + \rho^2 \le R^2$ sufficiently small, the $\hat{\imath}$ direction in $\mathrm{su}(2)$ is distinguished and

(3)
$$D_{B_0}\phi_0 = 4zdz\hat{\imath} + O(R^m),$$
$$F_{B_0} = 4\rho d\rho d\theta\hat{\imath} + O(R^{m-1}).$$

Letting

$$\beta(R) = \begin{cases} 0 & \text{for } R \le 2\varepsilon, \\ 1 & \text{for } R \ge 3\varepsilon, \end{cases}$$

we can therefore cut off the part of c_0 in the $\hat{\jmath}$, $\hat{k}$ directions with small loss of energy. Denote by $c_0^\varepsilon = (B_0^\varepsilon, \phi_0^\varepsilon)$ the resulting configuration. Now if $\sigma_\varepsilon : \mathbf{H}^3 \to \mathbf{H}^3$ is reflection in $R = \varepsilon$, we define c_γ, $0 \le \gamma \le \pi$, by

$$c_\gamma = \begin{cases} c_0^\varepsilon & \text{for } R \ge \varepsilon, \\ e^{-i\gamma}\sigma_\varepsilon^* c_0^\varepsilon e^{i\gamma} & \text{for } R \le \varepsilon. \end{cases}$$

Letting $\mathscr{C}^k$ denote the monopole sector with monopole number k, we see that this construction always gives a loop in $\mathscr{C}^0$. An argument of Taubes [T] shows that such a loop is *homotopically nontrivial*. Using elliptic theory, it is easy to obtain a *positive lower bound* on the energy of all loops homotopic to

c_γ. The *asymptotics of* (3) give us the important result that the Higgs energy $H(c_\gamma)$ is *strictly less* than $2H(c_0) = 8\pi m$.

4. Outline of the min-max argument

Let $h_\infty = \min_{d_\gamma \in \Lambda} \max_{0 \le \gamma \le 2\pi} H(d_\gamma)$. The results of the preceding section show that $0 < h_\infty < 8\pi m$. We choose a "good" minimizing sequence $\{c_i\} \in \mathscr{C}^0$ which converges in $L^2_{1,\text{loc}}$ to a critical point c_∞ of H. There are two potential problems we must contend with in order to show the existence of a nonminimal critical point in the limit. The lower bound eliminates the possibility that c_∞ has zero energy. hence, we would be finished if we knew that $c_\infty \in \mathscr{C}^0$. Now, for each i,

$$\frac{1}{4\pi m} \int_{|H|^3} \text{tr}(F_{A_i} \wedge D_{A_i} \phi_i) = 0$$

because each $c_i \in \mathscr{C}^0$. However, because of the "loc" convergence of the sequence, c_∞ need *not* be in C_0. Therefore, we have the possibility of "jumping" from the zero monopole sector to some other monopole sector in which c_∞ is then minimal. Fortunately, a result of Taubes [T] shows that the energy required to accomplish this must be at least twice the minimum energy. Therefore, the upper bound on energy implies $c_\infty \in \mathscr{C}^0$.

Moving back to S^4 we find

THEOREM 1. *For every integer $m \ge 2$, there is a nonminimal critical point of the Yang-Mills functional with $c_2 = 0$.*

REFERENCES

[A] M. F. Atiyah, *Magnetic monopoles in hyperbolic spaces*, Vector Bundles on Algebraic Varieties, Tata Inst. Fund. Res., Bombay, 1984, pp. 1–33.

[B] P. J. Braam, *Magnetic monopoles on three-manifolds*, J. Differential Geom. **30** (1989), 425–464.

[BL] J. P. Bourguignon and H. B. Lawson, Comm. Math. Phys. **79** (1981), 189–230.

[SS] L. M. Sibner and R. J. Sibner, *Classification of singular Sobolov connections by their holonomy*, Comm. Math. Phys. **144** (1992), 337–350.

[SSU] L. M. Sibner, R. J. Sibner, and K. Uhlenbeck, *Solutions to Yang-Mills equations that are not self-dual*, Proc. Nat. Acad. Sci. U.S.A. **86** (1989), 8610–8613.

[T] C. H. Taubes, *Existence of a non-minimal solution to the* SU(2) *Yang-Mills-Higgs equations on* R^3. I and II, Comm. Math. Phys. **86** (1982), 257–298, 299–320.

POLYTECHNIC UNIVERSITY

Proceedings of Symposia in Pure Mathematics
Volume **54** (1993), Part 2

Spectral Invariants of Pseudoconformal Manifolds

NANCY K. STANTON

I will survey some recent analogues for pseudoconformal, or CR, manifolds of some results in Riemannnian and conformal geometry. The proofs of these results are analogous to some of the proofs in the Riemannian and conformal cases, so I will begin by describing these.

Let M be a compact oriented Riemannian n-manifold and let $\Delta = d^*d$ be the Laplace-Beltrami operator on functions on M. A function f solves the heat equation on $M \times \mathbf{R}^+$ if $\frac{\partial f}{\partial t} + \Delta f = 0$. The initial value problem for the heat equation is the following. Given $f_0 \in \mathscr{C}(M)$, find $f(x, t)$ solving the heat equation with $\lim_{t \to 0} f(x, t) = f_0(x)$. The Laplace-Beltrami operator has a complete orthonormal set of eigenfunctions $\{\phi_i\}$ with eigenvalues λ_i of finite multiplicity. The solution of the initial value problem for the heat equation is obtained by integrating the initial value f_0 over M against the *heat kernel* $e(x, y, t) = \sum e^{-\lambda_i t}\phi_i(x)\phi_i(y)$. This kernel is smooth away from the diagonal of M at $t = 0$. Minakshisundaram and Pleijel [**MP**] showed that there is an asymptotic expansion

$$\sum e^{-\lambda_i t} \sim t^{-n/2} \sum_{j \geq 0} c_j t^j \quad \text{as } t \to 0^+.$$

McKean and Singer [**MS**] showed that c_j is the integral of a universal polynomial (depending only on n and j) in the curvature and its covariant derivatives. For example, $c_0 = \frac{\text{Vol } M}{(4\pi)^{n/2}}$ and $c_1 = c(n) \int_M R \, dV$ where R is the scalar curvature.

One proof of the McKean-Singer result uses pseudodifferential operators. Integration against $e(x, y, t-s)$ over $M \times [0, t]$ solves the inhomogeneous heat equation, so one can find the heat kernel by inverting the heat operator. Instead of the classical pseudodifferential calculus, one uses a variant in

1991 *Mathematics Subject Classification.* Primary 58G25, 32F20; Secondary 35N15.
Research supported in part by NSF grants.
This paper is in final form and no version will be submitted for publication elsewhere.

which τ, the dual variable to t, is a symbol of order 2. Then the principal symbol of $\frac{\partial}{\partial t} + \Delta$ is invertible, and all the classical pseudodifferential operator machinery can be applied to construct $(\frac{\partial}{\partial t} + \Delta)^{-1}$. In local coordinates, denote the metric by $ds^2 = \sum g_{jk} dx^j \otimes dx^k$, let $(g^{jk}) = (g_{jk})^{-1}$ and let $g = \det g_{jk}$. The symbol of $(\frac{\partial}{\partial t} + \Delta)^{-1}$ is a sum of fractions whose numerators are polynomials in g_{jk}, $g^{-1/2}$, derivatives of g_{jk} and ξ and whose denominators are powers of $\sum g^{jk} \xi_j \xi_k + i\tau$. If one works in Riemannian normal coordinates centered at x_0, then $g_{jk}(x_0) = \delta_{jk}$ and the coefficients in the Taylor expansion of g_{jk} about x_0 are polynomials in the curvature and its covariant derivatives. This fact allows one to give the geometric interpretation of the integrand for c_j.

In everything I said above, one can replace Δ by the *conformal Laplacian* $D = \Delta + \frac{n-2}{4(n-1)} R$ and the same results hold (with new coefficients c_j). The operator D is called the conformal Laplacian because it transforms nicely under conformal changes of metric; if the metric is multiplied by a conformal factor e^{2f}, then the new conformal Laplacian is $\tilde{D} = e^{-(n/2+1)f} D e^{(n/2-1)f}$. Parker-Rosenberg [PR] and Branson-Ørsted [BO] proved the following results.

THEOREM 1. *Suppose n is even and*

$$\operatorname{tr} e^{-tD} \sim t^{-n/2} \sum_{j \geq 0} c_j t^j \quad \text{as } t \to 0^+.$$

Then

(i) $c_{n/2}$, *the coefficient of t^0, is a conformal invariant of M;*

(ii) $c_{n/2-1}$, *the coefficient of t^{-1}, is the integral of a local conformal invariant p of weight -2, i.e., under a conformal change with conformal factor e^{2f}, $p\,dV \to e^{2f} p\,dV$ or, equivalently, $p \to e^{(-n+2)f} p$.*

Parker and Rosenberg prove these results by heat equation techniques. Basically, they use a ζ-function argument to prove (i). They show that

$$c_{n/2} = \operatorname{Res}_0 \int_0^1 t^{s-1} \operatorname{tr}(e^{-tD})\,dt.$$

Then they calculate the derivative with respect to a one-parameter conformal deformation and show that it is 0. To prove (ii) they identify the integrand for $c_{n/2-1}$ as a constant times the coefficient of the leading logarithmic singularity of a parametrix for D.

Let P be a self-adjoint operator which is bounded below and has eigenvalues of finite multiplicity. Suppose that the eigenvalues have no finite accumulation point and that some power of the nonzero eigenvalues (counted with multiplicities) is summable; this holds for example if P is elliptic. The

ζ-function of P is defined for $\mathrm{Re}\, s$ sufficiently large by

$$(1) \qquad \zeta(s, P) = \sum_{\lambda_i \neq 0} |\lambda_i|^{-s}$$

where the sum is over the nonzero eigenvalues of P, counted with multiplicities. The ζ-function $\zeta(s, D)$ is analytic for $\mathrm{Re}\, s > n/2$ and has a meromorphic continuation to the plane, which is regular at 0. Let ν be the number of negative eigenvalues of D. The *determinant* of D is defined by

$$(2) \qquad \det D = \begin{cases} (-1)^{\nu} e^{-\zeta'(0, D)}, & \ker D = 0, \\ 0, & \ker D \neq 0. \end{cases}$$

This terminology comes from the following fact. Let A be an $n \times n$ symmetric matrix. Then $\det A$ is given by (2), with D replaced by A. Parker and Rosenberg calculate the derivative at $\varepsilon = 0$ of a conformal variation $e^{2\varepsilon f} g$ of $\det D$. If n is even, they show that this derivative is

$$(3) \qquad -2 \det D \int f(x) C_{n/2}(x)\, dx$$

where $C_{n/2}$ is the integrand for $c_{n/2}$.

Now let M be a compact oriented strictly pseudoconvex CR manifold of dimension $2n + 1$. This means that there is a subbundle $T^{1,0}$ of the complexified tangent bundle CTM such that

(i) $T^{1,0} \cap \overline{T^{1,0}} = \{0\}$;

(ii) if Z, W are smooth sections of $T^{1,0}$, so is $[Z, W]$;

(iii) the Levi form (see below) is definite.

The boundary of any smooth bounded strictly convex domain in $\mathbf{C}^{n+1}$ is a compact strictly pseudoconvex CR manifold. Here, by a strictly convex domain I mean a convex domain which has a defining function with positive definite Hessian. Thus, the unit sphere S^{2n+1} and ellipsoids are examples of strictly pseudoconvex CR manifolds.

Let θ be a nonvanishing real one form on M which annihilates $T^{1,0}$. The *Levi form* determined by θ is the Hermitian form on $T^{1,0}$ given by

$$L_\theta(Z, W) = -i\, d\theta(Z, \overline{W}).$$

By changing the sign of θ if necessary, I can assume that L_θ is positive definite. Then θ is determined up to multiplication by a positive function e^{2f} and $L_{e^{2f}\theta} = e^{2f} L_\theta$, i.e., the conformal class of the Levi form is independent of the choice of θ. For this reason, E. Cartan referred to the geometry of CR manifolds as pseudoconformal geometry. Let T be the unique vector field satisfying $\theta(T) = 1$, $i(T)d\theta = 0$. The form θ determines a Hermitian metric on M as follows:

$$\langle Z, W \rangle = L_\theta(Z, W) \quad \text{for } Z, W \in T^{1,0};$$

$$T^{1,0} \perp \overline{T^{1,0}} \quad \text{and} \quad \text{conjugation is an isometry;}$$

$$T \perp T^{1,0} \oplus \overline{T^{1,0}} \quad \text{and} \quad |T| = 1.$$

The pair (M, θ) is called a *pseudo-Hermitian manifold*.

Let $H^* = \theta^\perp \subset T^*(M)$ and define

$$d_b = \pi \circ d : \mathscr{C}^\infty(M) \to H^*,$$

where π is the orthogonal projection onto H^*. The *sublaplacian* is

$$\Delta_b = d_b^* d_b.$$

A pseudo-Hermitian manifold (M, θ) has a canonical connection, the Tanaka-Webster connection [**T**, **W**]. Let

$$D_b = \left(2 + \frac{2}{n}\right)\Delta_b + R$$

where R is the scalar curvature of the connection. Jerison and Lee [**JL**] showed that D_b is a pseudoconformal analogue of the conformal Laplacian; if $\tilde{\theta} = r^{2/n}\theta$ with $r > 0$, then $\widetilde{D}_b r^{-1} = r^{-1-2/n}D_b$. The operator D_b is the linearization of the CR Yamabe equation. On $S^3 = \{|z|^2 + |w|^2 = 1\} \subset \mathbf{C}^2$ with $\theta = \frac{i}{2}(z\,d\bar{z} + w\,d\bar{w})$, $D_b = 4(-(Z\overline{Z} + \overline{Z}Z) + 1)$ where $Z = \sqrt{2}\left(\bar{z}\frac{\partial}{\partial w} - \overline{w}\frac{\partial}{\partial z}\right) \in T^{1,0}$. From this example, it is easy to see that D_b is not elliptic. There is no second order differention in the direction of T. However, it follows from Kohn's work [**FK**] that D_b is subelliptic with loss of one derivative and that D_b has a complete orthonormal set of eigenfunctions $\{\phi_i\}$ with eigenvalues of finite multiplicity.

THEOREM 2 [**BGS**]. $\operatorname{tr} e^{-tD_b} \sim t^{-n-1} \sum_{j \geq 0} k_j t^j$ *as* $t \to 0^+$.

(ii) $k_j = \int_M K_j(x)\,dV(x)$ *where* K_j *is a polynomial (depending only on n and j) in the components of the curvature and torsion of the Tanaka-Webster connection and their covariant derivatives.*

Note that the singularity in t is half a power worse than in the Riemannian case. Because T is obtained as a bracket of two vector fields, the problem behaves as if $\dim M = 2n + 2$.

The chief difficulty in proving this theorem is that D_b is not elliptic. One cannot use a classical pseudodifferential operator calculus. However, one can develop an analogue of the Beals-Greiner calculus [**BG**] which treats T and $\frac{\partial}{\partial t}$ as second order operators. The operator $\frac{\partial}{\partial t} + D_b$ is invertible in this calculus. Unfortunately, unlike the classical calculus, this calculus is not purely algebraic—it involves convolution on a nilpotent model group. However, one can read off the information required to prove the theorem.

THEOREM 3 [S]. k_{n+1} is an invariant of the CR structure on M, i.e., it is independent of the choice of θ.

I prove this using the ζ-function method of Parker-Rosenberg. To apply the method, I need to know that the kernel and the parametrix for $e^{-tD_b^\varepsilon}$ depend smoothly on ε, where D_b^ε is the pseudoconformal Laplacian for $\theta^\varepsilon = e^{2f\varepsilon}\theta$. A close look at the pseudodifferential operator calculus and the construction of the leading term of $(\frac{\partial}{\partial t} + D_b^\varepsilon)^{-1}$ in [BGS] shows that they are smooth in ε.

THEOREM 4 [S]. If $\widetilde{\theta} = r^2\theta$ then

$$\widetilde{K}_n(x)\, d\widetilde{V}(x) = r^2 K_n(x)\, dV(x)$$

or, equivalently,

$$\widetilde{K}_n(x) = r^{-2n} K_n(x).$$

Here, $\sim$ denotes quantities with respect to the pseudo-Hermitian structure $\widetilde{\theta}$.

To prove the Theorem I again follow Parker-Rosenberg. I identify K_n with the leading logarithmic singularity of a parametrix for D_b. To identify the integrand for $c_{n/2-1}$ with the logarithmic singularity of a parametrix for D, Parker and Rosenberg use the well-known fact that there is a parametrix for $\frac{\partial}{\partial t} + D$ of the form

$$(4) \qquad \left(\sum_{j=0}^{N} t^j u_j(x, y) + O(t^{N+1}) \right) (4\pi t)^{-n/2}\, e^{-r^2(x, y)/4t}.$$

There is no such simple form for the parametrix on a CR manifold. The key new ingredient is that, using results from [BGS], one can derive estimates which are a qualitative analogue of (4); these estimates are enough to make the proof work.

EXAMPLE 1. The sphere S^3 with $\theta = \frac{i}{2}\sum_{j=1}^{2} z^j\, d\bar{z}^j$. Here

$$\operatorname{tr} e^{-tD_b} = \frac{\pi^2}{256t^2} + O\left(\frac{1}{t^2} e^{-\pi^2/4t}\right) \qquad \text{as } t \to 0+,$$

so $K_1 = 0$ and $k_2 = 0$.

EXAMPLE 2. The sphere S^{2n+1} with $\theta = \frac{i}{2}\sum_{j=1}^{n+1} z^j\, d\bar{z}^j$. Here

$$\operatorname{tr} e^{-tD_b} = t^{-(n+1)} \sum_{j=0}^{n-1} k_j t^j + O\left(\frac{1}{t^{n+1}} e^{-1/t}\right) \qquad \text{as } t \to 0+,$$

so $K_n = 0$ and $k_{n+1} = 0$.

The results in Examples 1 and 2 are proved by techniques from analytic number theory; one knows the eigenvalues λ_i and their multiplicities. Of course, Theorem 4 and the fact that S^{2n+1} is pseudoconformally flat also give $K_n = 0$.

EXAMPLE 3 (Three dimensional CR manifolds). As in [BGS], one can use invariant theory to see that K_1 is a multiple of scalar curvature. By Theorem 4 the multiple must be 0. Using invariant theory together with a variation of the pseudo-Hermitian structure on the sphere, I can also show that $k_2 = 0$.

One can also define $\det D_b$, by formula (2) with D replaced by D_b. Recently, Lohrenz [L] calculated the variation of $\det D_b$ with respect to a pseudoconformal variation, and found that it is given by (3), where $C_{n/2}$ is replaced by K_{n+1} and D is replaced by D_b. Let χ be a finite dimensional unitary representation of $\pi_1(M)$ and let $L(\chi)$ be the associated flat vector bundle. Then, because the bundle is flat, d_b is well-defined on sections of $L(\chi)$. Because χ is unitary, the metric on M gives rise to one on $L(\chi)$-valued forms. Hence one can define $\Delta_{b,\chi}$ by $d_b^* d_b$, where the adjoint is with respect to the metric on $L(\chi)$-valued forms, and $D_{b,\chi} = (2 + \frac{2}{n})\Delta_{b,\chi} + R$. The integrand in (3) is locally determined, so it is independent of χ. Thus, we have the following theorem.

THEOREM 5 (LOHRENZ [L]). *If χ_1 and χ_2 are two m-dimensional representations of $\pi_1(M)$ satisfying $\ker D_{b,\chi_i} = 0$, $i = 1, 2$, then*

$$(5) \qquad a(M, \chi_1, \chi_2) = \frac{\det D_{b,\chi_1}}{\det D_{b,\chi_2}}$$

is a CR invariant of M, i.e., it is independent of the choice of pseudo-Hermitian structure.

EXAMPLE 4 (LOHRENZ [L]). Let $M = \mathbf{R}P^3 = S^3/\{\pm 1\}$; M inherits a pseudo-Hermitian structure from the one on S^3. Let $\chi_1 = 1$ be the trivial character of $\pi_1(M)$ and let χ_2 be the nontrivial character. Then $a(M, \chi_1, \chi_2) \neq 1$.

Lohrenz uses number-theoretic techniques to do the calculations required for Example 4. Note that $D_{b,\chi_i} > 0$, $i = 1, 2$, so χ_1 and χ_2 satisfy the hypothesis of Theorem 5. This example shows that the spectral invariant $a(M, \chi_1, \chi_2)$ is nontrivial.

REFERENCES

[BG]   R. Beals and P. C. Greiner, *Pseudodifferential operators associated to hyperplane bundles*, Conference on Linear Partial and Pseudo Differential Operators/Special Issue Rend. Sem. Mat. Torino (1983), 7–40.

[BGS] R. Beals, P. C. Greiner, and N. K. Stanton, *The heat equation on a CR manifold*, J. Differential Geom. **20** (1984), 343–387.

[BO]   T. P. Branson and B. Ørsted, *Conformal indices of Riemannian manifolds*, Compositio Math. **60** (1986), 261–293.

[FK] G. B. Folland and J. J. Kohn, *The Neumann problem for the Cauchy-Riemann complex*, Ann. of Math. Studies no. 75, Princeton Univ. Press, Princeton, New Jersey, 1972.

[JL] D. Jerison and J. M. Lee, *The Yamabe problem on CR manifolds*, J. Differential Geom. **25** (1987), 167–197.

[L] T. Lohrenz, *Determinants on CR manifolds*, preprint.

[MS] H. P. McKean, Jr. and I. M. Singer, *Curvature and the eigenvalues of the Laplacian*, J. Differential Geom. **1** (1967), 43–69.

[MP] S. Minakshisundaram and A. Pleijel, *Some properties of the eigenfunctions of the Laplace operator on Riemannian manifolds*, Canad. J. Math. **1** (1949), 242–256.

[PR] T. Parker and S. Rosenberg, *Invariants of conformal Laplacians*, J. Differential Geom. **25** (1987), 199–222.

[S] N. K. Stanton, *Spectral invariants of CR manifolds*, Michigan Math. J. **36** (1989), 267–288.

[T] N. Tanaka, *A differential geometric study on strongly pseudo-convex manifolds*, Kinokuniya Book-Store Co., Ltd., Tokyo, 1975.

[W] S. M. Webster, *Pseudo-hermitian structures on a real hypersurface*, J. Differential Geom. **13** (1978), 25–41.

UNIVERSITY OF NOTRE DAME
E-mail address: nancy@cartan.math.nd.edu

Proceedings of Symposia in Pure Mathematics
Volume **54** (1993), Part 2

L_2-Cohomology and Index Theory
of Noncompact Manifolds

MARK STERN

1. Introduction and preliminaries

The purpose of this paper is to survey aspects of recent work in the study of L_2-cohomology of projective varieties. We focus on two problems: (i) the realization of L_2-cohomology in the Fubini-Study metric as a topologically or algebraically defined cohomology, and (ii) the extension of index theory to quasiprojective varieties endowed with certain complete metrics. We present no proofs and no new theorems here but do formulate some results which are only implicit in earlier papers.

Let (X, g) be a smooth Riemannian manifold, and let $\Omega^{\cdot}(X)$ denote the smooth forms on X. Given any topological space X' containing X as an open dense subset, set $S = X' \backslash X$, and let $\Omega_c^{\cdot}(X')$ denote the subset of $\Omega^{\cdot}(X)$ consisting of those f such that $S \cup \text{support}(f)$ is compact in X'. Let $L_2^{\cdot}(X, g)$ denote the measurable forms on X which are square integrable with respect to the norm determined by the metric g. Set

$$D_{2,N}^{\cdot}(X, g) = \{f \in L_2^{\cdot}(X, g) : df \text{ exists in the weak sense and}$$
$$df \in L_2^{\cdot+1}(X, g)\}$$

and, for every topological space X' containing X as an open dense set, let

$$D_{2,D}^{\cdot}(X', g) = \{f \in D_{2,N}^{\cdot}(X, g) : \text{ there exists a sequence}$$
$$\{f_j\}_j \subset \Omega_c^{\cdot}(X') \text{ such that } (f_j, df_j) \xrightarrow{L_2} (f, df)\}.$$

We denote the cohomology of the complexes $(\Omega^{\cdot}(X), d)$, $(\Omega_c^{\cdot}(X'), d)$, $(D_{2,N}^{\cdot}(X), d)$, and $(D_{2,D}^{\cdot}(X'), d)$ by $H^{\cdot}(X)$, $H_c^{\cdot}(X')$, $H_{2,N}^{\cdot}(X, g)$, and

---

1991 *Mathematics Subject Classification.* Primary 58G12; Secondary 11F75.

Partially supported by a Sloan Foundation Fellowship, a National Science Foundation Post-doctoral Research Fellowship, and a Presidential Young Investigator Award.

This paper is in final form, and no version of it will be submitted for publication elsewhere.

$H^{\cdot}_{2,D}(X', g)$ respectively and call them (for $X = X'$ fixed) the de Rham cohomology, compactly supported cohomology, Neumann L_2-cohomology, and Dirichlet L_2-cohomology respectively. The de Rham and compactly supported cohomologies are well-known topological invariants. The L_2-cohomologies are not topological invariants but depend on the quasi-isometry class of the metric. For example, $H^0_{2,N}(X, g) = 0$ if g has infinite volume and is nonzero if the g has finite volume. When (X, g) is complete, a basic theorem of Gaffney implies that $D^{\cdot}_{2,N}(X, g) = D^{\cdot}_{2,D}(X, g)$. In this case, we will drop the N and D subscripts and refer simply to *the* L_2-cohomology. When (X, g) is a hermitian manifold, it is also useful to introduce the L_2-$\bar{\partial}$ complexes with coefficients in a Hermitian holomorphic vector bundle E. We denote these complexes by $D^{p,\cdot}_{2,N}(X, g, E)$, and $D^{p,\cdot}_{2,D}(X', g, E)$. These are defined as before, simply replacing d with $\bar{\partial}$. We will denote the corresponding cohomology groups $H^{p,\cdot}_{2,N}(X, E, g)$ and $H^{p,\cdot}_{2,D}(X', E, g)$ respectively. Once again these coincide for complete metrics, and we will delete the N and D subscripts when discussing the complete case. We will also suppress the E when it is trivial.

When X is compact, all smooth metrics are quasi-isometric, $\Omega^{\cdot}(X) = \Omega^{\cdot}_c(X)$, and the inclusion

$$(\Omega^{\cdot}(X), d) \hookrightarrow (D^{\cdot}_2(X, g), d)$$

induces an isomorphism on cohomology

$$(1) \qquad\qquad H^{\cdot}_2(X, g) = H^{\cdot}_{\mathrm{DR}}(X)(= H^{\cdot}_c(X)).$$

The realization of the topological invariant, the De Rham cohomology, as the metrically defined L_2-cohomology gives rise to powerful tools for analyzing its structure. These include (for X compact)

(2)

the Hodge (p, q) decomposition for Kähler manifolds: $H^j = \bigoplus_{p+q=j} H^{p,q}$,

(3)

the Atiyah-Singer index theorem and associated fixed point theorems,

(4)   Bochner type vanishing theorems.

In recent years there has been a great deal of interest in finding analogs of (1) and (2), and (3) for a larger category of spaces. In this note, we will discuss some of the recent work on this problem. See [7] for a discussion of earlier work. For simplicity, we will restrict our discussion to the category of projective varieties.

2. Projective varieties with the Fubini-Study metric

Let V be a projective variety, $\tilde{V}$ a smooth resolution, and $M \subset V$ the set of smooth points of V. Then the desire to extend (1) leads to the following question.

QUESTION 2.1. Given a "nice" metric on M, does there exist some topologically or algebraically defined cohomology $\mathscr{H}^{\cdot}$ on M, V or $\tilde{V}$ such that $\mathscr{H}^{\cdot} = H_{2,N}^{\cdot}(M, g)$?

The term "nice" is intentionally vague. One requirement might be that

N1. $H_{2,N}^{\cdot}(M, g)$ and $H_{2,D}^{\cdot}(M, g)$ are finite dimensional.

When this is the case, we may conclude that $H_{2,N}^{\cdot}(M, g)$ and $H_{2,D}^{\cdot}(M, g)$ may be represented by harmonic forms; hence, the Hodge star operator $*_g$ induces an isomorphism

$$(5) \qquad H_{2,N}^{p}(M, g) \cong H_{2,D}^{n-p}(M, g),$$

where $n = \dim_R M$. Two additional attributes a nice metric ideally should have are:

N2. naturality,

N3. computability of the associated cohomology groups.

Here "naturality" is a question of taste and "computability" a question of technology. In the context of projective varieties with no additional structure assumed, the most natural choice of metric on M is the metric g_V induced from some projective embedding of V. All such embeddings yield quasi-isometric metrics. If M is a locally symmetric space, then the invariant metric is also an important metric to study. By "computability," I mean that the metric definition of the L_2-cohomology actually implies some extra structure, for example, (2) and/or (3) above. One of the first examples of (2.1) is the conjecture of Cheeger, Goresky, and MacPherson. Let $\mathrm{IH}^{\cdot}(V)$ denote the intersection cohomology of V. We will not define intersection cohomology here (see [7]), but will merely note that it is a topological invariant of V which satisfies Poincaré duality.

CONJECTURE 2.2. $\mathrm{IH}^{\cdot}(V) = H_{2,N}^{\cdot}(M, g_V)$.

When g_V is conical, this was proved by Cheeger [6]. When $\dim_C M = 2$, this was studied in [12] and [18]. Recently Ohsawa has claimed the following strong result.

THEOREM 2.3 [20]. *Let V be a projective variety with isolated singularities; then* $\mathrm{IH}^{\cdot}(V) = H_{2,N}^{\cdot}(M, g_V)$.

Let N_ε be a cofinal system of neighborhoods in M about a singular point $o \in V$. This theorem is equivalent to proving

$$(6) \qquad H_{2,N}^{p}(N_\varepsilon, g_V) = 0, \quad \text{for } p \geq \dim_C V,$$

and

$$(7) \qquad H_{2,D}^{p}(N_\varepsilon \cup \{o\}, g_V) = 0, \quad \text{for } p \leq \dim_C V.$$

Equation (6) is equivalent to proving the following estimate for all smooth square integrable f satisfying Neumann boundary conditions:

$$(8) \qquad \|df\|^2 + \|d^* f\|^2 \geq c\|f\|^2.$$

The same estimate with Dirichlet boundary condition is equivalent to (7). The boundary condition includes an ideal boundary condition at o. Ohsawa's argument does not use this estimate directly, but instead considers as an intermediate step the analogous estimate for certain *complete* metrics for which he can apply the following lemma of Donnelly and Fefferman.

LEMMA 2.4 [9]. *Let* (Z, g) *be a complete Kähler manifold whose Kähler form* ω *is given by* $\omega = -i\partial\overline{\partial}F$, *for some function* F *with bounded gradient. Then there exists a positive constant* c *depending on* $\sup|dF|$ *such that for all* $f \in D_2^p(Z, g)$, $p \neq \dim_C Z$,

$$\|df\|^2 + \|d^*f\|^2 \geq c\|f\|^2.$$

Hence, $H_2^p(Z, g) = 0$, *for* $p \neq \dim_C Z$.

In order to prove (6) and (7), Ohsawa constructs on N_ε a family of complete Kähler metrics g_s, $s > 0$, whose Kähler forms are given by $-i\partial\overline{\partial}F_s$, with $\{F_s\}_{s>0}$ a family of functions with uniformly (g_s-) bounded gradients. In addition, one has the following norm comparison for some $A > 0$, and all measurable p-forms f with $p > \dim_C M$:

$$(9) \qquad\qquad\qquad \|f\|_s \leq A\|f\|_{g_V}.$$

Hence, $f \in D_{2,N}^p(X, g_V)$ implies $f \in D_{2,N}^p(X, g_s)$, for all s. As an immediate consequence of (2.4), we have

$$f = db_s,$$

for some $b_s \in D_{2,N}^{p-1}(X, g_s)$. The metrics g_s are constructed so that as s tends to 0 they converge on compact subsets to the given incomplete one. With a little analysis, one extracts a subsequence of the b_s which converge to some $b \in D_{2,N}^{p-1}(X, g_V)$, with

$$db = f.$$

This establishes the (local) vanishing of the cohomology above the middle dimension in the complete metric. Additional arguments are required to establish the desired vanishing in the middle degree.

Away from the middle dimension, it is possible to bypass the introduction of the complete metric in the proof of (6) and (7) by using a variant of (2.4) for the incomplete metric. Additional estimates are required, however, to make the formal arguments carry through.

This result goes a long way toward proving (2.2). One of the original motivations for studying this conjecture was the hope that it would yield a Hodge (p, q) decomposition for intersection cohomology. After some thought one realizes that this is not an immediate consequence of the conjecture. Recall that the usual derivation of the (p, q) decomposition for (X, g) compact Kähler says more or less that if $[f] \in H^{\cdot}(X)$ then we can find a harmonic representative h for $[f]$. We may decompose h into a sum of its (p, q)

components $h = \sum_{p,q} h_{p,q}$. Each $h_{p,q}$ is harmonic, and hence closed. Thus the decomposition descends to cohomology. In addition, because the $\bar{\partial}$ Laplacian is a scalar multiple of the d Laplacian each $h_{p,q}$ is also $\bar{\partial}$-closed. This gives a map (which is easily seen to be an isomorphism)

$$H_2^j = \bigoplus_{p+q=j} H_2^{p,q}.$$

This proof works for any complete Kähler metric with finite-dimensional L_2-cohomology. In the incomplete case, however, it is not even clear that $h_{p,q} \in D_{2,N}^{\cdot}(X, g)$ (or $D_{2,D}^{\cdot}(X, g)$). It is also not clear what candidate to expect for the (p, q) component of the L_2-cohomology. Observe that a consequence of (2.2) is the equality

$$H_{2,N}^p(M, g_V) = H_{2,D}^p(M, g_V).$$

This follows from (5) and the Poincaré duality for intersection cohomology. On the other hand, as we shall see later,

$$H_{2,N}^{p,q}(M, g_V) \neq H_{2,D}^{p,q}(M, g_V).$$

With such complications arising in the search for the correct Hodge components, we see that the simultaneous satisfaction of the requirement that a metric be natural and computable may require some relaxation of standards in the case of arbitrary projective varieties.

We are not yet able to relate the L_2-d-cohomology to the L_2-$\bar{\partial}$-cohomology, but we may study the L_2-$\bar{\partial}$-cohomology directly. Recall that in the category of smooth projective varieties X, the numbers $\dim H_2^{0,p}(X, g_V) = \dim H^{0,p}(X)$ are birational invariants; hence, the holomorphic Euler characteristic

$$\chi(X, \mathcal{O}) = \sum_p (-1)^p \dim H^{0,p}(X)$$

is also a birational invariant. In [14], MacPherson conjectured that if we define the L_2-holomorphic Euler characteristic to be

$$\chi_2(V, \mathcal{O}) = \sum_p (-1)^p \dim H_{2,D}^{0,p}(V, g_V),$$

then $\chi_2(V, \mathcal{O})$ is a birational invariant of the full category of projective varieties. This turns out to be true and follows from the following stronger result.

THEOREM 2.5 [22]. $\dim H_{2,D}^{0,p}(V, g_V)$ *is a birational invariant of projective varieties.*

In particular, we have

COROLLARY 2.6 [22]. $H_{2,D}^{0,p}(V, g_V) \cong H^{0,p}(\tilde{V})$.

Saper [23] has proved a related result for certain complete metrics on varieties with isolated singularities. See also [21], [5], and [11] for results in dimensions one and two.

It is also possible (but more difficult) to compute the Neumann $\bar{\partial}$-cohomology. Let Z denote the unreduced exceptional divisor in $\tilde{V}$ and assume that Z is supported along a divisor with normal crossings. The following theorem was conjectured by W. Pardon in [21].

THEOREM 2.7 [22]. *Suppose that the singularities of V are isolated, and* $\dim_C V \leq 2$. *Then* $H_{2,N}^{0,p}(V, g_V) \cong H^{0,p}(\tilde{V}; \mathscr{O}(Z - |Z|))$.

Similar theorems can be proved for higher dimensions. One sees immediately that the Neumann and Dirichlet $\bar{\partial}$-cohomologies do not coincide in general. This is contrary to what (2.2) implies for d-cohomology.

The tools used in the proof of (2.5) for the case $p \leq n - 2$ are similar to those used by Ohsawa in [19]. We also implicitly use a new cohomology comparison lemma which is well adapted to studying $H_{2,D}^{0,n-1}$ but which we expect to have wider application.

LEMMA 2.8 (Pardon-Stern). *Let* $(E^{\cdot}, D)$ *and* $(F^{\cdot}, D)$ *be two* L_2-*cohomology complexes. Suppose that* $F^{p-1} \subset E^{p-1}$, $F^p \subset E^p$, *and* $F^{p+1} \supset E^{p+1}$. *Suppose also that these inclusions are bounded in the* L_2-*norm (as opposed to the graph norm). Suppose that* $D: F^p \to F^{p+1}$, *and* $D: E^{p-1} \to E^p$, *have closed range. Then the inclusion*

$$H^p(F^p, D) \to H^p(E^p, D)$$

is a surjection.

This lemma provides an effective way to eliminate L_2 poles. It is especially well adapted to studying $H_{2,D}^{0,n-1}$ because all hermitian metrics agree on $(0, n)$ forms.

COROLLARY 2.9 (Pardon-Stern). *Let* X *be a complex manifold and* g_1 *and* g_2 *two Hermitian metrics on* X *with* $g_1 \geq cg_2$, *for some* $c > 0$. *Let* E *be a hermitian holomorphic vector bundle on* X. *Suppose also that* $H_{2,D}^{0,n-1}(X, g_2, E)$ *and* $H_{2,D}^{0,n}(X, g_1, E)$ *are finite dimensional. Then the inclusion*

$$H_{2,D}^{0,n-1}(X, g_1, E) \to H_{2,D}^{0,n-1}(X, g_2, E)$$

is a surjection.

3. Locally symmetric varieties

We see that the study of the L_2-cohomology of projective varieties with the natural incomplete metric is still in its infancy. We now consider varieties with complete metrics. The richest theory has been developed when M has the structure of a hermitian locally symmetric space of finite volume. So suppose that G is a semisimple Lie group, Γ an arithmetic subgroup, K a maximal compact subgroup, and $M = \Gamma \backslash G / K$ a Hermitian locally symmetric space. M may be viewed as the smooth points of a projective variety

V—the Baily-Borel-Satake compactification of M. The components F_P of the singular strata of V are also hermitian locally symmetric spaces, called boundary components. The boundary components are indexed by a set of representatives I of Γ-conjugacy classes of maximal rational parabolic subgroups. Let E be a flat vector bundle associated to an irreducible complex finite-dimensional representation of G. Each of these objects is equipped with a natural invariant metric.

The L_2-cohomology of M with coefficients in E is both computable and realizable as a topological invariant. We have the following theorem conjectured by Zucker.

THEOREM 3.1 ([25, 13]). $H_2^{\cdot}(M, E) \cong \mathrm{IH}^{\cdot}(V, E)$.

Here the L_2-cohomology with coefficients in E is defined in the obvious way, and $\mathrm{IH}^{\cdot}(V, E)$ denotes the intersection cohomology of V with coefficients in E.

The structure of the groups $\mathrm{IH}^{\cdot}(V, E)$ contains a great deal of arithmetic information. In particular, one wants to know the action of the Hecke operators on these spaces. We are led to study L_2-index theorems and associated fixed point formulas. We first recall some fixed point formulas in the compact case.

Let X be a smooth compact Riemannian manifold, and let $f\colon X \to X$ be a diffeomorphism with nondegenerate fixed points. Let f^* denote the induced map on $H^{\cdot}(X)$. Set

$$L(f, \tau^e) := \sum_i (-1)^i \operatorname{tr} f^*|_{d < H^i(X)} = \operatorname{tr} \tau^e f^*,$$

where τ^e denotes the involution $(-1)^{\mathrm{degree}}$. With this notation, the classical fixed point theorem is

THEOREM 3.2 (Lefschetz). $L(f, \tau^e) = \sum_p \operatorname{sign} \det(I - df_p) =: l(f, \tau^e)$.

The sum runs over the points p fixed by f. This formula was generalized in many ways by Atiyah, Bott, and Singer. The fundamental example arises as follows. Let τ_S be the involution defined on $\bigwedge^{\cdot} T^*X$ by $\tau_S = (-1)^{p(p-1)/2}*$ on p-forms. τ_S preserves the space $\mathscr{H}$ of harmonic representatives of the cohomology. It can be shown that the topological signature of the manifold is given by

$$\operatorname{Sign} X = \operatorname{tr} \tau_S|_{\mathscr{H}}.$$

Let f be an isometry. Then f induces a map $f^{\#}$ on $\mathscr{H}$. Set

$$L(f, \tau_S) = \operatorname{tr} \tau_S f^{\#},$$

the Lefschetz number associated to the signature complex. The generalization of (3.2) to this situation is given by the following theorem.

THEOREM 3.3 (Atiyah et al.). $L(f, \tau_S) = \sum_p \prod_j \cot \theta_j(p) =: l(f, \tau_S)$, where the $\theta_j(p)$ are a choice of rotation angles for df_p.

REMARK 3.4. This result can be extended to conformal maps [30].

There are variants of this theorem attached to any index theorem. One may remove the assumption that the fixed points are isolated. Then the local contribution $l(f, \tau_S)$ consists of certain complicated curvature integrals over the fixed submanifold.

Now we would like to extend these considerations to compute traces of the action of the Hecke operators on the L_2-cohomology of an arithmetic variety. Let G, Γ, and K be as above. Let $\alpha \in G$ be such that $\Gamma \alpha \Gamma$ can be written as a finite disjoint union

$$(10) \qquad\qquad \Gamma \alpha \Gamma = \bigcup_{i=1}^{N} \alpha_i \Gamma.$$

Define $T(\alpha) \colon L_2(\Gamma \backslash G) \to L_2(\Gamma \backslash G)$ by

$$T(\alpha) f(\Gamma x) = \sum_i f(\Gamma \alpha_i^{-1} x).$$

It is easy to check that this map is well defined and independent of the choice of α_i. $T(\alpha)$ extends to a bounded endomorphism of the space of L_2 E-valued forms which preserves the subspace of harmonic forms. Let $\alpha^{\#}$ denote the induced map on the harmonic forms. The Lefschetz numbers we wish to compute are

$$L(\alpha, \tau, E) := \operatorname{tr} \tau \alpha^{\#}, \quad \text{and} \quad L(\alpha, \tau^e, E) := \operatorname{tr} \tau^e \alpha^{\#},$$

where τ^e is as before, and τ is the natural extension of τ_S to E-valued forms. In particular, (see [15]) there exists an involution (possibly trivial) τ_E of E such that $\tau = \tau_S \tau_E$. With respect to this involution, E decomposes as a sum

$$E = E_+ \oplus E_-$$

of homogeneous eigenbundles.

Let d denote the exterior derivative for E-valued forms. Set $D = d + d^*$, and let D_+ (respectively D_e) denote the restriction of D to the $+1$ eigenspace of τ (respectively τ^e). When α is the identity, the Lefschetz numbers reduce to indices of elliptic operators:

$$L(1, \tau, E) = \operatorname{index} D_+, \quad \text{and} \quad L(1, \tau^e, E) = \operatorname{index} D_e.$$

Here the index is taken in the L_2 sense;

$$\operatorname{index} D_+ = \dim(\operatorname{Kernel} D_+ \cap L_2) - \dim(\operatorname{Kernel} D_+^* \cap L_2).$$

When M is real rank one, these indices (and more generally the index of any invariant Dirac operator) were first computed by Barbasch and Moscovici [3]. Their method can be described briefly as follows. Let $h_t(x, y)$ denote the

Schwarz kernel for the heat operator for D^2 on G/K. Then $\sum_{\gamma \in \Gamma} h_t(x, \gamma y)$ is the kernel for the heat operator e^{-tD^2} on M. The L_2 index is given by the trace of the restriction of τe^{-tD^2} to the discrete spectrum. This is then computed by applying the Selberg trace formula to $\sum_{\gamma \in \Gamma} h_t(x, \gamma y)$. Using related techniques and scattering theory, W. Muller [16, 17] extended these results to spaces of rational rank one. (See [4] for related work.)

In [26], we obtained an index theorem for signature operators on hermitian locally symmetric spaces of *arbitrary* rank and on equal rank locally symmetric spaces of rational rank one. (We assume Γ arithmetic). In real rank one we obtain the following expression for the index which corrects a minor error in [3]. (See [27, §6].) Let $M = \Gamma \backslash G / K$ be a real rank one, equal rank, locally symmetric space. The cusps are indexed by a set I of representatives of Γ-conjugacy classes of proper rational parabolic subgroups of G. For $P \in I$, let N_P denote the unipotent radical of P and let $\mathfrak{n}_P$ denote the Lie algebra of N_P. Let d_n denote the usual coboundary operator of $\mathfrak{n}_P$ with coefficients in E, and set $D_n = d_n + d_n^*$. (These are zero order operators—matrices.) Let $\tilde{\tau} = (-1)^{p(p-1)/2} *_n \tau_E$, where $*_n$ denotes the Hodge star operator for N_P. Let L denote the Hirzebruch L polynomial of M and let $\mathrm{ch}\, E_\pm$ denote the Chern character of $E_\pm$.

THEOREM 3.5 [26, Theorem 6.7].

$$\mathrm{Index}\, D_+ = \int_M (\mathrm{ch}\, E_+ - \mathrm{ch}\, E_-) \wedge L$$
$$+ \sum_{P \in I} \left(\mathrm{signature}(\tilde{\tau} D_n)/2 + \sum_p \varepsilon_p(G) \zeta(2p) c(N_P \cap \Gamma, E) \right),$$

where $\varepsilon_p(G) = 1$ *if G/K is the complex p-ball and is zero otherwise.*

Here $\zeta(s)$ denotes the Riemann zeta function, and $c(N_P \cap \Gamma, E)$ is an easily computed constant depending on E, certain Lie algebra data, and the lengths of the generators of the lattice $N_P \cap \Gamma$. When E is trivial, the term $\mathrm{signature}(\tilde{\tau} D_n)$ vanishes unless G/K is a complex p-ball for some p, and then admits the following cohomological interpretation. The cross-section of a cusp is a nilmanifold of the form $(\Gamma \cap N_P) \backslash N_P$. This nilmanifold is a circle bundle over a complex torus V_Γ. Let ω denote the Kähler form of V_Γ (which is also a multiple of the first Chern class of the circle bundle). Let Q denote the quadratic form defined on the cohomology of V_Γ by

$$Q(f) = \int_{V_\Gamma} f \wedge f \wedge \omega.$$

Then

$$\mathrm{signature}(\tilde{\tau} D_n) = - \mathrm{signature}\, Q.$$

We refer loosely to the terms signature($\tilde{\tau} D_n$)/2 and

$$\sum_p \varepsilon_p(G)\zeta(2p)c(N \cap \Gamma, E),$$

and their sum respectively as the eta term, the zeta term, and the defect term associated to the parabolic subgroup P. Recall that the eta invariant of a selfadjoint elliptic differential operator A with eigenvalues $\{\lambda\}$ is defined to be the value at $s = 0$ of the function

$$\eta(A, s) = \sum_\lambda \lambda|\lambda|^{-s}$$

$$(11) \qquad = \int_0^\infty \text{tr}\, A e^{-tA^2} t^{s-1/2}\, dt \left(\int_0^\infty e^{-t} t^{s-1/2}\, dt\right)^{-1}.$$

Substituting a finite rank operator for A yields $\eta(A, 0) = \text{signature}\, A$. The signature theorem of [26] for M of arbitrary rank has the following form.

$$(12) \qquad \text{Index}\, D_+ = \int_M (\text{ch}\, E_+ - \text{ch}\, E_-) \wedge L + \sum_{P \in I}(Z_P + \eta_P/2),$$

where the sum runs over the boundary components F_P of M, the term Z_P is a special value of a Sato-Shintani zeta function associated to F_P, which we will discuss in §4, and η_P is obtained by inserting into the expression (11) at $s = 0$, a certain differential operator $\hat{D}$ defined on a space of vector-valued differential forms on F_P. In the higher rank case, the operators $A e^{-tA^2}$ will not be trace class; so the trace must be defined by a limiting process. As an unexpected bonus, one can use the above formula to obtain rationality results for certain combinations of special values of Sato-Shintani zeta functions. Nonetheless, as these zeta functions do not yet have a nice geometric interpretation, we do not consider this theorem to be optimal. The Z_P do not appear in the index formula associated to τ^e.

This form of the index formula suffers from a severe defect; eta invariants are not usually computable. In [27], this defect is remedied. It is shown there that the computation of η_P can be reduced to the computation of the index of an elliptic operator (again of the form D_+ for some local system on F_P). Inductively, this allows one to complete the L_2 signature computation in terms of curvature date on boundary components and special values of zeta functions.

For M hermitian and G of the form $G = R_{k/Q}G'$ with G' an absolutely simple k group, k a totally real number field and $R_{k/Q}$ the restriction of scalars functor, all $\eta_P = 0$. Hence in this case, our computations greatly simplify, and the defect term is given solely in terms of the zeta terms and the Atiyah-Singer integrand. For example, consider the Hilbert modular varieties of dimension greater than 2, and suppose E is trivial. The L-polynomial vanishes on these rational rank one varieties and the Sato-Shintani zeta functions reduce to Shimizu L-functions. Hence, in the smooth case, we easily recapture the formula of Atiyah, Donnelly, and Singer [2] and Muller [16] for the signature of Hilbert modular varieties (and the associated conjecture of Hirzebruch). For M a Hilbert modular variety,

$$(13) \qquad \text{Signature}(M) = \sum_p L_p(0),$$

where $L_p(s)$ is the Shimizu L-function associated to the point boundary component F_p (see [2] for a definition of the Shimizu L-function).

We now return to the consideration of Hecke operators. As in the compact case, $L(\alpha, \tau^e, E)$ can be computed in many different ways. Using the trace formula, Arthur [1] has computed $L(\alpha, \tau^e, E)$ in terms of orbital integrals and other representation-theoretic type data. Using their intersection cohomology Lefschetz formula and (3.1), Goresky and MacPherson [10] have also computed $L(\alpha, \tau^e, E)$ in terms of more geometric data similar to that arising in the compact case. Their techniques do not seem to extend to the computation of the other Lefschetz numbers.

We will indicate how to compute the Lefschetz numbers in terms of fixed point data. To this end, define $l(\alpha, \tau, E)$ and $l(\alpha, \tau^e, E)$ to be the local data assigned to the fixed point sets by the same local recipe as in the compact case. For example, when E is trivial and the fixed points are isolated, $l(\alpha, \tau^e, E)$ is given by a sum over fixed points of ± 1. One needs to note, however, that $T(\alpha)$ does not arise from a diffeomorphism but from a correspondence. Hence, fixed point actually means the intersection of the correspondence with the diagonal in $M \times M$. In [28], we give a formula for the Lefschetz numbers in terms of special values of certain zeta functions and fixed point data on M and its boundary components. As in the case of the above index theorem, the first form of the Lefschetz theorem says that the Lefschetz numbers can be computed in terms of the local fixed point data, as in the compact case, plus a sum of defect terms associated to each boundary component. The defect terms can be written as a sum of a term $\eta_p(\alpha, \tau)$ (or $\eta_p(\alpha, \tau^e)$) analogous to the eta invariant term in (12) (merely insert $T(\alpha)$ into the trace) plus a special value of a Sato-Shintani zeta function $Z_p(\alpha)$. The latter term generically vanishes. In particular, if no α_i (in (10)) fixes a nonzero element of n_p then $Z_p(\alpha) = 0$. This gives us the following simplest case of the Lefschetz formula.

THEOREM 3.6 [28]. *Suppose that for some G' and some totally real number field $k \neq Q$, $G = R_{k|Q} G'$. Then*

$$L(\alpha, \tau^e, E) = l(\alpha, \tau^e, E),$$

and if no α_i (in (10)) fixes a nonzero element of n_p, $L(\alpha, \tau, E) = l(\alpha, \tau, E)$.

Rather than give the more general form of the theorem here, we first give the form of the contribution of a boundary component F_p under the restrictive assumption that M is real rank one, α normalizes Γ, and no α_i fixes a nonzero element of N_P, $P \in I$. We will then indicate the relation to the higher rank case. So suppose that M and α satisfy the preceding restrictive hypotheses. The component F_P corresponds to a maximal rational parabolic subgroup P with maximal rational split torus A. If no α_i is not conjugate to an element of P, then $\eta_P(\alpha, \tau) = \eta_P(\alpha, \tau^e) = 0$. If one α_i is conjugate to $\alpha' \in P$, let a denote the "A component" of α'. The induced metric on A determines an identification of A with $(-\infty, \infty)$. Let s denote the image of a under this map. Let $\{\mu\}$ denote the (finite set of) eigenvalues of $\tilde{\tau}D_n$ acting on $\bigwedge \mathfrak{n}^* \otimes E$ and write $\bigwedge \mathfrak{n}^* \otimes E = \bigoplus_\mu E_\mu$. Then

$$\eta_P(\alpha, \tau) = \text{(a universal constant)} \times \sum_\mu \text{sign}(\mu) e^{-|s|\mu}.$$

If we let ν denote the weights of A on $H^\cdot(\mathfrak{n}, E)$, then

$$\eta_P(\alpha, \tau^e) = \text{(a universal constant)} \times \sum_\nu \text{sign}(\nu) e^{-|s|\nu}.$$

Here the sign of a weight is $+1$ if it is greater than $1/2$ the sum of roots of A on $\mathfrak{n}$ and negative otherwise. For $G = \text{SL}(2, R)$, the classical Hecke operator T_p for p a prime corresponds to $T(\alpha)$ with

$$\alpha = \begin{pmatrix} p^{1/2} & 0 \\ 0 & p^{-1/2} \end{pmatrix},$$

and $e^{-|s|\nu} = p^n$, for $\nu = -n(\lambda_1 - \lambda_2)$ in the usual coordinates.

In the higher rank case, $\eta_P(\alpha, \tau)$ (and $\eta_P(\alpha, \tau^e)$) can be computed inductively in terms of a Lefschetz number for F_P (possibly of the identity Hecke operator) and a function of lower-rank eta invariant terms. The Hecke correspondence generated by α generates a sum of correspondences on F_P. Assume for simplicity of notation that these correspondences are generated by a single element α_P. The Lefschetz number for F_P is obtained by viewing $\text{sign}(\mu)$ as an involution on the space $E_{|\mu|} := E_\mu \oplus E_{-\mu}$ which extends τ to an involution $\tau_{|\mu|}$ on $E_{|\mu|}$-valued forms on F_P. Then

$$\eta_P(\alpha, \tau) = \text{constant} \times \sum_\mu L(\alpha_P, \tau_{|\mu|}, E_{|\mu|}) e^{-|s|\mu}$$

$$+ \text{a function of eta invariants on lower rank boundary}$$

$$\text{components}.$$

The lower rank terms can be evaluated inductively.

4. Zeta functions

We now recall briefly the basic facts about the zeta functions which Sato and Shintani associate to prehomogeneous vector spaces. Let H be a reductive complex linear algebraic group and (ϕ, V) a finite-dimensional rational representation of H on V of degree n. The triple (H, ϕ, V) is called a prehomogeneous vector space if there exists a proper algebraic subset S of V such that $V\backslash S$ is a single H orbit. Assume for simplicity that S is rationally irreducible. Then there exists a homogeneous polynomial B of degree d on V such that $S = B^{-1}(0)$. Let H_R and V_R denote the set of real points of H and V, and let

$$V_1 \cup \cdots \cup V_N = V_R\backslash V_R \cap S$$

be the decomposition of $V_R\backslash V_R \cap S$ into its connected components. Let L be a lattice in V_R, $L' = L\backslash L \cap S$, and Γ_H an arithmetic subgroup of H_R which preserves L. One may associate Dirichlet series to this data in the following manner. Set

$$\zeta_i(s, L) = \sum_{x \in \Gamma_H\backslash(L\cap V_i)} \mathrm{Vol}\{(\Gamma_H)_x\backslash(H_R^o)_x\}|B(x)|^{-s},$$

where for a group A, A_x denotes the subgroup of A which fixes x, and A^o denotes the connected component of the identity in A. Such Dirichlet series arise frequently in analytic number theory. For f a rapidly decreasing function on V_R, set

$$\Phi_i(f, s) = \int_{V_i} f(x)|B(x)|^s \, dx.$$

The Sato-Shintani zeta function associated to f and L is

$$Z(f, L, s) = \sum_i \zeta_i(s, L)\Phi_i(f, s - n/d).$$

These are introduced primarily as an intermediate tool for studying the Dirichlet series $\zeta_i(s, L)$.

We now relate the above definitions to the zeta function Z_P associated to a boundary component F_P when $\alpha = 1$ (see [28, §6]). In this case, V_R is the center U_P of N_P, $L = \Gamma \cap U_P$, and H_R is approximately the quotient of P by the kernel of the adjoint representation of P on U_P. The choice of f which is dictated to us by the solution of the Lefschetz problem is a polynomial times an exponential which leads to explicitly computable $\Phi_i(f, s)$; in fact, these can be computed in terms of gamma functions. Finally we have

(14) $$Z_P = Z(f, \Gamma \cap U_P, (\dim N - \dim V_R)/r),$$

where r is the real rank of H_R.

When F_P is a minimal boundary component and N_P is abelian, the Sato-Shintani zeta function is not absolutely convergent at $(\dim N - \dim V_R)/r$. It seems likely that if G is simple over R, then the zeta function for such P vanishes at this value. We have shown this [28, §6] for the rank two groups $G = \mathrm{SO}(p, 2)$ and $G = \mathrm{SU}(2, 2)$.

5. Complete Kähler manifolds

We now consider a useful model problem which points to a geometric interpretation of the Z_P discussed in the previous section.

Let V be a smooth compact Kähler manifold with Kähler form ω'. Let $\mathscr{D} = \bigcup_i \mathscr{D}_i \subset V$ be a divisor with normal crossings. Let σ_i be a holomorphic section of the Hermitian line bundle $[\mathscr{D}_i]$, with $\mathscr{D}_i = \sigma_i^{-1}(0)$, and let ν_i denote the first Chern class of the bundle. Equip $M = V - \mathscr{D}$ with the complete metric, g, of finite volume associated (for T large) to the Kähler form

$$\omega = T\omega' - \sum_j \partial\bar{\partial} \log\log^2 |\sigma_j|^2.$$

The function theory associated to such metrics has been studied at length in [8]. We have computed in [29] the L_2 indices of the standard elliptic complexes on (M, g). For example, we obtain the following theorem.

THEOREM 5.1 [29]. *The L_2 signature of (M, g) is given by*

$$\int_M L(TM) + \sum_I \int_{D_I} L(TD_I) \prod_i (L(\nu_i) - 1)/\nu_i,$$

where $D_I = \bigcap_{i \in I} D_i$ and $L(\nu_i)$ is the Hirzebruch L-polynomial.

Observe that

$$L(x) = 2 \sum_{k=0}^{\infty} (-1)^{k-1} \zeta(2k)(x/\pi)^{2k}.$$

In this model problem, the defect terms corresponding to Z_P are given by (products of) the $\zeta(2k)$ terms in this formula. It would be an interesting problem to express Z_P similarly in terms of characteristic classes associated to the divisors at ∞ of smooth toroidal compactifications of $\Gamma\backslash G/K$.

The proof of (5.1) combines the heat equation formalism with the parametrix method of computing indices. For example, again let $D = d + d^*$ and let D_+ denote the restriction of D to the $+1$ eigenspace of τ. We call D the signature operator of M. Recall that D_+ is Fredholm if and only if there exists a bounded operator Q such that

$$D_+ Q = I - S_1, \quad \text{and} \quad QD_+ = I - S_2,$$

with S_1 and S_2 trace class. Then

$$\operatorname{index} D_+ = \operatorname{tr} S_0 - \operatorname{tr} S_1,$$

where tr denotes trace. On a compact manifold, the operator

$$Q_t = \int_0^t De^{-sD^2}(1-\tau)\, ds/2$$

is a bounded operator and $S_i = e^{-tD^2}(1 + (-1)^i \tau)$, $i = 1, 2$, are trace class. This yields

$$\operatorname{index} D_+ = \operatorname{Trace} \tau e^{-tD^2}.$$

Thus the index can be computed from the small time asymptotics of the heat operator. On a noncompact manifold M, e^{-tD^2} is not generally trace class so this technique must be modified. Let $\{\rho_n\}_n$ be a family of smooth functions such that each $(1 - \rho_n)$ is supported on a compact set K_n, with $K_n \subset K_{n+1}$ and $\bigcup_n K_n = M$. Let Π denote the L_2 nullspace of D. When D_+ is Fredholm, we may define

$$Q = \int_0^t De^{-sD^2}(1 - \tau)\,ds/2 + \rho_n \int_t^\infty De^{-sD^2}(1 - \tau)\,ds/2.$$

This choice of Q gives the following expression for the index.

$$\text{index}\, D_+ = \text{tr}\,\tau(\rho_n\Pi + (1 - \rho_n)e^{-tD^2}) - \text{tr}[D, \rho_n]\int_t^\infty De^{-sD^2}(1 - \tau)\,ds/2.$$

The assumption that D_+ is Fredholm implies that

$$\lim_{n\to\infty} \text{tr}\,\tau\rho_n\Pi = 0.$$

To prove (5.1) we pick the sequence $\{\rho_n\}_n$ so that $t \sim$ the injectivity radius of the support of ρ_n. This enables us to compute the limits

$$(15) \qquad \lim_{n,\,1/t\to\infty} \text{tr}\,\tau(1 - \rho_n)e^{-tD^2},$$

and

$$\lim_{n,\,1/t\to\infty} \text{tr}[D, \rho_n]\int_t^\infty De^{-sD^2}(1 - \tau)\,ds/2$$

$$(16) \qquad = -\lim_{n,\,1/t\to\infty} \text{tr}[D, \rho_n]\int_t^\infty De^{-sD^2}\tau\,ds/2.$$

The first limit *appears* to be standard heat equation asymptotics, but it is not. The error terms arising in the usual local approximation to the trace of the heat kernel near a point x are $O(t^{N/2}/\text{inj}(x)^N)$, N large, where $\text{inj}(x)$ denotes the injectivity radius at x. Thus we only obtain the standard asymptotics when $t \ll \text{inj}(x)^2$. In the range where $t/\text{inj}(x)^2$ is not small, we must use a nonlocal approximation to the heat kernel in order to compute the trace (still for $t \to 0$). Locally M is a union of sets quasi-isometric to manifolds of the form $D_P^{*k} \times D^{n-k}$ where D is the unit disk with the usual metric and $D_P^* \approx ([0, \infty] \times S^1, dr^2 + e^{-2r}\,d\theta^2)$ is the punctured disk with the Poincaré metric. Essentially, the nonlocal approximation is constructed in the cases under consideration by Fourier expansion in the S^1 directions and separation of variables type arguments. It is this Fourier expansion which leads to the $\zeta(s)$ terms. For example, suppose that $\mathscr{D}$ has a single component, and let $\tilde{D}$ denote the signature operator of $\mathscr{D}$. Let γ denote the operation of multiplication by $\pi\nu/2$ (ν is the first Chern class of the normal bundle), and let p denote multiplication by $e^r\,dr \wedge d\theta$. Then one can show that in the limit as $n \to \infty$,

$$(17) \quad \text{tr}(1 - \rho_n)\tau e^{-tD^2} = \sum_{m\in\mathbf{Z}} \text{tr}(1 - \rho_n)\tau e^{-t\tilde{D}}e^{t\partial^2/\partial r^2}e^{-t(e^r m + ie^{-r}\gamma + ip)^2} + O(t).$$

Applying the Poisson summation formula to the expression on the right-hand side, integrating in r, and allowing t to tend to zero gives the formula in (5.1). In this case, (16) vanishes. The defect term (17) should be thought of as local at infinity since it arises from the small t asymptotics.

Computing the limit (16) is more delicate because it requires estimating large time behaviour of the heat operator. The computation requires three steps which I will only outline. For simplicity let us again assume that the divisor $\mathscr{D}$ has a single component. First, estimates show that the trace is unchanged in the limit if we replace De^{-sD^2} by its projection onto S^1 invariant forms. Second, we reduce the computation of the S^1-invariant trace to a computation involving heat operators on the cylinder $\mathbf{R} \times \mathscr{D}$. These operators are easily constructed using separation of variables. Finally, one can easily compute the trace in this simplified form. In the signature theorem, one proves vanishing of the corresponding trace for algebraic reasons. For the L_2 Riemann-Roch and L_2 Gauss-Bonnet theorems, the trace reduces to a constant times

$$\operatorname{tr} \tau^e e^{-t\widehat{D}^2},$$

where $\widehat{D} = \widetilde{D}$ for the Gauss-Bonnet theorem and is a multiple of the $\mathrm{spin_C}$ Dirac operator on $\mathscr{D}$ for the Riemann-Roch theorem. The trace is now computed in the L_2 space of the appropriate bundle on $\mathscr{D}$. The index theorem on compact manifolds now reduces this final trace to a curvature integral, completing the derivation of (5.1). Similar arguments can be used to compute the Lefschetz numbers of Hecke operators.

REFERENCES

1. J. Arthur, *The L^2-Lefschetz numbers of Hecke operators*, Invent. Math. **97** (1989), 257–290.
2. M. Atiyah and H. Donnelly, and I. Singer, *Eta invariants, signature defects of cusps, and values of L-functions*, Ann. of Math. (2) **118** (1983), 131–177.
3. D. Barbasch and H. Moscovici, *L^2-index and the Selberg trace formula*, J. Funct. Anal. **53** (1983), 151–201.
4. J. Bruning, *L^2-index theorems for certain complete manifolds*, J. Differential Geom. **32** (1990), 491–532.
5. J. Bruning, N. Peyerimhoff, and H. Schroder, *The $\bar{\partial}$-operator on algebraic curves*, Comm. Math. Phys. **129** (1990), 525–534.
6. J. Cheeger, *On the Hodge theory of Riemannian pseudomanifolds*, Proc. Sympos. Pure Math., vol. 36, Amer. Math. Soc., Providence, RI, 1980, pp. 91–146.
7. J. Cheeger, M. Goresky, and R. MacPherson, *L^2-cohomology and intersection homology for singular algebraic varieties*, Seminar in Differential Geometry (S.-T. Yau, ed.), Princeton Univ. Press, Princeton, NJ, 1981, pp. 303–340.
8. M. Cornalba and P. Griffiths, *Analytic cycles and vector bundles on non-compact algebraic varieties*, Invent. Math. **28** (1975), 1–106.
9. H. Donnelly and C. Fefferman, *L^2-cohomology and index theorem for the Bergman metric*, Ann. of Math. (2) **118** (1983), 593–618.
10. M. Goresky and R. MacPherson, *Lefschetz numbers of Hecke correspondences*, Comm. Math. Phys. **129** (1990), 525–534.

11. P. Haskell, L^2-*Dolbeault complexes on singular curves and surfaces*, Proc. Amer. Math. Soc. **107** (1989), 517–526.

12. W. C. Hsiang and V. Pati, L^2-*cohomology of normal algebraic surfaces*. I, Invent. Math. **81** (1985), 395–412.

13. E. Looijenga, L^2-*cohomology of locally symmetric varieties*, Compositio. Math. **67** (1988), 3–20.

14. R. MacPherson, *Global questions in the topology of singular spaces*, Proc. Internat. Congr. Math., Warsaw, 1983 PWN, Warsaw, 1984, pp. 213–235.

15. H. Moscovici, *The signature theorem with local coefficients of locally symmetric spaces*, Tôhoku Math. J. (2) **37** (1985), 513–522.

16. W. Muller, *Signature defects of cusps of Hilbert modular varieties and values of L-series at $s = 1$*, J. Differential Geom. **20** (1984), 55–119.

17. ____, *Manifolds with cusps of rank one: Spectral theory and L^2-index theorem*, Lecture Notes in Math., vol. 1244, Springer-Verlag, 1987.

18. M. Nagase, *Remarks on the L^2-cohomology of singular algebraic surfaces*, J. Math. Soc. Japan **41** (1989), 97–116.

19. T. Ohsawa, *Hodge spectral sequence on compact Kähler spaces*, Publ. Res. Inst. Math. Sci. **23** (1987), 613–625.

20. ____, *Cheeger-Goreski-MacPherson's conjecture for the varieties with isolated singularities*, Preprint.

21. W. Pardon, *The L^2-$\bar{\partial}$-cohomology of an algebraic surface*, Topology **28** (1989), 171–195.

22. W. Pardon and M. Stern, L^2-$\bar{\partial}$-*cohomology of complex projective varieties*, J. Amer. Math. Soc. **4** (1991), 603–621.

23. L. Saper, L^2-*cohomology of algebraic varieties with isolated singularities*, Preprint.

24. L. Saper and M. Stern, L_2-*cohomology of arithmetic varieties*, Proc. Nat. Acad. Sci. U.S.A. **84** (1987), 5516–5519.

25. ____, L_2-*cohomology of arithmetic varieties*, Ann. of Math. **132** (1990), 1–69.

26. M. Stern, L^2-*index theorems on locally symmetric spaces*, Invent. Math. **96** (1989), 231–282.

27. ____, *Eta invariants and hermitian locally symmetric spaces*, J. Differential Geom. **31** (1990), 771–789.

28. ____, *Lefschetz formulae for arithmetic varieties*, Preprint.

29. ____, *Index theory for certain complete Kähler manifolds*, J. Differential Geom. (to appear).

30. ____, *A fixed point theorem for conformal maps and the signature complex*, manuscript.

DUKE UNIVERSITY

Proceedings of Symposia in Pure Mathematics
Volume **54** (1993), Part 2

On Calabi-Yau Three-Folds Fibered Over Smooth Complex Surfaces

WING-WAH SUNG

ABSTRACT. Let X be a smooth complex projective three-fold with trivial canonical bundle and let $\pi\colon X \to S$ be a proper, surjective holomorphic map with connected fibers onto a smooth complex projective surface S. It is proved that the anticanonical bundle of S is numerically effective. In particular, if X is a Calabi-Yau three-fold and if $\pi\colon X \to S$ is a fibration of X over a smooth complex surface S, then S is among one of the following:

(i) $\mathscr{CP}^2$ or its blow-ups at no more than nine points;

(ii) $\mathscr{CP}^1 \times \mathscr{CP}^1$, the Hirzebruch surface Σ_2, or their blow-ups at no more than eight points.

0.1. Introduction. A Calabi-Yau three-fold is a smooth complex projective three-fold X with $\omega_X \cong \mathscr{O}_X$ and $h^0(X, \Omega^1_X) = 0$. Serre duality and symmetry of Hodge numbers imply that $h^0(X, \Omega^2_X) = 0$ as well. Recently such manifolds have attracted the attention of physicists due to developments in string theory. On the other hand, from the point of view of classification of smooth complex projective three-folds, not much is known about such class of manifolds. This article is an attempt in understanding those Calabi-Yau three-folds which admit fibrations over smooth compact complex surfaces.

Thus given a fibration $\pi\colon X \to S$ of a smooth complex projective three-fold X with $\omega_X \cong \mathscr{O}_X$ over a smooth complex projective surface S, one looks at the Leray spectral sequence for $R^i\pi_*(\pi^*[C])$, where $[C]$ is the line bundle associated to an irreducible curve C on S. By taking advantage of the fact that generic fibers of π are of dimension 1, this spectral sequence degenerates at E_3 level. On the other hand, as $c_1(X)$ vanishes and dimension $X = 3$, the Hirzebruch-Riemann-Roch formula takes a very simple form.

As a result of these, we arrive at an intersection formula $-C \cdot K_S = \frac{1}{12}\pi^*(c_1[C]) \cdot c_2(X)$. By virtue of the solution to Calabi conjecture by Yau [**10**], the right-hand side of this identity can be evaluated immediately by an

---

1991 *Mathematics Subject Classification.* Primary 53C55, 32J17.
This paper is in final form and no version will be submitted for publication elswhere.

appropriate representation of $c_2(X)$. In this way we arrive at the conclusion that $-C \cdot K_S$ is nonnegative. In other words, the anticanonical bundle of S is numerically effective.

On the other hand, given a fibration $\pi: X \to S$ of a Calabi-Yau three-fold X over a smooth compact complex surface S, one easily obtains that S is a projective surface with $q(S) = 0$. Moreover, by a Bochner type argument, one can show that $h^0(S, K_S^n) = 0$ for all positive integers n. Thus we conclude that S is a rational surface.

Armed with these two assertions, one can conclude easily that a minimal model of S is still a rational surface whose anticanonical bundle is numerically effective. It is now easy to determine what S should be by looking at the Enriques-Kodaira classification list. A complete list is given in the statement of the Main Theorem in §3.1.

This article is taken from part of the author's thesis at Brandeis University. I would like to express by gratitude to Professor Shing-Tung Yau for all his support, guidance and encouragement through all these years.

0.2. Notations.

fibration: a proper, surjective holomorphic map with connected fibers,

$\kappa(M)$: Kodaira dimension of a compact complex manifold M,

Ω_M^i: sheaf of germs of holomorphic sections of i-forms on a compact complex manifold M,

ω_M: sheaf of germs of holomorphic sections of n-forms on a compact complex manifold M of dimension n,

$\omega_{X/S}$: the relative canonical bundle $\omega_X \otimes \pi^* \omega_S^{-1}$ of a holomorphic map $\pi: X \to S$,

$R^i \pi_* \mathscr{F}$: theory ith higher direct image sheaf of a coherent sheaf $\mathscr{F}$ on X under π,

T_M': holomorphic tangent bundle of a complex manifold M,

Ω_M^1: holomorphic cotangent bundle of a complex manifold M,

$S^p V$: the pth symmetric power of a holomorphic vector bundle V,

$\Diamond$: end of proof of an assertion.

All varieties are defined over the field of complex numbers.

1. Rationality of the base

1.1. A Bochner type argument.

PROPOSITION 1.1.1. *Let M be a compact Kähler-Einstein manifold of dimension n. Let σ be a holomorphic tensor field on M of type (p, q), i.e. a global holomorphic section of $\bigotimes_p T_M' \otimes \bigotimes_q \Omega_M^1$. Then $\Delta \|\sigma\|^2 = \|\nabla \sigma\|^2 + c(q - p)\|\sigma\|^2$, where c is defined in $r_{i\bar{j}} = c g_{i\bar{j}}$.*

PROOF. Let $(g_{j\bar{j}})$ be a Kähler-Einstein metric on M and denote the induced metric on $\bigotimes_p T_M' \otimes \bigotimes_q \Omega_M^1$ by g_q^p. Locally σ is written as

$$\sigma^{i_1\cdots i_p}_{j_1\cdots j_q}\frac{\partial}{\partial z^{i_1}}\otimes\cdots\otimes\frac{\partial}{\partial z^{i_p}}\otimes dz^{j_1}\otimes\cdots\otimes dz^{j_q}$$

and

$$\begin{aligned}
\Delta\|\sigma\|^2 &= \Delta g^p_q(\sigma\otimes\overline{\sigma})\\
&= g^{k\bar{l}}\frac{\partial^2}{\partial z^k\partial\overline{z}^l}g^p_q(\sigma\otimes\overline{\sigma})\\
&= g^{k\bar{l}}g^p_q(\nabla^p_{q;k}\sigma\otimes\overline{\nabla^p_{q;l}\sigma}) + g^{k\bar{l}}g^p_q(\nabla^p_{q;\bar{l}}\nabla^p_{q;k}\sigma\otimes\overline{\sigma})\\
&= g^p_{q+1}(\nabla^p_q\sigma\otimes\overline{\nabla^p_q\sigma}) + Q(\sigma) = \|\nabla\sigma\|^2 + Q(\sigma),
\end{aligned}$$

where

$$Q(\sigma) = -\sum_{k=i_1}^{i_p}r^k_m\sigma^{i_1\cdots m\cdots i_p}_{j_1\cdots j_q}\sigma^{j_1\cdots j_q}_{i_1\cdots i_p} + \sum_{k=j_1}^{j_q}r^m_k\sigma^{i_1\cdots i_p}_{j_1\cdots m\cdots j_q}\sigma^{j_1\cdots j_q}_{i_1\cdots i_p},$$

$$r^i_j = g^{i\bar{h}}r_{j\bar{h}}\quad\text{and}\quad \sigma^{j_1\cdots j_q}_{i_1\cdots i_p} = g_{i_1\bar{k}_1}\cdots g_{i_p\bar{k}_p}{}^{j_1\bar{l}_1}\cdots g^{j_q\bar{l}_q}\overline{\sigma}^{k_1\cdots k_p}_{l_1\cdots l_q}.$$

Using $r_{i\bar{j}} = cg_{i\bar{j}}$, we have $r^i_j = c\delta^i_j$, so that

$$Q(\sigma) = (-pc + qc)\sigma^{i_1\cdots i_p}_{j_1\cdots j_q}\sigma^{j_1\cdots j_q}_{i_1\cdots i_p} = c(q-p)\|\sigma\|^2.$$

Result follows. $\Diamond$

COROLLARY 1.1.2. *Let M be a compact Kähler-Einstein manifold of dimension n. Then global holomorphic sections of $S^{mn}(T'_M)\otimes K^m_M$ and $S^{mn}(\Omega^1_M)$ $\otimes K^{-m}_M$ are parallel for all positive integers m.*

PROOF. Let σ be a global holomorphic section of $S^{mn}(T'_M)\otimes K^m_M$. Since $S^{mn}(T'_M)\otimes K^m_N = S^{mn}(T'_M)\otimes\bigotimes_m(\bigwedge^n\Omega^1_M)$, which in turn is a subbundle of $\bigotimes_{mn}T'_M\otimes\bigotimes_{mn}\Omega^1_M$, σ may be regarded as a holomorphic tensor field of type (mn, mn). By Proposition 1.1.1, $\Delta\|\sigma\|^2 = \|\nabla\sigma\|^2$, which is nonnegative. By Hopf's maximum principle [5], $\Delta\|\sigma\|^2$ is identically zero on M, so that $\nabla\sigma = 0$, i.e., σ is parallel. Similarly for $S^{mn}(\Omega^1_M)\otimes K^{-m}_M$. $\Diamond$

COROLLARY 1.1.3. *Let M be a compact Kähler manifold of dimension n with $c_1(M) = 0$. Then holomorphic tensor fields of type (p, q) on M are parallel for all nonnegative integers p and q.*

PROOF. By the solution of Calabi conjecture by Yau [10], M admits a Kähler-Einstein metric with $c = 0$. Result follows from same reasoning as in Corollary 1.1.2. $\Diamond$

Now let M be a compact Kähler manifold of dimension n with $c_1(M) = 0$. By works of Bogomolov, it is well known that the universal covering $\widetilde{M}$

of M is isomorphic to a product

$$\mathscr{C}^k \times \prod_i U_i \times \prod_j V_j,$$

where

(i) $\mathscr{C}^k$ is the unusual complex Euclidean space with the standard Kähler metric;

(ii) U_i's are simply-connected compact Kähler manifolds of odd complex dimension $u_i \geq 3$ with trivial canonical bundles and with irreducible holonomy groups $SU(u_i)$;

(iii) V_j's are simply-connected compact Kähler manifolds of even complex dimension v_j with trivial canonical bundles and with irreducible holonomy $Sp(\frac{v_j}{2})$.

COROLLARY 1.1.4. *Let X be a Calabi-Yau three-fold. Then* $h^0(X, \bigotimes_m \Omega^1_X)$ $= h^0(X, \bigotimes_m T'_X) = 0$ *for all positive integers* m.

PROOF. If σ were a nontrivial global holomorphic section of $\bigotimes_m \Omega^1_X$, consider its lift $\tilde{\sigma}$ to the universal cover of $\tilde{X}$. Since $\pi_1(X)$ is finite [**1**, §3, Proposition 2], $\tilde{X}$ does not contain Euclidean factors. On individual factors U_i and V_j of $\tilde{X}$, $\tilde{\sigma}$ is decomposed into holomorphic tensor fields of types $(0, q_i)$ and $(0, q_j)$, and therefore is parallel by Corollary 1.1.3 and hence identically zero by irreducible holonomy. Thus $\tilde{\sigma}$ is identically zero and so is σ.

Similarly for $\bigotimes_m T'_X$. ◇

1.2. Rationality of S.

PROPOSITION 1.2.1. *Let $\pi \colon X \to S$ be the fibration of a Calabi-Yau three-fold X over a smooth compact complex surface S. Then S is rational.*

PROOF. $q(S) = 0$ because $q(X) = 0$. We need only to prove that $h^0(S, K^n_S) = 0$ for all positive integers n.

If, on the contrary, that there were a nontrivial global holomorphic section σ of $K^n_S = \bigotimes_n(\bigwedge^2 \Omega^1_S)$ for some positive integer n, $\pi^*\sigma$ would then be a nontrivial global holomorphic section of $\bigotimes_n(\bigwedge^2 \Omega^1_X)$. As $\bigotimes_n(\bigwedge^2 \Omega^1_X)$ is a subbundle of $\bigotimes_{2n}(\Omega^1_X)$, $\pi^*\sigma$ would give a nontrivial global holomorphic section of $\bigotimes_{2n}(\Omega^1_X)$, which is impossible by Corollary 1.1.4.

Thus S is a rational. ◇

2. Numerical effectiveness of the anticanonical bundle of S

2.1. Spectral sequence computations.

PROPOSITION 2.1.1. *Let $\pi \colon X \to S$ be a fibration of a Calabi-Yau three-fold X over a smooth compact complex surface S. Then S has the following* *numerical invariants*:

(i) $h^{0,1}(S) = h^{1,0}(S) = 0$, $h^{0,2}(S) = h^{2,0}(S) = 0$;

(ii) $\mathscr{X}(\mathscr{O}_S) = 1$, $\mathscr{X}(\omega_S^n) = 1 + \frac{1}{2}n(n-1)c_1^2(S)$;

(iii) $c_1^2(S) + c_2(S) = 12$;

(iv) $b_1(S) = 0$, $b^+(S) = 1$, $b_2(S) = h^{1,1}(S)$.

PROOF. (i) We have $\pi_*\mathscr{O}_X \cong \mathscr{O}_S$ since π has connected fibers. Consider the exact sequence associated to the Leray spectral sequence of π,

$$0 \to H^1(S, \pi_*\mathscr{O}_X) \to H^1(X, \mathscr{O}_X) \to H^0(S, R^1\pi_*\mathscr{O}_X) \to \cdots.$$

Since $H^1(X, \mathscr{O}_X) = 0$, we have $H^1(S, \mathscr{O}_S) = H^1(S, \pi_*\mathscr{O}_X) = 0$, hence $h^{0,1}(S) = 0$. Also, $h^{i,0}(S) = \dim H^0(S, \Omega_S^i) = 0$ because X does not have nontrivial holomorphic one or two forms. Serre duality gives $h^{0,2}(S) = h^{2,0}(S) = 0$.

(ii) By definition,

$$\begin{aligned}\mathscr{X}(\mathscr{O}_S) &= 1 - h^1(S, \mathscr{O}_S) + h^2(S, \mathscr{O}_S) \\ &= 1 - h^{0,1}(S) + h^{0,2}(S) = 1.\end{aligned}$$

By Riemann-Roch,

$$\begin{aligned}\mathscr{X}(\omega_S^n) &= \mathscr{X}(\mathscr{O}_S) + \frac{1}{2}c_1(\omega_S^n)(c_1(\omega_S^n) + c_1(S)) \\ &= 1 + \frac{1}{2}n(n-1)c_1^2(S).\end{aligned}$$

(iii) Follows immediately from Noether's formula.

(iv) If $b_1(S)$ were odd, we would have $1 + b_1(S) = 2q(S) = 2h^{0,1}(S) = 0$, which is absurd. Thus $b_1(S)$ is even, and therefore $b_1(S) = 2q(S) = 0$ and $b^+(S) = 1 + 2p_g(S) = 1$. Also, $b_2(S) = h^{0,2}(S) + h^{1,1}(S) + h^{2,0}(S) = h^{1,1}(S)$.
◊

COROLLARY 2.1.2. *Let* $\pi: X \to S$ *be a fibration of a Calabi-Yau three-fold* X *over a smooth compact complex surface* S. *Then* S *is projective.*

PROOF. By $b^+(S) = 1$, $\exists \alpha \in H^2(S, \mathbb{Z})$ with $\alpha^2 > 0$. As $h^{0,1}(S) = h^{0,2}(S) = 0$, the first Chern class map $H^1(S, \mathscr{O}_S^*) \to H^2(S, \mathbb{Z})$ is an isomorphism. Thus there exists a holomorphic line bundle L on S with $c_1(L) = \alpha$, therefore $c_1^2(L) = \alpha^2 > 0$. Hence S is projective.   ◊

PROPOSITION 2.1.3. *Let* X *be a smooth projective three-fold with* $\omega_X \cong \mathscr{O}_X$ *and let* $\pi: X \to S$ *be a fibration of* X *over a smooth projective surface* S.

Then

(i)
$$R^i \pi_* \omega_X = \begin{cases} \mathscr{O}_S, & i = 0, \\ \omega_S, & i = 1, \\ 0, & i \geq 2, \end{cases}$$

(ii)
$$R^i \pi_* \omega_{X/S} = \begin{cases} \omega_S^{-1}, & i = 0, \\ \mathscr{O}_S, & i = 1, \\ 0, & i \geq 2. \end{cases}$$

PROOF. (i) $\pi_* \omega_X \cong \pi_* \mathscr{O}_X \cong \mathscr{O}_S$. The rest follows directly from Kollár [8, Theorem 2.1 and Proposition 7.6].

(ii) By projection formula,

$$R^i \pi_*(\omega_{X/S}) = R^i \pi_*(\omega_X \otimes \pi^* \omega_S^{-1}) = \omega_S^{-1} \otimes R^i \pi_* \omega_X$$

$$= \begin{cases} \omega_S^{-1}, & i = 0, \\ \omega_S^{-1} \otimes \omega_S \cong \mathscr{O}_S, & i = 1, \\ 0, & i \geq 2, \end{cases}$$

using results of (i). ◊

COROLLARY 2.1.4. *Let X be a smooth projective three-fold with $\omega_X \cong \mathscr{O}_X$ and let $\pi: X \to S$ be a fibration of X over a smooth projective surface S. Then for all $n \in Z$, we have*

$$H^0(X, \omega_{X/S}^n) \cong H^0(S, \omega_S^n),$$

$$H^1(X, \omega_{X/S}^n) \cong H^1(S, \omega_S^n) \otimes \mathrm{Ker}(d_2: H^0(S, \omega_S^{-n+1}) \to H^2(S, \omega_S^{-n})),$$

$$H^2(X, \omega_{X/S}^n) \cong H^1(S, \omega_S^{-n+1}) \otimes \frac{H^2(S, \omega_S^{-n})}{\mathrm{im}(d_2: H^0(S, \omega_S^{-n+1}) \to H^2(S, \omega_S^{-n}))},$$

$$H^3(X, \omega_{X/S}^n) \cong H^2(S, \omega_S^{-n+1}),$$

where d_2 is the differential on E_2 level.

PROOF. We apply the Leray spectral sequence whose E_2 terms are defined by $E_2^{p,q} = H^p(S, R^q \pi_* \omega_{X/S}^n) \Rightarrow H^{p+q}(X, \omega_{X/S}^n)$, $n \in Z$. Using formulas in Proposition 2.1.3, a direct calculation gives

$$R^q \pi_* \omega_{X/S}^n \cong \begin{cases} \omega_S^{-n}, & q = 0, \\ \omega_S^{-n+1}, & q = 1, \\ 0, & q \geq 2, \ \text{for all } n \text{ in } Z. \end{cases}$$

Thus $E_2^{p,q} = 0$ for all $q \geq 2$. Also, $\dim S = 2$ implies that $E_2^{p,q} = 0$ for all $p \geq 3$. Therefore the spectral sequence degenerates at E_3 level. In other words, $E_3^{p,q} \cong E_\infty^{p,q}$ and hence $H^i(X, \omega_{X/S}^n) = \bigoplus_{i=p+q} E_3^{p,q}$, $n \in Z$.

A straightforward computation shows that, for all $n \in Z$,

$$E_3^{0,0} \cong E_2^{0,0} \cong H^0(S, \omega_S^{-n}).$$

$$E_3^{1,0} \cong E_2^{1,0} \cong H^1(S, \omega_S^{-n}),$$

$$E_3^{0,1} \cong \text{Ker}(d_2 \colon H^0(S, \omega_S^{-n+1}) \to H^2(S, \omega_S^{-n})).$$

$$E_3^{2,0} \cong \frac{H^2(S, \omega_S^{-n})}{\text{im}(d_2 \colon H^0(S, \omega_S^{-n+1}) \to H^2(S, \omega_S^{-n}))},$$

$$E_3^{1,1} \cong E_2^{1,1} \cong H^1(S, \omega_S^{-n+1}),$$

$$E_3^{0,2} \cong 0.$$

$$E_3^{3,0} \cong 0,$$

$$E_3^{2,1} \cong H^2(S, \omega_S^{-n+1}),$$

$$E_3^{1,2} \cong 0,$$

$$E_3^{0,3} \cong 0.$$

The result follows immediately. $\Diamond$

COROLLARY 2.1.5. *Let* $\pi \colon X \to S$ *be a fibration of a Calabi-Yau three-fold over a smooth compact complex surface* S. *Then for all* $n \geq 0$, *we have*

$$H^0(X, \omega_{X/S}^n) \cong H^0(S, \omega_S^{-n}),$$

$$H^1(X, \omega_{X/S}^n) \cong H^1(S, \omega_S^{-n}) \oplus H^0(S, \omega_S^{-n+1}),$$

$$H^2(X, \omega_{X/S}^n) \cong H^1(S, \omega_S^{-n+1}),$$

$$H^3(X, \omega_{X/S}^n) \cong 0.$$

PROOF. By Corollary 2.1.2, S is projective. We can make use of Corollary 2.1.4. Serre duality and rationality of S give $H^2(S, \omega_S^{-n}) \cong H^0(S, \omega_S^{n+1})^v = 0$ and $H^2(S, \omega_S^{-n+1}) \cong H^0(S, \omega_S^n)^v = 0$, where "$v$" denotes the dual. The result now follows directly from Corollary 2.1.4. $\Diamond$

2.2. An intersection formula.

PROPOSITION 2.2.1. *Let* X *be a smooth projective three-fold with* $\omega_X \cong \mathscr{O}_X$ *and let* $\pi \colon X \to S$ *be a fibration of* X *over a smooth projective surface* S. *For any divisor* C *on* S, *we have* $-C \cdot K_S = \frac{1}{12}\pi^*(c_1[C]) \cdot c_2(X)$, *where* $[C]$ *is the holomorphic line bundle on* S *associated to the divisor* C.

PROOF. By Hirzebruch-Riemann-Roch on X,

$$\mathscr{X}(\pi^*[C]) = \{\text{ch}(\pi^*[C]) \cdot Td(X)\}_3,$$

where $\{*\}_3$ denotes evaluation of the degree 3 term of $*$ on the fundamental cycle $[X]$. As $c_1^3(\pi^*[C]) = \pi^*(c_1^3[C]) = 0$ and $c_1(X) = 0$, the right-hand side equals $\frac{1}{12}\pi^*(c_1[C]) \cdot c_2(X)$.

By definition, $\mathscr{X}(\pi^*[C]) = \sum_{i=0}^{3}(-1)^i h^i(X, \pi^*[C])$. To compute $h^i(X, \pi^*[C])$, we look at the Leray spectral sequence whose E_2 terms are

$$E_2^{p,q} = H^p(S, R^q \pi_*(\pi^*[C])) \Rightarrow H^{p+q}(X, \pi^*[C]).$$

Using $\omega_X \cong \mathscr{O}_X$ and formulas in Proposition 2.1.3, we find that

$$R^q \pi_*(\pi^*[C]) = \begin{cases} [C], & q = 0, \\ [C] \otimes \omega_S, & q = 1, \\ 0, & q \geq 2. \end{cases}$$

Therefore $E_2^{p,q} = 0$ for all $q \geq 2$. Also, $E_2^{p,q} = 0$ for all $p \geq 3$ since $\dim S = 2$. Hence the spectral sequence degenerates at E_3 level, and therefore $H^i(X, \pi^*[C]) \cong \bigoplus_{i=p+q} E_3^{p,q}$.

A straightforward computation gives

$$E_3^{0,0} \cong E_2^{0,0} \cong H^0(S, [C]).$$
$$E_3^{1,0} \cong H^1(S, [C]),$$
$$E_3^{0,1} \cong \operatorname{Ker}(d_2 : H^0(S, [C] \otimes \omega_S) \to H^2(S, [C])).$$
$$E_3^{2,0} \cong \frac{H^2(S, [C])}{\operatorname{im}(d_2 : H^0(S, [C] \otimes \omega_S) \to H^2(S, [C]))},$$
$$E_3^{1,1} \cong H^1(S, [C], \otimes \omega_S),$$
$$E_3^{0,2} \cong 0.$$
$$E_3^{3,0} \cong 0,$$
$$E_3^{2,1} \cong H^2(S, [C] \otimes \omega_S),$$
$$E_3^{1,2} \cong 0,$$
$$E_3^{0,3} = 0.$$

Therefore,

$$H^0(X, \pi^*[C]) \cong H^0(S, [C]),$$
$$H^1(X, \pi^*[C]) \cong H^1(S, [C]) \oplus \operatorname{Ker} d_2,$$
$$H^2(X, \pi^*[C]) \cong H^1(S, [C] \otimes \omega_S)$$
$$\oplus \frac{H^2(S, [C])}{\operatorname{im} d_2},$$
$$H^3(X, \pi^*[C]) \cong H^2(S, [C] \otimes \omega_S),$$

where $d_2 : H^0(S, [C] \otimes \omega_S) \to H^2(S, [C])$ is the differential on E_2 level.

Hence

$$
\begin{aligned}
\mathscr{X}(\pi^*[C]) &= \sum_{i=0}^{3}(-1)^i h^i(X,\pi^*[C]) \\
&= h^0(S,[C]) - \{h^1(S,[C]) + \dim(\operatorname{Ker} d_2)\} \\
&\quad + \{h^2(S,[C]\otimes\omega_S) + h^2(S,[C]) - \dim(\operatorname{Im} d_2)\} \\
&\quad - h^2(S,[C]\otimes\omega_S) \\
&= \{h^0(S,[C]) - h^1(S,[C]) + h^2(S,[C])\} \\
&\quad - \{\dim(\operatorname{Ker} d_2) + \dim(\operatorname{Im} d_2)\} \\
&\quad - \{h^2(S,[C]\otimes\omega_S) - h^1(S,[C]\otimes\omega_S)\} \\
&= \mathscr{X}(S,[C]) - \{h^2(S,[C]\otimes\omega_S) \\
&\qquad\qquad - h^1(S,[C]\otimes\omega_S) + h^0(S,[C]\otimes\omega_S)\} \\
&= \mathscr{X}(S,[C]) - \mathscr{X}(S,[C]\otimes\omega_S) \\
&= \mathscr{X}(S,[C]) - \mathscr{X}(S,[-C]) \quad \text{(by Serre duality)}.
\end{aligned}
$$

Now Riemann-Roch on S gives

$$
\begin{aligned}
\mathscr{X}(\pi^*[C]) &= \{\mathscr{X}(\mathcal{O}_S) + \tfrac{1}{2}(C^2 - C\cdot K_S)\} \\
&\quad - \{\mathscr{X}(\mathcal{O}_S) + \tfrac{1}{2}(C^2 + C\cdot K_S)\} \\
&= - C\cdot K_S.
\end{aligned}
$$

Thus

$$
-C\cdot K_S = \frac{1}{12}\pi^*(c_1[C])\cdot c_2(X). \quad \Diamond
$$

2.3. Numerical effectiveness of K_S^{-1} and some properties of S.

PROPOSITION 2.3.1. *Let X be a smooth projective three-fold with $\omega_X \cong \mathcal{O}_X$ and let $\pi\colon X \to S$ be a fibration of X over a smooth projective surface S. Then $-K_S$ is numerically effective.*

PROOF. By Proposition 2.2.1, we have

$$
-C\cdot K_S = \frac{1}{12}c_1([\pi^*C])\cdot c_2(X)
$$

for any irreducible curve C on S. It is enough to prove that the right-hand side of this identity is nonnegative.

Since the line bundle $[\pi^*C]$ comes from the divisor $D = \pi^*C$, $c_1[\pi^*C]$ is represented by the Poincaré dual η_D of the divisor D [4, p. 141]. Write $D = \sum_i a_i D_i$, where D_i's are irreducible components of D and $a_i \geq 0$ for all i since D is effective. We have $\eta_D = \sum_i a_i \eta_{D_i}$. Thus

$$-C \cdot K_S = \frac{1}{12} c_1([\pi^* C]) \cdot c_2(X) = \frac{1}{12} \int_X \eta_D \wedge c_2(X)$$

$$= \frac{1}{12} \sum_i a_i \int_X \eta_{D_i} \wedge c_2(X)$$

$$= \frac{1}{12} \sum_i a_i \int_{D_i} c_2(X) \quad \text{(by definition of Poincaré dual)}$$

$$= \frac{1}{12} \sum_i a_i \int_{D_i^*} c_2(X),$$

where D_i^* denotes the smooth part of the divisor D_i.

To evaluate $\int_{D_i^*} c_2(X)$, we restrict the expression for $c_2(X)$ to D_i^*. By a theorem of Chern [2], $c_2(X) = \frac{-1}{8\pi^2}(\Theta_j^j \wedge \Theta_k^k - \Theta_l^k \wedge \Theta_k^l)$, where $\Theta_l^k = R_{l\bar{k}p\bar{q}}\omega^p \wedge \overline{\omega}^q$ is the curvature given by a hermitian metric g_{ij} on X expressed in terms of a unitary coframe $(\omega^1, \omega^2, \omega^3)$.

As $c_1(X)$ vanishes, by the solution to Calabi conjecture by Yau [10], we may choose a Kähler-Einstein metric $(g_{i\bar{j}})$ on X with Ricci curvature $r_{p\bar{q}} = R_{j\bar{j}p\bar{q}} = 0$ for all p and q. Thus

$$\Theta_j^j = R_{i\bar{j}p\bar{q}}\omega^p \wedge \overline{\omega}^q = r_{p\bar{q}}\omega^p \wedge \overline{\omega}^q = 0.$$

Hence

$$c_2(X) = \frac{1}{8\pi^2}\Theta_l^k \wedge \Theta_k^l = \frac{1}{8\pi^2}(R_{l\bar{k}p\bar{q}} R_{k\bar{l}s\bar{t}})\omega^p \wedge \overline{\omega}^q \wedge \omega^s \wedge \overline{\omega}^t.$$

Denote $j: D \hookrightarrow X$ the inclusion. Since D_i^* is now a smooth complex submanifold of X, locally we may choose an adapted unitary coframe $(\omega^1, \omega^2, \omega^3)$ on X such that $(j^*\omega^1, j^*\omega^2)$ is a unitary coframe for the induced metric $(j^* g_{i\bar{j}})$ on D_i^* and $j^*\omega^3 = 0$.

Now $j^* c_2(X) = \frac{1}{8\pi^2} j^*(R_{l\bar{k}p\bar{q}} R_{k\bar{l}s\bar{t}}) j^*(\omega^p \wedge \overline{\omega}^q \wedge \omega^s \wedge \overline{\omega}^t)$. Because $j^*\omega^3 = 0$, the only terms survived are $\omega^1 \wedge \overline{\omega}^1 \wedge \omega^2 \wedge \overline{\omega}^2$, $\omega^1 \wedge \overline{\omega}^2 \wedge \omega^2 \wedge \overline{\omega}^1$, $\omega^2 \wedge \overline{\omega}^1 \wedge \omega^1 \wedge \overline{\omega}^2$ and $\omega^2 \wedge \overline{\omega}^2 \wedge \omega^1 \wedge \overline{\omega}^1$. Thus

$$j^* c_2(X) = \frac{1}{8\pi^2} j^*(R_{l\bar{k}1\bar{1}} R_{k\bar{l}2\bar{2}} - R_{l\bar{k}1\bar{2}} R_{k\bar{l}2\bar{1}} - R_{l\bar{k}2\bar{1}} R_{k\bar{l}1\bar{2}}$$

$$+ R_{l\bar{k}2\bar{2}} R_{k\bar{l}1\bar{1}}) j^*(\omega^1 \wedge \overline{\omega}^1 \wedge \omega^2 \wedge \overline{\omega}^2)$$

$$= \frac{1}{8\pi^2} j^*(-2R_{l\bar{k}1\bar{2}} R_{k\bar{l}2\bar{1}}) j^*(\omega^1 \wedge \overline{\omega}^1 \wedge \omega^2 \wedge \overline{\omega}^2).$$

$$(R_{l\bar{k}i\bar{i}} = R_{i\bar{i}l\bar{k}} = r_{l\bar{k}} = 0).$$

The associated $(1, 1)$ form of the Kähler-Einstein metric $(g_{i\bar{j}})$ on X is $\omega_X = \frac{\sqrt{-1}}{2}(\omega^1 \wedge \overline{\omega}^1 + \omega^2 \wedge \overline{\omega}^2 + \omega^3 \wedge \overline{\omega}^3)$, and the associated $(1, 1)$ form

of the induced metric on D_i^* is $\omega_{D_i^*} = j^*\omega_X = \frac{\sqrt{-1}}{2}j^*(\omega^1 \wedge \overline{\omega}^1 + \omega^2 \wedge \overline{\omega}^2)$.
Therefore

$$\text{volume form of } D_i^* = d\mu_{D_i^*} = \frac{1}{2!}\omega_{D_i^*} \wedge \omega_{D_i^*}$$

$$= \left(\frac{\sqrt{-1}}{2}\right)^2 j^*(\omega^1 \wedge \overline{\omega}^1 \wedge \omega^2 \wedge \overline{\omega}^2)$$

$$= -\frac{1}{4}j^*(\omega^1 \wedge \overline{\omega}^1 \wedge \omega^2 \wedge \overline{\omega}^2).$$

Thus

$$\int_{D_i^*} c_2(X) = \int_{D_i^*} j^* c_2(X)$$

$$= \frac{1}{8\pi^2}\int_{D_i^*}(-2R_{l\overline{k}1\overline{2}}R_{k\overline{l}2\overline{1}}) - (-4\,d\mu_{D_i^*})$$

$$= \frac{1}{\pi^2}\int_{D_i^*} R_{l\overline{k}1\overline{2}}\overline{R_{l\overline{k}1\overline{2}}}(d\mu_{D_i^*}) \qquad (R_{i\overline{j}k\overline{l}} = \overline{R_{j\overline{i}l\overline{k}}})$$

$$= \frac{1}{\pi^2}\int_{D_i^*}|R_{l\overline{k}1\overline{2}}|^2\,d\mu_{D_i^*} \geq 0.$$

Result follows. $\Diamond$

COROLLARY 2.3.2. *Let* $\pi: X \to S$ *be a fibration of a Calabi-Yau three-fold* X *over a smooth compact complex surface* S. *Then* $c_1^2(S) \geq 0$ *and* $h^0(S, K_S^{-1}) \geq 1$.

PROOF. By Corollary 2.1.2, S is projective. Thus Proposition 2.3.1 applies and enables us to conclude that $-K_S$ is numerically effective (nef). Hence $c_1^2(S) = (-K_S)(-K_S) \geq 0$ by Kleiman [6]. By Riemann-Roch,

$$\mathscr{X}(S, K_S^{-1}) = \mathscr{X}(O_S) + \tfrac{1}{2}(K_S^{-1}\cdot K_S^{-1} - K_S^{-1}\cdot K_S)$$
$$= 1 + c_1^2(S).$$

On the other hand, $\mathscr{X}(S, K_S^{-1}) = h^0(S, K_S^{-1}) - h^1(S, K_S^{-1}) + h^2(S, K_S^{-1})$. But $h^2(S, K_S^{-1}) = h^0(S, K_S^2) = 0$ since S is rational by Proposition 1.2.1. Therefore,

$$h^0(S, K_S^{-1}) = \mathscr{X}(S, K_S^{-1}) + h^1(S, K_S^{-1})$$
$$= 1 + c_1^2(S) + h^1(S, K_S^{-1}) \geq 1. \quad \Diamond$$

4. Numerical effectiveness of the anticanonical bundle of a minimal model S

3.1. Proof of main theorem.

PROPOSITION 3.1.1. *Let* $b: \widetilde{S} \to S$ *be a finite succession of blow-ups of a smooth compact complex surface* S. *If* $-K_{\widetilde{S}}$ *is nef, so is* $-K_S$.

PROOF. We can write

$$\widetilde{S} = S_m \xrightarrow{b_m} S_{m-1} \xrightarrow{b_{m-1}} \cdots \longrightarrow S_1 \xrightarrow{b_1} S_0 = S,$$

where $b = b_1 \circ \cdots \circ b_m$ and each b_i is a blow-up at a single point p_i of S_{i-1}.

It suffices to prove that $-K_{S_i}$ being nef implies that $-K_{S_{i-1}}$ is also nef. For simplicity we write p_i as p.

Let C be an irreducible curve on S_{i-1}. We separate into 2 cases:

(i) C does not pass through p: then $b_i^*(C) = b_1^{-1}(C)$ is still an irreducible curve on S_i, so that

$$0 \le b_i^*(C) = (-K_{S_i}) = b_i^*(C)(b_i^*(-K_{S_{i-1}}) - E_i)$$
$$= C \cdot (-K_{S_{i-1}}) - b_i^*(C) \cdot E_i = C \cdot (-K_{S_{i-1}}),$$

where E_i is the exceptional curve of b_i. Thus $-K_{S_{i-1}}$ is nef.

(ii) C passes through p: then $b_i^*(C) = \widehat{C} + mE_i$, where $\widehat{C}$ is the proper transform of C and $m = \mathrm{mult}_p(C) > 0$. Since $\widehat{C}$ is still an irreducible curve on S_i, we have

$$0 \le \widehat{C} \cdot (K_{S_i}) = (b_i^*(C) - mE_i)(b_i^*(-K_{S_{i-1}}) - E_i)$$
$$= C \cdot (-K_{S_{i-1}}) - m.$$

Thus

$$C \cdot (-K_{S_{i-1}}) \ge m > 0.$$

Hence $-K_{S_{i-1}}$ is still nef. $\Diamond$

PROPOSITION 3.1.2. *Let S be a minimal rational surface with $-K_S$ nef. Then S is either $\mathscr{C}\mathscr{P}^2$, $\mathscr{C}\mathscr{P}^1 \times \mathscr{C}\mathscr{P}^1$ or the Hirzebruch surface Σ_2.*

PROOF. All minimal rational surfaces are among $\mathscr{C}\mathscr{P}^2$ or Σ_n, $n = 0, 2, 3, \ldots$. $-K_{\mathscr{C}\mathscr{P}^2} = 3H$ is ample and hence nef. For Σ_n's, we have

$$-K_{\Sigma_n} = 2E_0 + (2-n)F, \qquad E_0^2 = n, \qquad E_0 \cdot F = 1, \qquad E_\infty \sim E_0 - nF,$$

where E_0, E_∞ and F are the zero-section, ∞-section and a fiber of the projection $p: \Sigma_n \to \mathscr{C}\mathscr{P}^1$.

For $-K_{\Sigma_n}$ to be nef,

$$0 \le (-K_{\Sigma_n}) \cdot E_0 = n + 2,$$
$$0 \le (-K_{\Sigma_n}) \cdot F = 2, \quad \text{and} \quad 0 \le (-K_{\Sigma_n}) \cdot E_\infty = 2 - n.$$

Therefore $n = 0, 1$ or 2. But Σ_1 is not minimal because it is $\mathscr{C}\mathscr{P}^2$ blown up at one point. We are left with $\Sigma_0 \cong \mathscr{C}\mathscr{P}^1 \times \mathscr{C}\mathscr{P}^1$ and Σ_2. $\Diamond$

MAIN THEOREM. *Let $\pi\colon X \to S$ be a fibration of a Calabi-Yau three-fold X over a smooth compact complex surface X. Then S is among one of the following:*

(i) *$\mathscr{C}\mathscr{P}^2$ or its blow-ups at no more than 9 points;*

(ii) *$\mathscr{C}\mathscr{P}^1 \times \mathscr{C}\mathscr{P}^1$, Σ_2, or their blow-ups at no more than 8 points.*

PROOF. By Proposition 1.2.1, Corollary 2.1.2, Proposition 2.3.1 and Corollary 2.3.2, S is a rational projective surface with $-K_S$ nef and $c_1^2(S) \geq 0$. Applying Proposition 3.1.1, we observe that a minimal model of S is a minimal rational surface whose anticanonical bundle is nef, and is therefore among $\mathscr{C}\mathscr{P}^2$, $\mathscr{C}\mathscr{P}^1 \times \mathscr{C}\mathscr{P}^1$ or Σ_2. Since $c_1^2(\mathscr{C}\mathscr{P}^2) = 9$ and $c_1^2(\mathscr{C}\mathscr{P}^1 \times \mathscr{C}\mathscr{P}^1) = c_1^2(\Sigma_2) = 8$, the result follows immediately. $\Diamond$

4. Structure of some fibrations

4.1. The associated fibration and structure of its fibers.

PROPOSITION 4.1.1. *Let $\pi\colon X \to S$ be a fibration of a Calabi-Yau three-fold X over a smooth compact complex surface S. Then π must have singular fibers.*

PROOF. Assume the contrary. Then all fibers of π are smooth elliptic curves by adjunction formula. We can define a map $j\colon S \to \mathscr{C}$ by invariant of the elliptic curve $\pi^{-1}(s)$, $s \in S$. Since all fibers are smooth, j is defined everywhere on S and is holomorphic, thus constant. Hence all fibers are isomorphic to each other. By Grauert-Fischer [3], π is a complex analytic fiber bundle. By $C_{m,n}$ for analytic fiber bundles [9],

$$\kappa(X) = \kappa(\text{fiber}) + \kappa(S),$$

where κ is the Kodaira dimension.. Thus $\kappa(S) = 0$, which is not true because S is rational by Proposition 1.2.1. $\Diamond$

PROPOSITION 4.1.2. *Let X be a Calabi-Yau three-fold, D a smooth divisor in X and $j\colon D \hookrightarrow X$ the inclusion. Then*

(i)
$$c_1(D) = -j^* c_1[D],$$
$$c_2(D) = j^* c_2(X) + j^* c_1^2[D],$$

where $[D]$ is the line bundle on X defined by the divisor D and $c_i(D)$ is the ith Chern class of D when D is regarded as a smooth complex manifold.

$$\frac{1}{6} j^* c_1^2[D] + \frac{1}{12} j^* c_2(X) = \frac{1}{6} c_1^3[D] + \frac{1}{12} c_1[D] \cdot c_2(X).$$

PROOF. (i) As D is smooth, by looking at the exact sequence $0 \to T_D \to T_{X|D} \to \mathscr{N}_{D/X} \to 0$ of holomorphic vector bundles on X, we have

$$c_1(T_{X|D}) = c_1(D) + c_1(\mathscr{N}_{D/X}),$$
$$c_2(T_{X|D}) = c_2(D) + c_1(D)c_1(\mathscr{N}_{D/X}).$$

Since $c_1(T_X) = 0$ and $\mathcal{N}_{D/X} \cong [D]|_D$, we have

$$c_1(D) = -c_1([D]|_D) = -j^* c_1[D],$$
$$c_2(D) = j^* c_2(X) - c_1(D)c_1(\mathcal{N}_{D/X}) = j^* c_2(X) + j^* c_1^2[D].$$

(ii) By looking at the exact sequence of sheaves

$$0 \longrightarrow \mathcal{O}_X \longrightarrow \mathcal{O}_X(D) \longrightarrow \mathcal{O}_D(D) \longrightarrow 0 \quad \text{on } X,$$

we have

$$\mathcal{X}(X, \mathcal{O}_{\mathcal{X}}(D)) = \mathcal{X}(X, \mathcal{O}_X) + \mathcal{X}(D, \mathcal{O}_D(D)).$$

Since X is a Calabi-Yau three-fold, $\mathcal{X}(X, \mathcal{O}_X) = 0$, therefore

(4.1.3) $\mathcal{X}(X, \mathcal{O}_{\mathcal{X}}(D)) = \mathcal{X}(D, \mathcal{O}_D(D)) = \mathcal{X}(D, [D]|_D).$

We compute both sides of (4.1.3) by Hirzebruch-Riemann-Roch.

$$\begin{aligned}
\text{R.H.S.} &= \{\text{ch}([D]|_D) \cdot Td(D)\}_2 \\
&= \frac{1}{12}(c_1^2(D) + c_2(D)) + \frac{1}{2}j^* c_1[D] \cdot c_1(D) + \frac{1}{2}j^* c_1^2[D] \\
&= \frac{1}{12}j^* c_2(X) + \frac{1}{6}j^* c_1^2[D], \quad \text{by formulas just proved in (i).} \\
\text{L.H.S.} &= \{\text{ch}([D]) \cdot Td(X)\}_3 \\
&= \frac{1}{6}c_1^3[D] + \frac{1}{12}c_1[D] \cdot c_2(X). \quad \text{Result follows.} \quad \Diamond
\end{aligned}$$

COROLLARY 4.1.4. *Let* $\pi: X \to S$ *be a fibration of a Calabi-Yau three-fold* X *over a smooth compact complex surface* S. *Let* C *be an irreducible curve on* S *and* $D = \pi^*(C)$ *the pull-back divisor in* X. *If* D *is smooth, then* $c_2(D) = -12K_S \cdot C$.

PROOF. Denote $i: C \hookrightarrow S$ the inclusion. We have the commutative diagram

$$\begin{array}{ccc}
D & \overset{j}{\hookrightarrow} & X \\
\downarrow{\pi} & & \downarrow{\pi} \\
C & \overset{i}{\hookrightarrow} & S
\end{array}$$

We have $[D] = \pi^*[C]$ as line bundles. Since D is assumed smooth, by (ii) of Proposition 4.1.2, we have

$$\frac{1}{6}j^* c_1^2[D] + \frac{1}{12}j^* c_2(X) = \frac{1}{6}c_1^3[D] + \frac{1}{12}c_1[D] \cdot c_2(X).$$

Now

$$\begin{aligned}
c_1^3[D] &= \pi^*(c_1^3[C]) = 0 \\
j^* c_1^2[D] &= j^* \pi^*(c_1^2[C]) = \pi^* i^*(c_1^2[C]) = 0.
\end{aligned}$$

Therefore

$$\frac{1}{12}j^*c_2(X) = \frac{1}{12}c_1[D] \cdot c_2(X) = \frac{1}{12}\pi^*(c_1[C]) \cdot c_2(X)$$
$$= -C \cdot K_S,$$

by the intersection formula in Proposition 2.2.1. Hence $j^*c_2(X) = -12C \cdot K_S$. By (i) of Proposition 4.1.2, $c_2(D) = j^*c_2(X) + j^*(c_1^2[D]) = j^*c_2(X)$. The result follows. $\Diamond$

DEFINITION 4.1.5. Let $\pi\colon X \to S$ be a fibration of a Calabi-Yau three-fold X over a smooth compact complex surface S. If S also fibers over $\mathscr{CP}^1$ via $p\colon S \to \mathscr{CP}^1$, the composite $\varphi = p \circ \pi\colon X \to \mathscr{CP}^1$ is still a fibration, called the *associated fibration* of π.

PROPOSITION 4.1.6. *Let $b\colon S \to \Sigma_n$ be a finite succession of blow-ups of a Hirzebruch surface Σ_n and $p\colon \Sigma_n \to \mathscr{CP}^1$ the projection map. Suppose $\pi\colon X \to S$ is a fibration of a Calabi-Yau three-fold X over S. Then generic fibers of the associated fibration $\varphi = p \circ b \circ \pi\colon X \to \mathscr{CP}^1$ are minimal algebraic elliptic $K3$ surfaces over $\mathscr{CP}^1$ without multiple fibers.*

PROOF.

$$X$$
$$\pi \downarrow$$
$$S \xrightarrow{\ h\ } \Sigma_n \xrightarrow{\ p\ } \mathscr{CP}^1$$

Generic fibers $X_u = \varphi^{-1}(u)$ of φ are smooth algebraic surfaces. By staying away from the finitely many exceptional curves of b, a generic point $u \in \mathscr{CP}^1$ has the property that $F = p^{-1}(u) \cong \mathscr{CP}^{-1}$, $F' = b^{-1}(F) \cong \mathscr{CP}^1$ and that $b(F') = F$.

By adjunction formula, $\omega_{X_u} \cong \mathscr{O}_{X_u}$, and therefore $\mathscr{X}(X_u) = \frac{1}{12}c_2(X_u)$ by Noether's formula. As $X_u = \pi^{-1}(F')$, Corollary 4.1.4 gives $c_2(X_u) = -12K_S \cdot F'$. By adjunction formula,

$$K_S \cdot F' = \left(b^*(K_{\Sigma_n}) + \sum_i E_i\right) \cdot b^*(F) = K_{\Sigma_n} \cdot F = -2,$$

where E_i's are exceptional curves of b. Therefore $c_2(X_u) = 24$, hence $\mathscr{X}(X_u) = 2$. But $p_g(X_u) = 1$ as $K_{X_u} \cong \mathscr{O}_{X_u}$, thus $q(X_u) = 0$.

Hence we have proved that X_u is an algebraic $K3$ surface fibered over $F' \cong \mathscr{CP}^1$. Again, using $K_{X_u} \cong \mathscr{O}_{X_u}$, then genus formula implies that X_u does not contain (-1) curves, and the adjunction formula implies that $\pi\colon X_u \to F' \cong \mathscr{CP}^1$ is an elliptic surface.

To prove that $\pi\colon X_u \to F' \cong \mathscr{CP}^1$ has no multiple fibers, we look at the canonical bundle formula for elliptic surfaces [7],

$$K_{X_u} = \pi^*(K_{\mathscr{CP}^1} - f) + \sum_\nu (m_\nu - 1)[P_\nu]$$

where $[P_\nu]$ is the line bundle on X_u associated to the multiple fiber P_ν, and f is a line bundle on $\mathscr{CP}^1$ with

$$c_1(f) = -p_g(X_u) + q(X_u) - 1 = -1 - 1 = -2.$$

Therefore $f \cong -2H$ on $\mathscr{CP}^1$ and

$$O_{X_u} \cong K_{X_u} = \pi^*(-2H + 2H) + \sum_\nu (m_\nu - 1)[P_\nu]$$
$$= \sum_\nu (m_\nu - 1)[P_\nu],$$

which is not possible if X_u contains multiple fibers. $\diamond$

REMARK 4.1.7. Bearing in mind that $\mathscr{CP}^2$ blown-up at 1 point is just Σ_1, all surfaces except $\mathscr{CP}^2$ on the list in the Main Theorem admit fibrations over $\mathscr{CP}^1$ and therefore we can talk about the associated fibrations. Proposition 4.1.6 just says that in such cases, "almost all" singular fibers of $\pi: X \to S$ are those which already occur on elliptic surfaces.

4.2. Higher direct images of dualizing sheaf.

PROPOSITION 4.2.1. Let X be a smooth projective three-fold with $\omega_X \cong \mathscr{O}_X$ and let $\pi: X \to S$ be a fibration X over a smooth projective surface S. Then $\pi_* c_2(X) = 12 c_1(S)$, where π_* is the push forward of cycles.

PROOF. We make use of the Grothendieck-Riemann-Roch formula $\mathrm{ch}(\pi_! \omega_X) \cdot Td(S) = \pi_*(\mathrm{ch}(\omega_X) \cdot Td(X))$, where $\pi_! \mathscr{F} = \sum_i (-1)^i R^i \pi_* \mathscr{F}$ for a coherent sheaf $\mathscr{F}$ on X.

As $\omega_X \cong \mathscr{O}_X$, $\mathrm{ch}(\omega_X) = 1$ and $Td(X) = 1 + \frac{1}{12} c_2(X)$. Therefore the right-hand side is equal to $\frac{1}{12} \pi_* c_2(X)$.

By the formula in Proposition 2.1.3, we have

$$\mathrm{ch}(\pi_! \omega_X) = \mathrm{ch}(\mathscr{O}_S - \omega_S) = \mathrm{ch}(\mathscr{O}_S) - \mathrm{ch}(\omega_S) = c_1(S) - \tfrac{1}{2} c_1^2(S).$$

Therefore the left hand side is equal to

$$\left(c_1(S) - \frac{1}{2} c_1^2(S) \right) \cdot Td(S)$$
$$= \left(c_1(S) - \frac{1}{2} c_1^2(S) \right) \left(1 + \frac{1}{2} c_1(S) + \frac{1}{12}(c_1^2(S) + c_2(S)) \right) = c_1(S).$$

Thus $c_1(S) = \frac{1}{12} \pi_* c_2(X)$, as asserted. $\diamond$

PROPOSITION 4.2.2. Let $b: S \to \Sigma_n$ be a finite succession of blow-ups of a Hirzebruch surface Σ_n and $p: \Sigma_n \to \mathscr{CP}^1$ the projection map. Suppose $\pi: X \to S$ is a fibration of a Calabi-Yau three-fold X over S and $\varphi = p \circ b \circ \pi: X \to \mathscr{CP}^1$ is its associated fibration. Then $R^1 \varphi_* \omega_X = 0$ and $R^2 \varphi_* \omega_X \cong \omega_{\mathscr{CP}^1}$.

PROOF. Write

$$X$$

$$\pi \downarrow$$

$$S = \Sigma_n^m \xrightarrow{b_m} \Sigma_n^{m-1} \xrightarrow{b_{m-1}} \cdots \longrightarrow \Sigma_n^1 \xrightarrow{b_1} \Sigma_n \xrightarrow{p} \mathscr{C}\mathscr{P}^1,$$

where $b_i \colon \Sigma_n^i \to \Sigma_n^{i-1}$ is a blow-up of Σ_n^{i-1} at a single point. We have $\varphi_* \omega_X \cong \varphi_* \mathscr{O}_X \cong \mathscr{O}_{\mathscr{C}\mathscr{P}^1}$.

By Kollár [8, Corollary 7.8],

$$R^1 \varphi_* \omega_X \cong R^1 \varphi_* \mathscr{O}_X \cong \operatorname{Hom}(R^1 \varphi_* \omega_{X/\mathscr{C}\mathscr{P}^1}, \mathscr{O}_{\mathscr{C}\mathscr{P}^1}),$$

which is reflexive on $\mathscr{C}\mathscr{P}^1$ and hence locally free. Thus

$$\begin{aligned}
R^1 \varphi_* \omega_X &\cong \operatorname{Hom}(R^1 \varphi_* \omega_{X/\mathscr{C}\mathscr{P}^1}, \mathscr{O}_{\mathscr{C}\mathscr{P}^1}) \\
&\cong \operatorname{Hom}(R^1 \varphi_* (\omega_X \otimes \varphi^* \omega_{\mathscr{C}\mathscr{P}^1}^{-1}), \mathscr{O}_{\mathscr{C}\mathscr{P}^1}) \\
&\cong \operatorname{Hom}(\omega_{\mathscr{C}\mathscr{P}^1}^{-1} \otimes R^1 \varphi_* \omega_X, \mathscr{O}_{\mathscr{C}\mathscr{P}^1}), \\
&\cong \omega_{\mathscr{C}\mathscr{P}^{-1}} \otimes (R^1 \varphi_* \omega_x)^v,
\end{aligned}$$

where "v" denotes the dual sheaf.

Also, $R^2 \varphi_* \omega_X \cong \omega_{\mathscr{C}\mathscr{P}^1}$ and $R^i \varphi_* \omega_X = 0$ $\forall i \geq 3$ by Kollár [8].

By Grothendieck-Riemann-Roch,

(4.2.3)
$$\begin{aligned}
\operatorname{ch}(\varphi_! \omega_X) \cdot Td(\mathscr{C}\mathscr{P}^1) &= \varphi_*(\operatorname{ch}(\omega_X) \cdot Td(X)) \\
&= \varphi_* \left(1 + \frac{1}{12} c_2(X)\right) \\
&= \frac{1}{12} \varphi_* c_2(X).
\end{aligned}$$

Now

$$\begin{aligned}
\operatorname{ch}(\varphi_! \omega_X) &= \operatorname{ch}(\varphi_* \omega_X) - \operatorname{ch}(R^1 \varphi_* \omega_X) + \operatorname{ch}(R^2 \varphi_* \omega_X) \\
&= 2 - r + c_1(\omega_{\mathscr{C}\mathscr{P}^1}) - c_1(R^1 \varphi_* \omega_X),
\end{aligned}$$

where $r = \operatorname{rank}(R^1 \varphi_* \omega_X)$.

But $R^1 \varphi_* \omega_X \cong \omega_{\mathscr{C}\mathscr{P}^1} \otimes (R^1 \varphi_* \omega_X)^v$, therefore $c_1(R^1 \varphi_* \omega_X) = \frac{r}{2} c_1(\omega_{\mathscr{C}\mathscr{P}^1})$. Hence $\operatorname{ch}(\varphi_! \omega_X) = 2 - r + \frac{r-2}{2} c_1(\mathscr{C}\mathscr{P}^1)$. Thus L.H.S. of (4.2.3) equals

$$\left(2 - r + \frac{r-2}{2} c_1(\mathscr{C}\mathscr{P}^1)\right) \left(1 + \frac{1}{2} c_1(\mathscr{C}\mathscr{P}^1)\right) = 2 - r.$$

Hence we have

$$2 - r = \frac{1}{12} \varphi_* c_2(X) = \frac{1}{12} (p \circ b_1 \circ \cdots \circ b_m)_* \pi_* c_2(X).$$

From Proposition 4.2.1, $\pi_* c_2(X) = 12 c_1(\Sigma_n^m)$. Therefore

(4.2.4)
$$(p \circ b_1 \circ \cdots \circ b_m)_* c_1(\Sigma_n^m) = 2 - r.$$

Now for each $1 \le i \le m$, $K_{\Sigma_n^i} = (b_i)^* K_{\Sigma_n^{i-1}} + E_i$, where E_i is the exceptional divisor of b_i, so that

$$(b_i)_* c_1(\Sigma_n^i) = (b_i)_* (b_i^* c_1(\Sigma_n^{i-1})) - (b_i)_* c_1(E_i)$$
$$= c_1(\Sigma_n^{i-1}),$$

where we have denoted $\Sigma_n^0 = \Sigma_n$.

Substitute this into (4.2.4) and simplify, we get $p_* c_1(\Sigma_n) = 2 - r$. Notice that $c_1(\Sigma_n) = 2c_1(E_0) + (2 - n)c_1(F)$, and therefore $p_* c_1(\Sigma_n) = 2p_* c_1(E_0) + (2 - n)p_* c_1(F) = 2$. Thus $2 = 2 - r$, which gives $r = 0$. Therefore $R^1 \varphi_* \omega_X$ is a locally free sheaf of rank 0. Result follows. $\diamond$

REFERENCES

1. A. Beauville, *Variétés kählériennes dont la première classe de Chern est nulle*, J. Differential Geom. **18** (1983), 755–782.
2. S. S. Chern, *Characteristic classes of hermitian manifolds*, Ann. of Math. (2) **47** (1946), 85–121.
3. W. Fischer and H. Grauert, *Lokal triviale Familien kompakter komplexer Mannigfaltigkeiten*, Nachr. Akad. Wiss. Goettingen, II. Math. Phys. Kl. (1965), 89–94.
4. P. Griffiths and J. Harris, *principles of algebraic geometry*, Wiley, NY, 1978.
5. E. Hopf, *Elementare Bemerkungen über die Lösungen partieller Differential-gleichungen zweiter Ordnung vom elliptischen Typus*, Sitzber. preuss. Akad. Wiss. Physik-math. Kl.**19** (1927), 147–152.
6. S. L. Kleiman, *Toward a numerical theory of ampleness*, Ann. of Math. (2) **84** (1966), 293–344.
7. K. Kodaira, *On the structure of compact complex analytic surfaces*. I, Amer. J. Math. **86** (1964), 751–798.
8. J. Kollár, *Higher direct images of dualizing sheaves*. I, Ann. of Math. (2) **123** (1986), 11–42.
9. K. Ueno and I. Nakamura, *An addition formula for Kodaira dimensions of analytic fiber bundles whose fiber are Moĭshezon manifolds*, J. Math. Soc. Japan **25** (1973), 363–371.
10. S. T. Yau, *Calabi's conjecture and some new results in algebraic geometry*, Proc. Nat. Acad. Sci. U.S.A. **74** (1977), 1789–1799.

BRANDEIS UNIVERSITY

Proceedings of Symposia in Pure Mathematics
Volume 54 (1993), Part 2

Degeneration of Kähler-Einstein Manifolds. I

GANG TIAN

Introduction

According to algebraic geometers, a degeneration of Kähler-Einstein manifolds is a holomorphic family $\pi : \mathscr{X} \mapsto \Delta$ with the following property: the fibers $X_t = \pi^{-1}(t)$ are smooth except for $t = 0$. By the famous Mumford's Semi-stable Reduction Theorem, after base change and birational modification, one can further assume that $\mathscr{X}$ is smooth and the central fiber X_0 is a reduced divisor with normal crossings. In this paper, we confine ourselves to the case: each X_t for $t \neq 0$ has ample canonical line bundle, in other words, the first Chern class is negative. Then each of these X_t admits a unique Kähler-Einstein metric $g_{E,t}$ with $\mathrm{Ric}(g_{E,t}) = -g_{E,t}$ (cf. [Y]). The basic problem is the limiting behavior of this family of canonical metrics as t goes to zero, for instance, does the family of metrics converge to a complete Kähler-Einstein metric on the regular part of X_0? The goal of this paper is to give a sufficient condition on X_0 under which the metrics $g_{E,t}$ converge to a metric g_∞ on $X_{0,\mathrm{reg}}$, where $X_{0,\mathrm{reg}}$ is the regular part of the central fiber. Of course, the metric g_∞ is a Kähler-Einstein one with $\mathrm{Ric}(g_\infty) = -g_\infty$. In case the relative canonical line bundle $K_{\mathscr{X}/\Delta}$ of $\mathscr{X}$ is numerically positive, H.Tsuji [Ts] proved such a convergence. However, the limit metric g_∞ can not be complete. It is often degenerate along some subvarieties. A line bundle L on Y is said to be numerically positive if $C_1(L)^n > 0$ and the degree of the restricted line bundle $L|_C$ for each holomorphic curve in Y is nonnegative, where n is the dimension of Y.

Our first main theorem is the following.

THEOREM 0.1. *Let* $\pi : \mathscr{X} \to \Delta$ *be a degeneration of Kähler-Einstein manifolds* $\{X_t, g_{E,t}\}$ *with* $\mathrm{Ric}(g_{E,t}) = -g_{E,t}$ *as above. Assume that the central fiber* X_0 *is the union of smooth hypersurfaces, say* $D_1, \ldots, D_m$,

1991 *Mathematics Subject Classification.* Primary 53B35.
This work is partially supported by NSF Grant DMS-8907648.
This paper is in final form and no version of it will be submitted for publication elsewhere.

with normal crossings and each line bundle $K_{D_i} + \sum_{j \neq i} D_j$ is ample on D_i, where $1 \leq i \leq m$. We further assume that no three of those divisors D_i $(1 \leq i \leq m)$ have nonempty intersection. Then the Kähler-Einstein metrics $g_{E,t}$ on X_t converge to a complete Kähler-Einstein metric on $X_0 \backslash \mathrm{Sing}(X_0)$ in the sense of Cheeger-Gromov: there are an exhaustion of compact subsets $F_\beta \subset\subset X_0 \backslash \mathrm{Sing}(X_0)$ and diffeomorphisms $\phi_{\beta,t}$ from F_β into X_t satisfying:

(1) $X_t \backslash \bigcup_{\beta=1}^{\infty} \phi_{\beta,t}(F_\beta)$ consists of finite union of submanifolds of real codimension 1;

(2) for each fixed β, $\phi_{\beta,t}^* g_{E,t}$ converge to $g_{E,0}$ on F_β in C^2-topology on the space of riemannian metrics as t goes to 0.

The author thinks that the ampleness condition in the above is almost necessary for the metrics g_t to converge to a nondegenerate metric. In general, if we do not require the nondegeneracy of the limit, the ampleness could be weaken to the bigness of divisors $K_{D_i \cap D_j} + \sum_{l \neq i,j} D_l$ on $D_i \cap D_j$. The later seems to be necessary for the Kähler-Einstein metric on the regular part of X_0 to be the limit of smooth Kähler-Einstein manifolds in the sense of Cheeger-Gromov. For instance, the author is able to prove: if $K_{D_i} + \sum_{l \neq i} D_l$ is numerical positive on D_i for each i, and the complete Kähler-Einstein metric on $X_0 \backslash \mathrm{Sing}(X_0)$ is the limit of a sequence of smooth Kähler-Einstein manifolds in the sense of Cheeger-Gromov, then for any i, j, the line bundle $K_{D_i \cap D_j} + \sum_{l \neq i,j} D_l$ is big on $D_i \cap D_j$. It will be discussed in a future paper. It is surely disirable to remove the extra assumption on the configuration of divisors D_i. Nevertheless, Theorem 1 can be generalized to the case that each X_t is quasiprojective (cf. §4). This allows us to deal with many cases where three of the divisors D_i might intersect.

As a corollary of Theorem 0.1, we estimate from above the restriction of the Peterson-Weil metric to the puctured disc for the above degeneration family.

THEOREM 0.2. *Let z be the euclidean coordinate of the disc in the complex line C^1. Then the restriction of the Peterson-Weil metric on the moduli space of complex structures to the degeneration $\pi: \mathcal{X} \to \Delta$ is bounded from above by a constant multiple of $\frac{|dz|^2}{|z|^2(-\log|z|^2)^3}$ on the punctured disc $\Delta \backslash \{0\}$.*

In particular, the Peterson-Weil metric is incomplete.

In fact, one should be able to prove the equivalence of the restriction of the Peterson-Weil metric and $\frac{|dz|^2}{|z|^2(-\log|z|^2)^3}$ on the punctured disc $\Delta \backslash \{0\}$. This has been proved for Riemann surfaces by Masur and Wolpert (cf. [**Ma**]). However, for higher dimensional Kähler manifolds, the proof would be more difficult and technical involving.

This paper is outlined as follows. In §1, we construct families of Kähler metrics on $\mathcal{X}$ with prescribed asympototic behavior near X_0. In §2, we prove Theorem 0.1 by using the estimates in [**Y**] and results from the previous section. Section 3 contains the proof of Theorem 0.2. In the last section, we discuss briefly the generalizations of Theorem 0.1.

1. Construction of families of Kähler metrics with asympototic behavior

Let $\pi : \mathscr{X} \to \Delta$ be a degeneration of Kähler-Einstein manifolds as in the introduction. In particular, the central fiber $X_0 = \pi^{-1}(0)$ is the union of smooth hypersurfaces $D_1, \ldots, D_m$ with normal crossings. For each i between 1 and m, we choose a neighborhood U_i of D_i such that $\overline{U}_i \cap \overline{U}_j = \varnothing$ whenever $\overline{D}_i \cap \overline{D}_j = \varnothing$, where $\overline{U}_i$, $\overline{U}_j$ denote the closures of U_i, U_j, respectively.

LEMMA 1.1. *There are defining sections S_i of D_i in $\mathscr{X}$ satisfying:*
  (1) $S_i = 1$ *outside U_i for each i*;
  (2) $S_1 \cdots S_m \equiv t$ *on $\mathscr{X}$*.

PROOF. The statement (1) simply follows from the fact that $D_i \subset U_i$. For (2), we only need to point out that the line bundle $\bigotimes_{i=1}^m [D_i] = [X_0]$ is trivial on $\mathscr{X}$, since $\mathscr{X}$ is smooth.

Fix hermitian metrics $\|\cdot\|_i$ of the line bundles $[D_i]$ with $\|S_i\|_i \equiv 1$ outside U_i, $1 \le i \le m$, and a relative volume form $\tilde{V}$ on $\mathscr{X}$. The later can be taken as the volume form of ω^n along fibers, where ω is a Kähler metric on $\mathscr{X}$ and n is the dimension of the fibers. Without losing the generality, we may assume $\mathscr{X} = \bigcup_{i=1}^m U_i$.

Let $\tilde{V}_i$ be the local represntation of the relative volume form $\tilde{V}$ on U_i, in particular, for each $t \in \Delta$, $\tilde{V}_i|_{X_t}$ is the volume form of $X_t \cap U_i$. Choose $U_i' \subset U_i$ be the neighborhoods of D_i such that $\overline{U}_i' \subset\subset U_i$. Denote by $U_{i_1 \cdots i_k}'$ the intersection $U_{i_1}' \cap \cdots \cap U_{i_k}'$ for each k-tuple $(i_1, \ldots, i_k)$, it is a neighborhood of $D_{i_1 \cdots i_k} = D_{i_1} \cap \cdots \cap D_{i_k}$.

LEMMA 1.2. *By shrinking $U_1', \ldots, U_m'$ if necessary, we can find smooth functions φ_i on $\mathscr{X}$ satisfying:*
  (1) $\prod_{j=1}^m e^{\varphi_i} \|\cdot\|_i \equiv 1$;
  (2) $e^{\varphi_i} \|S_i\|_i \equiv 1$ *outside U_i*, $i = 1, 2, \ldots, m$;
  (3) $e^{2\varphi_{i_j}} \|\cdot\|_{i_j}^2 \tilde{V}_{i_j} = e^{2\varphi_{i_l}} \|\cdot\|_{i_l}^2 \tilde{V}_{i_l}$ *on $U_{i_1, \ldots, i_k}'$ for $1 \le j, l \le k$.*

PROOF. Let $\tilde{k}$ be the largest integer with the property: $D_{i_1} \cap \cdots \cap D_{i_{\tilde{k}}} \ne \varnothing$ for some $\tilde{k}$-tuple $(i_1, \ldots, i_{\tilde{k}})$. Of course, $\tilde{k} \le n$. We define

$$X_{0,k} = \bigcup_{1 \le i_1 < \cdots < i_k \le m} \left(\bigcap_{j=1}^k D_{i_j} \right).$$

Then $X_{0,k} = \varnothing$ if $k > \tilde{k}$, and $X_{0,\tilde{k}}$ is a disjoint union of smooth submanifolds in X_0 of dimension $n - \tilde{k}$. Moreover, each connected component of $X_{0,\tilde{k}}$ is of form $D_{i_1} \cap \cdots \cap D_{i_k}$.

We will construct $\varphi_1, \dots, \varphi_m$ inductively on $\bigcup_{1 \le i_1 < \cdots < i_k \le m} U_{i_1 \cdots i_k}$ satisfying (1) and (2). First we assume $k = \tilde{k}$. We want to construct $\varphi_1, \dots, \varphi_m$ in the neighborhood $\bigcup_{1 \le i_1 < \cdots < i_k \le m} U_{i_1 \cdots i_k}$ of $X_{0,\tilde{k}}$. For this purpose, we pick up any $U_{i_1 \cdots i_k} \ne \varnothing$, say $U_{1 \cdots \tilde{k}}$ for simplicity. By adjunction formula, for any $1 \le i \le \tilde{k}$, $K_{D_i} + \sum_{j \ne i, j=1}^{\tilde{k}} D_j$ restricts to the canonical line bundle $K_{D_1 \cap \cdots \cap D_{\tilde{k}}}$ on $D_1 \cap \cdots \cap D_{\tilde{k}}$. Without losing generality, we may choose $U'_1, \dots, U'_m$ such that $D_{i_1} \cap \cdots \cap D_{i_k}$ are strong deformation retracts of $U'_{i_1} \cap \cdots \cap U'_{i_k}$. Therefore, the line bundles L_i $(i = 2, \dots, \tilde{k})$ are trivial in $U'_1 \cap \cdots \cap U'_{\tilde{k}}$, where we define L_i to be

$$\left(K_{D_i} + \sum_{j=1, j \ne i}^{\tilde{k}} D_j \right) - \left(K_{D_1} + \sum_{j=2}^{\tilde{k}} D_j \right).$$

In particular, it implies there are smooth functions $\psi_2, \dots, \psi_{\tilde{k}}$ in $U'_1 \cap \cdots \cap U'_{\tilde{k}}$ such that

$$\frac{\tilde{V}_i \| \cdot \|_i^2}{\tilde{V}_1 \| \cdot \|_1^2} = \frac{\tilde{V}_i \cdot \prod_{j=2}^{\tilde{k}} \| \cdot \|_j^2}{\tilde{V}_1 \cdot \prod_{j=1, j \ne i}^{\tilde{k}} \| \cdot \|_j^2} = e^{2\psi_i}, \qquad i = 2, \dots, \tilde{k}.$$

On the other hand, since each $[D_j]$ is trivial in $U'_1 \cap \cdots \cap U'_{\tilde{k}}$ for $j > \tilde{k}$ and $[X_0] = [D_1 + \cdots + D_m]$ is trivial, the product $\prod_{j=1}^m \| \cdot \|_j = e^{-\psi}$ for some function ψ in $U'_1 \cap \cdots \cap U'_{\tilde{k}}$. Now we define $\varphi_1 = (\psi - \sum_{j=2}^{\tilde{k}} \psi_j)/\tilde{k}$, $\varphi_i = \varphi_1 + \psi_i$ for $\tilde{k} \ge i \ge 2$ and $\varphi_i = 0$ for $i > \tilde{k}$. Similarly, we can define $\varphi_1, \dots, \varphi_m$ near the other components of $X_{0,\tilde{k}}$. Then we extend them to $\mathscr{X}$ such that φ_i is identically zero outside U_i. Next, we suppose that $\varphi_1, \dots, \varphi_m$ have been defined such that (1), (2) and (3) hold in the neighborhood of $X_{0,k+1}$ for $k + 1 \le \tilde{k}$, we need to modify them in a neighborhood of $X_{0,k}$ such that the properties (1), (2) and (3) hold near $X_{0,k}$. First we remark that $X_{0,k} \backslash X_{0,k+1}$ consists of disjoint submanifolds of dimension $n - k$ and each connected component is of form $(D_{i_1} \cap \cdots \cap D_{i_k}) \backslash (\bigcup_{j \ne i_l} D_j)$. It suffices to modify $\varphi_1, \dots, \varphi_m$ on each of these components. For simplicity, we assume $(i_1, \dots, i_k) = (1, 2, \dots, k)$. As above, we define

$$2\psi_i = \log(\tilde{V}_i e^{2\varphi_i} \| \cdot \|_i^2 / V_1 e^{2\varphi_1} \| \cdot \|_1^2) \quad \text{on } U'_1 \cap \cdots \cap U'_k, \ i = 2, \dots, k,$$

$$\psi = -\log \left(\prod_{j=1}^m e^{\varphi_j} \| \cdot \|_j \right)$$

then ψ_i, ψ are identically one in the neighborhood of $X_{0,k+1}$ by the hypothesis of the induction. Therefore, if we define $\tilde{\varphi}_1 = (\varphi - \sum_{j=2}^k \psi_j)/\tilde{k}$, $\tilde{\varphi}_2 = \tilde{\varphi}_1 + \psi_2, \dots, \tilde{\varphi}_k = \tilde{\varphi}_1 + \psi_k$, $\tilde{\varphi}_i = 0$ for $i \ge k + 1$, and replace the old $\varphi_1, \dots, \varphi_m$ by $\varphi_1 + \tilde{\varphi}_1, \dots, \varphi_m + \tilde{\varphi}_m$, we obtain the required functions in a neighborhood of $X_{0,k}$. The lemma then follows from induction.

REMARK. In fact, in this paper, we only need the special case of this lemma, namely, no more than two D_i have nonempty intersection. The proof for this special case is relatively simpler in notations.

From now on, we assume that no more than two divisors D_i intersect, for simplicity, we may assume $D_i \cap D_j = \varnothing$ if $|i - j| \geq 2$. By the above lemma, there are hermitian metrics $\| \cdot \|_i$ of the line bundles $[D_i]$ on $\mathscr{X}$ satisfying:

(1) $\|S_i\|_i \equiv 1$ on $\mathscr{X} \backslash U_i$, $i = 1, 2, \ldots, m$;
(2) $\|S_i\|_i \|S_{i+1}\|_{i+1} \equiv |t|$ on $U_i \cap U_{i+1}$, $1 \leq i \leq m - 1$;
(3) $\| \cdot \|_i^2 \widetilde{V}_i \equiv \| \cdot \|_{i+1}^2 \widetilde{V}_{i+1}$ on $U_i \cup U_{i+1}$, $1 \leq i \leq m - 1$.

Note that $\widetilde{V}_i$ are the local volume forms of a fixed one $\widetilde{V}$ on U_i. Without losing the generality, we may further assume that $\|S_i\|^2 \leq 3$ in $\mathscr{X}$ for all i. Now let us define a relative volume form V on $\mathscr{X} \backslash \mathrm{Sing}(X_0)$, i.e., $\mathscr{X} \backslash X_{0,1}$. For $t \in \Delta \backslash \{0\}$,

$$
V_{t,i} = \frac{\widetilde{V}_i}{\prod_{j=1,\, j\neq i}^m \|S_j\|_j^2}
$$
$$
\cdot \prod_{j=1,\, j\neq i}^m \left(\frac{\pi}{\log(\frac{|t|}{4})} \csc \left(\frac{\pi \log \left(\frac{\|S_j\|_j^2}{4} \right)}{2\log(\frac{|t|}{4})} \right) \right)^2 \quad \text{on } U_i \cap X_t
$$

and

$$
V_{0,i} = \frac{\widetilde{V}_i}{\prod_{j=1,\, j\neq i}^m \|S_j\|_j^2 (\log(\frac{\|S_j\|_j^2}{4}))^2} \quad \text{on } U_i \cap (X_0 \backslash \mathrm{Sing}(X_0)).
$$

In order to see that these volume forms can be glued together to give a global one V, we simply observe that on $U_i \cap U_{i+1} \cap X_t$ $(1 \leq i \leq m - 1)$,

$$
V_{t,i} = \frac{\widetilde{V}_i}{\|S_{i+1}\|_{i+1}^2} \left(\frac{\pi}{\log(\frac{|t|}{4})} \csc \left(\frac{-\pi \log 4}{2\log(\frac{|t|}{4})} \right) \right)^{2(m-2)}
$$
$$
\cdot \left(\frac{\pi}{\log(\frac{|t|}{4})} \csc \left(\frac{\pi \log(\frac{\|S_{i+1}\|_{i+1}^2}{4})}{2\log(\frac{|t|}{4})} \right) \right)^2
$$
$$
= \frac{\widetilde{V}_{i+1}}{\|S_i\|_i^2} \left(\frac{\pi}{\log(\frac{|t|}{4})} \csc \left(\frac{-\pi \log 4}{2\log(\frac{|t|}{4})} \right) \right)^{2(m-2)}
$$
$$
\cdot \left(\frac{\pi}{\log(\frac{|t|}{4})} \csc \left(\pi - \frac{\pi \log(\frac{\|S_i\|_i^2}{4})}{2\log(\frac{|t|}{4})} \right) \right)^2
$$
$$
= V_{t,i+1}.
$$

Define

$$\omega_t = \frac{\sqrt{-1}}{2\pi}\partial\bar{\partial}\log V_t \quad \text{on } X_t \ (t \neq 0),$$

$$\omega_0 = \frac{\sqrt{-1}}{2\pi}\partial\bar{\partial}\log V_0 \quad \text{on } X_0\backslash\operatorname{Sing}(X_0),$$

where $V_t(\cdot) = V(\cdot, t)$ for t in Δ. Simple computations show: for $t \neq 0$, in $U_j \cap X_t$,

$$\omega_t = -\operatorname{Ric}(\widetilde{V}_t) + \sum_{j\neq i}\operatorname{Ric}(\|\cdot\|_j)\left(1 + \frac{\pi}{\log(\frac{|t|}{4})}\operatorname{ctg}\left(\frac{\pi\log(\frac{\|S_j\|_j^2}{4})}{2\log(\frac{|t|}{4})}\right)\right)$$

$$+ \frac{\sqrt{-1}}{\pi}\sum_{j\neq i}\frac{DS_j\wedge\overline{DS_j}}{|S_j|^2}\left(\frac{\pi}{2\log(\frac{|t|}{4})}\operatorname{csc}\left(\frac{\pi\log(\frac{\|S_j\|_j^2}{4})}{2\log(\frac{|t|}{4})}\right)\right)^2$$

where $\operatorname{Ric}(\widetilde{V}_t)$, $\operatorname{Ric}(\|\cdot\|_j)$ denote the curvature tensor of the volume forms $\widetilde{V}_t = V|_{X_t}$ and the hermitian metrics $\|\cdot\|_j$ on D_j, respectively, DS_j denotes the covariant derivative of S_j with respect to $\|\cdot\|_j$. Also, in $D_j\backslash\operatorname{Sing}(X_0)$,

$$\omega_0 = -\operatorname{Ric}(\widetilde{V}_0) + \sum_{j\neq i}\operatorname{Ric}(\|\cdot\|_j)\left(1 + \frac{2}{\log(\frac{\|S_j\|_j^2}{4})}\right)$$

$$+ \frac{\sqrt{-1}}{2\pi}\sum_{j\neq i}\frac{DS_j\wedge\overline{DS_j}}{|S_j|^2}\left(\log\left(\frac{\|S_j\|_j^2}{4}\right)\right)^{-2}.$$

This follows

LEMMA 1.3. *Assume that for each $1 \leq i \leq m$, the line bundle $K_{D_i} + \sum_{j\neq i}D_j$ is ample on D_i. Then by choosing the volume form $\widetilde{V}$ properly, all ω_t, ω_0 are the Kähler forms of complete Kähler metrics g_t for t sufficiently small. Moreover, the Kähler metrics g_t converge to g_0 outside $\operatorname{Sing}(X_0)$ in the sense of Cheeger-Gromov: there are an exhaustion of compact subsets $F_\beta \subset\subset X_0\backslash\operatorname{Sing}(X_0)$ and diffeomorphisms $\phi_{\beta,t}$ from F_β into X_t satisfying:*

(1) $X_t\backslash\bigcup_{\beta=1}^\infty\phi_{\beta,t}(F_\beta)$ consists of finite union of submanifolds of real codimension 1;

(2) for each fixed β, $\phi_{\beta,t}^*g_t$ converge to g_0 on F_β in C^2-topology on the space of riemannian metrics as t goes to 0.

LEMMA 1.4. *Assume that $K_{D_i} + \sum_{j\neq i}D_j$ is ample on D_i for each i between 1 and m. Then the Kähler metrics g_t in the last lemma have uniformly bounded curvature tensors.*

PROOF. The boundedness of the curvature tensor of g_0 was known before (cf. [**TY**]). Let us prove the boundedness of $R(g_t)$ for $t \neq 0$ small, where $R(g_t)$ denotes the curvature tensor of g_t. It suffices to bound $R(g_t)$

near each intersection $D_i \cap D_{i+1}$, $(1 \leq i \leq m - 1)$. For a small $\delta > 0$, one can choose a finite covering of $\bigcup_{i=1}^{m-1} D_i \cap D_{i+1}$ by coordinate charts $\{(W_j; z_{j1}, \ldots, z_{jn+1})\}$ such that each W_j is defined by

$$|z_{j1}| < \delta, \ldots, |z_{jn+1}| < \delta, \qquad D_i \cap D_{i+1} \cap W_j = \{z_n = z_{n+1} = 0\},$$

and

$$W_j \cap X_t = \{(z_{j1}, \ldots, z_{jn+1}) | \, |z_{j\alpha}| < \delta$$
$$\text{for } \alpha = 1, 2, \ldots, n+1, \ z_{jn}z_{jn+1} = t\}.$$

Now we fix i and a coordinate chart $(W_j, z_{j1}, \ldots, z_{jn+1})$. For simplicity, we write W for W_j and z_α for $z_{j\alpha}$ $(\alpha = 1, 2, \ldots, n+1)$. By shrinking W if necessary, we may assume that $\|S_l\|_l = 1$ in W for $l \neq i, i+1$, and S_i, S_{i+1} are locally represented by functions z_n, z_{n+1} in W. This implies that if a, a' are the local representations of the hermitian metrics $\|\cdot\|_i$ and $\|\cdot\|_{i+1}$ in W, then $aa' \equiv 1$. By the definition of the metric g_t, its Kähler form is given in W by

$$\omega_t = \frac{\sqrt{-1}}{2\pi} \partial\overline{\partial} \log\left(\frac{b}{|z_n|^2} \left(\frac{\pi}{2\log(\frac{|t|}{4})} \csc\left(\frac{\pi\log(\frac{a|z_n|^2}{4})}{2\log(\frac{|t|}{4})} \right) \right)^2 \right)$$

where b is a smooth function of $z_1, \ldots, z_n, \frac{t}{z_n}$ with $\frac{\sqrt{-1}}{2\pi}\partial\overline{\partial}\log b$ positive definite. By exchanging z_n with z_{n+1} if necessary, we may assume that $|z_n| \geq \sqrt{|t|}$. Then $|\frac{t}{z_n}| \geq |z_n|$.

The straightforward computations show

$$\omega_t = \frac{\sqrt{-1}}{2\pi} \sum_{\alpha\beta=1}^{n+1} h_{\alpha\overline{\beta}} dz_\alpha \wedge d\overline{z}_\beta + \frac{\sqrt{-1}}{\pi}\partial\overline{\partial}\log$$

$$\cdot \left(\frac{\pi}{2\log(\frac{|t|}{4})} \csc\left(\frac{\pi\log(\frac{a|z_n|^2}{4})}{2\log(\frac{|t|}{4})} \right) \right)$$

$$= \frac{\sqrt{-1}}{2\pi} \sum_{\alpha\beta=1}^{n+1} h_{\alpha\overline{\beta}} dz_\alpha \wedge d\overline{z}_\beta$$

$$- \frac{\sqrt{-1}}{2\pi} \frac{\partial^2 \log a}{\partial z_\alpha \partial\overline{z}_\beta} \operatorname{ctg}\left(\frac{\pi\log(\frac{a|z_n|^2}{4})}{2\log(\frac{|t|}{4})} \right) \frac{\pi}{2\log(\frac{|t|}{4})} dz_\alpha \wedge d\overline{z}_\beta$$

$$+ \frac{\sqrt{-1}}{\pi} \left(\frac{\pi}{2\log(\frac{|t|}{4})} \csc\left(\frac{\pi\log(\frac{a|z_n|^2}{4})}{2\log(\frac{|t|}{4})} \right) \right)^2$$

$$\cdot \left(\frac{dz_n}{z_n} + \partial\log a \right) \wedge \overline{\left(\frac{dz_n}{z_n} + \partial\log a \right)}$$

where $h_{\alpha\bar{\beta}}$, $\frac{\partial^2 \log a}{\partial z_\alpha \partial \bar{z}_\beta}$ are smooth functions of $z_1, \bar{z}_1, \ldots, z_n, \bar{z}_n, t/z_n, \overline{t/z_n}$, and $\{h_{\alpha\bar{\beta}}\}$ is positive definite in W.

Then we have the following decomposition

$$\omega_t = \omega_{1t} + \omega_{2t} + \frac{\pi}{2\log(\frac{|t|}{4})} \operatorname{ctg}\left(\frac{\pi\log(\frac{a|z_n|^2}{4})}{2\log(\frac{|t|}{4})}\right)^2 \omega_{3t}$$

$$+ \left(\frac{\pi}{2\log(\frac{|t|}{4})} \csc\left(\frac{\pi\log(\frac{a|z_n|^2}{4})}{2\log(\frac{|t|}{4})}\right)\right)^2 \omega_{4t}$$

where

$$\omega_{1t} = \frac{\sqrt{-1}}{2\pi} \sum_{\alpha\beta=1}^{n-1} h_{\alpha\bar{\beta}}(z_1, \ldots, z_n, 0, 0)\, dz_\alpha d\bar{z}_\beta$$

$$+ \frac{\sqrt{-1}}{\pi}\left(\frac{\pi}{2\log(\frac{|t|}{4})} \csc\left(\frac{\pi\log(\frac{a|z_n|^2}{4})}{2\log(\frac{|t|}{4})}\right)\right)^2 \frac{dz_n \wedge d\bar{z}_n}{|z_n|^2}$$

and ω_{2t}, ω_{3t}, ω_{4t} are $(1,1)$-forms in $dz_\alpha, d\bar{z}_\beta, \frac{dz_n}{z_n}, \frac{\overline{dz_n}}{z_n}$ with coefficients being functions of form

$$u + \frac{t}{z_n}v + \frac{\overline{t}}{\overline{z_n}}v$$

where u, v are smooth functions of $z_1, \ldots, \bar{z}_n, \frac{t}{z_n}, \frac{\overline{t}}{\overline{z_n}}$. Using the fact that $|z_n| \geq \sqrt{|t|}$, one can easily check that $\omega_t - \omega_{1t}$ are uniformly bounded in C^3-topology with respect to the metric induced by ω_{1t}. On the other hand, the metric induced by ω_{1t} has bounded curvature tensor, so does g_t in W. The lemma is proved.

LEMMA 1.5. *There is a uniformly bounded function f on $\mathscr{X}\setminus \operatorname{Sing}(X_0)$ such that for $t \neq 0$,*

$$\operatorname{Ric}(g_t) + \omega_t = \frac{\sqrt{-1}}{2\pi}\partial\bar{\partial}f|_{X_t} \quad \text{on } X_t$$

while for $t = 0$,

$$\operatorname{Ric}(g_0) + \omega_0 = \frac{\sqrt{-1}}{2\pi}\partial\bar{\partial}f|_{X_0} \quad \text{on } X_0\setminus \operatorname{Sing}(X_0).$$

Moreover, the C^3-norms of $f|_{X_t}$ are uniformly bounded with respect to g_t for $t \in \Delta$.

PROOF. We define f by the formula

$$f|_{X_t} = -\log\frac{\omega_t^n}{V_t}.$$

It suffices to show that the C^2-norm of $f|_{X_t}$ are uniformly bounded with respect to g_t near each $D_i \cap D_{i+1}$ $(1 \le i \le m-1)$. Let $\{(W_j; z_{j1}, \ldots, z_{jn+1})\}$ be the finite covering of $\mathrm{Sing}(X_0)$ as in the proof of Lemma 1.3. Fix one of them, denote it by $(W; z_1, \ldots, z_{n+1})$, then using the notation in the proof of last lemma, we have $\omega_t^n = \omega_{1t}^n(1+\tilde{h})$ where $\tilde{h}$ has bounded C^3-norm with respect to the metric by g_t, and $\tilde{h} \to 0$ as $t \to 0$.

On the other hand, simple computations show

$$\frac{\omega_{1t}^n}{V_t} = \frac{c_n a \, \det(h_{\alpha\bar{\beta}})_{1 \le \alpha, \beta \le n-1}(z_1, \ldots, z_{n-1}, 0, 0)}{\tilde{V}_t}$$

is a positive function uniformly bounded away from zero, where c_n is a constant depending only on the dimension n. It follows the C^3-norm of $f|_{X_t}$ is uniformly bounded.

REMARK. The functions $f_t = f|_{X_t}$ converge to f_0 as (X_t, g_t) converge to $(X_{0,\mathrm{reg}}, g_0)$ in the sense of Lemma 1.3.

2. The proof of Theorem 0.1

In this section, we prove Theorem 0.1. Here all notations in the last section will be adopted. Recall that we constructed in last section a family of Kähler metrics g_t on X_t for $t \ne 0$ and g_0 on $X_0 \backslash \mathrm{Sing}(X_0)$ satisfying

(1) there is a uniform constant $C > 0$ such that

$$\sup_{X_t} \| \mathrm{Rm}(g_t) \|_{g_t} \le C \quad \text{for all } t \ne 0$$

and

$$\sup_{X_0 \backslash \mathrm{Sing}(X_0)} \| \mathrm{Rm}(g_0) \|_{g_0} \le C$$

where $\mathrm{Rm}(\cdot)$ denotes the sectional curvature tensor.

(2) There is an exhaustion of $X_0 \backslash \mathrm{Sing}(X_0)$ by compact sets $\{F_i\}$ such that for each i, there are diffeomorphisms $\phi_{it} : F_i \to X_t$ satisfying

$$\lim_{t \to 0} \sup_{F_i} \{ \| \phi_{it}^* g_t - g_0 \|_{g_0}, \| \nabla_{g_0}^k \phi_{it}^* g_t \|_{g_0} \} | 1 \le k \le 3 \} = 0$$

and $X_t \backslash \bigcup_{i=1}^{\infty} \phi_{it}(F_i)$ is a union of finitely many submanifolds of codimension 1, where ∇_{g_0} denotes the covariant derivative with respect to g_0.

(3) There is a smooth function f on $\mathscr{X} \backslash \mathrm{Sing}(X_0)$ such that for each t,

$$\mathrm{Ric}(g_t) + \omega_t = \frac{\sqrt{-1}}{2\pi} \partial \bar{\partial}(f|_{X_t}) \quad \text{on } X_t \backslash \mathrm{Sing}(X_t)$$

and if we put $f_t = f|_{X_t}$, then $\| f_t \|_{C^3(X_t, g_t)} \le C$. Now let $g_{E,t}$ be Kähler-Einstein metrics on $X_t \backslash \mathrm{Sing}(X_t)$. Note that $\mathrm{Sing}(X_t) = \varnothing$ for $t \ne 0$. Then there are smooth functions φ_t on $X_t \backslash \mathrm{Sing}(X_0)$ such that

$$\omega_{E,t} = \omega_t + \frac{\sqrt{-1}}{2\pi} \partial \bar{\partial} \varphi_t \quad \text{on } X_t \backslash \mathrm{Sing}(X_t)$$

where $\omega_{E,t}$ are Kähler forms of $g_{E,t}$. Furthermore, it follows from that $\text{Ric}(g_{E,t}) = -\omega_{E,t}$,

$$(2.1)_t \qquad\qquad \omega_{E,t}^n = e^{f_t + \varphi_t} \omega_t^n \quad \text{on} \quad X_t \backslash \text{Sing}(X_t).$$

The equations in $(2.1)_t$ are complex Monge-Ampere equations.

LEMMA 2.1. *There is a uniform constant C depending only on the C^0-norm of f such that $\sup_{X_t} |\varphi_t| \leq C$.*

PROOF. It follows from maximum principle and the uniform boundedness of f_t on X_t in (3).

LEMMA 2.2. *There is a uniform constant C, depending only on $\|\Delta_{g_t} f_t\|_{C^0(X_t)}$, the dimension n and the lower bound of the curvature tensor $\text{Rm}(g_t)$, such that $n + \Delta_{g_t} \varphi_t \leq C$.*

PROOF. It follows from the same computations as the second-order estimate in [Y].

COROLLARY 2.1. *There is a uniform constant C such that*

$$\sup_{X_t} \{\|\varphi_t\|, \|\nabla_{g_t}^k \varphi_t\| \mid 1 \leq k \leq 3\} \leq C.$$

PROOF. It can be proved by using the above lemma and the arguments in [Y] in deriving third-order estimate.

Now we conclude the proof of Theorem 0.1. Let $\{\varphi_{t_i}\}$ be any convergent subsequence in $C^{2,\frac{1}{2}}$-topology and φ_∞ be the limit on $X_0 \backslash \text{Sing}(X_0)$, then g_{E,t_i} will converge to a complete Kähler-Einstein metric $\tilde{g}_{E,0}$ outside $\text{Sing}(X_0)$. However, the Kähler-Einstein metric on $X_0 \backslash \text{Sing}(X_0)$ is unique, so $\tilde{g}_{E,0} = g_{E,0}$. Now, for any sequence $\{\varphi_{t_i}\}$, by Corollary 2.1, there is a convergent subsequence. This shows that φ_t converge to the unique smooth function φ_0 in $X_0 \backslash \text{Sing}(X_0)$, so $g_{E,t}$ converge to $g_{E,0}$ outside $\text{Sing}(X_0)$.

3. The Peterson-Weil metric is incomplete

In this section, we prove Theorem 0.2. Let z be the euclidean coordinate of $\Delta \subset C^1$, we need to show, in local coordinate z, the restriction of the Peterson-Weil metric to Δ^* is bounded from above by a constant multiple of $\frac{|dz|^2}{|z|^2(-\log|z|^2)^3}$. Note that the Poincaré metric on the punctured disc is $\frac{|dz|^2}{|z|^2(-\log|z|^2)^2}$. First let us recall the definition of the Peterson-Weil metric in this special case. Let $\pi : \mathscr{X} \to \Delta$ be the degenerating family of Kähler-Einstein manifolds as given in the introduction. For each t in Δ^*, the Kodaira-Spencer map ρ maps the tangent space $C = T_t\Delta^*$ into $H^1(X_t, T_{X_t})$, where T_{X_t} denotes the tangent sheaf of X_t. Denote by $\mathscr{G}_{PW}$ the Peterson-Weil metric on the moduli space of complex structure on X_t, and $\frac{\partial}{\partial t}$ the tangent vector of $T\Delta^*$. Then we have

(3.1) $$\mathscr{G}_{PW}\left(\frac{\partial}{\partial t}, \frac{\partial}{\partial t}\right)\Big|_{X_t} = \int_{X_t} \left\|H\rho\left(\frac{\partial}{\partial t}\right)\right\|^2_{g_{E,t}} dV_t$$

where dV_t is the volume form on X_t induced by the Kähler-Einstein metric $g_{E,t}$, and H denotes the harmonic projection.

In order to give a representation of $\rho(\frac{\partial}{\partial t})$, we need to construct proper diffeomorphisms $\phi_{tt'}$ between X_t and $X_{t'}$ for t, $t' \in \Delta^*$. Let U_i be the neighborhood of D_i and S_i are the defining sections of D_i as in the first section.

The following lemma is clearly true.

LEMMA 3.1. *By shrinking Δ is necessary, we may have a positive δ, independently of t in Δ, such that*

$$\{x \in \mathscr{X} \mid \|S_i\|_i(x) < 2\delta, \|S_{i+1}\|_{i+1}(x) < 2\delta\} \subseteq U_i \cap U_{i+1},$$
$$i = 1, 2, \dots, m-1$$

and there is a smooth family of diffeomorphisms $\psi_{tt'}$ from $X_t \backslash W$ onto $X_{t'} \backslash W$ satisfying

$$\psi_{tt'}(\partial(X_t \backslash W)) = \partial(X_{t'} \backslash W), \qquad \psi_{tt} = \mathrm{id}$$

where $W = \bigcup_{i=1}^{m-1}\{x \in \mathscr{X} \mid \|S_i\|_i < 2\delta, \|S_{i+1}\|_{i+1} < 2\delta\}$.

Denote by L_i the restriction of line bundle $[D_i]$ to $D_i \cap D_{i+1}$, then L_i^{-1} is the restriction of $[D_{i+1}]$ to $D_i \cap D_{i+1}$, where $1 \le i \le m-1$. The bundle $L_i \oplus L_i^{-1}$ is naturally identified with the normal bundle of $D_i \cap D_{i+1}$ in $\mathscr{X}$, so there are diffeomorphisms $f_i : B_i \to U_i \cap U_{i+1}$ such that the images are $\{x \in \mathscr{X} \mid \|S_i\|_i < 2\delta, \|S_{i+1}\|_{i+1} < 2\delta\}$, where B_i are neighborhoods of $D_i \cap D_{i+1}$ in the total spaces $L_i \oplus L_i^{-1}$ and we identify $D_i \cap D_{i+1}$ with the zero sections in $L_i \oplus L_i^{-1}$. The pull-backs $S_i \circ f_i$, $S_{i+1} \circ f_i$ are smooth functions on B_i and

$$|\bar{\partial}(S_i \circ f_i)|(x) = O(|x|), \qquad \text{for } x \text{ in } L_i,$$
$$|\bar{\partial}(S_{i+1} \circ f_i)|(x) = O(|x|), \qquad \text{for } x \text{ in } L_i^{-1},$$

where $|x|$ is the distance of x from zero section in L_i or L_i^{-1}. Note that $f_i^{-1}(X_t) \cap B_i = \{(S_i \circ f_i) \cdot (S_{i+1} \circ f_i) = t\}$. Clearly each point x in $f_i(B_i)$ is uniquely determined by $(\pi_i \circ f_i^{-1}(x), S_i(x), S_{i+1}(x))$, where $\pi_i : L_i \oplus L_i^{-1} \to D_i \cap D_{i+1}$ are the natural projections.

Now we define diffeomorphisms $\phi_{tt'}$ from X_t as follows. For any point x in X_t, if x is outside W, we put $\phi_{tt'}(x) = \psi_{tt'}(x)$. If x is in $\{\|S_i\|_i \le$

$\frac{3\delta}{2}$, $\|S_{i+1}\|_{i+1} \leq \frac{3\delta}{2}\}$, then we define $\phi_{tt'}(x)$ to be

(3.2)
$$\left(\pi_i \circ f_i^{-1}(x), \, S_i(x) e^{h\left(\frac{\log(\|S_i\|_i(x)/\delta)}{\log(|t|/\delta^2)} \right) \log \left| \frac{t'}{t} \right|}, \right.$$
$$\left. S_{i+1}(x) e^{h\left(\frac{\log(\|S_{i+1}\|_{i+1}(x)/\delta)}{\log |t/\delta^2|} \right) \log \left| \frac{t'}{t} \right|} \right)$$

where

$$h(u) = \frac{u^3}{3u^2 - 3u + 1}.$$

Observe that the function h satisfies the identity:

$$h(u) + h(1-u)$$
$$= \frac{u^3}{3u^2 - 3u + 1} + \frac{(1-u)^3}{3(1-u)^2 - 3(1-u) + 1} = 1$$

we have $\mathrm{Im}(\phi_{tt'}) \subset X_{t'}$.

Simple computation yields

(3.3)
$$\left\| \frac{\partial \phi_{tt'}}{\partial t'} \Big|_{X_t} \right\|_{C^2}(x) \leq \frac{C}{|t|(-\log|t|)^3}$$

for x in $X_t \backslash \bigcup_{i=1}^{m-1} \{\|S_i\|_i < \delta/2, \|S_{i+1}\|_{i+1} < \delta/2\}$ and in the domain where $\phi_{tt'}$ has been defined. Since $\phi_{tt}(x) = x$ whenever ϕ_{tt} has been defined, it is easy to extend $\phi_{tt'}$ across $W \backslash \bigcup_{i=1}^{m-1} \{\|S_i\|_i < 3\delta/2, \|S_{i+1}\|_{i+1} < 3\delta/2\}$ such that (3.3) holds for x in $X_t \backslash \bigcup_{i=1}^{m-1} \|S_i\|_i < \delta/2, \|S_{i+1}\|_{i+1} < \delta/2$.

To give a representation of $\rho(\frac{\partial}{\partial t})$, we choose a finite covering of $\mathscr{X}$ by open coordinate chart $\{(V_\alpha; z_{\alpha 1}, \ldots, z_{\alpha n+1})\}$ satisfying

(1) $V_\alpha = \{|z_{\alpha 1}| < \delta, \ldots, |z_{\alpha n+1}| < \delta\}$;
(2) if $V_\alpha \cap (D_i \cap D_{i+1}) \neq \varnothing$ for some i then $z_{\alpha n+1} = S_{i+1}$, $z_{\alpha n} = S_i$ on V_α;
(3) if $V_\alpha \cap (\bigcup_{i=1}^{m-1} D_i \cap D_{i+1}) = \varnothing$, then for each $i \leq m-1$, either $\|S_i\|_i(x) > \frac{\delta}{2}$ or $\|S_{i+1}\|_{i+1}(x) > \frac{\delta}{2}$, and $\pi(z_{\alpha 1}, \ldots, z_{\alpha n+1}) = z_{\alpha n+1}$.

Then for each α,

$$\rho\left(\frac{\partial}{\partial t} \right) \Big|_{X_t \cap V_\alpha} = \sum_{i=1}^n \bar{\partial} \left(\frac{\partial(z_{\alpha i} \circ \phi_{tt'})}{\partial t'} \Big|_{t=t'} \right) \frac{\partial}{\partial z_{\alpha i}} \Big|_{X_t \cap V_\alpha}.$$

It follows from (3.3) and Theorem 1 that

$$\left\| \rho\left(\frac{\partial}{\partial t} \right) \right\|_{g_{E,t}}^2(x) \leq \frac{C}{|t|^2 (\log|t|)^6}$$

for each x in $X_t \backslash \bigcup_{i=1}^{m-1} \{\|S_i\|_i < \delta/2, \|S_{i+1}\|_{i+1} < \delta/2\}$. It remains to estimate $\|\rho(\frac{\partial}{\partial t})\|_{g_{E,t}}^2(x)$ for x in one of $\{\|S_i\|_i < \delta/2, \|S_{i+1}\|_{i+1} < \delta/2\}$.

Put $(w_1, \ldots, w_n)(x) = (\pi_i \circ f_i^{-1}(x), z_{\alpha n}(x))$, then we have

$$\frac{\partial z_{\alpha i}}{\partial w_j} = \delta_{ij} + O\left(\min\left\{|z_{\alpha n}|, \left|\frac{t}{z_{\alpha n}}\right|\right\}\right),$$

$$\frac{\partial z_{\alpha i}}{\partial \overline{w}_j} = O\left(\min\left\{|z_{\alpha n}|, \left|\frac{t}{z_{\alpha n}}\right|\right\}\right),$$

therefore, on $V_\alpha \cap \{\|S_i\|_i < \delta/2, \|S_{i+1}\|_{i+1} < \delta/2\}$,

$$\rho\left(\frac{\partial}{\partial t}\right)\bigg|_{X_t} = \sum_{i=1}^n \overline{\partial}\left(\sum_{j=1}^n \frac{\partial z_{\alpha i}}{\partial w_j} \frac{\partial(w_j \circ \phi_{tt'})}{\partial t'}\bigg|_{t'=t}\right)\frac{\partial}{\partial z_{\alpha i}} + O\left(\frac{1}{|t|(-\log|t|)^3}\right)$$

$$= \sum_{i=1}^n \overline{\partial}\left(\frac{\partial z_{\alpha i}}{\partial w_n} \frac{\partial(w_n \circ \phi_{tt'})}{\partial t'}\bigg|_{tt'}\right)\frac{\partial}{\partial z_{\alpha i}} + O\left(\frac{1}{|t|(-\log|t|)^3}\right)$$

$$= \sum_{i=1}^n \left\{\overline{\partial}\left(\frac{\partial z_{\alpha i}}{\partial w_n}\right)\frac{\partial(w_n \circ \phi_{tt'})}{\partial t'}\bigg|_{t'=t}\right.$$

$$\left. + \frac{\partial z_{\alpha i}}{\partial w_n}\overline{\partial}\left(\frac{\partial(w_n \circ \phi_{tt'})}{\partial t'}\bigg|_{t'=t}\right)\right\}\frac{\partial}{\partial z_{\alpha i}}$$

$$+ O\left(\frac{1}{|t|(-\log|t|)^3}\right).$$

Using the definition (3.2) of $\phi_{tt'}$, we compute

$$(3.5) \qquad \frac{\partial(w_n \circ \phi_{tt'})}{\partial t'}\bigg|_{t'=t} = \frac{z_n}{2t}\, h\left(\frac{\log(a|z_n|^2/\delta)}{2\log(|t|/\delta^2)}\right)$$

where a is a positive function representing the hermitian metric $\|\cdot\|_i$ locally. Differentiating (3.5), we further compute

(3.6)

$$\overline{\partial}\left(\frac{\partial(w_n \circ \phi_{tt'})}{\partial t'}\bigg|_{tt'}\right)$$

$$= \frac{z_n}{16t\log(|t|/\delta^2)}\overline{\partial}$$

$$\cdot\left(\frac{(\log a + \log|z_n|^2/\delta)^3}{3\log^2(a|z_n|^2/\delta) - 6\log(a|z_n|^2/\delta)\log(|t|/\delta^2) + 4\log^2(|t|/\delta^2)}\right)$$

$$= \frac{z_n}{16t\log(|t|/\delta^2)}$$

$$\cdot\frac{3(\log(a|z_n|^2/\delta) - 2\log(|t|/\delta^2))^2(\log(a|z_n|^2/\delta))^2}{(3\log^2(a|z_n|^2/\delta) - 6\log(a|z_n|^2/\delta)\log(|t|/\delta^2) + 4\log^2(|t|/\delta^2))^2}$$

$$\cdot\left(\frac{\partial a}{a} + \frac{dz_n}{z_n}\right).$$

On the other hand, we have already known that the metric $g_{E,t}$ is equivalent to the metric on V_α given by its Kähler form

$$
(3.7) \quad \frac{\sqrt{-1}}{2\pi}\left(\sum_{i=1}^{n-1} dz_{\alpha i} \wedge d\bar{z}_{\alpha i} + \frac{dz_{\alpha n} \wedge d\bar{z}_{\alpha n}}{|z_n|^2} \right.
$$
$$
\left. \cdot \left(\frac{\pi}{2\log(\frac{|t|}{4})} \csc\left(\frac{\pi \log(\frac{a|z_n|^2}{4})}{2\log(\frac{|t|}{4})}\right)\right)^2 \right).
$$

Then it follows from (3.5), (3.6) and (3.7) that for x in $V_\alpha \cap \{\|S_i\|_i < \delta/2, \|S_{i+1}\|_{i+1} < \delta/2\}$,

$$
(3.8)
$$
$$
\left\| \rho\left(\frac{\partial}{\partial t}\right) \right\|_{g_{E,t}}^2 (x)
$$
$$
\leq \frac{C}{|t|^2 (\log|t|)^6} + \frac{C}{|t|^2} h^2 \left(\frac{\log(a|z_n|^2/\delta)}{2\log(|t|/\delta^2)}\right)
$$
$$
\cdot \left(\frac{\pi}{2\log|t|} \csc\left(\frac{\pi \log a|z_n|^2}{2\log|t|}\right)\right)^2
$$
$$
+ \frac{C}{|t|^2 (\log|t|)^2}
$$
$$
\times \frac{3\log^2(a|z_n|^2/\delta)(\log(a|z_n|^2/\delta) - 2\log|t|/\delta^2)^2}{(3\log^2(a|z_n|^2/\delta) - 6\log(a|z_n|^2/\delta)\log(|t|/\delta^2) + 4\log^2(|t|/\delta^2))^2}.
$$

By (3.1), (3.3), (3.8) and direct computations, we conclude the following estimate of the Peterson-Weil metric $\mathscr{G}_{PW}$,

$$
(3.9) \quad \mathscr{G}_{PW}\left(\frac{\partial}{\partial t}, \frac{\partial}{\partial t}\right)\bigg|_{X_t} \leq \frac{C}{|t|^2 (-\log|t|)^3}.
$$

So Theorem 0.2 is proved. In particular, the Peterson-Weil metric is incomplete.

4. Some generalizations

The previous discussions in §§1 and 2 can be generalized without much change to the degenerating families of quasi-projective manifolds. For simplicity, we only consider the following degeneration, that is, let $\pi: \mathscr{X} \to \Delta$ be a family of projective varieties with a divisor $\mathscr{E}$ satisfying:

(1) each $X_t = \pi^{-1}(t)$ is a union of smooth hypersurfaces in $\mathscr{X}$ with normal crossings;

(2) the divisor $\mathscr{E}$ consists of smooth components $E_1, \ldots, E_q$ with normal crossings such that the restrictions $d\pi|_{E_i}$ are surjective $(1 \leq i \leq q)$;

(3) for $t \neq 0$, $\mathrm{Sing}(X_t)$ are contained in $\mathscr{E} \cap X_t$;

(4) the divisor $\mathscr{E}$ intersects the central fiber X_0 and $\mathrm{Sing}(X_0)$ transversally. Put E_{it} to be $X_t \cap \mathscr{E}$ for $1 \le i \le q$ and $t \in \Delta$. Then each E_{it} is smooth. We further assume that for $t \ne 0$, the line bundle $K_{X_t} + \sum E_{it}$ is ample on X_t. It follows that there is a unique complete Kähler-Einstein metric $g_{E,t}$ on $X_t \backslash \mathscr{E}$ with $\mathrm{Ric}(g_{E,t}) = -g_{E,t}$ (cf. [CY, Ko]). The next theorem provides a sufficient condition on the convergence of this family of metrics $g_{E,t}$ as t goes to 0.

THEOREM 4.1. *Let* $\pi \colon \mathscr{X} \to \Delta$ *be the degenerating family with properties* (1)–(4) *given as above. Let* $D_1, \ldots, D_m$ *be the irreducible components. We assume that no three of* D_i *'s intersect outside* $\mathscr{E}$ *and each line bundle* $K_{D_i} + \sum_{j \ne i} D_j + \sum E_{j0}$ *is ample on* D_i, *where* $1 \le i \le m$. *Then the complete Kähler-Einstein metrics* $g_{E,t}$ *on* $X_t \backslash \mathscr{E}$ *converge to the unique Kähler-Einstein metric on* $X_0 \backslash (\mathrm{Sing}(X_0) \cup \mathscr{E})$ *in the sense of Cheeger-Gromov (cf. Theorem* 0.1).

The proof of this theorem is identical to that of Theorem 0.1. The only difference is the choice of the relative volume forms $\tilde{V}$ in constructing the family of Kähler metrics g_t on X_t with asympototic behavior. Here we should take $\tilde{V}$ to be of the form

$$\frac{V'}{\|f\|^2 (\log \|f\|^2)^2},$$

where V' is the smooth relative volume form on $\mathscr{X}$, f is the defining section of $\mathscr{E}$ and $\|\cdot\|$ is a proper hermitian metric of the line bundle $[\mathscr{E}]$ on $\mathscr{X}$. The details of the proof is omitted here.

This theorem can be used to deal with some cases of degenerations in which the central fiber has triple points.

REFERENCES

[CY] S. Y. Cheng and S. T. Yau, *Inequality between Chern numbers of singular Kähler surfaces and characterization of orbit space of discrete group of* $SU(2, 1)$, Contemporary Math. **49** (1986), 31–43.

[Ko] R. Kobayashi, *Einstein-Kähler metrics on open algebraic surfaces of general type*, Tôhoku Math. J. **37**(1985), 43–77.

[Ma] H. Masur, *The extension of the Weil-Petersson metric to the boundary of Teichmüller space*, Duke Math. J. **43** no. 3 (1976), 623–635.

[Ts] H. Tsuji, *An inequality of Chern numbers for open varieties*, Math. Ann **277**, no. 3 (1987), 483–487.

[TY] G. Tian and S. T. Yau, *Existence of Kähler-Einstein metrics on complete Kähler manifolds and their applications to algebraic geometry*, Math. Aspects of String Theory (San Diego, California, 1986), pp. 574–628, Adv. Ser. Math. Phys., 1, World Sci. Publishing, Singapore, 1987.

[Y] S. T. Yau, *On the Ricci curvature of a compact Kähler manifold and the complex Monge-Ampére equation.* I[*], Comm. Pure Appl. Math. **31** (1978), 339–441.

STATE UNIVERSITY OF NEW YORK

Proceedings of Symposia in Pure Mathematics
Volume **54** (1993), Part 2

Flat Connections on Products
of Determinant Bundles

YUE LIN L. TONG

1. Introduction

This is an announcement of some results of ongoing work on a problem posed by Beilinson and Schechtman [**BS**, §3]: the cancellation of anomalies in higher dimensions. Let X be complex manifold, and $E \to X$ a holomorphic vector bundle. The Atiyah algebra $\mathscr{A}_E$ is the Lie algebra sheaf of infinitesimal symmetries of E which is the extension of tangent sheaf $\mathscr{T}_X$ by $\mathrm{End}(E)$ (both given obvious Lie brackets),

$$0 \to \mathrm{End}(E) \to \mathscr{A}_E \to \mathscr{T}_X \to 0.$$

Let $\pi : X \to S$ be a proper smooth map of relative dimension n with fibers X_s, $s \in S$, and let $\lambda_E = \det R\pi_* E$ be the determinant bundle associated to the direct image sheaves. When $E = \omega^j$ (where $\omega = \omega_{X/S}$ is the relative canonical bundle) we write $\lambda_j = \lambda_E$.

Let $f_{n+1}^j(c_1, \ldots, c_n)$ be the term of total degree $n+1$ in the series $e^{-jc_1}\mathrm{Td}(c_1, \ldots, c_n)$ where Td is the Todd genus and $\deg c_i = i$. Thus $f_2^j(c_1) = \frac{1}{12}(6j^2 - 6j + 1)c_1^2$. Suppose $a_1, \ldots a_N$, $j_1, \ldots, j_N$ are integers such that $\sum_{i=1}^N a_i f_{n+1}^{j_i} = 0$. Then by Grothendieck Riemann Roch formula
(1)
$$c_1 \left(\bigotimes \lambda_{j_i}^{a_i} \right) = \sum_i a_i c_1(\lambda_{j_i})$$
$$= \sum_i a_i \int_{X_S} \{\mathrm{Td}(c_1(T_{X/S}), \ldots, c_n(T_{X/S}))\exp(-j_i c_1(T_{X/S}))\}_{n+1}$$
$$= 0.$$

1991 *Mathematics Subject Classification.* Primary 32C38, 14C40, 58G10.
Supported in part by NSF Grant No. DMS-8902243.
The detailed version of this paper has been submitted for publication elsewhere.

Cancellation of anomalies is to prove

(2) $\bigotimes_i \lambda_{j_i}^{a_i}$ has a canonical flat connection.

For example when $n = 1$, then $\lambda_j \otimes \lambda_1^{-(6j^2 - 6j + 1)}$ has a canonical flat connection. When $\pi : X \to S$ is a locally Kähler fibration a flat connection on the line bundle $\bigotimes \lambda_{j_i}^{a_i}$ is given by the connection associated to the products of Quillen metrics on λ_{j_i} (using the pointwise form level GRR proved by Bismut, Freed, Gillet, Soule [**BF**, **BGS**]). But this depends on the choice of metric on $T_{X/S}$, and does not yield a canonical one, at least not locally.

The statement (2) is equivalent to $\mathscr{A}_{\otimes \lambda_{j_i}^{a_i}} \cong \mathscr{A}_{\mathscr{O}_S}$ or written additively

(3) $\bigoplus a_i \mathscr{A}_{\lambda_{j_i}} \cong \mathscr{A}_{\mathscr{O}_S}.$

Let $\mathscr{A}_{E/S} \subset \mathscr{A}_E$ be the subalgebra of relative vector fields (tangent along fibers). $\mathscr{A}_{E,\pi} \subset \mathscr{A}_E$ the subalgebra of "projectable" vector fields and $\mathscr{A}_E^i$, $i = -1, 0$ the complex $\mathscr{A}_{E/S} \hookrightarrow \mathscr{A}_{E,\pi}$. There is a triangle in derived category [**BS**, §2.8], $\omega[n] \to \mathscr{K}_E^{\cdot} \to \mathscr{A}_E^{\cdot}$ such that when applied the direct image functor $(R^0 \pi)_*$ one gets $0 \to \mathscr{O}_S \to \mathscr{A}_{\lambda_E} \to \mathscr{T}_S \to 0$. the main point then is to find a concrete realization of $\mathscr{K}_E^{\cdot}$ in terms of which the structure of $\mathscr{A}_{\lambda_E}$ may be deduced via

(4) $R^0 \pi_*(\mathscr{K}_E^{\cdot}) \cong \mathscr{A}_{\lambda_E}.$

This is carried out for $n = 1$ in [**BS**]. In this case Beilinson and Schechtman prove much more than (3). Their model of $\mathscr{K}_E^{\cdot}$, the trace algebra, also contains a sheaf of Virasoro algebra and (3), (4) give the action of Virasoro algebra on λ_j. Furthermore via Tate residues and formal moduli spaces of triples (curve, point, parameter) this action has an embedding in Sato's universal Grassmannian. Parts of this beautiful picture were also observed by other authors in the papers [**ADKP**, **KNTY**, **K**]. Motivated partly by [**S**] I shall describe here a construction of $\mathscr{K}_E^{\cdot}$ and an approach to (3) for $n = 2$. The methods should no doubt go through for $n > 2$ as well, when the appropriate technical details are filled in. It is much less clear what happens to the rest of the above picture when $n > 1$. The Tate residues have a generalization found by Beilinson [**Be**], but so far appropriate generalization of Sato's Grassmannian seems to be missing [**Sa1,2**].

Finally Beilinson and Schechtman in [**BS**, §5] also take a C^∞ resolution of $\mathscr{K}_E^{\cdot}$ and represent $C^\infty(\mathscr{A}_{\lambda_E})$ by C^∞ objects on X via (4). In this way they describe a C^∞ connection on λ_E in terms of a relative parametrix, and they prove the pointwise GRR (for $n = 1$) using this connection. The techniques in [TT1] are suited to generalizing the relative parametrix. I hope to discuss this elsewhere.[1]

2. Construction of $\mathcal{K}_E^{\cdot}$ and some properties

In $\mathbb{C}^2 \times \mathbb{C}^2$ let $R = \mathcal{O}_{\mathbb{C}^2 \times \mathbb{C}^2}$ and $\xi_i = \zeta_i - z_i$, $i = 1, 2$ be the generators of $\mathcal{I}_\Delta$. A resolution of the coherent sheaf $R/\mathcal{I}_\Delta^2$ is given by the complex (cf. [S, §5]).

$$(5) \qquad K_{\cdot} = (K_2 \to K_1 \to K_0) = (R^2 \xrightarrow{d_2} R^3 \xrightarrow{d_1} R)$$

where

$$d_2 = \begin{pmatrix} \xi_2 & 0 \\ \xi_1 & -\xi_2 \\ 0 & -\xi_1 \end{pmatrix}, \qquad d_1 = (\xi_1^2, -\xi_1\xi_2, \xi_2^2).$$

Similar formulae are known for any dimensions and for $R/(\mathcal{I}_\Delta)^k$ by using the Eagon Northcott complex. The adjoint complex of (5) resolves the sheaf of holomorphic differential operators of order ≤ 1 on $\mathbb{C}^2$. Let $K^{-i} = \mathrm{Hom}(K_{2-i}, R)$ and $\Omega^{0,2}$ the holomorphic two forms with differentials in the second factor,

$$(6) \qquad 0 \to K^{-2} \otimes \Omega^{0,2} \xrightarrow{d_{-2}} K^{-1} \otimes \Omega^{0,2} \xrightarrow{d_{-1}} K^0 \otimes \Omega^{0,2} \xrightarrow{\mathcal{R}} \mathcal{D}^{\leq 1} \to 0$$

where given $a = (a_1, a_2) d\zeta_1 \wedge d\zeta_2 \in K^0 \otimes \Omega^{0,2}$,

$$(\mathcal{R}a)(f) = \mathrm{Res}_\Delta \left\{ \frac{a_1 f(\zeta)}{(\zeta_1 - z_1)^2(\zeta_2 - z_2)} d\zeta + \frac{a_2 f(\zeta)}{(\zeta_1 - z_1)(\zeta_2 - z_2)^2} d\zeta \right\}.$$

Let

$$D_i \varphi = \int_0^1 \frac{\partial \varphi}{\partial \zeta_i}(z, z + t(\zeta - z)) \, dt$$

and

$$t D_i \varphi = \int_0^1 t \frac{\partial \varphi}{\partial \zeta_i}(z, z + t(\zeta - z)) \, dt$$

then $\mathbb{C}$ linear chain homotopies of the complex $K^{\cdot}$ are given by

$$(7) \qquad P_0 = \begin{pmatrix} D_2 + \xi_1 D_1 D_2 & \xi_1 D_2^2 \\ \xi_1 D_1^2 & -\xi_2 D_2^2 \\ -\xi_2 D_1^2 & -D_1 - \xi_2 D_1 D_2 \end{pmatrix} : K^0 \to K^{-1}$$

$$P_{-1} = ((tD_1)^2, -2tD_1 tD_2, (tD_2)^2) : K^{-1} \to K^{-2}.$$

Let $\tilde{K}^{-2} = K^{-2}/\mathcal{I}_\Delta$, $\tilde{K}^{-1} = K^{-1}/\mathcal{I}_\Delta^3$, $\tilde{K}^0 = K^0/\mathcal{I}_\Delta^4$ then $\tilde{K}^{-2} \otimes \Omega^{0,2} = \omega$ the canonical sheaf on $\mathbb{C}^2$, and $\tilde{K}^i$ ($i = 0, -1$) is an extension (of length 2) of $\mathcal{D}(\mathcal{O})^{\leq 1}$ by ω. P_i are differential operators on $\tilde{K}^{\cdot}$.

---

[1]Note added in revision: Recently I have found a generalization of the Beilinson-Schechtman connection for determinant bundles in higher relative dimensions, and obtained a pointwise GRR without Kahler hypothesis. In particular the results described in §2, except the explicit calculation of twisted brackets, have been generalized to higher dimensions.

LEMMA 1. (i) $d_{-1}P_0 = 1$ on $\ker \mathscr{R}$.

(ii) $P_0 d_{-1} + d_{-2}P_{-1} = 1$, $P_{-1}d_{-2} = 1$.

(iii) $P_{-1}P_0 = 0$.

Now suppose $\pi: X \to S$ has relative dimension 2 and $E \to X$ a holomorphic vector bundle. Let $V \subset S$ be an open set and $\mathscr{U} = \{U_\alpha\}$ a Stein coordinate covering of $\pi^{-1}(V)$. We have local resolutions as in (5),

$$0 \to K_\cdot(U_\alpha \times_V U_\alpha) \to \mathscr{O}_{U_\alpha \times_V U_\alpha}/\mathscr{I}_\Delta^2 \to 0.$$

and their adjoint complexes with coefficients in $E \boxtimes E^*$ which resolves (here $E' = E^* \otimes \omega$),

$$(8) \qquad 0 \to K_\alpha^\cdot \otimes E \boxtimes E' \xrightarrow{\mathscr{R}} \mathscr{D}(E)_S^{\leq 1}|_{U_\alpha} \to 0$$

where $\mathscr{D}(E)_S$ is the sheaf of relative differential operators on E. From this one gets readily an exact sequence

$$(9) \qquad 0 \to \widetilde{K}_\alpha^\cdot(E) \xrightarrow{\mathscr{R}} \mathscr{A}_{E/S}|_{U_\alpha} \to 0$$

(cf. [BS, §2) where $\widetilde{K}_\alpha^{-2}(E) = \omega$.

These complexes are of course only locally defined. To "patch" them up we use the technique of twisted complexes [TT2, OTT]. This involves the construction of

$$a = a^{0,1} + a^{1,0} + a^{2,-1} \quad \text{where } a^{i,j} \in C^i(\mathscr{U} \times_V \mathscr{U}, \text{Hom}(K^\cdot, K^{\cdot+j}))$$

satisfying $\hat\delta a + a \cdot a = 0$. [TT2, §2]. Here $a_\alpha^{0,1}$ is the differential in $K_\alpha^\cdot$. To define $a_{\alpha\beta}^{1,0}$ let $z_\beta = \varphi_{\alpha\beta}(z_\alpha)$ be the coordinate transformation, and

$$A_{\alpha\beta}(z_\alpha, \zeta_\alpha) = \left(\int_0^1 \frac{\partial \varphi_i}{\partial z_j}(z_\alpha + t(\zeta_\alpha - z_\alpha)) dt \right) = (a_{ij})$$

then

$$(10) \qquad a_{\alpha\beta}^{1,0} = (\det A_{\alpha\beta}^{-1})A_{\alpha\beta}^{-1} : K_\beta^0 \to K_\alpha^0, \qquad a_{\alpha\beta}^{1,0} : K_\beta^{-1} \to K_\alpha^{-1}$$

is given by

$$\begin{pmatrix} a_{11}^2 & -2a_{11}a_{12} & a_{12}^2 \\ -a_{11}a_{21} & a_{11}a_{22} + a_{12}a_{21} & -a_{12}a_{22} \\ a_{21}^2 & -2a_{21}a_{22} & a_{22}^2 \end{pmatrix}$$

$$a_{\alpha\beta}^{1,0} = 1 : K_\beta^{-2} \to K_\alpha^{-2}.$$

LEMMA 2. $a_{\alpha\beta}^{1,0}$ is a chain map $K_\beta^\cdot \to K_\alpha^\cdot$. If $\varphi_{\alpha\beta}$ is affine then $a_{\alpha\beta}^{1,0}$ also commutes with the homotopy operators $(P_\cdot)_\beta$ and $(P_\cdot)_\alpha$.

Next $a_{\alpha\beta\gamma}^{2,-1}$ is a chain homotopy between $a_{\alpha\gamma}^{1,0}$ and $a_{\alpha\beta}^{1,0}a_{\beta\gamma}^{1,0}$. The construction of $a^{2,-1}$ depends on the homotopy operators $P_\cdot$, or their analogues for $K_\cdot$, the details here are similar to that of [TT2].

Let $(\tilde{K}^{\cdot}(E), a)$ be this "twisted" extension class of $\mathscr{A}_{E/S}$ by ω. We now define $\mathscr{K}_E^{\cdot}$ to be

(11)
$$\mathscr{K}_E^i \begin{cases} \mathscr{A}_{E,\pi}, & i = 0, \\ (\tilde{K}^{i+1}(E), a), & -2 \le i \le -1, \end{cases}$$

$\hat{\delta} + a$ defines a total differential on the bigraded complex $\bigoplus C^{\cdot}(\mathscr{U} \times_V \mathscr{U}, \mathscr{K}_E^{\cdot})$ whose zeroth cohomology is just $R^0 \pi_*(\mathscr{K}_E^{\cdot})|_V$. The Lie brackets on $\mathscr{D}(E)_S^{\le 1}$ or $\mathscr{A}_{E/S}$ can be first extended to a differential graded Lie bracket on $K^{\cdot} \otimes E \boxtimes E'$ or $\tilde{K}^{\cdot}(E)$ in (8) and (9), the bracket thus being compatible with $a^{0,1}$. To do this, $K_\alpha^{\cdot}$ can be identified with a local Čech complex built on the coordinate covering $\{\zeta_{\alpha_i} - z_{\alpha_i} \ne 0\}$. (Cf. [**TT1**, §7]) then $a^{0,1}$ is identified with sums of restriction maps. To have this bracket defined on cocycles for $\hat{\delta} + a$ we need to introduce correction terms and this leads to a twisted bracket. This descends to a bracket on $R^0 \pi_*(\mathscr{K}_E^{\cdot})$.

$\mathscr{K}_E^i$, $-2 \le i \le -1$, being a twisted extension class of $\mathscr{A}_{E/S}$ by ω, it defines a canonical element of

$$\text{Ext}^2((\pi \times \pi^{-1})(V); \Delta_* \mathscr{A}_{E/S}, \omega).$$

This element can also be calculated by starting with local splittings of $\mathscr{R}$ in (9) and completing to a total cocycle τ_E in a total complex using the operators $P_{\cdot}$. In the case $n = 1$, this is just the calculation in [**BS**, §3.1, p. 669]. The calculation of τ_E can be reduced to the absolute case $S = \text{pt.}$ where it is equivalent to the following question: given a vector field v acting on $E \to X$, it induces actions $v^i : H^i(X, E) \to H^i(X, E)$. What is a global cohomology formula for the Lefschetz number $L(X, E, v) = \Sigma (-1)^i \text{tr} \, v^i$? Local calculations with the operators $P_{\cdot}$ show that τ_E depends on three derivatives of transition functions. This makes it difficult to recognize a formula in Chern classes (in the component of τ_E with top Čech degree). This problem already occurs when $n = 1$, and Beilinson-Schechtman introduced a Čech coboundary term, or equivalently, a judicious choice of splitting in (9), to cancel out the third derivative term. Alternatively one can work in the Zariski topology and the sheaf of closed forms modulo exact forms. It appears to be a tedious matter to try to find the coboundaries or exact forms in general for $n > 1$. It turns out that nonetheless a simple formula exists in general for $L(X, E, v)$.

Let $(L_E)_\alpha = v_\alpha - i_{v_\alpha} d_\alpha$ be the endomorphism of $E|_{U_\alpha}$, where d_α is the local trivial holomorphic connection. Let $\theta(E)_{\alpha\beta}$ be the Atiyah obstruction cocycle, then $(L_E)_\alpha + \theta(E)_{\alpha\beta}$ is a cocycle in an appropriate Čech complex [**CL**]. Similarly we have $L_T + \theta(T)$ for the tangent bundle T.

PROPOSITION.

(12)
$$L(X, E, v) = \sum_{i+j=n+1} \text{Td}_i((L_T + \theta_T)) ch_j((L_E + \theta_E))[X].$$

Note that the formula involves homogeneous polynomials of total degree $n + 1$! If now $E = \omega^j$ then one gets exactly the f_{n+1}^j preceding (1). This is a key step to (3). For $n = 1$, the formula in (12) is equivalent to the formula in [**BS**, p. 669 preceding 3.1.1]. Various special cases of the formula (12) are known. If v has isolated nondegenerate zeros (12) can be derived from Atiyah Bott fixed point formula and techniques of Bott's vector field formula [**BV, Bt**]. If v is a Killing vector field preserving a Kähler metric (12) can be deduced from a formula of Bismut [**Bi**].

For $n = 2$ (12) may be obtained by first establishing a formula for the Lefschetz number of the one parameter family of maps e^{tv}, and then differentiate in t, evaluate at $t = 0$. For Lefschetz number of a map the theory developed in [**TT2,3, OB**] can be employed, except that we do not want to localize on the fixed point set. We have the Grothendieck dual class of $\Delta \hookrightarrow X \times X$ in $\text{Ext}^n(X \times X; \mathcal{O}_\Delta, \Omega^{0,n})$, and the Lefschetz number is given by its image in the maps: (Z is zero set of v),

$$\text{Ext}^n(X \times X; \mathcal{O}_\Delta, \Omega^{0,n}) \xrightarrow{(1, e^{tv})^*} \text{Ext}^n(X; \mathcal{O}_Z, \Omega_X^n) \to H^n(X, \Omega_X^n).$$

Finally a remark about the isomorphism (4). To directly construct this isomorphism it is necessary to define an action of $R^0\pi_*(\mathcal{K}_E^\cdot)$ on λ_E. In the case $n = 1$, this is defined via the Cousin resolution [**BS**, §2]. In our case $n > 1$ this can be defined by using techniques of a $\overline{\partial}$ parametrix [**TT1**]. Thus one uses $\overline{\partial}$ representative of cohomology along fibers, and to a total cocycle in $R^0\pi_*(\mathcal{K}_E^\cdot)$ one associates global C^∞ kernel functions which act on the cohomology.

REFERENCES

[ADKP] E. Arbarello, C. DeConcini, V. Kac and C. Procesi, *Moduli space of curves and representation theory*, Comment. Math. Phys. **117** (1988), 1-36.

[Be] A. Beilinson, *Residues and Adeles*, Funct. Anal. Appl. **14** (1) (1980), 34-35.

[BS] A. Beilinson, V. Schechtman, *Determinant bundles and Virasoro algebras*, Comment. Math. Phys. **118** (1988), 651-701.

[BV] N. Berline and M. Vergne, *Zeros d'un champ de vecteurs et classes caractéristiques equivariantes*, Duke Math. J. **50** (1983), 539-549.

[Bi] J.M. Bismut, *The infinitesimal Lefschetz formulas, A heat equation proof*, J. Funct. Anal. **62** (1985), 435-457.

[BF] J.M. Bismut and D.S. Freed, *The analysis of elliptic families*, Comment. Math. Phys. **106** (1986), 159-176; **107** (1986), 103–163.

[BGS] J.M. Bismut, H. Gillet and C. Soule, *Analytic torsion and holomorphic determinant bundles*, Comment. Math. Phys. **115** (1988), 49-126, 301-351.

[Bt] R. Bott, *Vector fields and characteritic numbers*, Michigan. Math. J. **14** (1967), 231-244.

[CL] J.B. Carrell and D.I. Lieberman, *Vector fields and Chern numbers*, Math. Ann. **225** (1977), 263-273.

[KNTY] N. Kawamoto, Y. Namikawa, A. Tsuchiya, and Y. Yamada, *Geometric realization of conformal field theory on Riemann surfaces*, Comment. Math. Phys. **116** (1988), 247-308.

[K] M.L. Kontsevich, *Virasoro algebra and Teichmüller spaces*, Fund. Anal. Appl. **21** (2) (1987), 78-79.

[OB] N.R. O'Brian, *Zeros of holomorphic vector fields and Grothendieck dualilty theory*, Trans. Amer. Math. Soc. **229** (1977), 289-306.

[OTT] N. R. O'Brian, D. Toledo, and Y. L. Tong, *A Grothendieck Riemann Roch formula for maps of complex manifolds*, Math. Ann. **271** (1985), 493-526.

[Sa1] M. Sato, *The KP hierarchy and infinite-dimensional Grassmann manaifolds*, Proc. Sympos. Pure Math., vol. 49, Amer. Math. Soc., Providence, R.I., 1989, **49** Part I, pp. 51–66.

[Sa2] _____, *$\mathscr{D}$-modules and nonlinear integrable systems*, Proc. JAMI Inaugural Conference, Suppl. to Amer. J. Math. (1989), 325-339.

[S] V.V. Schechtman, *Riemann Roch theorem after D. Toledo and Y.L. Tong*, Proc. of Winter School on Geometry and Physics, Rend. Cir. Mat. Palermo (2), Supp. 1989, no. 21.

[TT1] D.Toledo and Y.L. Tong, *A parametrix for $\bar{\partial}$ and Riemann Roch in Čech theory*, Topology **15** (1976), 273-301.

[TT2] _____, *Duality and intersection theory in complex manifolds*. I, Math. Ann. **237** (1978), 41-77.

[TT3] _____, *Duality and intersection theory in complex manifolds*. II, Ann. of Math. (2) **108** (1978), 519-538.

PURDUE UNIVERSITY

Proceedings of Symposia in Pure Mathematics
Volume **54** (1993), Part 2

Surfaces Riemanniennes à Singularités Simples

MARC TROYANOV

I. Description locale

Parmi les métriques singulières sur les surfaces, les plus simples sont celles ayant des singularités isolées. En dehors d'un ensemble discret, ces métriques sont donc lisses (disons de classe C^2). Nous ferons encore deux hypothèses qui sont naturelles d'un point de vue géométrique.

La première concerne la structure conforme. Si g est une métrique sur une surface S ayant une singularité isolée en p, et si U est un voisinage de p homéomorphe au disque, alors $U_0 := U - \{p\}$ possède une structure conforme bien définie (g étant lisse sur U_0, on peut appliquer le théorème de Korn-Lichtenstein). Par la classification des structures conformes sur l'anneau, on sait alors que U_0 est conformément équivalent à

$$A_\rho := \{z \in \mathbf{C} : \rho < |z| < 1\}$$

pour un certain $\rho \in [0, 1[$.

Nous supposerons toujours que U_0 est conformément équivalent à A_0, en d'autres termes; nous supposerons que la structure conforme de U_0 s'étend à U. En particulier, tout point de S (singulier ou non) possède un voisinage admettant des coordonnées (x, y) dans lesquelles la métrique s'écrit

$$g = \rho(x, y)(dx^2 + dy^2) = \rho(z)|dz|^2$$

($z = x + iy$) où ρ est une fonction positive et de classe C^2 en dehors des points singuliers. Ces coordonnées sont appelées "coordonnées isothermes."

CONTRE-EXEMPLE. Soit $\phi \geq 0$ une fonction sur $\mathbf{C}$ s'annulant exactement sur un ensemble Q connexe et de capacité positive (par exemple un ensemble contenant un segment). Notons $\mathbf{C}/Q$ l'espace obtenu en identifiant tous les points de Q et $g := \phi^2(z)|dz|^2$. Alors $(\mathbf{C}/Q, g)$ possède un point singulier $q = [Q]$ ne vérifiant pas l'hypothèse ci-dessus.

1991 *Mathematics Subject Classification.* Primary 53C20, 52A55.

This paper is in final form and no version of it will be submitted for publication elsewhere.

La seconde hypothèse concerne la courbure. Elle dit simplement que si K désigne la courbure de g et dA l'élément d'aire, alors

$$\int_{U_0} |K|\,dA < \infty.$$

Une classe importante de singularités vérifiant les deux hypothèses ci-dessus est donnée par les "singularités simples."

DÉFINITION. Soit S une surface de Riemann, une métrique strictement conforme g possède une *singularité simple* d'ordre β en $p \in S$ si on peut écrire localement

$$g = e^{2u(z)}|z|^{2\beta}|dz|^2$$

où β est un réel et u est une fonction vérifiant:

$$u \in L^1 \quad \text{et} \quad \Delta u \in L^1.$$

Dans cette définition, $z = x + iy$ est une coordonnée locale sur S définie dans un voisinage U de p et telle que $z(p) = 0$. L'espace L^1 est défini à partir de la mesure de Lebesgue $dx\,dy$ sur U. Le laplacien de u est défini au sens des distributions par $\Delta u = -\frac{\partial^2 u}{\partial x^2} - \frac{\partial^2 u}{\partial y^2}$.

Les singularités simples apparaissent naturellement dans de nombreux contextes comme nous le verrons au prochain paragraphe. Une premiere famille d'exemples est donnée par le résultat suivant du à MacOwen [**McO**, Appendice B]:

THÉORÈME. *Soit* $g = e^{2u}|dz|^2$ *une métrique conforme sur le disque unité* $D = \{z : |z| < 1\}$ *ayant une singularité en* 0. *Supposons la métrique lisse sur* $D^* = \{z : 0 < |z| < 1\}$.

S'il existe $\ell \in \mathbf{R}$ *et* $a, b > 0$ *tels que la courbure* K *de* g *satisfasse*

$$-b|z|^\ell \le K(z) \le -a|z|^\ell,$$

alors 0 *est une singularité simple pour* g.

Une singularité simple d'ordre $\beta < -1$ est toujours à distance infinie alors qu'une singularité simple d'ordre $\beta > -1$ est toujours à distance finie. Pour une singularité d'ordre -1, les deux cas peuvent se produire (cf. [**HT2**,§2.2]). Un *cusp* est une singularité simple à distance infinie admettant un voisinage d'aire finie. Un cusp est toujours d'ordre $\beta = -1$.

Une singularité simple d'ordre $\beta > -1$ est aussi appelée une *singularité conique* d'angle $\theta = 2\pi(\beta + 1)$. Une telle singularité peut en effet être approximée par un cône euclidien d'angle total θ. En particulier, lorsque la courbure de g reste bornée au voisinage d'une singularité conique, il existe une application exponentielle permettant de paramétriser un voisinage de la singularité conique par un voisinage du sommet de son cône tangent. En d'autres termes, il existe des coordonnées polaires. On montre de plus, que si la courbure est continue, alors ces coordonnées polaires sont de classe C^1 par rapport aux coordonnées isothermes (voir [T3]).

II. Description globale

Pour étudier une métrique ayant plusieurs singularités simples sur une surface, il est commode d'introduire la notion de diviseur:

DÉFINITIONS. Soit S une surface de Riemann. Un *diviseur* sur S est une somme formelle $\boldsymbol{\beta} = \sum_{i=1}^{n} \beta_i p_i$, le *support* de ce diviseur est l'ensemble supp$(\boldsymbol{\beta}) = \{p_i\}$. Une métrique conforme g sur S *représente le diviseur* $\boldsymbol{\beta}$ si elle est lisse en dehors de supp$(\boldsymbol{\beta})$ et si elle possède en p_i une singularité simple d'ordre β_i.

EXEMPLES. 1) La métrique $g = |dz|^2$ sur $\mathbf{C} \cup \infty$ représente le diviseur $\boldsymbol{\beta} = (-2) \cdot \infty$.

2) Plus généralement, la métrique $g = |z|^{2\alpha}|dz|^2$ sur $\mathbf{C} \cup \infty$ représente le diviseur $\boldsymbol{\beta} = \alpha \cdot 0 + (-2 - \alpha) \cdot \infty$.

3) Si $\omega = \varphi(z)dz$ est une différentielle méromorphe sur la surface de Riemann S, alors $g = |\omega|^2$ représente le diviseur div(ω).

4) Soit (S_1, g_1) une surface riemannienne lisse et $f : S \to S_1$ un revêtement ramifié. Alors $g := f^*(g_1)$ est une métrique sur S représentant le diviseur de ramification de f, à savoir $\boldsymbol{\beta} = \sum O_p(f) \cdot p$ (où $O_p(f)$ est l'ordre de ramification de f en p).

5) Si S est un polyèdre bidimensionnel (euclidien, hyperbolique ou sphérique) de sommets $p_1, p_2, \ldots, p_n$, alors la métrique induite par la réalisation géométrique de ce polyèdre représente le diviseur $\boldsymbol{\beta} = \sum \beta_i p_i$ où $2\pi(\beta_i + 1) = \theta_i$ est la somme des angles en p_i des faces incidentes à p_i.

6) Soit $(\widetilde{S}, \check{g})$ une surface riemannienne sur laquelle opère par isométries un groupe fini Γ tel que $S = \widetilde{S}/\Gamma$ est une surface sans bord. Alors S hérite d'une métrique g représentant le diviseur $\boldsymbol{\beta} = \sum \beta_i p_i$ où $\beta_i = (\frac{1}{n_i} - 1)$ si p_i est l'image d'un point $\tilde{p}_i \in \widetilde{S}$ tel que Γ ait en $\tilde{p}_i$ un stabilisateur d'ordre n_i. Cet exemple se généralise aux orbifolds.

Dans les exemples 4 à 6, tous les points singuliers sont des points coniques. Une source importante d'exemples (où aucun point singulier n'est conique) est donnée par le théorème d'Alfred Huber:

THÉORÈME DE HUBER (cf. [HT2, 1.1 et 2.9]). *Soit (S', g') une surface riemannienne complète de classe C^2 à courbure totale finie: $\int_{S'} |K| \, dA < \infty$.*

Alors il existe une surface de Riemann compacte S, un diviseur $\boldsymbol{\beta} = \sum_{i=1}^{n} \beta_i p_i$ sur S (tel que $\beta_i \leq -1$ pour tout i) et une métrique conforme g sur S représentant ce diviseur tels que (S', g') soit isométrique à $(S \setminus \text{supp}(\boldsymbol{\beta}), g)$.

Il est facile de voir, sous les hypothèses du théorème de Huber, que si l'on suppose de plus que l'aire de la surface est finie, alors $\beta_i = -1$ pour tout i.

III. Un peu de géométrie globale

Pour les surfaces riemanniennes compactes ayant des singularités simples, il existe une formule de Gauss-Bonnet due à R. Finn. Nous définissons pour cela la caractéristique d'Euler d'une surface S munie d'un diviseur $\boldsymbol{\beta} = \sum \beta_i p_i$ par $\chi(S, \boldsymbol{\beta}) := \chi(S) + \sum \beta_i$.

THÉORÈME (Formule de Gauss-Bonnet, cf. [HT2,2.8]). *Soit* (S, g) *une surface riemannienne compacte dont la métrique représente un diviseur* $\boldsymbol{\beta}$. *Alors la courbure totale de* (S, g) *est finie et l'on a*

$$\frac{1}{2\pi} \int_S K \, dA = \chi(S, \boldsymbol{\beta}).$$

Par exemple, si ω est une différentielle méromorphe sur une surface de Riemann compacte de genre γ, alors le degré de ω (= nombre de zéros - nombre de pôles) est égale à $2\gamma - 2$ (car $g = |\omega|^2$ est une métrique plate représentant le diviseur $\mathrm{div}(\omega)$). Une autre application de la formule de Gauss-Bonnet est la formule de Riemann-Hurwitz:

PROPOSITION (Formule de Riemann-Hurwitz). *Si* $f : S \to S_1$ *est un revêtement ramifié entre deux surfaces compactes sans bord, alors*

$$\chi(S) + \sum_p O_p(f) = n \, \chi(S_1) \, ,$$

où n *est le nombre de feuillets (le degré) du revêtement* f *et* $O_p(f)$ *est l'ordre de ramification de* f *en* p.

PREUVE. Choisissons une métrique lisse quelconque g_1 sur S_1, alors $g := f^*(g_1)$ est une métrique sur S représentant le diviseur de ramification $\boldsymbol{\beta} = \sum_p O_p(f) p$. Il est clair que l'on a

$$\int_S K \, dA = n \int_{S_1} K_1 \, dA_1 \, .$$

D'où le résultat, à partir de la formule de Gauss-Bonnet.   □

Le théorème de Huber, avec la formule de Gauss-Bonnet, est un raffinement de l'inégalité de Cohn-Vossen:

PROPOSITION (Cohn-Vossen). *Soit* (S', g') *une surface riemannienne complète de classe* C^2 *à courbure totale finie:* $\int_{S'} |K| \, dA < \infty$. *Alors on a*

$$\frac{1}{2\pi} \int_{S'} K \, dA \leq \chi(S').$$

De plus, lorsque (S', g) *est d'aire finie, alors on a égalité.*

En effet, le théorème de Huber nous dit que (S', g') admet une compactification (S, g) où g est une métrique représentant un diviseur $\boldsymbol{\beta} = \sum_{i=1}^n \beta_i p_i$

tel que $\beta_i \leq -1$ pour tout i. On a donc d'après la formule de Gauss-Bonnet,

$$\frac{1}{2\pi}\int_{S'} K' \, dA' = \frac{1}{2\pi}\int_S K \, dA = \chi(S) + \sum_{i=1}^n \beta_i$$

$$\leq \chi(S) - n = \chi(S').$$

Si l'aire de (S, g) est finie, alors on a $\beta_i = -1$ pour tout i et l'inégalité ci-dessus devient une égalité.

Observons toutefois qu'il convient de considérer la formule de Gauss-Bonnet comme un complément à l'inégalité de Cohn-Vossen et non l'inégalité de Cohn-Vossen comme un corollaire des théorèmes de Huber et Gauss-Bonnet. En effet la démonstration du théorème de Huber s'appuie sur l'inégalité de Cohn-Vossen.

Le resultat suivant montre que la difference entre $\chi(S')$ et $\chi(S, \boldsymbol{\beta})$ est une constante isoperimetrique:

THÉORÈME. *Soit* (S, g) *une surface riemannienne compacte dont la métrique* g *represente une diviseur* $\boldsymbol{\beta} = \sum_{i=1}^n \beta_i p_i$ *tel que* $\beta_i \leq -1$ *pour tout* i.

Si q *est un point de* $S' = S \setminus \{p_i\}$, *on note* $A(q, r)$ *l'aire de* $B_{q,r} := \{x \in S' : d(q, x) \leq r\}$ *et* $L(q, r)$ *la longueur de* $\partial B_{q,r}$. *Alors*

$$\lim_{r\to\infty} \frac{L^2(q, r)}{4\pi A(q, r)} = -\sum_{i=1}^n (\beta_i + 1) = \chi(S') - \chi(S, \boldsymbol{\beta}).$$

Ce théorème est dû à K. Shiohama [S], mais R. Finn avait obtenu un resultat partiel dans cette direction [F, §6, th. 10].

IV. Classification des métriques plates

Donnons à présent le théorème de classification des métriques plates à singularités simples sur une surface compacte.

THÉORÈME. *Soit* S *une surface de Riemann compacte munie d'un diviseur* $\boldsymbol{\beta} = \sum \beta_i p_i$. *Alors il existe une métrique plate conforme représentant* $\boldsymbol{\beta}$ *sur* S *si et seulement si* $\chi(S, \boldsymbol{\beta}) = 0$. *De plus cette métrique est unique à homothétie près.*

Ce théorème possède deux démonstrations (voir [T1] et [HT2, §7]). Nous présentons ici la démonstration contenue dans [HT2]:

PREUVE. Donnons-nous au voisinage de chacun des p_i une coordonnée z_i telle que $z_i(p_i) = 0$. Donnons-nous également une métrique conforme et lisse g_0 sur S telle que $g_0 = |dz_i|^2$ au voisinage de chacun des p_i.

Choisissons ensuite une fonction $\rho : S \to \mathbf{R}$ qui soit C^2 et positive sur $S \setminus \mathrm{supp}(\boldsymbol{\beta})$, et telle qu'au voisinage de p_i, $\rho(z) = |z_i|^{2\beta_i}$. Posons $g_1 := \rho \cdot g_0$; alors g_1 est une métrique conforme représentant $\boldsymbol{\beta}$.

Cherchons la métrique g sous la forme $g = e^{2u}g_1$. Si u est une fonction différentiable vérifiant l'équation

$$(\mathrm{E}_1) \qquad\qquad\qquad \Delta_1 u = -K_1$$

où Δ_1 et K_1 désignent le laplacien et la courbure de g_1, alors $g = e^{2u} g_1$ est une métrique conforme plate et représentant $\boldsymbol{\beta}$ sur S.

Comme Δ_1 est un opérateur singulier, on préfère écrire cette équation sous la forme

$$(\text{E}_0) \qquad\qquad\qquad \Delta_0 u = -\rho K_1$$

où Δ_0 est le laplacien de la métrique lisse g_0. Remarquons que K_1 est nulle ou voisinage des p_i et que ρ et K_1 sont de classe C^2 sur $S \setminus \text{supp}(\boldsymbol{\beta})$. Le second membre de (E_0) est donc une fonction de classe C^2.

On sait par conséquent que cette équation à une solution si et seulement si l'intégrale du second membre s'annule. Cela nous est garanti par Gauss-Bonnet :

$$\int_S K_1 \rho \, dA_0 = \int_S K_1 \, dA_1 = 2\pi\chi(S, \boldsymbol{\beta}) = 0.$$

On a donc montré l'existence. L'unicité découle du fait que si g et g' sont deux telles métriques, alors $g' = e^{2v} g$ pour une fonction *harmonique* v sur S. Cette fonction est constante (car S est compacte) et les deux métriques sont donc homothétiques.   □

REMARQUES. (1) A priori, la métrique donnée par la démonstration ci-dessus est seulement de classe $C^{1,\delta}$, mais comme cette métrique est de courbure nulle, elle est en fait de classe C^ω (en coordonnées isothermes) en dehors des singularités.

(2) Le théorème ci-dessus nous donne une démonstration élémentaire du théorème d'Uniformisation de Riemann pour la sphère:

THÉORÈME (d'uniformisation). *Soit S une surface de Riemann homéomorphe à une sphère. Alors S est conformément équivalente à $\mathbf{C} \cup \{\infty\}$.*

PREUVE. Choisissons un point $p \in S$ et considérons le diviseur $\boldsymbol{\beta} = (-2) \cdot p$. Observons que $\chi(S, \boldsymbol{\beta}) = 2 - 2 = 0$, le théorème précédent nous dit par conséquent qu'il existe une métrique conforme et plate g sur S représentant ce diviseur. Il est clair que (S, g) est isométrique (et donc conformément équivalente) à $(\mathbf{C} \cup \{\infty\}, |dz|^2)$.   □

Ce théorème d'uniformisation peut s'interpréter en disant qu'il existe sur S une métrique conforme (et lisse) de courbure constante (positive). Toutefois, il semble être beaucoup plus difficile de démontrer ce théorème en résolvant directement le problème de Berger-Nirenberg correspondant (i.e. en construisant directement une métrique conforme de courbure $+1$).

V. Le problème de M. S. Berger et L. Nirenberg
sur les surfaces à diviseurs

Le problème de Berger-Nirenberg est le suivant:

PROBLÈME I. *Soient S une surface de Riemann et $K : S \to \mathbf{R}$ une fonction sur cette surface. Existe-t-il une métrique conforme sur S dont la courbure est la fonction K? Cette métrique est-elle unique?*

Il est clair que ce problème est mal posé pour les surfaces ouvertes. On peut imaginer que ce problème est bien posé si l'on s'impose de n'étudier que des métriques complètes. Toutefois, il est facile de construire des familles $\{g_\lambda\}$ de métriques sur une surface de Riemann qui sont complètes, conformes, *de même courbure* et dont la géométrie à l'infini varie avec λ (voir l'exemple dans [HT1]).

Les discussions précédentes (en particulier le théorème de Huber) nous amènent à remplacer le problème de Berger-Nirenberg sur les surfaces ouvertes par un problème sur les surfaces compactes munies d'un diviseur.

PROBLÈME II. *Soit* $(S, \boldsymbol{\beta})$ *une surface de Riemann compacte munie d'un diviseur et d'une fonction* $K : S \to \mathbf{R}$. *Existe-t-il une métrique* g *sur* S *qui soit conforme, de courbure* K *et représentant* $\boldsymbol{\beta}$? *Cette métrique est-elle unique?*

Nous avons déjà répondu à cette question lorsque K est identiquement nulle.

Cette question est étudiée dans les articles [T2] (dans le cas des singularités coniques) et [HT2] dans le cas général. Les résultats peuvent êtres synthétisés sous une forme qui est similaire à la théorie classique dans le cas lisse (telle qu'on la trouve par exemple dans l'article de Jerry Kazdan et Frank Warner [KW]).

THÉORÈME. *Soient* $(S, \boldsymbol{\beta})$ *une surface de Riemann compacte avec un diviseur* $\boldsymbol{\beta} = \sum_{i=1}^{n} \beta_i p_i$, *et* $K : S \to \mathbf{R}$ *une fonction lisse. Supposons qu'il existe un nombre* $p > 1$ *tel que, au voisinage de chaque* p_i, $h_i(z) := (|z - p_i|^{2\beta_i} K(z))$ *est une fonction de* z *de classe* L^p. *De plus*

   a) *si* $\chi(S, \boldsymbol{\beta}) > 0$, *on suppose* $q \chi(S, \boldsymbol{\beta}) < 2$ *et* $\sup(K) > 0$ *(où* $\frac{1}{p} + \frac{1}{q} = 1$):
   b) *si* $\chi(S, \boldsymbol{\beta}) = 0$, *on suppose* $K \equiv 0$ *ou* $\sup(K) > 0$ *et* $\int_S K \, dA_0 < 0$ *(où* dA_0 *est l'élément d'aire d'une métrique conforme plate représentant* $\boldsymbol{\beta}$ *sur* S);
   c) *si* $\chi(S, \boldsymbol{\beta}) < 0$, *on suppose* $K \le 0$ *et* $K \not\equiv 0$.

Il existe alors une métrique conforme g *sur* S *qui représente le diviseur* $\boldsymbol{\beta}$ *et dont la courbure est* K. *Dans le cas* (c), *cette métrique est unique.*

Une idée très succincte de la démonstration est présentée dans [HT1] (voir [T2] et [HT2] pour les détails).

Certains cas particuliers de ce théorème ont étés obtenus précédemment par W. M. Ni, R. MacOwen et P. Aviles. Au début du siècle, Emile Picard avait déjà étudié le cas de la courbure -1 (voir [P]).

Les hypothèses du théorème précédent imposent une décroissance de K lorsqu'on s'approche des singularités d'ordre < -1. Le résultat suivant—uniquement valide en courbure non positive—n'impose pas un tel comportement.

THÉORÈME. *Soient* S *une surface de Riemann compacte et* g_1 *une métrique conforme représentant un diviseur* $\boldsymbol{\beta} = \sum_{i=1}^{n} \beta_i q_i$ *tel que* $\chi(S, \boldsymbol{\beta}) < 0$.

Soit $K : S \to \mathbf{R}$ *une fonction lisse non positive vérifiant*

(**P**) $$bK \leq K_1 \leq aK < 0$$

en dehors d'un compact de $S' := S \setminus \{q_i\}$ (K_1 *est la courbure de* g_1 *et* a *et* b *sont des constantes positives*).

Alors il existe une unique métrique g *sur* S *qui soit conforme, de courbure* K *et conformément quasi-isométriques à* g_1. *De plus,* g *représente* $\boldsymbol{\beta}$.

Voir [**HT2**, Théorème 8.1, 8.4] et [**McO**] pour la démonstration.

Ce théorème permet—par exemple—de construire des métriques à courbure (négative) prescrite ayant des cusps. Il admet aussi une généralisation aux surfaces non compactes (de type fini) ayant des bouts hyperboliques.

Nous terminons ce paragraphe sur deux resultats récents concernant les surfaces à diviseur dont la caractéristique d'Euler est négative. Soit $(S, \boldsymbol{\beta})$ une surface de Riemann compacte à diviseur dont la charactéristique d'Euler est négative, et soit $K : S \to \mathbf{R}$ une fonction sur S. Le premier résultat (dû à Tang Junjie) permet de résoudre le problème de Berger-Nirenberg (problème II) lorsque $\chi(S, \boldsymbol{\beta})$ est proche de 0. Le second résultat (dû à Dominique Hulin) permet de résoudre ce problème lorsque K est proche d'une fonction $K_1 \leq 0$.

THÉORÈME [**J**]. *Soit* S *une surface de Riemann compacte munie d'un diviseur* $\boldsymbol{\alpha} = \sum_{i=1}^{n} \alpha_i p_i$ *tel que* $\chi(S, \boldsymbol{\alpha}) = 0$. *Choisissons une métrique conforme plate* g_0 *représentant* $\boldsymbol{\alpha}$.

Donnons-nous sur S *un autre diviseur* $\boldsymbol{\beta} = \sum_{i=1}^{n} \beta_i p_i$ *ayant même support et tel que* $\chi(S, \boldsymbol{\beta}) < 0$. *Donnons-nous encore une fonction* $K : S \to \mathbf{R}$ *vérifiant* $K = O(|z - p_i|^{\ell_i}$ *(au voisinage de chaque* p_i) *où* $\ell_i > -2(1 + \alpha_i)$.

(**A**) *Si* $\beta_i < \alpha_i$ *pour tout* i, *alors une condition nécessaire à l'existence d'une métrique conforme* g *sur* S *représentant* $\boldsymbol{\beta}$ *est*

$$\int_S K \, dA_0 < 0,$$

où dA_0 *est l'élément d'aire de* g_0.

(**B**) *Il existe* $\epsilon > 0$ *(dépendant de* S, $\boldsymbol{\alpha}$ *et* K*) tel que si* $|\alpha_i - \beta_i| < \epsilon$ $(\forall i)$, *alors* $\int_S K \, dA_0 < 0$ *est une condition suffisante à l'existence d'une métrique conforme* g *représentant* $\boldsymbol{\beta}$.

THÉORÈME [**H**]. *Soit* S *une surface de Riemann compacte munie d'un diviseur* $\boldsymbol{\beta} = \sum_{i=1}^{n} \beta_i p_i$ *tel que* $\chi(S, \boldsymbol{\beta}) < 0$. *Soit* $K_1 : S \to \mathbf{R}$ *une fonction non positive (et non* $\equiv 0$) *et* $k : S \to \mathbf{R}$ *une fonction non négative. On suppose* K_1 *et* k *lisses et* $|z - p_i|^{2\beta_i} K_1(z)$, $|z - p_i|^{2\beta_i} k(z) \in L^p$ *pour un exposant* $p > 1$.

Alors il existe une constante $C > 0$ *(dépendant de* S, $\boldsymbol{\beta}$, k *et* K_1*) telle que si*

$$K_1 \leq K \leq K_1 + Ck,$$

alors il existe une métrique conforme de courbure K sur S et qui représente
$\boldsymbol{\beta}$.

La dépendence de la constante C en S, $\boldsymbol{\beta}$, k et K_1 est donnée dans [H, Théorème 6.1].

VI. Polyèdres sphériques

Convenons d'appeler "polyèdre sphérique" une surface riemannienne à singularités coniques homéomorphe à la sphère et dont la courbure est constante $K \equiv +1$.

Un théorème d'Aleksandrov (voir [A, §2 Chapitre XII, p. 457]) nous dit que si S est une telle surface, alors S peut être réalisée comme bord d'un polytope convexe de S^3 (muni de sa métrique standard) à condition que les angles θ_i des singularités vérifient $\theta_i < 2\pi$ (i.e. $-1 < \beta_i < 0$).

Le premier résultat classifie les polyèdres sphériques ayant moins de trois points coniques.

THÉORÈME [T4]. *Soit g une métrique sur la sphère S^2 représentant un diviseur $\boldsymbol{\beta} = \beta_1 p_1 + \beta_2 p_2$ et dont la courbure est constante $K \equiv +1$. Alors $\beta_1 = \beta_2$, de plus:*

i) *Si $\beta_i \notin \mathbb{N}$, alors les points p_1 et p_2 sont antipodaux (i.e. $d(p_1, p_2) = \pi$).*

ii) *Si $\beta_i \in \mathbb{N}$, alors (S, g) est isométrique à un revêtement de degré $\beta_1 + 1$ de la sphère standard (S^2, can) ramifié au dessus de deux points (et dont l'ordre de ramification est β_i).*

De plus, deux telles métriques sont isométriques si et seulement si leur singularités sont de même ordre et séparées par une même distance.

En particulier, un polyèdre sphérique ne peut pas avoir une seule singularité (on exprime parfois ce résultat en disant qu'une goutte d'eau n'admet pas de métrique à courbure constante).

Le dernier résultat, dû à Fang Luo et Gang Tian, classifie les diviseurs pouvant être représenté par un polyèdre sphérique ayant ou moins trois singularités et dont tous les angles sont $< 2\pi$.

THÉORÈME [LT]. *Soit $\boldsymbol{\beta} = \sum_{i=1}^n \beta_i p_i$ un diviseur sur $S^2 = \mathbf{C} \cup \{\infty\}$ tel que $-1 < \beta_i < 0$ ($\forall i$) et $n \geq 3$.*

Alors il existe une unique métrique conforme g à courbure constante $K \equiv +1$ représentant $\boldsymbol{\beta}$ si et seulement si

$$(C) \qquad 0 < 2 + \sum_{i=1}^n \beta_i < 2 + \min_i\{\beta_i\}.$$

La condition (C) s'exprime en fonctions des angles $\theta_i = 2\pi(\beta_i + 1)$ de la façon suivante:

$$(C') \qquad 0 < 4\pi + \sum_{i=1}^n (\theta_i - 2\pi) < 2\min_i\{\theta_i\}.$$

La premiere inégalité dans cette condition n'est autre que la formule de Gauss-Bonnet.

Observons que la condition (C') est similaire à la condition satisfaite par les angles $\varphi_1, \varphi_2, \ldots, \varphi_n$ d'un polygone convexe sphérique:

$$0 < 2\pi + \sum_{i=1}^{n} (\varphi_i - \pi) < 2 \min_i \{\varphi_i\}.$$

L'existence d'une métrique sphérique représentant β découle du premier théorème du paragraphe V ci-dessus (voir aussi [T2]). La nécessité de la condition (C), ainsi que l'unicité de la métrique ont été démontrés recemment par Feng Luo et Gang Tian [LT].

References

[A]   A. D. Alexandrov, *Die Innere Geometrie Der Konvexen Flächen*, Akademie Verlag, Berlin, 1955.

[F]   R. Finn, *On a class of conformal metrics, with application to differential geometry in the large*, Comment. Math. Helv. **40** (1965), 1–30.

[H]   D. Hulin, *Déformation conforme des surfaces non compactes à courbure négative*, J. Funct. Anal. (1992) (to appear).

[HT1] D. Hulin et M. Troyanov, *Sur la courbure des surfaces ouvertes*, C. R. Acad. Sci. Paris Sér. I Math. **310** (1990), 203–206.

[HT2] _____, *Prescribing curvature on open surfaces*, Math. Ann. **293** (1992), 277–315.

[J]   T. Junjie, *Prescribing curvature with negative total curvature on open Riemann surfaces*, Preprint, Northeastern Univ., 1991.

[KW]   J. Kazdan et F. Warner, *Curvature functions for compact 2-manifolds*, Ann. of Math. **99** (1974), 14–47.

[LT]   F. Luo & G. Tian, *Liouville equation and spherical convex polytopes*, Preprint, UCLA (1991).

[McO] R. McOwen, *Prescribed curvature and singularities of conformal metrics on Riemann surfaces*, Preprint, Northeastern Univ..

[P]   E. Picard, *De l'intégration de l'équation $\Delta u = e^u$ sur une surface de Riemann fermée*, Journal de Crelle **130** (1905), 243–258.

[S]   K. Shiohama, *Total curvature and minimal area of complete open surfaces,*, Proc. Amer. Math. Soc. **94** (1985), 310–316.

[T1]   M. Troyanov, *Les surfaces euclidiennes à singularités coniques*, Enseign. Math. **32** (1986), 79–94.

[T2]   _____, *Prescribing curvature on compact surfaces with conical singularities*, Trans. Amer. Math. Soc. **324** (1991), 793–821.

[T3]   _____, *Coordonnées polaires sur les surfaces riemanniennes singulières*, Ann. Inst. Fourier (Grenoble) **40** (1990), 913–937.

[T4]   _____, *Metrics of constant curvature on a sphere with two conical singularities*, Proc. Third Internat. Sympos. Differential Geom. (Peñiscola, 1988), Lecture Notes in Math., vol. 1410, Springer-Verlag.

UNIVERSITÉ DU QUÉBEC À MONTREAL, CANADA

Proceedings of Symposia in Pure Mathematics
Volume 54 (1993), Part 2

Remarks on Certain Higher-Dimensional Quasi-Fuchsian Domains

S.-T. YAU AND F. ZHENG

1

Let us begin with the uniformization theory of Riemann surfaces. Let G be $\mathrm{PSL}(2, \mathbb{C})$, acting on $\mathbb{P}^1$ as linear fractional transformations. For a discrete subgroup Γ of G, denote by:

$$\Omega^0(\Gamma) = \{x \in \mathbb{P}^1 | \exists \text{ neighbourhood } U \ni x \text{ s.t. } \gamma U \cap U = \varnothing \; \forall \gamma \in \Gamma \backslash \{1\}\},$$

$$\Omega(\Gamma) = \{x \in \mathbb{P}^1 | \exists \text{ neighbourhood } U \ni x \text{ s.t. } \gamma U \cap U = \varnothing$$
$$\text{for all but finitely many } \gamma \in \Gamma\}$$

$$\Lambda(\Gamma) = \{x \in \mathbb{P}^1 | \exists \text{ sequence } \{\gamma_n\} \subseteq \Gamma, \; x_0 \in \Omega^0(\Gamma) \text{ s.t.} \{\gamma_n(x_0)\} \to x\}.$$

When $\Omega^0(\Gamma)$ is nonempty, Γ is said to be a Kleinian group. For such a group, the limit set $\Lambda(\Gamma)$ either consists of at most two points, or is a perfect set; i.e., any point p in $\Lambda(\Gamma)$ is an accumulation point of $\Lambda(\Gamma) \backslash \{p\}$. $\mathbb{P}^1$ is the disjoint union of $\Omega(\Gamma)$ and $\Lambda(\Gamma)$, while $\Omega(\Gamma) \backslash \Omega^0(\Gamma)$ is a discrete subset in $\Omega(\Gamma)$. The quotient $\Omega(\Gamma)/\Gamma$ is a Riemann surface with at most countably many connected components, with possibly a discrete set of quotient singularities, while $\Omega^0(\Gamma)/\Gamma$ is smooth. By passing to a subgroup of finite index of Γ, one may always assume that $\Omega(\Gamma) = \Omega^0(\Gamma)$.

Recall that Γ is called a quasi-Fuchsian group if $\Omega(\Gamma) = \Omega^0(\Gamma)$ consists of exactly two connected components, both of which are invariant under Γ. In this case, $\Lambda(\Gamma)$ is a simple closed Jordan curve, while the quotient $\Omega(\Gamma)/\Gamma = \Sigma_1 \amalg \Sigma_2$ is the disjoint union of two smooth connected Riemann surfaces with Σ_1 homeomorphic to Σ_2.

By Bers's theorem on simultaneous uniformization, for any two homeomorphic hyperbolic Riemann surfaces Σ_1 and Σ_2, there always exists quasi-

1991 *Mathematics Subject Classification.* Primary 32M15.

The final version of this paper will be published elsewhere.

Research of the first author was supported by NSF grant NSF DMS 9206938.

Fuchsian group Γ such that $\Omega(\Gamma)/\Gamma = \Sigma_1 \amalg \Sigma_2$. (Furthermore, any orientation-reversing homeomorphism between Σ_1 and Σ_2 can be lifted to an involutive homeomorphism of $\mathbb{P}^1$ which commutes with each element in Γ.) Write $\Omega(\Gamma) = D \amalg D'$. Then D will not be linearly equivalent under $\mathrm{PSL}(2, \mathbb{C})$ to the unit disc in $\mathbb{C}$, if Σ_1 and Σ_2 are not biholomorphic to each other. In other words, there are lots of such "quasi-Fuchsian domains" other than the round discs, or equivalently, there are lots of quasi-Fuchsian groups which are not conjugate in $\mathrm{PSL}(2, \mathbb{C})$ to a subgroup of $\mathrm{PSL}(2, R)$. For more details see [**Ms**].

Suppose we want to generalize the one-dimensional theory to higher dimension. According to the view proposed by the first author in the Weyl Conference held in Duke University, a possible generalization is to consider subgroups of the birational group of an algebraic manifold. The reason is that, conjecturally, every compact algebraic manifold with no rational curves and with infinite fundamental group is birational to the quotient of a domain by a group of birational transformations, acting on the compact algebraic manifold containing the domain. Clearly one of the most interesting classes of algebraic manifolds is the Hermitian symmetric spaces.

We shall demonstrate in the next two sections that if we consider a discrete subgroup of the group of biregular transformations of the compact Hermitian symmetric space, and if the quotient is compact Kähler, then in fact the discrete group acts in a standard manner. No exotic actions can be created if the quotient is Kähler. However, there can be a rich class of non-Kähler quotient of domains in the Hermitian symmetric manifold by discrete subgroups of biregular transformations. They provide many interesting non-Kähler manifolds including class VII_0 surfaces. How shall we classify them?

2

From now on, X^n will be a compact complex manifold, with dimension $n \geq 2$, unless otherwise stated. Let $G = \mathrm{Aut}^0(X)$ be the identity component of the biholomorphism group of X.

DEFINITION. An open, simply connected subset Ω in X will be called a quasi-Fuchsian domain if there exists a subgroup Γ of G acting freely and properly discontinuously on Ω with the quotient space Ω/Γ being a compact Kähler manifold. (Ω, Γ) will be called a quasi-Fuchsian pair in X.

For example, let D be a bounded symmetric domain sitting in its compact dual X. For any uniform lattice Γ' in $\mathrm{Aut}(D)$, (D, Γ') becomes a quasi-Fuchsian pair in X.

DEFINITION. A quasi-Fuchsian pair (Ω, Γ) in X is said to be reducible, if there are quasi-Fuchsian pairs (Ω_i, Γ_i) in X_i ($i = 1, 2$), such that $X = X_1 \times X_2$, $\Omega = \Omega_1 \times \Omega_2$, and $\Gamma_1 \times \Gamma_2$ is a subgroup of finite index in Γ. In this case a finite cover of the quotient manifold Ω/Γ will be a product manifold.

(Ω, Γ) is said to be irreducible if it is not reducible.

Now we can state the result for the case when X is a hermitian symmetric space of compact type.

THEOREM. *Let X^n $(n \geq 2)$ be a compact hermitian symmetric space and let (Ω, Γ) be a quasi-Fuchsian pair in X where Ω is a bounded domain. Assume that none of the irreducible components of (Ω, Γ) are of dimension one, and Γ is Zariski dense in $G = \text{Aut}^0(X)$. Then (Ω, Γ) is conjugate to a standard quasi-Fuchsian pair in X; i.e., there exists $g \in G$ such that $g(\Omega)$ is the noncompact dual of X, while $g\Gamma g^{-1}$ is a uniform lattice in the automorphism group of $g(\Omega)$.*

At the time when we were studying this problem, Mok and Yeoung [M-Y] also studied a related question. They showed that the holomorphic projective structure on any compact complex manifold Y^n with $n \geq 2$ will be unique, if it exists. Let us now briefly indicate why this is the same as the rank one case of the above theorem.

Recall that a holomorphic projective structure on Y^n is simply a holomorphic immersion $\varphi: \widetilde{Y} \to \mathbb{P}^n$ such that, under this φ, the deck transformation group of Y becomes a subgroup $\text{PGL}(n+1, \mathbb{C})$. Here, $\widetilde{Y}$ is the universal covering space of Y. Now if (Ω, Γ) is a quasi-Fuchsian pair in $\mathbb{P}^n$ $(n \geq 2)$, the quotient space $Y = \Omega/\Gamma$ then admits a holomorphic projective structure from $\Omega \subseteq \mathbb{P}^n$. Using the holomorphic projective connection associated with this structure, one can show that the ratio of the first two Chern numbers of Y is equal to that ratio of $\mathbb{P}^n$. See Gunning [G], for example. Therefore the solution to the Calabi conjecture by the first author implies that Y is covered by the unit ball B^n in $\mathbb{C}^n$; i.e., $Y = B^n/\Gamma'$, with Γ' a uniform lattice in $\text{SU}(n, 1)$ and Γ' is intrinsically isomorphic to Γ. Note that (B^n, Γ') also gives a holomorphic projective structure on Y, and the uniqueness of such a structure simply says that (Ω, Γ) is conjugate to (B^n, Γ') in $\mathbb{P}^n$; i.e., there exists $g \in \text{PGL}(n+1, \mathbb{C})$ such that $g(\Omega) = B^n$, $g\Gamma g^{-1} = \Gamma'$.

3

Now let us assume that X^n $(n \geq 2)$ is a compact hermitian symmetric space other than $\mathbb{P}^n$, and (Ω, Γ) is a quasi-Fuchsian pair in it, with $Y = \Omega/\Gamma$ the quotient. By the theorem of Cheng-Yau [C-Y] and Mok-Yau [M-Y1], there exists a Kähler-Einstein metric with negative Ricci curvature on Y. By an argument due independent to the first author and to Kobayashi and Ochiai [K-O], one knows that (Y, ω) is locally hermitian symmetric; i.e., $Y = D/\Gamma'$, with D the noncompact dual of X. The key point in their argument is to demonstrate that the homology group H of (Y, ω) is reducible, and hence Berger's theorem can be applied to conclude that (Y, ω) is locally hermitian symmetric. One then constructs a holomorphic $(2, 2)$ tensor on X which is invariant under G, hence can be descended to Y. Then the Kodaira-Bochner identity implies that this tensor is parallel;

therefore the holonomy group H must be reducible. This construction is an algebraic one.

Let us give an alternative proof of the reducibility of H here, which is interesting by itself, and might have some other applications. For simplicity, let X^n be irreducible. Our goal here is to show that the holonomy group H of (Y, ω) cannot be the whole $U(n)$.

Write $X = G_C/K$, where G_C is a compact form of G, and K is the isotropy subgroup of G_C at a fixed base point $q \in \Omega \subseteq X$. Let $p = \pi(q)$, where $\pi: \Omega \to Y$ is the covering map. Then H and K become subgroups of $U(n)$. Of course it will be equal to K when (Ω, Γ) is a standard quasi-Fuchsian pair. The first observation here is:

$$(*) \qquad \begin{array}{l} \text{For any irreducible } U(n)\text{-representation } \rho \text{, any splitting pre-}\\ \text{served by } K \text{ is also preserved by } H. \end{array}$$

The reason for this is as follows. Let $V = \mathbb{C}^n$ be the standard $U(n)$-representation. For any nonnegative integers a, b, write $V^{a,b} = V^{\otimes a} \otimes (V^*)^{\otimes b}$, the tensor product representations of a-copies of V with b-copies of V^*, the dual representation of V. Suppose $V^{a,b} = V_1 \oplus V_2 \oplus \cdots \oplus V_r$ is a K-invariant splitting. Then the holomorphic vector bundle $T^{a,b}(X) = T_X^{\otimes a} \otimes (T_X^*)^{\otimes b}$ will split accordingly as $T^{a,b}(X) = E_1 \oplus E_2 \oplus \cdots \oplus E_r$, where T_X is the holomorphic tangent bundle of X. This splitting of $T^{a,b}(X)$ is also G-invariant, therefore can be restricted and descended down to Y:

$$(1) \qquad\qquad T^{a,b}(Y) = F_1 \oplus F_2 \oplus \cdots \oplus F_r.$$

Write $\omega^{a,b}$ for the induced metric on $T^{a,b}(Y)$ from (Y, ω). Then $(T^{a,b}(Y), \omega^{a,b})$ becomes a hermitian-Einstein bundle over (Y, ω). So the holomorphic splitting (1) will force the metric to split accordingly, and hence the holonomy group H will preserve (1).

Since any irreducible $U(n)$-representation ρ is an irreducible component of $V^{a,b}$ for certain a, b, this proves $(*)$.

Now let H_1 be the subgroup in $U(n)$ generated by H and K. Then $(*)$ is also satisfied by K and H_1. A simple modification of the classic Peter-Weyl theorem would imply that the commutator subgroup $[H_1, H_1]$ must be contained in K. Therefore K is a normal subgroup in H_1, and H_1 is an extension of K by an abelian group. Therefore H_1 cannot be equal to the whole $U(n)$.

4

Now we are ready to prove the theorem for the case $X^n \neq \mathbb{P}^n$.

Let $X, (\Omega, \Gamma), Y, G$ be as in the theorem, and assume $X^n \neq \mathbb{P}^n$. By the theorem mentioned above, one knows that Y is biholomorphic to D/Γ', with $D = G_0/K$ the noncompact dual of X, where $G = G_0^{\mathbb{C}}$, and Γ' is a uniform lattice in G_0.

Let Z be the center of G, $F = G/Z$. Since Γ' is isomorphic to Γ as abstract groups, we have a homomorphism $f: \Gamma' \to F$, which is the composition of $\Gamma' \overset{\cong}{\to} \Gamma \hookrightarrow G \to F$. B Margulis's super-rigidity theorem (Theorem $2'$ in [**Mr**]), there exists splitting $F = F_1 \times F_2$ as direct product, such that $\pi_1 \circ f(\Gamma')$ is relatively compact in F_1 (in $\mathbb{C}$-topology), while $\pi_2 \circ f: \Gamma' \to F_2$ can be extended to a rational morphism $\tilde{f}: G \to F_2$. Here $\pi_i: F \to F_i$ ($i = 1, 2$) are the projection maps.

Let us now prove that F_1 must be trivial: $F_1 = \{1\}$. Therefore $f: \Gamma' \to F$ is the restriction of a rational morphism $\tilde{f}: G \to F$.

Let $X = X_1 \times X_2$ be the symmetric space splitting corresponding to $F = F_1 \times F_2$. G is a finite extension of $G_1 \times G_2$. By lifting $Y = \Omega/\Gamma$ to a finite cover of it, one may suppose that Γ is actually contained in $G_1 \times G_2$; i.e., Γ preserves the tangent bundle splitting $T_X = T_{X_1} \oplus T_{X_2}$, so it descends down to a holomorphic splitting $T_Y = F_1 \oplus E_2$. Since (Y, ω) is Kähler-Einstein, this must be a metric splitting, and the de Rham theorem implies $\Gamma = \Gamma_1 \times \Gamma_2$, $\Omega = \Omega_1 \times \Omega_2$, (Ω_i, Γ_i) are quasi-Fuchsian pairs in X for $i = 1, 2$, and $Y = (\Omega_1/\Gamma_1) \times (\Omega_2/\Gamma_2)$. Now $\pi_1 \circ F(\Gamma') = \overline{\Gamma}_1 \subseteq F_1$ is the image of $\Gamma_1 \subseteq G_1$ under the projection map $G_1 \to F_1 = G_1/\text{Center}$. Note that since Γ_1 acts properly discontinuously on some nonempty set $\Omega_1 \subseteq X_1$, there are disjoint open sets $\{\mathcal{O}_\gamma : \gamma \in \Gamma_1\}$ in G_1 such that each $\mathcal{O}_\gamma$ contains only one element in Γ_1. That is, $\Gamma_1 \subseteq G_1$ is a discrete, hence closed, subgroup. When $\dim(G_1) > 0$, Γ_1 will be infinite; therefore it cannot be (relatively) compact. So we know that the (relative) compactness of $\overline{\Gamma}_1 \subseteq F_1$ implies that G_1 must be of dimension zero; i.e., $G_1 = \{1\}$.

Now modulo the finite center Z of G, the isomorphism f between Γ' and Γ is the restriction of a rational morphism φ from F to F. Since both Γ, Γ' are Zariski dense in G and F is semisimple and centerless, φ will be an inner automorphism of F. In other words, up to replacing by finite index subgroups if necessary, Γ and Γ' will be conjugate to each other in G; i.e., there is $g \in G$ such that $g \Gamma g^{-1} = \Gamma'$.

Now we want to compare the quasi-Fuchsian domains D and $\Omega' = g(\Omega)$ under the same quasi-Fuchsian group $\Gamma' = g\Gamma g^{-1}$. It would be sufficient to show that Ω' is actually a G_0-orbit in X.

Let $H = \{h \in G_0 : h\Gamma' h^{-1}$ is commensurable with $\Gamma'\}$. Assume that all the irreducible components of (Ω, Γ) are of rank greater than one (since the rank one case has already been dealt with by the work of Mok and Yeung [**M-Y**]. Then the arithmeticity [**Mr**] of Γ' implies that H will contain all the Γ'-rational points in G_0, and is dense in G_0.

For any $h \in H$, the group $\Gamma_h = \Gamma' \cap h\Gamma' h^{-1}$ acts invariantly and properly discontinuously on both Ω' and $h(\Omega')$, with compact quotients. Let $M = \Omega'/\Gamma_h$, $N = \Omega' \cup h(\Omega')/\Gamma_h$. If $\Omega' \cap h(\Omega') \neq \varnothing$, then N is connected. So $M \subseteq N$ and the compactness of M will force $M = N$. While the pull-back metrics from $M = N$ on Ω' and $\Omega' \cup h(\Omega')$ are both complete, this is only possible when $\Omega' = \Omega' \cup h(\Omega')$; hence $\Omega' \supset h(\Omega')$. By the same reason $\Omega' \subset h(\Omega')$, and therefore $\Omega' = h(\Omega')$. That is to say, for any element h in H, either $\Omega' \cap h(\Omega') = \varnothing$, or $\Omega' = h(\Omega')$. Since H is dense in G_0, this is also true for any $h \in G_0$, so by the connectedness of G_0 and the fact $1 \in G_0$, we know that $h(\Omega') = \Omega'$ for any $h \in G_0$. In other words, G_0 acts invariantly on Ω'. Therefore Ω' must be the union of some G_0-orbits in X.

By the work of Wolf [W], there are finitely many G_0-orbits in X. Among the maximal dimensional (or open) G_0-orbits in X, only one or two of them (depending on whether X is of tube type) do not contain compact subvarieties other than points, hence can be the universal covering space of Y (which is biholomorphic to D). In the case of two, there is an involution τ on X interchanging them, and this involution commutes with elements in G_0. By adding this involution if necessary, one gets

$$g_1 \Gamma g_1^{-1} = \Gamma', \quad g_1(\Omega) = D, \quad \text{for some } g_1 \in G,$$

where $g_1 = g$ as before, or $g_1 = \tau g$ in the case that there are two D-shape G_0 orbits in X and $g(\Omega)$ happens to be other than D.

This completes the proof of the argument.

REMARK. In the definition of quasi-Fuchsian pair, we required that the quotient space $Y = \Omega/\Gamma$ is a compact Kähler manifold. If we replace this by requiring Y to be a complete Kähler-Einstein manifold of negative Ricci curvature and of finite volume, we may call the pair (Ω, Γ) a 'nonuniform" quasi-Fuchsian pair, and it is interesting to study the rigidity of such pairs. It is easy to see that all the above arguments in §4 work for this case also. The only problem remains to prove that Y is locally Hermitian symmetric. The arguments will work if the invariants constructed in §3 are L^2 integrable.

In the theorem, we assume that Ω is bounded in order to get the Kähler-Einstein metric on Ω. If Ω is unbounded, there are other possibilities which include the case of $\Omega \simeq \mathbb{C}^n$. When $n = 2$, the only extra possibilities are $\Omega = \mathbb{C}^2$ or $\Omega = D \times \mathbb{C}P^1$ and the group Γ acts standardly. For $n > 2$, Mori's theory should be used to deal with similar situation.

REFERENCES

[C-Y] S. Y. Cheng and S. T. Yau, *On the existence of a complete Kähler metric on non-compact complex manifolds and the regularity of Fefferman's equation*, Comm. Pure Appl. Math. **33** (1980), 507–544.

[G] R. Gunning, *On uniformization of complex manifolds: the role of connections*, Princeton Univ. Press, Princeton, N.J., 1978.

[K-O] S. Kobayashi and T. Ochiai, *Holomorphic structures modeled after compact hermitian symmetric spaces*, Manifolds and Lie Groups (Notre Dame, Ind., 1980), Progr. Math., vol. 14, Birkhäuser, Boston, Mass., 1981, pp. 207–222.

[Mr] G. A. Margulis, *Arithmeticity of the irreducible lattices in the semi-simple groups of rank greater than 1*, Invent. Math. **76** (1984), 93–120.

[Ms] B. Maskit, *Kleinian groups*, Springer-Verlag, Berlin and New York, 1987.

[M-Y] N.-M. Mok and S.-K. Yeung, *Geometric realizations of uniformization of conjectures of hermitian locally symmetric manifolds*, preprint, 1990.

[M-Y1] N.-M. Mok and S.-T. Yau, *Completeness of the Kähler-Einstein metric on bounded domains and the curvature characterization of the domains of holomorphy by curvature conditions*, Proc. Symp. Pure Math. **39** (1983), 41–59.

[W] J. A. Wolf, *The action of a real semisimple Lie group on a complex flag manifold*, Bull. Amer. Math. Soc. **75** (1969), 1121–1237.

HARVARD UNIVERSITY

DUKE UNIVERSITY

Proceedings of Symposia in Pure Mathematics
Volume **54** (1993), Part 2

L^p-Cohomology: Banach Spaces and Homological Methods on Riemannian Manifolds

STEVEN ZUCKER

When I give a lecture on L^2-cohomology, somebody inevitably asks, "What about L^p for other values of p?" And if I talk about L^p-cohomology ($1 \leq p \leq \infty$), I am asked whether there is any reason for going beyond the case $p = 2$. Of course, the case of L^2-cohomology is the easiest one, as L^2 is a Hilbert space, its own dual by means of the inner product; it is the most important. The level of importance of $p \neq 2$ is not clear at this time. We mention, though, that treatment of general L^p-cohomology has already appeared in print, in [**9, 17, 23, 37**].

Through the early 1970s, there was a tendency to view L^2-methods as a technique for proving the vanishing of certain cohomology groups on manifolds that can be computed by differential forms (i.e., by the de Rham or Dolbeault theorem), without giving notice to the square-summable differential forms as a homological entity in its own right (that just happens to be sitting inside as a subcomplex). That this was an oversight, especially in the case of noncompact manifolds, is amply demonstrated by [**5, 32**], not to mention subsequent work.

We present the definition and general features of L^p-cohomology in §1. A useful fact is that both smooth and weakly-differentiable forms give rise to the same L^p-cohomology, stated as Theorem 1 in our (1.2). This is proved in [**5, §8**] for $p = 2$; here, we give an exposition of that, which makes it clear that the value of p is irrelevant.

Section 2 contains results (primarily on L^2-cohomology) in the case of Poincaré metrics (2.3), conical singularities (2.4), Fubini-Study metrics on projective varieties (2.5), and locally-symmetric varieties (2.7). The common bond is that these were obtained by local calculations at infinity, once the asymptotic properties of the metric were determined. In §3, we highlight

---

1991 *Mathematics Subject Classification*. Primary 55N33, 58A12, 58A14; Secondary 46E30.
This paper is in final form and no version will be submitted for publication elsewhere.

some recent methods that have been employed to circumvent such explicit calculation, notably from [19, 21].

1. Generalities

1.1. **Definitions.** Let M be a C^∞ manifold (without boundary, unless specified otherwise). Let $A^i(M)$ denote the space of C^∞ real-valued i-forms on M. Then $A^\cdot(M)$ is a cochain complex with exterior derivative d as differential. By the well-known theorem of de Rham, its cohomology is identified with the topological singular cohomology of M, $H^\cdot(M, \mathbb{R})$.

Although cohomology with any type of growth conditions can be defined analogously, we will restrict ourselves here to L^p-cohomology $(1 \leq p \leq \infty)$. *Suppose that a Riemannian metric has been specified on M*, and let dV_M be the associated volume density. A form on M is (in) L^p if

$$
(1) \quad
\begin{aligned}
\|\varphi\|_{(p)} &= \left(\int_M |\varphi(x)|^p \, dV_M(x) \right)^{1/p} < \infty, \quad \text{when } p < \infty, \\
\|\varphi\|_{(\infty)} &= \text{ess.}\sup\{|\varphi(x)| < \infty : x \in M\}, \quad \text{when } p = \infty.
\end{aligned}
$$

The following cochain complexes can be defined from L^p forms:

(2) $A_{(p)}^\cdot(M) = \{\varphi \in A^\cdot(M) : \varphi \text{ and } d\varphi \text{ are } L^p\}$,

(3) $\overline{A}_{(p)}^\cdot(M) = $ the completion of $A_{(p)}^\cdot(M)$ in the L^p graph-norm for d,

(4) $L_{(p)}^\cdot(M) = \{L^p \text{ forms } \varphi : \text{ the weak exterior derivative } d\varphi,$

$\qquad\qquad$ in the sense of distributions, is also $L^p\}$.

In (3) above, $\varphi \in \overline{A}_{(p)}^\cdot(M)$ means that there exists a sequence $\{\varphi_j\}$ in $A_{(p)}^\cdot(M)$ that converges to φ in L^p-norm, such that $\{d\varphi_j\}$ is Cauchy in L^p; one then puts $d\varphi = \lim_{j \to \infty} d\varphi_j$ (this is independent of the choice of $\{\varphi_j\}$). Clearly, there are inclusions

$$
(5) \qquad\qquad A_{(p)}^\cdot(M) \overset{\beta}{\hookrightarrow} \overline{A}_{(p)}^\cdot(M) \overset{\alpha}{\hookrightarrow} L_{(p)}^\cdot(M).
$$

It is simplest to proceed by making:

DEFINITION 1. The L^p-cohomology of M, denoted $H_{(p)}^\cdot(M)$, is the cohomology of the complex $A_{(p)}^\cdot(M)$. In other words,

$$
(6) \qquad\qquad H_{(p)}^i(M) \simeq Z_{(p)}^i(M) / B_{(p)}^i(M),
$$

where

$$
Z_{(p)}^i(M) = \{\varphi \in A_{(p)}^i(M) : d\varphi = 0\},
$$

$$
B_{(p)}^i(M) = \{\varphi \in A_{(p)}^i(M) : \varphi = d\eta \text{ for some } \eta \in A_{(2)}^{i-1}(M)\}.
$$

1.2. To gain more flexibility, we will verify the following:

PROPOSITION 1. *For any Riemannian manifold M, the mapping α in (5) is an isomorphism, and the mapping β induces an isomorphism on cohomology.*

It is convenient to introduce the following variant of Proposition 1, which actually contains it as a special case. Assume that M is the interior of a Riemannian manifold-with-boundary $\overline{M}$ (the boundary is permitted to have corners, or be empty). Let $A_{(p)}^{\cdot}(\overline{M})$ be the subcomplex of (2) consisting of forms that are smooth to the boundary, and define $\overline{A}_{(p)}^{\cdot}(\overline{M})$ in analogy with (3). We consider

$$(7) \qquad A_{(p)}^{\cdot}(\overline{M}) \overset{\overline{\beta}}{\hookrightarrow} \overline{A}_{(p)}^{\cdot}(\overline{M}) \overset{\overline{\alpha}}{\hookrightarrow} L_{(p)}^{\cdot}(M).$$

PROPOSITION 1$'$. *For any Riemannian manifold-with-boundary as above, the mapping $\overline{\alpha}$ in (7) is an isomorphism, and the mapping $\overline{\beta}$ induces an isomorphism on cohomology.*

It follows that we can work with complete normed spaces, and we have

COROLLARY. $H_{(p)}^{\cdot}(M)$ *is the cohomology of any of the complexes in* (7).

The material needed for proving Proposition 1$'$ can be found in an article of Cheeger [5] (see also [12]). All that one must do is to recognize that the arguments spelled out there for $p = 2$ carry over verbatim for all values of p. We will sketch the proof, but before doing so, we present some of the general features of L^p-cohomology.

1.3. **Basic properties.** First, L^p-cohomology is the same for quasi-isometric metrics, for they determine the same L^p complex. The inclusion $A_{(p)}^{\cdot}(M) \hookrightarrow A^{\cdot}(M)$ induces a mapping

$$(8) \qquad H_{(p)}^{\cdot}(M) \to H^{\cdot}(M),$$

it is an isomorphism whenever $\overline{M}$ is compact. The image of (8) is the space of cohomology classes admitting L^p representatives.

There is an important variant of the preceding. Suppose we are given a flat vector bundle (real or complex) on M, whose local system of horizontal sections is denoted $\mathbb{E}$. Then there is the de Rham complex $A^{\cdot}(M, \mathbb{E})$ of $\mathbb{E}$-valued forms on M. Moreover, given a metric (not necessarily flat) in the bundle—we then say that $\mathbb{E}$ is a *metrized local system*—one defines L^p-norms for $\mathbb{E}$-valued forms as in (1), and thereby $H_{(p)}^{\cdot}(M, \mathbb{E})$.

The case $p = 2$ is special, for the L^2 spaces are Hilbert spaces, and are thus self-dual. One can then try to represent $H^i_{(2)}(M)$ by

$$(9) \qquad\qquad h^i_{(2)}(M) = Z^i_{(2)}(M) \cap B^i_{(2)}(M)^{\perp}.$$

This is a subspace of the L^2 harmonic i-forms on M, and is the whole space if M is complete (see (2.6) for more on this). Harmonic forms are always smooth on $\overline{M}$. The most general assertion concerning (9) is the following version of the Hodge theorem:

$$(10) \qquad\qquad H^i_{(2)}(M) \simeq h^i_{(2)}(M) \oplus [\overline{B^i_{(2)}(M)}/B^i_{(2)}(M)],$$

where the "bar" denotes closure. By a little argument in functional analysis, the rightmost summand in (10) is either zero (when d has closed range), or is of infinite dimension otherwise. In particular:

PROPOSITION 2. *Let M be a complete Riemannian manifold, and suppose that $H^i_{(2)}(M)$ is finite dimensional. Then the latter is isomorphic to the space of L^2 harmonic i-forms on M.*

REMARK 1. (i) By what has been said earlier, the above contains the classical Hodge theorem, where M is compact (and thus the L^2-conditions are vacuous).

(ii) Another way of writing (10) is

$$Z^i_{(2)}(M)/\overline{B^i_{(2)}(M)} \simeq h^i_{(2)}(M).$$

We decidedly do *not* take the left-hand side as the definition of L^2-cohomology because we want to retain its homological nature. The significance of the Hodge theorem is that features of harmonic forms have implications in cohomology, e.g. the Hodge decomposition on a Kähler manifold (see [29]).

(iii) A nonlinear Hodge theory, where cohomology class representatives do not form a linear subspace, for L^p-cohomology $(p \neq 2)$ is given in [23].

1.4. We now begin to address the proof of Proposition 1'. Consider $M \subset \overline{M}$, and write $\partial M = \overline{M} - M$, $n = \dim M$.

DEFINITION 2. A *good covering* of $\overline{M}$ by cubes is a locally finite covering $\mathscr{U}$ of $\overline{M}$ by closed cubes $U \simeq \{t \in \mathbb{R} : 0 \leq t \leq 1\}^n$, such that

(i) For $U, U' \in \mathscr{U}$, $U \cap U'$ is relatively open in both U and U',

(ii) For $U \in \mathscr{U}$, if any closed face F of ∂U meets ∂M, then F is contained in ∂M.

(iii) The $(n-1)$-dimensional boundary face $t_j = 1$ is *not* contained in ∂M for any j.

The main theme of the proof of Proposition 1' is to do something on each member of a (countable) good covering of M by cubes, making sure that the process converges. To that end, we recall the construction of smoothing operators in $\mathbb{R}^n$. Let ψ be any nonnegative C^∞ function of compact support with $\int_{\mathbb{R}^n} \psi(t)\,dt = 1$, and put $\psi_\varepsilon(t) = \varepsilon^{-n}\psi(t/\varepsilon)$. Note that supp $\psi_\varepsilon = \varepsilon \cdot$ supp ψ, and $\int_{\mathbb{R}^n} \psi_\varepsilon(t)\,dt = 1$. The convolution operator on locally integrable functions

$$(11) \qquad\qquad \Psi_\varepsilon(f) = f * \psi_\varepsilon$$

has the property that it takes L^p functions to smooth L^p functions (all p), with operator norm $\|\Psi_\varepsilon\| = 1$, and for $f \in L^p$,

$$(12) \qquad\qquad \lim_{\varepsilon \downarrow 0} \Psi_\varepsilon(f) = f, \quad \text{(in } L^p\text{-norm)}.$$

One defines $\Psi_\varepsilon(\varphi)$, when φ is a differential form on $\mathbb{R}^n$, by expanding φ in terms of the standard coordinate differentials, and simply applying Ψ_ε to the coefficient functions. Then Ψ_ε commutes with d, and maps $L^{\cdot}_{(p)}(\mathbb{R}^n)$ into $A^{\cdot}_{(p)}(\mathbb{R}^n)$.

1.5. **Calculations on an interval.** We will also need a homotopy operator for Ψ_ε. Since the Poincaré Lemma underlies that, we begin there. For convenience, we exclude the case $p = \infty$ in (1).

Let $I = (0, 1)$ denote the unit interval in $\mathbb{R}$. Write $f = f(t)$ for functions on I, and $g = g(t)\,dt$ for one-forms, and note that for $p < \infty$,

$$(13) \qquad (\|f\|_{(p)})^p = \int_0^1 |f(t)|^p\,dt, \qquad \left(\|g\|_{(p)}\right)^p = \int_0^1 |g(t)|^p\,dt.$$

Define P_a and B_a by

$$(14) \qquad \begin{aligned} P_a f &= f(a) \text{ (constant function)}, \qquad P_a(g) = 0; \\ B_a f &= 0, \qquad B_a(g) = \int_a^t g(x)\,dx. \end{aligned}$$

It is clear that for smooth forms, $1 - P_a = dB_a + B_a d$. Then put

$$(15) \qquad \begin{aligned} Pf &= \int_I P_a f\,da, \quad \text{(constant function)}, \\ Bf &= \int_I (B_a f)\,da. \end{aligned}$$

We have, still, on $A^{\cdot}(I)$,

$$(16) \qquad\qquad 1 - P = dB + Bd.$$

But furthermore:

PROPOSITION 3. *P and B are bounded with respect to L^p-norm (any p), so extend continuously to $\overline{A}^{\cdot}_{(p)}(I)$.*

PROOF. If $f \in L^p$, then Hölder's inequality gives $|\int_I f| \leq \|f\|_{(p)}$, as I has unit length. The assertion follows easily.

COROLLARY. *For all* p, $H^0_{(p)}(I) \simeq \mathbb{R}$ *and* $H^1_{(p)}(I) = 0$.

PROOF. By (16), we see that a closed L^p form φ is L^p-cohomologous to $P\varphi$.

To move in the direction of proving Proposition 1′, we take as ψ (for (11)) a smooth function whose support is contained in $[-1, 0) \subset \mathbb{R}$. If f is a function on I, $\Psi_\varepsilon f$ is defined for $t \in (-\varepsilon, 1 - \varepsilon)$, and in particular is smooth at $t = 0$. Put $I_\varepsilon = (0, 1 - \varepsilon)$, and let

$$(17) \qquad\qquad A_\varepsilon = (1 - \Psi_\varepsilon)B|_{I_\varepsilon}.$$

We construe as understood the final restriction to I_ε. Since Ψ_ε and d commute, we get

$$dA_\varepsilon + A_\varepsilon d = (1 - \Psi_\varepsilon)(dB + Bd) = (1 - \Psi_\varepsilon)(1 - P)$$
$$= (1 - \Psi_\varepsilon) - (1 - \Psi_\varepsilon)P = 1 - \Psi_\varepsilon.$$

Thus, A_ε is essentially a homotopy operator for Ψ_ε.

1.6. **Cancellation of intervals.** In order to obtain versions of the results in (1.5) in several variables, we first prove

LEMMA. *For any Riemannian manifold* M,

$$H^{\cdot}_{(p)}(M \times I) \simeq H^{\cdot}_{(p)}(M).$$

PROOF. An element of $A^{\cdot}(M \times I)$ defines on the slice at $m \in M$ forms f and g on I (notation as in (1.5)), with values in $\bigwedge^{\cdot} T^*_{M, m}$. Define operators $\tilde{P}$ and $\tilde{B}$ from P and B, by extension of scalars (from $\mathbb{R}$ to $\bigwedge^{\cdot} T^*_{M, m}$). These are bounded with respect to all L^p-norms, and (16) gives $1 - \tilde{P} = d_I \tilde{B} + \tilde{B} d_I$, where d_I denotes partial exterior derivative along I. Since $\tilde{B}$ and d_M anticommute, we get

$$(18) \qquad\qquad 1 - \tilde{P} = d\tilde{B} + \tilde{B}d,$$

and our assertion follows (cf. corollary to Proposition 4).

REMARK 2. The above lemma is an instance of a "Künneth formula" for L^p-cohomology: under some fairly mild hypotheses, one expects to have for the Riemannian product of M_1 and M_2,

$$H^{\cdot}_{(p)}(M_1 \times M_2) \simeq H^{\cdot}_{(p)}(M_1) \otimes H^{\cdot}_{(p)}(M_2).$$

Sufficient conditions, when $p = 2$, are given in [**33**, (2.36)].

We can now show

PROPOSITION 4. (i) *There are operators* P *and* B *on* $A^{\cdot}(I^n)$, *bounded with respect to all* L^p-*norms, such that* P *is a projection onto the constant functions, and* $1 - P = dB + Bd$.

(ii) *For all* p, $H^i_{(p)}(I^n) = \begin{cases} \mathbb{R}, & \text{if } i = 0, \\ 0, & \text{otherwise.} \end{cases}$

PROOF. We wish to take $M = I^{n-1}$ in the above lemma, and iterate. Let P_j and B_j be the operators (15) for the jth factor of I. We allow ourselves to use the same symbols for their extensions to I^n (as in the proof of the above lemma, where, however, the notation $\tilde{P}$ and $\tilde{B}$ was introduced). Then take

$$(19) \qquad P = \prod_{1 \le j \le n} P_j, \qquad B = \sum_{j=1}^{n} B_j \prod_{k>j} P_k.$$

1.7. For the purpose of performing smoothing on I^n, let

$$(20) \qquad \psi(t) = \prod_{1 \le j \le n} \psi(t_j),$$

where ψ is, as in (1.5), supported in the interval $[-1, 0)$. The corresponding smoothing operator Ψ_ε produces forms on $(I_\varepsilon)^n$ that are smooth to the lower boundary (the faces on which some $t_j = 0$).

Next, let $f(t)$ be a C^∞ function on I, with support in I_ε, and put

$$(21) \qquad f.(t) = \prod_{1 \le j \le n} f(t_j).$$

The operators

$$(22) \qquad \begin{aligned} R_\varepsilon &= f.\Psi_\varepsilon + (1 - f.) - df. \wedge (1 - \Psi_\varepsilon)B, \\ H_\varepsilon &= f.(1 - \Psi_\varepsilon)B, \end{aligned}$$

where B is as in (19), extend in an obvious way to all of I^n, and one calculates

$$\begin{aligned} dH_\varepsilon + H_\varepsilon d &= df. \wedge (1 - \Psi_\varepsilon)B + f.(1 - \Psi_\varepsilon)dB + f.(1 - \Psi_\varepsilon)Bd \\ &= df. \wedge (1 - \Psi_\varepsilon)B + f.(1 - \Psi_\varepsilon)(1 - P) \\ &= df. \wedge (1 - \Psi_\varepsilon)B + f.(1 - \Psi_\varepsilon) = 1 - R_\varepsilon. \end{aligned}$$

In other words, H_ε is a homotopy operator for R_ε. One should observe that for all $\varphi \in L^\cdot_{(p)}(I^n)$, $R_\varepsilon\varphi$ is smooth on

$$(23) \qquad S = \{t \in I^n : f.(t) = 1\}.$$

1.8. We can now complete the proof of Proposition 1'. Let $\mathscr{U} = \{U_j : j \in N\}$ be a good covering of $\overline{M}$ by cubes; we will write $R^{(j)}_{\varepsilon_j}$, $f.^{(j)}$, $S^{(j)}$, etc. for the objects in (22), (23), etc. constructed for $U_j \simeq I^n$. We take $\{f.^{(j)}\}$

to be a partition of unity subordinate to $\mathscr{U}$, chosen such that

(24)
$$\bigcup_j S^{(j)} = \overline{M}.$$

The operators $R_{\varepsilon_j}^{(j)}$ and $H_{\varepsilon_j}^{(j)}$ extend naturally to $L_{(p)}^{\cdot}(M)$, as the identity and zero respectively in the complement of U_j. Put (cf. (19)) for any $\varepsilon = \{\varepsilon_j\}$,

(25)
$$\begin{aligned}
R_{\varepsilon} &= \cdots \circ R_{\varepsilon_2}^{(2)} \circ R_{\varepsilon_1}^{(1)}, \\
H_{\varepsilon} &= \sum_{j=1}^{\infty} H_{\varepsilon_j}^{(j)} \circ R_{\varepsilon_{j-1}}^{(j-1)} \circ \cdots \circ R_{\varepsilon_1}^{(1)}.
\end{aligned}$$

The above product and sum are actually finite on any compact subset of $\overline{M}$, and are thus well defined. One checks that if ε is suitably chosen, the operators $1 - R_{\varepsilon}$ and H_{ε} are bounded, and satisfy

$$1 - R_{\varepsilon} = dH_{\varepsilon} + H_{\varepsilon} d, \qquad R_{\varepsilon}(L_{(p)}^{\cdot}(M)) \subset A_{(p)}^{\cdot}(\overline{M}),$$

$$\| |\varphi - R_{\varepsilon}(\varphi)| \|_{(p)} \quad \text{and} \quad \| |d\varphi - R_{\varepsilon}(d\varphi)| \|_{(p)}$$

are arbitrarily small.

REMARK 3. Proposition 1.1' extends to metrized local systems. (The proof is left as an exercise.)

1.9. L^p **sheaves.** Let $M \subset \overline{M}$ be as in (1.2), and $\mathbb{E}$ a metrized local system on $\overline{M}$. It should be apparent that $U \mapsto L_{(p)}^{\cdot}(U, \mathbb{E})$ (U open in $\overline{M}$) defines a complex of presheaves on $\overline{M}$; likewise for $A_{(p)}^{\cdot}$. Let $\mathscr{L}_{(p)}^{\cdot}(\overline{M}, \mathbb{E})$ be the associated complex of sheaves. Proposition 5 implies immediately that this complex is a (fine) resolution of the sheaf $\mathbb{E}$ on $\overline{M}$. Thus,

(26)
$$H^{\cdot}(\overline{M}, \mathscr{L}_{(p)}^{\cdot}(\overline{M}, \mathbb{E})) \simeq H^{\cdot}(M, \mathbb{E}).$$

From the very nature of what a sheaf is, the reader may recognize that the global sections of $\mathscr{L}_{(p)}^{\cdot}(\overline{M}, \mathbb{E})$ consists precisely of the *locally L^p* forms φ whose derivatives are also locally L^p, so fails to give $L_{(p)}^{\cdot}(M, \mathbb{E})$ unless M is compact. To remedy this situation, one lets M^* be any compactification of $\overline{M}$, and defines $\mathscr{L}_{(p)}^{\cdot}(M^*, \mathbb{E})$ to be the complex of sheaves *on* M^* determined by the presheaf

$$U^* \mapsto L_{(p)}^{\cdot}(U, \mathbb{E}), \qquad (U^* \text{ open in } M^*, \quad U = U^* \cap M).$$

When $\mathbb{E} = \mathbb{R}$, we omit the coefficients. Note that each $\mathscr{L}_{(p)}^i(M^*, \mathbb{E})$ is an $\mathscr{L}_{(\infty)}^0(M^*)$-module. It is not hard to verify:

PROPOSITION 5. *If M^* admits a partition of unity consisting of functions in $L_{(\infty)}^0(M^*)$, then $\mathscr{L}_{(p)}^{\cdot}(M^*, \mathbb{E})$ is a complex of fine sheaves for any $\mathbb{E}$.*

Under its hypotheses, Proposition 6 provides an interpretation of L^p cohomology as the hypercohomology of a complex of sheaves on M^*; in other words, it comes from an object of the derived category, where one is less concerned about the particular complex than the cohomology it produces on the subsets of M^*. We assert

Main theme. Determine analogues of (26), i.e. the de Rham theorem, for L^p-cohomology (in interesting cases).

REMARK 4. One can define, in analogous manner, L^p *Dolbeault cohomology* and L^p sheaves for the $\bar{\partial}$-operator on a complex manifold. For a survey of this, see [29]; some recent work: [11, 24, 25]. I believe the roots of L^2-cohomology lie in the use of a priori L^2 estimates for $\bar{\partial}$ on pseudoconvex domains [16, 13] to imply the vanishing of ordinary Dolbeault cohomology.

2. Calculations

Most of the results involving L^2-cohomology have been obtained by explicit computations, using natural "coordinates" (compare Lemma of (1.6)). In this section, we will discuss some examples of this sort, most of which have been understood already for at least ten years.

2.1. **The half-line.** Having treated the Euclidean interval in (1.5), we advance to the case of the real half-line $\mathbb{R}^+$. It is instructive to see how one of the simplest Riemannian manifolds can be so badly-behaved with respect to L^2-cohomology. Moreover, the computations in this case feed into the more interesting examples that we consider later on.

PROPOSITION 6. (i) $h_{(2)}^{\cdot}(\mathbb{R}^+) = 0$,

(ii) $H_{(2)}^0(\mathbb{R}^+) = 0$,

(iii) $H_{(2)}^1(\mathbb{R}^+)$ *is infinite-dimensional.*

PROOF. (i) follows, for example, from the remark after (9); there are no nonzero L^2 harmonic functions (affine-linear here) on $\mathbb{R}^+$, and a harmonic one-form here is just a harmonic function times dt. The calculation of (ii) and (iii) begins as in (1.5). The zero function is the only L^2 constant. However, it is easy to see that an L^2 function on $\mathbb{R}^+$ need not have an L^2 antiderivative, e.g. $(t+1)^{-3/2}$. From this scant information, we can conclude that $B_{(2)}^1(\mathbb{R}^+)$ (see (10)) is not closed, and (iii) follows.

REMARK 5. Conclusions (ii) and (iii) hold for $H_{(p)}^{\cdot}(\mathbb{R}^+)$, for all p. See [37, (3.2)].

It is useful to treat also *weighted* L^p-cohomology, which can be viewed in the context of (1.3) as taking the tensor product with a trivial local system having nontrivial metrization. Explicitly, if w is a positive function on the

Riemannian manifold M, and $\mathbb{E}$ is a metrized local system thereon, one may consider the Banach spaces determined by the weighted notion of L^p,

$$(27) \qquad \|\varphi\|_{(p)}^p = \int_M |\varphi(x)|^p \, w(x) \, dV_M(x)$$

for any $1 \le p < \infty$ (contrast with (1)), and write $H_{(p)}^{\cdot}(M, \mathbb{E}; w)$ for the resulting cohomology groups.

We take as weight functions on $\mathbb{R}^+$, $w(t) = e^{-kt}$. The following is not hard to prove (see [37] again).

PROPOSITION 7. *For all* p,

$$H_{(p)}^0(\mathbb{R}^+; e^{-kt}) = \begin{cases} \mathbb{R}, & \text{if } k > 0, \\ 0, & \text{if } k < 0; \end{cases}$$

$$H_{(p)}^1(\mathbb{R}^+; e^{-kt}) = \quad 0, \quad \text{if } k \ne 0.$$

REMARK 6. The assertions of this section have analogues on $(\mathbb{R}^+)^m$ and $\mathbb{R}^m$.

2.2. **The Poincaré metric.** We consider metrics on a Riemann surface having isolated singularities asymptotic to that of the Poincaré metric of the punctured complex disc at the origin. Thus, we take as underlying manifold $M = S - \Sigma$, for Σ a discrete subset of the real surface S. It is easy to construct metrics on M of the sort described above; all are quasi-isometric when S is compact.

For our purposes, it is enough to study the punctured disc itself. In terms of the usual polar coordinates, the Poincaré metric is given by

$$(28) \qquad ds^2 = (r \log r)^{-2}(dr^2 + r^2 d\theta^2).$$

Making the change of variables

$$(29) \qquad u = \log|\log r| \qquad [du = (r \log r)^{-1} dr]$$

yields the formula

$$(30) \qquad ds^2 = du^2 + e^{-2u} d\theta^2,$$

which displays the Poincaré metric as the product of a circle warped by a half-line. Thus, $dV = e^{-u} du \, d\theta$.

REMARK 7. Keep in mind that we wish to stay clear of the singularities of (28) at $r = 1$. Thus we insist that (say) $r < \frac{1}{2}$; so the half-line above starts at $u = \log \log 2$.

For a function $f = f(\theta)$ (independent of u), one deduces from (30) that

$$(31) \qquad \|f\|_{(p)}^p = \iint |f(\theta)|^p e^{-u} du\, d\theta \, ;$$

similarly, for $g = g(\theta)d\theta$, $|d\theta|^p = e^{pu}$, so

$$(32) \qquad \|g\|_{(p)}^p = \iint |g(\theta)|^p e^{u(p-1)} du\, d\theta.$$

We see that such zero-forms are in L^p on Δ^* if and only if f is L^p on S^1, whereas any non-zero such one-form is not. This shows that

$$(33) \qquad H_{(p)}^0(\Delta^*, \mathbb{R}) = \mathbb{R} \quad (\text{any } 1 \le p < \infty),$$

and suggests.

PROBLEM. Investigate the assertion $H_{(p)}^1(\Delta^*, \mathbb{R}) = 0$.

For $p = 2$, the above vanishing is proved along the above lines in [33, (2.40)], which treats the general question of the L^2-cohomology of a warped product, as a variant of the Künneth formula (see Remark 2 of our (1.6)). The main issue is to construct homotopy operators for the inclusion of forms independent of du in the full L^p-complex, which is similar to what we did in §1. It seems like a good warm-up exercise to carry this out for all $p > 1$; I suspect that the L^1-cohomology, will go as in Proposition 7.

2.3. From here on, for security we restrict ourselves to the case of L^2-cohomology (i.e., $p = 2$). Let S now be a manifold of arbitrary dimension, Σ a union of smooth manifolds of codimension two that meet transversally, and $M = S - \Sigma$. An example of such is when S is a complex manifold, and Σ is a so-called divisor with normal crossings. At a point of Σ, M is locally diffeomorphic to

$$(34) \qquad (\Delta^*)^r \times I^s$$

for some integers $r > 0$ and s, and it is again easy to construct Riemannian metrics on M that are asymptotic to the product of Poincaré metrics on the Δ^* factors, and Euclidean on I^s. (In the complex analytic example, if S is Kähler one can arrange the metric on M to be Kählerian.) We call such a metric a *Poincaré metric on M, relative to S*. Then

PROPOSITION 8. *If M is given a Poincaré metric relative to S, $H_{(2)}^{\cdot}(M) \simeq H^{\cdot}(S, \mathbb{R})$.*

PROOF. With the help of the Künneth formula for L^2-cohomology, one sees that the space in (34) has all of its L^2-cohomology in degree 0. It follows that $\mathcal{L}_{(2)}^{\cdot}(S, \mathbb{R})$ is a fine resolution of the constant sheaf $\mathbb{R}_S$ on S.

REMARK 8. Proposition 9 includes the statement that certain forms singular along Σ represent cohomology classes on all of S, which can be applied to representing Chern classes.

While the above proposition is nice, its generalization to certain local systems whose metrization is asymptotically exponential in the coordinates of (34) is of greater interest. These are the ones underlying a *variation of Hodge structure*, about which we will say nothing here (see [31, §2]). The main result is

THEOREM 1 [4, 15]. *Let S be a compact complex manifold, Σ a divisor with normal crossings on S, and $M = S - \Sigma$. Endow M with a Poincaré metric relative to S. Let $\mathbb{E}$ be a local system on M underlying a variation of Hodge structure, metrized by the corresponding Hodge metric. Then*

$$(35) \qquad H^{\cdot}_{(2)}(M, \mathbb{E}) \simeq IH^{\cdot}(S, \mathbb{E}).$$

Here, the right-hand side denotes intersection (co)homology with middle perversity, as in [10].

REMARK 9. (i) The above is a consequence of the isomorphism in the derived category of $\mathscr{L}^{\cdot}_{(2)}(S, \mathbb{E})$ and the $\mathbb{E}$-valued intersection cochains $\mathscr{I}C^{\cdot}(S, \mathbb{E})$. Though this may sound rather fancy, the main issue in proving it is to verify that if U is a nice neighborhood of a point on the locus of r-fold crossings,

$$(36) \qquad H^i_{(2)}(U - \Sigma, \mathbb{E}) = 0 \quad \text{if } i \geq r,$$

as $\mathscr{I}C^{\cdot}(S, \mathbb{E})$ is characterized by certain axioms [10, (4.1), (6.1)]. (See [34, (3.1), (3.3), (3.13)] for more on this.)

(ii) In the case where S is a Riemann surface, (35) can be rewritten as

$$(37) \qquad H^{\cdot}_{(2)}(M, \mathbb{E}) \simeq H^{\cdot}(S, j_* \mathbb{E}),$$

where j denotes the inclusion of M in S. This was proved in [32]. (Note that we have for trivial coefficients, $j_* \mathbb{R}_M \simeq \mathbb{R}_S$.)

2.4. **Conical singularities.** Before presenting a significant generalization of (37), we first treat an important work in the development of modern L^2-cohomology, namely [5].

Let $M = \overset{\circ}{c}(N)$, the *metrical cone* on the Riemannian manifold N (where the vertex of the cone has been deleted). In other terms,

$$(38) \qquad M = I \times N, \quad \text{with metric } g_M = dt^2 + t^2 g_N.$$

This, too, is a warped product; but it is fundamentally different from the Poincaré metric, in that it is incomplete, with the vertex at finite distance.

PROPOSITION 9 [5, Lemma 3.4]. *Let N be a Riemannian manifold of dimension m. If $m = 2n + 1$ (odd), assume that $B^{n+1}_{(2)}(N)$ is closed. Then*

(i) $H^i_{(2)}(\overset{\circ}{c}(N)) \overset{\sim}{\to} H^i_{(2)}(N)$ if $i \le m/2$;

(ii) $H^i_{(2)}(\overset{\circ}{c}(N)) = 0$ if $i > m/2$.

The above is the expected outcome, by the sort of heuristic reasoning in (2.2), for (38) gives for an i-form φ on N, pulled back to $\overset{\circ}{c}(N)$,

$$(\|\varphi\|_{(2)})^2 = \int_N |\varphi(x)|^2 dV_N(x) \int_I t^{m-2i}\, dt.$$

A class of manifolds that are locally quasi-isometric to the product of a cone and a Euclidean cube, and satisfy (also locally) the hypothesis of Proposition 10, are the Riemannian pseudomanifolds (see [7, (3.4)] for the definition). One gets, by globalizing Proposition 10,

THEOREM 2 [5, Theorem 6.1]. Let X be a Riemannian pseudomanifold. Then $H^{\cdot}_{(2)}(X) \simeq IH^{\cdot}(X, \mathbb{R})$.

2.5. **The Cheeger-Goresky-MacPherson Conjecture.** Let X be an algebraic subvariety of $P^N(\mathbb{C})$. The natural (Fubini-Study) metric of projective space restricts to a Kähler metric on the regular locus M of X. Put $\Sigma = X - M$ (the singular locus of X). It is not hard to see that the asymptotics of the metric at any point of Σ depend only on the analytic germ of X at that point. It follows that the quasi-isometry class of the metric is an invariant of X, and is in particular independent of the projective embedding of X. Note, however, that these metrics are incomplete unless $M = X$.

When X has analytically conical singularities, it is easy to verify that X is a Riemannian pseudomanifold, so Theorem 2 gives

$$(39) \qquad\qquad H^{\cdot}_{(2)}(X) \simeq IH^{\cdot}(X).$$

There is the belief that this should remain true in general.

CONJECTURE 1 [7, §4]. Let X be a complex projective variety, endowed with a Fubini-Study metric. Then

(i) the isomorphism (39) holds, and

(ii) it imparts a Hodge decomposition to the intersection cohomology of X.

A proof of Conjecture 1 in the case of isolated singularities has recently been given in [21]. Earlier, the result had been proved for the cone on a variety [6], and assertion (i) was demonstrated for complex surfaces ([14, 20]), by calculating with metric asymptotics.

2.6. I want to digress a little about (ii) above and related matters, for it may help to dispel some of the mystery surrounding (9). A form is in $B^i_{(2)}(M)^\perp$ if and only if it is annihilated by the Hilbert space adjoint d^* of

d. By definition, $d^*\varphi$ is the L^2 form representing the linear functional

$$(40) \qquad\qquad \eta \mapsto \langle \varphi, d\eta \rangle,$$

when bounded. This operator has the formula $\pm * d *$; here "$*$" is the tensor that converts i-forms to $(m - i)$-forms on an m-dimensional oriented Riemannian manifold, characterized by the formula

$$(41) \qquad\qquad \varphi \wedge * \psi = \langle \varphi, \psi \rangle dV_M.$$

However, in contrast to (3), d^* can be described as the closure of $\pm * d *$ on smooth forms of compact support on M; thus, when $\overline{M}$ (of (1.2)) is compact, the domain of d^* consists, roughly, of forms that vanish on the boundary.

One introduces the *operator* Laplacian

$$(42) \qquad\qquad \Delta = dd^* + d^*d$$

(the negative of the classical operator Laplacian in the case of Euclidean space), whose domain is defined by the customary conventions of functional analysis. In particular,

$$(43) \qquad \varphi \in h_{(2)}^{\cdot}(M) \iff \Delta\varphi = 0 \iff d\varphi = 0 \quad\text{and}\quad d^*\varphi = 0.$$

On the other hand, φ is simply L^2 harmonic (as in (1.3)) if the single *second-order* differential expression $\Delta\varphi$ vanishes. When M is complete, the operator d is also the closure of its restriction to forms of compact support [8], and one can deduce that the same is true for Δ (see [35]). It follows that (43) is then equivalent to the assertion that φ is L^2 harmonic. One sees immediately

PROPOSITION 10. *Let M be a complete, oriented Riemannian manifold of dimension m. Then*

(i) $*$ *sets up an isomorphism* $h_{(2)}^i(M) \simeq h_{(2)}^{m-i}(M)$,

(ii) *If $H_{(2)}^i(M)$ and $H_{(2)}^{m-i}(M)$ are finite dimensional, $*$ induces an isomorphism*

$$H_{(2)}^i(M) \simeq H_{(2)}^{m-i}(M).$$

REMARK 10. (i) Likewise, on a complete manifold, any feature of harmonic forms that is determined from the formula for Δ passes to the L^2-cohomology (e.g., the Hodge decomposition).

(ii) A $*$-operator can be defined for $\mathbb{E}$-valued forms ($\mathbb{E}$ a metrized local system), taking its values in $\mathbb{E}^\vee$-valued forms. The analogue of Proposition 11 holds, not to mention (10).

One interprets Proposition 11 as asserting a "Poincaré duality" for L^2-cohomology, under hypotheses. The known Poincaré duality for intersection homology, which was a motivating force behind its conception, gave rise to expectations that statements like Conjecture 1 might hold.

The advantages of having a complete metric that are cited above has led to the idea of replacing the canonical Fubini-Study metric in (39) by constructed complete metrics (which fall into infinitely many quasi-isometry classes, though). This has been done by Saper [26, 27] (by determining metric asymptotics and computing with them) and Ohsawa (see our (3.2)).

We should point out that if X is a variety of complex dimension n having only isolated singularities, then one has a simple formula—only in this case—

$$IH^i(X) \simeq \begin{cases} H^i(X - \Sigma), & \text{if } i < n, \\ \text{im } \{H^n(X) \to H^n(X - \Sigma)\}, & \text{if } i = n, \\ H^i(X), & \text{if } i > n. \end{cases}$$

This is the global version of (36), in which the consequences of duality appear.

2.7. **The Zucker Conjecture.** An interesting class of complete Riemannian manifolds is provided by the arithmetic quotients of symmetric spaces. These are spaces of the form $M \simeq \Gamma \backslash D$, where $D = G/K$ is a symmetric space of noncompact type (G a semisimple Lie group, K a maximal compact subgroup), and Γ is a torsion-free arithmetically-defined subgroup of G. We take as Riemannian metric on M one that is induced by a G-invariant metric on D. It is easy to see that all such are quasi-isometric.

According to [3], M can be viewed as the interior of a compact manifold-with-corners $\hat{M}$, though *not* in the Riemannian sense, of course; the metric of M has good asymptotic descriptions at the boundary of $\hat{M}$, as multiply-warped products on the universal cover [2, (4.3)] (see also [33, (3.5)]). The space $\hat{M}$ admits a distinguished set of quotients, the *Satake compactifications* of M, determined from the restricted root structure of G (see [30, 37, §7]).

Any finite-dimensional representation E of G (a fortiori, of Γ) determines a local system $\mathbb{E}$ on M, whose associated vector bundle is just $\Gamma \backslash (D \times E)$. These come with a canonical metrization.

When D has a G-invariant complex structure, one says that D (or G, or M) is *Hermitian*. In this case, $\mathbb{E}$ underlies a variation of Hodge structure, and one of the Satake compactifications, which we denote M^*, can be realized as a normal, complex projective variety [1]; one then calls M a *locally-symmetric*, or *arithmetic*, *variety*, and M^* its *Baily-Borel Satake compactification*. On M^*, the asymptotics of the metric on M can be deduced from those on $\hat{M}$ (see [34, (3.19)]).

In [33], we had conjectured the following, on the basis of some examples:

THEOREM 3 [18, 28]. *Let M^* be the Baily-Borel Satake compactification of a locally-symmetric variety, and $\mathbb{E}$ the metrized local system associated to a finite-dimensional representation of the corresponding group G. Then*
$$H_{(2)}^{\cdot}(M, \mathbb{E}) \simeq IH^{\cdot}(M^*, \mathbb{E}).$$

The main point in the proof of Theorem 3 is to establish the vanishing (36), where "locus of r-fold crossings" is replaced by the more general "stratum of complex codimension r". For a summary of the two proofs, which are quite different from each other, and also the general issues involved, we refer the reader to [36]. The case of $G = SL_2$ is contained in the earlier [32] (see our (37)). For some further examples, discussed in the spirit of (2.2), see [37, §9].

REMARK 11. Note that one can also take $X = M^*$ in Conjecture 1 of (2.5), though the metric there is quite different. For $G = SU(n, 1)$, M^* has analytically conical singularities, so Conjecture 1 has been verified in that case. A comparison of the resulting Hodge decompositions is given in [35].

3. Averting calculations

In this section, we give a brief description of methods that may allow one to bypass the direct calculation of local L^2-cohomology from explicit metric asymptotics.

3.1. A priori estimates.
If one is interested in showing that an L^2-cohomology group vanishes (cf. Remark 9(i)) there is a very simple criterion.

PROPOSITION 11. One has $H^i_{(2)}(M) = 0$ if and only if there exists $C > 0$ such that for all i-forms φ in the domain of the Laplacian (42),

$$(44) \qquad \|\Delta\varphi\|_{(2)} \geq C \|\varphi\|_{(2)} .$$

REMARK 12. The inequality (44) says that Δ has a bounded (thus everywhere-definable) inverse, and is equivalent to an inequality of the form

$$(45) \qquad \|d\varphi\|^2 + \|d^*\varphi\|^2 \geq C \|\varphi\|^2 ,$$

where we are omitting the "sub-(2)".

In [28], this kind of reasoning is used for proving the necessary form of (36), although it is also possible to reformulate it differently (see [36, (4.6)]).

3.2. Work of Ohsawa.
Let $M = X - \Sigma$ be the regular locus of a variety, as in (2.5). The construction of complete Kähler metrics on M, mentioned in (2.6), is usually based on the selection of a suitable *potential* function ψ; specifically, the metric would have a Kähler form (which determines the metric, so we identify the two)

$$(46) \qquad \omega = \eta + \partial\overline{\partial}\psi ,$$

where η is a Kähler form "on X", and ψ is a smooth function on M with prescribed behavior along Σ. One arranges that $\partial\overline{\partial}\psi$ is asymptotically singularly positive, and the term η is then added to insure positivity away from Σ. (The Poincaré metrics of (2.4) and Saper's metrics are also of this type.)

One of the key points from [21] is that an estimate of the form

$$(47) \qquad \|d\varphi\|^2 + \|d^*\varphi\|^2 \geq C \|f\varphi\|^2 ,$$

where f is a function on M, is just about as useful as (45). Given that, the following perturbation technique plays a vital role. Suppose that ω_0 and ω_1 are two Kähler forms on M, with ω_1 dominating ω_0. Consider the family

$$(48) \qquad \omega_t = \omega_0 + t\omega_1, \qquad (0 \leq t \leq 1).$$

Of course, ω_t is quasi-isometric to ω_1 for $t \neq 0$, but the bound blows up as $t \to 0$. Under some conditions, it is possible to gain useful information about ω_0 as the limit. In fact, Ohsawa produces a family of estimates similar to (47) in the case where X is a variety with isolated singularities, ω_0 is Fubini-Study, and ω_1 is one of Saper's metrics [21, Proposition 10]; this case of Conjecture 1 is deduced therefrom.

REMARK 13. It is tempting to regard the preceding as the analytic analogue of the algebraic comparison methods of [35].

3.3. **Local "Hecke operators.".** The argument in [18] for Theorem 3 can be described in gross as the use of algebraic methods to reduce the proof to Theorem 1, which had already been verified by direct calculation. In [19], Looijenga and Rapoport found a simplification of the argument that bypasses Theorem 1 and moves it closer to arithmetic considerations. Here, we wish to present one result of theirs that illustrates how assertion (39) can be a consequence of fairly "soft" information.

Let X be an irreducible complex variety, with a stratification by subvarieties, and suppose that the dense stratum M has been given a Riemannian metric. For any lower-dimensional stratum S, let X_S denote the germ of X along S, and $\mathscr{I}_S$ the ideal sheaf of S in X. Also, fix an integer $q > 1$.

PROPOSITION 12. *In the above set-up, suppose that every X_S has an algebraic endomorphism f_S with the following properties:*

(a) $f_S^{-1}(S) = S$, *and f_S restricts to the identity on S,*

(b) f_S *is of finite degree $q^{c(S)}$, where $c(S)$ is the complex codimension of S in X,*

(c) $f_S^* \mathscr{I}_S = \mathscr{I}_S^q$,

(d) f_S *preserves the Riemannian metric.*
Then

(i) *M is complete,*

(ii) *$\mathscr{L}_{(2)}^{\cdot}(X)$ is a complex of fine sheaves,*

(iii) *$\mathscr{L}_{(2)}^{\cdot}(X)$ represents $\mathscr{I}C^{\cdot}(X)$.*

REMARK 14 (see [19, Theorem 3.2]). (i) There is a version of Proposition 13 for nonconstant coefficients.

(ii) Condition (c) can be relaxed a little.

(iii) There is a fourth conclusion in [19], concerning the action of f_S on local intersection cohomology, and the Hodge theoretic significance of its eigenspaces.

REFERENCES

1. W. Baily and A. Borel, *Compactification of arithmetic quotients of bounded symmetric domains*, Ann. of Math. (2) **84** (1966), 442–528.

2. A. Borel, *Stable real cohomology of arithmetic groups*, Ann. Sci. École Norm. Sup. **7** (1974), 235–272.

3. A. Borel and J.-P. Serre, *Corners and arithmetic groups*, Comm. Math. Helv. **48** (1973), 436–491.

4. E. Cattani, A. Kaplan, and W. Schmid, L^2 *and intersection cohomologies for a polarizable variation of Hodge structure*, Invent. Math. **87** (1987), 217–252.

5. J. Cheeger, *On the Hodge theory of Riemannian pseudomanifolds*, Proc. Sympos. Pure Math., vol. 36, Amer. Math. Soc., Providence, RI, 1980, pp. 91–146.

6. _____, *Hodge theory of complex cones*, Analyse et Topologie sur les Espaces Singuliers, Astérisque **101–102** (1983), 118–134.

7. J. Cheeger, M. Goresky, and R. MacPherson, L^2*-cohomology and intersection homology of singular algebraic varieties*, Yau, ed., Seminar on Differential Geometry, Annals of Mathematical Studies no. 102, Princeton Univ. Press, Princeton, NJ, 1982, 303–340.

8. M. Gaffney, *Hilbert space methods in the theory of harmonic integrals*, Trans. Amer. Math. Soc. **78** (1955), 426–444.

9. V. Gol'dshtein, V. Kuz'minov, and I. Shvedov, L^p*-cohomology of non-compact Riemannian manifolds*, Preprint, 1989/1990.

10. M. Goresky and R. MacPherson, *Intersection homology*. II, Invent. Math. **72** (1983), 77–129.

11. P. Haskell, L^2 *Dolbeault complexes on singular curves and surfaces*, Proc. Amer. Math. Soc. **107** (1989), 517–526.

12. L. Hörmander, *Weak and strong extensions of linear operators*, Comm. Pure Appl. Math. **14** (1961), 371–379.

13. _____, L^2 *estimates and existence theorems for the* $\bar{\partial}$ *operator*, Acta Math. **113** (1965), 89–152.

14. W.-C. Hsiang and V. Pati, L^2*-cohomology of normal algebraic surfaces*. I, Invent. Math. **81** (1985), 395–412.

15. M. Kashiwara and T. Kawai, *The Poincaré lemma for a variation of Hodge structure*, Publ. Res. Inst. Math. Sci. **23** (1987), 345–407.

16. J. Kohn, *Harmonic integrals on strongly pseudoconvex manifolds*. I, Ann. of Math. (2) **78** (1963), 112–148.

17. R. Lockhart, *Fredholm, Hodge and Liouville theorems on noncompact manifolds*, Trans. Amer. Math. Soc. **301** (1987), 1–35.

18. E. Looijenga, L^2*-cohomology of locally symmetric varieties*, Comp. Math. **67** (1988), 3–20.

19. E. Looijenga and M. Rapoport, *Weights in the local cohomology of a Baily-Borel compactification*, Proc. Sympos. Pure Math., vol. 53, Amer. Math. Soc., Providence, RI, 1991, pp. 223–260.

20. M. Nagase, *Remarks on the* L^2*-cohomology of singular algebraic surfaces*, J. Math. Soc. Japan **41** (1989), 97–116.

21. T. Ohsawa, *Cheeger-Goresky-MacPherson conjecture for the varieties with isolated singularities*, Math. Z. **206** (1991), 219–224.

22. _____, *On the L^2-cohomology groups of isolated singularities*, Adv. Stud. Pure Math. (to appear).

23. P. Pansu, Thèses.

24. W. Pardon, *The L_2-$\bar{\partial}$-cohomology of an algebraic surface*, Topology **28** (1989), 171–195.

25. W. Pardon and M. Stern, *L^2-$\bar{\partial}$-cohomology of complex projective varieties*, J. Amer. Math. Soc. (to appear).

26. L. Saper, *L_2-cohomology and intersection homology of certain varieties with isolated singularities*, Invent. Math. **82** (1985), 207–255.

27. _____, *L_2-cohomology of Kähler varieties with isolated singularities*, Preprint.

28. L. Saper and M. Stern, *L_2-cohomology of arithmetic varieties*, Ann. of Math. (2) **132** (1990), 1–69.

29. L. Saper and S. Zucker, *An introduction to L^2-cohomology*, Proc. Sympos. Pure Math., vol 52, Amer. Math. Soc., Providence, R.I., 1991, pp. 519–534.

30. I. Satake, *On compactifications of the quotient spaces for arithmetically defined discontinuous groups*, Ann. of Math. (2) **72** (1960), 555–580.

31. W. Schmid, *Variation of Hodge structure: the singularities of the period mapping*, Invent. Math. **22** (1973), 211–319.

32. S. Zucker, *Hodge theory with degenerating coefficients: L_2-cohomology in the Poincaré metric*, Ann. of Math. (2) **109** (1979), 415–476.

33. _____, *L^2-cohomology of warped products and arithmetic groups*, Invent. Math **70** (1982), 169–218.

34. _____, *L_2-cohomology and intersection homology of locally symmetric varieties*. II, Comp. Math. **59** (1986), 339–398.

35. _____, *The Hodge structures on the intersection homology of varieties with isolated singularities*, Duke Math. J. **55** (1987), 603–616.

36. _____, *L^2-cohomology and intersection homology of locally symmetric varieties*. III, Théorie de Hodge, Astérisque **179–180** (1989), 245–278.

37. _____, *L^p-cohomology and Satake compactifications*, (J. Noguchi and T. Ohsawa, (eds.)), Lecture Notes in Math., vol. 1468, Springer-Verlag, Berlin and New York, 1991, pp. 317–339.

THE JOHNS HOPKINS UNIVERSITY
E-mail address: sz@chow.mat.jhu.edu

Recent Titles in This Series

(*Continued from the front of this publication*)

(See the AMS catalog for earlier titles)

ISBN 0-8218-1495-8

9 780821 814956